DATA STRUCTURES USING C

REEMA THAREJA
Assistant Professor
Institute of Information Technology and Management
GGS IP University

OXFORD
UNIVERSITY PRESS

Oxford University Press is a department of the University of Oxford.
It furthers the University's objective of excellence in research, scholarship,
and education by publishing worldwide. Oxford is a registered trademark of
Oxford University Press in the UK and in certain other countries

Published in India by
Oxford University Press
YMCA Library Building, 1 Jai Singh Road, New Delhi 110001, India

First published in 2011
Second impression 2012

ISBN-13: 978-0-19-806544-9
ISBN-10: 0-19-806544-2

Typeset in Time New Roman
by Shubham Composers, New Delhi
Printed in India by Yash Printographics, Noida 201301

Dedicated to
my family
and
my uncle Mr B. L. Thareja

Preface

A data structure is defined as a group of data elements used for organizing and storing data. In order to be effective, data has to be organized in a manner that adds to the efficiency of an algorithm, and data structures such as stack, queue, linked list, heap, and tree provide different capabilities to organize data. While developing a program or an application, many developers find themselves more interested in the type of algorithm used rather than the type of data structure implemented. However, the choice of data structure used for a particular algorithm is always of the utmost importance.

Each data structure has its own unique properties and is constructed to suit various kinds of applications. Some of them are highly specialized to carry out specific tasks. For example, B-trees with their unique ability to organize indexes are well-suited for the implementation of databases. Similarly, stack, a linear data structure which provides 'last-in-first-out' access, is used to store and track the sequence of web pages while we browse the Internet. Specific data structures are essential components of many efficient algorithms, and make possible the management of large amounts of data, such as large databases and Internet indexing services.

C, as we all know, is the most popular programming language and is widespread among all the computer architectures. Therefore, it is not only logical but also fundamentally essential to start the introduction and implementation of various data structures through C. The course *data structures* is typically taught in the second or third semester of most engineering colleges and across most engineering disciplines in India. The aim of this course is to help students master the design and applications of various data structures and use them in writing effective programs.

About the Book

This book is aimed at serving as a textbook for undergraduate engineering students of computer science and postgraduate level courses of computer applications. The objective of this book is to introduce the concepts of data structures and apply these concepts in problem solving.

The book provides a thorough and comprehensive coverage of the fundamentals of data structures and the principles of algorithm analysis. The main focus has been to explain the principles required to select or design the data structure that will best solve the problem.

A structured approach is followed to explain the process of problem solving. A theoretical description of the problem is followed by the underlying technique. These are then ably supported by an example followed by an algorithm, and finally the corresponding program in C language.

The salient features of the book include:

- Explanation of the concepts using diagrams
- Numerous solved examples within the chapters
- Glossary of important terms at the end of each chapter
- Comprehensive exercises at the end of each chapter
- Practical implementation of the algorithms using tested C programs
- Objective type questions to enhance the analytical ability of the students
- Annexures to provide supplementary information to help generate further interest in the subject

The book is also useful as a reference and resource to young researchers working on efficient data storage and related applications, who will find it to be a helpful guide to the newly established techniques of a rapidly growing research field.

Organization of the Book

This book is organized into 16 chapters and a set of five annexures.

Chapter 1 provides an introduction to C, thereby offering an insight into the programming language that will be used to write programs in this book.

Chapter 2 deals with declaring, defining and calling functions. The chapter also discusses the storage classes as well as the variable scopes in C.

Chapter 3 gives an insight into the use of pointers, the knowledge of which will be helpful while writing C programs to implement different data structures.

Chapter 4 introduces data structure and its related terminology. The chapter explains how to calculate the time complexity which is a key concept to implement efficient storage structures of data.

From *Chapter 5* onwards, every chapter discusses individual data structures in detail.

Chapter 5 provides a detailed explanation of arrays that includes one-dimensional, two-dimensional and multi-dimensional arrays. Finally, the operations that can be performed on such arrays are also explained.

Chapter 6 unleashes the concept of strings which are better known as character arrays. The chapter not only focuses on reading and writing strings but also explains various operations that can be used to manipulate the character arrays.

Chapter 7 deals with structures. A structure is a basic programming constituent for implementing other data structures like linked list, trees, graphs, etc. So, this chapter introduces all the related concepts of structures. We will also read how structures can be used with pointers, arrays, and functions so that the interconnectivity between the programming techniques can be well understood.

Chapter 8 discusses different types of linked lists such as singly linked list, doubly linked list, circular linked list, doubly circular linked list, and header linked list. Linked list is a preferred data structure when it is required to allocate memory dynamically.

Chapter 9 introduces the concept of stacks and queues. In this chapter, the practical implementation of these data structures is shown using arrays as well as linked lists. The chapter also provides the applications of stacks and queues in the world of programming.

Chapter 10 focuses on binary trees, their traversal schemes and representation in memory. The chapter also discusses expression trees, tournament trees, and Huffman trees, all of which are variants of simple binary trees.

Chapter 11 broadens the discussion on trees taken up in *Chapter 10* by going one step ahead and discussing efficient binary trees. The chapter captures the details of binary search trees, AVL trees, B trees, B+ trees, M-way search trees, threaded binary trees, tries, splay trees, and red-black trees.

Chapter 12 introduces heaps. The chapter discusses three types of heaps, namely, binary heap, binomial heap, and Fibonacci heap. The chapter not only explains the operations on these data structures but also makes a comparison, thereby highlighting the key features of each structure.

Chapter 13 contains a detailed explanation of graphs as a data structure. It discusses the memory representation, traversal schemes, and applications of graphs in the real world.

Chapter 14 covers various sorting techniques. It gives the technique, complexity, example and program for different sorting techniques such as bubble sort, selection sort, heap sort, radix sort, shell sort, merge sort, and quick sort.

Chapter 15 deals with hashing and collision. It focuses on different methods of hashing and techniques to resolve collisions.

Chapter 16, the last chapter of the book, discusses the concept related to file organization. It explains the different ways in which files can be organized on the hard disk, the indexing techniques that can be used for fast retrieval of data, and the external sorting concepts that are used to sort large volumes of data.

The book also provides annexures to various chapters.

Annexure A (to Chapter 1) provides the ASCII codes of all printable as well as non-printable characters in the C language. Annexure B (to Chapter 7) introduces the concept of dynamic memory allocation in C programs. Annexure C (to Chapter 9) provides a brief on the stack abstract data type. Annexure D (to Chapter 11) discusses the practical implementation of a right-threaded binary tree and insertion in an AVL tree. Annexure E (to Chapter 13) helps to understand the application of graphs as a data structure by giving a C program for finding the minimum spanning tree.

Acknowledgements

The writing of this textbook was a mammoth task for which a lot of help was required from many people. Fortunately, I have had the fine support of my family, friends, and fellow members of the teaching staff at the Institute of Information Technology and Management (IITM).

My special thanks would always go to my father Mr Janak Raj Thareja and mother Mrs Usha Thareja, my brother Pallav and sisters Kimi and Rashi who were a source of abiding inspiration and divine blessings for me. I am especially thankful to my son Goransh who has been very patient and cooperative in letting me realize my dreams. My sincere thanks go to my uncle Mr B.L. Thareja for his inspiration and guidance in writing this book.

I would also like to thank my students and colleagues at IITM who had always been there to extend help while designing and testing the algorithms.

Finally, I would like to thank the editorial team at Oxford University Press for their help and support.

Reema Thareja

Contents

1 Introduction to C

Learning Objective

This book deals with the study of data structures through C. Before entering into a detailed analysis of data structures, it would be useful to familiarize ourselves with the basic knowledge of programming in C. Therefore, in this chapter we will learn about the various constructs of C such as data types, identifiers, constants, variables, operators, conditional statements, iterative statements, and storage classes of the variables.

1.1 INTRODUCTION

The programming language 'C' was developed in the early 1970s by Dennis Ritchie at Bell Laboratories. Although C was initially developed for writing system software, today it has become such a popular language that a variety of software programs are written using this language. The greatest advantage of using C for programming is that it can be easily used on different types of computers. Today, C is widely used with the UNIX operating system.

It is a good idea to learn C because it has been around for a long time which means that there is a lot of information available on it. Many other programming languages such as C++ and Java are also based on C which means that you will be able to learn them easily in the future.

1.1.1 Structure of a C Program

A C program contains one or more functions, where a function is defined as a group of C statements that are executed together. The statements in a C program are written in a logical sequence to perform a specific task. The `main()` function is the most important function and is a part of every C program. Rather, the execution of a C program begins at this function. Figure 1.1 shows the structure of a C program.

From the structure given in Fig. 1.1, we can conclude that a C program can have any number of functions depending on the tasks that have to be performed and each function can have any number of statements arranged according to specific meaningful sequence. Note that programmers can choose any name for the functions. It is not mandatory to write Function1, Function2, etc. but with an exception that every program must contain one function that has its name as `main()`.

1.2 KEYWORDS AND IDENTIFIERS

Every word in a C program is either a keyword or an identifier.

```
main()
{
        Statement 1;
        Statement 2;
        ............
        ............
        Statement N;
}
Function1()
{
        Statement 1;
        Statement 2;
        ............
        ............
        Statement N;
}
Function2()
{
        Statement 1;
        Statement 2;
        ............
        ............
        Statement N;
}
....................
....................
FunctionN()
{
        Statement 1;
        Statement 2;
        ............
        ............
        Statement N;
}
```

Figure 1.1 Structure of a C program

Keyword Like every computer language, C has a set of reserved words often known as keywords that cannot be used as an identifier. All keywords are basically a sequence of characters that have a fixed meaning. By convention, all keywords must be written in lower case (small) letters. Table 1.1 contains a list of keywords in C.

Identifier Identifiers are basically names given to program elements such as variables, arrays, and functions. Identifier may consist of an alphabet, digit, or an underscore.

Following are the rules for forming identifier names:

- It cannot include any special characters or punctuation marks (like #, $, ^, ?, ., etc) except the underscore "_"
- There cannot be two successive underscores
- Keywords cannot be used as identifiers
- The case of alphabetic characters that form the identifier name is significant. For example, 'FIRST' is different from 'first' and 'First'.
- The identifier name must begin with an alphabet or an underscore. However, use of underscore as the first character must be avoided because several complier-defined identifiers in the standard C library have underscore as their first character. So, inadvertently duplicated names may cause definition conflicts.
- Identifiers can be of any reasonable length. They should not contain more than 31 characters. (It can actually be longer than 31, but the compiler looks at only the first 31 characters of the name.)

Although not compulsory but it is a good practice to use meaningful identifier names. Examples of valid identifiers include:

```
roll_number, marks, name, emp_number, basic_pay, HRA, DA, dept_code
```

Examples of invalid identifiers include:

```
23_student, %marks, @name, #emp_number, basic.pay, -HRA, (DA), &dept_code, auto
```

Table 1.1 Keywords in C language

auto	break	case	char	const	continue	default	do
double	else	enum	extern	float	for	goto	if
int	long	register	return	short	signed	sizeof	static
struct	switch	typedef	union	unsigned	void	volatile	while

1.3 BASIC DATA TYPES IN C

C language provides very few basic data types. Table 1.2 lists the data type, their size, range, and usage for a C programmer.

Table 1.2 Basic data types in C

Data Type	Size in Bytes	Range	Use
char	1	-128 to 127	To store characters
int	2	-32768 to 32767	To store integer numbers
float	4	3.4E-38 to 3.4E+38	To store floating point numbers
double	8	1.7E-308 to 1.7E+308	To store big floating point numbers

Table 1.2 shows the basic data types. In addition to this, there are also variants of `int` and `float` data types. The `char` data type is one byte and is used to store single characters. Note that C does not provide any data type for storing text. This is because text is made up of individual characters.

You will be surprised to see that the range of `char` is given as –128 to 127. `char` is supposed to store characters not numbers, so why this range? The answer is that in the memory, characters are stored in their ASCII codes. For example, the character 'A' has the ASCII code 65. In memory we will not store 'A' but 65 (in binary number format). Table 1.3 shows the variants of basic data types in detail.

Table 1.3 Detailed list of data type

Data Type	Size in Bytes	Range
char	1	–128 to 127
unsigned char	1	0 to 255
signed char	1	–128 to 127
int	2	–32768 to 32767
unsigned int	2	0 to 65535
signed short int	2	–32768 to 32767
signed int	2	–32768 to 32767
short int	2	–32768 to 32767
unsigned short int	2	0 to 65535
long int	4	–2147483648 to 2147483647
unsigned long int	4	0 to 4294967295
signed long int	4	–2147483648 to 2147483647
float	4	3.4E–38 to 3.4E+38
double	8	1.7E–308 to 1.7E+308
long double	10	3.4E–4932 to 1.1E+4932

In Table 1.3, we have `unsigned char` and `signed char`. Do we have negative characters? No, then why do we have such data types? The answer is that we use `signed` and `unsigned char` to ensure portability of programs that store non-character data as `char`.

While the smaller data types take less memory, the larger data types incur a performance penalty. Although the data type we use for our variables does not have a big impact on the speed or memory usage of the application, but we should always try to use `int` unless there is a special need to use any other data type.

1.4 CONSTANTS AND VARIABLES

A variable is defined as a meaningful name given to the data storage location in the computer memory. When using a variable, we actually refer to the address of the memory where the data is stored. C language supports two basic kinds of variables.

1.4.1 Numeric Variables

Numeric variables can be used to store either integer values or floating point values. While an integer value is a whole number without a fraction part or decimal point in them, a floating point number, on the other hand, can have a decimal point in them.

Numeric values may also be associated with modifiers like `short`, `long`, `signed`, and `unsigned`. The difference between signed and unsigned numeric variables is that signed variables can be either negative or positive but unsigned variables can only be positive. Therefore, by using an unsigned variable we can increase the maximum positive range. When we omit the signed/unsigned modifier, C language automatically makes it a signed variable. To declare an unsigned variable, the unsigned modifier must be explicitly added during the declaration of the variable.

1.4.2 Character Variables

Character variables can include any letter from the alphabet or from the ASCII chart and numbers 0–9 that are put between single quotes. In C, a number that is put in single quotes is not the same as a number without them.

1.4.3 Declaring Variables

To declare a variable, specify the data type of the variable followed by its name. The data type indicates the kind of data that the variable will store. Variable names should always be meaningful and must reflect the purpose of their usage in the program. In C, variable declaration always ends with a semi-colon. For example,

```
int emp_num;
float salary;
char grade;
double balance_amount;
unsigned short int acc_no;
```

In C, variables can be declared at any place in the program but two things must be kept in mind. First, variables should be declared before using them. Second, variables should be declared closest to their first point of use so that the source code is easier to maintain.

1.4.4 Initializing Variables

While declaring the variables, we can also initialize them with some value. For example,

```
int emp_num = 7;
float salary = 5000;
char grade = 'A';
double balance_amount = 100000000;
```

1.4.5 Constants

Constants are identifiers whose value does not change. While variables can change their value at any time, constants can never change their value. Constants are used to define fixed values like PI or the charge on an electron so that their value does not get changed in the program even by mistake.

1.4.6 Declaring Constants

To declare a constant, precede the normal variable declaration with const keyword and assign it a value.

```
const float pi = 3.14;
```

1.5 WRITING THE FIRST C PROGRAM

To write a C program, we first need to write the code. For that, open a text editor. If you are a Windows user, you may use Notepad and if you prefer working on UNIX/Linux, you can use `emac` or `vi`. Once the text editor is opened on your screen, type the following statements.

```
#include<stdio.h>
int main()
{
printf("\n Welcome to the world of C ");
return 0;
}
```

`#include<stdio.h>` This is the first statement in our code that includes a file called `stdio.h`. This file has some in-built functions. So simply by including this file in our code, we can use these functions directly. `stdio` basically stands for Standard Input/Output which means it has functions for input and output of data like reading values from the keyboard and printing the results on the screen.

`int main()` `int` is the return value of the main function. After all the statements in the program have been written, the last statement of the program will return an integer value to the operating system. The concepts will be clear to us when we read this chapter in toto. So even if you do not understand certain things, do not worry.

`{ }` The two curly brackets are used to group all the related statements of the function main.

`printf("\n Welcome to the world of C ");` The printf function is defined in the `stdio.h` file and is used to print text on the screen. The message that has to be displayed on the screen is enclosed within double quotes and put inside brackets.

Table 1.4 Escape sequences

Escape Sequence	Purpose
\a	Audible signal
\b	Backspace
\t	Tab
\n	Newline
\v	Vertical tab
\f	New page\Clear screen
\r	Carriage return

The message is quoted because in C a text also known as a string or a sequence of characters is always put between inverted commas. The `\n` is an escape sequence and represents a newline character. It is used to print the message on a new line on the screen. Like the newline character, the other escape sequences supported by C language are shown in the Table 1.4.

`return 0;` This is a return command that is used to return the value 0 to the operating system to give an indication that there were no errors during the execution of the program.

Note Every statement in the main function ends with a semi-colon (;).

`first.c.` If you are a Windows user, then open the command prompt by clicking Start→Run and typing "command" and clicking Ok. Using the command prompt, change to the directory in which you had saved your file and then type:

```
C:\>tc first.c
```

In case you are working on UNIX/Linux operating system, then exit the text editor and type

```
$cc first.c -ofirst
```

The `-o` is for the output file name. If you leave out the `-o` then the file name `a.out` is used.

This command is used to compile your C program. If there are any mistakes in the program, then the compiler will tell you what mistake you have made and on which line the error has occurred. In case of errors, you need to re-open your `.c` file and correct the mistakes. However, if everything is right, then no error(s) will be reported and the compiler will create an `exe` file for your program. This `.exe` file can be directly run by typing

```
"first.exe" for Windows and "./first" for UNIX/Linux operating system.
```

When you run the `exe` file, the output of the program will be displayed on screen. That is,

```
Welcome to the world of C
```

Note The printf and return statements have been indented or moved away from the left side. This is done to make the code more readable.

Using Comments

Comments are a way of explaining what a program does. C supports two types of commenting.

- // is used to comment a single statement.
- /* is used to comment multiple statements. A /* is ended with */ and all statements that lie within these characters are commented.

Note that commented statements are not executed by the compiler. Rather, they are ignored by the compiler as they are simply added in the program to make the code understandable by the programmer as well as other people who read it. It is a good habit to always put a comment at the top of a program that tells you what the program does. This will help in defining the usage of the program the moment you open it.

Commented statements can be used anywhere in the program. You can also use comments in between your code to explain a piece of code that is complicated. The code given below shows the way in which we can make use of comments in our first program.

```
/* Author: Reema Thareja
   Description: To print "Welcome to the world of C" on the screen */
#include<stdio.h>
int main()
{
   printf("\n Welcome to the world of C "); //prints the message on screen
   return 0; // returns a value 0 to the operating system
}
```

1.6 HEADER FILES

When working with large projects, it is often desirable to separate certain subroutines from the `main()` of the program. There also may be a case that the same subroutine has to be used in different programs. In such a case, one option is to copy the code of the desired subroutine from one program to another. But copying the code is often tedious as well as error-prone and makes maintainability more difficult.

So, another option is to make subroutines and store them in a different file known as header file. The advantage of header files can be realized when:

- The programmer wants to use the same subroutines in different programs. For this, he simply has to compile the source code of the subroutines once, and then link it to the resulting object file in any other program in which the functionalities of these subroutine are required.
- The programmer wants to change or add subroutines, and have those changes be reflected in all other programs. For this, he just needs to change the source file for the subroutines, recompile its source code, and then recompile and re-link programs that use them. This way a huge amount of time can be saved as compared to editing the subroutines in every individual program that uses them.

Thus, we see that using a header file produces the same results as copying the header file into each source file that needs it. Also when a header file is included, the related declarations appear in only one place. If in future we need to modify the subroutines, we just need to make the changes in one place, and programs that include the header file will automatically use the new version when next recompiled. There is no need to find and change all the copies of the subroutine that have to be changed.

Conventionally, header file names ends with a 'dot h' (`.h`) extension and its name can use only letters, digits, dashes, and underscores. Although some standard header files are automatically available to C programmers but in addition to those header files, the programmer may have his own user-defined header files.

1.6.1 Standard Header Files

Till now, in our programs, we have used a function `printf()` defined in the `stdio.h` header file. Even in other programs that we will be writing, we will use many functions that are not written by us. For example, to use the `strcomp()` function that compares two strings, we will pass string arguments and retrieve the result. We do not know the details of how these functions work. Such functions that are provided by all C compilers are included in standard header files. Examples of these standard header files include:

- `string.h` : for string handling functions
- `stdlib.h` : for some miscellaneous functions
- `stdio.h` : for standardized input and output functions
- `math.h` : for mathematical functions
- `alloc.h` : for dynamic memory allocation
- `conio.h` : for clearing the screen

All the header files are referenced at the start of the source code file that uses one or more functions from that file.

1.7 INPUT/OUTPUT STATEMENT IN C

The most fundamental operation in a C program is to accept **input** values to the program from standard input device and **output** the data produced by the program to a standard output device. So far, we had been assigning values to variables using the assignment operator. For example,

```
int a = 3;
```

What if we want to assign value to variable `a` that is inputted from the user at run-time? This is done by using the `scanf` function that reads data from the keyboard. Similarly, for outputting results of the program, `printf` function is used that sends results out to a terminal. Like `printf` and `scanf`, there are different functions in C that can carry out the input-output operations. These functions are collectively known as standard Input/Output Library. A program that uses standard input/output functions must contain the statement

```
# include <stdio.h>
```

at the beginning of the program.

1.7.1 scanf()

The `scanf()` is used to read formatted data from the keyboard. The syntax of the `scanf()` can be given as,

```
scanf ("control string", arg1, arg2, arg3 ..........argn);
```

The control string specifies the type and format of the data that has to be obtained from the keyboard and stored in the memory locations pointed by the arguments, `arg1, arg2, ...,argn`. The prototype of the control string can be give as:

```
[=%[*][width][modifiers]type=], where
```

* is an optional argument that suppresses assignment of the input field. That is, it indicates that data should be read from the stream but ignored (not stored in the memory location).

width is an optional argument that specifies the maximum number of characters to be read. However, fewer characters will be read if the `scanf` function encounters a white space or an unconvertible character.

modifiers is an optional argument that can be **h, l,** or **L** for the data pointed by the corresponding additional arguments. Modifier **h** is used for `short int` or `unsigned short int`, **l** is used for `long int, unsigned long int,` or `double` values. Finally, **L** is used for `long double` data values.

type specifies the type of data that has to be read. It also indicates how this data is expected to be read from the user. The type specifiers for `scanf` function are given in Table 1.5.

Table 1.5 Type specifiers

Type	Qualifying Input
c	For single character
d	For decimal values
e,E,f,g,G	For floating point numbers
o	For Octal number
s	For a sequence of (string of) characters
u	For Unsigned decimal value
x,X	For Hexadecimal value

The `scanf` function ignores any blank spaces, tabs, and newlines entered by the user. The function simply returns the number of input fields successfully scanned and stored.

We will not read about functions in this chapter. So understanding `scanf` function in depth will be a bit difficult here, but for now just understand that the `scanf` function is used to store values in memory locations associated with variables. For this, the function should have the address of the variables. The address of the variable is denoted by an & sign followed by the name of the variable. Look at the following code that shows how we input values in variables of different data types.

```
int num;
scanf(" %d ", &num);
```

The `scanf` function reads an integer value (because the type specifier is `%d`) into the address or the memory location pointed by `num`.

```
float salary;
scanf(" %f ", &salary);
```

The `scanf` function reads a floating point number (because the type specifier is `%f`) into the address or the memory location pointed by `fnum`.

```
char ch;
scanf(" %c ", &ch);
```

The `scanf` function reads a single character (because the type specifier is `%c`) into the address or the memory location pointed by `ch`.

```
char str[10];
scanf(" %s ", str);
```

The `scanf` function reads a string or a sequence of characters (because the type specifier is `%s`) into the address or the memory location pointed by `str`. Note that in case of reading string, we do not use the `&` sign in the `scanf` function.

Look at the code given below which combines reading of all these variables of different data types in one single statement.

```
int num;
float fnum;
char ch;
char str[10];
scanf(" %d %f %c %s", &num, &fnum, &ch, str);
```

1.7.2 printf()

The `printf` function is used to display information required by the user and also prints the values of the variables. Its syntax can be given as:

```
printf ("conversion string", variable list);
```

After the control string, the function can have as many additional arguments as specified in the control string. The parameter control string in the `printf()` is nothing but a C string that contains text that has to be written on to the standard output device.

Note that there must be enough arguments because if there are not enough arguments, then the result will be completely unpredictable. However, if by mistake you specify more number of arguments, the excess arguments will simply be ignored. The prototype of the control string can be given as below:

```
%[flags][width][.precision][length]specifier
```

Each control string must begin with a % sign. After the % sign follows **Flags** which specifies output justification like decimal point, numerical sign, trailing zeros or octadecimal or hexadecimal prefixes. Table 1.6 shows the different types of flags with their descriptions.

Table 1.6 Flags in printf()

Flags	Description
-	Left-justify within the data given field width
+	Displays the data with its numeric sign (either + or -)
#	Used to provide additional specifiers like o, x, X, 0, 0x, or 0X for octal and hexa decimal values respectively for values different than zero
0	The number is left-padded with zeroes (0) instead of spaces

Width specifies the minimum number of characters to print after being padded with zeros or blank spaces.

Precision specifies the maximum number of characters to print.

- For integer specifiers (`d, i, o, u, x, X`): Precision flag specifies the minimum number of digits to be written. However, if the value to be written is shorter than this number, the result is padded with leading zeros. Otherwise, if the value is longer, it is not truncated.
- For character strings, precision specifies the maximum number of characters to be printed.

Length field can be explained as given in Table 1.7.

Table 1.7 Length field in printf()

Length	Description
h	When the argument is a short int or unsigned short int.
l	When the argument is a long int or unsigned long int for integer specifiers.
l	When the argument is a long double (used for floating point specifiers)

Specifier is used to define the type and the interpretation of the value of the corresponding argument:

The most simple `printf` statement is

```
printf ("Welcome to the world of C language");
```

the function when executed, prompts the message enclosed in the quotation to be displayed on the screen. The following code shows how we output values of variables of different data types.

```
int num;
scanf(" %d ", &num);
printf("%d", num);
```

The `printf` function prints an integer value (because the type specifier is `%d`) pointed by num on the screen.

```
float salary;
scanf(" %f ", &salary);
printf(".2%f", salary);
```

The `printf` function prints the floating point number (because the type specifier is `%f`) pointed by salary on the screen. Note that the control string specifies that only two digits after the decimal point must be displayed.

```
char ch;
scanf(" %c ", &ch);
printf("%c", ch);
```

The `printf` function prints a single character (because the type specifier is `%c`) pointed by `ch` on the screen.

```
char str[10];
scanf(" %s ", str);
printf("%s", str);
```

The `printf` function prints a string or a sequence of characters (because the type specifier is `%s`) pointed by `str` on the screen. The following code combines all these variables of different data types in one single statement.

```
int num;
char ch;
char str[10];
float fnum;
printf("\n Enter the values : ");
scanf("%d %f %c str = %s", &num, &fnum, &ch, str);
printf("\n num = %d \n fnum = %f \n ch = %c \n str = %s", num, fnum, ch, str);
```

Note that in the `printf` statement we have written `\n`. It is called the newline character and is used to `print` the following text on the new line. When the `printf` function gets executed, the following output will be generated.

```
Enter the values
2 3456.443 a abcde
num = 2
fnum = 3456.44
ch = a
str = abcde
```

1.8 OPERATORS IN C

The C language supports a lot of operators to be used in expressions. These operators can be categorized into the following major groups:

- Arithmetic operators
- Relational operators
- Equality operators
- Logical operators
- Unary operators
- Conditional operators
- Bitwise operators
- Assignment operators
- Comma operator
- Sizeof operator

In the following section, we will discuss all these operators.

1.8.1 Arithmetic Operators

Consider three variables declared as,

```
int a=9, b=3, result;
```

We will use these variables to explain arithmetic operators. Table 1.8 shows the arithmetic operators, their syntax, and usage in C language.

Table 1.8 Arithmetic operators

Operation	Operator	Syntax	Comment	Result
Multiply	*	a * b	result = a * b	27
Divide	/	a / b	result = a / b	3
Addition	+	a + b	result = a + b	12
Subtraction	-	a - b	result = a - b	6
Modulus	%	a % b	result = a % b	0

In Table 1.8, *a* and *b* (on which the operator is applied) are called **operands**. Arithmetic operators can be applied to any integer or floating point number. The addition, subtraction, and multiplication (+, –, and *) operators perform the usual arithmetic operations in C programs, so you are already familiar with these operators.

However, the operator % must be new to you. The modulus operator (%) finds the remainder of an integer division. This operator can be applied only to integer operands and cannot be used on float or double operands. Therefore, the code given below generates a compiler error.

```
#include<stdio.h>
#include<conio.h>
int main()
{
        float c = 20.0;
        printf("\n Result = %f", c % 5);
        return 0;
}
```

While performing modulo division, the sign of the result is always the sign of the first operand (the dividend). Therefore,

```
 16 %  3 =  1
-16 %  3 = -1
```

```
 16 % -3 =  1
-16 % -3 = -1
```

When both operands of the division operator (`/`) are integers, the division is performed as an integer division. Integer division always results in an integer result. So, the result is always rounded-off by ignoring the remainder. Therefore,

```
9/4 = 2  and  -9/4 = -3
```

From the above observation, we can conclude two things. If `op1` and `op2` are integers and the quotient is not an integer, then

- If `op1` and `op2` have the same sign then `op1/op2` is the largest integer less than the true quotient.
- If `op1` and `op2` have opposite signs then `op1/op2` is the smallest integer greater than the true quotient.

Note that it is not possible to divide any number by zero. This is an illegal operation that results in a run-time division-by-zero exception thereby terminating the program.

Except for modulus operator, all other arithmetic operators can accept a mix of integer and floating point numbers. If both operands are integers, the result will be an integer; if one or both operands are floating point numbers then the result would also be a floating point number.

All the arithmetic operators bind from left to right. As in mathematics, the multiplication, division, and modulus operators have higher precedence over the addition and subtraction operators. That is, if an arithmetic expression consists of a mix of operators, then multiplication, division, and modulus will be carried out first in a left to right order, before any addition and subtraction could be performed. For example,

```
  3 + 4 * 7
= 3 + 28
= 1
```

PROGRAMMING EXAMPLES

1. Write a program to perform addition, subtraction, division, integer division, multiplication, and modulo division on two integer numbers.

```
#include<stdio.h>
#include<conio.h>
int main()
{
        int num1, num2;
        int add_res=0, sub_res=0, mul_res=0, idiv_res=0, modiv_res=0;
        float fdiv_res=0.0;
        clrscr();
        printf("\n Enter the first number : ");
        scanf("%d", &num1);
        printf("\n Enter the second number : ");
        scanf("%d", &num2);
        add_res= num1+num2;
        sub_res=num1 - num2;
        mul_res = num1 * num2;
```

```
        idiv_res = num1/num2;
        modiv_res = num1%num2;
        fdiv_res = (float)num1/num2;
        printf("\n %d + %d = %d", num1, num2, add_res);
        printf("\n %d - %d = %d", num1, num2, sub_res);
        printf("\n %d × %d = %d", num1, num2, mul_res);
        printf("\n %d / %d = %d (Integer Division)", num1, num2, idiv_res);
        printf("\n %d %% %d = %d (Moduluo Division)", num1, num2, modiv_res);
        printf("\n %d / %d = %.2f (Normal Division)", num1, num2, fdiv_res);
        return 0;
}
```

2. Write a program to perform addition, subtraction, division, and multiplication on two floating point numbers.

```
#include<stdio.h>
#include<conio.h>
int main()
{
        float num1, num2;
        clrscr();
        printf("\n Enter the first number : ");
        scanf("%f", &num1);
        printf("\n Enter the second number : ");
        scanf("%f", &num2);
        printf("\n %f + %f = %f", num1, num2, num1 + num2);
        printf("\n %f - %f = %f", num1, num2, num1 - num2);
        printf("\n %f X %f = %f", num1, num2, num1 * num2);
        printf("\n %f / %f = %f ", num1, num2, num1 / num2);
        return 0;
}
```

3. Write a program to subtract two long integers.

```
#include<stdio.h>
#include<conio.h>
int main()
{
        long int num1= 1234567, num 2, diff=0;
        clrscr();
        printf("\n Enter the number : ");
        scanf("%ld", &num2);
        diff = num1 - num2;
        printf("\n Difference = %ld", diff);
        return 0;
}
```

1.8.2 Relational Operators

A relational operator, also known as a comparison operator, is an operator that compares two values. Expressions that contain relational operators are called *relational expressions*. Relational operators return true or false value, depending on whether the conditional relationship between the two operands holds or not.

For example, to test the expression, if x is less than y, relational operator < is used as x < y. This expression will return TRUE value if x is less than y; otherwise the value of the expression will be FALSE.

Relational operators can be used to determine the relationships between two operands. These relationships are illustrated in Table 1.9. These operators are evaluated from left to right.

Table 1.9 Relational operators

Operator	Meaning	Example
<	Less than	3 < 5 gives 1
>	Greater than	7 > 9 gives 0
>=	Less than or equal to	100 >= 100 gives 1
<=	Greater than equal to	50 >=100 gives 0

1.8.3 Equality Operators

C language supports two kinds of equality operators to compare their operands for strict equality or inequality. They are equal to (==) and not equal to (!=) operator. The equality operators have lower precedence than the relational operators.

The equal-to operator (==) returns **true** (1) if operands on both the side of the operator have the same value; otherwise, it returns **false** (0). On the contrary, the not-equal-to operator (!=) returns **true (1)** if the operands do not have the same value; else it returns **false (0)**. Table 1.10 summarizes equality operators.

Table 1.10 Equality operators

Operator	Meaning
==	Returns 1 if both operands are equal, 0 otherwise
!=	Returns 1 if operands do not have the same value, 0 otherwise

1.8.4 Logical Operators

C language supports three logical operators. They are Logical AND (&&), Logical OR (||), and Logical NOT (!). As in case of arithmetic expressions, the logical expressions are evaluated from left to right.

Logical AND (&&)

Logical AND operator is used to simultaneously evaluate two conditions or expressions with relational operators. If the expressions on both the sides (left and right side) of the logical operator are true, then the whole expression is true. The truth table of logical AND operator is given in Table 1.11. For example,

```
(a < b) && (b > c)
```

The expression to the left is (a < b) and that on the right is (b > c). The whole expression is true only if both expressions are true i.e., if b is greater than a and c.

Table 1.11 Truth table of logical AND

A	B	A && B
0	0	0
0	1	0
1	0	0
1	1	1

Logical OR (||)

Logical OR operator is used to simultaneously evaluate two conditions or expressions with relational operators. If one or both the expressions on the left side and right side of the logical operator is true, then the whole expression is true. The truth table of logical OR operator is given in Table 1.12. For example,

```
(a < b) || (b > c)
```

The expression to the left is `a < b` and that on the right is `b > c`. The whole expression is true if either `b` is greater than `a` or `b` is greater than `c`.

Table 1.12 Truth table of logical OR

A	B	A || B
0	0	0
0	1	1
1	0	1
1	1	1

Logical NOT (!)

The logical NOT operator takes a single expression and negates the value of the expression. That is, Logical NOT produces a zero if the expression evaluates to a non-zero value and produces a 1 if the expression produces a zero. In other words, it just reverses the value of the expression. The truth table of logical NOT operator is given in Table 1.13. For example,

Table 1.13 Truth table of logical NOT

A	! A
0	1
1	0

```
int a = 10, b;
b = !a;
```

Now the value of `b = 0`. This is because value of `a = 10`. `!a = 0`. The value of `!a` is assigned to `b`, hence, the result.

1.8.5 Unary Operators

Unary operators act on single operands. The C language supports three unary operators. They are unary minus, increment, and decrement operators.

Unary Minus (–)

Unary minus operator is strikingly different from the binary arithmetic operator that operates on two operands and subtracts the second operand from the first operand. When an operand is preceded by a minus sign, the unary operator negates its value. For example, if a number is positive then it becomes negative when preceded with a unary minus operator. Similarly, if the number is negative, it becomes positive after applying the unary minus operator. For example,

```
int a, b = 10;
a = -(b);
```

The result of this expression, is `a = -10`, because a variable *b* has a positive value. After applying unary minus operator (–) on the operand *b*, the value becomes –10, which indicates it as a negative value.

Increment Operator (++) and Decrement Operator (– –)

The increment operator is a unary operator that increases the value of its operand by 1. Similarly, the decrement operator decreases the value of its operand by 1. For example, – –x is equivalent to writing `x = x - 1`.

The increment/decrement operators have two variants: *prefix* or *postfix*. In a prefix expression (++x or --x), the operator is applied before an operand while in a postfix expression (x++ or x--), an operator is applied after an operand.

Therefore, an important point to note about unary increment and decrement operators is that ++x is not the same as x++. Similarly, --x is not the same as x--. Although, ++x and x++ both increment the value of x by 1 but in the former case, the value of x is returned before it is incremented. Whereas in the latter case, the value of x is returned after it is incremented. For example,

```
int x = 10, y;
y = x++;
```

is equivalent to writing

```
y = x;
x = x + 1;
```

whereas, y = ++x; is equivalent to writing

```
x = x + 1;
y = x;
```

The same principle applies to unary decrement operators. Note that unary operators have a higher precedence than the binary operators. And, if in an expression we have more than one unary operator then unlike arithmetic operators, they are evaluated from right to left.

1.8.6 Conditional Operator

The conditional operator also known as the ternary operator (?:) is just like an if...else statement that can be within expressions. The syntax of the conditional operator is:

```
exp1 ? exp2 : exp3
```

exp1 is evaluated first. If it is true, then exp2 is evaluated and becomes the result of the expression, otherwise *exp3* is evaluated and becomes the result of the expression. For example,

```
large = (a > b) ? a : b
```

The conditional operator is used to find largest of two given numbers. First exp1, that is a > b is evaluated. If a is greater than b, then large = a, else large = b. Hence, large is equal to either *a* or *b*, but not both.

Hence, conditional operator is used in certain situations, replacing if-else condition phrases. Conditional operators make the program code more compact, more readable, and safer to use as it is easier both to check and guarantee the arguments that are used for evaluation.

Conditional operator is also known as ternary operator as it is neither a unary nor a binary operator; it takes three operands.

1.8.7 Bitwise Operators

As the name suggests, bitwise operators are those operators that perform operations at the bit level. These operators include: bitwise AND, bitwise OR, bitwise XOR, and shift operators.

Bitwise AND

The bitwise AND operator (&) is a small version of the boolean AND (&&) as it performs operation on bits instead of bytes, chars, integers, etc. When we use the bitwise AND operator, the bit in the first operand is ANDed with the corresponding bit in the second operand. The truth table is the same as we had seen in logical AND operation. That is, the bitwise AND operator compares each bit of its first operand with the corresponding bit of its second operand. If both bits are 1, the corresponding bit in the result is 1 and 0 otherwise. For example,

```
10101010 & 01010101 = 00000000
```

In a C program, the & operator is used as follows:

```
int a = 10, b = 20, c = 0;
c = a&b;
```

Bitwise OR

The bitwise OR operator (|) is a small version of the boolean OR (||) as it performs operation on bits instead of bytes, chars, integers, etc. When we use the bitwise OR operator, the bit in the first operand is ORed with the corresponding bit in the second operand. The truth table is the same as we had seen in logical OR operation. That is, the bitwise-OR operator compares each bit of its first operand with the corresponding bit of its second operand. If one or both bits are 1, the corresponding bit in the result is 1 and 0 otherwise. For example,

```
10101010 & 01010101 = 11111111
```

In a C program, the | operator is used as follows:

```
int a = 10, b = 20, c = 0;
c = a|b
```

Bitwise XOR

The bitwise XOR operator (^) performs operation on individual bits of the operands. When we use the bitwise XOR operator, the bit in the first operand is XORed with the corresponding bit in the second operand. The truth table of bitwise XOR operator can be given as shown in Table 1.14. That is, the bitwise XOR operator compares each bit of its first operand with the corresponding bit of its second operand. If one of the bits is 1, the corresponding bit in the result is 1 and 0 otherwise. For example,

Table 1.14 Truth table of bitwise XOR

A	B	A ^ B
0	0	0
0	1	1
1	0	1
1	1	0

```
10101010 ^ 01010101 = 11111111
```

In a C program, the ^ operator is used as follows:

```
int a = 10, b = 20, c = 0;
c = a^b
```

Bitwise NOT (~)

The bitwise NOT or complement is a unary operation that performs logical negation on each bit of the operand. By performing negation of each bit, it actually produces the ones' complement of the given binary value. Bitwise NOT operator sets the bit to 1 if it was initially 0 and sets it to 0 if it was initially 1. For example,

```
10101011 = 01010100
```

Shift Operator

The language C supports two bitwise shift operators. They are shift left (<<) and shift right (>>). These operations are simple and are responsible for shifting bits to the left or to the right. The syntax for a shift operation can be given as

```
operand op num
```

where the bits in the operand are shifted left or right depending on the operator (left if the operator is << and right if the operator is >>) by number of places denoted by `num`. For example, if we have

```
x = 0001 1101
```

then `x << 1` produces `0011 1010`

When we apply a left shift, every bit in `x` is shifted to the left by one place. So, the MSB (most significant bit) of `x` is lost, the LSB (least significant bit) of `x` is set to 0. Therefore, if we have `x = 0001 1101`, then

```
x << 4 gives result = 11101000
```

If you observe carefully, you will notice that shifting once to the left multiplies the number by 2. Hence, multiple shifts of 1 to the left results in multiplying the number by 2 over and over again.

On the contrary, when we apply a right shift, every bit in `x` is shifted to the right by one place. So, the LSB of `x` is lost, the MSB of `x` is set to 0. For example, if we have `x` = 0001 1101, then

```
x >> 1 gives result = 0000 1110.
```

Similarly, if we have `x = 0001 1101`, then

```
x >> 4 gives result = 0000 0001.
```

If you observe carefully, you will notice that shifting once to the right divides the number by 2. Hence, multiple shifts of 1 to the right results in dividing the number by 2 over and over again.

1.8.8 Assignment Operators

In C language, the assignment operator is responsible for assigning values to the variables. While the equal sign (=) is the fundamental assignment operator, C language also supports other assignment operators that provide shorthand ways to represent common variable assignments.

When an equal sign is encountered in an expression, the compiler processes the statement on the right side of the sign and assigns the result to the variable on the left side. For example,

```
int x;
x = 10;
```

assigns the value 10 to variable `x`. If we have,

```
int x = 2, y = 3, sum = 0;
sum = x + y;
```

then `sum = 5`. The assignment operator has right-to-left associativity, so the expression

```
a = b = c = 10;
```

is evaluated as

```
(a = (b = (c = 10)));
```

First 10 is assigned to `c`, then the value of `c` is assigned to `b`. Finally, the value of `b` is assigned to `a`.

Other Assignment Operators

Table 1.15 contains a list of other assignment operators that are supported by C.

Table 1.15 Assignment operators

Operator	Syntax	Equivalent To	Meaning	Example
/=	variable /= expression	variable = variable / expression	Divides the value of a variable by the value of an expression and assigns the result to the variable.	float a=9.0; float b=3.0; a /= b;
\=	variable \= expression	variable = variable \ expression	Divides the value of a variable by the value of an expression and assigns the integer result to the variable.	int a= 9; int b = 3; a /= b;
*=	variable *= expression	variable = variable * expression	Multiplies the value of a variable by the value of an expression and assigns the result to the variable.	int a= 9; int b = 3; a *= b;
+=	variable += expression	variable = variable + expression	Adds the value of a variable to the value of an expression and assigns the result to the variable.	int a= 9; int b = 3; a += b;
-=	variable -= expression	variable = variable - expression	Subtracts the value of the expression from the value of the variable and assigns the result to the variable.	int a= 9; int b = 3; a -= b;
&=	variable &= expression	variable = variable & expression	Performs the bitwise AND with the value of variable and the value of the expression and assigns the result to the variable.	int a = 10; int b = 20; a &= b;
^=	variable ^= expression	variable = variable ^ expression	Performs the bitwise XOR with the value of variable and the value of the expression and assigns the result to the variable.	int a = 10; int b = 20; a ^= b;
<<=	variable <<= amount	variable = variable << amount	Performs an arithmetic left shift (amount times) on the value of a variable and assigns the result back to the variable.	int a= 9; int b = 3; a <<= b;
>>=	variable >>= amount	variable = variable >> amount	Performs an arithmetic right shift (amount times) on the value of a variable and assigns the result back to the variable.	int a= 9; int b = 3; a >>= b;

1.8.9 Comma Operator

The comma operator in C takes two operands. It works by evaluating the first and discarding its value, and then evaluates the second and returns the value as the result of the expression. Comma-separated operands when chained together are evaluated in left-to-right sequence with the right-most value yielding the result of the expression. Among all the operators, the comma operator has the lowest precedence.

Therefore, when a comma operator is used, the entire expression evaluates to the value of the right expression. For example, the following statement assigns the value of `b` to `x`, then increments `a`, and then increments `b`:

```
int a=2, b=3, x=0;
x = (++a, b+=a);
```

Now, the value of `x = 6`.

1.8.10 sizeof Operator

`sizeof` is a unary operator used to calculate the sizes of data types. This operator can be applied to all data types. When using this operator, the keyword `sizeof` is followed by a type name, variable, or expression. The operator returns the size of the variable, data type, or expression in bytes. That is, the `sizeof` operator is used to determine the amount of memory space that the variable/expression/data type will take.

When a type name is used, it is enclosed in parentheses, but in case of variable names and expressions, they can be specified with or without parentheses. A `sizeof` expression returns an unsigned value that specifies the size of the space in bytes required by the data type, variable, or expression. For example, `sizeof(char)` returns 1, that is the size of a character data type. If we have,

```
int a = 10;
unsigned int result;
result = sizeof(a);
```

then `result = 2`, that is, space required to store the variable `a` in memory. Since `a` is an integer, it requires 2 bytes of storage space.

1.8.11 Operator Precedence Chart

Table 1.16 lists the operators that C language supports in the order of their *precedence* (highest to lowest). The *associativity* indicates the order in which the operators of equal precedence in an expression are evaluated.

Table 1.16 Operators precedence chart

<table>
<tr><th>Operator</th><th>Associativity</th><th>Operator</th><th>Associativity</th></tr>
<tr><td>()
[]
.
>
++ --</td><td>left-to-right</td><td>^</td><td>left-to-right</td></tr>
<tr><td>++ --
+ -
! ~
(type)
*
&
sizeof</td><td>right-to-left</td><td>|</td><td>left-to-right</td></tr>
<tr><td>* / %</td><td>left-to-right</td><td>&&</td><td>left-to-right</td></tr>
<tr><td>+ -</td><td>left-to-right</td><td>||</td><td>left-to-right</td></tr>
<tr><td><< >></td><td>left-to-right</td><td>?:</td><td>right-to-left</td></tr>
<tr><td>< <=
> >=</td><td>left-to-right</td><td>=
+= -=
*= /=
%= &=
^= |=
<<= >>=</td><td>right-to-left</td></tr>
<tr><td>== !=</td><td>left-to-right</td><td>,</td><td>left-to-right</td></tr>
<tr><td>&</td><td>left-to-right</td><td></td><td></td></tr>
</table>

Note: If you do not recognize some of the operators in the chart, do not worry. We will discuss them later.

Examples of Expressions Using the Precedence Chart

If we have the following variable declaration,

```
int a = 0, b = 1, c = -1;
float x = 2.5, y = 0.0;
```

then,

```
(a) a && b = 0

(b) a < b && c < b = 1

(c) b + c || ! a
    = ( b + c) || (!a)
    = 1 ||1
    = 1

(d) x * 5 && 5 || ( b / c)
    = ((x * 5) && 5) || (b / c)
    = (1.25 && 5) || (1/-1)
    = 1

(e) a <= 10 && x >= 1 && b
    = ((a <= 10) && (x >= 1)) && b
    = 1 && 1 && 1
    = 1

(f) !x || !c || b + c
    = ((!x) || (!c)) || (b + c)
    = 0 || 0 || 0
    = 0

(g) x * y < a + b|| c
    = ((x * y) < (a + b)) || c
    = 0 < 1 || -1
    = 1

(h) (x > y) + !a || c++
    = ((x > y) + (!j)) || (c++)
    = 2.5 + 1 || 0
    = 1
```

PROGRAMMING EXAMPLES

4. Write a program to calculate the area of a circle.

```
#include<stdio.h>
#include<conio.h>
int main()
{
        float radius;
        double area;
        clrscr();
        printf("\n Enter the radius of the circle : ");
        scanf("%f", & radius);
        area = 3.14 * radius * radius;
        printf(" Area = %.2lf", area);
        return 0;
}
```

5. Write a program to print the ASCII value of a character.

```
#include<stdio.h>
#include<conio.h>
int main()
{
        char ch;
```

```
    clrscr();
    printf("\n Enter any character : ");
    scanf("%c", &ch);
    printf("\n The ascii value of %c is : %d",ch,ch);
    return 0;
}
```

6. Write a program to read a character in upper case and then print it in lower case.

```
#include<stdio.h>
#include<conio.h>
int main()
{
    char ch;
    clrscr();
    printf("\n Enter any character in upper case : ");
    scanf("%c", &ch);
    printf("\n The character in lower case is : %c", ch+32);
    return 0;
}
```

7. Write a program to swap two numbers using a temporary variable.

```
#include<stdio.h>
#include<conio.h>
int main()
{
    int num1, num2, temp;
    clrscr();
    printf("\n Enter the first number : ");
    scanf("%d",&num1;
    printf("\n Enter the second number : ");
    scanf("%d",&num2;
    temp = num1;
    num1=num2;
    num2=temp;
    printf("\n The first number is %d", num1);
    printf("\n The second number is %d", num2);
    return 0;
}
```

8. Write a program to swap two numbers without using a temporary variable.

```
#include<stdio.h>
#include<conio.h>
int main()
{
    int num1, num2;
    clrscr();
    printf("\n Enter the first number : ");
    scanf("%d",&num1;
```

```
        printf("\n Enter the second number : ");
        scanf("%d",&num2;
        num1 = num1 + num2;
        num2= num1 – num2;
        num1 = num1 – num2;
        printf("\n The first number is %d", num1);
        printf("\n The second number is %d", num2);
        return 0;
    }
```

9. Write a program that displays the size of every data type.

```
#include<stdio.h>
#include<conio.h>
int main()
{
    clrscr();
    printf("\n The size of short integer is : %d", sizeof(short int));
    printf("\n The size of unsigned integer is : %d", sizeof(unsigned int));
    printf("\n The size of signed integer is : %d", sizeof(signed int));
    printf("\n The size of integer is : %d", sizeof(int));
    printf("\n The size of long integer is : %d", sizeof(long int));
    printf("\n The size of character is : %d", sizeof(char));
    printf("\n The size of unsigned character is : %d", sizeof(unsigned char));
    printf("\n The size of signed character is : %d", sizeof(signed char));
    printf("\n The size of floating point number is : %d", sizeof(float));
    printf("\n The size of double number is : %d", sizeof(double));
    return 0;
}
```

1.9 TYPE CONVERSION AND TYPECASTING

Type conversion or typecasting of variables refers to changing a variable of one data type into another. While type conversion is done implicitly, casting has to be done explicitly by the programmer. We will discuss both of them here.

1.9.1 Type Conversion

Type conversion is done when the expression has variables of different data types. So to evaluate the expression, the data type is promoted from lower to higher level where the hierarchy of data types can be given as: `double`, `float`, `long`, `int`, `short`, and `char`. For example, type conversion is automatically done when we assign an integer value to a floating point variable. Consider the following code.

```
float x;
int y = 3;
x = y;
```

Now, `x = 3.0`, as automatically integer value is converted into its equivalent floating point representation.

1.9.2 Typecasting

Typecasting is also known as *forced conversion*. It is done when the value of a higher data type has to be converted into the value of a lower data type. For example, we need to explicitly typecast an integer variable into a floating point variable. The code to perform typecasting can be given as,

```
float salary = 10000.00;
int sal;
sal = (int) salary;
```

When floating point numbers are converted to integers, the digits after the decimal are truncated. Therefore, data is lost when floating point representations are converted to integral representations.

As we see in the code, typecasting can be done by placing the destination data type in parentheses followed by the variable name that has to be converted. Hence, we conclude that typecasting is done to make a variable of one data type to act like a variable of another type.

We can also typecast integer values to its character equivalent (as per ASCII code) and vice versa. Typecasting is also done in arithmetic operation to get correct result. For example, when dividing two integers, the result can be of floating type. Also when multiplying two integers the result can be of long int. So to get correct precision value, typecasting can be done. For instance,

```
int a = 500, b = 70 ;
float res;
res = (float) a/b;
```

PROGRAMMING EXAMPLES

10. Write a program to convert a floating point number into the corresponding integer.

```
#include<stdio.h>
#include<conio.h>
int main()
{
        float f_num;
        int i_num;
        clrscr();
        printf("\n Enter any floating point number: ");
        scanf("%f", &f_num);
        i_num = (int)f_num;
        printf("\n The integer variant of %f is = %d", f_num, i_num);
        return 0;
}
```

11. Write a program to convert an integer into the corresponding floating point number.

```
#include<stdio.h>
#include<conio.h>
int main()
{
        float f_num;
        int i_num;
        clrscr();
```

```
    printf("\n Enter any integer: ");
    scanf("%d", &i_num);
    f_num = (float)i_num;
    printf("\n The floating point variant of %d is = %f", i_num, f_num);
    return 0;
}
```

1.10 DECISION CONTROL STATEMENTS

Till now we know that the code in the C program is executed sequentially from the first line of the program to its last line. That is, the second statement is executed after the first, the third statement is executed after the second, so on and so forth.

Although this is true, but in some cases we want only selected statements to be executed. Such type of conditional processing extends the usefulness of programs. It allows the programmers to build logic that determine which statements of the code should be executed and which should be ignored.

The language C supports decision control statements that can alter the flow of a sequence of instructions. These statements help to jump from one part of the program to another depending on whether a particular condition is satisfied or not. These decision control statements include:

(a) `if` statement, (b) `if-else` statement, (c) `if-else-if` statement, and (d) `switch-case` statement.

1.10.1 If Statement

`if` statement is the simplest form of decision control statements that is frequently used in decision making. The general form of a simple `if` statement is shown in Fig. 1.2.

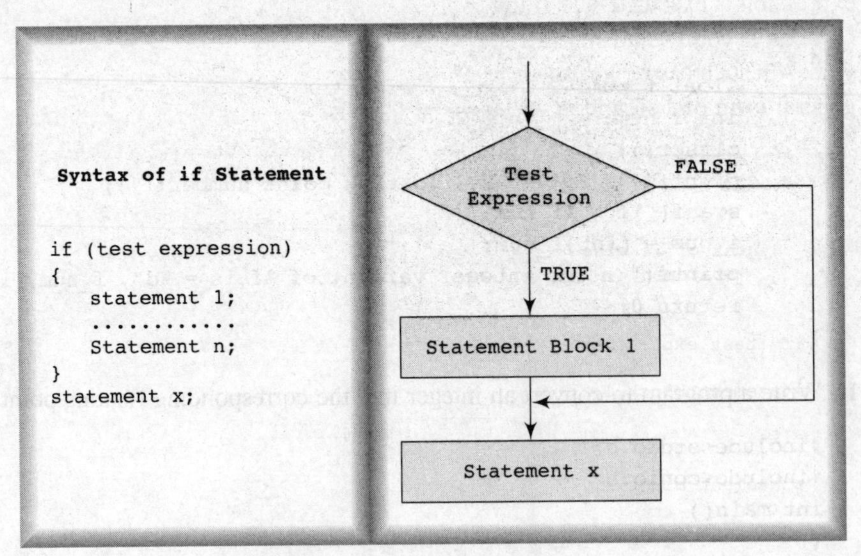

Figure 1.2 `if` statement construct

The `if` structure may include 1 statement or *n* statements enclosed within curly brackets. First the test expression is evaluated. If the test expression is true, the statements of the `if` block (statement 1 to *n*) are executed, otherwise these statements will be skipped and the execution will jump to `statement x`.

The statement in an `if` construct is any valid C language statement and the test expression is any valid C language expression that may include logical operators. Note that there is no semi-colon after the test expression. This is because the condition and statement should be put together as a single statement.

```
#include<stdio.h>
int main()
{
    int x=10;
    if (x>0) x++;
    printf("\n x = %d", x);
    return 0;
}
```

In the above code, we take a variable `x` and initialize it to 10. In the test expression, we check if the value of `x` is greater than 0. If the test expression evaluates to true, then the value of `x` is incremented. After that, the value of `x` is printed on the screen. The output of this program is

```
x = 1
```

Observe that the `printf` statement will be executed even if the test expression is false.

In case the statement block contains only one statement, putting curly brackets becomes optional. If there are more than one statement in the statement block, putting curly brackets becomes mandatory.

1.10.2 If-Else Statement

We have studied that using `if` statement plays a vital role in conditional branching. Its usage is very simple. The test expression is evaluated, if the result is true, the statement(s) followed by the expression is executed, else if the expression is false, the statement is skipped by the compiler.

What if you want a separate set of statements to be executed if the expression returns a zero value? In such cases, we use an `if-else` statement rather than using a simple `if` statement. The general form of a simple `if-else` statement is shown in Fig. 1.3.

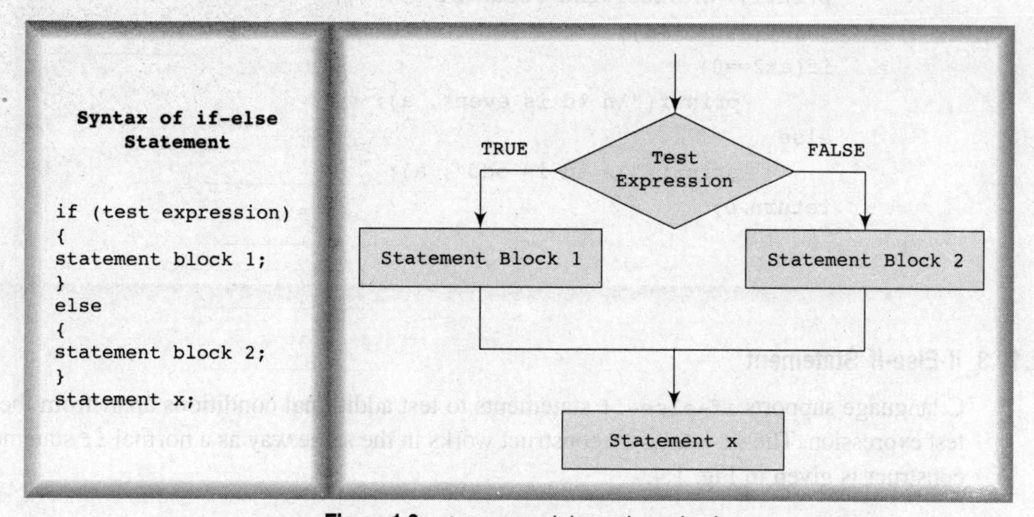

Figure 1.3 `if-else` statement construct

In the above syntax, we have written statement block. A statement block may include one or more statements. According to the `if-else` construct, first the test expression is evaluated. If the expression is true, statement block 1 is executed and statement block 2 is skipped. Otherwise, if the expression is false, statement block 2 is executed and statement block 1 is ignored. Now in any case after the statement block 1 or 2 gets executed, the control will pass to statement `x`. Therefore, statement `x` is executed in every case.

PROGRAMMING EXAMPLES

12. Write a program to find the larger of two numbers.

```
#include<stdio.h>
main()
{
        int a, b, large;
        printf("\n Enter the value of a and b : ");
        scanf("%d %d", &a, &b);
        if(a>b)
                large = a;
        else
                large = b;
        printf("\n LARGE = %d", large);
        return 0;
}
```

13. Write a program to find whether a number is even or odd.

```
#include<stdio.h>
main()
{
        int a;
        printf("\n Enter the value of a : ");
        scanf("%d", &a);
        if(a%2==0)
                printf("\n %d is even", a);
        else
                printf("\n %d is odd", a);
        return 0;
}
```

1.10.3 If-Else-If Statement

C language supports `if-else-if` statements to test additional conditions apart from the initial test expression. The `if-else-if` construct works in the same way as a normal `if` statement. Its construct is given in Fig. 1.4.

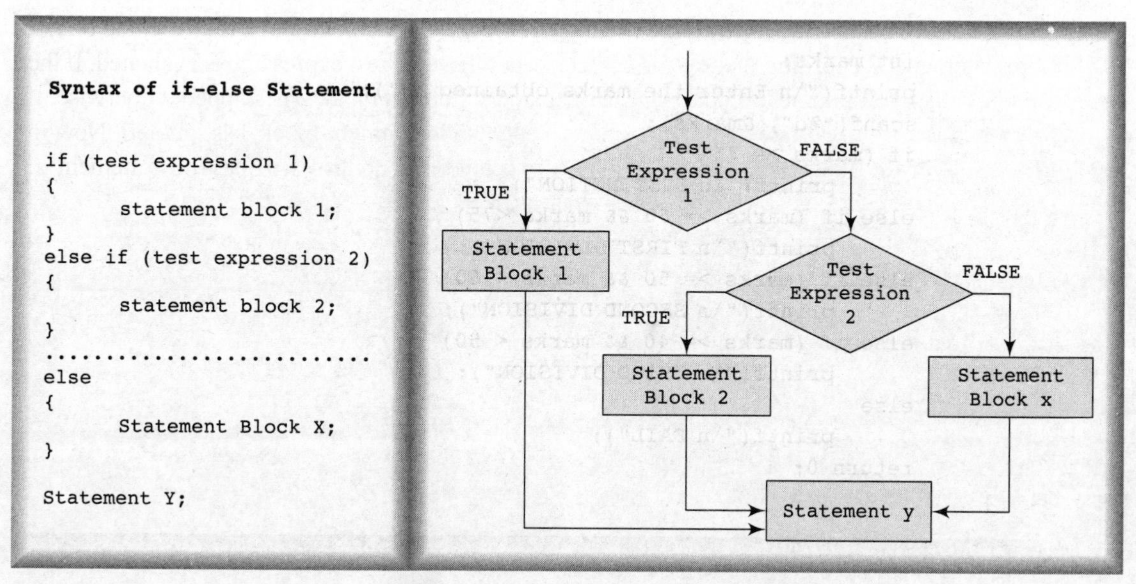

Figure 1.4 `if-else-if` statement construct

Note that it is not necessary that every `if` statement should have an `else` block as C supports simple `if` statements. After the first test expression or the first `if` branch, the programmer can have as many `else-if` branches as he wants depending on the expressions that have to be tested. For example, the following code tests whether a number entered by the user is negative, positive, or equal to zero.

```
#include<stdio.h>
main()
{
        int num;
        printf("\n Enter any number : ");
        scanf("%d", &num);
        if(num==0)
                printf("\n The value is equal to zero");
        else if(num>0)
                printf("\n The number is positive");
        else
                printf("\n The number is negative");
        return 0;
}
```

Note that if the first test expression evaluates a true value, then the rest of the statements in the code will be ignored and after executing the printf statement that displays. 'The value is equal to zero', the control will jump to return 0 statement. Consider the code given below which shows usage of the `if-else-if` statement.

PROGRAMMING EXAMPLE

14. Write a program to display the result.

```
#include<stdio.h>
main()
```

```
{
        int marks;
        printf("\n Enter the marks obtained : ");
        scanf("%d", &marks);
        if (marks >= 75)
                printf("\n DISTINCTION");
        else if (marks >= 60 && marks <75)
                printf("\n FIRST DIVISION");
        else if (marks >= 50 && marks < 60)
                printf("\n SECOND DIVISION");
        else if (marks >= 40 && marks < 50)
                printf("\n THIRD DIVISION");
        else
                printf("\n FAIL");
        return 0;
}
```

1.10.4 Switch-Case Statement

A `switch-case` statement is a multi-way decision statement that is a simplified version of an `if-else` block that evaluates only one variable. The general form of a `switch` statement is shown in Fig. 1.5.

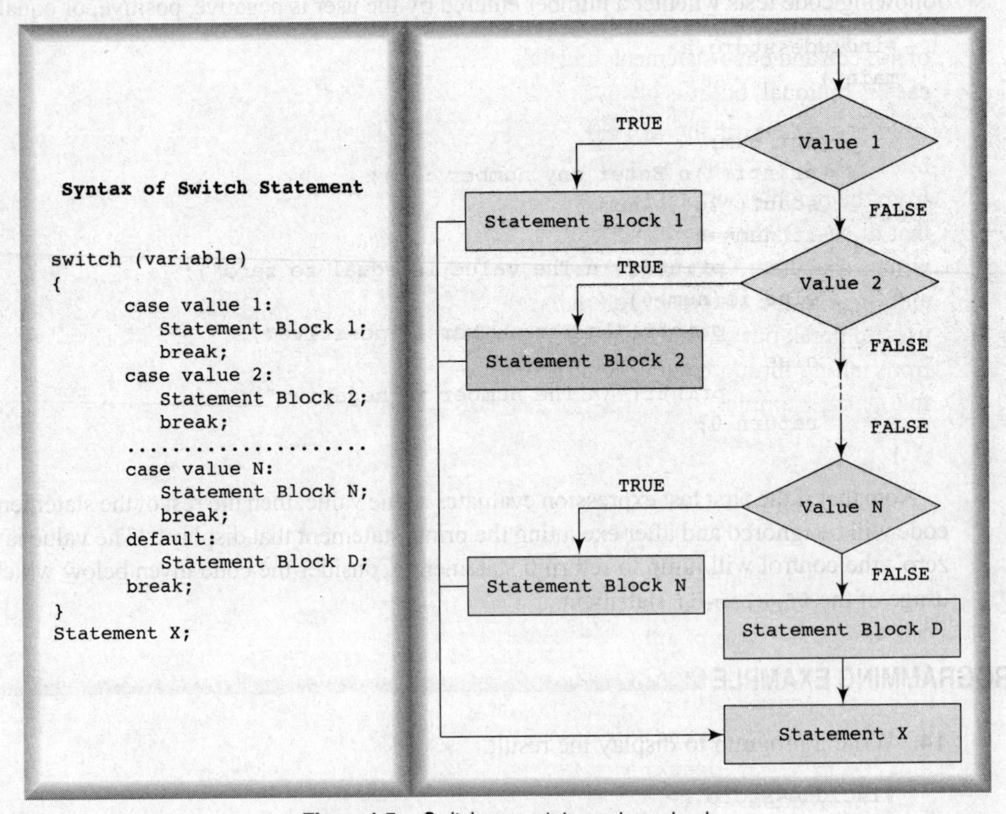

Figure 1.5 Switch-case statement construct

Table 1.17 Switch and if-else construct

switch statement	if-else statement
`switch(x) {`	`if(expl) {`
`case 1: // do this`	`// do this`
`case 2: // do this`	`} else if(exp2) {`
`case 3: // do this`	`// do this`
`....`	`} else if(exp3) {`
`default:`	`// do this`
`//do this`	`}`
`}`	

Table 1.17 compares the general form of a switch statement with that of an `if-else` statement.

The power of nested `if-else` statement's lies in the fact that it can evaluate more than one expression in a single logical structure. `switch` statements are mostly used in two situations:

- When there is only one variable to evaluate in the expression
- When many conditions are being tested for

When there are many conditions to test, using the `if` and `else-if` constructs becomes complicated and confusing. Therefore, `switch case` statements are often used as an alternative to long `if` statements that compare a variable to several 'integral' values (integral values are those values that can be expressed as an integer, such as the value of a `char`). `Switch` statements are also used to handle the input given by the user.

We have already seen the syntax of the `switch` statement. The `switch case` statement compares the value of the variable given in the `switch` statement with the value of each case statement that follows. When the value of the switch and the case statement matches, the statement block of that particular case is executed.

Did you notice the keyword `default` in the syntax of the `switch case` statement? Default is also a case that is executed when the value of the variable does not match with any of the values of the `case` statement. That is, default case is executed when there is no match found between the values of `switch` and `case` statements and thus there are no statements to be executed. Although, the default case is optional, but it is always recommended to include it as it handles any unexpected cases.

In the syntax of the `switch-case` statement, we have used another keyword `break`. The `break` statement must be used at the end of each case because if it is not used, then all the cases from the one met will be executed. For example, if the value of `switch` statement matched with that of case 2, then all the statements in case 2 as well as the rest of the cases including default will be executed. The `break` statement tells the compiler to jump out of the switch case statement and execute the statement following the `switch-case` construct. Thus, the keyword `break` is used to break out of the case statements. It indicates the end of a case and prevents the program from falling through and executing the code in all the rest of the case statements. Consider the following example.

```
switch(grade)
{
        case 'O':
                printf("\n Outstanding");
                break;
        case 'A':
                printf("\n Excellent");
                break;
        case 'B':
                printf("\n Good");
                break;
        case 'C':
```

```
            printf("\n Fair");
            break;
        case 'F':
            printf("\n Fail");
            break;
        default:
            printf("\n Invalid Grade");
            break;
    }
```

PROGRAMMING EXAMPLE

15. Write a program to determine whether an entered character is a vowel or not.

```
#include<stdio.h>
int main()
{
    char ch;
    printf("\n Enter any character : ");
    scanf("%c", &ch);
    switch(ch)
    {
        case 'A':
        case 'a':
            printf("\n % c is VOWEL", ch);
            break;
        case 'E':
        case 'e':
            printf("\n % c is VOWEL", ch);
            printf("\n VOWEL");
            break;
        case 'I':
        case 'i':
            printf("\n % c is VOWEL", ch);
            break;
        case 'O':
        case 'o':
            printf("\n % c is VOWEL", ch);
            break;
        case 'U':
        case 'u':
            printf("\n % c is VOWEL", ch);
            break;
        default: printf("%c is not a vowel", ch);
    }
    return 0;
}
```

Note that there is no break statement after case A, so if the character A is entered then control will execute the statements given in case a. For example, consider a simple calculator program that can be used to add, multiply, subtract, and divide two integers.

Advantages of using a Switch Case Statement

`switch case` statement is preferred by programmers due to the following reasons:

- Easy to debug
- Easy to read and understand
- Ease of maintenance as compared with its equivalent `if-else` statements
- Like `if-else` statements, `switch` statements can also be nested
- Execute faster than its equivalent `if-else` construct

PROGRAMMING EXAMPLES

16. Write a program to find whether a given year is a leap year or not.

```
#include<stdio.h>
#include<conio.h>
int main()
{
        int year;
        clrscr();
        printf("\n Enter any year : ");
        scanf("%d",&year);
        if((year%4 = = 0)&&((year%100!=0)||(year% 400==0)))
                printf("\n Leap Year");
        else
                printf("\n Not A Leap Year");
        return 0;
}
```

17. Write a program to find the greatest among three numbers.

```
#include<stdio.h>
#include<conio.h>
int main()
{
        int num1, num2, num3, big=0;
        clrscr();
        printf("\n Enter the first number : ");
        scanf("%d", &num1);
        printf("\n Enter the second number : ");
        scanf("%d", &num2);
        printf("\n Enter the third number : ");
        scanf("%d", &num3);
        if(num1>num2)
        {
```

```
        if(num1>num3)
            printf("\n %d is greater than %d and %d", num1, num2, num3);
        else
            printf("\n %d is greater than %d and %d", num3, num1, num2);
    }
    else if(num2>num3)
        printf("\n %d is greater than %d and %d", num2, num1, num3);
    else
        printf("\n %d is greater than %d and %d", num3, num1, num2);
    return 0;
}
```

18. Write a program to input three numbers and then find the largest of them using && operator.

```
#include<stdio.h>
#include<conio.h>
int main()
{
    int num1, num2, num3;
    clrscr();
    printf("\n Enter the first number : ");
    scanf("%d", &num1);
    printf("\n Enter the second number : ");
    scanf("%d", &num2);
    printf("\n Enter the third number : ");
    scanf("%d", &num3);
    if(num1>num2 && num1>num3)
        printf("\n %d is the largest number", num1);
    if(num2>num1 && num2>num3)
        printf("\n %d is the largest number", num2);
    else
        printf("\n %d is the largest number", num3);
    getch();
    return 0;
}
```

19. Write a program to find the largest of two numbers using ternary operator.

```
#include<stdio.h>
#include<conio.h>
int main()
{
    int num1, num2, large;
    clrscr();
    printf("\n Enter the first number : ");
    scanf("%d", &num1);
    printf("\n Enter the second number : ");
    scanf("%d", &num2);
    large = num1>num2?num1:num2;
```

```
        printf("\n The largest number is : %d", large);
        return 0;
}
```

20. Write a program to find the largest of three numbers using ternary operator.

```
#include<stdio.h>
#include<conio.h>
int main()
{
        int num1, num2, num3, large;
        clrscr();
        printf("\n Enter the first number : ");
        scanf("%d", &num1);
        printf("\n Enter the second number : ");
        scanf("%d", &num2);
        printf("\n Enter the third number : ");
        scanf("%d", &num3);
        large = num1>num2?(num1>num3?num1:num3):(num2>num3?num2:num3);
        printf("\n The largest number is : %d", large);
        retun 0;
}
```

21. Write a program to take input from the user and then check whether it is a number or a character. If it is a character, determine whether it is in upper case or lower case.

```
#include<stdio.h>
#include<conio.h>
int main()
{
        char ch;
        clrscr();
        printf("\n Enter any character : ");
        scanf("%c", &ch);
        if(ch >='A' && ch<='Z')
                printf("\n Upper case character was entered");
        if(ch >='a' && ch<='z')
                printf("\n Lower case character was entered");
        else if(ch>='0' && ch<='9')
                printf("\n You entered a number");
        return 0;
}
```

22. Write a program to enter any character. If the entered character is in lower case, then convert it into upper case and if it is a lower case character, then convert it into upper case.

```
#include<stdio.h>
#include<conio.h>
int main()
{
```

```
        char ch;
        clrscr();
        printf("\n Enter any character : ");
        scanf("%c", &ch);
        if(ch >='A' && ch<='Z')
            printf("\n The entered character was in upper case. In lower
case it is : %c", (ch+32));
        else
            printf("\n The entered character was in lower case. In upper
case it is : %c", (ch-32));
        return 0;
}
```

23. Write a program to enter the marks of a student in four subjects. Then calculate the total, aggregate and display the grade obtained by the student.

```
#include<stdio.h>
#include<conio.h>
{
        int marks1, marks2, marks3, marks4, total = 0;
        float avg =0.0;
        clrscr();
        printf("\n Enter the marks in Mathematics : ");
        scanf("%d", &marks1);
        printf("\n Enter the marks in Science : ");
        scanf("%d", &marks2);
        printf("\n Enter the marks in Social Science : ");
        scanf("%d", &marks3);
        printf(\n Enter the marks in Computers : ");
        scanf("%d", &marks4);
        total = marks1 + marks2 + marks3 + marks4;
        avg = (float) total/4;
        printf("\n Total = %d", total);
        printf("\n Aggregate = %f", avg);
        if(avg >= 75)
            printf("\n DISTINCTION");
        if(avg>=60 && avg<75)
            printf("\n FIRST DIVISION");
        if(avg>=50 && avg<60)
            printf("\n SECOND DIVISION");
        if(avg>=40 && avg<50)
            printf("\n THIRD DIVISION");
        else
            printf("\n FAIL");
        return 0;
}
```

24. Write a program to enter a number from 1–7 and display the corresponding day of the week using switch case statement.

```
#include<stdio.h>
#include<conio.h>
int main()
{
    int day;
    clrscr();
    printf("\n Enter any number from 1 to 7 : ");
    scanf("%d",&day);
    switch(day)
    {
        case 1 : printf("\n SUNDAY");
            break;
        case 2 : printf("\n MONDAY");
            break;
        case 3 : printf("\n TUESDAY");
            break;
        case 4 : printf("\n WEDNESDAY");
            break;
        case 5 : printf("\n THURSDAY");
            break;
        case 6 : printf("\n FRIDAY");
            break;
        case 7 : printf("\n SATURDAY");
            break;
        default: printf("\n Wrong Number");
    }
    return 0;
}
```

1.11 ITERATIVE STATEMENTS

Language C supports three types of iterative statements also known as looping statements. They are:

- While loop
- Do-while loop
- For loop

Iterative statements are used to repeat the execution of a sequence of statements, depending on the value of an integer expression. In this section, we will discuss all these statements.

1.11.1 While Loop

The `while` loop provides a mechanism to repeat one or more statements while a particular condition is true. Figure 1.6 shows the syntax and general form of representation of a while loop:

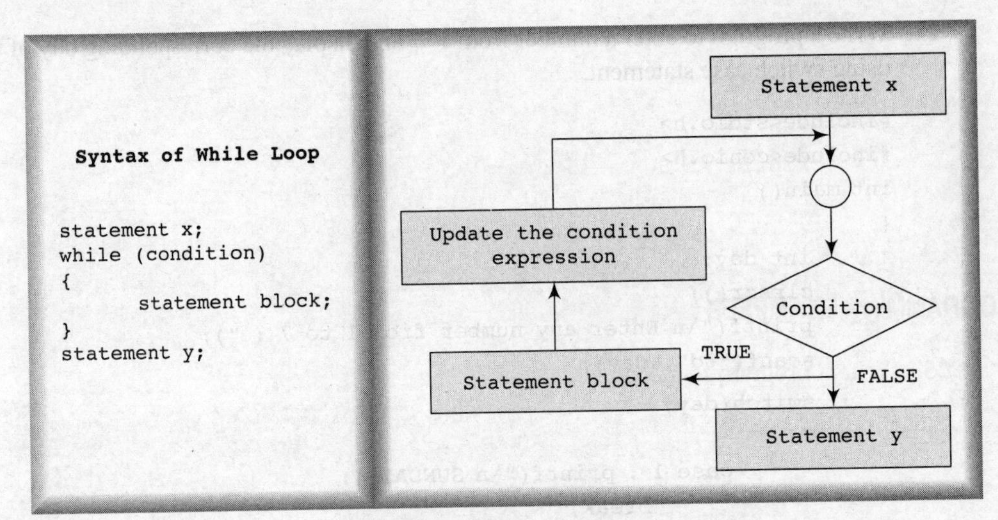

Figure 1.6 While loop construct

Note that in the `while` loop, the condition is tested before any of the statements in the statement block is executed. If the condition is true, only then the statements will be executed, otherwise if the condition is false, the control will jump to `statement y`, that is the immediate statement outside the `while` loop block.

In the flow diagram of Figure 1.6, it is clear that we need to constantly update the condition of the `while` loop. It is this condition which determines when the loop will end. The `while` loop will execute as long as the condition is true. Note that if the condition is never updated and the condition never becomes false, then the computer will run into an infinite loop which is never desirable. For example, the following code prints the first 10 numbers using a `while` loop.

```
#include<stdio.h>
int main()
{
        int i = 0;
        while(i<=10)
        {
                printf("\n %d", i);
                i = i + 1;      // condition updated
        }
        getch();
        return 0;
}
```

Note that initially `i = 0` and is less than 10, i.e., the condition is true, so in the `while` loop the value of `i` is printed and the condition is updated so that with every execution of the loop, the condition becomes more approachable. The same code can be modified to calculate the sum of first 10 numbers, as follows:

```
#include<stdio.h>
int main()
{
        int i = 0, sum = 0;
        while(i<=10)
```

```
        {
            sum = sum + i;
            i = i + 1;        // condition updated
        }
        printf("\n SUM = %d", sum);
        return 0;
}
```

PROGRAMMING EXAMPLES

25. Write a program to calculate the sum of numbers from *m* to *n*.

```
#include<stdio.h>
int main()
{
        int n, m, sum =0;
        clrscr();
        printf("\n Enter the value of m : ");
        scanf("%d", &m);
        printf("\n Enter the value of n : ");
        scanf("%d", &n);
        while(m<=n)
        {
                sum = sum + m;
                m = m + 1;
        }
        printf("\n The sum of numbers from %d to %d = %d", m, n, sum);
        return 0;
}
```

26. Write a program to display the largest of 10 numbers using ternary operator.

```
#include<stdio.h>
#include<conio.h>
int main()
{
        int i=0, large = -1, num;
        clrscr();
        while(i<=10)
        {
                printf("\n Enter the number : ");
                scanf("%d",&num);
                large = num>large?num:large;
                i++;
        }
        printf("\n The largest of ten numbers entered is : %d", large);
        return 0;
}
```

27. Write a program to read the numbers until –1 is encountered. Also count the negative, positive, and zeroes entered by the user.

```
#include<stdio.h>
#include<conio.h>
int main()
{
        int num;
        int negatives=0, positives=0, zeroes=0;
        clrscr();
        printf("\n Enter -1 to exit….");
        printf("\n\n Enter any number : ");
        scanf("%d",&num);
        while(num != -1)
        {
                if(num>0)
                        positives++;
                else if(num<0)
                        negatives++;
                else
                        zeroes++;
                printf("\n\n Enter any number : ");
                scanf("%d",&num);
        }
        printf("\n Count of positive numbers entered = %d", positives);
        printf("\n Count of negative numbers entered = %d", negatives);
        printf("\n Count of zeroes entered = %d", zeroes);
        getch();
        return 0;
}
```

Now look at the code given below which hangs the computer in an infinite loop. The code given below is supposed to calculate average of first 10 numbers, but since the condition never becomes false, the output will not be generated and the intended task will not be performed.

```
#include<stdio.h>
int main()
{
        int i = 0, sum =0;
        float avg = 0.0;
        while(i<=10)
        {
                sum = sum + i;
        }
        avg = (float)sum/10;
        printf("\n The sum of first 10 numbers = %d", sum);
```

```
        printf("\n The average of first 10 numbers = %f", avg);
        return 0;
    }
```

1.11.2 Do-while Loop

The `do-while` loop is similar to the `while` loop. The only difference is that in a `do-while` loop, the test condition is tested at the end of the loop. Now that the test condition is tested at the end, this clearly means that the body of the loop gets executed at least one time (even if the condition is false). Figure 1.7 shows the syntax and the general form of representation of a `do-while` loop.

Note that the test condition is enclosed in parentheses and followed by a semi-colon. The statements in the statement block are encapsulated within a curly bracket. The curly bracket is optional if there is only one statement in the body of the `do-while` loop.

Like the `while` loop, the `do-while` loop continues to execute while a condition is true. There is no choice whether to execute the loop or not because the loop will be executed at least once irrespective of whether the condition is true or false. Hence, entry in the loop is automatic, there is only a choice to continue it further or not. The `do-while` loop will continue to execute while the condition is true and when the condition becomes false, the control will jump to statement following the `do-while` loop.

The major disadvantage of using a `do-while` loop is that it always executes at least once, so even if the user enters some invalid data, the loop will execute. However, `do-while` loops are widely used to print a list of options for a menu-driven program. For example, consider the following code.

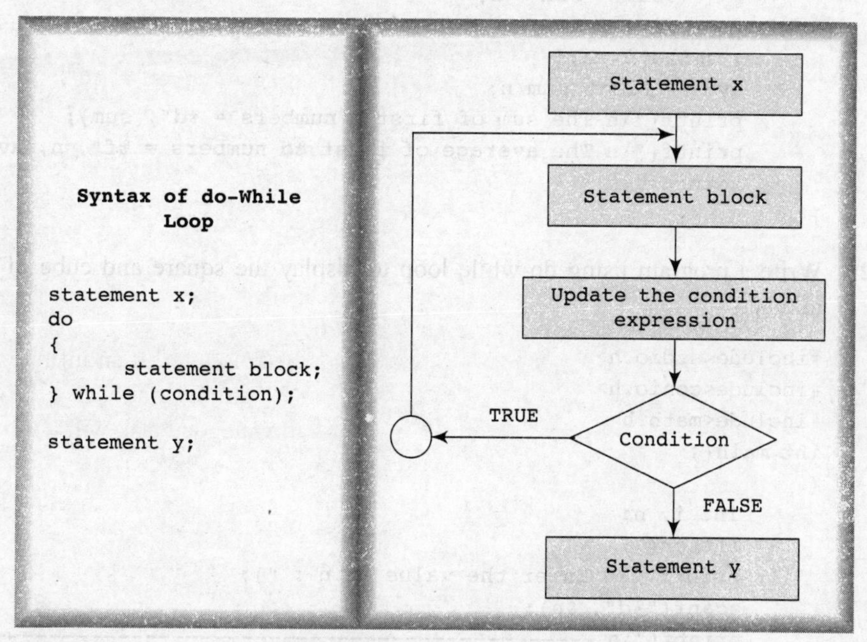

Figure 1.7 `Do-while` construct

```
#include<stdio.h>
int main()
{
```

```
    int i = 0;
    do
    {
        printf("\n %d", i);
        i = i + 1;
    } while(i<=10);
    return 0;
}
```

What do you think will be the output? Yes, the code will print numbers from 0 –11 and not till 10.

PROGRAMMING EXAMPLES

28. Write a program to calculate the average of first *n* numbers.

```
#include<stdio.h>
int main()
{
    int n, i = 0, sum =0;
    float avg = 0.0;
    printf("\n Enter the value of n : ");
    scanf("%d", &n);
    do
    {
        sum = sum + i;
        i = i + 1;
    } while(i<=n);
    avg = (float)sum/n;
    printf("\n The sum of first n numbers = %d", sum);
    printf("\n The average of first %d numbers = %f", n, avg);
    return 0;
}
```

29. Write a program using do-while loop to display the square and cube of first *n* natural numbers.

```
#include<stdio.h>
#include<conio.h>
#include<math.h>
int main()
{
    int i, n;
    clrscr();
    printf("\n Enter the value of n : ");
    scanf("%d", &n);
    printf("\n ----------------------------------------------");
    i=1;
    do
    {
        printf("\n | \t %d \t | \t %d \t | \t %ld \t |", i, pow(i,2),
pow(i,3));
```

```
        i++;
    } while(i<=n);
    printf("\n -------------------------------------------------------");
    return 0;
}
```

30. Write a program to list all the leap years from 1900 to 2100.

```
#include<stdio.h>
#include<conio.h>
int main()
{
    int m=1900, n=2100;
    clrscr();
    do
    {
        if(i%4 == 0)
            printf("\n %d is a leap year",m);
        else
            printf("\n %d is not a leap year", m);
        m = m+1;
    }while(m<=n);
    return 0;
}
```

31. Write a program to read a character until a * is encountered. Also count the number of upper case, lower case, and numbers entered by the users.

```
#include<stdio.h>
#include<conio.h>
int main()
{
    char ch;
    int lowers = 0, uppers = 0, numbers = 0;
    clrscr();
    printf("\n Enter any character : ");
    scanf("%c, &ch);
    do
    {
        if(ch >='A' && ch<='Z')
            uppers++;
        if(ch >='a' && ch<='z')
            lowers++;
        if(ch >='0' && ch<='9')
            numbers++;
        printf("\n Enter another character. Enter * to exit.");
        scanf("%c", &ch);
    } while(ch != '*');
    printf("\n Total count of lower case characters entered = %d", lowers);
    printf("\n Total count of upper case characters entered = %d", uppers);
    printf("\n Total count of numbers entered = %d", numbers);
```

```
        return 0;
}
```

32. Write a program to read the numbers until –1 is encountered. Also calculate the sum and mean of all positive numbers entered and the sum and mean of all negative numbers entered separately.

```
#include<stdio.h>
#include<conio.h>
int main()
{
        int num;
        int sum_negatives=0, sum_positives=0;
        int positives = 0, negatives = 0;
        float mean_positives = 0.0, mean_negatives = 0.0;
        clrscr();
        printf("\n Enter -1 to exit….");
        printf("\n\n Enter any number : ");
        scanf("%d",&num);
        do
        {
                if(num>0)
                {
                        sum_positives += num;
                        positives++;
                }
                else if(num<0)
                {
                        sum_negatives += num;
                        negatives++;
                }
                printf("\n\n Enter any number : ");
                scanf("%d",&num);
        } while(num != -1);
        mean_positives = (float)sum_positives/positives;
        mean_negatives = (float)mean_neagtives/negatives;
        printf("\n Sum of all positive numbers entered = %d", sum_positives);
        printf("\n Mean of all positive numbers entered = %f", mean_positives);
        printf("\n Sum of all negative numbers entered = %d", sum_negatives);
        printf("\n Mean of all negative numbers entered = %f", mean_negatives);
        return 0;
}
```

1.11.3 For Loop

Like the `while` and `do-while` loop, the `for` loop provides a mechanism to repeat a task until a particular condition is true. The syntax and general form of a `for` loop is given in Fig. 1.8.

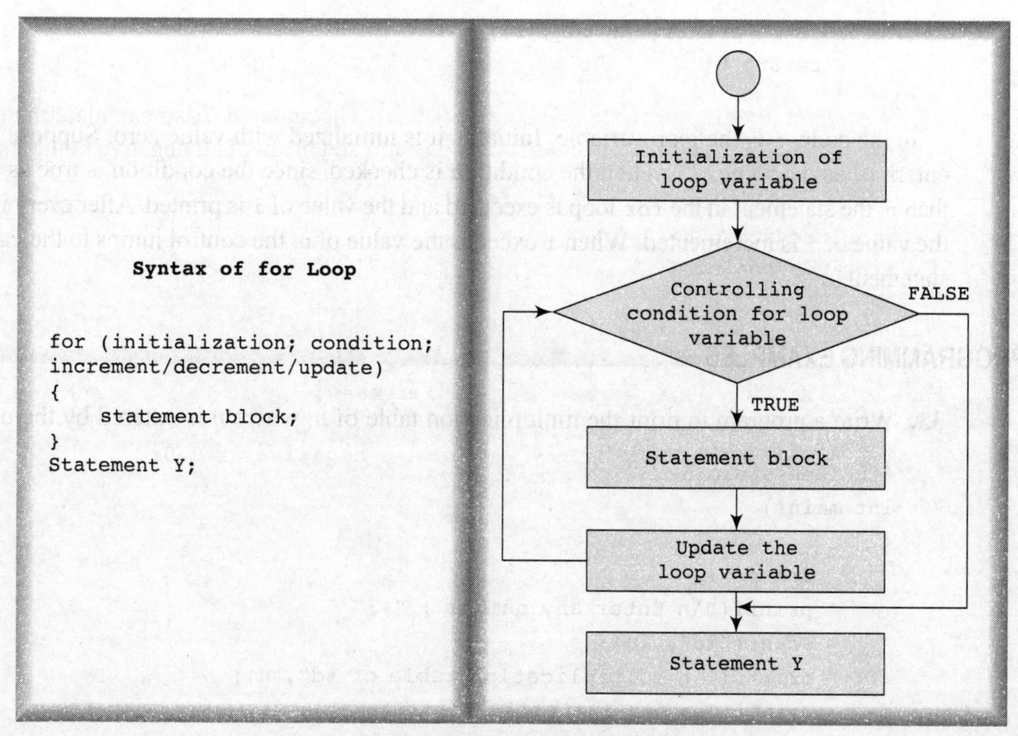

Figure 1.8 `for` loop construct

When a `for` loop is used, the loop variable is initialized only once. With every iteration of the loop, the value of the loop variable is updated and the condition is checked. If the condition is true, the statement block of the loop is executed else, the statements comprising the statement block of the `for` loop are skipped and the control jumps to the immediate statement following the `for` loop body.

In the syntax of the `for` loop, initialization of the loop variable allows the programmer to give it a value. Second, the condition specifies that while the conditional expression is true, the loop should continue to repeat itself. Every iteration of the loop must make the condition approachable. So, with every iteration, the loop variable must be updated. Updating the loop variable may include incrementing the loop variable, decrementing the loop variable or setting it to some other value like, `i +=2`, where `i` is the loop variable.

Note that every section of the `for loop` is separated from the other with a semi-colon. It is possible that one of the sections may be empty, though the semi-colons still have to be there. However, if the condition is empty, it is evaluated as true and the loop will repeat until something else stops it.

The `for` loop is widely used to execute a single or a group of statements for a limited number of times. The following code shows how to print the first *n* numbers using a `for` loop.

```
#include<stdio.h>
int main()
{
        int i, n;
        printf("\n Enter the value of n :");
        scanf("%d", &n);
        for(i=0;i<=n;i++))
```

```
            printf("\n %d", i);
        return 0;
    }
```

In the code, `i` is the loop variable. Initially, it is initialized with value zero. Suppose the user enters 10 as the value of `n`. Then the condition is checked, since the condition is true as `i` is less than `n`, the statement in the `for` loop is executed and the value of `i` is printed. After every iteration, the value of `i` is incremented. When `i` exceeds the value of `n`, the control jumps to the `return 0` statement.

PROGRAMMING EXAMPLES

33. Write a program to print the multiplication table of *n*, where *n* is entered by the user.

```
#include<stdio.h>
int main()
{
    int n, i;
    printf("\n Enter any number : ");
    scanf("%d", &n);
    printf("\n Multiplication table of %d", n);
    printf("\n ************************");
    for(i=0;i<20;i++)
        printf("\n %d × %d = %d", num, i, (n * i));
    return 0;
}
```

34. Write a program to print numbers from *m* to *n*.

```
#include<stdio.h>
#include<conio.h>
int main()
{
    int n, m;
    printf("\n Enter the value of m : ");
    scanf("%d", &m);
    printf("\n Enter the value of n : ");
    scanf("%d", &n);
    for(i=m;i<=n;i++)
        printf("\n %d", m);
    return 0;
}
```

35. Write a program using for loop to print all the numbers from *m* to *n*, thereby classifying them as even or odd.

```
#include<stdio.h>
#include<conio.h>
int main()
```

```
{
        int i, m, n;
        clrscr();
        printf("\n Enter the value of m : ");
        scanf("%d", &m);
        printf("\n Enter the value n : ");
        scanf("%d", &n);
        for(i=m;i<=n;i++)
        {
                if(i%2 == 0)
                        printf("\n %d is even",i);
                else
                        printf("\n %d is odd", i);
        }
        return 0;
}
```

36. Write a program using for loop to calculate the average of first *n* natural numbers.

```
#include<stdio.h>
#include<conio.h>
int main()
{
        int n, i, sum =0;
        float avg = 0.0;
        clrscr();
        printf("\n Enter the value of n : ");
        scanf("%d", &n);
        for(i=0;i<=n;i++)
                sum = sum + i;
        avg = (float)sum/n;
        printf("\n The sum of first n natural numbers = %d", sum);
        printf("\n The average of first n natural numbers = %f", avg);
        return 0;
}
```

37. Write a program using for loop to calculate the factorial of a number.

```
#include<stdio.h>
#include<conio.h>
int main()
{
        int fact = 1, num;
        clrscr();
        printf("\n Enter the number : ");
        scanf("%d",&num);
        if(num == 0)
```

```
                fact = 1;
        else
        {
                for(i=1; i<=num;i++)
                        fact = fact * num;
        }
        printf("\n Factorial of %d is %d : ", num, fact);
        return 0;
}
```

38. Write a program to classify a given number as prime or composite.

```
#include<stdio.h>
#include<conio.h>
int main()
{
        int flag = 0, i, num;
        clrscr();
        printf("\n Enter any number : ");
        scanf("%d", &num);
        for(i=2; i<num/2;i++)
        {
                if(num%i == 0)
                {
                        flag =1;
                        break;
                }
        }
        if(flag == 1)
                printf("\n %d is not a prime number", num);
        else
                printf("\n %d is a prime number", num);
        return 0;
}
```

39. Write a program using do-while loop to read the numbers until –1 is encountered. Also count the number of prime numbers and composite numbers entered by the user.

```
#include<stdio.h>
#include<conio.h>
int main()
{
        int num, i;
        int primes=0, composites=0, flag=0;
        clrscr();
        printf("\n Enter -1 to exit….");
        printf("\n\n Enter any number : ");
```

```
        scanf("%d",&num);
        do
        {
                for(i=2;i<=num%2;i++)
                {
                        if(num%i==0)
                        {
                                flag=1;
                                break;
                        }
                }
                if(flag==0)
                        primes++;
                else
                        composites++:
                flag=0;
                printf("\n\n Enter any number : ");
                scanf("%d",&num);
        } while(num != -1);
        printf("\n Count of prime numbers entered = %d", primes);
        printf("\n Count of composite numbers entered = %d", composites);
        return 0;
}
```

40. Write a program to calculate pow(x,n).

```
#include<stdio.h>
#include<conio.h>
int main()
{
        int i, num, n;
        long int result =1;
        clrscr();
        printf("\n Enter the number : ");
        scanf("%d", &num);
        printf("\n Till which power to calculate : ");
        scanf("%d", &n);
        for(i=1;i<=n;i++)
                result = result * num;
        printf("\n pow(%d, %d) = %ld", num, n, result);
        return 0;
}
```

41. Write a program to print the reverse of a number.

```
#include<stdio.h>
#include<conio.h>
```

```
int main()
{
        int num, temp;
        clrscr();
        printf("\n Enter the number : ");
        scanf("%d", &num);
        printf("\n The reversed number is : ");
        while(num != 0)
        {
                temp = num%10;
                printf("%d",temp);
                num = num/10;
        }
        return 0;
}
```

42. Write a program to enter a number and then calculate the sum of its digits.

```
#include<stdio.h>
#include<conio.h>
int main()
{
        int num, temp, sumofdigits = 0;
        clrscr();
        printf("\n Enter the number : ");
        scanf("%d", &num);
        while(num != 0)
        {
                temp = num%10;
                sumofdigits += temp;
                num = num/10;
        }
        printf("\n The sum of digits = %d", sumofdigits);
        return 0;
}
```

43. Write a program to enter a decimal number. Calculate and display the binary equivalent of this number.

```
#include<stdio.h>
#include<conio.h>
#include<math.h>
int main()
{
        int decimal_num, remainder, binary_num = 0, i =0;
        clrscr();
        printf("\n Enter the decimal number : ");
```

```
        scanf("%d", &decimal_num);
        while(decimal_num != 0)
        {
                remainder = decimal_num%2;
                binary_num += remainder*pow(10,i);
                decimal_num = decimal_num/2;
                i++;
        }
        printf("\n The binary equivalent is = %d", binary_num);
        return 0;
}
```

44. Write a program to enter a decimal number. Calculate and display the octal equivalent of this number.

```
#include<stdio.h>
#include<conio.h>
#include<math.h>
int main()
{
        int decimal_num, remainder, octal_num= 0, i =0;
        clrscr();
        printf("\n Enter the decimal number : ");
        scanf("%d", &decimal_num);
        while(decimal_num != 0)
        {
                remainder = decimal_num%8;
                octal_num += remainder*pow(10,i);
                decimal_num = decimal_num/8;
                i++;
        }
        printf("\n The octal equivalent = %d",octal_num);
        return 0;
}
```

45. Write a program to enter a decimal number. Calculate and display the hexadecimal equivalent of this number.

```
#include<stdio.h>
#include<conio.h>
#include<math.h>
int main()
{
        int decimal_num, hex_num =0 , i =0, remainder;
        clrscr();
        printf("\n Enter the decimal number : ");
        scanf("%d", &decimal_num);
        while(decimal_num != 0)
```

```
        {
                remainder = decimal_num%16;
                hex_num += remainder*pow(10,i);
                decimal_num = decimal_num/16;
                i++;
        }
        printf("\n The hexa decimal equivalent = %d", hex_num);
        return 0;
}
```

46. Write a program to enter a binary number. Calculate and display the decimal equivalent of this number.

```
#include<stdio.h>
#include<conio.h>
#include<math.h>
int main()
{
        int decimal_num = 0, remainder, binary_num, i =0;
        clrscr();
        printf("\n Enter the binary number : ");
        scanf("%d", &binary_num);
        while(binary_num != 0)
        {
                remainder = binary_num%10;
                decimal_num += remainder*pow(2,i);
                binary_num = binary_num/10;
                i++;
        }
        printf("\n The decimal equivalent is = %d", decimal_num);
        return 0;
}
```

47. Write a program to enter an octal number. Calculate and display the decimal equivalent of this number.

```
#include<stdio.h>
#include<conio.h>
#include<math.h>
int main()
{
        int decimal_num= 0, remainder, octal_num, i =0;
        clrscr();
        printf("\n Enter the octal number : ");
        scanf("%d", &octal_num);
        while(octal_num != 0)
        {
```

```
            remainder = octal_num%10;
            decimal_num += remainder*pow(8,i);
            octal_num = octal_num/10;
            i++;
      }
      printf("\n The decimal equivalent is = %d", decimal_num);
      return 0;
}
```

48. Write a program to enter a hexadecimal number. Calculate and display the decimal equivalent of this number.

```
#include<stdio.h>
#include<conio.h>
#include<math.h>
int main()
{
      int decimal_num= 0, remainder, hex_num, i =0;
      clrscr();
      printf("\n Enter the hexadecimal number : ");
      scanf("%d", &hex_num);
      while(hex_num != 0)
      {
            remainder = hex_num%10;
            decimal_num += remainder*pow(16,i);
            hex_num = hex_num/10;
            i++;
      }
      printf("\n The decimal equivalent is = %d", decimal_num);
      return 0;
}
```

49. Write a program to calculate GCD of two numbers.

```
#include<stdio.h>
#include<conio.h>
int main()
{
      int num1, num2, temp;
      int dividend, divisor, remainder;
      clrscr();
      printf("\n Enter the first number : ");
      scanf("%d", &num1);
      printf("\n Enter the second number : ");
      scanf("%d", &num2);
      if(num1>num2)
      {
```

```
            dividend = num1;
            divisor = num2;
        }
        else
        {
            dividend = num2;
            divisor = num1;
        }
        while(divisor)
        {
            remainder = dividend%divisor;
            dividend = divisor;
            divisor = remainder;
        }
        printf("\n GCD of %d and %d is = %d", num1, num2, dividend);
        return 0;
}
```

50. Write a program to sum the series $1+\frac{1}{2}+\frac{1}{3}+...+\frac{1}{n}$

```
#include<stdio.h>
#include<conio.h>
main()
{
        int n;
        float sum=0.0, a, i;
        clrscr();
        printf("\n Enter the value of n : ");
        scanf("%d", &n);
        for(i=1.0;i<=n;i++)
        {   a=(float)1/i;
            sum = sum +a;
        }
        printf("\n The sum of series 1/1 + 1/2 + .... + 1/%d = %f",n,sum);
        return 0;
}
```

51. Write a program to sum the series $\frac{1}{1^2}+\frac{1}{2^2}+...+\frac{1}{n^2}$

```
#include<stdio.h>
#include<math.h>
#include<conio.h>
main()
{
        int n;
```

```
        float sum=0.0, a, i;
        clrscr();
        printf("\n Enter the value of n : ");
        scanf("%d", &n);
        for(i=1.0;i<=n;i++)
        {       a=(float)1.0/pow(i,2);
                sum = sum +a;
        }
        printf("\n The sum of series 1/1² + 1/ 2² + …. 1/n² = %f",sum);
        return 0;
}
```

52. Write a program to sum the series $\frac{1}{2}+\frac{2}{3}+...+\frac{n}{n+1}$

```
#include<stdio.h>
#include<conio.h>
main()
{
        int n;
        float sum=0.0, a, i;
        clrscr();
        printf("\n Enter the value of n : ");
        scanf("%d", &n);
        for(i=1.0;i<=n;i++)
        {       a=(float)i/(i+1);
                sum = sum +a;
        }
        printf("\n The sum of series 1/2 + 2/3 + .... + %d/%d = %f",n,n+1,sum);
        return 0;
}
```

53. Write a program to sum the series $\frac{1}{1}+\frac{2^2}{2}+\frac{3^3}{3}+...$

```
#include<stdio.h>
#include<conio.h>
#include<math.h>
main()
{
        int n, NUM;
        float i,sum=0.0;
        clrscr();
        printf("\n Enter the value of n : ");
        scanf("%d", &n);
        for(i=1.0;i<=n;i++)
        {
```

```
        NUM = pow(i,i);
        sum += (float)NUM/i;
        }
        printf("\n 1/1 + 4/2 + 27/3 + .... = %f", sum);
        return 0;
}
```

54. Write a program to calculate the sum of cubes of numbers from 1 to *n*.

```
#include<stdio.h>
#include<conio.h>
#include<math.h>
main()
{
        int i, n;
        int term, sum = 0;
        clrscr();
        printf("\n ENter the value of n : ");
        scanf("%d", &n);
        for(i=1;i<=n;i++)
        {
                term = pow(i,3);
                sum += term;
        }
        printf("\n 1³ + 2³ + 3³ + ....... = %d", sum);
        return 0;
}
```

55. Write a program to sum the squares of even numbers and cubes of odd numbers.

```
#include<stdio.h>
#include<conio.h>
#include<math.h>
main()
{
        int i, n;
        int term, sum = 0;
        clrscr();
        printf("\n ENter the value of n : ");
        scanf("%d", &n);
        for(i=1;i<=n;i++)
        {
                if(i%2 == 0)
                        term = pow(i,2);
                else
                        term = pow(i,3);
                sum += term;
        }
        printf("\n sum of the series = %d", sum);
```

```
        return 0;
    }
```

56. Write a program to find whether the given number is an armstrong number or not.

```
#include<stdio.h>
#include<conio.h>
#include<math.h>
main()
{
        int num, sum=0, r, n;
        clrscr();
        printf("\n Enter the number : ");
        scanf("%d", &num);
        n=num;
        while(n>0)
        {
                r=n%10;
                sum += pow(r,3);
                n=n/10;
        }
        if(sum==num)
                printf("\n %d is an armstrong number", num);
        else
                printf("\n %d is not an armstrong number", num);
        return 0;
}
```

57. Write a program to print the multiplication table.

```
#include<stdio.h>
#include<conio.h>
int main()
{
        int i, j;
        clrscr();
        printf("\n\t\t Multiplication table", );
        printf("\n *******************************************************");
        for(i=1;i<=20;i++)
        {
                printf("\n %d", i);
                for(j=1;j<=20;j++)
                        printf("\t %d", (i*j));
        }
        getch();
        return 0;
}
```

1.12 BREAK AND CONTINUE STATEMENT

In C, the `break` statement is used to terminate the execution of the nearest enclosing loop in which it appears. We have already seen its usage in the switch statement. The break statement is widely used with `for` loop, `while` loop, and `do-while` loop. When compiler encounters a `break` statement, the control passes to the statement that follows the loop in which the `break` statement appears. Its syntax is quite simple, just type keyword `break` followed by a semi-colon.

```
break;
```

In a `switch` statement, if the `break` statement is missing, then every case from the matched case label to the end of the `switch`, including the default, is executed. The example given below shows the manner in which `break` statement is used to terminate the statement in which it is embedded.

```
#include<stdio.h>
int main()
{
        int i = 0;
        while(i<=10)
        {
                if (i==5)
                        break;
                printf("\n %d", i);
                i = i + 1;
        }
        return 0;
}
```

Note that the code is meant to print first 10 numbers using a `while` loop, but it will actually print only numbers from 0 to 4. As soon as `i` becomes equal to 5, the `break` statement is executed and the control jumps to the statement following the `while` loop.

Hence, the `break` statement is used to exit a loop from any point within its body, bypassing its normal termination expression. When the `break` statement is encountered inside a loop, the loop is immediately terminated, and program control is passed to the next statement following the loop.

Continue Statement

Like the `break` statement, the `continue` statement can only appear in the body of a loop. When the compiler encounters a `continue` statement, then the rest of the statements in the loop are skipped and the control is unconditionally transferred to the loop-continuation portion of the nearest enclosing loop. Its syntax is quite simple, just type keyword `continue` followed by a semi-colon.

```
continue;
```

Again like the `break` statement, the continue statement cannot be used without an enclosing `for`, `while`, or `do` statement. When the `continue` statement is encountered in the `while` loop and in the `do-while` loop, the control is transferred to the code that tests the controlling expression. However, if placed with a `for` loop, the continue statement causes a branch to the code that updates the loop variable. For example, consider the following code.

```
#include<stdio.h>
int main()
```

```
    {
        int i;
        for(i=0; i<= 10; i++)
        {
            if (i==5)
                continue;
            printf("\t %d", i);
            i = i + 1;
        }
        return 0;
    }
```

Note that the code is meant to print numbers from 0 to 10. But as soon as i becomes equal to 5, the continue statement is encountered, so rest of the statements in the `for` loop are skipped and the control passes to the expression that increments the value of `i`. The output of this program would thus be:

```
0 1 2 3 4 6 7 8 9 10
```

Note There is no 5 because it could not be printed, as continue caused an early incrementation of i and skipping of statement that printed the value of i on screen.

Hence, we conclude that the `continue` statement is somewhat the opposite of the `break` statement. It forces the next iteration of the loop to take place, skipping any code in between itself and the test condition of the loop. It is generally used to restart a statement sequence when an error occurs.

SUMMARY

- C was developed in the early 1970s by Dennis Ritchie at Bell Laboratories.
- A function is defined as a group of C statements that are executed together. The execution of a C program begins at this function.
- Every word in a C program is either a keyword or an identifier. C has a set of reserved words often known as keywords that cannot be used as an identifier.
- A variable is defined as a meaningful name given to the data storage location in the computer memory. When using a variable, we actually refer to the address of the memory where the data is stored.
- The difference between signed and unsigned numeric variables is that signed variables can be either negative or positive but unsigned variables can only be positive. By default, C makes a signed variable.
- Return 0 returns the value 0 to the operating system to give an indication that there were no errors during the execution of the program.
- The conditional operator or the ternary (?:) is just like an if .. else statement that can be within expressions. Conditional operator is also known as ternary operator as it is neither a unary nor a binary operator; it takes three operands.
- The bitwise NOT or complement produces the ones' complement of the given binary value.
- The comma operator evaluates the first expression and discards its value, and then evaluates the second and returns the value as the result of the expression.
- Size of is a unary operator used to calculate the sizes of data types. This operator can be applied to all data types.
- While type conversion is done implicitly, casting has to be done explicitly by the programmer. Type-casting is done when the value of a higher data type has to be converted in to the value of a lower data type.
- C supports decision control statements that can alter the flow of a sequence of instructions. A switch case statement is a multi-way decision

statement that is a simplified version of an if-else block that evaluates only one variable.

- Default is also a case that is executed when the value of the variable does not match with any of the values of the case statement.
- Iterative statements are used to repeat the execution of a list of statements, depending on the value of an integer expression.
- The break statement is used to terminate the execution of the nearest enclosing loop in which it appears.
- When the compiler encounters a continue statement then the rest of the statements in the loop are skipped and the control is unconditionally transferred to the loop-continuation portion of the nearest enclosing loop.

GLOSSARY

ANSI C American National Standards Institute's definition of the C programming language. It is the same as the ISO definition.

Constant A value that cannot be changed.

Data type Definition of the data. For example, int, char, float.

Escape sequence Control codes that comprises of combinations of a backslash followed by letters or digits which represent non-printing characters.

Expression A sequence of operators and operands that may yield a single value as the result of its computation.

Executable program Program which will run in the environment of the operating system or within an appropriate run time environment.

Floating point number Number that comprises of a decimal place and exponent.

Format specification A string which controls the manner in which input or output of values has to be done.

Identifier The names used to refer to stored data values as in case of constants, variables or functions.

Integer A number that has no fractional part.

Keyword A word which has a predefined meaning to a 'C' compiler and therefore must not be used for any other purpose.

Library file The file which comprises of compiled versions of commonly used functions that can be linked to an object file to make an executable pro-gram.

Library function A function whose source code is stored in the external library file.

Linker The tool that connects object code and libraries to form a complete, executable program.

Operator precedence The order in which operators are applied to operands during the evaluation of an expression.

Program A text file that contains the source code to be compiled.

Reserved word (keyword) A word which has a predefined meaning to a 'C' compiler. Such words cannot be used for any other purpose.

Source code A text file that contains the source code to be compiled.

Statement A simple statement in C language that is followed by a semi-colon.

Variable An identifier (and storage) for a data type. The value of a variable may change as the program runs.

EXERCISES

Review Questions

1. Differentiate between declaration and definition.
2. How is memory reserved using a declaration statement?
3. What does the data type of a variable signify?
4. Discuss the structure of a C program?
5. What do you understand by identifiers and keywords?
6. Write a short note on basic data types that the C language supports.
7. Why do we need signed and unsigned char?
8. Explain the terms variables and constants? How many types of variables are supported by C?
9. Why do we include <stdio.h> in our programs?
10. What are header files? Explain their significance.

11. Write a short note on operators available in C language.
12. Draw the operator precedence chart.
13. Evaluate the expression: `(x > y) + ++a || !c`
14. Differentiate between typecasting and type conversion.
15. What are decision control statements? Explain in detail.
16. Write a short note on the iterative statements that C language supports.
17. When will you prefer to work with a switch statement?
18. What are header files? Why are they important? Can we write a C program without using any header file?
19. Write a program to read an integer. Display the value of that integer in decimal, octal, and hexadecimal notation.
20. Write short notes on printf and scanf functions.
21. Explain the utility of `#define` and `#include` statements.
22. Write a program that prints a floating point value in exponential format with the following specifications:
    (a) correct to two decimal places;
    (b) correct to four decimal places; and
    (c) correct to eight decimal places.

Programming Exercises

1. Write a program to read 10 integers. Display these numbers by printing three numbers in a line separated by commas.
2. Write a program to print the count of even numbers between 1-200. Also print their sum.
3. Write a program to count the number of vowels in a text.
4. Write a program to read the address of a user. Display the result by breaking it in multiple lines.
5. Write a program to read two floating point numbers. Add these numbers and assign the result to an integer. Finally, display the value of all the three variables.
6. Write a program to read a floating point number. Display the rightmost digit of the integral part of the number.
7. Write a program to calculate simple interest and compound interest.
8. Write a program to calculate salary of an employee given his basic pay (to be entered by the user), HRA = 10% of the basic pay, TA = 5% of basic pay. Define HRA and TA as constants and use them to calculate the salary of the employee.
9. Write a program to prepare a grocery bill. Enter the name of the items purchased, quantity in which it is purchased, and its price per unit. Then display the bill in the following format.

```
************** B I L L **************
   Item   Quantity   Price   Amount
_____________________________________

_____________________________________
      Total Amount to be paid
_____________________________________
```

10. Write a C program using printf statement to print BYE in the following format.

```
BBB     Y   Y    EEEE
B   B    Y Y     E
BBB       Y      EEEE
B   B     Y      E
BBB       Y      EEEE
```

Multiple Choice Questions

1. The operator which compares two values is
   (a) Assignment (b) Relational
   (c) Unary (d) Equal
2. Which operator is used to simultaneously evaluate two expressions with relational operators?
   (a) AND (b) OR
   (c) NOT (d) All of these
3. Ternary operator operates on how many operands?
   (a) 1 (b) 2 (c) 3 (d) 4
4. Which operator produces the ones' complement of the given binary value.
   (a) Logical AND (b) Bitwise AND
   (c) Logical OR (d) Bitwise NOT
5. Which operator has the lowest precedence?
   (a) Sizeof (b) Unary
   (c) Assignment (d) Comma

6. Short integer has which conversion character associated with it?
   (a) `%c` (b) `%d` (c) `%e`
   (d) `%f` (e) `%hd`
7. Which of the following is not a character constant?
   (a) `'A'` (b) `'bb'` (c) `"A"`
   (d) `' '` (e) `'*'`
8. Which of the following is not a floating point constant?
   (a) `20` (b) `–4.5` (c) `'a'`
   (d) `"1"` (e) `pi`
9. Identify the invalid variable names.
   (a) Initial.Name (b) A+B
   (c) $amt (d) Floats
   (e) 1st_row (f) Col-Amt
   (g) Col Amt
10. Which operator cannot be used with float operands?
    (a) `+` (b) `-` (c) `%`
    (d) `*` (e) `/`
11. Identify the erroneous expression.
    (a) `X=y=2, 4;` (b) `res = ++a * 5;`
    (c) `res = /4;` (d) `res = a++ -b *2`

True or False

1. We can have only one function in a C program.
2. Keywords are case sensitive.
3. Variable 'first' is the same as 'First'.
4. Signed variables can increase the maximum positive range.
5. Commented statements are not executed by the compiler.
6. Equality operators have higher precedence than the relational operators.
7. Shifting once to the left multiplies the number by 2.
8. Decision control statements are used to repeat the execution of a list of statements.
9. Equality operators have higher precedence than the relational operators.
10. Shifting once to the left multiplies the number by 2.
11. `printf("%d", scanf("%d", &num));` is a valid C statement.
12. 1,234 is a valid integer constant.
13. A `printf` statement can generate only one line of output.
14. `stdio.h` is used to store the source code of the program.
15. The closing brace of `main()` is the logical end of the program.
16. The declaration section gives instructions to the computer.
17. Any valid printable ACII character can be used for a variable name.
18. Declaration of variables can be done anywhere in the program.
19. Underscore can be used anywhere in the variable name.
20. amt is same as AMT in C.
21. `void` is a data type in C.
22. `scanf` can be used to read only one value at a time.
23. All arithmetic operators have same precedence.
24. The modulus operator can be used only with integers.
25. The expression containing all integer operands is called an integer expression.

Fill in the Blanks

1. C was developed by ______.
2. ______ is a group of C statements that are executed together.
3. The execution of a C program begins at ______.
4. In the memory, characters are stored as ______.
5. `return 0` returns 0 to the ______.
6. ______ finds the remainder of an integer division.
7. ______ operator reverses the value of the expression.
8. `sizeof` is a ______ operator used to calculate the sizes of data types.
9. ______ is also known as forced conversion.
10. `scanf()` returns ______.
11. ______ is executed when the value of the variable does not match with any of the values of the case statement.
12. ______ function prints data on the monitor.
13. ______ establishes the original value for a variable.
14. Character constants are quoted using ______.

15. A C program ends with a ______.
16. ______ file contains mathematical functions.
17. ______ causes the cursor to move to the next line.
18. Floating point values denote ______ values by default.
19. A variable can be made constant by declaring it with the qualifier ______ at the time of initialization.
20. The sign of the result is positive in modulo division if ______.
21. Associativity of operators defines ______.
22. ______ can be used to change the order of evaluation expressions.
23. ______ operator returns the number of bytes occupied by the operand.
24. The ______ specification is used to read/write a short integer.
25. The ______ specification is used to read/write a hexadecimal integer.
26. To print the data left-justified, ______ specification is used.

Practice Questions

1. Find errors in the following declaration statements:
   (a) `Int n;`
   (b) `float a b;`
   (c) `double = a, b;`
   (d) `complex a b;`
   (e) `a,b : INTEGER`
   (f) `long int a;b;`
2. Find error in the following code:
```
int a = 9;
float y = 2.0;
a = b % a;
printf("%d", a);
```
3. Find error in the following scanf statement:
```
scanf("%d%f", &marks, &avg);
```

Find the Output of the Following Programs

1.
```
#include<stdio.h>
main()
{
  int x=3, y=5, z=7;
  int a, b;
  a = x * 2 + y / 5 - z * y;
  b = ++x * (y - 3) / 2 - z++ * y;
  printf("\n a = %d", a);
  printf("\n b = %d", b);
  return 0;
}
```
2.
```
#include<stdio.h>
main()
{
  int a;
  printf("\n %d", 1/3 + 1/3);
  printf("\n %f", 1.0/3.0 + 1.0/3.0);
  a = 15/10.0 + 3/2;
  printf("\n %d", a);
  return 0;
}
```
3.
```
#include<stdio.h>
main()
{
  int a, b =3;
  char c = 'A';
  a = b + c;
  printf("\n a = %d", a);
  return 0;
}
```
4.
```
#include<stdio.h>
int main()
{
  int a = 4);
  printf("\n %d ", 10 + a++);
  printf("\n %d ", 10 + ++a);
  return 0;
}
```
5.
```
#include<stdio.h>
main()
{
  int a = 4, b =5, c= 6;
  a = b == c;
  printf("\n a = %d ", a);
  return 0;
}
```
6.
```
#include<stdio.h>
main()
{
  int a = 4, b = 5;
  printf("\n %d ", (a > b)? a: b);
  return 0;
}
```

7.
```
#include<stdio.h>
#include<conio.h>
main()
{
  int a=1, b=2, c=3, d=4, e=5, res;
  clrscr();
  res = a + b /c - d * e;
  printf("\n The value of the
  expression a + b /c - d * e = %d",res);
  res = (a + b) /c - d * e;
  printf("\n The value of the expression
  (a + b) /c - d * e = %d",res);
  res = a + ( b / (c -d)) * e;
  printf("\n The value of the
  expression a +( b / (c - d)) * e
  = %d",res);
  return 0;
}
```

8.
```
#include<stdio.h>
int main()
{
  int a = 4, b = 12, c= -3, res;
  res = a > b && a < c;
  printf("\n %d ", res);
  res = a == c || a < b;
  printf("\n %d ", res);
  res = b >10 b && c < 0 || a > 0;
  printf("\n %d ", res);
  res = (a/2.0 == 0.0 && b/2.0 != 0.0)
  || c < 0.0;
  printf("\n %d ", res);
  return 0;
}
```

9.
```
#include<stdio.h>
int main()
{
  int a = 20, b = 5, result;
  float c = 20.0, d= 5.0;
  printf("\n 10 + a / 4 * b = %d", 10
  + a / 4 * b);
  printf("\n c / d * b + a % b = %d",
  c / d * b + a % b);
  return 0;
}
```

10.
```
#include<stdio.h>
int main()
{
  int a, b;
  printf("\n a = %d \t b = %d \t a + b
  = %d", a, b, a+b);
  return 0;
}
```

11.
```
#include<stdio.h>
int main()
{
  printf("\n %d", 'F');
  return 0;
}
```

12.
```
#include<stdio.h>
int main()
{
  int n = -2;
  printf("\n n = %d", -n);
  return 0;
}
```

13.
```
#include<stdio.h>
int main()
{
  int n = 2;
  n = !n;
  printf("\n n = %d", n);
  return 0;
}
```

14.
```
#include<stdio.h>
int main()
{
  int a = 2, b = 3, c, d;
  c = a++;
  d = ++b;
  printf("\n c = %d d = %d", c, d);
  return 0;
}
```

15.
```
#include<stdio.h>
int main()
{
  int a = 2, b = 3;
  printf("\n %d", ++(a - b));
  return 0;
}
```

16.
```
#include<stdio.h>
int main()
{
  int a = 2, b = 3, c, d;
  a++;
  ++b;
  printf("\n a = %d b = %d", a, b);
```

```
    return 0;
  }
```

17.
```
#include<stdio.h>
int main()
{
    int a = 2, b = 3;
    printf("\n %d", ++a - b);
    return 0;
}
```

18.
```
#include<stdio.h>
int main()
{
    int a = 2, b = 3;
    printf("\n a * b = %d", a*b);
    printf("\n a / b = %d", a/b);
    printf("\n a % b = %d", a%b);
    printf("\n a && b = %d", a*&&b);
    return 0;
}
```

19.
```
#include<stdio.h>
int main()
{
    int a = 2;
    a = a + 3*a++;
    printf("\n a= %d", a);
    return 0;
}
```

20.
```
#include<stdio.h>
int main()
{
    int a = 2,
    b = 3, c=4;
    a=b==c;
    printf("\n a = %d",a);
    return 0;
}
```

21.
```
#include<stdio.h>
int main()
{
    int result;
    result = 3 + 5 - 1 * 17 % -13;
    printf("%d", result);
    result = 3 * 2 + ( 15 / 4 % 7);
    printf("%d", result);
    result = 18 / 9 / 3 * 2 * 3 * 5 % 10
    / 4;
    printf("%d", result);
```

```
    return 0;
}
```

22.
```
#include<stdio.h>
int main()
{
    int num = 070;
    printf("\n num = %d", num);
    printf("\n num = %o", num);
    printf("\n num = %x", num);
    return 0;
}
```

23.
```
#include<stdio.h>
int main()
{
    int n = 2;
    printf("\n %d %d %d", n++, n, ++n);
    return 0;
}
```

Hint: *The arguments in the function call will be pushed from left to right. However, these arguments are evaluated from right to left.*

24.
```
#include<stdio.h>
int main()
{
    printf("\n  %40.27s Welcome to C
    programming");
    printf("\n  %40.20s Welcome to C
    programming");
    printf("\n  %40.14s Welcome to C
    programming");
    printf("\n  %-40.27s Welcome to C
    programming");
    printf("\n  %-40.20s Welcome to C
    programming");
    printf("\n  %-40.14s Welcome to C
    programming");
    return 0;
}
```

25.
```
#include<stdio.h>
main()
{
    int a = -21, b = 3;
    printf("\n %d", a/b + 10);
    b = -b;
    printf("\n %d", a/b + 10);
    return 0;
}
```

26.
```
#include<stdio.h>
main()
{
  int a;
  float b;
  printf("\n Enter any four digit number
  : ");
  scanf("%2d", &a);
  printf("\n Enter any floating point
  number : ");
  scanf("%f", &b);
  printf("\n The numbers are : %d and
  %f", a, b);
  return 0;
}
```

27.
```
#include<stdio.h>
main()
{
  char a, b, c;
  printf("\n Enter three characters :
  ");
  scanf("%c %c %c", &a, &b, &c);
  a++; b++; C==;
  printf("\n a = %c b = %c and d = %c",
  a, b, c);
  return 0;
}
```

28.
```
#include<stdio.h>
{
  int x=10, y=20, res;
  res = y++ + x++;
  res += ++y + ++x;
  printf("\n x = %d y = %d RESULT =
  %d", x,y, res");
  return 0;
}
```

29.
```
#include<stdio.h>
{
  int x=10, y=20, res;
  res = x+++b;
  printf("\n x = %d y = %d RESULT =
  %d", x,y, res");
  return 0;
}
```

ASCII Table and Description

Dec	Hx	Oct	Char	Dec	Hx	Oct	Html	Chr	Dec	Hx	Oct	Html	Chr	Dec	Hx	Oct	Html	Chr
0	0	000	NUL (null)	32	20	040		Space	64	40	100	@	@	96	60	140	`	`
1	1	001	SOH (start of heading)	33	21	041	!	!	65	41	101	A	A	97	61	141	a	a
2	2	002	STX (start of text)	34	22	042	"	"	66	42	102	B	B	98	62	142	b	b
3	3	003	ETX (end of text)	35	23	043	#	#	67	43	103	C	C	99	63	143	c	c
4	4	004	EOT (end of transmission)	36	24	044	$	$	68	44	104	D	D	100	64	144	d	d
5	5	005	ENQ (enquiry)	37	25	045	%	%	69	45	105	E	E	101	65	145	e	e
6	6	006	ACK (acknowledge)	38	26	046	&	&	70	46	106	F	F	102	66	146	f	f
7	7	007	BEL (bell)	39	27	047	'	'	71	47	107	G	G	103	67	147	g	g
8	8	010	BS (backspace)	40	28	050	(	(	72	48	110	H	H	104	68	150	h	h
9	9	011	TAB (horizontal tab)	41	29	051	)	)	73	49	111	I	I	105	69	151	i	i
10	A	012	LF (NL line feed, new line)	42	2A	052	*	*	74	4A	112	J	J	106	6A	152	j	j
11	B	013	VT (vertical tab)	43	2B	053	+	+	75	4B	113	K	K	107	6B	153	k	k
12	C	014	FF (NP form feed, new page)	44	2C	054	,	,	76	4C	114	L	L	108	6C	154	l	l
13	D	015	CR (carriage return)	45	2D	055	-	-	77	4D	115	M	M	109	6D	155	m	m
14	E	016	SO (shift out)	46	2E	056	.	.	78	4E	116	N	N	110	6E	156	n	n
15	F	017	SI (shift in)	47	2F	057	/	/	79	4F	117	O	O	111	6F	157	o	o
16	10	020	DLE (data link escape)	48	30	060	0	0	80	50	120	P	P	112	70	160	p	p
17	11	021	DC1 (device control 1)	49	31	061	1	1	81	51	121	Q	Q	113	71	161	q	q
18	12	022	DC2 (device control 2)	50	32	062	2	2	82	52	122	R	R	114	72	162	r	r
19	13	023	DC3 (device control 3)	51	33	063	3	3	83	53	123	S	S	115	73	163	s	s
20	14	024	DC4 (device control 4)	52	34	064	4	4	84	54	124	T	T	116	74	164	t	t
21	15	025	NAK (negative acknowledge)	53	35	065	5	5	85	55	125	U	U	117	75	165	u	u
22	16	026	SYN (synchronous idle)	54	36	066	6	6	86	56	126	V	V	118	76	166	v	v
23	17	027	ETB (end of trans. block)	55	37	067	7	7	87	57	127	W	W	119	77	167	w	w
24	18	030	CAN (cancel)	56	38	070	8	8	88	58	130	X	X	120	78	170	x	x
25	19	031	EM (end of medium)	57	39	071	9	9	89	59	131	Y	Y	121	79	171	y	y
26	1A	032	SUB (substitute)	58	3A	072	:	:	90	5A	132	Z	Z	122	7A	172	z	z
27	1B	033	ESC (escape)	59	3B	073	;	;	91	5B	133	[	[	123	7B	173	{	{
28	1C	034	FS (file separator)	60	3C	074	<	<	92	5C	134	\	\	124	7C	174			|
29	1D	035	GS (group separator)	61	3D	075	=	=	93	5D	135	]	]	125	7D	175	}	}
30	1E	036	RS (record separator)	62	3E	076	>	>	94	5E	136	^	^	126	7E	176	~	~
31	1F	037	US (unit separator)	63	3F	077	?	?	95	5F	137	_	_	127	7F	177		DEL

Source: www.LookupTables.com

2 Functions

Learning Objective

Till now, we have used the word 'function' several times and we know that `main()`, `printf()`, and `scanf()` are the functions that we have been using in every program. In this chapter, we will explore in a little more depth what functions are and why they are important.

2.1 INTRODUCTION

C enables its programmers to break up a program into segments commonly known as *functions*, each of which can be written more or less independently of the others. Every function in the program is supposed to perform a well-defined task. Therefore, the program code of one function is completely insulated from the other functions.

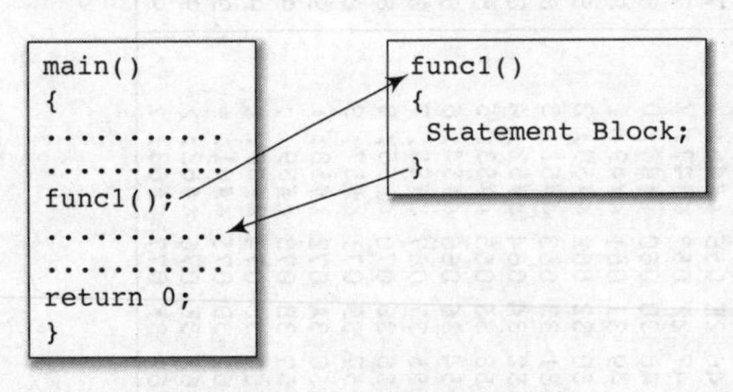

Figure 2.1 main() calls func1()

Every function interfaces to the outside world in terms of how information is transferred to it and how results generated by the function are transmitted back from it. This interface is basically specified by the function name. For example, look at Fig. 2.1 which explains how the `main()` function calls another function to perform a well-defined task.

In the figure, we see that `main()` calls a function named `func1()`. Therefore, `main()` is known as the *calling function* and `func1()` is known as the *called function*. The moment the compiler encounters a function call, instead of executing the next statement in the calling function, the control jumps to the statements that are a part of the called function. After the called function is executed, the control is returned to the calling program.

The `main()` function can call as many functions as it wants and as many times as it wants. For example, a function call placed within a `for` loop, `while` loop, or `do-while` loop may call the same function multiple times until the condition holds true.

Not only `main()`, any function can call any other function. For example, look at Fig. 2.2 which shows one function calling another, and the other function in turn calling some other function.

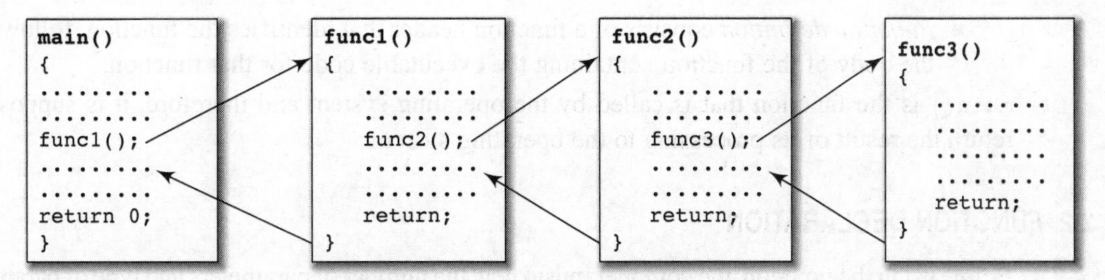

Figure 2.2 Function calling another function

Therefore, we see that every function encapsulates a set of operations and when called, it returns information to the calling program.

2.1.1 Why are Functions Needed?

Let us analyse the reasons why segmenting a program into manageable chunks is an important aspect of programming.

- Dividing the program into separate well-defined functions facilitates each function to be written and tested separately. This simplifies the process of getting the total program to work.
- Understanding, coding, and testing multiple separate functions are far easier than doing the same for one huge function.
- If a big program has to be developed without the use of any function other than `main()`, then there will be countless lines in the `main()` and maintaining that program will be a big mess.
- All the libraries in C contain a set of functions that the programmers are free to use in their programs. These functions have been pre-written and pre-tested, so the programmers use them without worrying about their code details. This speeds up program development, by allowing the programmer to concentrate only on the code that he has to write.
- Like C libraries, programmers can also make their functions and use them from different point in the main program or any other program that needs its functionalities.
- When a big program is broken into comparatively smaller functions, then different programmers working on that project can divide the workload by writing different functions.

2.1.2 Using Functions

A function can be compared to a *black box* that takes in inputs, processes it, and then outputs the result. However, we may also have a function that does not take any inputs at all, or a function that does not return anything at all. While using functions, we will be using the following terminologies:

- A function *f* that uses another function *g* is known as the *calling function,* and *g* is known as the *called function*.
- The inputs that the function takes are known as *arguments.*
- When a called function returns some result back to the calling function, it is said to *return* that result.
- The calling function may or may not pass *parameters* to the called function. If the called function accepts arguments, the calling function will pass parameters, else not.
- *Function declaration* is a declaration statement that identifies a function with its name, a list of arguments that it accepts, and the type of data it returns.

- *Function definition* consists of a function header that identifies the function, followed by the body of the function containing the executable code for that function.

`main()` is the function that is called by the operating system and therefore, it is supposed to return the result of its processing to the operating system.

2.2 FUNCTION DECLARATION

Before using the function, the compiler must know the number of parameters and type of parameters that the function expects to receive and the data type of value that it will return to the calling program. Placing the function declaration statement prior to its use enables the compiler to make a check on the arguments used while calling that function.

The general format for declaring a function that accepts some arguments and returns some value as result can be given as:

```
return_data_type function_name(data_type variable1, data_type variable2,..);
```

Here, **`function_name`** is a valid name for the function. Naming a function follows the same rules that are followed while naming the variables. A function should have a meaningful name that must clarify the task that the function will perform.

`return_data_type` specifies the data type of the value that will be returned to the calling function as a result of the processing performed by the called function.

`(data_type variable1, data_type variable2, ...)` is a list of variables of specified data types. These variables are passed from the calling function to the called function. They are also known as arguments or parameters that the called function accept to perform its task. Table 2.1 shows a few examples of valid function declarations in C.

Table 2.1 Valid function declarations

Function Declaration	Use of the Function
Return data type `char convert_to_uppercase (char ch);`	Converts a character to upper case. The function receives a character as an argument, converts it into upper case, and returns the converted character back to the calling program.
Function name `float avg (int a, int b);`	Calculates the average of two numbers *a* and *b* received as arguments. The function returns a floating point value.
`int find_large (int a, int b, int c);` Data type of variable	Finds the largest of the three numbers *a*, *b*, and *c* received as arguments. An integer value which is the largest number of the three numbers is returned to the calling function.
`double multiply (float a, float b);` Variable 1	Multiplies two floating point numbers *a* and *b* that are received as arguments.
`void swap (int a, int b);`	Swaps or interchanges the value of integer variables *a* and *b* received as arguments. The function returns no value, therefore the data type is void.
`void print(void);`	The function is used to print information on screen. The function neither accepts nor returns any value. Therefore, the return type is void and the argument list contains void.

The following points are to be noted while declaring a function:

- After the declaration of every function, there is a semi-colon. If the semi-colon is missing, the compiler will generate an error message.
- The name of function is global. Therefore, the declared function can be called from any point in the program.
- No function can be declared within the body of another function.

Note A function having void as its return type cannot return any value. Similarly, a function having void as its parameter list cannot accept any value.

2.3 FUNCTION DEFINITION

When a function is defined, space is allocated for that function in the memory. A function definition comprises of two parts:

- Function header
- Function body

The syntax of a function definition can be given as:

```
return_data_type function_name(data_type variable1, data_type variable2,..)
{
        .............
        statements
        .............
        return( variable);
}
```

Note that the number of arguments and the order of arguments in the function header must be the same as that given in the function declaration statement.

While **return_data_type function_name(data_type variable1, data_type variable2,...)** is known as the function header, the rest of the portion comprising of program statements within the curly brackets { } is the function body which contains the code to perform the specific task.

Note that the function header is the same as that of the function declaration. The only difference between the two is that a function header is not followed by a semi-colon.

2.4 FUNCTION CALL

The function call statement invokes the function. When a function is invoked, the compiler jumps to the called function to execute the statements that are a part of that function. Once the called function is executed, the program control passes back to the calling function. A function call statement has the following syntax:

```
function_name(variable1, variable2, ...);
```

The following points are to be noted while calling a function:

- Function name, the number, and the type of arguments in the function call must be same as that given in the function declaration and the function header of the function definition.

- Names (and not the types) of variables in function declaration, function call, and header of function definition may vary.
- Arguments may be passed in the form of expressions to the called function. In such a case, arguments are first evaluated and converted to the type of formal parameter and then the body of the function gets executed.
- If the return type of the function is not void, then the value returned by the called function may be assigned to some variable as given below.

```
variable_name = function_name(variable1, variable2, ...);
```

Let us now try some programs using functions.

PROGRAMMING EXAMPLES

1. Write a program to add two integers using functions.

```
#include<stdio.h>
int sum(int a, int b); // FUNCTION DECLARATION
int main()
{
        int num1, num2, total = 0;
        printf("\n Enter the first number : ");
        scanf("%d", &num1);
        printf("\n Enter the second number : ");
        scanf("%d", &num2);
        total = sum(num1, num2);         // FUNCTION CALL
        printf("\n Total = %d", total);
        return 0;
}
// FUNCTION DEFNITION
int sum ( int a, int b) // FUNCTION HEADER
{                       // FUNCTION BODY
        int result;
        result = a + b;
        return result;
}
```

2. Write a program to find the largest of three integers using functions.

```
#include<stdio.h>
int greater( int a, int b, int c);
int main()
{
        int num1, num2, num3, large;
        printf("\n Enter the first number : ");
        scanf("%d", &num1);
        printf("\n Enter the second number : ");
        scanf("%d", &num2);
```

```
    printf("\n Enter the third number : ");
    scanf("%d", &num3);
    large = greater(num1, num2, num3);
    printf("\n Largest number = %d", large);
    return 0;
}
int greater( int a, int b, int c)
{
    if(a>b && a>c)
        return a;
    if(b>a && b>c)
        return b;
    else
        return c;
}
```

3. Write a program to calculate the area of a circle using function.

```
#include<stdio.h>
float cal_area(float r);
int main()
{
    float area, radius;
    printf("\n Enter the radius of the circle : ");
    scanf("%f", &radius);
    area = cal_area(radius);
    printf("\n Area of the circle with radius %f = %f", radius, area);
    return 0;
}
float cal_area(float radius)
{
    return (3.14 * radius * raidus);
}
```

4. Write a program to find whether a number is even or odd using functions.

```
#include<stdio.h>
int evenodd(int);
int main()
{
    int num, flag;
    printf("\n Enter the number : ");
    scanf("%d", &num);
    flag = evenodd(num);
    if (flag == 1)
        printf("\n %d is EVEN", num);
```

```
        else
                printf("\n %d is ODD", num);
        return 0;
}
int evenodd(int a)
{
        if(a%2 == 0)
                return 1;
        else
                return 0;
}
```

2.5 PASSING PARAMETERS TO THE FUNCTION

There are two ways in which arguments or parameters can be passed to the called function.

Call by value The values of the variables are passed by the calling function to the called function. All the programs that we have written so far call the function using the call by value method of passing parameters.

Call by reference The address of the variables are passed by the calling function to the called function.

2.5.1 Call by Value

Till now, we had been calling the functions and passing arguments to them using the call-by-value method. In this method, the called function creates new variables to store the value of the arguments passed to it. Therefore, the called function uses a copy of the actual arguments to perform its intended task.

If the called function is supposed to modify the value of the parameters passed to it, then the change will be reflected only in the called function. In the calling function, no change will be made to the value of the variables. This is because all the changes were made to the copy of the variables and not on the actual variables. To understand this concept, consider the code given below. The function add accepts an integer variable num and adds 10 to it. In the calling function, the value of `num = 2`. In `add()`, the value of `num` is modified to 20 but in the calling function, the change is not reflected.

```
#include<stdio.h>
void add( int n);
int main()
{
   int num = 2;
   printf("\n The value of num before calling the function = %d", num);
   add(num);
   printf("\n The value of num after calling the function = %d", num);
   return 0;
}
```

```
void add(int n)
{
        n = n + 10;
        printf("\n The value of num in the called function = %d", n);
}
```

Output:

```
The value of num before calling the function = 2
The value of num in the called function = 20
The value of num after calling the function = 2
```

Note that since the called function uses a copy of num, the value of num in the calling function remains untouched. This concept is clearly explained in Fig. 2.3.

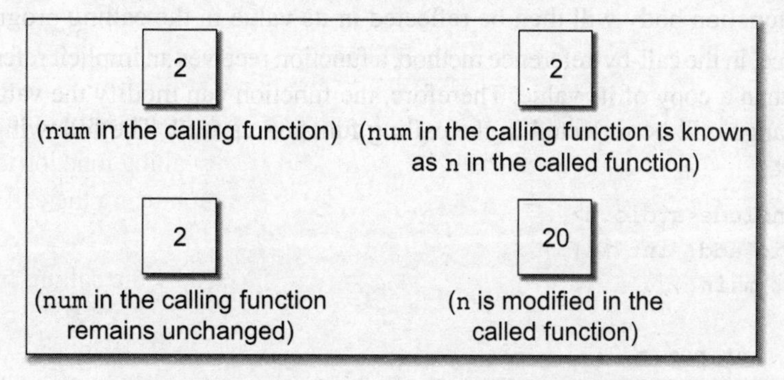

Figure 2.3 Call by value method of argument passing

Following are the points to remember while passing arguments to a function using the call-by-value method:

- When arguments are passed by value, the called function creates new variables of the same data type as the arguments passed to it.
- The values of the arguments passed by the function are copied into the newly created variables.
- Arguments are called by value when the function does not need to modify the values of the original variables in the calling program.
- Values of the variables in the calling functions remain unaffected when the arguments are passed using the call-by-value technique.

Therefore, the call-by-value method of passing arguments to a function must be used only in two cases:

- When the called function does not modify the value of the actual parameter; it simply uses the value of the parameters to perform its task.
- When you want the called function to modify the value of the variables only temporarily and not permanently. Although the called function may modify the value of the variables, these variables remain unchanged in the calling program.

Pros and Cons

The biggest advantage of using the call-by-value technique is that, arguments can be passed as variables, literals, or expressions, while its main drawback is that copying data consumes additional storage space. In addition, it can take a lot of time to copy, thereby resulting in performance penalty, especially if the function is called many times.

2.5.2 Call by Reference

When the calling function passes arguments to the called function using the call-by-value method, the only way to return the modified value of the argument to the caller is explicitly using the return statement. The better option, when a function can modify the value of the argument, is to pass arguments using the call-by-reference technique. In this method, we declare the function parameters as references rather than normal variables. When this is done, any changes made by the function to the arguments it received are visible by the calling program.

To indicate that an argument is passed using call by reference, an ampersand sign (&) is placed after the type in the parameter list. In this way, the changes made to that parameter in the called function body will then be reflected in its value in the calling program.

Hence, in the call-by-reference method, a function receives an implicit reference to the argument, rather than a copy of its value. Therefore, the function can modify the value of the variable and that change will be reflected in the calling function as well. The following code illustrates this concept.

```
#include<stdio.h>
void add( int *n);
int main()
{
    int num = 2;
    printf("\n The value of num before calling the function = %d", num);
    add(&num);
    printf("\n The value of num after calling the function = %d", num);
    return 0;
}
void add( int *n)
{
    *n = *n + 10;
    printf("\n The value of num in the called function = %d", *n);
}
```

Output:

```
The value of num before calling the function = 2
The value of num in the called function = 20
The value of num after calling the function = 20
```

Advantages

The advantages of using the call-by-reference technique of passing arguments include:

- Since arguments are not copied into the new variables, it provides greater time and space-efficiency.
- The function can change the value of the argument and the change is reflected in the caller function.
- A function can return only one value. In case we need to return multiple values, pass those arguments by reference, so that the modified values are visible in the calling function.

Disadvantages

However, the drawback of using this technique is that when an argument is passed using call by reference, it becomes difficult to tell whether that argument is meant for input, output, or both.

Consider the code given below which swaps the value of two integers. Note the value of integers in the calling function and called function.

```
//This function swaps the value of two variables
#include<stdio.h>
void swap_call_by_val(int, int);
void swap_call_by_ref(int *, int *);
void main()
{
    int a=1, b=2, c=3, d=4;
    printf("\n In main(), a = %d and b = %d", a, b);
    swap_call_val(a, b);
    printf("\n In main(), a = %d and b = %d", a, b);
    printf("\n\n In main(), c = %d and d = %d", c, d);
    swap_call_ref(&c, &d);
    printf("\n In main(), c = %d and d = %d", c, d);
    return 0;
}
void swap_call_by_val(int a, int b)
{
    int temp;
    temp = a;
    a = b;
    b = temp;
    printf("\n In function (Call By Value Method) - a = %d and b = %d", a, b);
}
void swap_call_by_refl(int *c, int *d)
{
    int temp;
    temp = *c;
    *c = *d;
    *d = temp;
    printf("\n In function (Call By Reference Method) - c = %d and d = %d",
    *c, *d);
}
```

Output:

```
In main(), a = 1 and b = 2
In function (Call By Value Method) - a = 2 and b = 1
In main(), a = 1 and b = 2
In main(), c = 3 and d = 4
In function (Call By Reference Method) - c = 4 and d = 3
In main(), c = 4 and d = 3
```

2.6 SCOPE OF VARIABLES

In C, all constants and variables have a defined *scope*. By scope, we mean the accessibility and visibility of the variables at different points in the program. A variable or a constant in C has four types of scope: *block*, *function*, *file*, and *program scope*.

2.6.1 Block Scope

We have studied that a statement block is a group of statements enclosed within an opening and closing curly brackets ({ }). If a variable is declared within a statement block, then as soon as the control exits that block, the variable will cease to exist. Such a variable, also known as a *local variable*, is said to have a block scope.

So far, we had been using local variables. For example, if we declare an integer `x` inside a function, then that variable is unknown to the rest of the program outside (that is, outside that function).

A program may also contain a nested block, like a `while` loop inside `main()`. If an integer `x` is declared and initialized in the `main()`, and then re-declared and re-initialized in the `while` loop, then the integer variable `x` will occupy different memory slots and are considered as different variables. Look at the code given below which reveals this concept.

```
#include <stdio.h>
int main()
{
    int x = 10;
    int i;
    printf("\n The value of x outside the while loop is %d", x);
    while (i<3)
    {
        int x = i;
        printf("\n The value of x inside the while loop is %d", x);
        i++;
    }
    printf("\n The value of x outside the while loop is %d", x);
    return 0;
}
```

Output:

```
The value of x outside the while loop is 10
The value of x inside the while loop is 0
The value of x inside the while loop is 1
The value of x inside the while loop is 2
The value of x outside the while loop is 10
```

You may get an error message while executing this code. This is because some C compilers make it mandatory to declare all the variables first before you do anything with them. That is, they permit declaration of variables right after the curly bracket of `main()` starts.

2.6.2 Function Scope

Function scope is applicable only with `goto` label names. That is, the programmer cannot have the same label name inside a function. Since, using `goto` statements these days is not recommended as it is not considered to be a good programming practice, we will neither discuss the `goto` statement nor its function scope.

2.6.3 Program Scope

If you want the functions to be able to access some variables which are not passed to them as arguments, then declare those variables outside any function block. Such variables are commonly known as *global variables*. Hence, global variables are those variables that can be accessed from any point in the program. Look at the code given below.

```
#include<stdio.h>
int x = 10;
void print();
int main()
{
        printf("\n The value of x in the main() = %d", x);
        int x = 2;
        printf("\n The value of local variable x in the main() = %d", x);
        print();
        return 0;
}
void print()
{
        printf("\n The value of x in the print() = %d", x);
}
```

Output:

```
The value of x in the main() = 10
The value of local variable x in the main() = 2
The value of x in the print() = 10
```

In the above code, we see that local variables overwrite the value of global variables. In big programs, using global variables is not recommended unless it is very important to use them because there is a big risk of confusing them with any local variables of the same name.

2.6.4 File Scope

When a global variable is accessible until the end of the file, the variable is said to have *file scope*. To allow a variable to have file scope, declare that variable with the `static` keyword before specifying its data type as follows:

```
static int x = 10;
```

A global static variable can be used anywhere from the file in which it is declared, but it is not accessible by any other files. Such variables are useful when the programmer writes his own header files.

2.7 STORAGE CLASSES

The storage class of a variable defines the scope (visibility) and lifetime of variables and/or the functions declared within a C program. In addition to this, the storage class gives the following information about the variable or the function:

- The storage class of a function or a variable determines the part of memory where storage space will be allocated for that variable or function (whether the variable/function will be stored in a register or in RAM).
- It specifies how long the storage allocation will continue to exist for that function or variable.
- It specifies the scope of the variable or function. That is, the storage class indicates the part of the C program in which the variable name is visible, or the part in which it is accessible. In other words, whether the variable/function can be referenced throughout the program or only within the function, block, or source file where it has been defined.
- It specifies whether the variable or function has internal, external, or no linkage.
- It specifies whether the variable will be automatically initialized to zero or to any indeterminate value.

C supports four storage classes: *automatic*, *register*, *external*, and *static*.

2.7.1 auto Storage Class

The auto storage class specifier is used to explicitly declare a variable with *automatic storage*. It is the default storage class for variables declared inside a block. For example, if we write

```
auto int x;
```

then `x` is an integer that has automatic storage. It is deleted when the block in which `x` was declared exits.

`auto` storage class can be used for variables declared in a block or for function parameters. However, since the variable names or names of function parameters by default have automatic storage, the `auto` storage class specifier is therefore treated as redundant while declaring data. The following points are to be noted while declaring variables with `auto` storage class:

- They should be declared at the start of the program's block (Right after the opening curly bracket {).
- Memory for the variable is automatically allocated upon entry to a block and freed automatically upon exit from that block.
- The scope of the variable is local to the block in which it is declared. These variables may be declared within a nested block.
- Every time the block (in which the automatic variable is declared) is entered, the value of the variable declared with initializers are initialized.

2.7.2 register Storage Class

When a variable is declared using `register` as its storage class, it is stored in a CPU register instead of RAM. Since the variable is stored in RAM, the maximum size of the variable is equal to the register size. One drawback of using a `register` variable is that they cannot be operated using the unary `&` operator because it does not have a memory location associated with it. A `register` variable is declared in the following manner:

```
register int x;
```

`register` variables are used when programmers need quick access to the variable. Note that it is not always necessary that the `register` variable will be stored in the register. Rather, the `register` variable might be stored in a register depending on the hardware and implementation restrictions.

Hence, programmers can only suggest the compiler to store those variables in the registers which are used repeatedly or whose access times are critical. However, for the compiler, it is not an obligation to always accept such requests. In case the compiler rejects the request to store the variable in the register, the variable is treated as having the storage class specifier `auto`.

Like `auto` variables, `register` variables also have automatic storage duration. That is, each time a block is entered, the storage for `register` variables defined in that block are accessible and the moment that block is exited, the variables become no longer accessible to use.

2.7.3 extern Storage Class

`extern` is used to give a reference of a global variable that is visible to all the program files. When a variable is declared using the `extern` keyword, then that variable cannot be initialized, as it simply points to the variable name at a storage location that has been previously defined.

When there are multiple files in a program and you need to use a particular function or variable in a file apart from which it is declared, then use the keyword `extern`. To declare a variable `x` as `extern`, write

```
extern int x;
```

External variables may be declared outside any function in a source code file as any other variable is declared. Usually, external variables are declared and defined at the beginning of a source file.

Memory is allocated for external variables when the program begins execution, and remains allocated until the program terminates. External variables may be initialized while they are declared. However, the initializer must be a constant expression. The compiler will initialize its value only once during the compile time. In case the `extern` variable is not initialized, it will be initialized to zero by default.

External variables have global scope, that is, these variables are visible and accessible from all the functions in the program. However, if any function has a local variable with the same name and types as that of the global or `extern` variable, then references to the name will access the local variable rather than the `extern` variable. Hence, `extern` variables are overwritten by local variables. Let us look at a program code in which we will use the `extern` keyword.

```
// FILE 1.C
#include<stdio.h>
#include<FILE2.C>// Programmer's own header file
int x;
void print(void);
int main()
{
        x = 10;
        print();
        return 0;
}
// END OF FILE1.C
```

```
// FILE2.C
#include<stdio.h>
extern int x;
void print()
{
        printf("\n x = %d", x);
}
main()
{
        // Statements
}
// END OF FILE2.C
```

Note that in the program, we have used two files: `File1` and `File2`. `File1` has declared a global variable `x`. `File1` also includes `File2` which has a print function that uses the external variable `x` to print its value on the screen.

2.7.4 static Storage Class

While `auto` is the default storage class for all local variables, `static` is the default storage class for global variables. `static` variables have a lifetime over the entire program. That is, memory for the `static` variables is allocated when the program begins running and is freed when the program terminates. To declare an integer `x` as `static`, write

```
static int x = 10;
```

`static` variables are accessible from all the functions in this source file. `static` variables, when defined within a function, are initialized at the runtime. The difference between an `auto` variable and a `static` variable is that the `static` variable, when defined within a function, is not re-initialized when the function is called again and again. It is initialized just once and further calls to the function share the value of the `static` variable. Hence, the `static` variable inside a function retains its value during various calls.

`static` variables are automatically initialized to zero when memory is allocated for them. `static` storage class can be specified for `auto` as well as external variables. For example, we can write,

```
static extern int x;
```

Although `static` automatic variables exist even after the block in which they are defined terminates, but their scope is local to the block in which they are defined. `static` variables can be initialized while they are being declared. But this initialization is done only once at the compile time when memory is being allocated for the `static` variable. Also, the value with which the `static` variable is initialized must be a constant expression.

Look at the following code which clearly differentiates between a `static` variable and a normal variable.

```
#include<stdio.h>
void print(void);
int main()
{
        printf("\n First call of print()");
        print();
```

```
        printf("\n\n Second call of print()");
        print();
        printf("\n\n Third call of print()");
        print();
        return 0;
}
void print()
{
        static int x;
        int y = 0;
        printf("\n Static integer variable, x = %d", x);
        printf("\n Integer variable, y = %d", y);
        x++;
        y++;
}
```

Output:

```
First call of print()
Static integer variable, x = 0
Integer variable, y = 0
Second call of print()
Static integer variable, x = 1
Integer variable, y = 0
Third call of print()
Static integer variable, x = 2
Integer variable, y = 0
```

2.8 RECURSIVE FUNCTIONS

A recursive function is defined as a function that calls itself to solve a smaller version of its task until a final call is made which does not require a call to itself. Every recursive solution has two major cases:

Base case The problem is simple enough to be solved directly without making any further calls to the same function.

Recursive case First the problem at hand is divided into simpler sub-parts. Second, the function calls itself but with sub-parts of the problem obtained in the first step. Third, the result is obtained by combining the solutions of simpler sub-parts.

Therefore, recursion is defining large and complex problems in terms of a smaller and more easily solvable problem. In recursive function, a complicated problem is defined in terms of simpler problems and the simplest problem is given explicitly.

To understand recursive functions, let us take an example of calculating the factorial of a number. To calculate `n!`, what we do is multiply the number `n` with the factorial of `(n-1)`. In other words, `n! = n × (n-1)!`

Let us say, we need to find the value of 5!

$$5! = 5 \times 4 \times 3 \times 2 \times 1 = 120$$

This can be written as

`5! = 5 × 4!` where

`4! = 4 × 3!`

Therefore,

`5! = 5 × 4 × 3!`

Similarly, we can also write,

`5! = 5 × 4 × 3 × 2!`

Expanding further,

`5! = 5 × 4 × 3 × 2 × 1!`

We know, `1! = 1`. Therefore, the series of problem and its solution can be given as shown in Fig. 2.4.

PROBLEM	SOLUTION
5!	5 × 4 × 3 × 2 × 1!
= 5 × 4!	= 5 × 4 × 3 × 2 × 1
= 5 × 4 × 3!	= 5 × 4 × 3 × 2
= 5 × 4 × 3 × 2!	= 5 × 4 × 6
= 5 × 4 × 3 × 2 × 1!	= 5 × 24
	= 120

Figure 2.4 Recursive factorial function

Now if you look at the problem carefully, you can see that we can write a recursive function to calculate the factorial of a number. Note that we have said that every recursive function must have a base case and a recursive case. For the factorial function,

- **Base case** is when `n = 1`, because if `n = 1`, the result is known to be 1 as `1! = 1`.
- **Recursive case** of the factorial function will call itself but with a smaller value of n, this case can be given as:
  `factorial(n) = n × factorial (n-1)`

Look at the following code which calculates the factorial of a number recursively.

```
#include<stdio.h>
int Fact(int);
main()
{
        int num;
        printf("\n Enter the number : ");
        scanf("%d", &num);
        printf("\n Factorial of %d = %d", num, Fact(num));
        return 0;
}
int Fact(int n)
{
        if(n==1)
                retrun 1;
```

```
        return (n * Fact(n-1));
}
```

From the above example, let us analyse the basic steps of a recursive program.

Step 1: Specify the base case which will stop the function from making a call to itself.

Step 2: Check to see whether the current value being processed matches with the value of the base case. If yes, process and return the value.

Step 3: Divide the problem into a smaller or simpler sub-problem.

Step 4: Call the function on the sub-problem.

Step 5: Combine the results of the sub-problems.

Step 6: Return the result of the entire problem.

2.8.1 Greatest Common Divisor

The greatest common divisor (GCD) of two numbers (integers) is the largest integer that divides both the numbers. We can find the GCD of two numbers recursively by using the Euclid's algorithm that states:

$$\texttt{GCD(a, b)} = \begin{cases} \texttt{b, if b divides a} \\ \texttt{GCD (b, a mod b), otherwise} \end{cases}$$

GCD can be implemented as a recursive function because if `b` does not divide `a`, then we call the same function (GCD) with another set of parameters that are smaller and simpler than the original ones. (Here we assume that `a > b`. However if `a < b`, then interchange `a` and `b` in the formula given above).

Working

```
Assume a = 62 and b = 8
    GCD( 62, 8)
            rem = 62 % 8 = 6
            GCD(8, 6)
                    rem = 8 % 6 = 2
                    GCD(6, 2)
                            rem = 6 % 2 = 0
    Return 2
```

PROGRAMMING EXAMPLE

5. Write a program to calculate the GCD of two numbers using recursive functions.

```
#include<stdio.h>
int GCD(int, int);
main()
{
        int num1, num2, res;
        printf("\n Enter the two numbers : ");
        scanf("%d %d", &num1, &num2);
        res = GCD(num1, num2);
```

```
        printf("\n GCD of %d and %d = %d", num1, num2, res);
        return 0;
}
int GCD(int x, int y)
{
        int rem;
        rem = x%y;
        if(rem==0)
                return y;
        else
                return (GCD(y, rem));
}
```

2.8.2 Finding Exponents

We can find a solution to find exponent of a number using recursion. To find x^y, the base case would be when y=0, as we know that any number raised to the power 0 is 1. Therefore, the general formula to find x^y can be given as:

$$EXP(x, y) = \begin{cases} 1, \text{ if } y == 0 \\ x*EXP(x^{y-1}), \text{ otherwise} \end{cases}$$

Working

```
exp_rec(2, 4) = 2 * exp_rec(2, 3)
        exp_rec(2, 3) = 2 * exp_rec(2, 2)
                exp_rec(2, 2) - 2 * exp_rec(2, 1)
                        exp_rec(2, 1) = 2 * exp_rec(2, 0)
                                exp_rec(2, 0) = 1
                        exp_rec(2, 1) = 2 * 1 = 2
                exp_rec(2, 2) = 2 * 2 = 4
        exp_rec(2, 3) = 2 * 4 = 8
exp_rec(2, 4) = 2 * 8 = 16
```

PROGRAMMING EXAMPLE

6. Write a program to calculate exp (x, y) using recursive functions.

```
#include<stdio.h>
int exp_rec(int, int);
main()
{
        int num1, num2, res;
        printf("\n Enter the two numbers : ");
        scanf("%d %d", &num1, &num2);
        res = exp_rec(num1, num2);
        return 0;
}
```

```
int exp_rec(int x, int y)
{
        if( y==0)
                return 1;
        else
                return (x * exp_rec(x, y-1));
}
```

2.8.3 Fibonacci Series

The Fibonacci series can be given as:

0 1 1 2 3 5 8 13 21 34 55......

That is, the third term of the series is the sum of the first and second terms. On similar grounds, the fourth term is the sum of second and third terms, so on and so forth. Now we will design a recursive solution to find the nth term of the Fibonacci series. The general formula to do so can be given as:

```
FIB(n) = { 1, if n <= 2
         { FIB (n – 1) + FIB (n – 2), otherwise
```

Working

If n = 7.

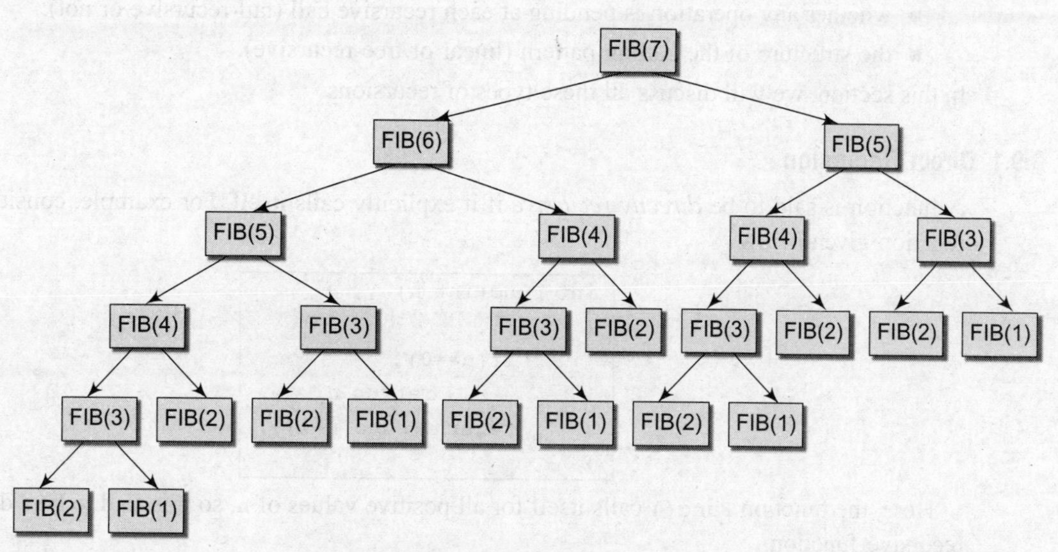

Figure 2.5 Recursion structure of Fib function

PROGRAMMING EXAMPLE

7. Write a program to print the Fibonacci series using recursion.

```
#include<stdio.h>
int Fibonacci(int);
```

```
main()
{
        int n;
        printf("\n Enter the number of terms in the series : ");
        scanf("%d", &n);
        for(i=0;i<n;i++)
                printf("\n Fibonacci (%d) = %d", i, Fibonacci(i));
        return 0;
}
int Fibonacci(int num)
{
        if(num <= 2)
                return 1;
        return ( Fibonacci (num - 1) + Fibonacci(num - 2));
}
```

2.9 TYPES OF RECURSION

Recursion is a technique that breaks a problem into one or more sub-problems that are similar to the original problem. Any recursive function can be characterized based on:

- whether the function calls itself directly or indirectly (direct or indirect recursion).
- whether any operation is pending at each recursive call (tail-recursive or not).
- the structure of the calling pattern (linear or tree-recursive).

In this section, we will discuss all these types of recursions.

2.9.1 Direct Recursion

A function is said to be *directly recursive* if it explicitly calls itself. For example, consider the function given below.

```
int Func(int n)
{
        if(n==0)
                retrun n;
        return (Func(n-1));
}
```

Here, the function `Func()` calls itself for all positive values of `n`, so it is said to be a directly recursive function.

2.9.2 Indirect Recursion

A function is said to be *indirectly recursive* if it contains a call to another function which ultimately calls it. Look at the functions given below. These two functions are indirectly recursive as they both call each other.

int Func1(int n)	int Func2(int x)
{	{
if(n==0)	return Func1(x-1);
return n;	}
return Func2(n);	
}	

2.9.3 Tail Recursion

A recursive function is said to be *tail recursive* if no operations are pending to be performed when the recursive function returns to its caller. That is, when the called function returns, the returned value is immediately returned from the calling function. Tail recursive functions are highly desirable because they are much more efficient to use as in their case, the amount of information that has to be stored on the system stack is independent of the number of recursive calls.

```
int Fact( int n)
{
    if(n==1)
        retrun 1;
    return (n * Fact(n-1));
}
```

For example, the factorial function that we have written is a non-tail-recursive function, because there is a pending operation of multiplication to be performed on return from each recursive call.

Whenever there is a pending operation to be performed, the function becomes non-tail-recursive. In such a non-tail recursive function, information about each pending operation must be stored so the amount of information directly depends on the number of calls. However, the same factorial function can be written in a tail-recursive manner as shown below.

int Fact(n)	int Fact1(int n, int res)
{	{
return Fact1(n, 1);	if (n==1)
}	return res;
	return Fact1(n-1, n*res);
	}

In the code, `Fact1` function preserves the syntax of the `Fact(n)`. Here the recursion occurs in the `Fact1` function and not in `Fact` function. Carefully observe that `Fact` has no pending operation to be performed on return from recursive calls. The value computed by the recursive call is simply returned without any modification. So in this case, the amount of information to be stored on the system stack is constant (just the value of `n` and `res` needs to be stored) and is independent of the number of recursive calls.

Converting Recursive Functions to Tail Recursive

A non-tail recursive function is converted into a tail recursive function by using an 'auxiliary' parameter as we did in case of the Factorial function. The auxiliary parameter is used to form the result. When we use such a parameter, the pending operation is incorporated into the auxiliary parameter so that the recursive call no longer has a pending operation. We generally use an auxiliary function while using the auxiliary parameter. This is done to keep the syntax clean and to hide the fact that auxiliary parameters are needed.

2.9.4 Linear and Tree Recursion

Recursive functions can also be characterized depending on the way in which the recursion grows– in a linear fashion or forming a tree structure.

In simple words, a recursive function is said to be *linearly recursive* when no pending operation involves another recursive call to the function. For example, the factorial function is linearly recursive as the pending operation involves only multiplication to be performed and does not involve another call to fact.

On the contrary, a recursive function is said to be *tree recursive* (or *non-linearly recursive*) if the pending operation makes another recursive call to the function. For example, the Fibonacci function `Fib` in which the pending operations recursively calls the `Fib` function.

```
int Fibonacci(int num)
{
   if(num <= 2)
      return 1;
   return (Fibonacci (num — 1) + Fibonacci(num — 2));
}
```

Observe the series of function calls. When the function returns, the pending operation in turn calls the function.

```
Fibonacci(7) = Fibonacci(6) + Fibonacci(5)
Fibonacci(6) = Fibonacci(5) + Fibonacci(4)
Fibonacci(5) = Fibonacci(4) + Fibonacci(3)
Fibonacci(4) = Fibonacci(3) + Fibonacci(2)
Fibonacci(3) = Fibonacci(2) + Fibonacci(1)
```

Now we have,

```
Fibonacci(3) = 1 + 1 = 2
Fibonacci(4) = 2 + 1 = 3
Fibonacci(5) = 3 + 2 = 5
Fibonacci(6) = 3 + 5 = 8
Fibonacci(7) = 5 + 8 = 13
```

2.10 RECURSION VS ITERATION

Recursion is more of a top-down approach to problem-solving in which the original problem is divided into smaller sub-problems. On the contrary, iteration follows a bottom-up approach that begins with what is known and then constructing the solution step-by-step.

Recursion is an excellent way of solving complex problems especially when the problem can be defined in recursive terms. For such problems, a recursive code can be written and modified in a much simpler and clearer manner.

However, recursive solutions are not always the best solutions. In some cases, recursive programs may require substantial amount of runtime overhead. Therefore, when implementing a recursive solution, there is a trade-off involved between the time spent in constructing and maintaining the program and the cost incurred in running time and memory space required for the execution of the program.

Whenever, a recursive function is called, some amount of overhead in the form of a runtime stack is always involved. Before jumping to the function with a smaller parameter, the original parameters, the local variables, and the return address of the calling function are all stored on the system stack. Therefore, while using recursion, a lot of time is needed to first push all the information on the stack when function is called and then time is again involved in retrieving the information stored on the stack once the control passes back to the calling function.

To conclude, one must use recursion only to find solution to a problem for which no obvious iterative solution is known. To summarize the concept of recursion, let us briefly discuss the pros and cons of recursion.

Advantages

The advantages of using a recursive program include:

- Recursive solutions often tend to be shorter and simpler than non-recursive ones.
- Code is clearer and easier to use.
- Recursion represents the original formula to solve a problem.
- Follows a divide-and-conquer technique to solve problems.
- In some (limited) instances, recursion may be more efficient.

Disadvantages

The drawbacks of using a recursive program include:

- For some programmers and readers, recursion is a difficult concept.
- Recursion is implemented using system stack. If the stack space on the system is limited, recursion to a deeper level will be difficult to implement.
- Aborting a recursive process in midstream is slow and sometimes nasty.
- Using a recursive function takes more memory and time to execute as compared to its non-recursive counterpart.
- It is difficult to find bugs, particularly when using global variables.

Conclusion The advantages of recursion pays off for the extra overhead involved in terms of time and space required.

2.11 TOWER OF HANOI

Tower of Hanoi is one of the main applications of recursion. It says, 'if you can solve `n-1` cases, then you can easily solve the nth case.'

Figure 2.6 shows three rings mounted on pole `A`. The problem is to move all these rings from pole `A` to pole `C` while maintaining the same order. The main issue is that the smaller disk must always come above the larger disk.

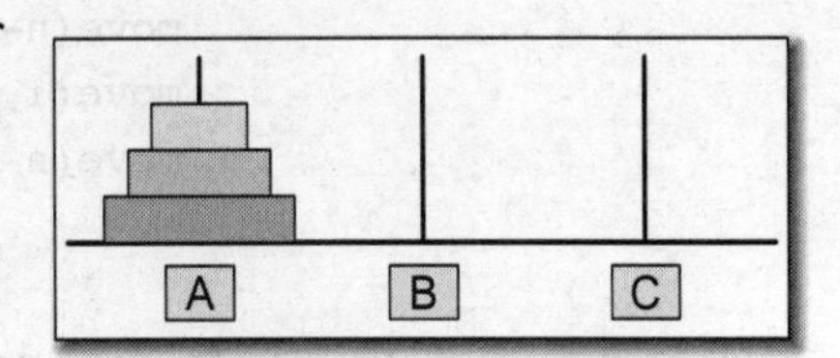

Figure 2.6 Tower of Hanoi

We will do this using a spare pole. In our case, `A` is the source pole, `C` is the destination pole, and `B` is the spare pole. To transfer all the three rings from `A` to `C`, we will first shift the upper two rings (`n-1 rings`) from the source pole to the spare pole. That is, move first two rings from pole `A` to `B` as given in Fig. 2.7.

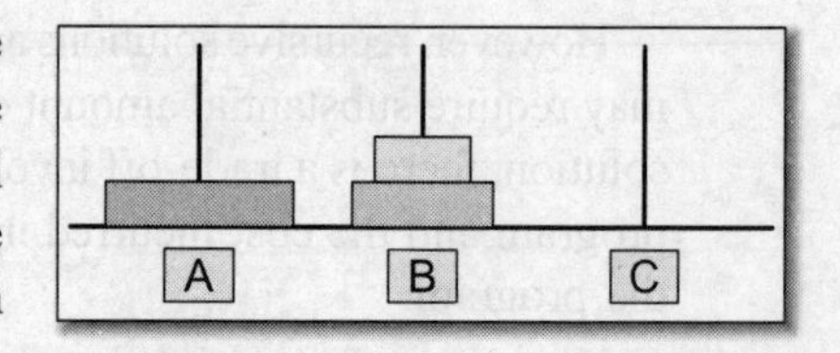

Figure 2.7 Move rings from A to B

Now that `n-1` rings have been removed from pole `A`, the source pole, the `n`th ring can be easily moved from the source pole `(A)` to the destination pole `(C)`. Figure 2.8 shows this step.

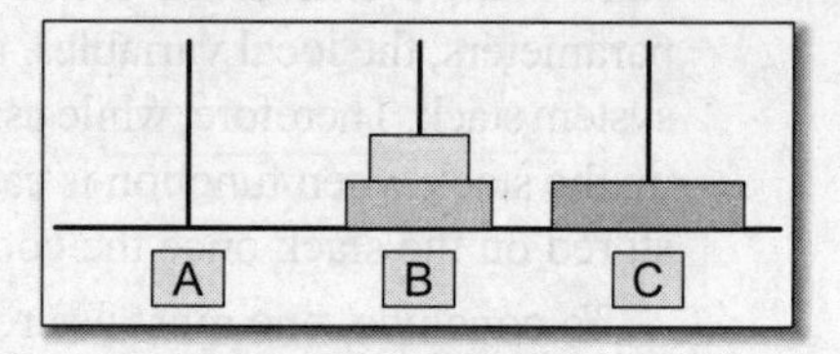

Figure 2.8 Move ring from A to C

The final step is to move the `n-1` rings from the spare pole `(B)` to the destination pole `(C)`. This is shown in Fig. 2.9.

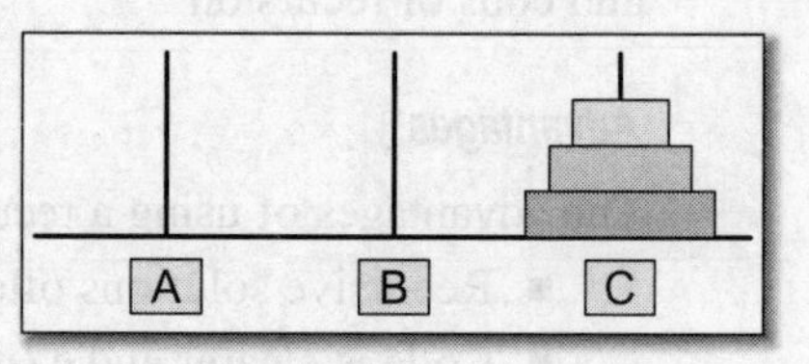

Figure 2.9 Move rings from B to C

To summarize, the solution to our problem of moving `n` rings from `A` to `C` using `B` as a spare can be given as:

Base case: if `n=1`

- Move the ring from `A` to `C` using `B` as a spare

Recursive case:

- Move `n-1` rings from `A` to `B` using `C` as the spare
- Move the one ring left on `A` to `C` using `B` as the spare
- Move `n-1` rings from `B` to `C` using `A` as the spare

Look at the following code which implements the solution of the Tower of Hanoi problem.

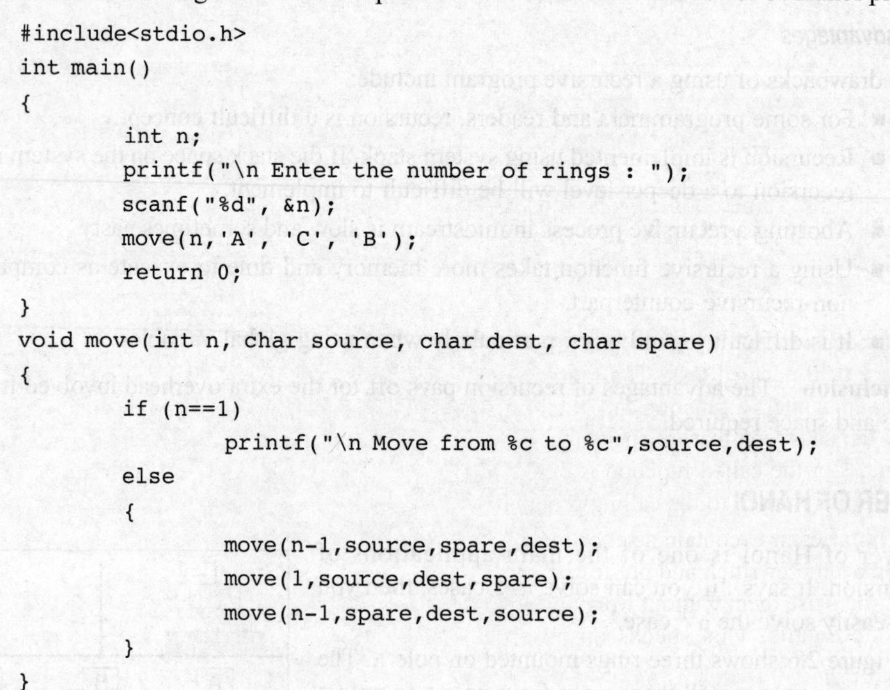

```
#include<stdio.h>
int main()
{
        int n;
        printf("\n Enter the number of rings : ");
        scanf("%d", &n);
        move(n,'A', 'C', 'B');
        return 0;
}
void move(int n, char source, char dest, char spare)
{
        if (n==1)
                printf("\n Move from %c to %c",source,dest);
        else
        {
                move(n-1,source,spare,dest);
                move(1,source,dest,spare);
                move(n-1,spare,dest,source);
        }
}
```

Let us look at the Tower of Hanoi problem in detail using the given program. Figure 2.10 explains the working of the program using 1, then 2, and finally, 3 rings.

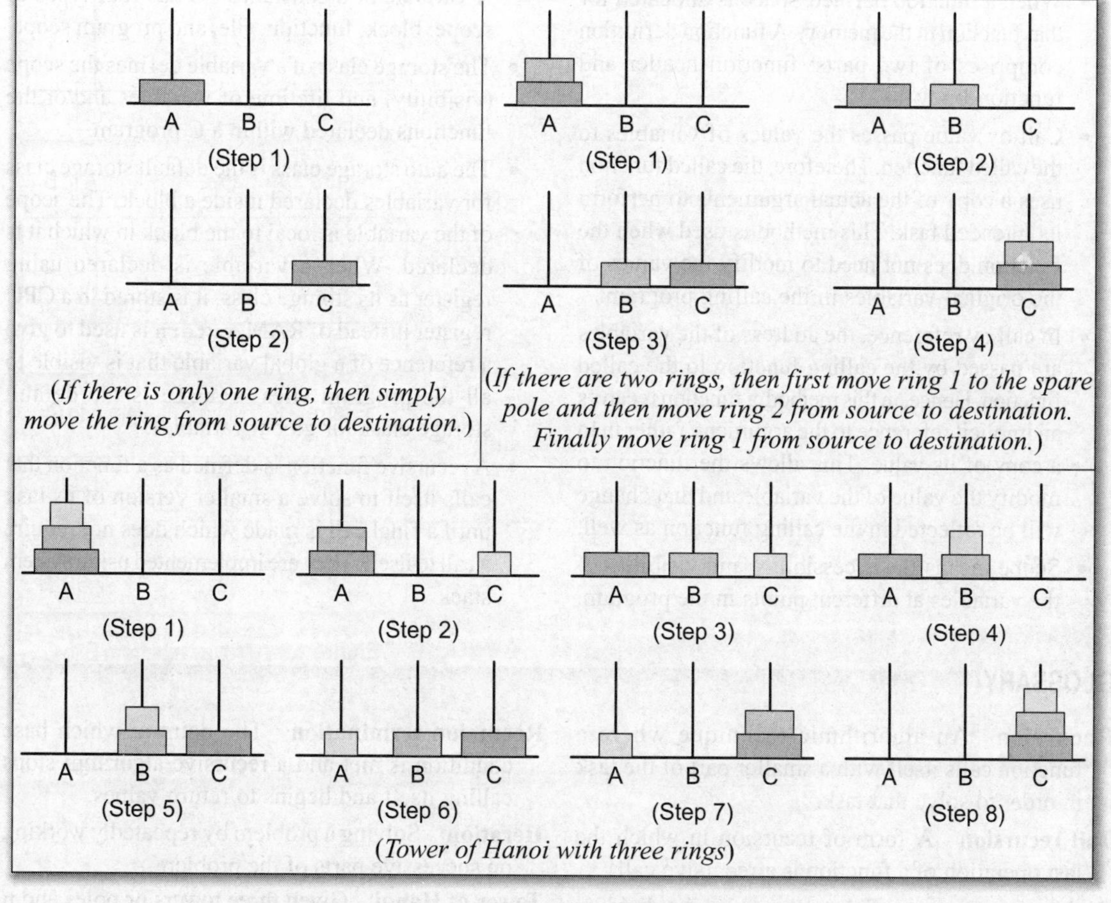

Figure 2.10 Working of Tower of Hanoi with one, two, and three rings

Hence, the same concept can be extended to move four or more poles.

SUMMARY

- Every function in the program is supposed to perform a well-defined task. The moment the compiler encounters a function call, the control jumps to the statements that are a part of the called function. After the called function is executed, the control is returned to the calling program.
- All the libraries in C contain a set of functions that have been prewritten and pre-tested, so the programmers use them without worrying about their code details. This speeds up program development.
- While function declaration statement identifies a function with its name, a list of arguments that it accepts and the type of data it returns, the function definition, on the other hand, consists of a function header that identifies the function, followed by the body of the function containing the executable code for that function.
- `main()` is the function that is called by the operating system and therefore, it is supposed to return the result of its processing to the operating system.
- Placing the function declaration statement prior to its use enables the compiler to make a check on the arguments used while calling that function.
- A function having `void` as its return type cannot return any value. Similarly, a function having void as its parameter list cannot accept any value.

- When a function defined, space is allocated for that function in the memory. A function definition comprises of two parts: function header and function body.
- Call by value passes the values of variables to the called function. Therefore, the called function uses a copy of the actual arguments to perform its intended task. This method is used when the function does not need to modify the values of the original variables in the calling program.
- In call by reference, the address of the variables are passed by the calling function to the called function. Hence, in this method, a function receives an implicit reference to the argument, rather than a copy of its value. This allows the function to modify the value of the variable and that change will be reflected in the calling function as well.
- Scope means the accessibility and visibility of the variables at different points in the program. A variable or a constant in C has four types of scope: block, function, file, and program scope.
- The storage class of a variable defines the scope (visibility) and lifetime of variables and/or the functions declared within a C program.
- The auto storage class is the default storage class for variables declared inside a block. The scope of the variable is local to the block in which it is declared. When a variable is declared using register as its storage class, it is stored in a CPU register instead of RAM. `extern` is used to give a reference of a global variable that is visible to all the program files. `static` is the default storage class for global variables.
- A recursive function is defined as a function that calls itself to solve a smaller version of its task until a final call is made which does not require a call to itself. They are implemented using system stack.

GLOSSARY

Recursion An algorithmic technique where a function calls itself with a smaller part of the task in order to solve that task.

Tail recursion A form of recursion in which the last operation of a function is a recursive call.

Divide and conquer Solving a problem by dividing it into two or more smaller instances. Each of these smaller instances is recursively solved, and the solutions are combined to produce a solution for the original problem.

Recursion tree Technique to analyse the complexity of an algorithm by diagramming the recursive function calls.

Recursion termination The point at which base condition is met and a recursive algorithm stops calling itself and begins to return values.

Iteration Solving a problem by repeatedly working on successive parts of the problem.

Tower of Hanoi Given three towers or poles and n disks of decreasing sizes, move the disks from one pole to one by one without putting a larger disk on a smaller one.

Argument A value passed to the called function by the calling function.

Block A sequence of definitions, declarations, and statements, enclosed within curly brackets {}.

EXERCISES

Review Questions

1. Define a function. Why are they needed?
2. Explain the concept of making function calls.
3. Differentiate between function declaration and function definition.
4. How many types of storage classes does C language support? Why do we need different types of such classes?
5. What are the features of each storage class.
6. Explain the concept of recursive functions with example.
7. Differentiate between an iterative function and a recursive function. Which one will you prefer to use in what circumstances?

8. Write a program to calculate factorial of a number using recursion. Also write a non-recursive procedure to do the same job.
9. Explain the tower of Hanoi problem.
10. Differentiate between call by value and call by reference using suitable examples.
11. What do you understand by scope of a variable? Explain in detail with suitable examples.
12. Why is function declaration statement placed prior to function definition?
13. Write a function to reverse a string using recursion.

Programming Exercises

1. Write a program to concatenate two strings using recursion.
2. Write a program to read an integer number. Print the reverse of this number using recursion.
3. Write a program to compute `F(x, y)`, where

   `F(x, y) = F(x-y, y) + 1 if y ≤ x`

   and `F(x, y) = 0 if x<y`
4. Write a program to compute `F(n, r)` where `F(n, r)` can be recursively defined as:

   `F(n, r) = F(n-1, r) + F(n-1, r-1)`
5. Write a program to compute `Lambda(n)` for all positive values of `n` where `Lambda(n)` can be recursively defined as:

   `Lambda(n) = Lambda(n/2) + 1 if n>1`

   `and Lambda(n) = 0 if n =1`
6. Write a program to compute `F(M, N)` where `F(M, N)` can be recursively defined as:

   `F(M,N) = 1 if M=0 or M=N=1`

   `and F(M,N) = F(M-1,N) + F(M-1, N-1), otherwise`
7. Write a menu driven program to add, subtract, multiply, and divide two integers using functions.
8. Write a program to find biggest of three integers using functions.
9. Write a program to calculate the area of a triangle using functions.
10. Write a program to find whether a number is even or odd using functions.
11. Write a program to illustrate call by value technique of passing arguments to a function.
12. Write a program to illustrate call by reference technique of passing arguments to a function.
13. Write a program to swap two integers using call by value method of passing arguments to a function.
14. Write a program to swap two integers using call by reference method of passing arguments to a function.
15. Write a program to calculate factorial of a number with recursion.
16. Write a program to calculate factorial of a number without recursion.
17. Write a program to reverse a string using recursion.
18. Write a program to reverse a string without using recursion.
19. Write a program to calculate GCD of two numbers using recursion.
20. Write a program to calculate GCD of two numbers without using recursion.
21. Write a program to calculate exp(x, y) using recursion.
22. Write a program to calculate exp(x, y) without using recursion.
23. Write a program to print the Fibonacci series using recursion.
24. Write a program to print the Fibonacci series without using recursion.

Multiple Choice Questions

1. The function that is invoked is known as
   (a) Calling function
   (b) Caller function
   (c) Called function
   (d) Invoking function
2. Function declaration statement identifies a function with its
   (a) Name
   (b) Arguments
   (c) Data type of return value
   (d) All of these
3. Which return type cannot return any value to the caller?
   (a) int (b) float
   (c) void (d) double
4. Memory is allocated for a function when the function is
   (a) declared (b) defined
   (c) called (d) returned

5. Which keyword allows a variable to have file scope?
   (a) auto (b) static
   (c) register (d) extern
6. The default storage class of global variables is
   (a) auto (b) static
   (c) register (d) extern

True or False

1. The calling function must pass parameters to the called function.
2. Function header is used to identify the function.
3. The name of function is global.
4. No function can be declared within the body of another function.
5. The function call statement invokes the function.
6. Auto variables are stored inside CPU registers.
7. Extern variables are initialized by default.
8. The default storage class of local variables is extern.
9. Recursion follows a divide-and-conquer technique to solve problems.
10. Local variables overwrite the value of global variables.

Fill in the Blanks

1. ______ provides an interface to use the function.
2. After the function is executed, the control passes back to the ______.
3. A function that uses another function is known as the ______.
4. The inputs that the function takes are known as ______.
5. `Main()` is called by the ______.
6. Function definition consist of ______ and ______.
7. In ______ method, address of the variable is passed by the calling function to the called function.
8. Function scope is applicable only with ______.
9. Recursive functions are implemented using ______.
10. ______ function is defined as a function that calls itself.

3 Pointers

Learning Objective

In this chapter, we will learn about a special variable known as the pointer variable. Pointer variables are used to access the memory directly using memory addresses. In this chapter, we will use variables that point to all types of variables, be it `int, float, double,` or `char`. Pointers also provide a handy way to access arrays and strings. Moreover, we may also have function pointers in our C programs. Last but not the least, the most powerful use of pointers is to dynamically allocate memory for variables. We will learn the basics of pointers in this chapter and then in the subsequent chapters, we will be using them to implement our data structures.

3.1 INTRODUCTION

Every variable in C has a name and a value associated with it. When a variable is declared, a specific block of memory within the computer is allocated to hold the value of that variable. The size of the allocated block depends on the data type. Let us write a program to find the size of various data types on your system. (Note that the size of integer may vary from one system to another. On 32-bit systems, an integer variable is allocated 4 bytes while on 16-bit systems, it is allocated 2 bytes).

```
#include<stdio.h>
int main()
{
    printf("\n The size of short integer is : %d", sizeof(short int));
    printf("\n The size of unsigned integer is : %d", sizeof(unsigned int));
    printf("\n The size of signed integer is : %d", sizeof(signed int));
    printf("\n The size of integer is : %d", sizeof(int));
    printf("\n The size of long integer is : %d", sizeof(long int));
    printf("\n The size of character is : %d", sizeof(char));
    printf("\n The size of unsigned character is : %d", sizeof(unsigned char));
    printf("\n The size of signed character is : %d", sizeof(signed char));
    printf("\n The size of floating point number is : %d", sizeof(float));
    printf("\n The size of double number is : %d", sizeof(double));
    return 0;
}
```

Consider the following statement.

```
int x = 10;
```

When this statement executes, the compiler sets aside 2 bytes of memory to hold the value 10. It also sets up a symbol table in which it adds the symbol x and the relative address in the memory where those 2 bytes were set aside.

Thus, every variable in C has a value and also a memory location (commonly known as *address*) associated with it. Some texts use the term `rvalue` and `lvalue` for the value and the address of the variable, respectively.

The `rvalue` appears on the right side of the assignment statement (10 in the above statement) and cannot be used on the left side of the assignment statement. Therefore, writing `10 = k;` is illegal. If we write,

```
int x, y;
x = 10;
y = x;
```

then, we have two integer variables `x` and `y`. The compiler reserves memory for the integer variable `x` and stores the `rvalue` 10 in it. When we say `y=x`, then `x` is interpreted as its `rvalue` (since it is on the right hand side of the assignment operator =). Therefore, here `x` refers to the value stored at the memory location set aside for `x`, in this case 10. After this statement is executed, the `rvalue` of `y` is also 10.

You must be wondering why we are discussing addresses and `lvalues`. Actually pointers are nothing but memory addresses. A pointer is a variable that contains the memory location of another variable. Therefore, a pointer is a variable that represents the location of a data item, such as a variable or an array element. Pointers are frequently used in C, as they have a number of useful applications. These applications include:

- Pointers are used to pass information back and forth between a function and its reference point.
- Pointers enable the programmers to return multiple data items from a function via function arguments.
- Pointers provide an alternate way to access the individual elements of an array.
- Pointers are used to pass arrays and strings as function arguments. We will discuss this in subsequent chapters.
- Pointers are used to create complex data structures, such as trees, linked list, linked stack, linked queue, and graphs.
- Pointers are used for the dynamic memory allocation of a variable.

3.2 DECLARING POINTER VARIABLES

The general syntax of declaring pointer variables can be given as below.

```
data_type *ptr_name;
```

Here, `data_type` is the data type of the value that the pointer will point to. For example,

```
int *pnum;
char *pch;
float *pfnum;
```

In each of the above statements, a pointer variable is declared to point to a variable of the specified data type. Although all these pointers, `pnum`, `pch`, and `pfnum` point to different data types, they will occupy the same amount of space in the memory. But how much space they will occupy will depend on the platform where the code is going to run. Now let us declare an integer pointer variable and start using it in our program code.

```
int x= 10;
int *ptr;
ptr = &x;
```

In the above statement, `ptr` is the name of the pointer variable. The `*` informs the compiler that `ptr` is a pointer variable and the `int` specifies that it will store the address of an integer variable. An integer pointer variable, therefore, 'point to' an integer variable. In the last statement, `ptr` is assigned the address of `x`. The `&` operator retrieves the `lvalue` (address) of `x`, and copies that to the contents of the pointer `ptr`. Consider the memory cells given in Fig. 3.1.

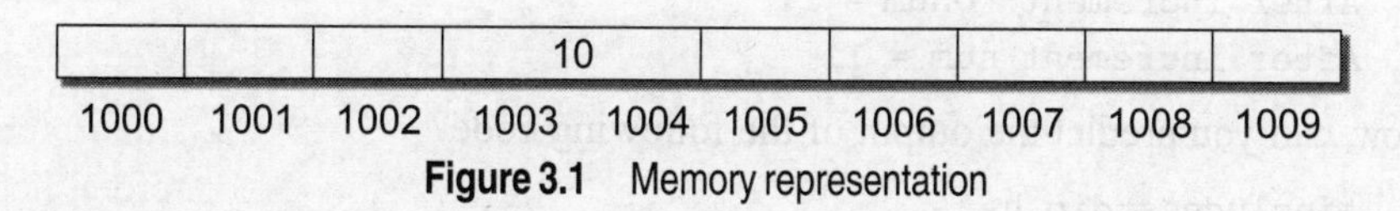

Figure 3.1 Memory representation

Now, since `x` is an integer variable, it will be allocated 2 bytes. Assuming that the compiler assigns it memory locations 1003 and 1004, we say the value of `x = 10` and the address of `x` (written as `&x`) is equal to 1003, that is the starting address of `x` in the memory. When we write, `ptr = &x`, then `ptr = 1003`.

We can 'dereference' a pointer, *i.e.*, we can refer to the value of the variable to which it points by using unary the `*` operator as in `*ptr`. That is, `*ptr = 10`, since 10 is the value of `x`. Look at the following code which shows the use of a pointer variable.

```
#include<stdio.h>
int main()
{
	int num, *pnum;
	pnum = &num;
	printf("\n Enter the number : ");
	scanf("%d", &num);
	printf("\n The number that was entered is : %d", *pnum);
	return 0;
}
```

Output:

```
Enter the number : 10
The number that was entered is : 10
```

What will be the value of `*(&num)`? It is equivalent to simply writing `num`. We can also assign values to variables using the pointer variables and modify their value. The following code shows this.

```
#include<stdio.h>
int main()
{
	int num, *pnum;
	pnum = &num;
```

```
        *pnum = 10;
        printf("\n *pnum = %d", *pnum);
        printf("\n num = %d", num);
        *pnum = *pnum + 1;              // increments the value of num
        printf("\n After increment *pnum = %d", *pnum);
        printf("\n After increment num = %d", num);
        return 0;
    }
```

Output:

```
*pnum = 10
num = 10
After increment *pnum = 11
After increment num = 11
```

Now, can you predict the output of the following code?

```
#include<stdio.h>
int main()
{
    int num, *pnum1, *pnum2;
    pnum1 = &num;
    *pnum1 = 10;
    pnum2 = pnum1;
    printf("\n Value of num using all three variables (num, *pnum1, *pnum2)
= %d %d %d", num, *pnum1, *pnum2);
    printf("\n Address of num using all three variables (&num, pnum1, pnum2)
= %x %x %x", num, pnum1, pnum2);
    return 0;
}
```

While the first `printf` statement will print the value of num, the second `printf` statement will print the address of num. These are just three different ways to refer to the value and address of the same variable. Note that any number of pointers can point to the same address.

One thing to remember always is that, the data type of the pointer variable and the variable whose address it will store must both be of the same type. Therefore, the following code is not valid.

```
int x = 10;
float y = 2.0;
int *px;
float *py;
px = &y; //INVALID
py = &x; //INVALID
```

3.3 POINTER EXPRESSIONS AND POINTER ARITHMETIC

Like other variables, pointer variables can also be used in expressions. For example, if `ptr1` and `ptr2` are pointers, then the following statements are valid.

```
int num1 = 2, num2 = 3, sum = 0, mul = 0, div = 1;
int *ptr1, *ptr2;
ptr1 = &num1;
ptr2 = &num2;
sum = *ptr1 + *ptr2;
mul = sum * (*ptr1);
*ptr2 += 1;
div = 9 + (*ptr1)/(*ptr2) - 30;
```

In C, the programmer may add integers to or subtract integers from pointers as well as subtract one pointer from the other. We can also use shorthand operators with the pointer variables as we use them with other variables.

C also allows comparing pointers by using relational operators in the expressions. For example, `p1 > p2`, `p1 == p2` and `p1! = p2` are all valid in C.

While using pointers, unary increment `(++)` and decrement `(--)` operators have greater precedence than the dereference operator `(*)`. But both these operators have a special behaviour when used as a suffix. In that case, the expression is evaluated with the value it had before being increased. Therefore, the expression `*ptr++` is equivalent to `*(ptr++)`, as ++ has greater operator precedence than `*`. Therefore, the expression will increase the value of `ptr` so that it now points to the next element. This means that the statement `*ptr++` does not do the intended task. Therefore, to increment the value of the variable whose address is stored in `ptr`, you should write `(*ptr)++`. Consider another C statement:

```
int num1=2, num2=3;
int *p = &num1, *q=&num2;
*p++ = *q++;
```

What will `*p++ = *q++` do? As ++ has a higher precedence than `*`, the value assigned to `*p` is `*q` before both `p` and `q` are incremented. So the statement is equivalent to writing:

```
*p = *q;
++p; ++q;
```

PROGRAMMING EXAMPLES

1. Write a program to print Hello world using pointers.

```
#include<stdio.h>
int main()
{
        char *ch = "Hello world";
        printf("%s", ch);
        return 0;
}
```

2. Write a program to print a character. Also print its ASCII value and rewrite the character in upper case.

```
#include<stdio.h>
int main()
```

```
{
        int ch, *pch;
        pch = &ch;
        printf("\n Enter the character : ");
        scanf("%c", &ch);
        printf("\n The character that was entered is : %c", *pch);
        printf("\n The ascii value of the character is : %d", *pch);
        printf("\n The character in upper case is : %c", *pch - 32);
        return 0;
}
```

3. Write a program to add two floating point numbers. The result should contain only two digits after the decimal.

```
#include<stdio.h>
int main()
{
        float num1, num2, sum = 0.0;
        float *pnum1, *pnum2, *psum;
        *pnum1 = &num1;
        *pnum2 = &num2;
        *psum = &sum;
        printf("\n Enter the two numbers : ");
        scanf("%f %f", pnum1, pnum2); // pnum1 = &num1;
        *psum = *pnum1 + *pnum2;
        printf("\n %f + %f = %.2f", *pnum1, *pnum2, *psum);
        return 0;
}
```

4. Write a program to calculate the area of a circle.

```
#include<stdio.h>
#include<conio.h>
int main()
{
        double radius, area = 0.0;
        double *pradius, *parea;
        pradius = &radius;
        parea = &area;
        printf("\n Enter the radius of the circle : ");
        scanf("%lf", pradius);
        *parea = 3.14 * (*pradius) * (*pradius);
        printf("\n The area of the circle with radius = %lf %lf", *pradius,
*parea);
        return 0;
}
```

5. Write a program to convert a floating point number into an integer.

```
#include<stdio.h>
int main()
```

```
{
        float fnum, *pfnum;
        int num, *pnum;
        pfnum = &fnum;
        pnum= &num;
        printf("\n Enter the floating point number : ");
        scanf("%f", &fnum);
        *pnum = (int)*pfnum;
        printf("\n The integer equivalent of %f = %d", *pfnum, *pnum);
        return 0;
}
```

6. Write a program to find the biggest of three numbers.

```
#include<stdio.h>
int main()
{
        int num1, num2, num3;
        int *pnum1, *pnum2, *pnum3;
        pnum1 = &num1;
        pnum2 = &num2;
        pnum3 = &num3;
        printf("\n Enter the first number : ");
        scanf("%d", pnum1);
        printf("\n Enter the second number : ");
        scanf("%d", pnum2);
        printf("\n Enter the third number : ");
        scanf("%d", pnum3);
        if(*pnum1 > *pnum2 && *pnum1 > *pnum3)
                printf("\n %d is the largest number", *pnum1);
        if(*pnum2 > *pnum1 && *pnum2 > *pnum3)
                printf("\n %d is the largest number", *pnum2);
        else
                printf("\n %d is the largest number", *pnum3);
        return 0;
}
```

7. Write a program to enter a character and then determine whether it a vowel or not.

```
#include<stdio.h>
int main()
{
        char ch, *pch;
        pch = &ch;
        printf("\n Enter any character : ");
        scanf("%c", pch);
        if(*pch ='a' || *pch =='e' || *pch=='i' || *pch=='o' || *pch=='u'
        || *pch=='A' || *pch=='E' || *pch=='I' || *pch=='O' || *pch=='U' )
```

```
            printf("\n %c is a VOWEL", ch);
        else
            printf("\n %c is not a vowel");
        return 0;
    }
```

8. Write a program to test whether a number is positive, negative, or equal to zero.

```
#include<stdio.h>
int main()
{
        int num, *pnum;
        pnum = &num;
        printf("\n Enter any number : ");
        scanf("%d", pnum);
        if(*pnum>0)
            printf("\n The number is positive");
        else
        {
            if(*pnum<0)
                printf("\n The number is negative");
            else
                printf("\n The number is equal to zero");
        }
        return 0;
}
```

9. Write a program to take an input from the user and then check whether it is a number or a character. If it is a character, determine whether it is in upper case or lower case.

```
#include<stdio.h>
int main()
{
        char ch, *pch;
        pch = &ch;
        printf("\n Enter any character : ");
        scanf("%c", pch);
        if(*pch >= 'A' && *pch <= 'Z')
            printf("\n Upper case character was entered");
        if(*pch >='a' && *pch <='z')
            printf("\n Lower case character was entered");
        else if(*pch >= '0' && *pch <= '9')
            printf("\n You entered a number");
        return 0;
}
```

10. Write a program to display the sum and average of numbers from *m* to *n*.

```
#include<stdio.h>
int main()
{
	int num, *pnum;
	int m, *pm;
	int n, *pn;
	int sum = 0, *psum;
	float avg, *pavg;
	pn = &n;
	pm = &m;
	pnum = &num;
	psum = &sum;
	pavg = &avg;
	printf("\n Enter the starting and ending limit of the numbers to be
summed : %d %d", pm, pn);
	while( *pm <= *pn)
	{
		*psum = *psum + *pm;
		*pm = *pm + 1;
	}
	printf("\n Sum of numbers = %d", *psum);
	*pavg = *psum /(float)(*pm - *pn);
	printf("\n Average of numbers = %d", *pavg);
	return 0;
}
```

11. Write a program to print all even numbers from *m* to *n*.

```
#include<stdio.h>
#include<conio.h>
int main()
{
	int m, *pm;
	int n, *pn;
	pn = &n;
	pm = &m;
	printf("\n Enter the starting and ending limit of the numbers : %d
%d", pm, pn);
	while(*pm <= *pn)
	{
		if(*pm %2 == 0)
			printf("\n %d is even", *pm);
		else
			printf("\n %d is odd", *pm);
		*pm = *pm + 1;
```

```
    }
    printf("\n Sum of numbers = %d", *psum);
    return 0;
}
```

12. Write a program to read numbers until –1 is entered. Also display whether the number is prime or composite.

```
#include<stdio.h>
int main()
{
    int num, *pnum;
    int i, flag = 0;
    pnum = &num;
    printf("\n ***** ENTER -1 TO EXIT ******");
    printf("\n Enter any number : ");
    scanf("%d", pnum);
    while( *pnum != -1)
    {
        if(*pnum == 1)
            printf("\n %d is neither prime nor composite",*pnum);
        else if(*pnum == 2)
            printf("\n %d is prime", *pnum);
        else
        {
            for(i=2; i<*pnum/2; i++)
            {
                if( *pnum/i == 0)
                flag =1;
            }
            if( flag == 0)
                printf("\n %d is prime", *pnum);
            else
                printf("\n %d is composite", *pnum);
        }
        flag = 0;
        printf("\n Enter any number : ");
        scanf("%d", pnum);
    }
    return 0;
}
```

3.4 NULL POINTERS

So far, we have studied that a pointer variable is a pointer to some other variable of the same data type. However, in some cases, we may prefer to have a *null pointer* which is a special pointer value that is known not to point anywhere. This means that a `NULL` pointer does not point to any valid memory address.

To declare a null pointer, you may use the predefined constant `NULL` which is defined in several standard header files including `<stdio.h>`, `<stdlib.h>`, and `<string.h>`. After including any of these files in your program, just write

```
int *ptr = NULL;
```

You can always check whether a given pointer variable stores the address of some variable or contains a null by writing,

```
if (ptr == NULL)
{
        Statement block;
}
```

You may also initialize a pointer as a null pointer by using a constant 0, like

```
int *ptr,
ptr = 0;
```

This is a valid statement in C, as even `NULL` which is a preprocessor macro, typically has the value, or replacement text, 0. However, to avoid ambiguity, it is always better to use `NULL` to declare a null pointer. A function that returns pointer values can return a null pointer when it is unable to perform its task.

Null pointers are used in situations if one of the pointers in the program points somewhere some of the time, but not all the time. In such situations, it is always better to set it to a null pointer when it does not point anywhere, and to test to see if it is a null pointer before using it.

3.5 GENERIC POINTERS

A generic pointer is a pointer variable that has *void* as its data type. The *void pointer*, or the generic pointer, is a special type of pointer that can be pointed at variables of any data type. It is declared like a normal pointer variable but using the `void` keyword as the pointer's data type. For example,

```
void *ptr;
```

In C, since you cannot have a variable of type `void`, the void pointer will therefore not point to any data and thus, cannot be dereferenced. You need to cast a void pointer (generic pointer) to another kind of pointer before using it.

Generic pointers are often used when you want a pointer to point to data of different types at different times. For example, take a look at the following code.

```
#include<stdio.h>
int main()
{
      int x=10;
      char ch = 'A';
      void *gp;
      gp = &x;
```

```
        printf("\n Generic pointer points to the integer value = %d", *(int*)gp);
        gp = &ch;
        printf("\n Generic pointer now points to the character %c", *(char*)gp);
        return 0;
    }
```

Output:

```
Generic pointer points to the integer value = 10
Generic pointer now points to the character = A
```

It is always recommended to avoid using void pointers unless absolutely necessary, as they effectively allow you to avoid type checking.

3.6 PASSING ARGUMENTS TO FUNCTION USING POINTERS

In Chapter 2, we have seen the call-by-value method of passing parameters to a function. Using the call-by-value method, it is impossible to modify the actual parameters in the call when you pass them to a function. Furthermore, the incoming arguments to a function are treated as local variables in the function and those local variables get a *copy* of the values passed from their caller.

Pointers provide a mechanism to modify the data declared in one function using the code written in another function. In other words, if the data is declared in `func1()` and we want to write the code in `func2()` that modifies the data in `func1()`, then we must pass the addresses of the variables we want to change.

The calling function sends the addresses of the variables and the called function must declare those incoming arguments as pointers. In order to modify the variables sent by the caller, the called function must dereference the pointers that were passed to it. Thus, passing pointers to a function avoids the overhead of copying the data from one function to another.

PROGRAMMING EXAMPLES

13. Write a program to add two integers using functions.

```
#include<stdio.h>
int main()
{
        int num1, num2, total;
        printf("\n Enter the first number : ");
        scanf("%d", &num1);
        printf("\n Enter the second number : ");
        scanf("%d", &num2);
        sum(&num1, &num2, &total);
        printf("\n Total = %d", total);
        getch();
        return 0;
}
void sum (int *a, int *b, int *t)
```

```
{
        *t = *a + *b;
}
```

14. Write a program to find the greatest of three integers using functions.

```
#include<stdio.h>
void greater( int *a, int *b, int *c, int *large);
int main()
{
        int num1, num2, num3, large;
        printf("\n Enter the first number : ");
        scanf("%d", &num1);
        printf("\n Enter the second number : ");
        scanf("%d", &num2);
        printf("\n Enter the third number : ");
        scanf("%d", &num3);
        greater(&num1, &num2, &num3, &large);
        return 0;
}
void greater(int *a, int *b, int *c, int *large)
{
        if(*a > *b && *a > *c)
                *large = *a;
        if( *b > *a && *b > *c)
                *large = *b;
        else
                *large = *c;
        printf("\n Largest number = %d", *large);
}
```

15. Write a program to calculate the area of a triangle.

```
#include<stdio.h>
void read(float *b, float *h);
void calculate_area( float *b, float *h, float *a)
int main()
{
        float base, height, area;
        read(&base, &height);
        calculate_area(&base, &height, &area);
        printf("\n Area of the triangle with base %f and height %f = %f",
base, height, area);
        return 0;
}
void read(float *b, float *h)
{
        printf("\n Enter the base of the triangle : ");
```

```
        scanf("%f", b);
        printf("\n Enter the height of the triangle : ");
        scanf("%f", h);
}
void calculate_area( float *b, float *h, float *a)
{
        *a = 0.5 * (*b) * (*h);
}
```

3.7 POINTER TO FUNCTION

C allows operations with pointers to functions. This is a useful technique for passing a function as an argument to another function. In order to declare a pointer to a function, we have to declare it as a prototype of the function except that the name of the function is enclosed between parentheses `()` and an asterisk `(*)` is inserted before the name.

```
/* pointer to function returning int */
int (*func)(int a, float b);
```

Because of precedence, if you do not put the function name within parenthesis, you will end up in declaring a function returning a pointer.

```
/* function returning pointer to int */
int *func(int a, float b);
```

If we have declared a pointer to the function, then that pointer can be assigned to the address of the right sort of function just by using its name. Like in the case of an array, a function name is turned into an address when it is used in an expression.

When a pointer to a function is declared, it can be called using one of the following two forms:

```
(*func)(1,2);
```

Or

```
func(1,2);
```

Look at the following program which uses a pointer to a function.

```
#include <stdio.h>
void print(int n);
main()
{
    void (*fp)(int);
    fp = print;
    (*fp)(10);
    fp(20);  // same as above
    return 0;
}
void print(int value)
{
```

```
    printf("\n %d", value);
}
```

Let us declare a function pointer and initialize it to NULL.

```
int (*fp)(int) = NULL;
```

3.7.1 Comparing Function Pointers

Comparison-operators like == and != can be used the same way as usual. Consider the following code which checks if `fp` actually contains the address of the function `print(int)`.

```
if(fp >0){ // check if initialized
    if(fp == print)
        printf("\n Pointer points to Print");
else
    printf("\n Pointer not initialized!");
```

3.7.2 Passing a Function Pointer as an Argument to a Function

A function pointer can be passed as a function's calling argument. This is in fact necessary if you want to pass a pointer to a callback function. The following code shows how to pass a pointer to a function which returns an `int` and accepts two `int` values.

```
int Add (int a, int b)
{
    printf("\n ADD");
    return a+b;
}

void PassPtr(int (*fp)(int a, int b))
{
    int result = (*fp)(2, 3); // call using function pointer
    printf("\n RESULT = %d", result);
}
void Pass_A_Function_Pointer()
{
    printf("\n Executing 'Pass_A_Function_Pointer'");
    PassPtr(&Add);
}
```

3.8 POINTER TO POINTERS

In C, you are also allowed to use pointers that point to pointers. The pointers in turn point to data (or even to other pointers). To declare pointers to pointers, just add an asterisk * for each level of reference.

For example, suppose we have:

```
int x=10;
int *px, **ppx;
px = &x;
ppx = &px;
```

Let us assume, the memory location of these variables are as shown in Fig. 3.2.

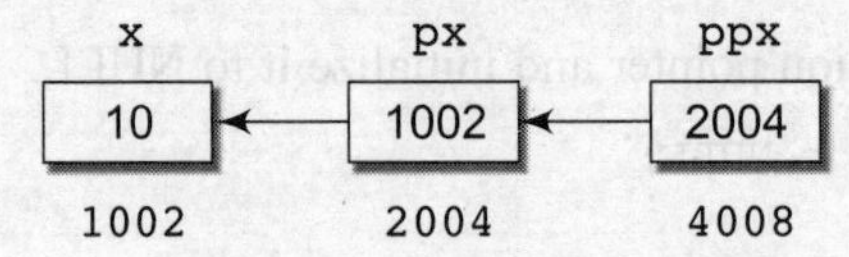

Figure 3.2 Pointer to pointer

Now if we write,

```
printf("\n %d", **ppx);
```

Then, it would print 10, the value of `x`.

SUMMARY

- Pointers are nothing but memory addresses. A pointer is a variable that contains the memory address of another variable.
- The `&` operator retrieves the `lvalue` (address) of the variable. We can 'dereference' a pointer, *i.e.*, refer to the value of the variable to which it points, by using the unary `*` operator.
- Unary increment and decrement operators have greater precedence than the dereference operator `*`.
- Null pointer is a special pointer value that is known not to point anywhere. This means that a `NULL` pointer does not point to any valid memory address. To declare a null pointer, you may use the predefined constant `NULL`. You may also initialize a pointer as a null pointer by using a constant 0.
- A generic pointer is a pointer variable that has `void` as its data type. The generic pointer can be pointed at variables of any data type.

GLOSSARY

Alias A reference (usually in the form of a pointer) to an object which is also known via other references that may include its own name, or other pointers.

Pointer Variable that stores addresses.

Dereference To look up a value referred to. Usually, the 'value referred to' is the value pointed to by a pointer. Therefore, 'dereference a pointer' means to see what it points to. In C, a pointer is dereference either using the unary `*` operator or the array subscripting operator `[]`.

Function pointer A pointer to any function type.

Lvalue An expression that appears on the left-hand side of an assignment operator, hence, something that can perhaps be assigned to. An `lvalue` specifies something that has a location, as opposed to a transient value.

Null pointer A pointer value which is not the address of any object or function. A null pointer points to nothing.

Null pointer constant An integral constant expression with the value 0 (or such an expression cast to void `*`) that represents a null pointer.

Rvalue An expression that appears on the right-hand side of an assignment operator. Generally, `rvalue` can participate in an expression or be assigned to some other variable.

EXERCISES

Review Questions

1. Explain the difference between a null pointer and a void pointer.
2. Define pointers.
3. Write a short note on pointers.

4. Compare pointer and array name.
5. Explain the result of the following code:

```
int num1 = 2, num2 = 3;
int *p = &num1, *q = &num2;
*p++ = *q++;
```

6. What do you understand by a null pointer?
7. How are generic pointers different from other pointer variables?
8. What do you understand by the term 'pointer to a function'?
9. Can we have an array of function pointers? If yes, illustrate with the help of a suitable example.
10. Write a short note on pointers to pointers.
11. Differentiate between a function returning pointer to `int` and a pointer to function returning `int`.
12. Write a program that illustrates passing of character arrays as an argument to a function (use pointers).
13. Differentiate between pointer to constants and constant to pointers.
14. What is a void pointer?

Programming Exercises

1. Write a program to access the records of students in a class using pointers to structure.
2. Write a program to print 'Programming in C is Fun' using pointers.
3. Write a program to read a character and print it. Also, print its ASCII value. If the character is in lower case, print it in upper case and vice versa. Repeat the process until a '`*`' is entered.
4. Write a program to add three floating point numbers. The result should contain only two digits after the decimal.
5. Write a program to calculate the circumference and area of a circle.
6. Write a program to convert an `integer` into a floating point number and vice versa.
7. Write a program to find the biggest of four numbers.
8. Write a program to enter a character and then, determine whether it a vowel or not. Use switch case statement.
9. Write a program to test whether a number is prime or composite.
10. Write a program to take input from the user and then check whether it is a number or a character. If it is a character, determine whether it is in upper case or lower case. Also print its ASCII value.
11. Write a program to display the sum and average of numbers from 1 to `n`. Use `while` loop.
12. Write a program to print all odd numbers from `m` to `n`.
13. Write a program to print all prime numbers from `m` to `n`.
14. Write a program to read numbers until −1 is entered. Also display whether the number is prime or composite.
15. Write a program to add two floating point numbers using pointers and functions.
16. Write a program to calculate the area of a triangle.

Multiple Choice Questions

1. `*` signifies a
   (a) Referencing operator
   (b) Dereferencing operator
   (c) Address operator
   (d) None of these
2. `*(&num)` is equivalent to writing
   (a) `&num` (b) `*num`
   (c) `num` (d) None of these
3. Pointers are used to create complex data structures like
   (a) trees (b) linked list
   (c) stack (d) queue
   (e) all of these
4. Which operator retrieves the `lvalue` of a variable?
   (a) `&` (b) `*`
   (c) `->` (d) None of these
5. Which operator is used to dereference a pointer?
   (a) `&` (b) `*`
   (c) `->` (d) None of these

True or False

1. A pointer is a variable.
2. The `&` operator retrieves the `lvalue` of the variable.
3. Unary increment and decrement operators have greater precedence than the dereference operator.
4. The generic pointer can be pointed at variables of any data type.
5. A function pointer cannot be passed as a function's calling argument.

6. On 32-bit systems, an integer variable is allocated 4 bytes.
7. `Lvalues` cannot be used on the left side of an assignment statement.
8. Pointers provide an alternate way to access individual elements of an array.
9. Pointer is a variable that represents the contents of a data item.
10. Unary increment and decrement operators have greater precedence than the dereference operator.

Fill in the Blanks

1. Size of character pointer is ______.
2. Allocating memory at runtime is known as ______.
3. A pointer to a pointer stores ______ of another ______ variable.
4. ______ pointer does not point to any valid memory address.
5. The size of memory allocated for a variable depends on its ______.
6. On 16-bit systems, an integer variable is allocated ______ bytes.
7. The ______ appears on the right side of the assignment statement.
8. Pointers are nothing but ______.
9. ______ enable programmers to return multiple values from a function via function arguments.
10. The ______ operator informs the compiler that the variable is a pointer variable.

4 Introduction to Data Structures

Learning Objective

In the last chapter, we discussed the basics of programming in C and learnt how to write, debug, and run simple programs. Our aim has been to design good programs, where a good program is defined as a program that

- runs correctly and efficiently,
- is easy to read and understand,
- is easy to debug, and
- is easy to modify.

A program should, no doubt, give correct results. But along with that, it should run efficiently. A program is said to be efficient when it executes in the minimum time with minimum memory space. Before going into the details of efficient programming, let us look at some of the key terms that we will frequently come across while discussing data structures.

4.1 INTRODUCTION

A *data structure* is an arrangement of data, either in the computer's memory or on the disk storage. Some common examples of data structures include arrays, linked lists, queues, stacks, binary trees, and hash tables. Data structures are widely applied in areas such as compiler design, operating system, statistical analysis package, DBMS, numerical analysis, simulation, artificial intelligence, and graphics.

Data structures play an important role in DBMS. The major data structures used in the network data model are graphs. In the same way, hierarchical data model uses trees and RDBMS uses arrays.

Algorithms are used to manipulate the data contained in these data structures, as in *searching* and *sorting*. While working with data structures, algorithms are used to insert new data, search for a specified item, and delete a specific item.

4.2 TYPES OF DATA STRUCTURES

C has a variety of data structures. In this section, we will just introduce them. We will discuss them in detail in subsequent chapters.

Arrays

An array is a collection of similar data elements. These data elements have the same data type. The elements of the array are stored in consecutive memory locations and are referenced by an *index* (also known as the *subscript*).

Arrays are declared using the following syntax.

```
type name[size];
```

For example,

```
int marks[10];
```

The above statement declares an array `marks` that contains 10 elements. In C, the array index starts from zero. This means that the array marks will contain 10 elements in all. The first element will be stored in `marks[0]`, second element in `marks[1]`, so on and so forth. Therefore, the last element, that is the 10th element, will be stored in `marks[9]`. In the memory, the array will be stored as shown in Fig. 4.1.

1st element	2nd element	3rd element	4th element	5th element	6th element	7th element	8th element	9th element	10th element
marks[0]	marks[1]	marks[2]	marks[3]	marks[4]	marks[5]	marks[6]	marks[7]	marks[8]	marks[9]

Figure 4.1 Memory representation of an array of 10 elements

Arrays have the following limitations:

- Arrays are of fixed size.
- Data elements are stored in continuous memory locations which may not be always available.
- Adding and removing of elements is problematic because of shifting the elements from their positions.

However, these limitations can be solved by using linked lists. We will discuss more about arrays in Chapter 5.

Linked Lists

Linked list is a very flexible, dynamic data structure in which the elements can be added to or deleted from anywhere at will. In contrast to using static arrays, a programmer need not worry about how many elements will be stored in the linked list. This feature enables the programmers to write robust programs which require less maintenance.

In a linked list, each element (is called a *node*) is allocated space as it is added to the list. Every node in the list points to the next node in the list. Therefore, in a linked list, every node contains the following two types of information:

- The value of the node or any other data that corresponds to that node, and
- A pointer or link to the next node in the list.

The last node in the list contains a `NULL` pointer to indicate that it is the end or *tail* of the list. Since the memory for a node is dynamically allocated when it is added to the list, the total number of nodes that may be added to a list is limited only by the amount of memory available. Figure 4.2 shows a linked list of seven nodes.

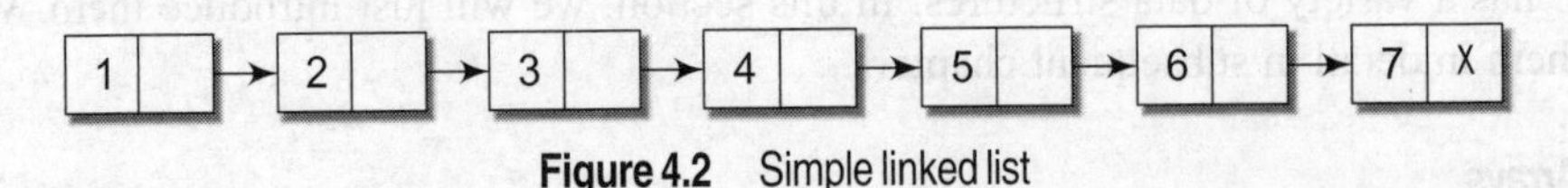

Figure 4.2 Simple linked list

Note

Advantage: Provides quick insert and delete operations.

Disadvantage: Slow search operation and requires more memory space.

Stacks

In the computer's memory, stacks can be represented as a linear array. Every stack has a variable `TOP` associated with it. `Top` is used to store the address of the topmost element of the stack. It is this position from where the element will be added or deleted. There is another variable `MAX`, which will be used to store the maximum number of elements that the stack can store.

If `TOP = NULL`, then it indicates that the stack is empty and if `TOP = MAX`, then the stack is full.

A	AB	ABC	ABCD	ABCDE					
0	1	2	3	TOP = 4	5	6	7	8	9

Figure 4.3 Stack

In Fig. 4.3, `TOP = 4`, so insertions and deletions will be done at this position. Here, the stack can store a maximum of 10 elements where the indices range from 0–9. In the above stack, five more elements can still be stored.

A stack has three basic operations: `push`, `pop`, and `peep`. The `push` operation adds an element to the top of the stack. The `pop` operation removes the element from the top of the stack. Finally, the `peep` operation returns the value of the topmost element of the stack (without deleting it).

However, before inserting an element in the stack, we must check for overflow conditions. An overflow occurs when we try to insert an element into a stack that is already full.

Similarly, before deleting an element from the stack, we must check for underflow conditions. An underflow condition occurs when we try to delete an element from a stack that is already empty. If `TOP = –1`, it indicates that there is no element in the stack.

Note *Advantage:* Last-in, first-out (LIFO) access.
Disadvantage: Slow access to other elements.

Queue

A queue is a FIFO (First-In First-Out) data structure in which the element that was inserted first is the first one to be taken out. The elements in a queue are added at one end called the `rear` and removed from the other end called the `front`. Like stacks, queues can be implemented either by using arrays or linked lists.

Every queue will have `front` and `rear` variables that will point to the position from where deletions and insertions can be done, respectively. Consider a queue given in Fig. 4.4.

12	9	7	18	14	36				
0	1	2	3	4	5	6	7	8	9

Figure 4.4 Queue

Here, `front = 0` and `rear = 5`. If we want to add one more value in the list, say, if we want to add another element with the value 45, then the `rear` would be incremented by 1 and the value would be stored at the position pointed by the `rear`. The queue, after the addition, would be as shown in Fig. 4.5.

Here, `front = 0` and `rear = 6`. Everytime a new element has to be added, we will repeat the same procedure.

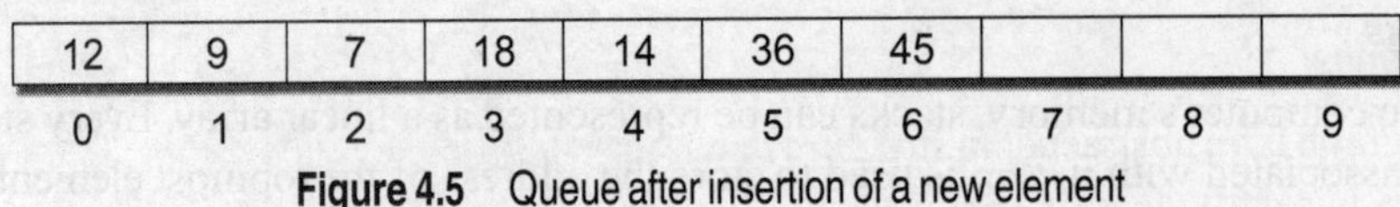

12	9	7	18	14	36	45			
0	1	2	3	4	5	6	7	8	9

Figure 4.5 Queue after insertion of a new element

Now, if we want to delete an element from the queue, then the value of the `front` will be incremented. Deletions are done only from this end of the queue. The queue after the deletion will be as shown in Fig. 4.6.

	Front					Rear			
	9	7	18	14	36	45			
0	1	2	3	4	5	6	7	8	9

Figure 4.6 Queue after deletion of an element

However, before inserting an element in the queue, we must check for overflow conditions. An overflow occurs when we try to insert an element into a queue that is already full. A queue is full when `Rear = MAX – 1`, where `MAX` is the size of the queue, that is `MAX` specifies the maximum number of elements in the queue. Note that we have written `MAX – 1` because the index starts from 0.

Similarly, before deleting an element from the queue, we must check for underflow conditions. An underflow condition occurs when we try to delete an element from a queue that is already empty. If `front = -1` and `rear = -1`, then there is no element in the queue. We will discuss stacks and queues in Chapter 9.

Note *Advantage:* Provides first-in, first-out (FIFO) data access.
Disadvantage: Slow access to other items.

Trees

A binary tree is a data structure which is defined as a collection of elements called the nodes. Every node contains a left pointer, a right pointer, and a data element. Every binary tree has a root element pointed by a 'root' pointer. The root element is the topmost node in the tree. If `root = NULL`, then the tree is empty.

Figure 4.7 shows a binary tree. If the root node `R` is not `NULL`, then the two trees T_1 and T_2 are called the left and right subtrees of `R`. If T_1 is non-empty, then T_1 is said to be the left successor of `R`. Likewise, if T_2 is non-empty, then it is called the right successor of `R`.

In Fig. 4.7, node 2 is the left successor and node 3 is the right successor of the root node 1. Note that the left subtree of the root node consists of the nodes, 2, 4, 5, 8, and 9. Similarly, the right subtree of the root node consists of the nodes, 3, 6, 7, 10, 11, and 12. We will discuss trees in detail in Chapter 10.

Figure 4.7 Binary tree

Note *Advantage:* Provides quick search, insert, and delete operations.
Disadvantage: Complicated deletion algorithm.

Graphs

A graph is an abstract data structure that is used to implement the graph concept from mathematics. It is basically a collection of *vertices* (also called *nodes*) and *edges* that connect these vertices. A graph is often viewed as a generalization of the tree structure, where instead of a having a purely parent-to-child relationship between tree nodes, any kind of complex relationships between the nodes can be represented.

In a tree structure, the nodes can have many children but only one parent, a graph on the other hand relaxes all such kinds of restrictions. Figure 4.8 shows a graph with five nodes.

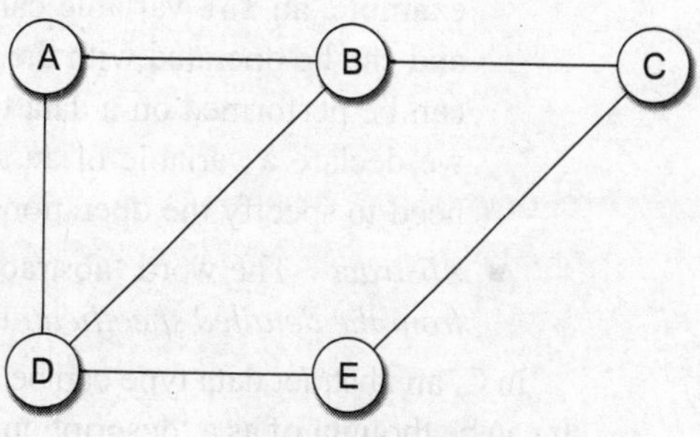

Figure 4.8 Graph

Every node in the graph may represent a city and the edges connecting the nodes could represent the roads. A graph can also be used to represent a computer network where the nodes are workstations and the edges are the network connections. Graphs have so many applications in computer science and mathematics that several algorithms have been written to perform the standard graph operations, such as searching the graph and finding the shortest path between the nodes of a graph.

Note that unlike trees, graphs do not have any root node. Rather, every node in the graph can be connected with another node in the graph. When two nodes are connected via an edge, the two nodes are known as *neighbours*. For example, in Fig. 4.8, node A has two neighbors: B and D. We will discuss graphs in detail in Chapter 13.

Advantage: Best models real-world situations.
Disadvantage: Some algorithms are slow and very complex.

Linear and Nonlinear Data Structures

Now that we have a little idea about data structures, we can easily classify them into two distinct types: *linear* and *nonlinear* data structures. If the elements of a data structure are stored sequentially, then it is a linear data structure. In linear data structures, we can traverse either forward or backward from a node. Examples include arrays, stacks, queues, and linked lists.

Linear data structures can be represented in the memory in two different ways. One way is to have the linear relationship between the elements by means of sequential memory locations. Such linear structures are called arrays. The second way is to have the linear relationship between the elements represented by means of links. Such linear data structures are called linked list.

However, if the elements of a data structure are not stored in a sequential order, then it is a nonlinear data structure. It branches to more than one node and cannot be traversed in a single run. Examples include trees and graphs.

4.3 ABSTRACT DATA TYPE

An *abstract data type* (ADT) is the way we look at a data structure, focusing on what it does and ignoring how it does its job. For example, stacks and queues are perfect examples of an abstract data type. We can implement both these ADTs using an array or a linked list. This demonstrates the 'abstract' nature of stacks and queues.

To further understand the meaning of an abstract data type, we will break the term into 'data type' and 'abstract', and then discuss their meanings.

- *Data type* Data type of a variable is the set of values that the variable may take. We have already read the basic data types in C that include `int, char, float,` and `double`. When we talk about a primitive type (built-in data type), we actually consider two things: a data item with certain characteristics and the permissible operations on that data. For example, an `int` variable can contain any whole-number value from `-32768` to `32767` and can be operated with the operators +, –, *, and /. In other words, the operations that can be performed on a data type are an inseparable part of its identity. Therefore, when we declare a variable of an abstract data type (for example, stack or a queue), we also need to specify the operations that can be performed on them.
- *Abstract* The word 'abstract' in the context of data structures means *considered apart from the detailed specifications or implementation*.

In C, an abstract data type can be a structure considered without regard to its implementation. It can be thought of as a 'description' of the data in the structure with a list of operations that can be performed on the data within that structure.

The end-user is not concerned about the details of how the methods carry out their tasks. They are only aware of the methods that are available to them and are only concerned about calling those methods and getting the results. They are not concerned about how they work.

For example, when we use a stack or a queue, the user is concerned only with the type of data and the operations that can be performed on it. Therefore, the fundamentals of how the data is stored should be invisible to the user. They should not know how the methods work or what structures are being used to store the data. They should just know that to work with stacks, they have `push()` and `pop()` functions available to them. Using these functions, they can manipulate the data (add or delete) stored on the stack.

Implementation of the Stack ADT

Let us take an example and see the implementation of the stack ADT in C. Figure 4.9 shows an imperative style interface of the stack implementation. This code can be implemented in a `stack.h` file. Note that in the code, we have not discussed whether the stack will be represented as an array or a linked list.

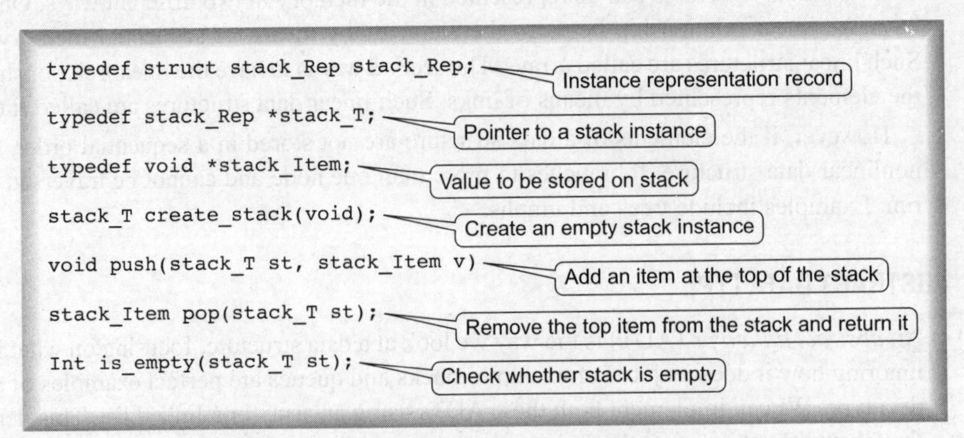

Figure 4.9 `Stack.h` to be used in abstract stack implementation

Figure 4.10 shows the code that uses the abstract stack implementation.

```
#include <stack.h>             /* Include the stack interface. */

stack_T s = create_stack();    /* Create a stack */

int val = 09;                  /* Any arbitrary value to be stored on the stack */

push(s, &val);                 /* Push address of val on stack */

void *v = pop(s);              /* Delete the value on top of the stack and return it */

if (is_empty(s))               /* Take action if the stack is empty */
{...}
```

Figure 4.10 Implementation of abstract stack using `stack.h`

Advantage of using ADT

In the real world, programs *evolve* as a result of new requirements or constraints, so a modification to a program commonly requires a change in one or more of its data structures. For example, if you want to add a new field to a student's record to keep track of more information about each student, then it will be better to replace an array with a linked structure to improve the program's efficiency. In such a scenario, rewriting every procedure that uses the changed structure is not desirable. Therefore, a better alternative is to *separate* the use of a data structure from the details of its implementation. This is the principle underlying the use of abstract data types.

4.4 ALGORITHM

The typical meaning of algorithm is 'a formally defined procedure for performing some calculation'. If a procedure is formally defined, then it must be implemented using some formal language, and such a language is known as a *programming language*. In general terms, an algorithm provides a blueprint to write a program to solve a particular problem. It is considered to be an effective procedure for solving a problem in finite number of steps. That is, a well-defined algorithm always provides an answer and is guaranteed to terminate.

Algorithms are mainly used to achieve *software reuse*. Once we have an idea or a blueprint of a solution, we can implement it in any high-level language like C, C++, or Java.

An algorithm is basically a set of instructions that solve a problem. It is not uncommon to have multiple algorithms to tackle the same problem, but the choice of a particular algorithm must depend on the time and space complexity of the algorithm. In this section, we will discuss how we can analyse algorithms to determine which one is the most efficient. First, let us look at a few examples of algorithms.

4.5 KEY FEATURES OF AN ALGORITHM

An algorithm has a finite number of steps. Some steps may involve decision-making and repetition. Broadly speaking, an algorithm exhibits the following three key features: (a) Sequence, (b) Decision, and (c) Repetition.

Sequence

By sequence, we mean that each step of an algorithm is executed in a specified order. Let us write an algorithm to add two numbers. This algorithm performs the steps in a purely sequential order, as shown in Fig. 4.11.

```
Step 1: Input the first number as A
Step 2: Input the second number as B
Step 3: SET SUM = A+B
Step 4: PRINT SUM
Step 5: END
```

Figure 4.11 Algorithm to add two numbers

Decision

Decision statements are used when the outcome of a process depends on some condition. For example, if `x = y`, then print `EQUAL`. So the general form of `IF` construct can be given as:

`IF` *condition* `then` *process*

A condition in this context is any statement that may evaluate either to a true value or a false value. In the above example, a variable `x` can either be equal to `y` or not equal to `y`. However, it cannot be both true and false. If the condition is true, then the process is executed.

A decision statement can also be stated in the following manner:

```
IF condition
    Then process1
ELSE process2
```

This form is popularly known as the `if-else` construct. Here, if the condition is true, then `process1` is executed, else `process2` is executed. Figure 4.12 shows an algorithm to check if two numbers are equal.

```
Step 1: Input the first number as A
Step 2: Input the second number as B
Step 3: IF A = B
            Then PRINT "EQUAL"
        ELSE
            PRINT "NOT EQUAL"
Step 4: END
```

Figure 4.12 Algorithm to test for equality of two numbers

Repetition

Repetition, which involves executing one or more steps for a number of times, can be implemented using constructs such as `while`, `do-while`, and `for` loops. These loops execute one or more steps until some condition is true. Figure 4.13 shows an algorithm that prints the first 10 natural numbers.

```
Step 1: [INITIALIZE] SET I = 0, N = 10
Step 2: Repeat Step while I<=N
Step 3: PRINT I
Step 4: END
```

Figure 4.13 Algorithm to print the first 10 natural numbers

PROGRAMMING EXAMPLES

1. Write an algorithm for swapping two values.

```
Step 1: Input first number as A
Step 2: Input second number as B
Step 3: SET TEMP = A
Step 4: SET A = B
Step 5: SET B = TEMP
Step 6: PRINT A, B
Step 7: END
```

2. Write an algorithm to find the larger of two numbers.

```
Step 1: Input first number as A
Step 2: Input second number as B
Step 3: IF A>B
             then PRINT A
        ELSE
             IF A<B
                    then PRINT B
             ELSE
                    PRINT "The numbers are equal"
Step 4: END
```

3. Write an algorithm to find whether a number is even or odd.

```
Step 1: Input the first number as A
Step 2: IF A%2 =0
             Then Print "EVEN"
        ELSE
             PRINT "ODD"
Step 3: END
```

4. Write an algorithm to print the grade obtained by a student using the following rules.

```
Step 1: Enter the Marks obtained as M
Step 2: IF M>75
             then PRINT O
Step 3: IF M>=60 AND M<75
             then PRINT A
Step 4: IF M>=50 AND M<60
             then PRINT B
Step 5: IF M>=40 AND M<50
             then PRINT C
        ELSE
             PRINT D
Step 6: END
```

Marks	Grade
Above 75	O
60-75	A
50-59	B
40-49	C
Less than 40	D

5. Write an algorithm to find the sum of first N natural numbers.

```
Step 1: Input N
Step 2: SET I = 0, SUM = 0
Step 3: Repeat Step 3 and 4 while I <= N
Step 4:      SET SUM = SUM + I
             SET I = I + 1
Step 5: PRINT SUM
Step 6: END
```

4.6 TIME AND SPACE COMPLEXITY

To analyse an algorithm means determining the amount of resources (such as time and storage) needed to execute it. Algorithms are generally designed to work with an arbitrary number of inputs, so the efficiency or complexity of an algorithm is stated in terms of time and space complexity.

The *time complexity* of an algorithm is basically the running time of a program, as a function of the input size. On similar grounds, the *space complexity* of an algorithm is the amount of computer memory that is required during the program execution, as a function of the input size.

In other words, the number of machine instructions which a program executes during its execution is called its time complexity. This number is primarily dependant on the size of the program's input and the algorithm used.

Generally, the space needed by a program depends on the following two parts:

- *Fixed part*, that varies from problem to problem. It includes the space needed for storing instructions, constants, variables, and structured variables (like arrays and structures).
- *Variable part*, that varies from program to program. It includes the space needed for recursion stack, and for structured variables that are allocated space dynamically during the runtime of a program.

However, running time requirements are more critical than memory requirements. Therefore, do not worry about memory requirements. In this section, we will concentrate on the running time efficiency of algorithms.

4.6.1 Worst Case, Average Case, Best Case, and Amortized Time Complexity

Worst-case running time This denotes the behaviour of the algorithm with respect to the worst-possible case of the input instance. The worst-case running time of an algorithm is an upper bound on the running time for any input. Therefore, having the knowledge of worst-case running time gives us an assurance that the algorithm will never go beyond this time limit.

Average-case running time The average-case running time of an algorithm is an estimate of the running time for an 'average' input. It specifies the expected behaviour of the algorithm when the input is randomly drawn from a given distribution. Average-case running time assumes that all inputs of a given size are equally likely.

Amortized running time While calculating the amortized running time, the time required to perform a sequence of (related) operations is averaged over all the operations performed. Amortized analysis guarantees the average performance of each operation in the worst case.

Best-case running time The term 'best-case performance' is used to analyse an algorithm under optimal conditions. For example, the best case for a simple linear search on an array occurs when the desired element is the first in the list. However, while developing and choosing an algorithm to solve a problem, we hardly base our decision on the best-case performance. It is always recommended to improve the average performance and the worst-case performance of an algorithm.

4.6.2 Time-Space Trade-off

The best algorithm to solve a particular problem at hand is, no doubt, the one that requires less memory space and takes less time to complete its execution. But practically, designing such an ideal algorithm is not a trivial task. There can be more than one algorithm to solve a particular problem. One may require less memory space, while the other may require less CPU time to execute. Thus, it is not uncommon to sacrifice one thing for the other. Hence, there exists a time-space trade-off among algorithms.

So, if space is a big constraint, then one might choose a program that takes less space at the cost of more CPU time. On the contrary, if time is a major constraint, then one might choose a program that takes minimum time to execute at the cost of more space.

4.6.3 Expressing Time and Space Complexity

The time and space complexity can be expressed using a function `f(n)` where `n` is the input size for a given instance of the problem being solved. Expressing the complexity is required when:

- We want to predict the rate of growth of complexity as the size of the problem increases.
- There are multiple algorithms that find a solution to a given problem and we need to find the algorithm that is most efficient.

The most widely used notation to express this function `f(n)` is the Big-Oh notation. It provides the upper bound for the complexity.

4.6.4 Algorithm Efficiency

If a function is linear (without any loops or recursions), the efficiency of that algorithm or the running time of that algorithm can be given as the number of instructions it contains. However, if an algorithm contains certain loops or recursive functions, then the efficiency of that algorithm may vary depending on the number of loops and the running time of each loop in the algorithm.

The efficiency of an algorithm is expressed in terms of the number of elements that has to be processed. So, if `n` is the number of elements, then the efficiency can be stated as:

```
f(n) = efficiency
```

Let us consider different cases in which loops determine the efficiency of an algorithm.

Linear Loops

To calculate the efficiency of an algorithm that has a single loop, we need to first determine the number of times the statements in the loop will be executed. This is because the number of iterations is directly proportionate to the loop factor. Higher the loop factor, more is the number of iterations. For example, consider the loop given below:

```
for(i=0;i<100;i++)
      statement block;
```

Here, 100 is the loop factor. We have already said that efficiency is directly proportional to the number of iterations. Hence, the general formula in the case of liner loops may be given as

```
f(n) = n
```

However calculating efficiency is not as simple as is shown in the above example. Consider the loop given below:

```
for(i=0;i<100;i+=2)
      statement block;
```

Here, the number of iterations is just half the number of the loop factor. So, here the efficiency can be given as

```
f(n) = n/2
```

Logarithmic Loops

We have seen that in linear loops, the loop update either adds or subtracts. However, in logarithmic loops, the loop-controlling variable is either multiplied or divided during each iteration of the loop. For example, look at the loops given below:

```
for(i=1;i<100;i*=2)                 for(i=0;i<100;i/=2)
        statement block;                    statement block;
```

Consider the first `for` loop in which the loop-controlling variable `i` is multiplied by 2. After each iteration of the loop, the loop will be executed only 10 times and not 100 times. Therefore, putting this analysis in general terms, we can conclude that the efficiency of loops in which iterations divide/multiply the loop-controlling variables, can be given as

```
f(n) = log n
```

Nested Loops

Loops that contain loops are known as *nested loops*. In order to analyse nested loops, we need to determine the number of iterations each loop completes. The total is then obtained as the product of the number of iterations in the inner loop and the number of iterations in the outer loop.

```
Total no. of   No. of iterations    No. of iterations
iterations   = in inner loop      * in outer loop
```

In this case, we analyse the efficiency of the algorithm based on whether it is a linear logarithmic, quadratic, or dependant quadratic nested loop.

Linear logarithmic loop Consider the following code in which the loop-controlling variable of the inner loop is multiplied after each iteration. The number of iterations in the inner loop is `log 10`. This inner loop is controlled by an outer loop which iterates 10 times. Therefore, according to the formula, the number of iterations for this code can be given as `10 log 10`.

```
for(i=0;i<10;i++)
        for(j=1; j<10;j*=2)
                statement block;
```

In more general terms, the efficiency of such loops can be given as `f(n) = n log n`.

Quadratic loop In a quadratic loop, the number of iterations in the inner loop is equal to the number of iterations in the outer loop. Consider the following code in which the outer loop executes 10 times and for each iteration of the outer loop, the inner loop also executes 10 times. Therefore, the efficiency here is 100.

```
for(i=0;i<10;i++)
        for(j=1; j<10;j++)

                statement block;
```

The generalized formula for quadratic loop can be given as, `f(n)` = n^2.

Dependent quadratic loop In a dependent quadratic loop, the number of iterations in the inner loop is dependant on the outer loop. Consider the code given below which shows such an example.

```
for(i=0;i<10;i++)
        for(j=1; j<i;j++)
                statement block;
```

In this code, the inner loop will execute just once in the first iteration, twice in the second iteration, thrice in the third iteration, so on and so forth. In this way, the number of iterations can be calculated as

```
1 + 2 + 3 + 4 + 5 + … + 9 + 10 = 55
```

If we calculate the average of this loop (55/10 = 5.5), we will observe that it is equal to the number of iterations in the outer loop (10) plus 1 divided by 2. In general terms, the inner loop iterates `(n + 1)/2` times. Therefore, the efficiency of such a code can be given as

```
f(n) = n (n + 1)/2
```

4.7 BIG-OH NOTATION

In today's era of massive advancement in computer technology, we are hardly concerned about the efficiency of algorithms. Rather, we are more interested in knowing the generic order of the magnitude of the algorithm. If we have two different algorithms to solve the same problem where one algorithm executes in 10 iterations and the other in 20 iterations, the difference between the two algorithms is not much. However, if the first algorithm executes in 10 iterations and the other in 1000 iterations, then it is a matter of concern.

We have seen that the number of statements executed in the function for n elements of the data is a function of the number of elements, expressed as `f(n)`. Even if the equation derived for a function may be complex, a dominant factor in the equation is sufficient to determine the order of the magnitude of the result and hence, the efficiency of the algorithm. This factor is the `Big-Oh`, as in `on the order of`, and is expressed as `O(n)`.

The Big-Oh notation, where the `O` stands for 'order of', is concerned with what happens for very large values of `n`. For example, if a sorting algorithm performs n^2 operations to sort just `n` elements, then that algorithm would be described as an $O(n^2)$ algorithm.

When expressing complexity using the Big-Oh notation, constant multipliers are ignored. So, an `O(4n)` algorithm is equivalent to `O(n)`, which is how it should be written.

If `f(n)` and `g(n)` are the functions defined on a positive integer number `n`, then

```
f(n) = O(g(n))
```

That is, `f of n is big-oh of g of n` if and only if positive constants `c` and `n` exist, such that $f(n) \le Cg(n) \le n$. It means that for large amounts of data, `f(n)` will grow no more than a constant factor than `g(n)`. Hence, `g` provides an upper bound. Note that here, `C` is a constant which depends on various factors like:

- the programming language used,
- the quality of the compiler or interpreter,
- the CPU speed,
- the size of the main memory and the access time to it,
- the knowledge of the programmer, and
- the algorithm itself, which may require simple but also time-consuming machine instructions.

Now let us look at some examples of `g(n)` and `f(n)`. Table 4.1 shows the relationship between `g(n)` and `f(n)`. Note that the constant values will be ignored because the main purpose of the Big-Oh notation is to analyse the algorithm in a general fashion, so the anomalies that appear for small input sizes are simply ignored.

Table 4.1 Example of `f(n)` and `g(n)`

g(n)	f(n) = O(g(n))
10	O(1)
$2n^3 + 1$	$O(n^3)$
$3n^2 + 5$	$O(n^2)$
$2n^3 + 3n^2 + 5n - 10$	$O(n^3)$

Categories of Algorithms

According to the Big-Oh notation, we have five different categories of algorithms:

- Constant time algorithm; running time complexity given as $O(1)$
- Linear time algorithm; running time complexity given as $O(n)$
- Logarithmic time algorithm; running time complexity given as $O(\log n)$
- Polynomial time algorithm; running time complexity given as $O(n^k)$ where $k > 1$
- Exponential time algorithm; running time complexity given as $O(2^n)$

Table 4.2 shows the number of operations that would be performed for various values of n.

Table 4.2 Number of operations for different functions of n

n	O(1)	O(log n)	O(n)	O(n log n)	$O(n^2)$	$O(n^3)$
1	1	1	1	1	1	1
2	1	1	2	2	4	8
4	1	2	4	8	16	64
8	1	3	8	24	64	512
16	1	4	16	64	256	4,096

Hence, the Big-Oh notation provides a convenient way to express the worst-case scenario for a given algorithm, although it can also be used to express the average case. The Big-Oh notation is derived from `f(n)` using the following steps:

Step 1: Set the coefficient of each term to 1.

Step 2: Rank each term starting from the lowest to the highest. Then, keep the largest term in the function and discard the rest. For example, look at the terms given below that are ranked from lowest to highest.

$$\log n,\ n \log n,\ n^2,\ n^3,\ \ldots\ n^k\ \ldots\ 2^n\ \ldots\ n!$$

Example: Calculate the Big-Oh notation for the function.

$$f(n) = \frac{n\,(n + 1)}{2}$$

The function can be expanded as $\frac{1}{2}n^2 + \frac{1}{2}n$

Step 1: Set the coefficient of each term to 1, so now we have $n^2 + n$.

Step 2: Keep the largest term and discard the rest, so discard n and the Big-Oh notation can be given as $O(f(n)) = O(n^2)$.

Limitations of Big-Oh Notation

There are certain limitations with the Big-Oh notation of expressing the complexity of algorithms. These limitations include:

- Many algorithms are simply too hard to analyse mathematically.
- There may not be sufficient information to calculate the behaviour of the algorithm in the average case.
- Big-Oh analysis only tells us how the algorithm grows with the size of the problem, not how efficient it is, as it does not consider the programming effort.
- It ignores important constants. For example, if one algorithm takes $O(n^2)$ time to execute and the other takes $O(100000n^2)$ time to execute, then as per Big-Oh, both algorithms have equal time complexity. In real-time systems, this may be a serious consideration.

NP-Complete

In computational complexity theory, the complexity class NP-C (Non-deterministic Polynomial time–Complete) is a class of problems that exhibits the following two properties:

- A set of problems is said to be NP if any given solution to the problem can be verified quickly in polynomial time.
- If the problem can be solved quickly in polynomial time, then it implies that every problem in NP can be solved in polynomial time.

Even if a given solution to a problem can be verified quickly, there is no known efficient way to locate a solution in the first place. Therefore, it is not wrong to say that solutions to NP-complete problems may not even be known.

If we try to solve a problem using any known algorithm available today, we will see that the time required to solve the problem will increase very quickly as the size of the problem grows. Consequently, the time required to solve even moderately large versions of many of these problems easily runs into billions or trillions of years. Therefore, determining whether or not it is possible to solve these problems quickly is one of the key issues in computer science today.

A problem p *in NP is also there in NPC if and only if every other problem in NP can be transformed into* p *in polynomial time.* Since the ability to quickly verify the solutions to a problem (NP) seems to correlate with the ability to quickly solve that problem (P), NP-complete problems are generally studied. However, it is never known whether every problem in NP can be quickly solved, which is called the P = NP problem. *But if any single problem in NP-complete can be solved quickly, then every problem in NP can also be solved quickly.* This statement always hold true because the definition of an NP-complete problem states that every problem in NP must be quickly reducible to every problem in NP-complete (that is, it can be reduced in polynomial time). Hence, NP-complete problems are more difficult to solve than NP problems. To conclude, a problem C is NP-complete if:

- C is in NP, and
- Every problem in NP is reducible to C in polynomial time.

SUMMARY

- A data structure is an arrangement of data, either in the computer's memory or on the disk storage.
- There are two types of data structures: linear and non-linear data structures. If the elements of a data structure are stored sequentially, then it is a linear data structure. However, if the elements of a data structure are not stored in sequential order, then it is a non-linear data structure.
- An Abstract Data Type is the way we look at a data structure, focusing on what it does and ignoring how it does its job.
- Top is used to store the address of the topmost element of the stack. It is this position from where an element will be added or deleted.
- A queue is a FIFO (First-In First-Out) data structure in which the element that was inserted first is the first one to be taken out. The elements in a queue are added at one end called the rear and removed from the other end called the front.
- The root element is the topmost node in the tree. If root = NULL, then the tree is empty.
- A graph is often viewed as a generalization of the tree structure, where instead of having a purely parent-to-child relationship, any kind of complex relationship, between the nodes can be represented.
- An algorithm is basically a set of instructions that solve a problem.
- The time complexity of an algorithm is basically the running time of a program as a function of the input size. On similar grounds, the space complexity of an algorithm is the amount of computer memory required during the program execution as a function of the input size.
- The worst-case running time of an algorithm is an upper bound on the running time for any input.
- Average-case running time specifies the expected behaviour of the algorithm when the input is randomly drawn from a given distribution.
- Amortized analysis guarantees the average performance of each operation in the worst case.

GLOSSARY

Abstract data type A set of the data values and associated operations that are not dependant on any particular implementation.

Algorithm A computable set of steps that are performed to achieve a desired result.

Asymptotic space complexity The limiting behaviour of memory usage of an algorithm when the size of the problem goes to infinity. This is usually denoted in Big-Oh notation.

Asymptotic time complexity The limiting behaviour of the time needed to execute an algorithm when the size of the problem goes to infinity. This is usually denoted in Big-Oh notation.

Average case Average case is related with the mathematical average of all cases.

Best case Best case is related with the situation or input for which an algorithm or data structure takes the least time or resources.

Worst case Worst case is related with the situation or input that forces an algorithm or data structure to take the maximum time or resources.

Big-Oh notation A theoretical measure of the execution of an algorithm which is used to express the time or memory needed, given the problem size n, which is usually the number of items.

Complexity Minimum amount of resources, such as memory, time, messages, etc. needed to solve a problem or execute an algorithm.

Computable A function that can be computed by an algorithm.

Data structure An organization of data in the memory, for better algorithm efficiency. Examples of data structure include queue, stack, linked list, heap, and tree.

Lower bound A function or growth rate below which solving a problem is not possible.

NP The complexity class of problems for which the solutions can be checked by an algorithm whose runtime is polynomial in the size of the input. However, this the does not require or imply that the solution can be found quickly. It only means that any claimed solution can be verified quickly.

NP-complete The complexity class of problems for which the solutions can be checked for correctness by an algorithm whose runtime is polynomial in the size of the input (that is, it is NP) and no other NP problem is more than a polynomial factor harder. In other words, a problem is NP-complete if the solutions can be verified quickly, and a quick algorithm to solve this problem can be used to solve all other NP problems quickly.

NP-hard The complexity class of problems which are intrinsically harder than those that can be solved by a non-deterministic Turing machine in polynomial time.

Runtime The time needed to execute an algorithm.

Polynomial time An algorithm has a polynomial time when its execution time `m(n)` is no more than a polynomial function of the problem size `n`. More formally, $m(n) = O(n^k)$ where `k` is a constant.

Linear function A linear function is written as: $f(x)=c_1x + c_0$ where C_1 and C_0 are constants. While calculating the complexity, the measure of computation `m(n)` is bound by a linear function of the problem size `n`. More formally, `m(n) = O(n)`.

Logarithmic function A function that is a constant times the logarithm of the argument: `f(x)= c log x`. While calculating the complexity theory, the measure of computation `m(n)` is bound by a logarithmic function of the problem size `n`. More formally, `m(n) = O(log n)`.

EXERCISES

Review Questions

1. Explain the features of a good program.
2. Define data structures. Give some examples.
3. In how many ways can you categorize a data structure? Explain each of them.
4. Discuss the applications of data structures.
5. Define an array with a suitable example.
6. What is a linked list?
7. Compare linked list with an array.
8. Write a short note on stacks.
9. Write a short note on abstract data type.
10. Explain the different types of data structures. Also discuss their merits and demerits.
11. Define an algorithm. Explain its features with the help of suitable examples.
12. Define a queue.
13. Write a brief note on trees as a data structure.
14. What do you understand by a graph?
15. Explain the criteria that you will keep in mind while choosing an appropriate algorithm to solve a particular problem.
16. What do you understand by time-and-space trade-off?
17. What do you understand by the efficiency of an algorithm?
18. How will you express the time complexity of a given algorithm?
19. Discuss the significance and limitations of the Big-Oh notation.
20. Differentiate between linear and non-linear data structure.
21. Discuss the best case, worst case, average case, and amortized time complexity of an algorithm.
22. Categorize algorithms based on their running time complexity.
23. What do you understand by the NP-complete problem?

Multiple Choice Questions

1. Which data structure is defined as a collection of similar data elements?
   (a) Arrays (b) Linked lists
   (c) Trees (d) Graphs
2. The data structure used in hierarchical data model is
   (a) Array (b) Linked list
   (c) Tree (d) Graph
3. The data structure used in RDBMS is
   (a) Array (b) Linked list
   (c) Tree (d) Graph
4. In a stack, insertion is done at
   (a) Top (b) Front
   (c) Rear (d) Mid

5. The position in a queue from which an element is deleted is called as
   (a) Top (b) Front
   (c) Rear (d) Mid
6. Which data structure has fixed size?
   (a) Arrays (b) Linked lists
   (c) Trees (d) Graphs
7. If `TOP = MAX`, then that the stack is
   (a) Empty
   (b) Full
   (c) Contains some data
   (d) None of these
8. Which among the following is a LIFO data structure?
   (a) Stacks (b) Linked lists
   (c) Queues (d) Graphs
9. Which data structure is used to represent complex relationships between the nodes?
   (a) Arrays (b) Linked lists
   (c) Trees (d) Graphs
10. Examples of linear data structure include
   (a) Arrays (b) Stacks
   (c) Queue (d) Linked list
   (e) All of these
11. The running time complexity of a linear time algorithm is given as
   (a) `O(1)` (b) `O(n)`
   (c) `O(n log n)` (d) `O(log n)`
   (e) $O(n^2)$

True or False

1. Trees and graphs are the examples of linear data structures.
2. Queue is a FIFO data structure.
3. Trees can represent any kind of complex relationship between the nodes.
4. The average-case running time of an algorithm is an upper bound on the running time for any input.
5. Array is an abstract data type.
6. Array elements are stored in continuous memory locations.
7. The `pop` operation adds an element to the top of a stack.
8. Graphs have a purely parent-to-child relationship between their nodes.
9. Trees and graphs are the examples of linear data structures.
10. The worst-case running time of an algorithm is a lower bound on the running time for any input.

Fill in the Blanks

1. ______ is an arrangement of data either in the computer's memory or on the disk storage.
2. ______ are used to manipulate the data contained in various data structures.
3. In ______, the elements of a data structure are stored sequentially.
4. ______ of a variable specifies the set of values that the variable can take.
5. In a queue, a new element is added at ______ position.
6. A tree is empty if ______.
7. Abstract means ______.
8. The time complexity of an algorithm is the running time given as a function of ______.
9. ______ analysis guarantees the average performance of each operation in the worst case.
10. The elements of an array are referenced by an ______.
11. ______ is used to store the address of the topmost element of a stack.
12. The ______ operation returns the value of the topmost element of a stack.
13. An overflow occurs when ______.
14. ______ is a FIFO data structure.
15. The elements in a queue are added at ______ and removed from ______.
16. If the elements of a data structure are stored sequentially, then it is a ______.
17. ______ is basically a set of instructions that solve a problem.
18. The number of machine instructions that a program executes during its execution is called its ______.
19. ______ specifies the expected behaviour of an algorithm when an input is randomly drawn from a given distribution.
20. The running time complexity of a constant time algorithm is given as ______.

5 Arrays

Learning Objective

In this chapter, we will discuss arrays. An array is a user-defined data type that stores related information together. However, all the information stored in an array belongs to the same data type. So, in this chapter, we will learn how arrays are defined, declared, initialized, accessed, etc. We will also discuss the different operations that can be performed on array elements and the different types of arrays such as two-dimensional arrays, multi-dimensional arrays, and sparse matrices.

5.1 INTRODUCTION

Suppose, we take a situation in which we have 20 students in a class and we are asked to write a program that reads and prints the marks of all the 20 students. In this program, we will need 20 integer variables with different names, as shown in Fig. 5.1.

Marks1	Marks5	Marks9	Marks13	Marks17
Marks2	Marks6	Marks10	Marks14	Marks18
Marks3	Marks7	Marks11	Marks15	Marks19
Marks4	Marks8	Marks12	Marks16	Marks20

Figure 5.1 Twenty variables

Now, to read the values of these twenty variables, we must have twenty read statements. Similarly, to print the value of these variables, we need 20 write statements. If it is just a matter of 20 variables, then it might be acceptable for the user to follow this approach. But imagine, would it be possible to follow this approach if we had to read and print the marks of students

- in the entire course (say 100 students)
- in the entire college (say 500 students)
- in the entire university (say 10,000 students)

The answer is no, definitely not. To process a large amount of data, we need a data structure known as *array*.

An array is a collection of similar data elements. These data elements have the same data type. The elements of the array are stored in consecutive memory locations and are referenced by an `index` (also known as the *subscript*). The subscript indicates an ordinal number of the element, counting from the beginning of the array.

5.2 DECLARATION OF ARRAYS

We have already seen that every variable must be declared before it is used. The same concept holds true for array variables. An array must be declared before being used. Declaring an array means specifying the following:

- *Data type*– what kind of values it can store, for example, `int`, `char`, `float`, `double`.
- *Name*– to identify the array.
- *Size*– the maximum number of values that the array can hold.

Arrays are declared using the following syntax.

```
type name[size];
```

Here, the type can be either `int`, `float`, `double`, `char`, or any other valid data type. The number within brackets indicates the size of the array, i.e. the maximum number of elements that can be stored in the array. For example, if we write,

```
int marks[10];
```

then the statement declares `marks` to be an array containing 10 elements. In C, the array index starts from zero. This means that the array marks will contain 10 elements in all. The first element will be stored in `marks[0]`, second element in `marks[1]`, and so on. Therefore, the last element, that is the 10th element, will be stored in `marks[9]`. Note that 0, 1, 2, 3 written within square brackets are the subscripts. In the memory, the array will be stored as shown in Fig. 5.2.

1st element	2nd element	3rd element	4th element	5th element	6th element	7th element	8th element	9th element	10th element
marks[0]	marks[1]	marks[2]	marks[3]	marks[4]	marks[5]	marks[6]	marks[7]	marks[8]	marks[9]

Figure 5.2 Memory representation of an array of 10 elements

Figure 5.3 shows how different type of arrays are declared.

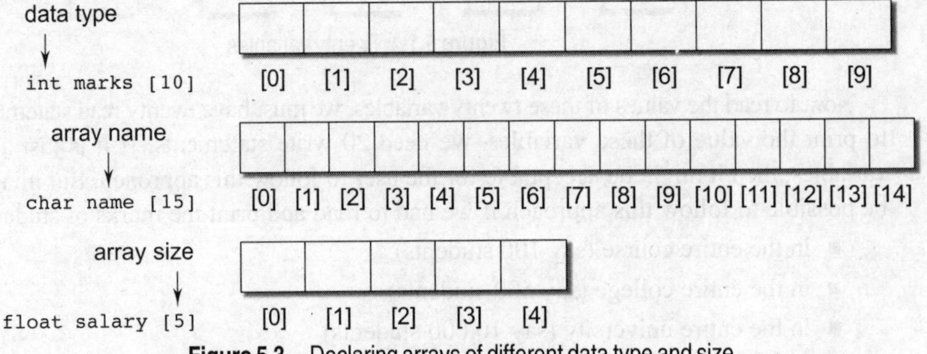

Figure 5.3 Declaring arrays of different data type and size

5.3 ACCESSING ARRAY ELEMENTS

Storing the related data items in a single array enables the programmers to develop concise and efficient programs. But there is no single operation that can operate on all the elements of an array. To access all the elements, you must use a loop. That is, we can access all the elements of an array by varying the value of the subscript into the array. But note that the subscript must be an integral value or an expression that evaluates to an integral value. As shown in Fig. 5.2, the first element of the array `marks[10]` can be accessed by writing `marks[0]`. Now to process all the elements of the array, we use a loop as shown in Fig. 5.4.

```
// Set each element of the array to -1
int i, marks[10];
for(i=0;i<10;i++)
        marks[i] = -1;
```

Figure 5.4 Code to initialize each element of the array to –1

Figure 5.5 shows the result of the code shown in Fig. 5.4. The code accesses every individual element of the array and sets each value to –1. In the `for` loop, first the value of `marks[0]` is set to –1, then the value of the index `(i)` is incremented and the next value, that is, `marks[1]` is set to –1. The procedure continues until all the 10 elements of the array are set to –1.

– 1	– 1	– 1	– 1	– 1	– 1	– 1	– 1	– 1	– 1
[0]	[1]	[2]	[3]	[4]	[5]	[6]	[7]	[8]	[9]

Figure 5.5 Array `marks` after executing the code given in Fig. 5.4

Note There is no single statement that can read, access, or print all the elements of an array. To do this, we have to use a loop to execute the same statement with different index values.

Calculating the address of array elements You must be wondering how C gets to know where an individual element of the array is located in the memory. The answer is that the array name is a symbolic reference for the address to the first byte of the array. When we use the array name, we are actually referring to the first byte of the array.

The subscript or the index represents the offset from the beginning of the array to the element being referenced. That is, with just the array name and the index, C can calculate the address of any element in the array.

Since an array stores all its data elements in consecutive memory location, storing just the base address, that is the address of the first element in the array, is sufficient. The address of other data elements can simply be calculated using the base address. The formula to perform this calculation is,

```
Address of data element, A[k] = BA(A) + w( k - lower_bound)
```

Here, `A` is the array, `k` is the index of the element of which we have to calculate the address, `BA` is the base address of the array `A`, and `w` is the word size of one element in memory, for example, size of `int` is 2.

Example 5.1: Given an array `int marks[] = {99,67,78,56,88,90,34,85}`, calculate the address of `marks[4]` if the `base address = 1000`.

Solution

99	67	78	56	**88**	90	34	85
marks[0]	Marks[1]	Marks[2]	Marks[3]	**Marks[4]**	Marks[5]	Marks[6]	Marks[7]
1000	1002	1004	1006	**1008**	1010	1012	1014

We know that storing an integer value requires 2 bytes, therefore here, the word size is 2 bytes.

```
Marks[4] = 1000 + 2(4 - 0)
         = 1000 + 2(4) = 1008
```

5.4 STORING VALUES IN ARRAYS

When we declare and define an array, we are just allocating space for its elements; no values are stored in the array. There are three ways to store values in an array. First, to initialize the array elements; second, to input values for every individual element; third, to assign values to the elements. This is shown in the Fig. 5.6.

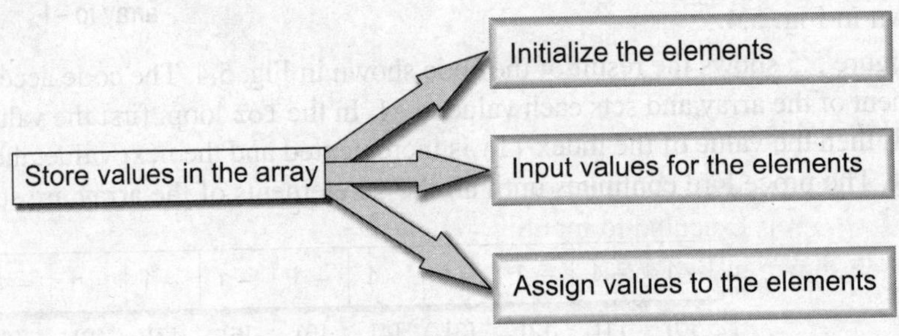

Figure 5.6 Storing values in an array

5.4.1 Initialization of Arrays

The elements of an array can also be initialized at the time of declaration, just as every other variable. When an array is initialized, we need to provide a value for every element in the array. Arrays are initialized by writing,

```
type array_name[size]={list of values};
```

Note that the values are written with curly brackets and every value is separated by a comma. It is a compiler error to specify more values than there are elements in the array. When we write,

```
int marks[5]={90, 82, 78, 95, 88};
```

An array with the name `marks` is declared that has enough space to store five elements. The first element, that is, `marks[0]` is assigned with the value 90. Similarly, the second element of the array, that is `marks[1]`, has been assigned 82, and so on. This is shown in Fig. 5.7.

marks[0]	90
marks[1]	82
marks[2]	78
marks[3]	95
marks[4]	88

Figure 5.7 Initialization of array `Marks[5]`

While initializing the array at the time of declaration, the programmer may omit the size of the array. For example,

```
int marks[]= { 98, 97, 90};
```

The above statement is absolutely legal. Here, the compiler will allocate enough space for all the initialized elements. Note that if the number of values provided is less than the number of elements in the array, the un-assigned elements are filled with zeros. Figure 5.8 shows the initialization of arrays.

```
int marks [5] = {90, 45, 67, 85, 78};
```

90	45	67	85	78
[0]	[1]	[2]	[3]	[4]

```
int marks [5] = {90, 45};
```

90	45	0	0	0
[0]	[1]	[2]	[3]	[4]

Rest of the elements are filled with 0's

```
int marks [] = {90, 45, 72, 81, 63, 54};
```

90	45	72	81	63	54
[0]	[1]	[2]	[3]	[4]	[5]

```
int marks [5] = {0};
```

0	0	0	0	0	0
[0]	[1]	[2]	[3]	[4]	[5]

Figure 5.8 Initialization of array elements

5.4.2 Inputting Values

An array can be filled by inputting values from the keyboard. In this method, a `while/do-while` or a `for` loop is executed to input the value for each element of the array. For example, look at the code shown in Fig. 5.9.

In the code, we start at the index `i` at 0 and input the value for the first element of the array. Since the array can have 10 elements, we must input values for elements whose index varies from 0 to 9. Therefore, in the `for loop`, we test for the condition `(i < 10)` which means the number of elements in the array.

```
int i, marks[10];
for(i=0;i<10;i++)
        scanf("%d", &marks[i]);
```

Figure 5.9 Code for inputting each element of the array

5.4.3 Assigning Values

The third way to assign values to individual elements of the array is by using the assignment operator. Any value that evaluates to an appropriate data type as that of the array can be assigned to the individual array element. A simple assignment statement can be written as

```
marks[3] = 100;
```

Here, 100 is assigned to the fourth element of the array which is specified as `marks[3]`.

Note that we cannot assign one array to another array, even if the two arrays have the same type and size. To copy an array, you must copy the value of every element of the first array into the elements of the second array. Figure 5.10 illustrates the code to copy an array.

```
int i, arr1[10], arr2[10];
for(i=0;i<10;i++)
        arr2[i] = arr1[i];
```

Figure 5.10 Code to copy an array at the individual element level

In Fig. 5.10, the code accesses each element of the first array and simultaneously assigns its value to the corresponding element of the second array. Finally, the index value `i` is incremented to access the next element in succession. Therefore, when this code is executed, `arr2[0] = arr1[0]`, `arr2[1] = arr1[1]`, `arr2[2] = arr1[2]`, and so on.

We can also use a loop to assign a pattern of values to the array elements. For example, if we want to fill an array with even integers (starting from 0), then we will write the code as shown in Fig. 5.11.

```
// Fill an array with even number
int i, arr[10];
for(i=0;i<10;i++)
      arr[i] = i*2;
```

Figure 5.11 Code for filling an array with even numbers

In the code, we assign to each element a value equal to twice of its index, where the index starts from 0. So after executing this code, we will have `arr[0] = 0`, `arr[1] = 2`, `arr[2] = 4`, and so on.

5.5 CALCULATING THE LENGTH OF AN ARRAY

The length of an array is given by the number of elements stored in it. The general formula to calculate the length of the array is,

```
Length = upper_bound - lower_bound + 1
```

where `upper_bound` is the index of the last element and `lower_bound` is the index of the first element in the array.

EXAMPLE 5.2: Let `Age` be an array of integers such that

```
Age[0] = 2, Age[1] = 5, Age[2] = 3, Age[3] = 1, Age[4] = 7
```

Show the memory representation of the array and calculate its length.

Solution

The memory representation of the array `Age` is given as below.

2	5	3	1	7
Age [0]	Age [1]	Age [2]	Age [3]	Age [4]

`Length = upper_bound - lower_bound + 1`

Here, `lower_bound = 0`, `upper_bound = 4`

Therefore, `length` = 4 – 0 + 1 = 5

5.6 OPERATIONS ON ARRAYS

There are a number of operations that can be preformed on arrays. These operations include:

- Traversal
- Insertion
- Searching
- Deletion
- Merging
- Sorting

We will discuss all these operations in detail in this section.

5.6.1 Traversal

Traversing the array element means accessing each and every element of the array for a specific purpose. We have already seen this in Section 5.3. This is just a re-visit of the topic.

If `A` is an array of homogeneous data elements, then traversing the data elements can include printing every element, counting the total number of elements, or performing any processing on these elements. Since, array is a linear data structure (because all its elements forms a sequence), traversing its elements is very simple and straightforward. The algorithm for array traversal is given in Fig. 5.12.

```
Step 1: [Initialization] Set I = lower_bound
Step 2: Repeat steps 3 to 4 while I <= upper_bound
Step 3:      Apply Process to A[I]
Step 4:      Set  I = I + 1
          [End of Loop]
Step 5: Exit
```

Figure 5.12 Algorithm for array traversal

In step 1, we initialize the index to the `lower bound` of the array. In step 2, a `while loop` is executed. Steps 3 and 4 form a part of the loop. Step 3 processes the individual array element as specified by the array name and index value. Step 4 increments the index value so that the next array element could be processed. The `while loop` in step 2 is executed until all the elements in the array are processed, *i.e.*, until `I` is less than or equal to the `upper bound` of the array.

EXAMPLE 5.3: Assume that there is an array `Marks[]` such that the index of the array specifies the roll number of a student and the value of a particular element denotes the marks obtained by the student. For example, if it is given as `Marks[4] = 78`, then the student whose roll number is 4 has obtained 78 marks in the examination. Now, write an algorithm to:

(a) Find the total number of students who have secured 80 or more marks.

(b) Print the roll number and marks of all the students who have got distinction.

Solution

(a)

```
Step 1: [Initialization] Set Count = 0, Marks[0] = – 1
Step 2: Repeat for I = lower_bound to upper_bound
            If Marks[I] >= 80, then : Set Count = Count + 1
     [End of Loop]
Step 3: Exit
```

(b)

```
Step 1: Repeat for I = lower_bound to upper_bound
               If Marks[I] >= 75, Write: I , Marks[I]
          [End of Loop]
Step 2: Exit
```

PROGRAMMING EXAMPLES

1. Write a program to read and display *n* numbers using an array.

```
#include<stdio.h>
#include<conio.h>
int main()
{
        int i=0, n, arr[20];
```

```
    clrscr();
    printf("\n Enter the number of elements : ");
    scanf("%d", &n);
    for(i=0;i<n;i++)
    {
        printf("\n arr[%d] = ", i);
        scanf("%d",&arr[i]);
    }
    printf("\n The array elements are ");
    for(i=0;i<n;i++)
        printf("arr[%d] = %d\t", i, arr[i]);
    return 0;
}
```

2. Write a program to find the mean of *n* numbers using arrays.

```
#include<stdio.h>
#include<conio.h>
int main()
{
    int i, n, arr[20], sum =0;
    float mean = 0.0;
    clrscr();
    printf("\n Enter the number of elements in the array : ");
    scanf("%d", &n);
    for(i=0;i<n;i++)
    {
        printf("\n arr[%d] = ", i);
        scanf("%d",&arr[i]);
    }
    for(i=0;i<n;i++)
        sum += arr[i];
    mean = (float)sum/n;
    printf("\n The sum of the array elements = %d", sum);
    printf("\n The mean of the array elements = %f", mean);
    return 0;
}
```

3. Write a program to find the largest of *n* numbers using arrays.

```
#include<stdio.h>
#include<conio.h>
int main()
{
    int i, n, arr[20], large =-1111;
    clrscr();
```

```
        printf("\n Enter the number of elements in the array : ");
        scanf("%d", &n);
        for(i=0;i<n;i++)
                scanf("%d",&arr[i]);
        for(i=0;i<n;i++)
        {
                if(arr[i]>large)
                        large = arr[i];
        }
        printf("\n The largest of number is : %d", large);
        return 0;
}
```

4. Write a program to print the position of the smallest number of *n* numbers using arrays.

```
#include<stdio.h>
#include<conio.h>
int main()
{
        int i, n, arr[20], small =1234, pos = -1;
        clrscr();
        printf("\n Enter the number of elements in the array : ");
        scanf("%d", &n);
        for(i=0;i<n;i++)
                scanf("%d",&arr[i]);
        for(i=0;i<n;i++)
        {
                if(arr[i]<small)
                {
                        small = arr[i];
                        pos = i;
                }
        }
        printf("\n The smallest of element is : %d", small);
        printf("\n The position of the smallest number in the array is : %d",
pos);
        return 0;
}
```

5. Write a program to interchange the largest and the smallest number in an array.

```
#include<stdio.h>
#include<conio.h>
int main()
{
        int i, n, arr[20], temp;
```

```
        int small =1234, small_pos =0;
        int large = -1234, large_pos = 0;
        clrscr();
        printf("\n Enter the number of elements in the array : ");
        scanf("%d", &n);
        for(i=0;i<n;i++)
        {
                printf("\n Enter the %d th element : ",i);
                scanf("%d",&arr[i]);
        }
        for(i=0;i<n;i++)
        {
                if(arr[i]<small)
                {
                        small = arr[i];
                        small_pos = i;
                }
                if(arr[i]>large)
                {
                        large = arr[i];
                        large_pos = i;
                }
        }
        printf("\n The smallest of these numbers is : %d", small);
        printf("\n The position of the smallest number in the array is :
%d",small_pos);
        printf("\n The largest of these numbers is : %d", large);
        printf("\n The position of the largest number in the array is :
%d",large_pos);
        temp = arr[large_pos];
        arr[large_pos] =arr[small_pos];
        arr[small_pos] = temp;
        printf("\n The new array is : ");
        for(i=0;i<n;i++)
                printf(" \n arr[%d] = %d ", i, arr[i]);
        return 0;
}
```

6. Write a program to find the second largest number using an array of *n* numbers.

```
#include<stdio.h>
#include<conio.h>
int main()
{
        int i, n, arr[20], large =-1111, second_largest = —12345;
```

```
        clrscr();
        printf("\n Enter the number of elements in the array : ");
        scanf("%d", &n);
        printf("\n Enter the elements");
        for(i=0;i<n;i++)
                scanf("%d",&arr[i]);
        for(i=0;i<n;i++)
        {
                if(arr[i]>large)
                        large = arr[i];
        }
        for(i=0;i<n;i++)
        {
                if(arr[i] != large)
                {
                        if(arr[i]>second_large)
                                second_large = arr[i];
                }
        }
        printf("\n The numbers you entered are : ");
        for(i=0;i<n;i++)
                printf("%d", arr[i]);
        printf("\n The largest of these numbers is : %d",large);
        printf("\n The second largest of these numbers is : %d",second_large);
        return 0;
}
```

7. Write a program to enter *n* number of digits. Form a number using these digits.

```
#include<stdio.h>
#include<conio.h>
#include<math.h>
int main()
{
        int number=0, digit[10], numofdigits,i;
        clrscr();
        printf("\n Enter the number of digits : ");
        scanf("%d", &numofdigits);
        for(i=0;i<numofdigits;i++)
        {
                printf("\n Enter the %d th digit : "i);
                scanf("%d", &digit[i]);
        }
        i=0;
```

```
        while(i<numofdigits)
        {
                number = number + digit[i] * pow(10,i);
                i++;
        }
        printf("\n The number is : %d", number);
        return 0;
}
```

8. Write a program to find whether the array of integers contain a duplicate number.

```
#include<stdio.h>
#include<conio.h>
int main()
{
  int array1[10], i, n, j;
  clrscr();
  printf("\n Enter the size of the array : ");
  scanf("%d", &n);
  for(i=0;i<n;i++)
  {
        printf(" array [%d] = ", i);
        scanf("%d", &array1[i]);
  }
  for(i=0;i<n;i++)
  {
        for(j= i+1;j<n;j++)
        {
              if(array1[i] == array1[j] && i!=j)
                    printf("\n Duplicate number %d found at location %d and %d",
array1[i], i, j);
        }
  }
  return 0;
}
```

5.6.2 Insertion

Inserting an element in the array means adding a new data element in an already existing array. If the element has to be inserted at the end of the existing array, then the task of inserting is quite simple. We just have to add 1 to the `upper_bound` and assign the value. Here, we assume that the memory space allocated for the array is still available. For example, if an array is declared to contain 10 elements, but currently it has only 8 elements, then obviously there is space to accommodate two more elements. But if it already has 10 elements, then we will not be able to add another element to it.

```
Step 1: Set upper_bound = upper_bound + 1
Step 2: Set A[upper_bound] = VAL
Step 3: EXIT
```

Figure 5.13 Algorithm to append a new element to an existing array

Figure 5.13 shows an algorithm to insert a new element to the end of the array. In step 1 of the array, we increment the value of the upper_bound. In step 2, the new value is stored at the position pointed by the upper_bound. For example, let us assume an array has been declared as

```
int marks[60];
```

The array is declared to store the marks of all the students in a class. Now, suppose there are 54 students and a new student comes and is asked to take the same test. The marks of this new student would be stored in marks[55]. Assuming that the student secured 68 marks, we will assign the value as

```
marks[55] = 68;
```

However, if we have to insert an element in the middle of the array, then this is not a trivial task. On an average, we might have to move as much as half of the elements from its position in order to accommodate space for the new element.

For example, consider an array whose elements are arranged in ascending order. Now, if a new element has to be added, it will have to be added probably somewhere in the middle of the array. To do this, we first find the location where the new element will be inserted and then move all the elements (that have a greater value than that of the new element) one space to the right so that space can be created to store the new value.

EXAMPLE 5.4: Data[] is an array that is declared as int Data[20]; and contains the following values:

```
Data[] = {12, 23, 34, 45, 56, 67, 78, 89, 90, 100};
```

(a) Calculate the length of the array.
(b) Find out the upper_bound and lower_bound.
(c) Produce the memory representation of the array.
(d) If a new data element with the value 75 has to be inserted, find its position.
(e) Insert a new data element and produce the memory representation after the insertion.

Solution

(a) length of the array = number of elements
Therefore, length of the array = 10
(b) By default, lower_bound = 0 and upper_bound = 9
(c)

12	23	34	45	56	67	78	89	90	100
Data[0]	Data[1]	Data[2]	Data[3]	Data[4]	Data[5]	Data[6]	Data[7]	Data[8]	Data[9]

(d) Since the elements of the array are stored in ascending order, the new data element will be stored after 67, *i.e.*, at the 6th location. So, all the array elements from the 6th position will be moved one space towards the right to accommodate the new value.
(e)

12	23	34	45	56	67	75	78	89	90	100
Data[0]	Data[1]	Data[2]	Data[3]	Data[4]	Data[5]	Data[6]	Data[7]	Data[8]	Data[9]	Data[10]

Algorithm to Insert an Element in the Middle of an Array

The algorithm `INSERT` will be declared as `INSERT (A, N, POS, VAL)`. The arguments are:

(a) `A`, the array in which the element has to be inserted,

(b) `N`, the number of elements in the array,

(c) `POS`, the position at which the element has to be inserted, and

(d) `VAL`, the value that has to be inserted.

In the algorithm given in Fig. 5.14, in step 1, we first initialize I with the total number of elements in the array. In step 2, a `while` loop is executed which will move all the elements having an index greater than `POS` one space towards right to create space for the new element. In step 5, we increment the total number of elements in the array by 1 and finally in step 6, the new value is inserted at the desired position.

```
Step 1: [INITIALIZATION] SET I = N
Step 2: Repeat Steps 3 and 4 while I >= POS
Step 3:         SET A[I + 1] = A[I]
Step 4:         SET I = I – 1
              [End of Loop]
Step 5: SET N = N + 1
Step 6: SET A[POS] = VAL
Step 7: EXIT
```

Figure 5.14 Algorithm to insert an element in the middle of an array

Now, let us visualize this algorithm by taking an example.

Initial `Data[]` is given as below.

45	23	34	12	56	20
Data[0]	Data[1]	Data[2]	Data[3]	Data[4]	Data[5]

Calling `INSERT (Data, 6, 3, 100)` will lead to the following processing in the array:

45	23	34	12	56	20	20
Data[0]	Data[1]	Data[2]	Data[3]	Data[4]	Data[5]	Data[6]

45	23	34	12	56	56	20
Data[0]	Data[1]	Data[2]	Data[3]	Data[4]	Data[5]	Data[6]

45	23	34	12	12	56	20
Data[0]	Data[1]	Data[2]	Data[3]	Data[4]	Data[5]	Data[6]

45	23	34	100	12	56	20
Data[0]	Data[1]	Data[2]	Data[3]	Data[4]	Data[5]	Data[6]

PROGRAMMING EXAMPLES

9. Write a program to insert a number at a given location in an array.

```
#include<stdio.h>
#include<conio.h>
int main()
{
```

```
        int i, n, num, pos, arr[10];
        clrscr();
        printf("\n Enter the number of elements in the array : ");
        scanf("%d", &n);
        for(i=0;i<n;i++)
        {
                printf("\n Arr[%d] = : ", i);
                scanf("%d", &arr[i]);
        }
        printf("\n Enter the number to be inserted : ");
        scanf("%d", &num);
        printf("\n Enter the position at which the number has to be added : ");
        scanf("%d", &pos);
        for(i=n;i>=pos;i–)
                arr[i+1] = arr[i];
        arr[pos] = num;
        printf("\n The array after insertion of %d is : ", num);
        for(i=0;i<n+1;i++)
                printf("\n Arr[%d] = %d", i, arr[i]);
        getch();
        return 0;
}
```

10. Write a program to insert a number in an array that is already sorted in ascending order.

```
#include<stdio.h>
#include<conio.h>
int main()
{
        int i, n, j, num, arr[10];
        clrscr();
        printf("\n Enter the number of elements in the array : ");
        scanf("%d", &n);
        for(i=0;i<n;i++)
        {
                printf("\n Arr[%d] = ", i);
                scanf("%d", &arr[i]);
        }
        printf("\n Enter the number to be inserted : ");
        scanf("%d", &num);
        for(i=0;i<n;i++)
        {
                if(arr[i] > num)
                {
                        for(j = n; j<=i; j–)
```

```
                    arr[j+1] = arr[j];
                arr[i] = num;
                break;
            }
        }
        printf("\n The array after insertion of %d is : ", num);
        for(i=0;i<n+1;i++)
            printf("\n Arr[%d] = %d", i, arr[i]);
        getch();
        return 0;
}
```

5.6.3 Deletion

Deleting an element from an array means removing a data element from an already existing array. If the element has to be deleted from the end of the existing array, then the task of deletion is quite simple. We just have to subtract 1 from the upper_bound. Figure 5.15 shows an algorithm to delete an element from the end of an array.

```
Step 1: Set upper_bound = upper_bound - 1
Step 2: EXIT
```

Figure 5.15 Algorithm to delete the last element of an array

For example, if we have an array that is declared as:

`int marks[];`

The array is declared to store the marks of all the students in the class. Now, suppose there are 54 students and the student with roll number 54 leaves the course. The score of this student was therefore stored in `marks[54]`. We just have to decrement the `upper_bound`. Subtracting 1 from the `upper_bound` will indicate that there are 53 valid data in the array.

However, if we have to delete an element from the middle of an array, then it is not a trivial task. On an average, we might have to move as much as half of the elements from their position in order to occupy the space of the deleted element.

For example, consider an array whose elements are arranged in ascending order. Now, suppose an element has to be deleted, probably from somewhere in the middle of the array. To do this, we first find the location from where the element has to be deleted and then move all the elements (having a greater value than that of the element) one space towards left so that the space vacated by the deleted element can be occupied by rest of the elements.

EXAMPLE 5.5: `Data[]` is an array that is declared as `int Data[10];` and contains the following values:

`Data[] = { 12, 23, 34, 45, 56, 67, 78, 89, 90, 100};`

(a) Calculate the length of the array.

(b) Find out the `upper_bound` and `lower_bound`.

(c) Show the memory representation of the array.

(d) If a data element with value 56 has to be deleted, find its position.

(e) Delete a data element and show the memory representation after the deletion.

Solution

(a) `length of the array = number of elements`

Therefore, length of the array = 10

(b) By default, `lower_bound = 0` (it can be set to any value) and `upper_bound = 9`

(c)

12	23	34	45	56	67	78	89	90	100
Data[0]	Data[1]	Data[2]	Data[3]	Data[4]	Data[5]	Data[6]	Data[7]	Data[8]	Data[9]

(d) Since the elements of the array are stored in ascending order, we will compare the value that has to be deleted with the value of every element in the array. As soon as `VAL = Data[I]`, where `I` is the index or subscript of the array, we will get the position from which the element has to be deleted. For example, if we see this array, here `VAL = 56`. `Data[0] = 12` which is not equal to 56. We will continue to compare and finally get the value of `POS = 4`.

(e)

12	23	34	45	56	67	78	89	90	100
Data[0]	Data[1]	Data[2]	Data[3]	Data[4]	Data[5]	Data[6]	Data[7]	Data[8]	Data[9]

12	23	34	45	67	67	78	89	90	100
Data[0]	Data[1]	Data[2]	Data[3]	Data[4]	Data[5]	Data[6]	Data[7]	Data[8]	Data[9]

12	23	34	45	67	78	78	89	90	100
Data[0]	Data[1]	Data[2]	Data[3]	Data[4]	Data[5]	Data[6]	Data[7]	Data[8]	Data[9]

12	23	34	45	67	78	89	89	90	100
Data[0]	Data[1]	Data[2]	Data[3]	Data[4]	Data[5]	Data[6]	Data[7]	Data[8]	Data[9]

12	23	34	45	67	78	89	90	90	100
Data[0]	Data[1]	Data[2]	Data[3]	Data[4]	Data[5]	Data[6]	Data[7]	Data[8]	Data[9]

12	23	34	45	67	78	89	90	100	100
Data[0]	Data[1]	Data[2]	Data[3]	Data[4]	Data[5]	Data[6]	Data[7]	Data[8]	Data[9]

12	23	34	45	67	78	89	90	100
Data[0]	Data[1]	Data[2]	Data[3]	Data[4]	Data[5]	Data[6]	Data[7]	Data[8]

Algorithm to Delete an Element from the Middle of an Array

The algorithm `DELETE` will be declared as `DELETE (A, N, POS)`. The arguments are:

(a) `A`, the array from which the element has to be deleted,

(b) `N`, the number of elements in the array, and

(c) `POS`, the position from which the element has to be deleted.

Figure 5.16 shows the algorithm in which we first initialize `I` with the position from which the element has to be deleted. In step 2, a `while` loop is executed which will move all the elements having an index greater than `POS` one space towards left to occupy the space vacated by the deleted element. When we say that we are deleting an element, actually we are overwriting the element with the value of its successive element. In step 5, we decrement the total number of elements in the array by 1.

```
Step 1: [INITIALIZATION] SET I = POS
Step 2: Repeat Steps 3 and 4
                while I <= N - 1
Step 3:         SET A[I] = A[I + 1]
Step 4:         SET I = I + 1
           [End of Loop]
Step 5: SET N = N - 1
Step 6: EXIT
```

Figure 5.16 Algorithm to delete an element from the middle of an array

Now, let us visualize this algorithm by taking an example given in Fig. 5.17. Calling `DELETE (Data, 6, 2)` will lead to the following processing in the array.

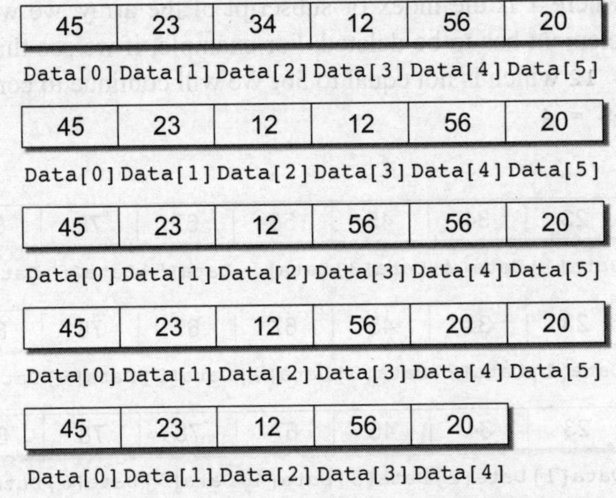

Figure 5.17 Deleting elements from the array

PROGRAMMING EXAMPLE

11. Write a program to delete a number from a given location in an array.

```
#include<stdio.h>
#include<conio.h>
int main()
{
    int i, n, pos, arr[10];
    clrscr();
    printf("\n Enter the size of the array : ");
    scanf("%d", &n);
    for(i=0;i<n;i++)
    {
        printf("\n Arr[%d] = ", i);
        scanf("%d", &arr[i]);
    }
```

```
        printf("\nEnter the position from which the number has to be deleted
: ");
        scanf("%d", &pos);
        for(i=pos; i<n;i++)
             array1[i] = arr [i+1];
        n--;
        printf("\n The array after deletion is : ");
        for(i=0;i<n;i++)
             printf("\n Arr[%d] = %d", i, arr[i]);
        getch();
        return 0;
}
```

12. Write a program to delete a number from an array that is already sorted in ascending order.

```
#include<stdio.h>
#include<conio.h>
int main()
{
        int i, n, j, num, arr[10];
        clrscr();
        printf("\n Enter the number of elements in the array : ");
        scanf("%d", &n);
        for(i=0;i<n;i++)
        {
             printf("\n Arr[%d] = ", i);
             scanf("%d", &arr[i]);
        }
        printf("\n Enter the number to be deleted : ");
        scanf("%d", &num);
        for(i=0;i<n;i++)
        {
             if(arr[i] == num)
             {
                  for(j=i; j<n;i++)
                       arr[j] = arr[j+1];
             }
        }
        printf("\n The array after deletion is : ");
        for(i=0;i<n-1;i++)
             printf("\n Arr[%d] = %d", i, arr[i]);
        getch();
        return 0;
}
```

5.6.4 Merging

Merging two arrays in a third array means first copying the contents of the first array into the third array and then copying the contents of the second array into the third array. Hence, the merged array contains the contents of the first array followed by the contents of the second array.

If the arrays are unsorted, then merging the arrays is very simple, as one just needs to copy the contents of one array into another. But merging is not a trivial task when the two arrays are sorted and the merged array also needs to be sorted. Let us first discuss the merge operation on unsorted arrays. This operation is shown in Fig 5.18.

Array 1-	90	56	89	77	69							
Array 2-	45	88	76	99	12	58	81					
Array 3-	90	56	89	77	69	45	88	76	99	12	58	81

Figure 5.18 Merging of two unsorted arrays

PROGRAMMING EXAMPLE

13. Write a program to merge two unsorted arrays.

```
#include<stdio.h>
#include<conio.h>
main()
{
        int arr1[10], arr2[10], arr3[20];
        int i, n1, n2, m, index=0;
        clrscr();
        printf("\n Enter the number of elements in array1 : ");
        scanf("%d", &n1);
        printf("\n\n Enter the Elements of the first array");
        printf("\n ************************************");
        for(i=0;i<n1;i++)
        {
                printf("\n Arr1[%d] = ", i);
                scanf("%d", &arr1[i]);
        }
        printf("\n Enter the number of elements in array2 : ");
        scanf("%d", &n2);
        printf("\n\n Enter the Elements of the second array");
        printf("\n ************************************");
        for(i=0;i<n2;i++)
        {
                printf("\n Arr2[%d] = ", i);
                scanf("%d", &arr2[i]);
        }
        m = n1+n2;
        for(i=0;i<n1;i++)
```

```
        {
            arr3[index] = arr1[i];
            index++;
        }
        for(i=0;i<n2;i++)
        {
            arr3[index] = arr2[i];
            index++;
        }
        printf("\n\n The merged array is");
        printf("\n ********************");
        for(i=0;i<m;i++)
            printf("\n Arr[%d] = %d", i, arr3[i]);
        getch();
        return 0;
}
```

If we have two sorted arrays and the resultant merged array also needs to be a sorted one, then the task of merging the arrays becomes a little difficult. The task of merging can be explained using Fig. 5.19.

Array 1-	20	30	40	50	60							
Array 2-	15	22	31	45	56	62	78					
Array 3-	15	20	22	30	31	40	45	50	56	60	62	78

Figure 5.19 Merging of two sorted arrays

Figure 5.19 shows how the merged array is formed using two sorted arrays. Here, we first compare the 1st element of `array1` with the 1st element of `array2`, put the smaller element in the merged array. Since, 20 > 15, we put 15 as the first element in the merged array. We then compare the 2nd element of the second array with the 1st element of the first array. Since 20 < 22, now 20 is stored as the second element of the merged array. Next, the 2nd element of the first array is compared with the 2nd element of the second array. Since, 30 > 22, we store 22 as the third element of the merged array. Now, we will compare the 2nd element of the first array with the 3rd element of the second array. Because, 30 < 31, we store 30 as the 4th element of the merged array. This procedure will be repeated until elements of both the arrays are placed in the right location in the merged array.

PROGRAMMING EXAMPLE

14. Write a program to merge two sorted arrays.

```
#include<stdio.h>
#include<conio.h>
```

```
main()
{
        int arr1[10], arr2[10], arr3[20];
        int i, n1, n2, m, index=0;
        int index_first = 0, index_second = 0;
        clrscr();
        printf("\n Enter the number of elements in array1 : ");
        scanf("%d", &n1);
        printf("\n\n Enter the Elements of the first array");
        printf("\n ***********************************");
        for(i=0;i<n1;i++)
        {
                printf("\n Arr1[%d] = ", i);
                scanf("%d", &arr1[i]);
        }
        printf("\n Enter the number of elements in array2 : ");
        scanf("%d", &n2);
        printf("\n\n Enter the Elements of the second array");
        printf("\n ************************************");
        for(i=0;i<n2;i++)
        {
                printf("\n Arr2[%d] = ", i);
                scanf("%d", &arr2[i]);
        }
        m = n1+n2;
        while(index_first < n1 && index_second < n2)
        {
                if(arr1[index_first]<arr2[index_second])
                {
                        arr3[index] = arr1[index_first];
                        index_first++;
                }
                else
                {
                        arr3[index] = arr2[index_second];
                        index_second++;
                }
                index++;
        }
        // if elements of the first array is over and the second array has
           some elements
        if(index_first == n1)
        {
                while(index_second<n2)
                {
                        arr3[index] = arr2[index_second];
```

```
                index_second++;
                index++;
            }
        }
        //if elements of the second array is over and the first array has
           some elements
        else if(index_second == n2)
        {
            while(index_first<n1)
            {
                arr3[index] = arr1[index_first];
                index_first++;
                index++;
            }
        }
        printf("\n\n The contents of the merged array are-");
        printf("\n ***********************************");
        for(i=0;i<m;i++)
            printf("\n Arr[%d] = %d", i, arr3[i]);
        getch();
        return 0;
}
```

5.6.5 Searching the Array Elements

Searching means to find whether a particular value is present in the array or not. If the value is present in the array, then searching is said to be successful and the searching process gives the location of that value in the array. Otherwise, if the value is not present in the array, the searching process displays the appropriate message and in this case searching is said to be unsuccessful.

There are two popular methods for searching the array elements: *linear search* and *binary search*. The algorithm that should be used depends entirely on how the values are organized in the array. For example, if the elements of the array are arranged in ascending order, then binary search should be used, as it is more efficient for sorted list in terms of complexity. We will discuss these two methods in detail in this section.

Linear Search

Linear search, also called as *sequential search*, is a very simple method used for searching an array for a particular value. It works by comparing every element of the array one by one in a sequence until a match is found. Linear search is mostly used to search an unordered list of elements (array in which data elements are not sorted). For example, if an array `A[]` is declared and initialized as,

```
int A[] = { 10, 8, 2, 7, 3, 4, 9, 1, 6, 5};
```

and `VAL = 7`, then searching means to find out whether the value '7' is present in the array or not. If yes, then it returns to the position of its occurrence. Here, `POS = 4` (index starting from 0).

Thus, we see that linear search executes in `O(n)` time where `n` is the number of elements in the array. Obviously, the best case of linear search is when `VAL` is equal to the first element of the array. In this case, only one comparison will be made.

Likewise, the worst case will happen when either `VAL` is not present in the array or it is equal to the last element of the array. In both the cases, `n` comparisons will have to be made. However, the performance of the linear search algorithm can be improved by using a sorted array. Figure 5.20 shows the algorithm for linear search.

```
LINEAR_SEARCH(A, N, VAL, POS)
Step 1: [INITIALIZE] SET POS = -1
Step 2: [INITIALIZE] SET I = 0
Step 3:      Repeat Step 4 while I<N
Step 4:           IF A[I] = VAL, then
                       SET POS = I
                       PRINT POS
                       Go to Step 6
             [END OF IF]
        [END OF LOOP]
Step 5: PRINT "Value Not Present In The Array"
Step 6: EXIT
```

Figure 5.20 Algorithm for Linear Search

In steps 1 and 2 of the algorithm, we initialize the value of `POS` and `I`. In step 3, a `while` loop is executed that would be executed until `I` is less than `N` (total number of elements in the array). In step 4, a check is made to see if a match is found between the current array element and the `VAL`. If a match is found, then the position of the array element is printed, else the value of `I` is incremented to match the next element with `VAL`. However, if all the arrays elements have been compared with `VAL` and no match is found, then it means that the `VAL` is not present in the array.

PROGRAMMING EXAMPLE

15. Write a program to search an element in an array using the linear search technique.

```
#include<stdio.h>
#include<conio.h>
main()
{
          int arr[10], num, i, n, found = 0, pos = -1;
          clrscr();
          printf("\n Enter the number of elements in the array : ");
          scanf("%d", &n);
          printf("\n Enter the elements -");
          for(i=0;i<n;i++)
          {
                printf("\n arr[%d] = ", i);
                scanf("%d", &arr[i]);
          }
```

```
        printf("\n Enter the number that has to be searched : ");
        scanf("%d", &num);
        for(i=0;i<n;i++)
        {
                if(arr[i] == num)
                {
                        found =1;
                        pos=i;
                        printf("\n %d is found in the array at position = %d",
    num, i);
                        break;
                }
        }
        if (found == 0)
                printf("\n %d DOES NOT EXIST in the array");
        getch();
        return 0;
    }
```

Binary Search

Binary search is a searching algorithm that works efficiently with a sorted list. The algorithm finds the position of a particular element in the array. The mechanism of binary search can be better understood by an analogy of a telephone directory. When we are searching for a particular name in the directory, we first open the directory from the middle and then decide whether to look for the name in the first part of the directory or in the second part of the directory. Again, we open some page in the middle and the whole process is repeated until we finally find the right name.

Take another analogy. How do we find words in a dictionary? We first open the dictionary somewhere in the middle. Then, we compare the first word on that page with the desired word whose meaning we are looking for. If the desired word comes before the word that appear on the page, we look in the first half of the dictionary, else we look in the second half. Again, we open a page in the first half of the dictionary. Compare the first word on that page with the desired word and repeat the same procedure until we finally get the word. The same mechanism is applied in the binary search.

Now, let us consider how this mechanism is applied to search for a value in a sorted array. Given an array that is declared and initialized as

```
int A[] = {0, 1, 2, 3, 4, 5, 6, 7, 8, 9, 10};
```

and `VAL = 9`, the algorithm will proceed in the following manner.

```
BEG = 0, END = 10, MID = (0 + 10)/2 = 5
```

Now, `VAL = 9` and `A[MID] = A[5] = 5`

`A[5]` is less than `VAL`, therefore, we now search for the value in the later half of the array. So, we change the values of `BEG` and `MID`.

Now, `BEG = MID + 1 = 6, END = 10, MID = (6 + 10)/2 =16/2 = 8`

`VAL = 9` and `A[MID] = A[8] = 8`

`A[8]` is less than `VAL`, therefore, we now search for the value in the later half of the array. So, again we change the values of `BEG` and `MID`.

Now, `BEG = MID + 1 = 9, END = 10, MID = (9 + 10)/2 = 9`

Now, `VAL = 9` and `A[MID] = 9`.

In this algorithm, we see that `BEG` and `END` are the beginning and ending positions of the segment that we are looking to search for the element. `MID` is calculated as `(BEG + END)/2`. Initially, `BEG = lower_bound` and `END = upper_bound`. The algorithm will terminate when `A[MID] = VAL`. When the algorithm ends, we will set `POS = MID`. `POS` is the position at which the value is present in the array.

However, if `VAL` is not equal to `A[MID]`, then the values of `BEG`, `END`, and `MID` will be changed depending on whether the `VAL` is smaller or greater than `A[MID]`.

(a) If `VAL < A[MID]`, then `VAL` will be present in the left segment of the array. So, the value of `END` will be changed as `END = MID – 1`.

(b) If `VAL > A[MID]`, then `VAL` will be present in the right segment of the array. So, the value of `BEG` will be changed as `BEG = MID + 1`.

Finally, if `VAL` is not present in the array, then eventually, `END` will be less than `BEG`. When it happens, the algorithm should terminate, as it will indicate that the element is not present in the array and the search will be unsuccessful. Let us consider another example.

```
int A[] = {0,1, 2, 3, 4, 5, 6, 7, 8, 9, 10}; and VAL = 2
Step 1: BEG = 0, END = 10, MID = 5
        A[MID] > VAL
Step 2: BEG = 0, END = MID - 1 = 4, MID = 2
        A[MID] = VAL
```

Figure 5.21 shows the algorithm for binary search.

```
BINARY_SEARCH(A, lower_bound, upper_bound, VAL, POS)
Step 1: [INITIALIZE] SET BEG = lower_bound, END = upper_bound, POS = - 1
Step 2: Repeat Step 3 and Step 4 while BEG <= END
Step 3:         SET MID = (BEG + END)/2
Step 4:         IF A[MID] = VAL, then
                      POS = MID
                      PRINT POS
                      Go to Step 6
                IF A[MID] > VAL then;
                      SET END = MID - 1
                ELSE
                      SET BEG = MID + 1
                [END OF IF]
        [END OF LOOP]
Step 5: IF POS = -1, then
            PRINTF "VAL IS NOT PRESENT IN THE ARRAY"
        [END OF IF]
Step 6: EXIT
```

Figure 5.21 Algorithm for binary search

In step 1, we initialize the value of variables, BEG, END, and POS. In step 2, a while loop is executed until BEG is less than or equal to END. In step 3, the value of MID is calculated. In step 4, we check if the value of MID is equal to VAL (item to be searched in the array). If a match is found, then the value of POS is printed and the algorithm exits. However, if a match is not found, then if the value of A[MID] is greater than VAL, the value of END is modified, otherwise if A[MID] is greater than VAL, then the value of BEG is altered. In step 5, if the value of POS = –1, then the VAL is not present in the array and an appropriate message is printed on the screen before the algorithm exits.

The complexity of the binary search algorithm can be expressed as f(n), where n is the number of elements in the array. The complexity of the algorithm is calculated depending on the number of comparisons that are made. In the binary search algorithm, we see that with each comparison, the size of the segment where search has to be made is reduced to half. Thus, we can say that, in order to locate a particular VAL in the array, the total number of comparisons that will be made is given as

$$2^{f(n)} > n \text{ or } f(n) = \log_2 n$$

PROGRAMMING EXAMPLE

16. Write a program for Binary Search.

```
#include<stdio.h>
#include<conio.h>
main()
{
        int arr[10], num, i, n, pos = -1, beg, end mid, found =0;
        clrscr();
        printf("\n Enter the number of elements in the array : ");
        scanf("%d", &n);
        printf("\n Enter the elements -");
        for(i=0;i<n;i++)
        {
                printf("\n arr[%d] = ", i);
                scanf("%d", &arr[i]);
        }
        printf("\n Enter the number that has to be searched : ");
        scanf("%d", &num);
        beg = 0, end = n-1;
        while(beg<=end)
        {
                mid = (beg + end)/2;
                if (arr[mid] == num)
                {
                        printf("\n %d is present in the array at position = %d",
num, mid);
```

```
                found =1;
                break;
            }
            if (arr[mid]>num)
                end = mid-1;
            else if (arr[mid] < num)
                beg = mid+1;
        }
        if (beg > end && found == 0)
            printf("\n %d DOES NOT EXIST IN THE ARRAY, num");
        getch();
        return 0;
    }
```

5.7 ONE-DIMENSIONAL ARRAYS FOR INTER-FUNCTION COMMUNICATION

Like variables of other data types, we can also pass an array to a function. In some situations, you may want to pass individual elements of the array; while in other situations, you may want to pass the entire array. In this section, we will discuss both the cases. Look at Fig. 5.22 which will help you understand the concept.

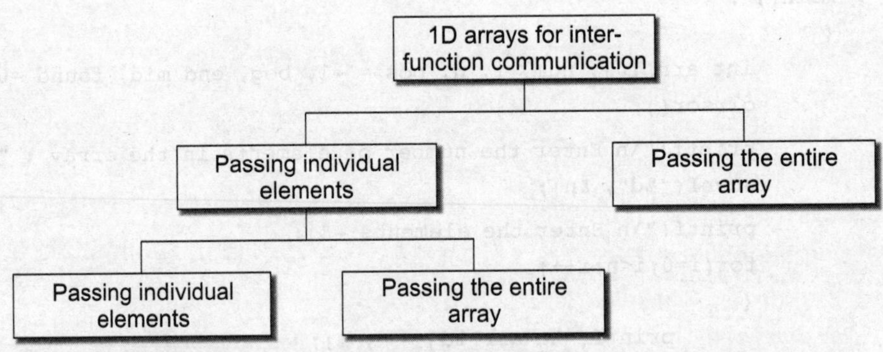

Figure 5.22 One dimensional arrays for inter-function communication

5.7.1 Passing Individual Elements

The individual elements of an array can be passed to a function either by passing their addresses or data values.

Passing Data Values

Individual elements can be passed in the same manner as we pass variables of any other data type. The condition is just that the data type of the array element must match the type of the function parameter. Look at Fig. 5.23(a) which shows the code to pass an individual array element by passing the data value.

Calling function

```
main()
{
        int arr[5] ={1, 2, 3, 4, 5};
        func(arr[3]);
}
```

Called function

```
void func(int num)
{
        printf("%d", num);
}
```

Figure 5.23 (a) Passing individual array elements to function

In the above example, only one element of the array is passed to the called function. This is done by using the index expression. So, `arr[3]` actually evaluates to a single integer value. The called function hardly bothers whether a normal integer variable is passed to it or an array value is passed.

Passing Addresses

Like ordinary variables, we can pass the address of an individual array element by preceding the address operator to the element's indexed reference. Therefore, to pass the address of the fourth element of the array to the called function, we write `&arr[3]`.

However, in the called function, the value of the array element must be accessed using the indirection `(*)` operator. Look at the code shown in Fig. 5.23(b).

Calling function

```
main()
{
        int arr[5] ={1, 2, 3, 4, 5};
        func(&arr[3]);
}
```

Called function

```
void func(int *num)
{
        printf("%d", num);
}
```

Figure 5.23 (b) Passing individual array elements to a function

5.7.2 Passing the Entire Array

We have discussed that in C, the array name refers to the first byte of the array in the memory. The address of the rest of the elements in the array can be calculated using the array name and the index value of the element. Therefore, when we need to pass an entire array to a function, we can simply pass the name of the array. Figure 5.24 illustrates the code which passes the entire array to the called function.

Calling function

```
main()
{
        int arr[5] ={1, 2, 3, 4, 5};
        func(arr);
}
```

Called function

```
void func(int arr[5])
{
        int 1;
        for(i=0;i<5;i++)
                printf("%d", arr[i]);
}
```

Figure 5.24 Passing entire array to a function

Note that in case we want the called function to make no changes to the array, the array must be received as a constant array by the called function. This prevents any type of unintentional modifications of the array elements. To declare an array as a constant array, simply add the keyword `const` before the data type of the array.

PROGRAMMING EXAMPLES

17. Write a program to read and print an array of *n* numbers.

```
#include<stdio.h>
#include<conio.h>
void read_array( int arr[], int);
void display_array( int arr[], int);
int main()
{
    int num[10], n;
    clrscr();
    printf("\n Enter the size of the array : ");
    scanf("%d", &n);
    read_array(num, n);
    display_array(num, n);
    getch();
    return 0;
}
void read_array( int arr[10], int n)
{
    int i;
    for(i=0;i<n;i++)
    {
        printf("\n array[%d] = ", i);
        scanf("%d", &arr[i]);
    }
}
void display_array( int arr[10], int n)
{
    for(i=0;i<n;i++)
        printf("\n array[%d] = %d", i, arr[i]);
}
```

18. Write a program to read an array of *n* numbers, then find out the smallest number.

```
#include<stdio.h>
#include<conio.h>
void read_array( int arr[], int);
int find_small(int arr[], n);
int main()
{
```

```
        int num[10], n, smallest;
        clrscr();
        printf("\n Enter the size of the array : ");
        scanf("%d", &n);
        read_array(num, n);
        smallest = find_small(num, n);
        printf("\n The smallest number in the array is = %d", smallest);
        getch();
        return 0;
}
void read_array( int arr[10], int n)
{
        int i;
        for(i=0;i<n;i++)
        {
                printf("\n array[%d] = ", i);
                scanf("%d", &arr[i]);
        }
}
int find_small(int arr[10], int n)
{
        int i, small = 1234567;
        for(i=0;i<n;i++)
        {
                if(arr[i] < small)
                        small = arr[i];
        }
        return small;
}
```

19. Write a program to merge two integer arrays. Also display the merged array in reverse order.

```
#include<stdio.h>
#include<conio.h>
void read_array(int arr1[], int);
void display_array(int arr3[], int);
void merge_array(int arr3[], int, int arr1[], int arr2[], int );
void reverse_array(int my_array[], int);
int main()
{
        int arr1[10], arr2[10], arr3[20], n, m, t;
        clrscr();
        printf("\n Enter the size of the first array : ");
        scanf("%d", &m);
        read_array(arr1, m);
        printf("\n Enter the size of the second array : ");
        scanf("%d", &n);
```

```
        read_array(arr2, n);
        t = m + n;
        merge_array(arr3, t, arr1, m, arr2, n);
        printf("\n The merged array is : ");
        display_array(arr3, t);
        printf("\n The merged array in reverse order is : ");
        reverse_array(arr3, t);
        getch();
        return 0;
}
void read_array( int my_array[10], int n)
{
        int i;
        for(i=0;i<n;i++)
        {
                printf("\n array[%d] = ", i);
                scanf("%d", &my_array[i]);
        }
}
void merge_array(int my_array3[], int t, int my_array1[], int m, int
my_array2[], int n)
{
        int i, j=0;
        for(i=0; i<n; i++)
        {
                my_array3[j] = my_array1[i];
                j++;
        }
        for(i=0; i<n; i++)
        {
                my_array3[j] = my_array2[i];
                j++;
        }
}
void display_array( int my_array[], int n)
{
        int i;
        for(i=0;i<n;i++)
                printf("\n array[%d] = %d", i, my_array[i]);
}
void reverse_array(int my_array[], int m)
{
        int i, j;
        for(i=m-1, j=0;i>=0;i–, j++)
                printf("\n array[%d] = %d", j, my_array[i]);
}
```

20. Write a program to interchange the biggest and the smallest number in an array.

```
#include<stdio.h>
#include<conio.h>
void read_array(int my_array[], int);
void display_array(int my_array[], int);
void interchange(int arr[], int);
int find_biggest_pos( int my_array[10], int n);
int find_smallest_pos( int my_array[10], int n);
int main()
{
	int arr[10], n;
	clrscr();
	printf("\n Enter the size of the array : ");
	scanf("%d", &n);
	read_array(arr, n);
	display_array(arr, n);
	interchange(arr, n);
	display_array(arr,n);
	getch();
	return 0;
}
void read_array(int my_array[10], int n)
{
	int i;
	for(i=0;i<n;i++)
	{
		printf("\n array[%d] = ", i);
		scanf("%d", &my_array[i]);
	}
}
void display_array( int my_array[10], int n)
{
	int i;
	for(i=0;i<n;i++)
		printf("\n array[%d] = %d", i, my_array[i]);
}
void interchange(int my_array[10], int n)
{
	int temp, big_pos, small_pos;
	big_pos = find_biggest_pos(my_arr, n);
	small_pos = find_smallest_pos(my_arr,n);
	temp = my_array[big_pos];
	my_array[big_pos] = my_array[small_pos];
	my_array[small_pos] = temp;
}
int find_biggest_pos(int my_array[10], int n)
{
```

```
        int i, large = -123456, pos=-1;
        for(i=0;i<n;i++)
        {
                if (my_array[i] > large)
                {
                        large = my_array[i];
                        pos=i;
                }
        }
        return pos;
}
int find_smallest(int my_array[10], int n)
{
        int i, small = 123456, pos=-1;
        for(i=0;i<n;i++)
        {
                if (my_array[i] < small)
                {
                        small = my_array[i];
                        pos=i;
                }
        }
        return pos;
}
```

5.8 POINTERS AND ARRAYS

The concept of array is bound to the concept of pointers. An array occupies consecutive memory locations. For example, if we have an array declared as

```
int arr[5] = {1, 2, 3, 4, 5};
```

then in the memory, it would be as shown in Fig. 5.25. Now, let us use a pointer variable as given in the following statement:

```
int *ptr;
ptr = &arr[0];
```

Here, `ptr` is made to point to the first element of the array. Similarly, writing `ptr = &arr[2]` makes `ptr` point to the third element of the array having the index 2. Look at Fig. 5.26, which shows `ptr` pointing to the third element of the array.

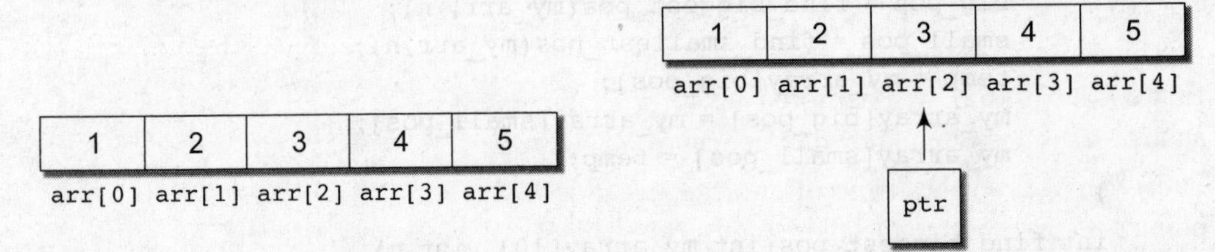

Figure 5.25 Memory representation of `arr[]`

Figure 5.26 `ptr` pointing to the third element of the array

If the pointer variable `ptr` holds the address of the first element in the array, then the address of the successive element can be calculated by writing, `ptr++`.

```
int *ptr = &arr[0];
ptr++;
printf("\n The value of the second element of the array is %d", *ptr);
```

The `printf()` statement will print the value 2 because after being incremented, `ptr` points to the next location. One point to note here is that if `x` is an integer variable, then `x++` adds 1 to the value of `x`. But `ptr` is a pointer variable, so when we write `ptr + i`, then adding `i` gives a pointer that points `i` elements further along an array than the original pointer.

Since `++ptr` and `ptr++` are both equivalent to `ptr + 1`, incrementing a pointer using the unary ++ operator increments the address it stores by the amount given by `sizeof(type)`, where `type` is the data type of the variable it points to (*i.e.*, 2 for an integer).

If `ptr` originally points to `arr[2]`, then `ptr++` will make it point to the next element, i.e., `arr[3]`. This is clearly shown in Fig. 5.27.

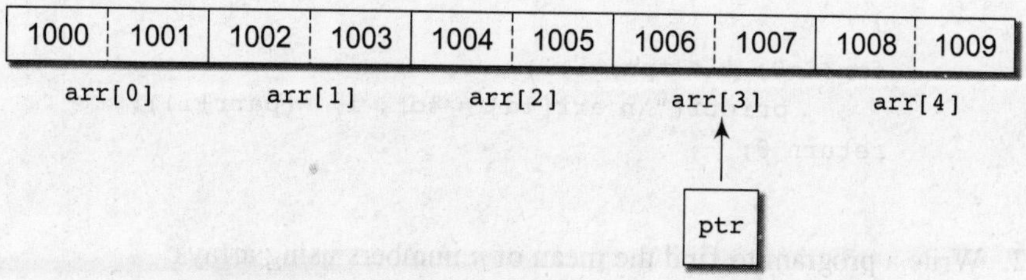

Figure 5.27 `ptr` pointing to the fourth element of the array

Note that had it been a character array, every byte in the memory would have been used to store an individual character. `ptr++` would then add only 1 byte to the address of `ptr`.

While using pointers, an expression like `arr[i]` is equivalent to writing `*(arr+i)`. If `arr` is the array name, then the compiler implicitly takes

```
arr = &arr[0]
```

To print the value of the third element of the array, we can straightaway use the expression `*(arr+2)`. Note that `arr[i] = *(arr + i)`.

Many beginners get confused by assuming the array name as a pointer. For example, we can write,

```
ptr = arr;  // that is ptr = &arr[0]
```

while the following statement is incorrect.

```
arr = ptr;
```

This is because while `ptr` is a variable, `arr` is a constant. That is, the location at which the first element of `arr` will be stored cannot be changed once `arr[]` has been declared. Therefore, an array name is often known to be a constant pointer.

To summarize, the name of an array is equivalent to the address of its first element, as a pointer is equivalent to the address of the element that it points to. Therefore, arrays and pointers use the same concept.

PROGRAMMING EXAMPLES

21. Write a program to read and display an array of *n* integers.

```
#include<stdio.h>
int main()
{
        int i, n;
        int arr[10], *parr, *pn;
        parr = arr;
        pn = &n;
        printf("\n Enter the number of elements : ");
        scanf("%d", pn);
        for(i=0; i < *pn; i++)
        {
                printf("\n Enter the %dth element : ", i);
                scanf("%d", parr+i);
        }
        for(i=0; i < *pn; i++)
                printf("\n arr[%d] = %d", i, *(parr+i));
        return 0;
}
```

22. Write a program to find the mean of *n* numbers using arrays.

```
#include<stdio.h>
int main()
{
        int i, n, arr[20], sum =0;
        int *pn, *parr, *psum;
        float mean = 0.0, *pmean;
        pn = &n;
        psum = &sum;
        pmean = &mean;
        parr = &arr;
        printf("\n Enter the number of elements in the array : ");
        scanf("%d", pn);
        for(i=0; i < *pn; i++)
        {
                printf("\n Enter the number: ");
                scanf("%d", parr + i);
        }
        for(i=0; i < *pn; i++)
                *psum += *(arr + i);
        *pmean = (float)(*psum)/(*pn);
        printf("\n The numbers you entered are : ");
        for(i=0; i < *pn; i++)
```

```
            printf("%d", *(arr + i));
        printf("\n The sum of these numbers is : %d", *psum);
        printf("\n The mean of these numbers is : %f", *pmean);
        return 0;
}
```

23. Write a program to find the largest of *n* numbers using arrays. Also display its position.

```
#include<stdio.h>
int main()
{
        int i, n, arr[20], large = -1111, pos = 0;
        int *pn, *parr, *plarge, *ppos;
        clrscr();
        pn = &n;
        parr = arr;
        plarge = &large;
        ppos = &pos;
        printf("\n Enter the number of elements in the array : ");
        scanf("%d", pn);
        for(i=0; i < *pn;i++)
        {
              printf("\n Enter the number: ");
              scanf("%d", arr+i);
        }
        for(i=0; i < *pn; i++)
        {
              if( *(arr+i) > *plarge)
              {
                    *plarge = *(arr+i);
                    *ppos = i;
              }
        }
        printf("\n The numbers you entered are : ");
        for(i=0;i<*pn;i++)
              printf("%d", *(arr+i));
        printf("\n The largest of these numbers is : %d", *plarge);
        printf("\n The position of largest numbers in the array is : %d",
*ppos);
        return 0;
}
```

24. Write a program to read and print an array of *n* numbers, then find out the smallest number. Also print the position of the smallest number.

```
#include<stdio.h>
void read_array(int *arr, int n);
```

```
void print_array( int *arr, int n);
void find_small(int *arr, int n, int *small, int *pos);
int main()
{
     int num[10], n, smallest, pos;
     printf("\n Enter the size of the array : ");
     scanf("%d", &n);
     read_array(num, n);
     find_small(num, n, &smallest, &pos);
     printf("\n The smallest number in the array is = %d", smallest);
     return 0;
}
void read_array(int *arr, int n)
{
     int i;
     for(i=0;i<n;i++)
     {
          printf("\n array[%d] = ", i);
          scanf("%d", arr+i);
     }
}
void print_array(int *arr, int n)
{
     int i;
     for(i=0;i<n;i++)
          printf("\n array[%d] = %d", i, arr[i]);
}
void find_small(int *arr, int n, int *small, int *pos)
{
     for(i=0;i<n;i++)
     {
          if(*(arr+i]) < *small)
          {
               *small = *(arr+i);
               *pos = i;
          }
     }
}
```

5.9 ARRAY OF POINTERS

An array of pointers can be declared as

```
int *ptr[10]
```

The statement declares an array of 10 pointers where each of the pointer points to an integer variable. For example, look at the code given below.

```
int *ptr[10];
int p=1, q=2, r=3, s=4, t=5;
ptr[0]=&p;
ptr[1]=&q;
ptr[2]=&r;
ptr[3]=&s;
ptr[4]=&t
```

What will be the output of the following statement?

```
printf("\n %d", *ptr[3]);
```

The output will be 4 because `ptr[3]` stores the address of the integer variable `s` and `*ptr[3]` will therefore print the value of `s` which is 4.

5.10 TWO-DIMENSIONAL ARRAYS

Till now, we have only discussed one-dimensional arrays. One-dimensional arrays are organized linearly in only one direction. But at times, we need to store data in the form of matrices or tables. Here, the concept of single-dimension arrays is extended to incorporate two-dimensional data structures. A two-dimensional array is specified using two subscripts where one subscript denotes the row and the other denotes the column. The C compiler treats a two-dimensional array as an array of one-dimensional arrays. Figure 5.28 shows a two-dimensional array which can be viewed as array of arrays.

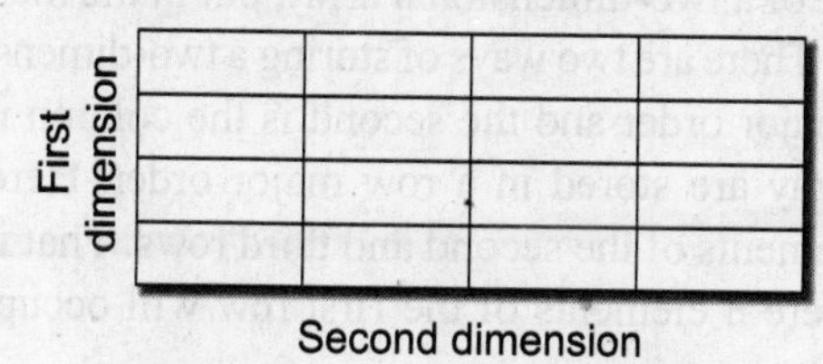

Figure 5.28 Two-dimensional array

5.10.1 Declaration of Two-dimensional Arrays

Any array must be declared before being used. The declaration statement tells the compiler the name of the array, the data type of each element in the array, and the size of each dimension. A two-dimensional array is declared as:

```
data_type array_name[row_size][column_size];
```

Therefore, a two-dimensional `m` × `n` array is an array that contains `m*n` data elements and each element is accessed using two subscripts, `i` and `j` where `i <= m` and `j <= n`.

For example, if we want to store the marks obtained by 3 students in 5 different subjects, we can declare a two-dimensional array as:

```
int marks[3][5]
```

A two-dimensional array called `marks` is declared that has `m(3)` rows and `n(5)` columns. The first element of the array is denoted by `marks[0][0]`, the second element as `marks[0][1]`, and so on. Here, `marks[0][0]` stores the marks obtained by the first student in the first subject, `marks[1][0]` stores the marks obtained by the second student in the first subject.

The pictorial form of a two-dimensional array is given in Fig. 5.29.

Rows / Columns	Col 0	Col 1	Col 2	Col 3	Col 4
Row 0	Marks[0][0]	Marks[0][1]	Marks[0][2]	Marks[0][3]	Marks[0][4]
Row 1	Marks[1][0]	Marks[1][1]	Marks[1][2]	Marks[1][3]	Marks[1][4]
Row 2	Marks[2][0]	Marks[2][1]	Marks[2][2]	Marks[2][3]	Marks[2][4]

Figure 5.29 Two-dimensional array

Hence, we see that a 2D array is treated as a collection of 1D arrays. The elements of the 2D array in turn comprises of 1D array (the rows). To understand this, we can also see the representation of a two-dimensional array as shown in Fig. 5.30.

Marks[0] –	Marks[0]	Marks[1]	Marks[2]	Marks[3]	Marks[4]
Marks[1] –	Marks[0]	Marks[1]	Marks[2]	Marks[3]	Marks[4]
Marks[2] –	Marks[0]	Marks[1]	Marks[2]	Marks[3]	Marks[4]

Figure 5.30 Representation of two-dimensional array `Marks[3][5]`

Although we have shown a rectangular picture of a two-dimensional array, but in the memory, these elements actually will be stored sequentially. There are two ways of storing a two-dimensional array in the memory. The first way is the row major order and the second is the column major order. Let us see how the elements of a 2D array are stored in a row major order. Here, the elements of the first row are stored before the elements of the second and third rows. That is, the elements of the array are stored row by row where n elements of the first row will occupy the first nth locations. This is illustrated in Fig. 5.31.

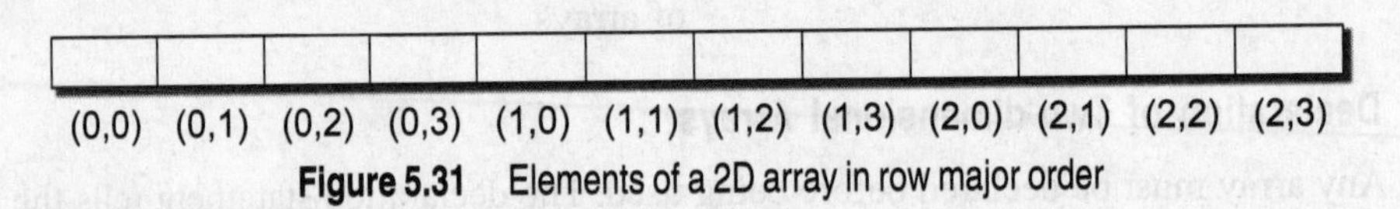

Figure 5.31 Elements of a 2D array in row major order

However, when we store the elements in a column major order, the elements of the first column are stored before the elements of the second and third column. That is, the elements of the array are stored column by column where n elements of the first column will occupy the first nth locations. This is illustrated in Fig. 5.32.

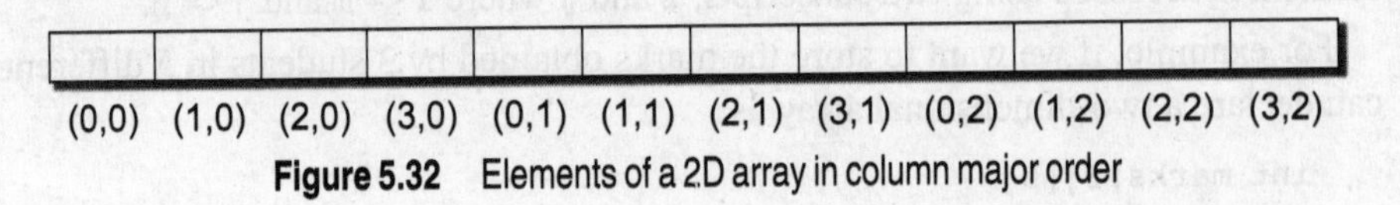

Figure 5.32 Elements of a 2D array in column major order

In one-dimensional arrays, we have seen that the computer does not keep track of the address of every element in the array. It stores only the address of the first element and calculates the address of other elements from the base address (address of the first element). Same is the case with a two-dimensional array. Here also, the computer stores the base address and the address of the other elements is calculated using the following formula.

If the array elements are stored in column major order,

```
Address(A[I][J] = Base_Address + w{M ( J - 1) + (I - 1)}
```

And if the array elements are stored in row major order,

```
Address(A[I][J] = Base_Address + w{N ( I - 1) + (J - 1)}
```

where w is the number of words stored per memory location, m is the number of columns, n is the number of rows, and I and J are the subscripts of the array element.

EXAMPLE 5.6: Consider a 20 × 5 two-dimensional array `Marks` which has its base address = 1000 and the number of words per memory location of the array = 2. Now compute the address of the element, `Marks[18, 4]` assuming that the elements are stored in row major order.

Solution

```
Address(A[I][J] = Base_Address + w{N (I - 1) + (J - 1)}
Address(Marks[18,4]) = 1000 + 2 {5(18 - 1) + (4 - 1)}
                     = 1000 + 2 {5(17) + 3}
                     = 1000 + 2 (88)
                     = 1000 + 176 = 1176
```

5.10.2 Initialization of Two-dimensional Arrays

Like in the case of other variables, declaring a two-dimensional array only reserves space for the array in the memory. No values are stored in it. A two-dimensional array is initialized in the same way as a single-dimensional array is initialized. For example,

```
int marks[2][3]={90, 87, 78, 68, 62, 71};
```

Note that the initialization of a two-dimensional array is done row by row. The above statement can also be written as:

```
int marks[2][3]={{90,87,78},{68, 62, 71}};
```

The above two-dimensional array has two rows and three columns. Here, the elements in the first row are initialized first and then the elements of the second row are initialized.

Therefore, `marks[0][0] = 90` `marks[0][1] = 87` `marks[0][2] = 78`
`marks[1][0] = 68` `marks[1][1] = 62` `marks[1][2] = 71`

In the above example, each row is defined as a one-dimensional array of three elements that are enclosed in braces. Note that the commas are used to separate the elements in the row as well as to separate the elements of two rows.

In case of one-dimensional arrays, we have discussed that if the array is completely initialized, we may omit the size of the array. The same concept can be applied to a two-dimensional array, except that only the size of the first dimension can be omitted. Therefore, the declaration statement given below is valid.

```
int marks[][3]={{90,87,78},{68, 62, 71}};
```

In order to initialize the entire two-dimensional array to zeros, simply specify the first value as zero. That is,

```
int marks[2][3] = {0};
```

5.10.3 Accessing the Elements

The elements in a 2D array are stored in contiguous memory locations. While accessing the elements, remember that the last subscript varies most rapidly whereas the first varies least rapidly.

In case on one-dimensional arrays, we used a single `for` loop to vary the index `i` in every pass, so that all the elements could be scanned. Similarly, since the two-dimensional array contains two subscripts, we will use two `for` loops to scan the elements. The first `for` loop will loop for each row in the 2D array and the second `for` loop will scan individual columns for every row in the array. Look at the code given below which prints the elements of a 2D array on the screen.

PROGRAMMING EXAMPLE

25. Write a program to print the elements of a 2D array.

```
#include<stdio.h>
#include<conio.h>
main()
{
     int arr[2][2] = {12, 34, 56,32};
     int i, j;
     for(i=0;i<2;i++)
     {
          printf("\n");
          for(j=0;j<2;j++)
               printf("%d\t", arr[i][j]);
     }
     return 0;
}
```

Output:

```
12   34
56   32
```

The individual elements of a two-dimensional array can be initialized using the assignment operator as shown here.

```
marks[1][2] = 79;
```

or

```
marks[1][2] = marks[1][1] + 10;
```

5.11 OPERATIONS ON 2D ARRAYS

We can perform the following operations on a two-dimensional array:

- **Transpose** Transpose of a $m \times n$ matrix A is given as a $n \times m$ matrix B, where $B_{i,j} = A_{j,i}$
- **Sum** Two matrices that are compatible with each other can be added together, thereby storing the result in the third matrix. Two matrices are said to be compatible when they

have the same number of rows and columns. The elements of two matrices can be added by writing:

$C_{i,j} = A_{i,j} + B_{i,j}$

- **Difference** Two matrices that are compatible with each other can be subtracted, thereby storing the result in the third matrix. Two matrices are said to be compatible when they have the same number of rows and columns. The elements of two matrices can be subtracted by writing:

  $C_{i,j} = A_{i,j} - B_{i,j}$

- **Product** Two matrices can be multiplied with each other if the number of columns in the first matrix is equal to the number of rows in the second matrix. Therefore, `m×n` matrix `A` can be multiplied with a `p×q` matrix if `n=q`. The elements of two matrices can be multiplied by writing:

  $C_{i,j} = \Sigma A_{i,k}B_{j,k}$, for $k=1$ to $k<n$

PROGRAMMING EXAMPLES

26. Write a program to read and display a 3 × 3 matrix.

```
#include<stdio.h>
#include<conio.h>
int main()
{
    int i, j, mat[3][3];
    clrscr();
    printf("\n Enter the elements of the matrix ");
    printf("\n **************************");
    for(i=0;i<3;i++)
    {
        for(j=0;j<3;j++)
        {
            printf("\n mat[%d][%d] = ", i, j);
            scanf("%d",&mat[i][j]);
        }
    }
    printf("\n The elements of the matrix are ");
    printf("\n **************************");
    for(i=0;i<3;i++)
    {
        printf("\n");
        for(j=0;j<3;j++)
            printf("\t mat[%d][%d] = %d",i, j, mat[i][j]);
    }
    return 0;
}
```

27. Write a program to transpose a 3×3 matrix.

```
#include<stdio.h>
#include<conio.h>
int main()
{
    int i, j, mat[3][3], transposed_mat[3][3];
    clrscr();
    printf("\n Enter the elements of the matrix ");
    printf("\n **************************");
    for(i=0;i<3;i++)
    {
        for(j=0;j<3;j++)
        {
            printf("\n mat[%d][%d] = ",i, j);
            scanf("%d", &mat[i][j]);
        }
    }
    printf("\n The elements of the matrix are ");
    printf("\n **************************");
    for(i=0;i<3;i++)
    {
        printf("\n");
        for(j=0;j<3;j++)
            printf("\t mat[%d][%d] = %d", i, j, mat[i][j]);
    }
    for(i=0;i<3;i++)
    {
        for(j=0;j<3;j++)
            transposed_mat[i][j] = mat[j][i];
    }
    printf("\n The elements of the transposed matrix are ");
    printf("\n **************************");
    for(i=0;i<3;i++)
    {
        printf("\n");
        for(j=0;j<3;j++)
            printf("\t transposed_mat[%d][%d] = %d",i, j, transposed_
mat[i][j]);
    }
    return 0;
}
```

28. Write a program to input two $m \times n$ matrices and then calculate the sum of their corresponding elements and store it in a third $m \times n$ matrix.

```
#include<stdio.h>
#include<conio.h>
```

```
int main()
{
        int i, j;
        int rows1, cols1, rows2, cols2, rows_sum, cols_sum;
        int mat1[5][5], mat2[5][5], sum[5][5];
        clrscr();
        printf("\n Enter the numbers of rows in the first matrix : ");
        scanf("%d",&rows1);
        printf("\n Enter the numbers of columns in the first matrix : ");
        scanf("%d",&cols1);
        printf("\n Enter the numbers of rows in the second matrix : ");
        scanf("%d",&rows2);
        printf("\n Enter the numbers of columns in the second matrix : ");
        scanf("%d",&cols2);
        if(rows1 != rows2 || cols1 != cols2)
        {
                printf("\n The number of rows and columns of both the matrices
must be equal");
                getch();
                exit();
        }
        rows_sum = rows1;
        cols_sum = cols1;
        printf("\n Enter the elements of the first matrix ");
        printf("\n **************************");
        for(i=0;i<rows1;i++)
        {
                for(j=0;j<cols1;j++)
                {
                        printf("\n mat1[%d][%d] = ",i,j);
                        scanf("%d",&mat1[i][j]);
                }
        }
        printf("\n Enter the elements of the second matrix ");
        printf("\n **************************");
        for(i=0;i<rows2;i++)
        {
                for(j=0;j<cols2;j++)
                {
                        printf("\n mat2[%d][%d] = ",i,j);
                        scanf("%d",&mat2[i][j]);
                }
        }
        for(i=0;i<rows_sum;i++)
```

```
        {
              for(j=0;j<cols_sum,j++)
                    sum[i][j] = mat1[i][j] + mat2[i][j];
        }
        printf("\n The elements of the resultant matrix are ");
        printf("\n ***************************");
        for(i=0;i<rows_sum;i++)
        {
              printf("\n");
              for(j=0;j<cols_sum;j++)
                    printf("\t %d", sum[i][j]);
        }
        return 0;
  }
```

29. Write a program to multiply two m × n matrices.

```
  #include<stdio.h>
  #include<conio.h>
  int main()
  {
        int i, j, k;
        int rows1, cols1, rows2, cols2, res_rows, res_cols;
        int mat1[5][5], mat2[5][5], res[5][5];
        clrscr();
        printf("\n Enter the numbers of rows in the first matrix : ");
        scanf("%d",&rows1);
        printf("\n Enter the numbers of columns in the first matrix : ");
        scanf("%d",&cols1);
        printf("\n Enter the numbers of rows in the second matrix : ");
        scanf("%d",&rows2);
        printf("\n Enter the numbers of columns in the second matrix : ");
        scanf("%d",&cols2);
        if(cols1 != rows2)
        {
              printf("\n The number of columns in the first matrix must be
  equal to the number of rows in the second matrix");
              getch();
              exit();
        }
        res_rows = rows1;
        res_cols = cols2;
        printf("\n Enter the elements of the first matrix ");
        printf("\n ***************************");
        for(i=0;i<rows1;i++)
```

```
    {
        for(j=0;j<cols1;j++)
        {
            printf("\n mat1[%d][%d] = ",i,j);
            scanf("%d",&mat1[i][j]);
        }
    }
    printf("\n Enter the elements of the second matrix ");
    printf("\n ************************");
    for(i=0;i<rows2;i++)
    {
        for(j=0;j<cols2;j++)
        {
            printf("\n mat2[%d][%d] = ",i,j);
            scanf("%d",&mat2[i][j]);
        }
    }
    for(i=0;i<res_rows;i++)
    {
        j=0;
        for(j=0;j<res_cols;j++)
        {
            res[i][j]=0;
            for(k=0; k<res_cols;k++)
            res[i][j] += mat1[i][k] * mat2[k][j];
        }
    }
    printf("\n The elements of the product matrix are ");
    printf("\n *************************");
    for(i=0;i<res_rows;i++)
    {
        printf("\n");
        for(j=0;j<res_cols;j++)
            printf("\t res[%d][%d] = %d",i,j, res[i][j]);
    }
    return 0;
}
```

5.12 2D ARRAYS FOR INTER-FUNCTION COMMUNICATION

There are three ways of passing parts of a two-dimensional array to a function. First, we can pass individual elements of the array. This is exactly the same as passing an element of a one-dimensional array. Second, we can pass a single row of the two-dimensional array. This is equivalent to passing the entire one-dimensional array to a function that has already been discussed in the previous section. Third, we can pass the entire two-dimensional array to the function. Figure 5.33 shows the three ways of using two-dimensional arrays for inter-function communication.

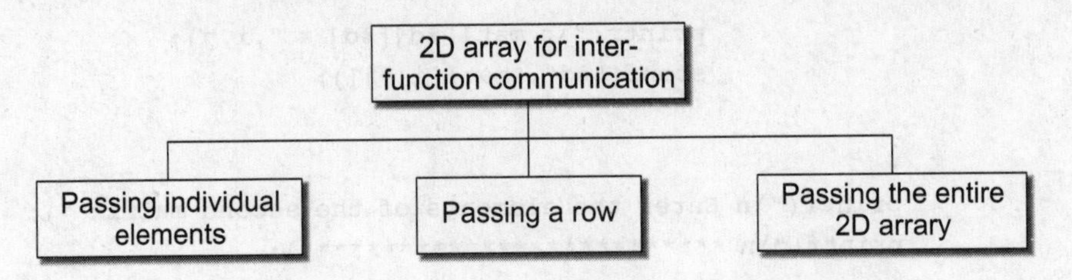

Figure 5.33 2D arrays for inter-function communication

5.12.1 Passing a Row

A row of a two-dimensional array can be passed by indexing the array name with the row number. Look at Fig. 5.34 which illustrates how a single row of a two-dimensional array is passed to the called function.

Calling function

```
main()
{
    int arr[2][3] = ({1, 2, 3}, {4, 5, 6});
    func(arr[1]);
}
```

Called function

```
void func(int arr[])
{
    int i;
    for(i=0;i<5;i++)
        printf("%d", arr[i] * 10);
}
```

Figure 5.34 Passing 2D arrays for inter-function communication

5.12.2 Passing the Entire 2D Array

To pass a two-dimensional array to a function, we use the array name as the actual parameter (The same we did in case of a 1D array). However, the parameter in the called function must indicate that the array has two dimensions.

PROGRAMMING EXAMPLES

30. Write a menu-driven program to read and display an m × n matrix. Also find the sum, transpose, and product of two m × n matrices.

```
#include<stdio.h>
#include<conio.h>
void read_matrix(int mat[][], int, int);
void sum_matrix(int mat1[][], int mat2[][], int, int);
void mul_matrix(int mat1[][], int mat2[][], int, int);
void transpose_matrix(int mat2[][], int, int);
void display_matrix(int mat[5][5], int r, int c);
int main()
{
        int option, row, col;
```

```
        int mat1[5][5], mat2[5][5];
        clrscr();
        do
        {
              printf("\n ******* MAIN MENU ********");
              printf("\n 1. Read the two matrices");
              printf("\n 2. Add the matrices");
              printf("\n 3. Multiply the matrices");
              printf("\n 4. Transpose the matrix");
              printf("\n 5. EXIT");
              printf("\n\n Enter your option : ");
              scanf("%d", &option);
              switch(option)
              {
                    case 1:
                          printf("\n Enter the number of rows and columns of
the matrix : ");
                          scanf("%d %d", &row, &col);
                          read_matrix(mat1, row, col);
                          printf("\n Enter the second matrix : ");
                          read_matrix(mat2, row, col);
                          break;
                    case 2:
                          sum_matrix(mat1, mat2, row, col);
                          break;
                    case 3:
                          if(col == row)
                                mul_matrix(mat1, mat2, row, col);
                          else
                                printf("\n To multiply two matrices, number
of columns in the first matrix must be equal to number of rows in the second
matrix");
                          break;
                    case 4:
                          transpose_matrix(mat1, row, col);
                          transpose_matrix(mat2, row, col);
                          break;
              }
        }while(option != 5);
        getch();
        return 0;
}
void read_matrix(int mat[5][5], int r, int c)
{
        int i, j;
        for(i=0;i<r;i++)
```

```
        {
            for(j=0;j<c;j++)
            {
                printf("\n mat[%d][%d] = ");
                scanf("%d", &mat[i][j]);
            }
        }
}
void sum_matrix(int mat1[5][5], mat2[5][5], int r, int c)
{
        int i, j, sum[5][5];
        for(i=0;i<r;i++)
        {
            for(j=0;j<c;j++)
                sum[i][j] = mat1[i][j] + mat2[i][j];
        }
        display_matrix(sum, r, c);
}
void mul_matrix(int mat1[5][5], mat2[5][5], int r, int c)
{
        int i, j, k, prod[5][5];
        for(i=0;i<r;i++)
        {
            for(j=0;j<c;j++)
            {
                prod[i][j] = 0;
                for(k=0;k<col;k++)
                    prod[i][j] += mat1[i][k] * mat2[k][j];
            }
        }
        display_matrix(prod, r, c);
}
void transpose_matrix(int mat[5][5], int r, int c)
{
        int i, j, tp_mat[5][5];
        for(i=0;i<r;i++)
        {
            for(j=0;j<c;j++)
                tp_mat[i][j] = mat[j][i];
        }
        display_matrix(tap_mat, r, c);
}
void display_matrix(int mat[5][5], int r, int c)
{
```

```
        int i, j;
        for(i=0;i<r;i++)
        {
                printf("\n");
                for(j=0;j<c;j++)
                        printf("\t mat[%d][%d] = %d", mat[i][j]);
        }
}
```

5.13 POINTERS AND TWO-DIMENSIONAL ARRAYS

The elements of a two-dimensional array are stored in contiguous memory locations. A two-dimensional array is not the same as an array of pointers to one-dimensional arrays. To declare a pointer to a two-dimensional array, you may write

```
int **ptr
```

Here, `int **ptr` is an array of pointers (to one-dimensional arrays), while `int mat[5][5]` is a 2D array. They are not the same type, and are not interchangeable.

Consider a two-dimensional array declared as

```
int mat[5][5];
```

Individual elements of the array can be accessed using either:

`mat[i][j]` or `*(*(mat + i) + j)`

To understand better, let us replace `*(multi + row)` with `X`, so the expression `*(*(mat + i) + j)` becomes `*(X + col)`

Using pointer arithmetic, we know that the address pointed to by (*i.e.*, value of) `X + col + 1` must be greater than the address `X + col` by an amount equal to `sizeof(int)`.

Since `mat` is a two-dimensional array, we know that in the expression `multi + row` as used above, `multi + row + 1` must increase by an amount equal to that needed to 'point to' the next row, which in this case, would be an amount equal to `COLS * sizeof(int)`.

Thus, in case of a two-dimensional array, in order to evaluate an expression (for a row major 2D array), we must know a total of four values:

1. The address of the first element of the array, which is given by the name of the array, *i.e.*, `mat` in our case.
2. The size of the type of the elements of the array `i`, *i.e.*, the size of integers in our case.
3. The specific index value for the row.
4. The specific index value for the column.

Note that

```
int (*ptr)[10];
```

declares `ptr` to be a pointer to an array of 10 integers. This is different from

```
int *ptr[10];
```

which would make `ptr` the name of an array of 10 pointers to type `int`. You must be wondering how a pointer arithmetic works, if you have an array of pointers. For example:

```
int * arr[10];
int ** ptr = arr;
```

In this case, `arr` has type `int**`. Since all the pointers have the same size, the address of `ptr + i` can be calculated as:

```
addr(ptr + i) = addr(ptr) + [sizeof(int *) * i]
              = addr(ptr) + [2 * i]
```

Note that since `arr` has type `int **`,

```
arr[0] = & arr[0][0],
arr[1] = & arr[1][0],
```

and in general,

```
arr[i] = & arr[i][0].
```

According to the pointer arithmetic, `arr + i = &arr[i]`. Yet, this skips an entire row of 5 elements. That is, it skips 10 byes (5 elements each of 2 bytes size). Therefore, if **`arr`** is address 1000, then `arr + 2` is address 1010.

To conclude, a two-dimensional array is not the same as an array of pointers to 1D arrays. Actually a two-dimensional array is declared as:

```
int (*ptr)[10];
```

Here `ptr` is a pointer to an array of 10 elements. The parentheses are not optional. In the absence of these parentheses, `ptr` becomes an array of 10 pointers, not a pointer to an array of 10 integers.

PROGRAMMING EXAMPLE

31. Write a program to read and display a 3 × 3 matrix.

```
#include<stdio.h>
int main()
{
     int i, j, mat[3][3];
     clrscr();
     printf("\n Enter the elements of the matrix ");
     printf("\n **************************");
     for(i=0;i<3;i++)
     {
          for(j=0;j<3;j++)
          {
               printf("\n mat[%d][%d] = ",i,j);
               scanf("%d", (mat + i*3)+j));
          }
     }
```

```
    printf("\n The elements of the matrix are ");
    printf("\n ************************");
    for(i=0;i<3;i++)
    {
        printf("\n");
        for(j=0;j<3;j++)
            printf("\t mat[%d][%d] = %d",i,j, *((mat + i*3) +j));
    }
    return 0;
}
```

5.14 ARRAY OF FUNCTION POINTERS

When an array of function pointers is created, the appropriate function is selected using an index. The code given below shows how to define and use an array of function pointers in C.

Step 1: Use the keyword `typedef` so that `fp` can be used as a type.

```
typedef int (*fp)(int, int);
```

Step2: Define the array and initialize each element to `NULL`. This can be done in two ways.

// with 10 pointers to functions which return an `int` and take two `ints`

First way: `fp funcArr[10] = {NULL};`

Second way: `int (*funcArr[10])(int, int) = {NULL};`

Step 3: Assign the function's address – Add and Subtract.

```
funcArr1[0] = funcArr2[1] = &Add;
funcArr[0] = &Add;
funcArr[1] = &Subtract;
```

Step 4: Call the function using an index to address the function pointer.

```
printf("%d\n", funcArr[1](2, 3));      // short form
printf("%d\n", (*funcArr[0])(2, 3));   // correct way
```

PROGRAMMING EXAMPLE

32. Write a program that uses an array of function pointers.

```
#include <stdio.h>
int sum(int a, int b);
int subtract(int a, int b);
int mul(int a, int b);
int div(int a, int b);

int (*fp[4]) (int a, int b);
int main(void)
{
```

```
    int result;
    int num1, num2, op;
    fp[0] = sum;
    fp[1] = subtract;
    fp[2] = mul;
    fp[3] = div;
    printf("\n Enter the numbers: ");
    scanf("%d %d", &num1, &num2);
do
{
    printf("\n 0: Add \n 1: Subtract \n 2: Multiply \n 3: Divide \n 4. EXIT");
    printf("Enter the operation: ");
    scanf("%d", &op);
    result = (*fp[op]) (num1, num2);
    printf("%d", result);
} while(op!=4);
    return 0;
}
int sum(int a, int b)
{
    return a + b;
}
int subtract(int a, int b)
{
    return a - b;
}
int mul(int a, int b)
{
    return a * b;
}
int div(int a, int b)
{
    if(b)
        return a / b;
    else
        return 0;
}
```

5.15 MULTI-DIMENSIONAL ARRAYS

A multi-dimensional array in simple terms is an array of arrays. As we have one index in a one-dimensional array, two indices in a two-dimensional array, in the same way, we have n indices in an n-dimensional array or multi-dimensional array. Conversely, an n-dimensional array is specified using n indices. An n-dimensional $m_1 \times m_2 \times m_3 \times \ldots m_n$ array is a collection of $m_1 * m_2 * m_3 * \ldots * m_n$ elements. In a multi-dimensional array, a particular element is specified by using n subscripts as `A[I1][I2][I3]...[In]`, where

$$I_1 <= M_1,\ I_2 <= M_2,\ I_3 <= M_3,\ \ldots\ I_n <= M_n$$

A multi-dimensional array can contain as many indices as needed and the requirement of memory increases with the number of indices used. However, in practice, we hardly use more than three indices in any program. Figure 5.35 shows a three- dimensional array. The array has three pages, four rows, and two columns.

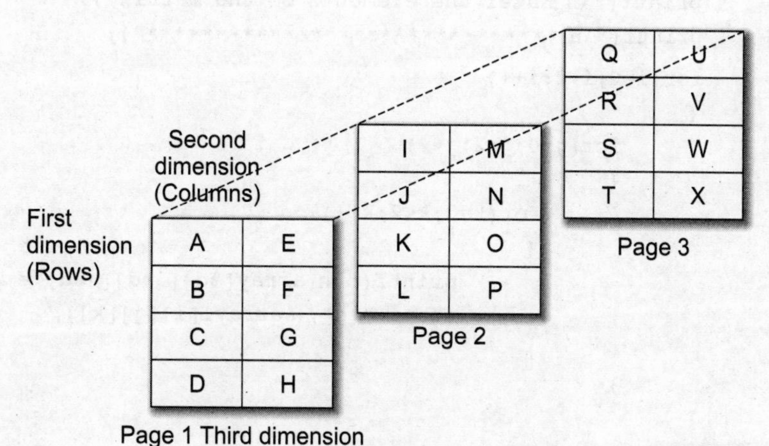

Figure 5.35 Three-dimensional array

Note A multi-dimensional array is declared and initialized the same way we declare and initialize one- and two-dimensional arrays.

EXAMPLE 5.7: Consider a three-dimensional array defined as `int A[2][3][2]`. Calculate the number of elements in the array. Also, show the memory representation of the array in the row major order and the column major order.

Solution

A three-dimensional array consists of pages. Each page, in turn, contains m rows and n columns.

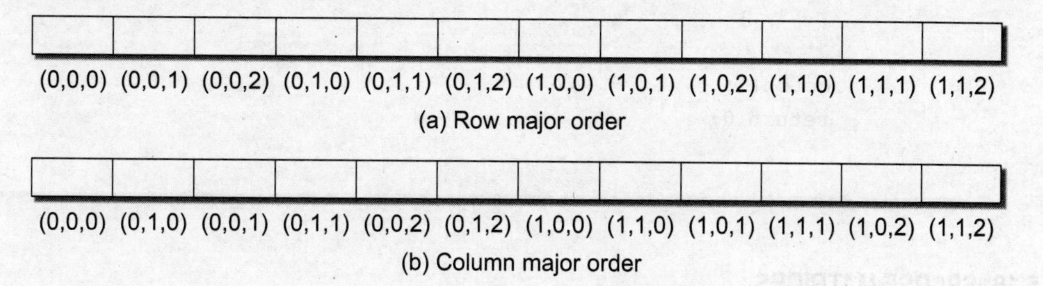

The three dimension array will contain $2 \times 3 \times 2 = 12$ elements

PROGRAMMING EXAMPLE

33. Write a program to read and display a $2 \times 2 \times 2$ array.

```
#include<stdio.h>
#include<conio.h>
int main()
{
        int array1[3][3][3], i, j, k;
        clrscr();
        printf("\n Enter the elements of the matrix");
        printf("\n ******************************");
        for(i=0;i<2;i++)
        {
                for(j=0;j<2;j++)
                {
                        for(k=0;k<2;k++)
                        {
                                printf("\n array[%d][ %d][ %d] = ", i, j, k);
                                scanf("%d", &array1[i][j][k]);
                        }
                }
        }
        printf("\n The matrix is : ");
        printf("\n *********************************");
        for(i=0;i<2;i++)
        {
                printf("\n\n");
                for(j=0;j<2;j++)
                {
                        printf("\n");
                        for(k=0;k<2;k++)
                                printf("\t array[%d][ %d][ %d] = %d", i, j, k,
array1[i][j][k]);
                }
        }
        getch();
        return 0;
}
```

5.16 SPARSE MATRICES

Sparse matrix is a matrix that has many elements with a value zero. In order to efficiently utilize the memory, specialized algorithms and data structures that take advantage of the sparse structure should be used. If we apply the operations using standard matrix structures and algorithms to sparse matrices, then the execution will slow down and the matrix will consume large amounts of memory. Sparse data can be easily compressed, which in turn can significantly reduce memory usage.

Basically, there are two types of sparse matrices. In the first type of sparse matrix, all elements above the main diagonal have a value zero. This type of sparse matrix is also called a **(lower) triagonal matrix** because if you see it pictorially, all the elements with a non-zero value appear below the diagonal. In a lower triangular matrix, $A_{i,j} = 0$ where $i < j$. An $n \times n$ lower-triangular matrix A has one non-zero element in the first row, two non-zero element in the second row and likewise, n non-zero elements in the nth row. Look at Fig. 5.36 which shows a lower-triangular matrix.

$$\begin{bmatrix} 1 & & & & \\ 5 & 3 & & & \\ 2 & 7 & -1 & & \\ 3 & 1 & 4 & 2 & \\ -9 & 2 & -8 & 1 & 7 \end{bmatrix}$$

Figure 5.36 Lower-triangular matrix

To store a lower-triangular matrix efficiently in the memory, we can use a one- dimensional array which stores only non-zero elements. The mapping between a two-dimensional matrix and a one-dimensional array can be done in any one of the following ways:

(a) **Row-wise mapping**– Here, the contents of array `A[ ]` will be {1, 5, 3, 2, 7, –1, 3, 1, 4, 2, –9, 2, –8, 1, 7}

(b) **Column-wise mapping**– Here, the contents of array `A[ ]` will be {1, 5, 2, 3, –9, 3, 7, 1, 2, –1, 4, –8, 2, 1, 7}

In an **upper-triangular matrix**, $A_{i,j} = 0$ where $i > j$. An $n \times n$ upper-triangular matrix A has n non-zero elements in the first row, $n-1$ non-zero elements in the second row and likewise, one non-zero elements in the nth row. Look at Fig. 5.37 which shows an upper-triangular matrix.

$$\begin{bmatrix} 1 & 2 & 3 & 4 & 5 \\ & 3 & 6 & 7 & 8 \\ & & -1 & 9 & 1 \\ & & & 9 & 3 \\ & & & & 7 \end{bmatrix}$$

Figure 5.37 Upper-triangular matrix

In the second variant of a sparse matrix, elements with a non-zero value can appear only on the diagonal or immediately above or below the diagonal. This type of matrix is also called a *tri-diagonal matrix*. Hence in a tridiagonal matrix, $A_{i,j} = 0$, where $|i - j| > 1$. If we look at it mathematically, we realize that in a tridiagonal matrix, if elements are present on

(a) the main diagonal, it contains non-zero elements for $i=j$. In all, there will be n elements.

(b) below the main diagonal, it contains non-zero elements for $i=j+1$. In all, there will be $n-1$ elements.

(c) above the main diagonal, it contains non-zero elements for $i=j-1$. In all, there will be $n-1$ elements.

Figure 5.38 shows a tri-diagonal matrix. To store a tri-diagonal matrix efficiently in the memory, we can use a one-dimensional array that stores only non-zero elements. The mapping between a two-dimensional matrix and a one-dimensional array can be done in any one of the following ways:

$$\begin{bmatrix} 4 & 1 & & & & \\ 5 & 1 & 2 & & & \\ & 9 & 3 & 1 & & \\ & & 4 & 2 & 2 & \\ & & & 5 & 1 & 9 \\ & & & & 8 & 7 \end{bmatrix}$$

Figure 5.38 Tri-diagonal matrix

(a) Row-wise mapping– Here, the contents of array `A[ ]` will be {4, 1, 5, 1, 2, 9, 3, 1, 4, 2, 2, 5, 1, 9, 8, 7}

(b) Column-wise mapping– Here, the contents of array `A[ ]` will be {4, 5, 1, 1, 9, 2, 3, 4, 1, 2, 5, 2, 1, 8, 9, 7}

(c) Diagonal-wise mapping– Here, the contents of array `A[ ]` will be {5, 9, 4, 5, 8, 4, 1, 3, 2, 1, 7, 1, 2, 1, 2, 9}

GLOSSARY

Array An array is a collection of similar data elements.

Array index Location of an item in an array.

Matrix A two-dimensional array in which the first index is the row and the second index is the column.

Lower-triangular matrix A matrix A_{ij} forms a lower triangular when $i \geq j$.

Upper-triangular matrix A matrix A_{ij} forms an upper triangular when $i \leq j$.

Sparse matrix A matrix that has relatively few non-zero elements.

Linear search It searches an array by checking the elements one at a time.

Binary search It searches a sorted array by repeatedly dividing the search interval in half. Begin with an interval covering the whole array. If the value of the search key is less than the item in the middle of the interval, narrow down the interval to the lower half. Otherwise, narrow it to the upper half. Repeatedly check until the value is found or the interval is empty.

One-dimensional array An array with one dimension or one subscript/index.

Two-dimensional array An array with two dimensions or two subscripts/indices.

Three-dimensional array An array with three dimensions or three subscripts/indices.

K-dimensional array An array with exactly k orthogonal axes or k dimensions.

Rectangular matrix An $n \times m$ matrix whose size may not be the same in both the dimensions.

EXERCISES

Review Questions

1. Why are arrays needed?
2. How is an array represented in the memory?
3. How is a two-dimensional array represented in the memory?
4. What is the use of multi-dimensional arrays?
5. Explain sparse matrix.
6. How are pointers used on two-dimensional arrays?
7. Why does storing of sparse matrices need extra consideration? How are sparse matrices stored efficiently in the computer's memory?
8. For an array declared as `int arr[50]`, calculate the address of `arr[35]`, if `Base(arr) = 1000` and `w = 2`.
9. Consider a 10×5 two-dimensional array `Marks` having its base address as 2000 and the number of words per memory location of the array is 2. Now, compute the address of the element, `Marks[8, 5]`, assuming that the elements are stored in row major order.
10. Which technique of searching an element in the array do you prefer to use and in which situation?
11. Explain the concept of array of function pointers with the help of a program.
12. How are arrays related to pointers?
13. Briefly explain the concept of array of pointers.
14. How can one-dimensional arrays be used for inter-process communication?
15. How are two-dimensional arrays represented in the main memory?
16. Consider a 10×10 two-dimensional array which has base address = 1000 and the number of words per memory location of the array = 2. Now, compute the address of the element `arr[8][5]` assuming that the elements are stored in row major order. Then calculate the same assuming the elements are stored in column major order.
17. Consider an array `MARKS[20][5]` which stores the marks obtained by 20 students in 5 subjects. Now write a program to:
    (a) find the average marks obtained in each subject.
    (b) find the average marks obtained by every student.
    (c) find the number of students who have scored below 50 in their average.
    (d) display the scores obtained by every student.

18. Consider the array given below:

Name[0]	Adam
Name[1]	Charles
Name[2]	Dicken
Name[3]	Esha
Name[4]	Georgia
Name[5]	Hillary
Name[6]	Mishael

(a) How many elements would be moved if the name Andrew has to be added in it?
(i) 7 (ii) 4 (iii) 5 (iv) 6

(b) How many elements would be moved if the name Esha has to be deleted from it?
(i) 3 (ii) 4 (iii) 5 (iv) 6

(c) How many comparisons need to be made to search for the name Hillary?
(i) 3 (ii) 4 (iii) 5 (iv) 6

Programming Exercises

1. Write a program to delete all duplicate elements from an array.

2. Write a program that reads an array of 100 integers. Display all the pairs of elements whose sum is 50.

3. Write a program to interchange the largest element with the smallest element.

4. Write a program to interchange the second element with the second last element.

5. Write a program that calculates the sum of squares of all the elements.

6. Write a program to compute the sum and mean of the elements of a two-dimensional array.

7. Write a program to read and display a square (using functions).

8. Write a program that computes the sum of the elements that are stored on the main diagonal of a matrix using pointers.

9. Write a program to add two 3×3 matrix using pointers.

10. Write a program to transpose a $m \times n$ matrix.

11. Write a program that computes the product of the elements that are stored on the diagonal above the main diagonal.

12. Write a program to count the total number of non-zero elements in a two- dimensional array.

13. Write a program to read and display an array of n integers.

14. Write a program to read and print an array of n numbers using functions.

15. Write a program to find the mean of n numbers using arrays.

16. Write a program to find the largest of n numbers using arrays. Also, display its position.

17. Write a program to read and print an array of n numbers, then find out the smallest number. Also, print the position of the smallest number.

18. Write a program to input the elements of a two-dimensional array. Then from this array, make two arrays – one that stores all odd elements of the two-dimensional array and the other stores all even elements of the array.

19. Write a program using functions to find out the largest number in an array of 10 floating point numbers.

20. Write a program to merge two integer arrays. Also, display the merged array in reverse order.

21. Write a program using pointers to interchange the second biggest and the second smallest number in an array.

22. Write a program to insert a new name in the string array `STUD[][]`, assuming that names are sorted alphabetically.

23. Write a program to delete a name in the string array `STUD[][]`, assuming that names are sorted alphabetically.

24. Write a program to read and display a 3×3 matrix.

25. Write a program to multiply two $m \times n$ matrices.

26. Write a menu driven program to read and display a $p \times q \times r$ matrix. Also, find the sum, transpose, and product of the two $p \times q \times r$ matrices.

27. Write a program to input two $m \times n$ matrices and then calculate the sum of their corresponding elements in the first matrix.

28. Write a program to eliminate duplicate entries in an array.

29. Write a menu driven program that reads, displays, adds, subtracts, transposes, and multiplies two matrices.

30. Write a program that reads a matrix and displays the sum of its diagonal elements.

31. Write a program using the functions to print the elements of a 2D array.
32. Write a program that reads a matrix and displays the sum of the elements above the main diagonal. (*Hint: Calculate the sum of elements array* `A` *where* A_{ij} *and* `i<j`)
33. Write a program that reads a matrix and displays the sum of the elements below the main diagonal. (*Hint: Calculate the sum of elements array* `A` *where* A_{ij} *and* `i>j`)
34. Write a program that reads a square matrix of size `n×n`. Write a function `int isUpperTriangular(int a[][], int n)` that returns 1 if the matrix is upper triangular. (*Hint: Array* `A` *is upper triangular if* A_{ij}= `0` *for* `i>j`)
35. Write a program that reads a square matrix of size `n×n`. Write a function `int isLowerTriangular (int a[][], int n)` that returns 1 if the matrix is lower triangular. (*Hint: Array* `A` *is lower triangular if* A_{ij}= `0` *for* `i<j`)
36. Write a program that reads a square matrix of size `n×n`. Write a function `int isSymmetric(int a[][], int n)` that returns 1 if the matrix is upper triangular. (*Hint: Array* `A` *is symmetric if* A_{ij}= A_{ji} *for all values of* `i` *and* `j`)
37. Write a program to calculate `XA + YB` where `A` and `B` are matrices and `X` = 2 and `Y` = 3.
38. A company has an array `YEAR[]` such that `YEAR[I]` contains the number of employees born in year `I`. Write a program to print the years in which no employee was born. Also, display the names of the employees who will complete their 50th year in 2010.
39. Write a program to sort 10 integers using an array of pointers.
40. Write a program to illustrate the use of a pointer that points to a 2D array.
41. Write a program to read and display 10 floating point numbers using an array.
42. Write a program to find the mean of 10 floating point numbers using arrays.
43. Write a program to find the largest of 10 floating point numbers using arrays.
44. Write a program for linear search.
45. Write a program for binary search.
46. Write a program to print the position of the smallest number of n numbers using arrays.
47. Write a program to interchange the largest and the smallest number in the array.
48. Write a program to find the second biggest number using an array of `n` numbers.
49. Write a program to enter a number and break it into `n` number of digits.
50. Write a program to delete all the duplicate entries from an array of `n` integers.
51. Write a program to insert a number at a given location in an array.
52. Write a program to insert a number in an array that is already sorted in the ascending order.
53. Write a program to delete a number from a given location in an array.
54. Write a program to delete a number from an array that is already sorted in the ascending order.
55. Write a program to merge two unsorted arrays.
56. Write a program to merge two sorted arrays.
57. Write a menu driven program to read and display an `m×n` matrix. Also find the sum, transpose, and product of two `m×n` matrices.
58. Write a program to read and display a `2×2×2` array.
59. Consider an array YEAR in which a company stores the birth year of its frequent customers. Now write a program to
    (a) Find the year in which no customer was born
    (b) Find the year in which maximum customers were born
    (c) Find the year in which lowest number of customers were born
    (d) Find number of customers who will be 25 at the end of the current year

Multiple Choice Questions

1. If an array is declared as `int arr[50]`, how many elements can it hold?
   (a) 49 (b) 50 (c) 51 (d) 0
2. If an array is declared as `int arr[5][5]`, how many elements can the store?
   (a) 5 (b) 25 (c) 10 (d) 0

3. The worst case complexity is ______ when compared with the average case complexity of a binary search algorithm.
   (a) Equal (b) Greater
   (c) Less (d) None of these
4. The complexity of binary search algorithm is
   (a) `O(n)` (b) $O(n^2)$
   (c) `O(n log n)` (d) `O(log n)`
5. In linear search, where `VAL` is equal to the first element of the array, which case it is?
   (a) Worst case (b) Average case
   (c) Best case (d) Amortized case
6. Given an integer array `arr[]`; the `i`th element can be accessed by writing
   (a) `*(arr+i)` (b) `*(i + arr)`
   (c) `arr[i]` (d) All of these

True or False

1. An array is used to refer multiple memory locations having the same name.
2. An array name can be used as a pointer.
3. An array need not be declared before being used.
4. A loop is used to access all the elements of the array.
5. An array stores all its data elements in non-consecutive memory location.
6. To copy an array, you must copy the value of every element of the first array into the element of the second array.
7. Lower bound is the index of the last element in the array.
8. Merged array contains contents of the first array followed by the contents of the second array.
9. Binary search is also called sequential search.
10. Linear search is performed on a sorted array.
11. It is possible to pass the entire array as a function argument.
12. `arr[i]` is equivalent to writing `*(arr+i)`.
13. Array name is equivalent to the address of its last element.
14. `mat[i][j]` is equivalent to `*(*(mat + i) + j)`.

Fill in the blanks

1. An array is a ______.
2. Every element is accessed using a ______.
3. The elements of an array are stored in ______ memory locations.
4. An *n*-dimensional array contains ______ subscripts.
5. Name of the array acts as a ______.
6. Declaring an array means specifying the ______, ______, and ______.
7. The subscript or the index represents the offset from the beginning of the array to ______.
8. ______ is the address of the first element in the array.
9. Length of the array is given by the number of ______.
10. ______ means accessing each element of the array for a specific purpose.
11. Performance of the linear search algorithm can be improved by using a ______.
12. The complexity of linear search algorithm is ______.
13. An array occupies ______ memory locations.
14. An array name is often known to be a ______ pointer.
15. A multi-dimensional array, in simple terms, is an ______.

6 Strings

Learning Objective

In this chapter, we will discuss the array of characters commonly known as strings. We will see how arrays are stored, declared, initialized, and accessed. We will also learn how different operations can be performed on strings.

6.1 INTRODUCTION

These days, computers are widely used for word processing applications like creating, inserting, updating, and modifying textual data. Besides this, we need to search for a particular pattern within a text, delete it, or replace it with another pattern. So, there is a lot that we, as users, do to manipulate the textual data.

In C, a string is a null-terminated character array. This means that after the last character, a `null` character (`'\0'`) is stored to signify the end of the character array. For example, if we write

```
char str[] = "HELLO";
```

then we are declaring a character array that has five usable characters, namely, `H`, `E`, `L`, `L`, and `O`. Apart from these characters, a `null` character (`'\0'`) is stored at the end of the string. So, the internal representation of the string becomes `HELLO'\0'`. Note that to store a string of length 5, we need 5 + 1 locations (1 extra for the `null` character). The name of the character array (or the string) is a pointer to the beginning of the string. Figure 6.1 shows the difference between character storage and string storage.

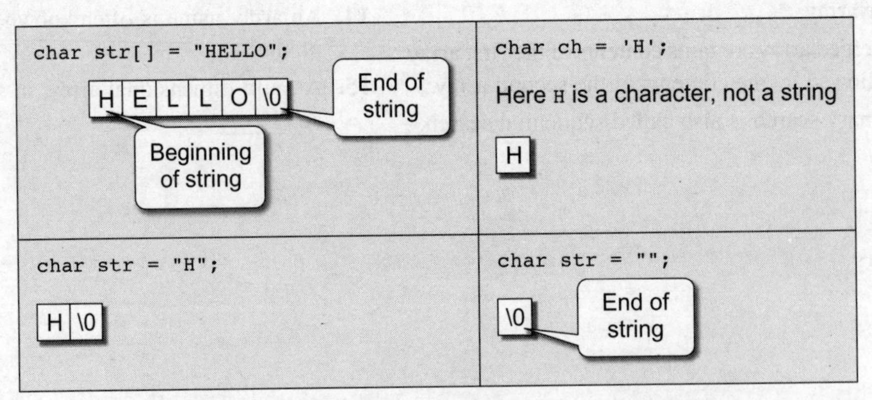

Figure 6.1 Difference between character storage and string storage

Suppose, we had declared `str` as

```
char str[5] = "HELLO";
```

then, the `null` character will not be appended automatically to the character array. This is because `str` can hold only 5 characters and the characters in `HELLO` have already filled the space allocated to it.

We use subscripts (also known as index) to access the elements of a character array. The subscript starts with a zero (0). All the characters of a string array are stored in successive memory locations. Figure 6.2 shows how `str[]` is stored in the memory.

Thus, in simple terms, a string is a sequence of characters. In Fig. 6.2, 1000, 1001, 1002, etc. are the memory addresses of individual characters. For simplicity, the figure shows that H is stored at memory location 1000 but in reality, the ASCII codes of characters are stored in the memory and not the character itself. So, at address 1000, 72 will be stored, as the ASCII code for `H` is 72.

str[0]	1000	H
str[1]	1001	E
str[2]	1002	L
str[3]	1003	L
str[4]	1004	O
str[5]	1005	\0

Figure 6.2 Memory representation of a character array

The statement

```
char str[] = "HELLO";
```

declares a constant string, as we have assigned a value to it while declaring the string. However, the general form of declaring a string is

```
char str[size];
```

When we declare the string like this, we can store `size-1` characters in the array because the last character would be the null character. For example, `char mesg[100];` can store a maximum 99 usable characters.

Till now, we have only seen one way of initializing strings. The other way to initialize a string is to initialize it as an array of characters. For example,

```
char str[] = {'H', 'E', 'L', 'L', 'O', '\0'};
```

Note that in this example, we have added the `null` character.

6.1.1 Reading Strings

If we declare a string by writing

```
char str[100];
```

then, `str` can be read by the user in three ways:

1. using `scanf` function,
2. using `gets()` function, and
3. using `getchar()` function repeatedly.

The string can be read using `scanf()` by writing

```
scanf("%s", str);
```

Although the syntax of using `scanf()` function is well-known and easy to use, but the main pitfall with this function is that the function terminates as soon as it finds a blank space. For example, if the user enters `Hello World`, the `str` will contain only `Hello`. This is because the moment a blank space is given, the string gets terminated by the `scanf()` function. You may also specify a field width to indicate the maximum number of characters that can be read. Remember that extra characters are left unconsumed in the input buffer.

Note that unlike `integer`, `float`, and `characters`, `%s` format does not require the ampersand before the variable `str`.

The next method of reading a string is by using the `gets()` function. The string can be read by writing

```
gets(str);
```

`gets()` is a simple function that overcomes the drawbacks of the `scanf()` function. The `gets()` function takes the starting address of the string which will hold the input. The string inputted using `gets()` is automatically terminated with a `null` character.

Last but not the least, the string can also be read by calling the `getchar()` repeatedly to read a sequence of single characters (unless a terminating character is entered) and simultaneously storing it in a character array as shown below.

```
i=0;
getchar(ch);
while(ch != '*')
{
        str[i] = ch;
        i++;
        getchar(ch);
}
str[i] = '\0';
```

Note that in this method, you have to deliberately append the characters with a `null` character. The other two functions automatically do this.

6.1.2 Writing Strings

The string can be displayed on screen using the following three ways:

1. using `printf()` function,
2. using `puts()` function, and
3. using `putchar()` function repeatedly.

The string can be displayed using `printf()` by writing

```
printf("%s", str);
```

Like the `scanf()`, `printf()` also terminates the string as soon as it finds a blank space. For example, if we declare `str[] = Hello World`, then `printf()` will print only `Hello`. This is because the moment a blank space is encountered, the `printf()` is terminated. Note that we use the conversion character `s` to output a string. We may also use width and precision specifications along with `&s` (as discussed in Chapter 1). The width specifies the minimum output field width. If the string is short, extra space is either left padded or right padded. The precision specifies the maximum number of characters to be displayed. A negative width left pads short string rather than the default right justification. If the string is long, the extra characters are truncated. For example,

```
printf("%5.3s", str);
```

would print only the first three characters in a total field of five characters. Also these three characters are right justified in the allocated width.

The next method of writing a string is by using `puts()` function. The string can be displayed by writing

```
puts(str);
```

`puts()` is a simple function that overcomes the drawbacks of the `printf()`.

Last but not the least, the string can also be written by calling the `putchar()` function repeatedly to print a sequence of single characters.

```
i=0;
while(str[i] != '\0*)
{
        putchar(str[i]);
        i++;
}
```

6.2 STRING OPERATIONS

In this section, we will learn the different operations that are performed on character arrays.

6.2.1 Length

The number of characters in a string constitutes the length of the string. For example, `LENGTH("C PROGRAMMING IS FUN")` will return 20. Note that even blank spaces are counted as characters in the string.

`LENGTH('0') = 0` and `LENGTH('') = 0`, because both the strings do not contain any character. Since the ASCII code of `'\0'` is zero, both the strings are empty and thus of zero length. Figure 6.3 shows an algorithm that calculates the length of a string.

```
Step 1: [INITIALIZE] SET I = 0
Step 2: Repeat Step 3 while STR[I] != '\0'
Step 3:             SET I = I + 1
        [END OF LOOP]
Step 4:         SET LENGTH = I
Step 5: END
```

Figure 6.3 Algorithm to calculate length of a string

In this algorithm, `I` is used as an index into the string `STR`. To traverse each and every character of `STR`, we increment the value of `I`. Once we encounter a `null` character, the control jumps out of the `while` loop and the length is initialized with the value of `I`. This is because the number of characters in the string constitutes its length.

Note There is a library function `strlen(s1)` that returns the length of `s1`. It is defined in `string.h`

PROGRAMMING EXAMPLE

1. Write a program to find the length of a string.

```
#include<stdio.h>
#include<conio.h>
int main()
{
    char str[100], i = 0, length;
```

```
    clrscr();
    printf("\n Enter the string : ");
    gets(str)
    while(str[i] != '\0')
        i++;
    length = i;
    printf("\n The length of the string is : %d", length);
    getch();
}
```

6.2.2 Converting Characters of a String into Upper Case

We have already discussed that only ASCII codes are stored in the memory, instead of the real value. The ASCII code for `A-Z` varies from 65 to 91 and the ASCII code for `a-z` ranges from 97 to 123. So, if we have to convert a lower case character into upper case, we just need to subtract 32 from the ASCII value of the character. Figure 6.4 shows an algorithm that converts the characters of a string into upper case.

Note There is a library function `toupper()` that converts a character into upper case. It is defined in `ctype.h`

```
Step 1: [Initialize] SET I=0
Step 2: Repeat Step 3 while STR[I] != '\0'
Step 3:         IF STR[1] > 'a' AND STR[I] < 'z'
                        SET Upperstr[I] = STR[I] - 32
                ELSE
                        SET Upperstr[I] = STR[I]
        [END OF IF]
        [END OF LOOP]
Step 4: SET Upperstr[I] = '\0'
Step 5: EXIT
```

Figure 6.4 Algorithm to convert characters of a string into upper case

When all the characters have been traversed, a `null` character is appended to the `Upperstr` (as done in Step 4).

PROGRAMMING EXAMPLE

2. Write a program to convert the characters of a string into upper case.

```
#include<stdio.h>
#include<conio.h>
int main()
{
    char str[100], upper_str [100];
    int i=0;
    clrscr();
```

```
        printf("\n Enter the string :");
        gets(str);
        while(str[i] != '\0')
        {
                if(str[i]>='a' && str[i]<='z')
                        upper_str[j] = str[i] - 32;
                else
                        upper_str[i] = str[i];
                i++;
        }
        Upper_str[i] = '\0';
        printf("\n The string converted into upper case is : ");
        puts(upper_str);
        return 0;
}
```

6.2.3 Converting Characters of a String Into Lower Case

Figure 6.5 shows an algorithm that converts characters of a string into lower case.

Note There is a library function `tolower()` that converts a character into lower case. It is defined in `ctype.h`

In the algorithm, we initialize `I` to zero. Using `I` as the index of `STR`, we traverse each character from Step 2 to 3. If the character is already in lower case, then it is copied in the `Lowerstr` string, else the upper case character is converted into lower case by adding 32 to its ASCII value. The lower case character is then stored in `Lowerstr`. Finally, when all the characters have been traversed, a `null` character is appended to the `Lowerstr` (as done in Step 4).

```
Step 1: [Initialize] SET I=0
Step 2: Repeat Step 3 while STR[I] != '\0'
Step 3:      IF STR[1] > 'A' AND STR[I] < 'Z'
                  SET Lowerstr[I] = STR[I] + 32
             ELSE
                  SET Lowerstr[I] = STR[I]
        [END OF IF]
```

Figure 6.5 Algorithm to convert characters of a string into lower case

PROGRAMMING EXAMPLE

3. Write a program to convert the characters of a string into lower case.

```
#include<stdio.h>
#include<conio.h>
int main()
{
        char str[100], lower_str [100];
        int i=0;
        clrscr();
        printf("\n Enter the string :");
        gets(str);
```

```
        while(str[i] != '\0')
        {
            if(str[i]>='A' && str[i]<='Z')
                lower_str[j] = str[i] + 32;
            else
                lower_str[i] = str[i];
            i++;
        }
        lower_str[i] = '\0';
        printf("\n The string converted into upper case is : ");
        puts(lower_str);
        return 0;
    }
```

6.2.4 Concatenating Two Strings to form a New String

If S_1 and S_2 are two strings, then the concatenation operation produces a string which contains the characters of S_1, followed by the characters of S_2. Figure 6.6 shows an algorithm that concatenates two strings.

```
Step 1. Initialize I =0 and J=0
Step 2. Repeat step 3 to 4 while I <= LENGTH(str1)
Step 3.       SET new_str[J] = str1[I]
Step 4.       Set I =I+1 and J=J+1
        [END of step2]
Step 5. SET I=0
Step 6. Repeat step 6 to 7 while I <= LENGTH(str2)
Step 7.       SET new_str[J] = str1[I]
Step 8.       Set I =I+1 and J=J+1
       [END of step5]
Step 9.    SET new_str[J] = '\0'
Step 10. EXIT
```

Figure 6.6 Algorithm to concatenate two strings

In this algorithm, we first initialize the two counters `I` and `J` to zero. To concatenate the strings, we have to copy the contents of the first string followed by the contents of the second string in a third string, `new_str`. Steps 2 to 4 copy the contents of the first string in the `new_str`. Likewise, Steps 6 to 8 copy the contents of the second string in the `new_str`. After the contents have been copied, a `null` character is appended at the end of the `new_str`.

PROGRAMMING EXAMPLE

4. Write a program to concatenate two strings.

```
#include<stdio.h>
#include<conio.h>
int main()
{
        char str1[100], str2[100], str3[100];
        int i=0, j=0;
        clrscr();

        printf("\n Enter the first string : ");
        gets(str1);
```

```
        printf("\n Enter the second string : ");
        gets(str2);
        while(str1[i] != '\0')
        {
                str3[j] = str1[i];
                i++;
                j++;
        }
        i=0;
        while(str2[i] != '\0')
        {
                str3[j] = str2[i];
                i++;
                j++;
        }
        str3[j] = '\0';
        printf("\n The concatenated string is : ");
        puts(str3);
        getch();
        return 0;
}
```

6.2.5 Appending

Appending one string to another string involves copying the contents of the source string at the end of the destination string. For example, if `S1` and `S2` are two strings, then appending `S1` to `S2` means we have to add the contents of `S1` to `S2`. So, `S1` is the source string and `S2` is the destination string. The appending operation would leave the source string `S1` unchanged and the destination string `S2 = S2 + S1`. Figure 6.7 shows an algorithm that appends two strings.

Note There is a library function `strcat(s1, s2)` that concatenates `s2` to `s1`. It is defined in `string.h`

```
Step 1: [Initialize] SET I =0 and J=0
Step 2: Repeat Step 3 while Dest_Str[I] != '\0'
Step 3:  SET I = I + 1
         [END OF LOOP]
Step 4: Repeat Step 5 to 7 while Source_Str[J] !='\0'
Step 5:  Dest_Str[I] = Source_Str[J]
Step 6:  SET I = I + 1
Step 7:  SET J = J + 1
         [END OF LOOP]
Step 8: SET Dest_Str[J] = '\0'
Step 9: EXIT
```

Figure 6.7 Algorithm to append two strings

In this algorithm, we first traverse through the destination string to reach its end, that is, reach the position where a `null` character is encountered. The characters of the source string are then copied into the destination string starting from that position. Finally, a null character is added to terminate the destination string.

PROGRAMMING EXAMPLE

5. Write a program to append a string to another string.

```
#include<stdio.h>
#include<conio.h>
main()
{
        char Dest_Str[100], Source_Str[50];
        int i=0, j=0;
        clrscr();
        printf("\n Enter the Source String : ");
        gets(Source_Str);
        printf("\n Enter the Destination String : ");
        gets(Dest_Str);
        while(Dest_Str[i] != '\0')
                i++;
        while(Source_Str[j] != '\0')
        {
                Dest_Str[i] = Source_Str[j];
                i++;
                j++;
        }
        Dest_Str[i] = '\0';
        printf("\n After appending, the destination string is : ");
        puts(Dest_Str);
        getch();
        return 0;
}
```

6.2.6 Comparing Two Strings

If S1 and S2 are two strings, then comparing the two strings will give either of the following results:

(a) S1 and S2 are equal,

(b) S1>S2, when in dictionary order, S1 will come after S2, and

(c) S1<S2, when in dictionary order, S1 precedes S2.

To compare the two strings, each and every character is compared from both the strings. If all the characters are the same, then the two strings are said to be equal. Figure 6.8 shows an algorithm that compares two strings.

Note There is a library function `strcmp(s1, s2)` that concatenates s2 to s1. It is defined in `string.h`

```
Step 1: [Initialize] SET I=0, SAME =0
Step 2: SET Len1 = Length(STR1), Len2 = Length(STR2)
Step 3: IF len1 != len2, then
            Write "Strings Are Not Equal"
        ELSE
            Repeat while I<Len1
                  IF STR1[I] == STR2[I]
                        SET I = I + 1
                  ELSE
                        Go to Step 4
                  [END OF IF]
            [END OF LOOP]
            IF I = Len1, then
                  SET SAME =1
                  Write "Strings are equal"
            [END OF IF]
Step 4:     IF SAME = 0, then
            IF STR1[I] > STR2[I], then
                  Write "String1 is greater than String2"
            ELSE IF STR1[I] < STR2[I], then
                  Write "String2 is greater than String1"
            [END OF IF]
        [END OF IF]
Step 5: EXIT
```

Figure 6.8 Algorithm to compare two strings

In this algorithm, we first check whether the two strings are of the same length. If not, then there is no point in moving ahead, as it straight away means that the two strings are not the same. However, if the two strings are of the same length, then we compare character by character to check if all the characters are same. If yes, then the variable `SAME` is set to 1. Else, if `SAME = 0`, then we check which string precedes the other in the dictionary order and prints the corresponding message.

PROGRAMMING EXAMPLE

6. Write a program to compare two strings.

```
#include<stdio.h>
#include<conio.h>
main()
{
      char str1[50], str2[50];
      int i=0, len1=0, len2=0, same =0;
      clrscr();
      printf("\n Enter the first string : ");
      gets(str1);
      printf("\n Enter the second string : ");
      gets(str2);
      len1 = strlen(str1);
      len2 = strlen(str2);
      if(len1 == len2)
      {
```

```
            while(i<len1)
            {
                    if(str1[i] == str2[i])
                            i++;
                    else break;
            }
            if(i==len1)
             {
                    same=1;
                    printf("\n The two strings are equal");
             }
        }
        if(len1!=len2)
            printf("\n The two strings are not equal");
        if(same == 0)
        {
            if(str1[i]>str2[i])
                    printf("\n String1 is greater than string2");
            else if(str1[i]<str2[i])
                    printf("\n String2 is greater than string1");
        }
        getch();
        return 0;
    }
```

6.2.7 Reversing a String

If S1 = "HELLO", then reverse of S1 = "OLLEH". To reverse a string, we just need to swap the first character with the last, second character with the second last character, and so on. Figure 6.9 shows an algorithm that reverses a string.

Note There is a library function strrev(s1) that reverses all the characters in the string except the null character. It is defined in string.h

In Step 1, I is initialized to zero and J is initialized to the length of the string STR. In Step 2, a while loop is executed until all the characters of the string are accessed. In Step 4, we swap the ith character of STR with its jth character. (As a result, the first character of STR will be replaced with its last character, the second character will be replaced with the second last character of STR, and so on.). In Step 4, the value of I is incremented and J is decremented to traverse STR in the forward and backward direction respectively.

```
Step 1: [Initialize] SET I=0, J= Length(STR)
Step 2: Repeat Step 3 and 4 while I < J
Step 3:      SWAP(STR(I), STR(J))
Step 4:      SET I = I + 1, J = J – 1
        [END OF LOOP]
Step 5: EXIT
```

Figure 6.9 Algorithm to reverse a string

PROGRAMMING EXAMPLE

7. Write a program to reverse a given string.

```
#include<stdio.h>
#include<conio.h>
int main()
{
        char str[100], reverse_str[100], temp;
        int i=0, j=0;
        clrscr();
        printf("\n Enter the string : ");
        gets(str);
        j = strlen(str);
        while(i < j)
        {
                temp = str[j];
                str[j] = str[i];
                str[i] = temp;
                i++;
                j–;
        }
        printf("\n The reversed string is : ");
        puts(str);
        getch();
        return 0;
}
```

6.2.8 Extracting a Substring from the Left

In order to extract a substring from its main string, we need to copy the content of the string starting from the first position to the `nth` position, where `n` is the number of characters to be extracted.

For example, if `S1 = "Hello World"`, then `Substr_Left(S1, 7) = Hello W`.

Figure 6.10 shows an algorithm that extracts the first `n` characters from a string.

In Step 1, we initialize the index variable `I` with zero. In Step 2, a `while` loop is executed until all the characters of `STR` have been accessed and `I` is less than `N`. In Step 3, the `I`th character of `STR` is copied in the `I`th character of `Substr`. In Step 4, the value of `I` is incremented to access the next character in `STR`. In Step 5, the `Substr` is appended with a null character.

```
Step 1: [Initialize] SET I=0
Step 2: Repeat Step 3 while STR[I] != '\0' AND I<N
Step 3:        SET Substr[I] = STR[I]
Step 4:        SET I = I + 1
        [END OF LOOP]
Step 5: SET Substr[I] ='\0'
Step 6: EXIT
```

Figure 6.10 Algorithm to extract the first n characters from a string

PROGRAMMING EXAMPLE

8. Write a program to extract the first N characters of a string.

```
#include<stdio.h>
#include<conio.h>
int main()
{
	char str[100], substr[100];
	int i=0, n;
	clrscr();
	printf("\n Enter the string : ");
	gets(str);
	printf("\n Enter the number of characters to be copied : ");
	scanf("%d", &n);
	i = 0;
	while(str[i] != '\0' && i <n)
	{
		substr[i] = str[i];
		i++;
	}
	substr [i] = '\0';
	printf("\n The substring is : ");
	puts(substr);
	getch();
	return 0;
}
```

6.2.9 Extracting a Substring from the Right

In order to extract a substring from the right side of the main string, we need to first calculate the position. For example, if `S1 = "Hello World"` and we have to copy seven characters starting from the right, then we have to actually start extracting characters from the 5th position. This is calculated by, `total number of characters - n + 1`.

For example, if `S1 = "Hello World"`, then `Substr_Right(S1, 7) = o World`.

Figure 6.11 shows an algorithm that extracts n characters from the right of a string.

```
Step 1: [Initialize] SET I=0, J = Length(STR) - N + 1
Step 2: Repeat Step 3 while STR[J] != '\0'
Step 3:     SET Substr[I] = STR[J]
Step 4:     SET I = I + 1, J = J + 1
        [END OF LOOP]
Step 5: SET Substr[I] ='\0'
Step 6: EXIT
```

Figure 6.11 Algorithm to extract N characters from the right of a string

In Step 1, we initialize the index variable `I` to zero and `J` to `Length(STR) – N + 1`, so that `J` points to the character from which the string has to be copied in the substring. In Step 2, a `while` loop is executed until the `null` character in `STR` is accessed. In Step 3, the `J`th character of `STR` is copied in the `I`th character of `Substr`. In Step 4, the value of `I` and `J` is incremented. In Step 5, the `Substr` is appended with a `null` character.

PROGRAMMING EXAMPLE

9. Write a program to extract the first N characters of a string from the right side.

```
#include<stdio.h>
#include<conio.h>
int main()
{
    char str[100], substr[100];
    int i=0, j=0, n;
    clrscr();
    printf("\n Enter the string : ");
    gets(str);
    printf("\n Enter the number of characters to be copied : ");
    scanf("%d", &n);
    j = strlen(str) – n + 1;
    while(str[j] != '\0')
    {
        substr[i] = str[j];
        i++, j++;
    }
    substr [i] = '\0';
    printf("\n The substring is : ");
    puts(substr);
    getch();
    return 0;
}
```

6.2.10 Extracting a Substring from the Middle

To extract a substring from the middle of a given string, we need to have data on the following three parameters:

1. the main string,
2. the position of the first character of the substring in the given string, and
3. the maximum number of characters/length of the substring.

For example, if we have a string

```
str[] = "Welcome to the world of programming";
```

then,

```
SUBSTRING(str, 15, 5) = world
```

```
Step 1: [INITIALIZE] Set I=M, J = 0
Step 2: Repeat steps 3 to 6
            while str[I] != '\0' and N>=0
Step 3:   SET substr[J] = str[I]
Step 4:   SET I = I + 1
Step 5:   SET J = J + 1
Step 6:   SET N = N - 1
    [END of loop]
Step 7: SET substr[J] = '\0'
Step 8: EXIT
```

Figure 6.12 Algorithm to extract a substring from the middle of a string

Figure 6.12 shows an algorithm that extracts a substring from the middle of a string.

In this algorithm, we initialize a loop counter `I` to `M`, that is, the position from which the characters have to be copied. Steps 3 to 6 are repeated until `N` characters have been copied. With every character copied, we decrement the value of `N`. The characters of the string are copied into another string called the `substr`. At the end, a null character is appended to the `substr` to terminate the string.

PROGRAMMING EXAMPLE

10. Write a program to extract a substring from the middle of a given string.

```
#include<stdio.h>
#include<conio.h>
int main()
{
        char str[100], substr[100];
        int i=0, j=0, n, m;
        clrscr();
        printf("\n Enter the main string : ");
        gets(str);
        printf("\n Enter the position from which to start the substring: ");
        scanf("%d", &m);
        printf("\n Enter the length of the substring: ");
        scanf("%d", &n);
        i=m;
        while(str[i] != '\0' && n >= 0)
        {
                substr[j] = str[i];
                i++;
                j++;
                n–;
        }
        substr[j] = '\0';
        printf("\n The substring is : ");
        puts(str);
        getch();
        return 0;
}
```

6.2.11 Insertion

The insertion operation inserts a string `S` in the main text `T` at the kth position. The general syntax of this operation is `INSERT(text, position, string)`. For example, `INSERT("XYZXYZ", 3, "AAA") = "XYZAAAXYZ"`

Figure 6.13 shows an algorithm to insert a string in a given text at the specified position.

```
Step 1: [INITIALIZE] SET I=0, J=0 and K=0
Step 2: Repeat steps 3 to 4 while text[I] != '0'
Step 3: IF I = pos, then
          Repeat while str[K] != '\0'
          new_str[j] = str[k]
          SET J=J+1
          SET K = K+1
        [END OF INNER LOOP]
        ELSE
          new_str[[J] = text[I]
          SET J = J+1
Step 4: SET I = I+1
        [END OF OUTER LOOP]
Step 5: SET new_str[J] = '\0'
Step 6: EXIT
```

Figure 6.13 Algorithm to insert a string in a given text at the specified position

This algorithm first initializes the indexes into the string to zero. From Steps 3 to 6, the contents of `new_str` are built. If `I` is exactly equal to the position at which the substring has to be inserted, then the inner loop copies the contents of the substring into the `new_str`. Otherwise, the contents of the text are copied into it.

PROGRAMMING EXAMPLE

11. Write a program to insert a string in the main text.

```
#include<stdio.h>
#include<conio.h>
main()
{
        char text[100], str[20], ins_text[100];
        int i=0, j=0, k=0,pos;
        clrscr();
        printf("\n Enter the main text : ");
        gets(text);
        printf("\n Enter the string to be inserted : ");
        gets(str);
        printf("\n Enter the position at which the string has to be inserted: ");
        scanf("%d", &pos);
        while(text[i]! = '\0')
        {
                if(i==pos)
```

```
        {
             while(str[k] != '\0')
             {
                  ins_text[j] = str[k];
                  j++;
                  k++;
             }
        }
        else
        {
             ins_text[j] = text[i];
             j++;
        }
        i++;
    }
    ins_text[j] = '\0';
    printf("\n The new string is : ");
    puts(ins_text);
    getch();
    return 0;
}
```

6.2.12 Indexing

This operation returns the position in the string where the string pattern first occurs. For example,

```
INDEX("Welcome to the world of programming", "world") = 15
```

However, if the pattern does not exist in the string, the `INDEX` function returns 0. Figure 6.14 shows an algorithm to find the index of the first occurrence of a string within a given text.

```
Step 1: [Initialize] SET I=0 and MAX = LENGTH(text)-LENGTH(str)+1
Step 2: Repeat Steps 3 to 6 while I <= MAX
Step 3:    Repeat step 4 for K = 0 To Length(str)
Step 4:           IF str[K] != text[I + K], then GOTO step 6
           [END of inner loop]
Step 5:    SET INDEX = I. Goto step 8
Step 6:    SET I = I+1
        [END OF OUTER LOOP]
Step 7: SET INDEX = -1
Step 8: EXIT
```

Figure 6.14 Algorithm to find the index of the first occurrence of a string within a given text

In this algorithm, `MAX` is initialized to `LENGTH(text) - Length(str) + 1`. For example, if a text contains `'Welcome To Programming'` and the string contains `'World'`, in the main text, we will at the most look for 22 – 5 + 1 = 18 characters because after that, there is no scope left for the string to be present in the text.

Steps 3 to 6 are repeated until each and every character of the text has been checked for the occurrence of the string within it. In the inner loop in Step 3, we check the `n` characters of string with the `n` characters of text to find if the characters are same. If it is not the case, then we move to Step 6, where `I` is incremented. If the string is found, then the index is initialized with `I`, else it is set to –1. For example, if

```
TEXT = WELCOME TO THE WORLD
STRING = COME
```

In the first pass of the inner loop, we will compare `WORLD` with `WELC` character by character. As soon as `E` and `O` do not match, the control will move to Step 6 and then `ELCO` will be compared with `WORLD`. In the next pass, `COME` will be compared with `COME`.

We will write the programming code of indexing operation in the operations that follow.

6.2.13 Deletion

The deletion operation deletes a substring from a given text. We write it as `DELETE(text, position, length)`. For example,

```
DELETE("ABCDXXXABCD", 5, 3) = "ABCDABCD"
```

Figure 6.15 shows an algorithm to delete a substring from a given text.

In this algorithm, we first initialize the indexes to zero. Steps 3 to 6 are repeated until all the characters of the text are scanned. If `I` is exactly equal to `M` (the position from which deletion has to be done), then the index of the text is incremented and `N` is decremented. `N` is the number of characters that have to be deleted starting from position `M`. However, if `I` is not equal to `M`, then the characters of the text are simply copied into the `new_str`.

```
Step 1: [INITIALIZE] SET I=0 and J=0
Step 2: Repeat steps 3 to 6 while text[I] != '\0'
Step 3:  IF I=M, then
              Repeat Step 4 while N>=0
                    SET I = I+1
                    SET N = N – 1
              [END of inner loop]
        [END OF IF]
Step 4: SET new_str[J] = text[I]
Step 5: SET J = J + 1
Step 6: SET I = I + 1
        [END of outer loop]
Step 7: SET new_str[J] = '\0'
Step 8: EXIT
```

Figure 6.15 Algorithm to delete a substring from a text

Delete operation can also be used to delete a string from a given text. To do this, we first find out the first index at which the string occurs in the text. Then, we delete `n` number of characters from the text, where `n` is the number of characters in the string.

PROGRAMMING EXAMPLE

12. Write a program to delete a substring from a text.

```
#include<stdio.h>
#include<conio.h>
main()
{
      char text[200], str[20], new_text[200];
      int i=0, j=0, found=0, k, n=0, copy_loop=0;
```

```
        clrscr();
        printf("\n Enter the main text : ");
        gets(text);
        fflush(stdin);
        printf("\n Enter the string to be deleted : ");
        gets(str);
        fflush(stdin);
        while(text[i]!='\0')
        {
                j=0, found=0, k=i;
                  while(text[k]==str[j] && str[j]!='\0')
                  {
                        k++;
                        j++;
                  }
                  if(str[j]=='\0')
                        copy_loop=k;
                  new_text[n] = text[copy_loop];
                  i++;
                  copy_loop++;
                  n++;
                }
        new_str[n]='\0';
        printf("\n The new string is : ");
        puts(new_str);
        getch();
        return 0;
}
```

6.2.14 Replacement

The replacement operation is used to replace the pattern P_1 by another pattern P_2. This is done by writing REPLACE(text, pattern$_1$, pattern$_2$). For example,

("AAABBBCCC", "BBB", "X") = AAAXCCC

("AAABBBCCC", "X", "YYY")= AAABBBCC

Note that in the second example, there is no change as X does not appear in the text. Figure 6.16 shows an algorithm to replace a pattern P_1 with another pattern P_2 in the text.

```
Step 1: [INITIALIZE] SET Pos = INDEX(TEXT, P1)
Step 2: SET TEXT = DELETE(TEXT, Pos, LENGTH(P1))
Step 3: INSERT(TEXT, Pos, P2)
Step 4: EXIT
```

Figure 6.16 Algorithm to replace a pattern P_1 with another pattern P_2 in the text

The algorithm is very simple, where we first find the position pos, at which the pattern occurs in the text, then delete the existing pattern from that position and insert a new pattern there.

PROGRAMMING EXAMPLE

13. Write a to program replace a pattern with another pattern in the text.

```
#include<stdio.h>
#include<conio.h>
main()
{
        char str[200], pat[20], new_str[200], rep_pat[100];
        int i=0, j=0, k, n=0, copy_loop=0, rep_index=0;
        clrscr();
        printf("\n Enter the string : ");
        gets(str);
        fflush(stdin);
        printf("\n Enter the pattern: ");
        gets(pat);
        fflush(stdin);
        printf("\n Enter the replace pattern : ");
        gets(rep_pat);
        while(str[i]!='\0')
        {
                j=0,k=i;
                while(str[k]==pat[j] && pat[j]!='\0')
                {
                        k++;
                        j++;
                }
                if(pat[j]=='\0')
                {
                        copy_loop=k;
                        while(rep_pat[rep_index] !='\0')
                        {
                         new_str[n] = rep_pat[rep_index];
                         rep_index++;
                         n++;
                        }
                }
                new_str[n] = str[copy_loop];
                i++;
                copy_loop++;
                n++;
        }
        new_str[n]='\0';
        printf("\n The new string is : ");
        puts(new_str);
        getch();
        return 0;
}
```

6.3 ARRAY OF STRINGS

Till now, we have seen that a string is an array of characters. For example, if we say `char name[] = "Mohan"`, then the name is a string (character array) that has five characters.

Now, suppose that there are 20 students in a class and we need a string that stores the names of all the 20 students. How can this be done? Here, we need a string of strings or an array of strings. Such an array of strings would store 20 individual strings. An array of strings is declared as

```
char names[20][30];
```

Here, the first index will specify how many strings are needed and the second index will specify the length of every individual string. So here, we allocate space for 20 names where each name can be a maximum 30 characters long.

Let us see the memory representation of an array of strings. If we have an array declared as

```
char name[5][10] = {"Ram", "Mohan", "Shyam", "Hari", "Gopal"};
```

then in the memory, the array will be stored as shown in Fig. 6.17.

Name[0]	R	A	M	'\0'						
Name[1]	M	O	H	A	N	'\0'				
Name[2]	S	H	Y	A	M	'\0'				
Name[3]	H	A	R	I	'\0'					
Name[4]	G	O	A	A	L	'\0'				

Figure 6.17 Memory representation of a 2D character array

By declaring the array names, we allocate 60 bytes. But the actual memory occupied is 28 bytes. Thus, we see that more than half of the memory allocated lies wasted. Figure 6.18 shows an algorithm to process individual string from an array of strings.

In step 1, we initialize the index variable `I` to zero. In step 2, a `while` loop is executed until all the strings in the array are accessed. In step 3, each individual string is processed.

```
Step 1: [Initialize] SET I=0
Step 2: Repeat step 3 while I< N
Step 3:      Apply Process to NAMES[I]
          [END OF LOOP]
Step 4: EXIT
```

Figure 6.18 Algorithm to process individual string from an array of strings

PROGRAMMING EXAMPLES

14. Write a program to read and print the names of *n* students of a class.

```
#include<stdio.h>
#include<conio.h>
main()
{
        char names[5][10];
        int i, n;
        clrscr();
        printf("\n Enter the number of students : ");
        scanf("%d", &n);
```

```
    for(i=0;i<n;i++)
    {
        printf("\n Enter the name of %dth student : ", i+1);
        gets(names[i]);
    }
    printf("\n Names of the students are : \n");
    for(i=0;i<n;i++)
        puts(names[i]);
    getch();
    return 0;
}
```

15. Write a program to sort the names of students.

```
#include<stdio.h>
#include<conio.h>
main()
{
    char names[5][10], temp[10];
    int i, n, j;
    clrscr();
    printf("\n Enter the number of students : ");
    scanf("%d", &n);
    for(i=0;i<n;i++)
    {
        printf("\n Enter the name of the %dth student : ", i+1);
        fflush(stdin);
        gets(names[i]);
    }
    for(i=0;i<n;i++)
    {
        for(j=0;j<n-i-1;j++)
        {
            if(strcmp(names[j], names[j+1])>0)
            {
                strcpy(temp, names[j]);
                strcpy(names[j], names[j+1]);
                strcpy(names[j+1], temp);
            }
        }
    }
    printf("\n Names of the students as stored in alphabetical order are
: ");
    for(i=0;i<n;i++)
        puts(names[i]);
    getch();
    return 0;
}
```

16. Write a program to read a sentence until a '.' is entered.

```
#include<stdio.h>
#include<conio.h>
int main()
{
    char str[1000];
    int i=0;
    clrscr();
    printf("\n Enter . to end");
    printf("\n **************");
    printf("\n Enter the sentence : ");
    scanf("%c", &str[i]);
    while(ch[i] != '.')
    {
        i++;
        scanf("%c", &str[i]);
    }
    str[i] = '\0';
    printf("\n The sentence is : ");
    i=0;
    while(str[i] != '\0')
    {
        printf("%c", str[i]);
        i++;
    }
    return 0;
}
```

17. Write a program to read a line until a new line.

```
#include<stdio.h>
#include<conio.h>
int main()
{
    char str[1000];
    int i=0;
    clrscr();
    printf("\n Enter a new line character to end");
    printf("\n **************");
    printf("\n Enter the line of text : ");
    scanf("%c", &str[i]);
    while(i == 79 && str[i] != '\n')
    {
        i++;
        scanf("%c", &str[i]);
    }
```

```
        str[i] = '\0';
        printf("\n The text is : ");
        i=0;
        while(str[i] != '\0')
        {
               printf("%c", str[i]);
               i++;
        }
        return 0;
}
```

18. Write a program to read and print the text until a `'*'` is encountered. Also count the number of characters in the text entered.

```
#include<stdio.h>
#include<conio.h>
int main()
{
       char str[1000];
       int i=0;
       clrscr();
       printf("\n Enter * to end");
       printf("\n **************");
       printf("\n Enter the text : ");
       scanf("%c", &str[i]);
       while(str[i] != '*')
       {
              i++;
              scanf("%c", &str[i]);
       }
       str[i] = '\0';
       printf("\n The text is : ");
       i=0;
       while(str[i] != '\0')
       {
              printf("%c", str[i]);
              i++;
       }
       printf("\n The count of characters is : %d",i-1);
       return 0;
}
```

19. Write a program to read a sentence and the number of words in the sentence.

```
#include<stdio.h>
#include<conio.h>
int main()
```

```
{
        char str[1000];
        int i=0, count=1;
        clrscr();
        printf("\n Enter the sentence : ");
        gets(str);
        while(str[i] != '\0')
        {
                if(str[i] == ' ' && str[i+1] != ' ')
                        count++;
                i++;
        }
        printf("\n The total count of words is : %d", count);
        return 0;
}
```

20. Write a program to read multiple lines of text until a '*' is entered. Also count the number of characters, words, and lines in the text.

```
#include<stdio.h>
#include<conio.h>
int main()
{
        char str[1000];
        int i=0, word_count = 1, line_count =1, char_count = 1;
        clrscr();
        printf("\n Enter a new line character to end");
        printf("\n **************");
        printf("\n Enter the text : ");
        scanf("%c", &str[i]);
        while(str[i] != '*')
        {
                i++;
                scanf("%c", &str[i]);
        }
        str[i] = '\0';
        i=0;
        while(str[i] != '\0')
        {
                if(str[i] == '\n' || i==79)
                        line_count++;
                if(str[i] == ' ' &&str[i+1] != ' ')
                        word_count++;
                char_count++;
                i++;
        }
```

```
        printf("\n The total count of words is : %d", word_count);
        printf("\n The total count of lines is : %d", line_count);
        printf("\n The total count of characters is : %d", char_count);
        return 0;
}
```

21. Write a program to copy the first *n* characters of a string in another string.

```
#include<stdio.h>
#include<conio.h>
int main()
{
        char str[1000], copy_str{1000];
        int i=0, n;
        clrscr();
        printf("\n Enter the string : ");
        gets(str);
        printf("\n Enter the number of characters to be copied : ");
        scanf("%d", &n);
        i = 0;
        while(str[i] != '\0' && i <n)
        {
                copy_str[i] = str[i];
                i++;
        }
        copy_str[i] = '\0';
        printf("\n The copied text is : ");
        puts(copy_str);
        return 0;
}
```

22. Write a program to copy *n* characters of a string from the mth position in another string.

```
#include<stdio.h>
#include<conio.h>
int main()
{
        char str[1000], copy_str{1000];
        int i=0, j=0, m, n;
        clrscr();
        printf("\n Enter the text : ");
        gets(str);
        printf("\n Enter the position from which to start : ");
        scanf("%d", &m);
        printf("\n Enter the number of characters to be copied : ");
        scanf("%d", &n);
```

```
    i = m;
    while(str[i] != '\0' && n>0)
    {
        copy_str[j] = str[i];
        i++;
        j++;
        n–;
    }
    copy_str[j] = '\0';
    printf("\n The copied text is : ");
    puts(copy_str);
    return 0;
}
```

23. Write a program to copy the last *n* characters of a character array in another character array.

```
#include<stdio.h>
#include<conio.h>
int main()
{
    char str[1000], copy_str{1000];
    int i=0, j=0, n;
    clrscr();
    printf("\n Enter the string : ");
    gets(str);
    printf("\n Enter the number of characters to be copied (from the
end): ");
    scanf("%d", &n);
    i = strlen(str) -n+1;
    while(str[i] != '\0')
    {
        copy_str[j] = str[i];
        i++;
        j++;
    }
    copy_str[j] = '\0';
    printf("\n The copied text is : ");
    i=0;
    while(copy_str[i] != '\0')
    {
        printf("%c", copy_str[i]);
        i++;
    }
    return 0;
}
```

24. Write a program to enter a text that has commas. Replace all the commas with semi-colons and then display the text.

```
#include<stdio.h>
#include<conio.h>
int main()
{
        char str[1000], copy_str[1000];
        int i=0;
        clrscr();
        printf("\n Enter the text : ");
        gets(str);
        while(str[i] != '\0')
        {
                if(str[i] ==',')
                        copy_str[i] = ';';
                else
                        copy_str[i] = str[i];
                i++;
        }
        copy_str[i] = '\0';
        printf("\n The copied text is : ");
        i=0;
        while(copy_str[i] != '\0')
        {
                printf("%c", copy_str[i]);
                i++;
        }
        return 0;
}
```

25. Write a program to enter a text that contains multiple lines. Rewrite this text by printing line numbers before the text where the line starts.

```
#include<stdio.h>
#include<conio.h>
int main()
{
        char str[1000];
        int i=0, linecount = 1;
        clrscr();
        printf("\n Enter the text : ");
        gets(str);
        while(str[i] != '\0')
        {
                if(linecount == 1 && i == 0)
                        printf("\n %d\t", linecount);
```

```
            if(str[i] == '\n')
            {
                linecount++;
                printf("\n %d\t", linecount);
            }
            printf("%c", str[i]);
            i++;
        }
        return 0;
    }
```

26. Write a program to enter a text that contains multiple lines. Display the n lines of text starting from the mth line.

```
#include<stdio.h>
#include<conio.h>
int main()
{
    char str[1000];
    int i=0, m, n, linecount = 0;
    clrscr();
    printf("\n Enter the text : ");
    gets(str);
    printf("\n Enter the line number from which to copy : ");
    scanf("%d", &m);
    printf("\n Enter the line number till which to copy : ");
    scanf("%d", &n);
    i=0,
    while(str[i] != '\0')
    {
        if(linecount == m )
        {
            j = i;
            while(n>0)
            {
                printf("%c", str[j]);
                j++;
                if(str[j] == '\n')
                {
                    n--;
                    linecount++;
                    printf("%d \t", linecount);
                }
            }
        }
        else
```

```
                {
                        i++;
                        if(str[i] == '\n')
                                linecount++;
                }
        }
        getch();
        return 0;
}
```

27. Write a program to enter a text. Then enter a pattern and count the number of times the pattern is repeated in the text.

```
#include<stdio.h>
#include<conio.h>
main()
{
        char str[200], pat[20];
        int i=0, j=0, found=0, k, count=0;
        clrscr();
        printf("\n Enter the string : ");
        gets(str);
        printf("\n Enter the pattern: ");
        gets(pat);
         while(str[i]!='\0')
         {
                j=0, k=i;
                while(str[k]==pat[j] && pat[j]!='\0')
                {
                        k++;
                        j++;
                }
                if(pat[j]=='\0')
                {
                        found=1;
                        count++;
                }
                i++;
         }
        if(found==1)
                printf("\n PATTERN FOUND %d TIMES", count);
        else
                printf("\n PATTERN NOT FOUND");
        return 0;
}
```

28. Write a program to enter a text and a pattern. If the pattern exists in the text, then delete the text and display it.

```
#include<stdio.h>
#include<conio.h>
main()
{
      char str[200], pat[20], new_str[200];
      int i=0, j=0, found=0, k, n=0, copy_loop=0;
      clrscr();
      printf("\n Enter the string : ");
      gets(str);
      fflush(stdin);
      printf("\n Enter the pattern: ");
      gets(pat);
      fflush(stdin);
      while(str[i]!='\0')
      {
            j=0, found= 0, k=i;
            while(str[k]==pat[j] && pat[j]!='\0')
             {
                  k++;
                  j++;
              }
              if(pat[j]=='\0')
                  copy_loop=k;
              new_str[n] = str[copy_loop];
              i++;
              copy_loop++;
              n++;
       }
      new_str[n]='\0';
      printf("\n The new string is : ");
      puts(new_str);
      return 0;
}
```

29. Write a program to enter a text and a pattern. If the pattern exists in the text, then replace it with another pattern and then display the text.

```
#include<stdio.h>
#include<conio.h>
main()
{
      char str[200], pat[20], new_str[200], rep_pat[100];
      int i=0, j=0, k, n=0, copy_loop=0, rep_index=0;
      clrscr();
      printf("\n Enter the string : ");
```

```
        gets(str);
        fflush(stdin);
        printf("\n Enter the pattern: ");
        gets(pat);
        fflush(stdin);
        printf("\n ENter the replace pattern : ");
        gets(rep_pat);
        while(str[i]!='\0')
        {
                j=0,k=i;
                while(str[k]==pat[j] && pat[j]!='\0')
                {
                      k++;
                      j++;
                }
                if(pat[j]=='\0')
                {
                        copy_loop=k;
                        while(rep_pat[rep_index] !='\0')
                        {
                             new_str[n] = rep_pat[rep_index];
                             rep_index++;
                             n++;
                        }
                }
                new_str[n] = str[copy_loop];
                i++;
                copy_loop++;
                n++;
        }
        new_str[n]='\0';
        printf("\n The new string is : ");
        puts(new_str);
        return 0;
}
```

30. Write a program to find whether a string is a palindrome or not.

```
#include<stdio.h>
#include<conio.h>
int main()
{
        char str[100];
        int i = 0, j, length;
        clrscr();
        printf("\n Enter the string : ");
        scanf("%d", &str[i]);
```

```
    while(str[i] != '*')
    {
        i++:
        scanf("%c", str[i]);
    }
    str[i] = '\0';
    length = i;
    i=0;
    j = length - 1;
    while(i <= length/2)
    {
        if(str[i] == str[j])
        {
            i++;
            j—;
        }
        else
            break;
    }
    if(i>=j)
        printf("\n PALINDROME");
    else
        printf("\n NOT A PALINDROME");
    return 0;
}
```

6.4 POINTERS AND STRINGS

In C, strings are treated as array of characters that are terminated with a binary zero character (written as '\0'). Consider, for example,

```
char str[10];
str[0] = 'H';
str[1] = 'i';
str[2] = '!':
str[3] = '\0';
```

C provides two alternate ways of declaring and initializing a string. First, you may write,

```
char str[10] = {'H', 'i', '!', '\0',};
```

But, this also takes more typing than is convenient. So, C permits,

```
char str[10] = "Hi!";
```

When the double quotes are used, a `null` character ('\0') is automatically appended to the end of the string.

When a string is declared like this, the compiler sets aside a contiguous block of the memory, i.e., 10 bytes long, to hold characters and initializes its first four characters as `Hi!\0`.

Now, consider the following program that prints a text.

```
#include<stdio.h>
int main()
{
        char str[] = "Oxford";
        char *pstr;
        pstr = str;
        printf("\n The string is : ");
        while(*pstr != '\0')
        {
                printf("%c", *pstr);
                pstr++;
        }
        return 0;
}
```

In this program, we declare a character pointer `*pstr` to show the string on the screen. We then point the pointer `pstr` at `str`. Then, we print each character of the string in the `while` loop. Instead of using the `while` loop, we could straightaway use the function `puts()`, like

```
puts(pstr);
```

Consider here that the function prototype for `puts()` is:

```
int puts(const char *s);
```

Here the `const` modifier is used to assure the user that the function will not modify the contents pointed to by the source pointer. Note that the address of the string is passed to the function as an argument.

Consider another program that reads a string and then scans each character to count the number of upper and lower case characters entered.

```
#include<stdio.h>
int main()
{
        char str[100], *pstr;
        int upper = 0, lower = 0;
        printf("\n Enter the string : ");
        gets(str);
        pstr = str;
        while(*pstr != '\0')
        {
                if(*pstr >= 'A' && *pstr <= 'Z')
                                upper++;
                else if(*pstr >= 'a' && *pstr <= 'z')
                        lower++;
                pstr++;
        }
        printf("\n Total number of upper case characters = %d", upper);
```

```
        printf("\n Total number of lower case characters = %d", lower);
        return 0;
}
```

PROGRAMMING EXAMPLES

31. Write a program to read and print a text. Also, count the number of characters, words, and lines in the text.

```
#include<stdio.h>
int main()
{
        char str[100], *pstr;
        int chars = 0, lines=0, words=0;
        pstr=str;
        printf("\n Enter the string : ");
        gets(str);
        pstr = str;
        while( *pstr != '\0')
        {
                if(*pstr == '\n')
                        lines++;
                if(*pstr == ' ' && *(pstr + 1) != ' ')
                        words++;
                chars++;
                ptstr++;
        }
        printf("\n The string is : ");
        puts(str);
        printf("\n Number of characters = %d", chars);
        printf("\n Number of lines = %d", lines);
        printf("\n Number of words = %d", words);
        return 0;
}
```

32. Write a program to copy a character array in another character array.

```
#include<stdio.h>
int main()
{
        char str[100], copy_str[100];
        char *pstr, *pcopy_str;
        int i=0;
        pstr = str;
        pcopy_str = copy_str;
        printf("\n Enter the string : ");
        gets(str);
```

```
	pstr= str;
	while( *pstr != '\0')
	{
		*pcopy_str = *pstr;
		pstr++, pcopy_str++;
	}
	*pcopy_str = '\0';
	printf("\n The copied text is : ");
	pcopy_str = copy_str;
	while(*pcopy_str != '\0')
	{
		printf("%c", *pcopy_str);
		pcopy_str++;
	}
	return 0;
}
```

33. Write a program to copy n characters of a character array from the mth position in another character array.

```
#include<stdio.h>
int main()
{
	char str[100], copy_str[100];
	char *pstr, *pcopy_str;
	int m, n, i=0;
	pstr = str;
	pcopy_str = copy_str;
	printf("\n Enter the position from which to start : ");
	scanf("%d", &m);
	printf("\n Enter the number of characters to be copied : ");
	scanf("%d", &n);
	pstr = pstr + m - 1;
	i=0;
	while(*pstr != '\0' && i <n)
	{
		*pcopy_str = *pstr;
		pcopy_str++;
		pstr++;
		i++;
	}
	*pcopy_str = '\0';
	printf("\n The copied text is : ");
	puts(copy_str);
	return 0;
}
```

34. Write a program to copy the last *n* characters of a character array in another character array. Also, convert the lower case letters into upper case while copying.

```
#include<stdio.h>
#include<string.h>
int main()
{
        char str100], copy_str[100];
        char *pstr, *pcopy_str;
        int i=0, n;
        pstr = str;
        pcopy_str = copy_str;
        printf("\n Enter the string:");
        gets(str);
        printf("\n Enter the number of characters to be copied (from the
end): ");
        scanf("%d", &n);
        pstr = pstr + len(str) – n;
        while(*pstr != '\0')
        {
                *pcopy_str = *pstr - 32;
                pstr++; pcopy_str++;
        }
        *pcopy_str = '\0';
        printf("\n The copied text is : ");
        puts(copy_str);
        return 0;
}
```

35. Write a program to read a text, delete all the semi-colons it has and finally replace all '.' with a ','.

```
#include<stdio.h>
int main()
{
        char str[100], copy_str[100];
        char *pstr, *pcopy_str;
        pstr = str;
        pcopy_str = copy_str;
        printf("\n Enter the string : ");
        gets(str);
        pstr = str;
        while( *pstr != '\0')
        {
                if( *pstr != ';')
                { }   // do nothing
                if ( *pstr == '.')
```

```
                *pcopy_str = ',';
            else
                *pcopy_str = *pstr;
            pstr++; pcopy_str++;
        }
        *pcopy_str = '\0';
        printf("\n The new text is : ");
        pcopy_str = copy_str;
        while( *pcopy_str != '\0')
        {
            printf("%c", *pcopy_str);
            pcopy_str++;
        }
        return 0;
    }
```

36. Write a program to reverse a string.

```
#include<stdio.h>
int main()
{
    char str[100], copy_str[100];
    char *pstr, *pcopy_str;
    pstr = str;
    pcopy_str = copy_str;
    printf("\n Enter * to end");
    printf("\n **************");
    printf("\n Enter the string : ");
    scanf("%c", pstr);
    while( *pstr != '*')
    {
        pstr++;
        scanf("%c", pstr);
    }
    *pstr = '\0';
    pstr--;
    while (pstr >= str)
    {
        *pcopy_str = *pstr;
        pcopy_str++;
        pstr–;
    }
    *pcopy_str = '\0';
    printf("\n The new text is : ");
    pcopy_str = copy_str;
```

```
        while(*pcopy_str != '\0')
        {
              printf("%c", *pcopy_str);
              pcopy_str++;
        }
        return 0;
  }
```

37. Write a program to concatenate two strings.

```
  #include<stdio.h>
  int main()
  {
        char str1[100], str2[100], copy_str[2000];
        char *pstr, *pcopy_str;
        clrscr();
        pstr = str1;
        pcopy_str = copy_str;
        printf("\n Enter the first string : ");
        gets(str1);
        printf("\n Enter the second string : ");
        gets(str2);
        pstr=str1;
        while(*pstr != '\0')
        {
              *pcopy_str = *pstr;
              pcopy_str++, pstr++;
        }
        pstr=str2;
        while( *pstr != '\0')
        {
              *pcopy_str = *pstr;
              pcopy_str++, pstr++;
        }
        *pcopy_str = '\0';
        printf("\n The new text is : ");
        pcopy_str = copy_str;
        while(*pcopy_str != '\0')
        {
              printf("%c", *pcopy_str);
              pcopy_str++;
        }
        return 0;
  }
```

38. Write a menu-driven program to read a string, display the string, merge two strings, copy *n* characters from the *m*th position, and calculate the length of the string.

```
#include<stdio.h>
void read_str( char *my_str);
void display( char *my_str);
void merge_str(char *my_str1, char *my_str2, char *my_str3);
void copy_str( char my_str[], int m, int n);
int cal_len(char my_str[]);
void count(char my_str[]);
void count_wlc(char my_str[]);
void replace_str(char *my_str);
int main()
{
        char str1[100], str2[100], merged_str[200], copy_str[100];
        int option, m, n, length=0;
        do
        {
                printf("\n 1. Enter the string");
                printf("\n 2. Display the string");
                printf("\n 3. Merge two strings");
                printf("\n 4. Copy n characters from mth position");
                printf("\n 5. Calculate length of the string");
                printf("\n 6. Count the number of upper case, lower case,
numbers and special characters");
                printf("\n 7. Count the number of words, lines and characters");
                printf("\n 8. Replace , with ;");
                printf("\n 9. EXIT");
                printf("\n\n Enter your option : ");
                scanf("%d", &option);
                switch(option)
                {
                        case 1:
                                fflush(stdin);
                                read_str(str1);
                                break;
                        case 2:
                                display_str(str1);
                                break;
                        case 3:
                                read_str(str2);
                                merge_str(str1, str2, merged_str);
                                break;
                        case 4:
                                printf("\n Enter the position from which to copy
the text : ");
                                scanf("%d" &m);
```

```
                    printf("\n Enter the number of characters to be
copied : ");
                    scanf("%d", &n);
                    copy_str(str1, m, n);
                    break;
                case 5:
                    length = cal_len(str1);
                    printf("\n The length of the string is : %d",
length);
                    break;
                case 6:
                    count(str1);
                    break;
                case 7:
                    count_wlc(str1);
                    break;
                case 8:
                    replace_str(str1);
                    break;
            }
        }while (option != 9);
        return 0;
}
void read_str( char *my_str)
{
        fflush(stdin);
        printf("\n Enter the string : ");
        gets(my_str);
}
void display_str(char *my_str)
{
        printf("\n The string is : ");
        while(*my_str != '\0')
        {
            printf("%c", *my_str);
            my_str++;
        }
}
void merge_str(char *my_str1, char *my_str2, char *my_str3)
{
        strcpy(my_str3, my_str1);
        strcat(my_str3, my_str2);
        display_str(my_str3);
```

```
}
void copy_str(char my_str[], int m, int n)
{
      int i=0;
      char *prt;
      printf("\n The copied string is:");
      while (i < n || my_str[m] != '\0')
      {
            *pstr = my_str[m];
            m++, i++;
            printf("%c", *pstr);
      }
}
int cal_len(char *my_str[])
{
      char *str = my_str;
      int len=0;
      while (*str != '\0')
      {
            str++;
            len++;
      }
      return len;
}
void count (char my_str[])
{
      char *pstr = my_str;
      int upper_case = 0, lower_case = 0, numbers = 0, spcl_char =0;
      while (*pstr != '\0')
      {
            if (*pstr >= 'A' && *pstr <= 'Z')
                  upper_case++;
            if(*pstr >= 'a' && *pstr <= 'z')
                  lower_case++;
            if (*pstr >= '0' && *pstr <= '9')
                  numbers++;
            else
                  spcl_char++; pstr++;
                  printf("\n Upper case character = %d", upper_case);
                  printf("\n Lower case character = %d", lower_case);
                  printf("\n Number = %d", numbers);
                  printf("\n Special characters = %d", spcl_char);
      }
}
```

```
void count_wlc(char mystr[])
{
        char *pstr = my_str;
        int words = 0, lines =0, characters = 0;
        while(*pstr != '\0')
        {
                if (*pstr == '\n')
                        lines++;
                if ( *pstr == ' ' && *(pstr + 1) != ' ')
                        words++;
                        character++;
                        pstr++;
}
        printf("\n Number of words = %d", words);
        printf("\n Number of lines = %d", lines);
        printf("\n Number of characters = %d", characters);
}
void replace_str(char my_str[])
{
        char *pstr = my_str;
        while (*pstr == '\0')
        {
                if (*pstr != ',')
                        *pstr = ';'
                pstr++;
        }
        display_str(my_str);
}
```

SUMMARY

- A string is a null-terminated character array.
- A string is terminated with a `null` character (`'\0'`) to signify the end of the character array.
- A string is a character array from which individual characters can be accessed using a subscript that starts from zero.
- All the characters of a string array are stored in successive memory locations.
- A string can be read by the user using three ways: using `scanf()` function, using `gets()` function, or using `getchar()` function repeatedly.
- The `scanf()` function terminates as soon as it finds a blank space.
- The `gets()` function takes the starting address of the string which will hold the input. The string inputted using `gets()` is automatically terminated with a `null` character.
- The string can also be read by calling `getchar()` repeatedly to read a sequence of single characters.
- A string can be displayed on the screen using three ways: using `printf` function, using `puts()` function, or using `putchar()` function repeatedly.

- In `printf()`, the width specifies the minimum output field width. If the string is short, extra space is either left padded or right padded. The precision specifies the maximum number of characters to be displayed. A negative width left-pads a short string rather than the default right justification. If the string is long, the extra characters are truncated.
- The number of characters in a string constitutes the length of the string.
- Appending one string to another string involves copying the contents of the source string at the end of the destination string. There is a library function `strcat(s1, s2)` that concatenates `s2` to `s1`. It is defined in `string.h`.
- To extract a substring from a given string requires information regarding three things: the main string, the position of the first character of the substring in the given string, and the maximum number of characters/length of the substring.
- Index operation returns the position in the string where the string pattern first occurs.
- Replacement operation is used to replace the pattern P_1 by another pattern P_2.

GLOSSARY

String matching The problem of finding occurrences of a pattern string within another string.

Brute force string matching algorithm An algorithm to find a string within another string by trying each position one at a time.

Substring A string `u` is a substring of a string `v` if `v` = `v' uv"` for some prefix `v'` and suffix `v"`.

String An array of characters.

EXERCISES

Review Questions

1. What are strings? Discuss some of the operations that can be performed on strings.
2. Explain how strings are represented in the main memory.
3. Explain string operations.
4. How are strings read from the standard input device? Explain the different functions used to perform the string input operation.
5. Explain how strings can be displayed on the screen.
6. Explain the syntax of `printf()` and `scanf()`.
7. Write a short note on string operations.
8. List all the substrings that can be formed from the string 'ABCD'.
9. What do you understand by pattern matching? Give an algorithm for it.
10. Write a short note on array of strings.
11. How is an array of strings represented in the memory?
12. Explain with an example how an array of strings is stored in the main memory.
13. Explain how pointers and strings are related to each other with the help of a suitable program.
14. If the Substring function is given as `SUBSTRING (string, position, length)`, then find `S(5, 9)` if `S = "Welcome to world of C Programming"`
15. If the Index function is given as `INDEX(text, pattern)`, then find `index(T, P)` where `T = "Welcome to world of C Programming"` and `P = "of"`

Programming Exercises

1. Write a program in which a string is passed as an argument to a function.
2. Write a program to concatenate the first n characters of a string with another string.
3. Write a program that compares the first n characters of one string with the first n characters of another string.
4. Write a program that removes leading and trailing spaces from a string.
5. Write a program that replaces a given character with another character in the string.
6. Write a program to display the given string array in reverse order.

7. Write a program to count the number of characters, words, and lines in a given text.
8. Write a program to count the number of digits, upper case and lower case characters, and special characters in a given string.
9. Write a program to count the total number of occurrences of a character in a string.
10. Write a program to accept a text. Count and display the number of times the word 'the' appears in the text.
11. Write a program to count the total number of occurrences of a word in the text.
12. Write a program to find the last instance of occurrence of a sub-string within a string.
13. Write a program to insert a substring in the middle of a given string.
14. Write a program to input an array of strings. Then, reverse the string in the format shown below.

    `"HAPPY BIRTHDAY TO YOU"` should be displayed as `"YOU TO BIRTHDAY HAPPY"`
15. Write a program to append a given string in the following format.

    `"GOOD MORNING MORNING GOOD"`
16. Write a program to trim a string.
17. Write a program to input a text of at least two paragraphs. Interchange the first and second paragraphs and then re-display the text on the screen.
18. Write a program to input a text of at least two paragraphs. Construct an array PAR such that PAR[I] contains the location of the Ith paragraph in the text.
19. Write a program to find the length of a string.
20. Write a program to convert the characters of a string into upper case.
21. Write a program to convert the characters of a string into lower case.
22. Write a program to concatenate two strings.
23. Write a program to append a string to another string.
24. Write a program to compare two strings.
25. Write a program to reverse a given string.
26. Write a program to extract the first N characters of a string.
27. Write a program to extract a substring from a given string.
28. Write a program to insert a string in the main text.
29. Write a program to delete a substring from a text.
30. Write a menu-driven program to read a string, display the string, merge two strings, copy `n` characters from the `m`th position, and calculate the length of the string.
31. Write a program to concatenate two strings.
32. Write a program to reverse a string.
33. Write a program to read a text, delete all the semi-colons it has, and finally replace all `'.'` with a `','`.
34. Write a program to copy the last `n` characters of a character array in another character array. Also, convert the lower case letters into upper case while copying.
35. Write a program to copy `n` characters of a character array from the `m`th position in another character array.
36. Write a program to copy a character array in another character array.
37. Write a program to read and print a text. Also, count the number of characters, words, and lines in the text.
38. Write a program to replace a pattern with another pattern in the text.
39. Write a program to read and print the names of `n` students of a class.
40. Write a program to sort the names of students.
41. Write a program to read a sentence until a `'.'` is entered.
42. Write a program to read a line until a newline.
43. Write a program to read and print the text until a `'*'` is encountered. Also count the number of characters in the text entered.
44. Write a program to read a sentence and count the number of words in the sentence.
45. Write a program to read multiple lines of a text until a `'*'` is entered. Then, count the number of characters, words, and lines in the text
46. Write a program to copy the first `n` characters of a string in another string.
47. Write a program to copy `n` characters of a string from the `m`th position in another string.
48. Write a program to copy the last `n` characters of a character array in another character array.
49. Write a program to enter a text that has commas. Replace all the commas with semi- colons and then display the text.

50. Write a program to enter a text that contains multiple lines. Rewrite this text by printing line numbers before the text of the line starts.

51. Write a program to enter a text that contains multiple lines. Display the n lines of text starting from the mth line.

52. Write a program to count the number of times a pattern is repeated in a text.

53. Write a program to check whether a pattern exists in a text. If it does, delete the pattern and display it.

54. Write a program to replace a pattern with another pattern and display the text.

55. Write a program to find whether a string is a palindrome or not.

Multiple Choice Questions

1. `LENGTH('0') =`
   (a) – 1 (b) 0
   (c) 1 (d) None of these

2. ASCII code for `a-z` ranges from
   (a) 0–26 (b) 35–81
   (c) 97–123 (d) None of these

3. `Insert("XXXYYYZZZ", 1, "PPP") =`
   (a) PPPXXXYYYZZZ
   (b) XPPPXXYYYZZZ
   (c) XXXYYYZZZPPP

4. `Delete("XXXYYYZZZ", 4,3) =`
   (a) XXYZ (b) XXXYYZZ
   (c) XXXYZZ

5. If `str[] = "Welcome to the world of programming"`, then `SUBSTRING(str, 15, 5) =`
   (a) world (b) programming
   (c) welcome (d) none of these

6. `strcat()` is defined in which header file?
   (a) `ctype.h` (b) `stdio.h`
   (c) `string.h` (d) `math.h`

7. A string can be read using which functions?
   (a) `gets()` (b) `scanf()`
   (c) `getchar()` (d) all of these

8. `Replace("XXXYYYZZZ", "XY", "AB") =`
   (a) XXABYYZZZ (b) XABYYYZZZ
   (c) ABXXXYYYZZ

True or False

1. A string `Hello World` can be read using `scanf()`.
2. A string when read using `scanf()` needs an ampersand character.
3. The `gets()` function takes the starting address of a string which will hold the input.
4. `tolower()` is defined in `ctype.h` header file.
5. If S_1 and S_2 are two strings, then the concatenation operation produces a string which contains the characters of S_2 followed by the characters of S_1.
6. Appending one string to another string involves copying the contents of the source string at the end of the destination string.
7. `S1<S2`, when in dictionary order, `S1` precedes `S2`.
8. If `S1 = "GOOD MORNING"`, then `Substr_Right (S1, 5) = MORNING.`
9. `Replace ("AAABBBCCC", "X", "YYY")= AAABBBCC.`

Fill in the blanks

1. Strings are ______.
2. Every string is terminated with a ______.
3. If a string is given as `'AB CD'`, the length of this string is ______.
4. The subscript of a string starts with ______.
5. The characters of a string are stored in ______ memory locations.
6. `char mesg[100];` can store a maximum of ______ characters.
7. ______ function terminates as soon as it finds a blank space.
8. `LENGTH('') =` ______.
9. The ASCII code for `A-Z` varies from ______.
10. `toupper()` is used to ______.
11. `S1>S2` means ______.
12. The function to reverse a string is ______.
13. If `S1 = "GOOD MORNING"`, then `Substr_Left (S1, 7) =` ______.
14. `INDEX("Welcome to the world of programming", "world") =` ______.
15. ______ returns the position in the string where the string pattern first occurs.

7 Structures

Learning Objective

A structure extends the concept of arrays by storing related information of heterogeneous data types together under a single umbrella. In this chapter, we will see how a structure is defined, declared, and accessed using C.

7.1 INTRODUCTION

A structure is in many ways similar to a record. It is basically a user-defined data type that can store related information (even of different data types) together, while an array can store only entities of same data type.

A structure is therefore a collection of variables under a single name. The variables within a structure are of different data types and each has a name that is used to select it from the structure.

7.1.1 Structure Declaration

A structure is declared using the keyword `struct`, followed by a structure name. All the variables are declared within the structure. A structure type is generally defined by using the following syntax.

```
struct struct-name
{
  data_type var-name;
  data_type var-name;
  ...
};
```

For example, if we have to define a structure for a student, then what will be its related information? The answer is `roll_number`, `name`, `course`, and `fees`. This structure can be defined as:

```
struct student
{
  int r_no;
  char name[20];
  char course[20];
  float fees;
};
```

Now, the structure has become a user-defined data type. The structure definition however does not allocate any memory or consume storage space. It just gives a template that conveys to the C compiler how the structure is laid out in the memory and also gives the details of member names. Like any other data type, memory is allocated for the structure when we declare a variable of the structure. For example, we can define a variable of student by writing

```
struct student stud1;
```

7.1.2 `Typedef` Declarations

When we precede a `struct` name with the keyword `typedef`, then the `struct` becomes a new type. It is used to make the construct shorter with more meaningful names for types that are already defined by C or for types that you have declared. With a `typedef` declaration, it becomes a synonym for the type. For example,

```
typedef struct student
{
  int r_no;
  char name[20];
  char course[20];
  float fees;
};
```

Now that you have preceded the structure's name with the keyword `typedef`, `student` becomes a new data type. Therefore, now you can straightaway declare the variables of this new data type as you declare the variables of type `int`, `float`, `char`, `double`, etc. To declare a variable of structure `student`, you write `student stud1`. Note that we have not written `struct student stud1`.

7.1.3 Initialization of Structures

A structure can be initialized in the same way as other data types are initialized. Initializing a structure means assigning some constants to the members of the structure. When the user does not explicitly initialize the structure, then C automatically does it. For `int` and `float` members, the values are initialized to zero, and `char` and `string` members are initialized to `'\0'` by default (in the absence of any initialization done by the user).

The initializers are enclosed in braces and are separated by commas. However, care must be taken to see that the initializers match their corresponding types in the structure definition.

The general syntax to initialize a structure variable is as follows.

```
struct struct_name
{
       data_type member_name1;
       data_type member_name2;
       data_type member_name3;
       .......................................
}struct_var = {constant1, constant2, constant 3,...};
```

OR

```
struct struct_name
{
        data_type member_name1;
        data_type member_name2;
        data_type member_name3;
        ..................................
};
struct struct_name struct_var = {constant1, constant2, constant 3,...};
```

For example, we can initialize a student structure by writing,

```
struct student
{
   int r_no;
   char name[20];
   char course[20];
   float fees;
}stud1 = {01, "Rahul", "BCA", 45000};
```

Or, by writing,

```
struct student stud1 = {01, "Rahul", "BCA", 45000};
```

Figure 7.1 illustrates how the values will be assigned to individual fields of the structure.

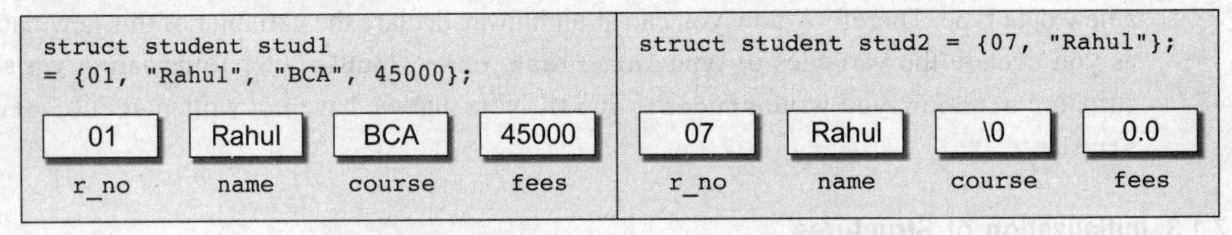

Figure 7.1 Assigning values to structure elements

7.1.4 Accessing the Members of a Structure

Each member of a structure can be used just like a normal variable, but its name will be a little longer. A structure member variable is generally accessed using a `'.'` (dot operator). The syntax of accessing a structure or a member of a structure can be given as:

```
struct_var.member_name
```

The dot operator is used to select a particular member of the structure. For example, to assign value to the individual data members of a structure variable `Rahul`, we may write

```
stud1.r_no = 01;
stud1.name = "Rahul";
stud1.course = "BCA";
stud1.fees = 45000;
```

Memory is allocated only when we declare the variables of the structure. In other words, the memory is allocated only when we instantiate the structure. In the absence of any variable, structure definition is just a template that will be used to reserve memory when a variable of type `struct` will be declared.

Once the variables of a structure are defined, we can perform a few operations on them. For example, we can use the assignment operator (=) to assign the values of one variable to another.

7.1.5 Copying Structures

We can assign a structure to another structure of the same type. For example, if we have two structure variables `stud1` and `stud2` of type `struct student` given as

```
struct student stud1 = {01, "Rahul", "BCA", 45000};
struct student stud2;
```

then, to assign one structure variable to another, we will write,

```
stud2 = stud1;
```

PROGRAMMING EXAMPLES

1. Write a program using structures to read and display the information about a student.

```
#include<stdio.h>
#include<conio.h>
int main()
{
        struct student
        {
           int roll_no;
           char name[80];
           float fees;
           char DOB[80];
        };
        struct student stud1;
        clrscr();
        printf("\n Enter the roll number : ");
        scanf("%d", &stud1.roll_no);
        printf("\n Enter the name : ");
        scanf("%s", stud1.name);
        printf("\n Enter the fees : ");
        scanf("%f", &stud1.fees);
        printf("\n Enter the DOB : ");
        scanf("%s", stud1.DOB);
        printf("\n ********STUDENT'S DETAILS *******");
        printf("\n ROLL No. = %d", stud1.roll_no);
        printf("\n NAME. = %s", stud1.name);
        printf("\n ROLL No. = %f", stud1.fees);
        printf("\n ROLL No. = %s", stud1.DOB);
        getch();
        return 0;
}
```

2. Write a program to find the largest of three numbers using structures.

```
#include<stdio.h>
#include<conio.h>
int main()
{
        struct numbers
        {
          int a, b, c;
          int largest;
        };
        struct numbers num;
        clrscr();
        printf("\n Enter the three numbers : ");
        scanf("%d %d %d", &num.a, &num.b, &num.c);
        if (num.a > num.b && num.a > num.c)
               num.largest = num.a;
        if (num.b > num.a && num.b > num.c)
               num.largest = num.b;
        if (num.c > num.a && num.c > num.b)
               num.largest = num.c;
        printf("\n The largest number is : %d", num.largest);
        getch();
        return 0;
}
```

3. Write a program to read, display, add, and subtract two complex numbers.

```
#include<stdio.h>
#include<conio.h>
int main()
{
        typedef struct complex
        {
          int real;
          int imag;
        }COMPLEX;
        COMPLEX c1, c2, sum_c, sub_c;
        int option;
        clrscr();
        do
        {
          printf("\n ******** MAIN MENU *********");
          printf("\n 1. Read the complex numbers ");
          printf("\n 2. Display the complex numbers");
          printf("\n 3. Add the complex numbers");
          printf("\n 4.Subtract the complex numbers");
```

```
            printf("\n 5. EXIT");
            printf("\n Enter your option : ");
            scanf("%d", &option);
            switch(option)
            {
               case 1:
                     printf("\n Enter the real and imaginary parts of the
first complex number : ");
                     scanf("%d %d", &c1.real, &c1.imag);
                     printf("\n Enter the real and imaginary parts of the
second complex number : ");
                     scanf("%d %d", &c2.real, &c2.imag);
                     break;
               case 2:
                     printf("\n The first complex number is : %d %di", c1.real,
c1.imag);
                     printf("\n The second complex number is : %d %di",
c2.real, c2.imag);
                     break;
               case 3:
                     sum_c.real = c1.real + c2.real;
                     sum_c.imag = c1.real + c2.imag;
                     printf("\n The sum of two complex numbers is : %d %di",
sum_c.real, sum_c.imag);
                     break;
               case 4:
                     sub_c.real = c1.real - c2.real;
                     sub_c.imag = c1.real - c2.imag;
                     printf("\n The difference between two complex numbers is
: %d %di", sub_c.real, sub_c.imag);
                     break;
               }
         }while(option != 5);
         getch();
         return 0;
}
```

4. Write a program to enter two points and then calculate the distance between them.

```
#include<stdio.h>
#include<conio.h>
#include<math.h>
int main()
{
  typedef struct point
  {
    int x, y;
  }POINT;
```

```
    Point p1, p2;
    float distance;
    clrscr();
    printf("\n Enter the coordinates of the first point : ");
    scanf("%d %d", &p1.x, &p1.y);
    printf("\n Enter the coordinates of the second point : ");
    scanf("%d %d", &p2.x, &p2.y);
    distance = sqrt((pow(p1.x - p2.x), 2) + (pow(p1.y - p2.y), 2));
    printf("\n The coordinates of the first point are : %dx %dy", p1.x, p1.y);
    printf("\n The coordinates of the second point are : %dx %dy", p2.x,
p2.y);
    printf("\n Distance between p1 and p2 = %f", distance);
    getch();
    return 0;
}
```

7.2 NESTED STRUCTURES

A structure can be placed within another structure. That is, a structure may contain another structure as its member. Such a structure that contains another structure as its member is called a *nested structure.*

Let us now see how we declare nested structures. Although it is possible to declare a nested structure with one declaration, but declaring it in this way is not recommended. The easier and clearer way is to declare the structures separately and then group them in the high-level structure. When you do this, take care to check that nesting must be done from inside-out (from lowest level to the most inclusive level). That is, declare the inner-most structure, then the next level structure, working towards the outer (most inclusive) structure.

```
typedef struct
{
    char first_name[20];
    char mid_name[20];
    char last_name[20];
}NAME;
typedef struct
{
    int dd;
    int mm;
    int yy;
}DATE;
struct student
{
    int r_no;
    NAME name;
```

```
    char course[20];
    Date DOB;
    float fees;
};
```

In the given example, we see that the structure `student` contains two other structures, `NAME` and `DATE`. Both these structures have their own fields. The structure `NAME` has three fields, `first_name`, `mid_name`, and `last_name`. The structure `DATE` also has three fields, `dd`, `mm`, and `yy`, which specify the day, month, and year of the date. Now, to assign value to the structure fields, we will write,

```
struct student stud1;
stud1.name.first_name = "Janak";
stud1.name.mid_name = "Raj";
stud1.name.last_name = "Thareja";
stud1.course = "BCA";
stud1.DOB.dd = 15;
stud1.DOB.mm = 09;
stud1.DOB.yy = 1990;
stud1.fees = 45000;
```

Note that in case of nested structures, we use the dot operator in conjunction with the structure variables to access the members of the innermost as well as the outermost structures. The use of nested structures is illustrated in the next program.

PROGRAMMING EXAMPLE

5. Write a program to read and display the information of a student using a nested structure.

```
#include<stdio.h>
#include<conio.h>
int main()
{
    struct DOB
    {
        int day;
        int month;
        int year;
    };
    struct student
    {
        int roll_no;
        char name[100];
        float fees;
        struct DOB date;
    };
    struct student stud1;
    clrscr();
```

```
        printf("\n Enter the roll number : ");
        scanf("%d", &stud1.roll_no);
        printf("\n Enter the name : ");
        scanf("%s", stud1.name);
        printf("\n Enter the fees : ");
        scanf("%f", &stud1.fees);
        printf("\n Enter the DOB : ");
        scanf("%d %d %d", &stud1.date.day, &stud1.date.month, &stud1.date.year);
        printf("\n ********STUDENT'S DETAILS *******");
        printf("\n ROLL No. = %d", stud1.roll_no);
        printf("\n NAME. = %s", stud1.name);
        printf("\n FEES. = %f", stud1.fees);
        printf("\n DOB = %d - %d - %d", stud1.date.day, stud1.date.month,
stud1.date.year);
        getch();
        return 0;
}
```

7.3 ARRAYS OF STRUCTURES

We have discussed how to declare a structure and assign values to its data members. Now, we will see how an array of structures is declared. For this purpose, let us first analyse where we would need an array of structures.

In a class, we do not have just one student. But there may be at least 30 students. So, the same definition of the structure can be used for all the 30 students. This would be possible when we make an array of structures. An array of structures is declared in the same way as we declare an array of a built-in data type.

Another example where an array of structures is desirable is in case of an organization. An organization has a number of employees. So, defining a separate structure for every employee is not a viable solution. So, here we can have a common structure definition for all the employees. This can again be done by declaring an array of structure employee.

The general syntax for declaring an array of structures can be given as,

```
struct struct_name
{
   data_type member_name1;
   data_type member_name2;
   data_type member_name3;
   ....................................
};
struct struct_name struct_var[index];
```

Consider the given structure definition.

```
struct student
{
   int r_no;
```

```
    char name[20];
    char course[20];
    float fees;
  };
```

A student array can be declared by writing,

```
struct student stud[30];
```

Now, to assign values to the ith student of the class, we will write,

```
stud[i].r_no = 09;
stud[i].name = "RASHI";
stud[i].course = "MCA";
stud[i].fees = 60000;
```

PROGRAMMING EXAMPLES

6. Write a program to read and display the information of all the students in a class.

```
#include<stdio.h>
#include<conio.h>
int main()
{
      struct student
      {
         int roll_no;
         char name[80];
         float fees;
         char DOB[80];
      };
      struct student stud[50];
      int n, i;
      clrscr();
      printf("\n Enter the number of students : ");
      scanf("%d", &n);
      for(i=0;i<n;i++)
      {
         printf("\n Enter the roll number : ");
         scanf("%d", &stud[i].roll_no);
         printf("\n Enter the name : ");
         scanf("%s", stud[i].name);
         printf("\n Enter the fees : ");
         scanf("%f", stud[i].fees);
         printf("\n Enter the DOB : ");
         scanf("%s", stud[i].DOB);
      }
      for(i=0;i<n;i++)
      {
```

```
        printf("\n ********DETAILS OF %dth STUDENT*******", i+1);
        printf("\n ROLL No. = %d", stud[i].roll_no);
        printf("\n NAME. = %s", stud[i].name);
        printf("\n FEES = %f", stud[i].fees);
        printf("\n DOB = %s", stud[i].DOB);
    }
    getch();
    return 0;
}
```

7. Write a program to read and display the information of all the students in a class. Then edit the details of the ith student and redisplay the entire information.

```
#include<stdio.h>
#include<conio.h>
int main()
{
    struct student
    {
        int roll_no;
        char name[80];
        float fees;
        char DOB[80];
    };
    struct student stud[50];
    int n, i, rolno, new_rolno;
    float new_fees;
    char new_DOB[80], new_name[80];
    clrscr();
    printf("\n Enter the number of students : ");
    scanf("%d", &n);
    for(i=0;i<n;i++)
    {
        printf("\n Enter the roll number : ");
        scanf("%d", &stud[i].roll_no);
        printf("\n Enter the name : ");
        scanf("%s", stud[i].name);
        printf("\n Enter the fees : ");
        scanf("%f", stud[i].fees);
        printf("\n Enter the DOB : ");
        scanf("%s", stud[i].DOB);
    }
    for(i=0;i<n;i++)
    {
        printf("\n *******DETAILS OF %dth STUDENT*******", i+1);
        printf("\n ROLL No. = %d", stud[i].roll_no);
        printf("\n NAME. = %s", stud[i].name);
```

```
        printf("\n FEES. = %f", stud[i].fees);
        printf("\n DATE OF BIRTH = %s", stud[i].DOB);
    }
    printf("\n Enter the rolno of the student whose record has to be
edited : ");
    scanf("%d", &rolno);
    printf("\n Enter the new roll number : ");
    scanf("%d", &new_rolno);
    printf("\n Enter the new name : "):
    scanf("%s", new_name);
    printf("\n Enter the new fees : ");
    scanf("%f", &new_fees);
    printf("\n Enter the new date of birth : ");
    scanf("%s", new_DOB);
    stud[rolno].roll_no = new_rolno;
    strcpy(stud[rolno].name, new_name);
    stud[rolno].fees = new_fees;
    stud[rolno].DOB, new_DOB);
    for(i=0;i<n;i++)
    {
        printf("\n ********DETAILS OF %dth STUDENT*******", i+1);
        printf("\n ROLL No. = %d", stud[i].roll_no);
        printf("\n NAME. = %s", stud[i].name);
        printf("\n FEES. = %f", stud[i].fees);
        printf("\n DATE OF BIRTH = %s", stud[i].DOB);
    }
    getch();
    return 0;
}
```

7.4 STRUCTURES AND FUNCTIONS

For structures to be fully useful, we must have a mechanism to pass them to functions and return them. A function may access the members of a structure in three ways as shown in Fig. 7.2.

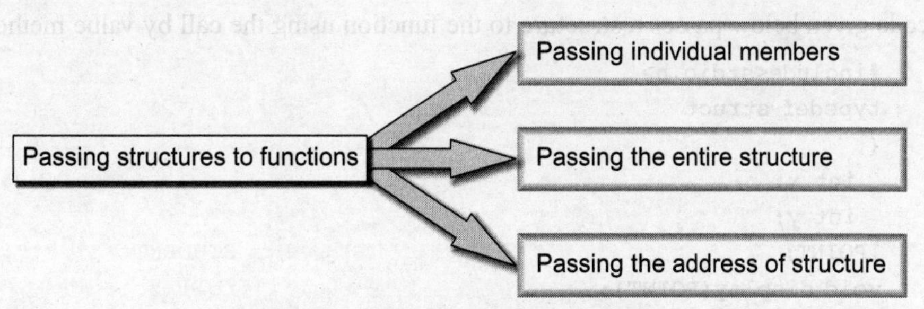

Figure 7.2 Passing structures to a function

7.4.1 Passing Individual Members

To pass any individual member of a structure to a function, we must use the direct selection operator to refer to the individual members for the actual parameters. The called program does not know if the two variables are ordinary variables or structure members. Look at the code given below which illustrates this concept.

```
#include<stdio.h>
typedef struct
{
  int x;
  int y;
}POINT;
void display(int, int);
main()
{
  POINT p1 = {2, 3};
  display(p1.x, p1.y);
  return 0;
}
void display(int a, int b)
{
  printf("%d %d", a, b);
}
```

7.4.2 Passing the Entire Structure

Just like any other variable, we can pass an entire structure as a function argument. When a structure is passed as an argument, it is passed using the call by value method. That is, a copy of each member of the structure is made. No doubt, this is a very inefficient method especially when the structure is very big or the function is called frequently. Therefore, in such a situation, passing and working with pointers may be more efficient.

The general syntax for passing a structure to a function and returning a structure can be given as,

```
struct struct_name func_name(struct struct_name struct_var);
```

The above syntax can vary as per the requirement. For example, in some situations, we may want a function to receive a structure but return a void or the value of some other data type. The code given below passes a structure to the function using the call by value method.

```
#include<stdio.h>
typedef struct
{
  int x;
  int y;
}POINT;
void display(POINT);
main()
```

```
{
  POINT p1 = {2, 3};
  display(p1);
  return 0;
}
void display(POINT p)
{
  printf("%d %d", p.x, p.y);
}
```

PROGRAMMING EXAMPLES

8. Write a program to read, display, add, and subtract two distances. Distance must be defined using kms and meters.

```
#include<stdio.h>
#include<conio.h>
typedef struct distance
{
  int kms;
  int meters;
}DISTANCE;
DISTANCE d1, d2, d3, d4;
DISTANCE add_distance (DISTANCE, DISTANCE);
DISTANCE subtract_distance (DISTANCE, DISTANCE);
int main()
{
  int option;
  clrscr();
  do
  {
       printf("\n ******** MAIN MENU *********");
       printf("\n 1. Read the distances ");
       printf("\n 2. Display the distances");
       printf("\n 3. Add the distances");
       printf("\n 4. Subtract the distances");
       printf("\n 5. EXIT");
       printf("\n Enter your option : ");
       scanf("%d", &option);
       switch(option)
       {
            case 1:
                   printf("\n Enter the kms and meters in the first distance : ");
                   scanf("%d %d", &d1.kms, &d1.meters);
                   printf("\n Enter the kms and meters in the second distance : ");
                   scanf("%d %d", &d2.kms, &d2.meters);
                   break;
```

```
			case 2:
				printf("\n The first distance is : %dkms %dmeters", d1.kms,
	d1.meters);
				printf("\n The second distance is : %dkms %dmeters",
	d2.kms, d2.meters);
				break;
			case 3:
				d3 = add_distances(d1, d2);
				printf("\n The sum of two distances is : %dkms %dmeters",
	d3.kms, d3.meters);
				break;
			case 4:
				d4 = subtract_distances(d1, d2);
				printf("\n The difference between two distances is :
	%dkms %dmeters", d4.kms, d4.meters);
				break;
		}
	}while(option != 5);
	getch();
	return 0;
}
DISTANCE add_distance(DISTANCE d1, DISTANCE d2)
{
	DISTANCE sum;
	sum.meters = d1.meters + d2.meters;
	sum.kms = d1.kms + d2.kms;
	while (sum.meters > 1000)
	{
		sum.meters = sum.meters % 1000;
		sum.kms += 1;
	}
	return sum;
}
DISTANCE subtract_distance(DISTANCE d1, DISTANCE d2)
{
	DISTANCE sub;
	if(d1.kms > d2.kms)
	{
		sub.meters = d1.meters - d2.meters;
		sub.kms = d1.kms - d2.kms;
	}
	else
	{
		sub.meters = d2.meters - d1.meters;
		sub.kms = d2.kms - d1.kms;
```

```
        }
        if(sub.meters < 0)
        {
            sub.kms = sum.kms - 1;
            sub.meters = sum.meters + 1000;
        }
        return sub;
}
```

9. Write a program that uses a structure called date that is passed to an isLeapYear function to determine if the year is a leap year.

```
#include <stdio.h>
struct date
{
        int day;
        int month;
        int year;
};
int isLeapYear(struct date d);
int main()
{
        struct date d;
        printf("Enter the date (eg: 05/06/1978): ");
        scanf("%d/%d/%d", &d.day, &d.month, &d.year);
        if (isLeapYear(d) == 0)
                printf("\n %d is not a leap year", d.year);
        else
                printf("\n %d is a leap year", d.year);
        return 0;
}
int isLeapYear(struct date d)
{
        if ((d.year % 4 == 0 && d.year % 100 != 0) || d.year % 400 == 0)
                return 1;
        else
                return 0;
}
```

10. Write a program using pointer to structure to initialize the members in a structure. Use functions to print the student's information.

```
#include<stdio.h>
#include<conio.h>
struct student
{
```

```
    int r_no;
    char name[20];
    char course[20];
    float fees;
  };
  void display(struct student *);
  main()
  {
    struct student *ptr_stud1;
    struct student stud1 = {01, "Rahul", "BCA", 45000};
    clrscr();
    ptr_stud1 = &stud1;
    display(ptr_stud1);
    return 0);
  }
  void display( struct student *ptr_stud1)
  {
    printf("\n DETAILS OF STUDENT");
    printf("\n ---------------------------------");
    printf("\n ROLL NUMBER = %d", ptr_stud1->r_no);
    printf("\n NAME = ", puts(ptr_stud1->name));
    printf("\n COURSE = ", puts(ptr_stud1->course));
    printf("\n FEES = %f", ptr_stud1->fees);
  }
```

11. Write a program to read, display, add, and subtract two times. Time must be defined using hour, minutes, and seconds.

```
#include<stdio.h>
#include<conio.h>
typedef struct
{
  int hr;
  int min;
  int sec;
}TIME;
TIME t1, t2, t3;
TIME subtract_time(TIME, TIME);
TIME add_time(TIME, TIME);
int main()
{
  int option;
  clrscr();
  do
  {
```

```
        printf("\n ******** MAIN MENU *********");
        printf("\n 1. Read time");
        printf("\n 2. Display time");
        printf("\n 3. Add two time");
        printf("\n 4.Subtract two time");
        printf("\n 5. EXIT");
        printf("\n Enter your option : ");
        scanf("%d", &option);
        switch(option)
        {
               case 1:
                      printf("\n Enter the hrs, mins, and secs of the first
time : ");
                      scanf("%d %d %d", &t1.hr, &t1.min, &t1.sec);
                      printf("\n Enter the hrs, mins and secs of the second
time : ");
                      scanf("%d %d %d %d", &t2.hr, &t2.min, &t2.sec);
                      break;
               case 2:
                      printf("\n The first time is : %dhr %dmin %dsec", t1.hr,
t1.min, t1.sec);
                      printf("\n The second time is : %dhr %dmin %dsec", t2.hr,
t2.min, t2.sec);
                      break;
               case 3:
                      t3 = add_time(t1, t2);
                      printf("\n The sum of two times is : %dhr %dmin %dsec",
t3.hr, t3.min, t3.sec);
                      break;
               case 4:
                      t4 = subtract_time(t1, t2);
                      printf("\n The difference of two times is : %dhr %dmin
%dsec", t4.hr, t4.min, t4.sec);
                      break;
        }
   }while(option != 5);
   getch();
   return 0;
}
TIME add_time(TIME t1, TIME t2)
{
        TIME sum;
        sum.sec = t1.sec + t2.sec;
        while(sum.sec > 60)
        {
```

```
                sum.sec -= 60;
                sum.min++;
            }
            sum.min = t1.min + t2.min;
            while(sum.min>60)
            {
                sum.min -= 60;
                sum.hr++;
            }
            sum.hr = t1.hr + t2.hr;
            return sum;
        }
        TIME subtract_time(TIME t1, TIME t2)
        {
            TIME sub;
            if( t1.hr > t2.hr)
            {
                if(t1.sec < t2.sec)
                {
                    t1.sec += 60;
                    t1.min–;
                }
                sub.sec = t1.sec - t2.sec;
                if(t1.min < t2.min)
                {
                    t1.min += 60;
                    t1.hr–;
                }
                sub.min = t1.min - t2.min;
                sub.hr = t1.hr - t2.hr;
            }
            else
            {
                if(t2.sec < t1.sec)
                {
                    t2.sec+=60;
                    t2.min–;
                }
                sub.sec = t2.sec - t1.sec;
                if(t2.min < t1.min)
                {
                    t2.min+=60;
                    t2.hr–;
                }
                sub.min = t2.min - t1.min;
```

```
                sub.hr = t2.hr - t1.hr;
        }
        return sub;
}
```

12. Write a program to read, display, add, and subtract two heights. Height must be defined using feet and inches.

```
#include<stdio.h>
#include<conio.h>
typedef struct
{
  int ft;
  int inch;
}HEIGHT;
HEIGHT h1, h2, h3;
HEIGHT add_height(HEIGHT, HEIGHT)
HEIGHT subtract_height(HEIGHT, HEIGHT)
int main()
{
  int option;
  clrscr();
  do
  {
       printf("\n ******** MAIN MENU *********");
       printf("\n 1. Read height ");
       printf("\n 2. Display height");
       printf("\n 3. Add two heights");
       printf("\n 4.Subtract two heights");
       printf("\n 5. EXIT");
       printf("\n Enter your option : ");
       scanf("%d", &option);
       switch(option)
       {
              case 1:
                     printf("\n Enter the feet and inches of the first height
: ");
                     scanf("%d %d", &h1.ft, &h1.inch);
                     printf("\n Enter the feet and inches of the second height
: ");
                     scanf("%d %d", &h2.ft, &h2.inch);
                     break;
              case 2:
                     printf("\n The first height is : %dft %dinch %dsec",
h1.ft, h1.inch);
                     printf("\n The second height is : %dft %dinch %dsec",
h2.ft, h2.inch);
                     break;
```

```
                    case 3:
                            h3 = add_height(h1, h2);
                            printf("\n The sum of two heights is : %dft %dinch",
        h3.ft, h3.inch);
                            break;
                    case 4:
                            h3 = subtract_height(h1, h2);
                            printf("\n The difference of two heights is : %dft
        %dinch", h3.ft, h3.inch);
                            break;
                }
          }while(option != 5);
          getch();
          return 0;
        }
        HEIGHT add_height(HEIGHT h1, HEIGHT h2)
        {
                HEIGHT sum;
                sum.inch = h1.inch + h2.inch;
                while(sum.inch > 12)
                {
                        sum.inch -= 12;
                        sum.ft++;
                }
                sum.ft = h1.ft + h2.ft;
                return sum;
        }
        HEIGHT subtract_height(HEIGHT h1, HEIGHT h2)
        {
                HEIGHT sub;
                if( h1.ft > h2.ft)
                {
                        if(h1.inch < h2.inch)
                        {
                                h1.inch += 12;
                                h1.ft–;
                        }
                        sub.inch = h1.inch - h2.inch;
                        sub.ft = h1.ft - h2.ft;
                }
                else
                {
                        if(h2.inch < h1.inch)
```

```
            {
                H2.inch += 12;
                H2.ft—;
            }
            sub.inch = h2.inch - h1.inch;
            sub.ft = h2.ft - h1.ft;
        }
        return sub;
    }
```

7.4.3 Passing Structures through Pointers

Passing large structures to functions using the call by value method is very inefficient. Therefore, it is preferred to pass structures through pointers. It is possible to create a pointer to almost any type in C, including the user-defined types. It is extremely common to create pointers to structures. Like in other cases, a pointer to a structure is never itself a structure, but merely a variable that holds the address of a structure. The syntax to declare a pointer to a structure can be given as,

```
struct struct_name
{
        data_type member_name1;
        data_type member_name2;
        data_type member_name3;
        ....................................
}*ptr;
```

Or,

```
struct struct_name *ptr;
```

For our `student` structure, we can declare a pointer variable by writing,

```
struct student *ptr_stud, stud;
```

The next thing to be done is to assign the address of `stud` to the pointer using the address operator (`&`), as we would do in case of any other pointer. So to assign the address, we will write,

```
ptr_stud = &stud;
```

One way to access the members of a structure is to write,

```
/* get the structure, then select a member */
(*ptr_stud).roll_no;
```

Since parentheses have a higher precedence than `*`, writing this statement would work well. But this statement is not easy to work with, especially for a beginner. So, C introduces a new operator to do the same task. This operator is known as 'pointing-to' operator (–>). Here, it is used as:

```
/* the roll_no in the structure ptr_stud points to */
ptr_stud->roll_no = 01;
```

This statement is far easier than its alternative.

PROGRAMMING EXAMPLES

13. Write a program using pointer to structure to initialize the members in a structure.

```
#include<stdio.h>
#include<conio.h>
struct student
{
  int r_no;
  char name[20];
  char course[20];
  float fees;
};
main()
{
  struct student stud1, stud2, *ptr_stud1, *ptr_stud2;
  clrscr();
  ptr_stud1 = &stud1;
  ptr_stud2 = &stud2;
   ptr_stud1 -> r_no = 01;
   strcpy(ptr_stud1 -> name, "Rahul");
   strcpy(ptr_stud1 -> course, "BCA");
   ptr_stud1 -> fees = 45000;
   printf("\n Enter the details of the second student :");
   printf("\n Enter the Roll Number =");
   scanf("%d", &ptr_stud2 -> r_no);
   printf("\n Enter the Name = );
   gets(ptr_stud2 -> name));
   printf("\n Enter the Course = ");
   gets(ptr_stud2 -> course));
   printf("\n Enter the Fees = );
   scanf("%f", &ptr_stud2 -> fees);
   printf("\n DETAILS OF FIRST STUDENT");
   printf("\n ---------------------------------");
   printf("\n ROLL NUMBER = %d", ptr_stud1 -> r_no);
   printf("\n NAME = ", puts(ptr_stud1 -> name));
   printf("\n COURSE = ", puts(ptr_stud1 -> course));
   printf("\n FEES = %f", ptr_stud1 -> fees);
   printf("\n\n\n\n DETAILS OF SECOND STUDENT");
   printf("\n ---------------------------------");
   printf("\n ROLL NUMBER = %d", ptr_stud2 -> r_no);
   printf("\n NAME = ", puts(ptr_stud2 -> name));
   printf("\n COURSE = ", puts(ptr_stud2 -> course));
   printf("\n FEES = %f", ptr_stud2 -> fees);
   return 0;
  }
```

14. Write a program using pointer to structure to initialize the members in the structure using an alternative technique.

```
#include<stdio.h>
#include<conio.h>
struct student
{
    int r_no;
    char name[20];
    char course[20];
    float fees;
};
main()
{
    struct student *ptr_stud1;
    struct student stud1 = {01, "Rahul", "BCA", 45000};
    clrscr();
    ptr_stud1 = &stud1;
    printf("\n DETAILS OF STUDENT");
    printf("\n -----------------------------------");
    printf("\n ROLL NUMBER = %d", ptr_stud1->r_no);
    printf("\n NAME = ", puts(ptr_stud1->name));
    printf("\n COURSE = ", puts(ptr_stud1->course));
    printf("\n FEES = %f", ptr_stud1->fees);
    return 0;
}
```

7.5 SELF-REFERENTIAL STRUCTURES

Self-referential structures are those structures that contain a reference to data of its same type. That is, a self-referential structure, in addition to other data, contains a pointer to a data that is of the same type as that of the structure. For example, consider the structure node given below.

```
struct node
{
    int val;
    struct node *next;
};
```

Here, the structure node will contain two types of data, an integer `val` and a pointer `next`. You must be wondering why we need such a structure. Actually, self-referential structure is the foundation of other data structures. We will be using them throughout this book and their purpose will be clearer to you when we discuss linked lists, trees, and graphs.

SUMMARY

- Structure is basically a user-defined data type that can store related information (even of different data types) together.
- A structure is declared using the keyword `struct`, followed by a structure name. The structure definition however does not allocate any memory or consume storage space. It just gives a template that conveys to the C compiler how the structure is laid out in the memory and gives details of the member names.
- When we precede a `struct` name with the keyword `typedef`, then the `struct` becomes a new type.
- When the user does not explicitly initialize the structure, then C automatically does it. For `int` and `float` members, the values are initialized to zero and `char` and `string` members are initialized to `'\0'` by default.
- A structure member variable is generally accessed using a `'.'` (dot operator).
- A structure can be placed within another structure. That is, a structure may contain another structure as its member. Such a structure is called a nested structure.
- Self-referential structures are those structures that contain a reference to data of its same type. That is, a self-referential structure, in addition to other data, contains a pointer to a data that is of the same type as that of the structure.

GLOSSARY

Structure It is a user-defined data type that can store related information (even of different data types) together.

`Typedef` declaration When we precede a `struct` name with the keyword `typedef`, then the `struct` becomes a new type. It is used to make the construct shorter with more meaningful names for types already defined by C or for types that you have declared.

Structure initialization Initializing a structure means assigning some constants to the members of the structure. When the user does not explicitly initialize the structure, then C automatically does that. For `int` and `float` members, the values are initialized to zero and `char` and `string` members are initialized to `'\0'` by default.

Copying structures Assigning a structure to another structure of the same type.

Nested structure A structure placed within another structure. That is, a structure that contains another structure as its member.

Self-referential structures Structures that contain a reference to data of its same type. That is, a self-referential structure, in addition to other data, contains a pointer to a data that is of the same type as that of the structure.

EXERCISES

Review Questions

1. Declare a structure that represents the following hierarchical information.
   - (a) Student
   - (b) Roll Number
   - (c) Name
     - (i) First name
     - (ii) Middle Name
     - (iii) Last Name
   - (d) Sex
   - (e) Date of Birth
     - (i) Day
     - (ii) Month
     - (iii) Year
   - (f) Marks
     - (i) English
     - (ii) Mathematics
     - (iii) Computer Science

2. Define a structure to store the name, an array `marks[]` which stores the marks of five different subjects, and a character grade. Write a program to display the details of the student whose name is entered by the user. Use the structure definition of the first question to make an array of students. Display the name of the students who have secured less than 40% of the aggregate.
3. Modify Question 2 to print each student's average marks and the class average (that includes average of all the student's marks).
4. Make an array of students as illustrated in Question 1 and write a program to display the details of the student with the given Date of Birth.
5. Make an array of student as illustrated in Question 1 and write a program to delete the record of the student with the given last name.
6. What is the advantage of using structures?
7. Structure declaration reserves memory for the structure. Comment on this statement with valid justifications.
8. Differentiate between a structure and an array.
9. Write a short note on structures and inter-process communication.
10. Explain the utility of the keyword `typedef` in structures.
11. Explain with an example how structures are initialized.
12. Is it possible to create an array of structures? Explain with the help of an example.

Programming Exercises

1. Write a program using structures to read and display the information about a student.
2. Write a program to find the largest of three numbers using structures.
3. Write a program to enter two points and then calculate the distance between them.
4. Write a program to read, display, add, and subtract two complex numbers.
5. Write a program to read and display the information of a student using nested structure.
6. Write a program to read and display the information of all the students in a class.
7. Write a program to read and display the information of all the students in a class. Then, edit the details of the ith student and redisplay the entire information.
8. Write a program to read, display, add, and subtract two distances. Distance must be defined using kms and meters.
9. Write a program to read, display, add, and subtract two heights. Height must be defined using feet and inches.
10. Write a program to read, display, add, and subtract two times. Time must be defined using hour, minutes, and seconds.
11. Write a program that uses a structure called `Date` that has is passed to an `isLeapYear` function to determine if the year is a leap year.
12. Write a program using pointer to structure to initialize the members in a structure. Use functions to print the student's information.
13. Write a program to create a structure with the information given below. Then, read and print the data.

    Employee[10]

    (a) Emp_Id
    (b) Name
        (i) First Name
        (ii) Middle Name
        (iii) Last Name
    (c) Address
        (i) Area
        (ii) City
        (iii) State
    (d) Age
    (e) Salary
    (f) Designation

Multiple Choice Questions

1. A data structure that can store related information together is called

   (a) Array (b) String
   (c) Structure (d) All of these

2. A data structure that can store related information of different data types together is called

   (a) Array (b) String
   (c) Structure (d) All of these

3. Memory for a structure is allocated at the time of
   (a) Structure definition
   (b) Structure variable declaration
   (c) Structure declaration
4. A structure member variable is generally accessed using
   (a) Address operator
   (b) Dot operator
   (c) Comma operator
   (d) Ternary operator
5. A structure that can be placed within another structure is known as
   (a) Self-referential structure
   (b) Nested structure
   (c) Parallel structure

True or False

1. Structures contain related information but all those information has the same data type.
2. Structure declaration reserves memory for the structure.
3. Initializing a structure means assigning some constants to the members of the structure.
4. When the user does not explicitly initialize the structure, then C automatically does it.
5. The dereference operator is used to select a particular member of the structure.
6. Memory is allocated for a structure only when we declare its variables.
7. A nested structure contains another structure as its member.

Fill in the blanks

1. Structure is a ______ data type.
2. A structure is same as that of ______.
3. ______ contains related information of the same data type.
4. ______ contains related information of the same or different data type(s).
5. ______ is just a template that will be used to reserve memory when a variable of type `struct` is declared.
6. A ______ is a collection of variables under a single name.
7. A structure is declared using the keyword `struct` followed by a ______.
8. When we precede a `struct` name with ______, then the `struct` becomes a new type.
9. For `int` and `float` structure members, the values are initialized to ______.
10. `char` and `string` structure members are initialized to ______ by default.
11. A structure member variable is generally accessed using a ______.
12. A structure that can be placed within another structure is called a ______.
13. ______ structures contain a reference to data of its same type.

Annexure B Memory Allocation in C Programs

C supports three kinds of memory allocation through the variables in C programs:

Static allocation When we declare a static or global variable, static allocation is done. Each static or global variable is allocated a fixed size of memory space. The number of bytes reserved for the variable cannot change during the execution of the program. Till now, we have been using this technique to define variables, arrays, and pointers.

Automatic allocation When we declare an automatic variable, such as a function argument or a local variable, automatic memory allocation is done. The space for an automatic variable is allocated when the compound statement containing the declaration is entered, and is freed when that compound statement is exited.

Dynamic allocation A third important kind of memory allocation is known as *dynamic allocation.* In this section, we will discuss dynamic memory allocation using pointers.

MEMORY USAGE

Before jumping into dynamic memory allocation, let us first understand how memory is used. Conceptually, memory is divided into two parts: *program memory* and *data memory*.

Program memory consists of memory used for the `main()` and other called functions in the program. Data memory on the other hand consists of memory needed for permanent definitions such as global data, local data, constants and dynamic memory data. The way in which C handles the need of memory is a function of the operating system and the compiler.

When a program is being executed, its `main()` and all other functions are always kept in the memory. However, the local variables of the function are available in the memory only when they are active. In Chapter 2, when we had studied recursive functions, we had seen that system stack is used to store a single copy of the function and multiple copies of the local variables.

Apart from the stack, we also have a memory allocation known as heap. Heap memory is unused memory allocated to the program and available to be assigned during its execution. When we dynamically allocate memory for variables, heap acts as a memory pool from which memory is allocated to those variables. However, this is just a conceptual view of memory and its implementation is entirely in the hands of system designers.

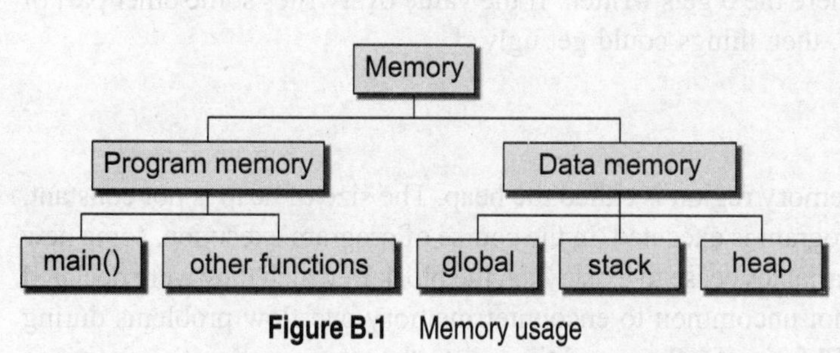

Figure B.1 Memory usage

DYNAMIC MEMORY ALLOCATION

The process of allocating memory to the variables during the execution of a program or at runtime is known as *dynamic memory allocation*. C language has four library routines which allow this function.

Till now, whenever we needed an array, we had declared a static array of fixed size, say,

```
int arr[100];
```

When this statement is executed, consecutive space for 100 integers is allocated. It is not uncommon that we may be using only 10% or 20% of the allocated space, thereby wasting rest of the space. To overcome this problem and to utilize the memory efficiently, C language provides a mechanism of dynamically allocating memory so that only the amount of memory that is actually required is reserved. We reserve space only at the runtime for the variables that are actually required. Dynamic memory allocation gives the best performance in situations in which we do not know the memory requirements in advance.

C provides four library routines to automatically allocate memory at the runtime. These routines are shown in Table B.1.

Table B.1 Memory allocation/de-allocation functions

Function	Task
malloc	Allocates memory and returns a pointer to the first byte of allocated space.
calloc	Allocates space for an array of elements and initializes them to zero. Like malloc(), calloc() also returns a pointer to the memory.
free	Frees previously allocated memory.
realloc	Alters the size of previously allocated memory.

When we have to dynamically allocate memory in our programs, then pointers are the only way to go. When we use `malloc()` for dynamic memory allocation, then we need to manage the memory allocated for variables manually. If you miss to keep a track of the memory which `malloc` has given you, then it is not uncommon to accidentally use a pointer which points 'nowhere', thereby giving unpleasant results. That is, if you assign a value to the location pointed to by a pointer

```
*ptr = 0;
```

and if `ptr` points 'nowhere', it can be construed to point somewhere, just not where you wanted it to, and that 'somewhere' is where the 0 gets written. If the value overwrites some other part of your program or any other data, then things could get ugly.

MEMORY ALLOCATION PROCESS

In computer science, the free memory region is called the heap. The size of heap is not constant, as it keeps changing when the program is executed. In the course of program execution, some new variables are created and some variables cease to exist when the block in which they were declared is exited. For this reason, it is not uncommon to encounter memory overflow problems during the dynamic allocation process. When an overflow condition exists, the memory allocation functions mentioned above will return a null pointer.

ALLOCATING A BLOCK OF MEMORY

Let us see how memory is allocated using `malloc()`. `malloc` is declared in `<stdlib.h>`, so we include that header in any program that calls `malloc`. It reserves a block of memory of specified size and returns a pointer of type `void`. It means we can assign it to any type of pointer. The general syntax of `malloc()` is

```
ptr=(cast-type*)malloc(byte-size);
```

where, `ptr` is a pointer of type `cast-type`. The `malloc()` returns a pointer (of `cast-type`) to an area of memory with the size `byte-size`. For example,

```
arr=(int*)malloc(10*sizeof(int));
```

This statement is used to dynamically allocate memory equivalent to 10 times the area of `int` bytes. On successful execution of the statement, the space is reserved and and the address of the first byte of memory allocated is assigned to the pointer `arr` of type `int`.

`Calloc()` is another function that reserves memory at runtime. It is normally used to request multiple blocks of storage each of the same size and then sets all bytes to zero. `Calloc` stands for contiguous memory allocation and is primarily used to allocate memory for arrays. The syntax of `calloc()` can be given as

```
ptr=(cast-type*) calloc(n,elem-size);
```

The above statement allocates contiguous space for `n` blocks each of elements size `bytes`. The only difference between `malloc()` and `calloc()` is that when we use `calloc()`, all bytes are initialized to zero. `Calloc()` returns a pointer to the first byte of the allocated region.

Note that when we allocate memory using `malloc()` or `calloc()`, a null pointer will be returned if there is not enough space in the system to allocate. A null pointer points definitively nowhere. It is a 'not a pointer' marker; therefore, it is not a pointer you can use. Thus, whenever you allocate memory using `malloc()` or `calloc()`, you must check the returned pointer before using it. If the program receives a null pointer, it should at the very least print an error message and exit, or perhaps figure out some way of proceeding without the memory it asked for. But in any case, the program cannot go on to use the null pointer it got back from `malloc()` or `calloc()`.

A call to `malloc`, with an error check, typically looks something like this:

```
int *ip = malloc(100 * sizeof(int));
if(ip == NULL)
        {
        printf("out of memory\n");
        exit or return
        }
```

PROGRAMMING EXAMPLE

1. Write a program to read and display the values of an integer array. Allocate space dynamically for the array.

```
#include<stdio.h>
#include<conio.h>
main()
{
        int i, n;
```

```
        int *arr;
        clrscr();
        printf("\n Enter the number of elements in the array : ");
        scanf("%d", &n);
        arr = (int *)malloc(n*sizeof(int));
        if(arr == NULL)
        {
                printf("\n Memory Allocation Failed");
                exit(0);
        }
        for(i=0;i<n;i++)
        {
                printf("\n Enter the %dth value of the array : ", i);
                scanf("%d", &arr[i]);
        }
        printf("\n The array contains the following values - \n");
        for(i=0;i<n;i++)
                printf("%d", arr[i]); // another way is to write *(arr+i)
        return 0;
}
```

RELEASING THE USED SPACE

When a variable is allocated space during compile time, the memory used by that variable is automatically released by the system in accordance with its storage class. But when we dynamically allocate memory, it is our responsibility to release the space when it is not required. This is even more important when the storage space is limited. Therefore, if we no longer need the data stored in a particular block of memory and we do not intend to use that block for storing any other information, then as a good programming practice, we must release that block of memory for future use, using the free function. The general syntax of `free()` is

```
free(ptr);
```

where `ptr` is a pointer that has been created by using `malloc()` or `calloc()`.

ALTERING THE SIZE OF ALLOCATED MEMORY

At times, the memory allocated by `calloc()` or `malloc()` might be insufficient or in excess. In both the situations, we can always use `realloc()` to change the memory size already allocated by `calloc()` and `malloc()`. This process is called *reallocation of memory*. The general syntax for `realloc()` can be given as

```
ptr=realloc(ptr,newsize);
```

`Realloc()` allocates new memory space of size specified by `newsize` to the pointer variable `ptr`. It returns a pointer to the first byte of the memory block. The allocated new block may or may not be at the same region. Thus, we see that `realloc()` takes two arguments. The first is the pointer referencing the memory and the second is the total number of bytes you want to reallocate.

If you pass zero as the second argument, it will be equivalent to calling `free()`. Like `malloc()` and `calloc()`, `realloc` returns a void pointer if successful, else a `null` pointer is returned.

If `realloc()` was able to make the old block of memory bigger, it returns the same pointer. Otherwise, if `realloc()` has to go elsewhere to get enough contiguous memory, then it returns a pointer to the new memory, after copying your old data there. However, if `realloc()` can't find enough memory to satisfy the new request at all, it returns a null pointer. So you must check the pointer returned by `realloc()` is not a null pointer.

```
/*Example program for reallocation*/
#include<stdio.h>
#include<stdlib.h>
define NULL 0
main()
{
        char *str;
        /*Allocating memory*/
str = (char *)malloc(10);
if(str==NULL)
{
        printf("\n Memory could not be allocated");
        exit(1);
}
strcpy(str,"Oxford");
printf("\n STR = %s", str);
/*Reallocation*/
str = (char *)realloc(str,20);
if(str==NULL)
{
        printf("\n Memory could not be reallocated");
        exit(1);
}
printf("\n STR size modified.\n");
printf("\n STR = %s\n", str);
strcpy(str,"Oxford University");
printf("\n STR = %s", str);
/*freeing memory*/
free(str);
        return 0;
}
```

Note that with `realloc()`, you can allocate more bytes without losing your data.

PROGRAMMING EXAMPLE

2. Write a program using pointer to structure to initialize the members in a structure. Allocate memory for the structure using `malloc()`.

```
#include<stdio.h>
#include<conio.h>
```

```
struct student
{
        int r_no;
        char name[20];
        char course[20];
        float fees;
};
main()
{
        struct student *ptr_stud;
        int i, n;
        clrscr();
        printf("\n Enter the number of students : ");
        scanf("%d", &n);
        ptr_stud = (struct student*)malloc(n*sizeof(struct student));
        if(ptr_stud == NULL)
        {
                printf("\n Memory Allocation Failed");
                exit(0);
        }
        for(i=0;i<n;i++)
        {
                printf("\n Enter the details of the %dth student :", i);
                printf("\n Enter the Roll Number =");
                scanf("%d", &ptr_stud[i]->r_no);
                printf("\n Enter the Name = ");
                gets(ptr_stud[i]->name));
                printf("\n Enter the Course = ");
                gets(ptr_stud[i]->course));
                printf("\n Enter the Fees = ");
                scanf("%f", &ptr_stud[i]->fees);
        }
        for(i=0;i<n;i++)
        {
                printf("\n DETAILS OF THE %dth STUDENT", i);
                printf("\n ---------------------------------------");
                printf("\n ROLL NUMBER = %d", ptr_stud[i]->r_no);
                printf("\n NAME = ", puts(ptr_stud[i]->name));
                printf("\n COURSE = ", puts(ptr_stud[i]->course));
                printf("\n FEES = %f", ptr_stud[i]->fees);
        }
        return 0;
}
```

DRAWBACK OF POINTERS

Although pointers are very useful in C, they are not free from limitations. If used incorrectly, pointers can lead to bugs that are difficult to unearth. For example, if you use a pointer to read a memory location but that pointer is pointing to an incorrect location, then you may end up reading a wrong value. An erroneous input always leads to an erroneous output, therefore however efficient your program code may be, the output will always be disastrous. Same is the case when writing a value to a particular memory location.

Consider a scenario in which the program code is supposed to read the account balance of a customer, add new amount to it, and then re-write the modified value to that location. If the pointer is pointing to the account balance of some other customer, then the account balance of the wrong customer will be updated.

Let us try to find some common errors when using pointers.

```
int x, *px;
x=10;
*px = 20;
```

ERROR: un-initialized pointer. `px` is pointing to an unknown memory location. Hence it will overwrite that location's contents and store 20 in it.

```
int x, *px;
x=10;
px = x;
```

ERROR: it should be `px = &x;`

```
int x=10, y=20, *px, *py;
px = &x, py = &y;
if(px<py)          // it should be *px and *py
        printf("\n px is less than py");
else
        printf("\n py is less than px");
```

8 Linked Lists

Learning Objective

A linked list is a data structure created using a self-referential structure. In this chapter, we will learn about different types of linked lists and the techniques applied to insert and delete nodes in a linked list.

8.1 INTRODUCTION

A linked list, in simple terms, is a linear collection of data elements. These data elements are called *nodes*. Linked list is a data structure which in turn can be used to implement other data structures. Thus, it acts as a building block to implement data structures like stacks, queues, and their variations. A linked list can be perceived as a train or a sequence of nodes in which each node contains one or more data fields and a pointer to the next node.

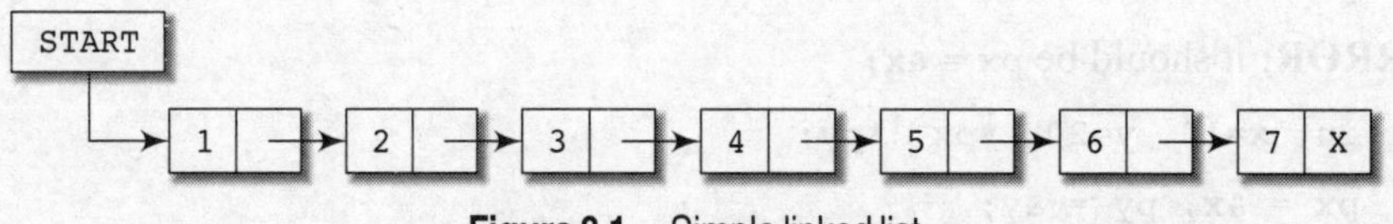

Figure 8.1 Simple linked list

In Fig. 8.1, we see a linked list in which every node contains two parts, an integer and a pointer to the next node. The left part of the node which contains data may include a simple data type, an array, or a structure. The right part of the node contains a pointer to the next node (or address of the next node in sequence). The last node will have no next node connected to it, so it will store a special value called NULL. In Fig. 8.1, the NULL pointer is represented by X. But when we do the programming, we usually define NULL as –1. Hence, a NULL pointer denotes the end of the list. Since in a linked list, every node contains a pointer to another node which is of the same type, it is also called a *self-referential data type*.

Linked lists contain a pointer variable START that stores the address of the first node in the list. We can traverse the entire list using a single pointer variable called START. The START node will contain the address of the first node; the next part of the first node will in turn store the address of its succeeding node. Using this technique, the individual nodes of the list will form a chain of nodes. If START = NULL, then the linked list is empty and contains no nodes.

In C, we will implement a linked list using the following code:

```
struct node
{
  int data;
```

```
    struct node *next;
  };
```

Let us see how a linked list is maintained in the memory. In order to form a linked list, we need a structure called *node* which has two fields, DATA and NEXT. DATA will store the information part and NEXT will store the address of the next node in sequence. Consider Fig. 8.2.

START = 1

	Data	Next
1	H	4
2		
3		
4	E	7
5		
6		
7	L	8
8	L	10
9		
10	O	–1

Figure 8.2 START pointing to the first element of the linked list in the memory

In the figure, we see that the variable START is used to store the address of the first node. Here, in this example, START = 1, so the first data is stored at address 1, which is H. The corresponding NEXT stores the address of the next node, which is 4. So, we will look at address 4 to fetch the next data item. The second data element obtained from address 4 is E. Again, we see the corresponding NEXT to go to the next node. From the entry in the NEXT, we get the next address, that is 7, and fetch L as the data. We repeat this procedure until we reach a position where the NEXT entry contains –1 or NULL, as this would denote the end of the linked list. When we traverse DATA and NEXT in this manner, we finally see that the linked list in the above example stores characters that when put together forms the word HELLO.

Note that the figure shows a chunk of memory locations whose address ranges from 1 to 10. The shaded portion contains data for other applications. Remember that the nodes of a linked list need not be in consecutive memory locations. In our example, the nodes for the linked list are stored at addresses 1, 4, 7, 8, and 10.

Let us take another example to see how two linked lists are maintained together in the computer's memory. For example, the students of Class XI of Science group are asked to choose between Biology and Computer Science. Now, we will maintain two linked lists, one for each subject. That is, the first linked list will contain the roll numbers of all the students who have opted for Biology and the second list will contain the roll numbers of students who have chosen Computer Science.

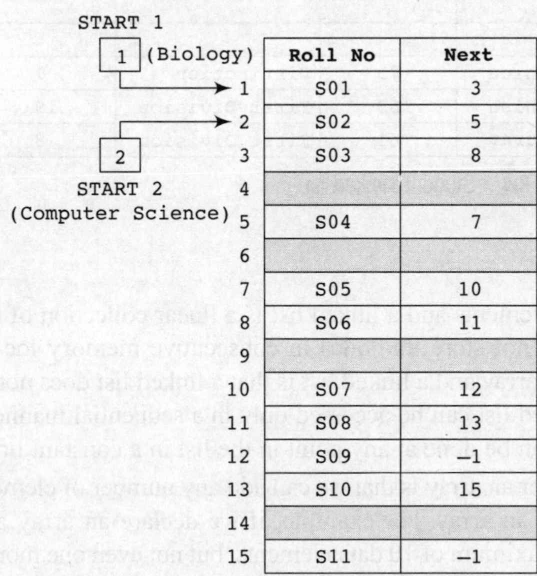

	Roll No	Next
1	S01	3
2	S02	5
3	S03	8
4		
5	S04	7
6		
7	S05	10
8	S06	11
9		
10	S07	12
11	S08	13
12	S09	–1
13	S10	15
14		
15	S11	–1

Figure 8.3 Two linked lists are simultaneously maintained in the memory

Now, look at Fig. 8.3. Two different linked lists are simultaneously maintained in the memory. There is no ambiguity in traversing through the list because each list maintains a separate Start pointer, which gives the address of the first node of their respective linked lists. The rest of the nodes are reached by looking at the value stored in the NEXT.

By looking at the figure, we can conclude that roll number of the students who have opted for Biology are S01, S03, S06, S08, S10, and S11. Similarly, roll number of the students who chose Computer Science are S02, S04, S05, S07, and S09.

We have already said that the `DATA` part of a node may contain just a single data item, an array, or a structure. Let us take an example to see how a structure is maintained in a linked list that is stored in the memory.

Consider a scenario in which the roll number, name, aggregate, and grade of students are stored using linked lists. Now, we will see how the `NEXT` pointer is used to store the data alphabetically. This is shown in Fig. 8.4.

	Roll No	Name	Aggregate	Grade	Next
1	S01	Ram	78	Distinction	6
2	S02	Shyam	64	First Division	7
3					
4	S03	Mohit	89	Outstanding	17
5					
6	S04	Rohit	77	Distinction	14
7	S05	Varun	86	Outstanding	10
8	S06	Karan	65	First Division	12
9					
10	S07	Veena	54	Second Division	–1
11	S08	Meera	67	First Division	4
12	S09	Krish	45	Third Division	13
13	S10	Kusum	91	Outstanding	11
14	S11	Silky	72	First Division	2
15					
16					
17	S12	Monica	75	Distinction	1
18	S13	Ashish	63	First Division	19
19	S14	Gaurav	61	First Division	8

START = 18

Figure 8.4 Student's linked list

8.2 LINKED LIST VERSUS ARRAYS

An array is a linear collection of data elements and a linked list is a linear collection of nodes. But unlike an array, a linked list does not store its nodes in consecutive memory locations. Another point of difference between an array and a linked list is that a linked list does not allow random access of data. Nodes in a linked list can be accessed only in a sequential manner. But like an array, insertions and deletions can be done at any point in the list in a constant time.

Another advantage of a linked list over an array is that we can add any number of elements in the list. This is not possible in case of an array. For example, if we declare an array as `int marks[10]`, then the array can store a maximum of 10 data elements, but not even one more than that. There is no such restriction in case of a linked list.

Thus, linked lists provide an efficient way of storing related data and perform basic operations such as insertion, deletion, and updating of information at the cost of extra space required for storing the address of next nodes.

8.3 MEMORY ALLOCATION AND DE-ALLOCATION FOR A LINKED LIST

We have seen how a linked list is represented in the memory. If we want to add a node to an already existing linked list in the memory, we first find free space in the memory and then use it to store the information. For example, consider the linked list given in Fig. 8.5. The linked list contains the roll number of students, marks obtained by them in Biology, and finally a `NEXT` field which stores the address of the next node in sequence. Now, if a new student joins the class and is asked to appear for the same test that the other students had taken, then the new student's marks should also be recorded in the linked list. For this purpose, we find a free space and store the information there. If, the grey shaded portion shows free space, then we see that we have 4 locations of memory available. We can use any one of them to store our data. This is illustrated in Fig. 8.5 (a) and (b).

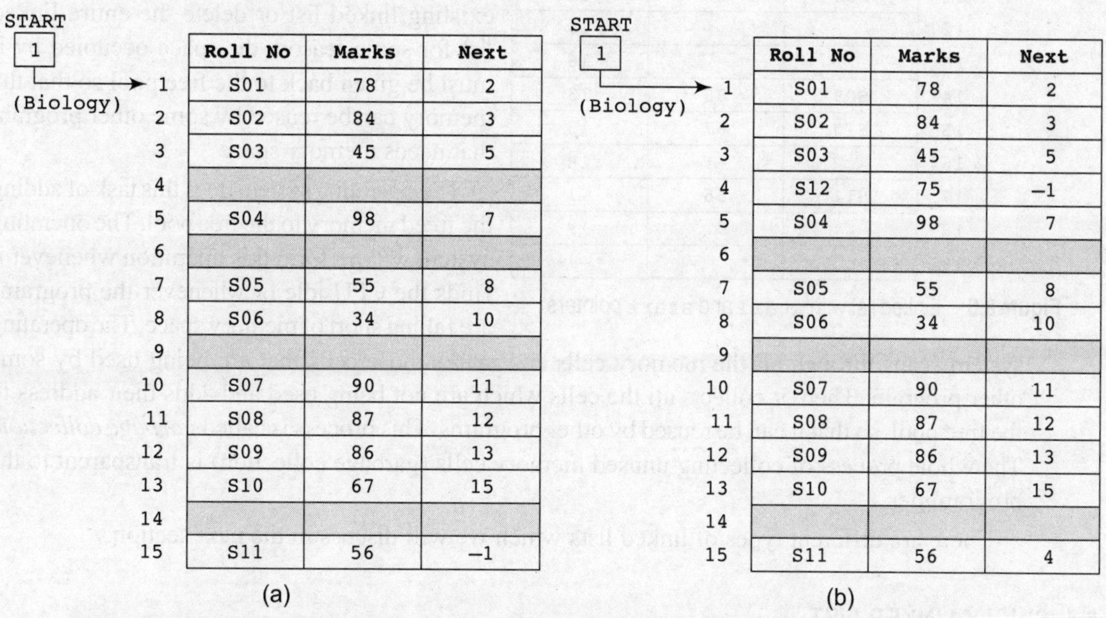

(a)

	Roll No	Marks	Next
1	S01	78	2
2	S02	84	3
3	S03	45	5
4			
5	S04	98	7
6			
7	S05	55	8
8	S06	34	10
9			
10	S07	90	11
11	S08	87	12
12	S09	86	13
13	S10	67	15
14			
15	S11	56	–1

(b)

	Roll No	Marks	Next
1	S01	78	2
2	S02	84	3
3	S03	45	5
4	S12	75	–1
5	S04	98	7
6			
7	S05	55	8
8	S06	34	10
9			
10	S07	90	11
11	S08	87	12
12	S09	86	13
13	S10	67	15
14			
15	S11	56	4

Figure 8.5 (a) Marks obtained by students in biology (b) linked list after the new student's record has been added

Now, the big question is which part of the memory is available and which part is occupied? When we delete a node from a linked list, then who changes the status of the memory occupied by it from occupied to available? The answer is, the operating system. Discussing the mechanism of how the operating system does all this is out of the scope of this book. So, in simple language, we can say that the computer does it on its own without any intervention from the user or the programmer. As a programmer, you just have to take care of the code to perform insertions and deletions in the list.

However, let us briefly discuss the basic concept behind it. The computer maintains a list of all the free memory cells. This list of available space is called the *free pool*.

We have seen that every linked list has a pointer variable `START` which stores the address of the first node of the list. Likewise, for the free pool (which is a linked list of all the free memory cells), we have a pointer variable `AVAIL` which stores the address of the first free space. Let us revisit the memory representation of the linked list storing all the student's marks in Biology.

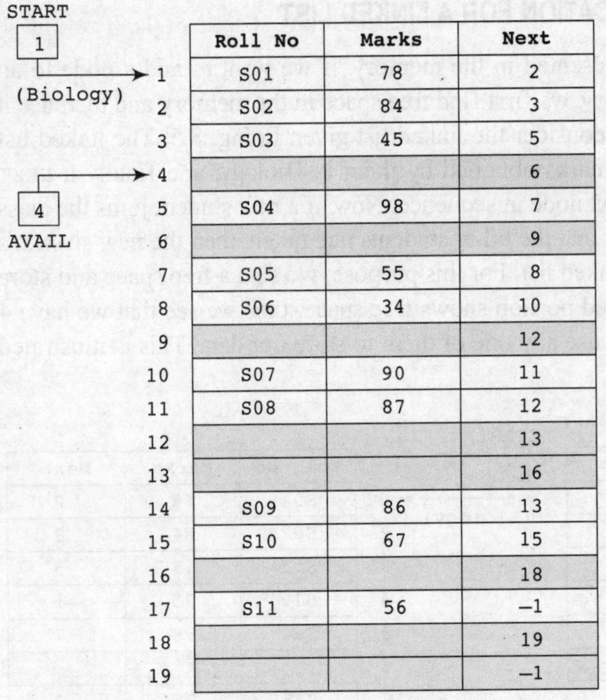

	Roll No	Marks	Next
1	S01	78	2
2	S02	84	3
3	S03	45	5
4			6
5	S04	98	7
6			9
7	S05	55	8
8	S06	34	10
9			12
10	S07	90	11
11	S08	87	12
12			13
13			16
14	S09	86	13
15	S10	67	15
16			18
17	S11	56	–1
18			19
19			–1

Figure 8.6 Linked list with `avail` and `start` pointers

Now, when a new student's record has to be added, the memory address pointed by `AVAIL` will be taken and used to store the desired information. After the insertion, the next available free space's address will be stored in `AVAIL`. For example, in Fig. 8.6, when the first free memory space is utilized for inserting the new node, the `AVAIL` will be set to contain address 6.

This was all about inserting a new node in an already existing linked list. Now, we will discuss deleting a node or the entire linked list. When we delete a particular node from an existing linked list or delete the entire linked list for some reason, the space occupied by it must be given back to the free pool so that the memory can be reused by some other program that needs memory space.

The operating system does this task of adding the freed memory to the free pool. The operating system will perform this operation whenever it finds the CPU idle or whenever the programs are falling short of memory space. The operating system scans through all the memory cells and marks those cells that are being used by some other program. Then, it collects all the cells which are not being used and adds their address to the free pool, so that it can be reused by other programs. This process is called *garbage collection*. The whole process of collecting unused memory cells (garbage collection) is transparent to the programmer.

There are different types of linked lists which we will discuss in the next section.

8.4 SINGLY LINKED LIST

A singly linked list is the simplest type of linked list in which every node contains some data and a pointer to the next node of the same data type. By saying that the node contains a pointer to the next node, we mean that the node stores the address of the next node in sequence. A singly linked list allows traversal of data only in one way. Figure 8.7 shows a singly linked list.

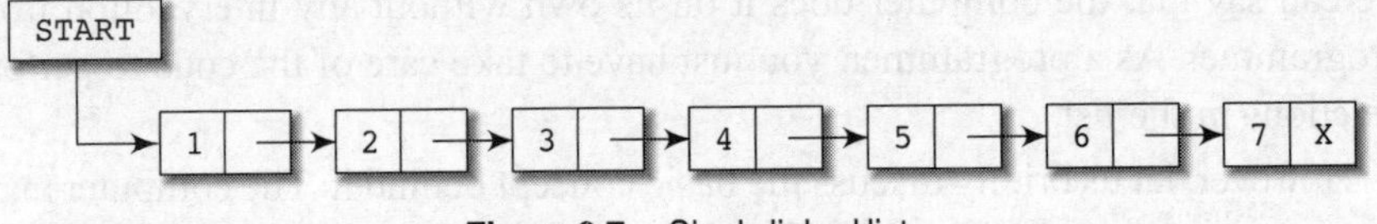

Figure 8.7 Singly linked list

8.4.1 Traversing a Singly Linked List

Traversing a linked list means accessing the nodes of the list in order to perform some processing on them. Remember, a linked list always contains a pointer variable `START` which stores the

```
Step 1: [INITIALIZE] SET PTR = START
Step 2: Repeat Steps 3 and 4 while PTR != NULL
Step 3:          Apply Process to PTR -> DATA
Step 4:          SET PTR = PTR -> NEXT
        [END OF LOOP]
Step 5: EXIT
```

Figure 8.8 Algorithm for traversing a linked list

address of the first node of the list. End of the list is marked by storing `NULL` or –1 in the `NEXT` field of the last node. For traversing the linked list, we also make use of another pointer variable `PTR` which points to the node that is currently being accessed. The algorithm to traverse a linked list is shown in Fig. 8.8.

In this algorithm, we first initialize `PTR` with the address of `START`. So now, `PTR` points to the first node of the linked list. Then in Step 2, a `while` loop is executed which is repeated till `PTR` processes the last node, that is until it encounters `NULL`. In Step 3, we apply the process (for example, `print`) to the current node, that is, the node pointed by `PTR`. In Step 4, we move to the next node by making the `PTR` variable point to the node whose address is stored in the `NEXT` field. The algorithm to print the information stored in each node of the linked list is shown in Fig. 8.9.

```
Step 1: [INITIALIZE] SET PTR = START
Step 2: Repeat Steps 3 and 4 while PTR != NULL
Step 3:          Write PTR -> DATA
Step 4:          SET PTR = PTR -> NEXT
        [END OF LOOP]
Step 5: EXIT
```

Figure 8.9 Algorithm to print each node of the linked list

Let us now write an algorithm to count the number of nodes in the linked list. To do this, we will traverse each and every node of the list and while traversing every individual node, we will increment the counter by 1. Once we reach `NULL`, that is, when all the nodes of the linked list have been traversed, the final value of the counter will be displayed. Figure 8.10 shows the algorithm to print the number of nodes in the linked list.

```
Step 1: [INITIALIZE] SET Count = 0
Step 2: [INITIALIZE] SET PTR = START
Step 3: Repeat Steps 4 and 5 while PTR != NULL
Step 4:          SET Count = Count + 1
Step 5:          SET PTR = PTR -> NEXT
        [END OF LOOP]
Step 6: EXIT
```

Figure 8.10 Algorithm to print the number of nodes in the linked list

8.4.2 Searching a Linked List

Searching a linked list means to find a particular element in the linked list. As already discussed, a linked list consists of a node which is divided into two parts, the information part and the next part. So obviously, searching means finding whether a given value is present in the information part of the node or not. If it is present, the searching algorithm returns the address of the node that contains the value.

There are two variants of searching an algorithm. One algorithm is used to search for a value in a sorted linked list, while the other works for an unsorted linked list. Let us first look at the algorithm that is used to search for a value in an unsorted linked list. Figure 8.11 shows the algorithm to search an unsorted linked list.

```
Step 1: [INITIALIZE] SET PTR = START
Step 2: Repeat Steps 3 while PTR != NULL
Step 3:       IF VAL = PTR -> DATA
                   SET POS = PTR
                   Go To Step 5
              ELSE
                   SET PTR = PTR -> NEXT
              [END OF IF]
        [END OF LOOP]
Step 4: SET POS = NULL
Step 5: EXIT
```

Figure 8.11 Algorithm to search an unsorted linked list

In Step 1, we initialize a pointer variable PTR with START that contains the address of the first node. In Step 2, a while loop is executed which will compare every node's DATA with the VAL for which the search is being made. If the search is successful, that is, the VAL has been found, then the address of that node is stored in POS and the control jumps to the last statement of the algorithm. However, if the search is unsuccessful, POS is set to NULL which indicates that the VAL is not present in the linked list.

Consider the linked list shown in Fig. 8.12. If we have VAL = 4, then the flow of the algorithm can be explained as shown in the figure.

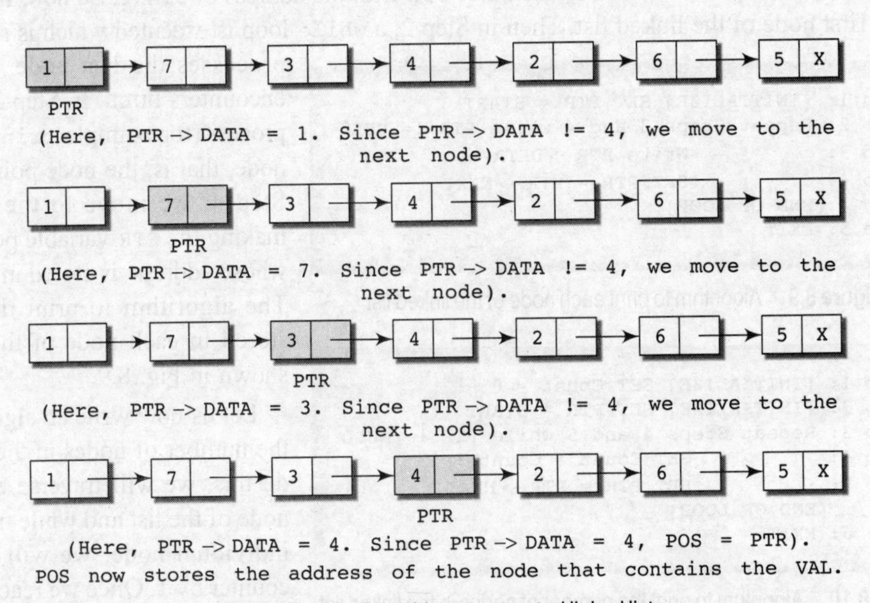

Figure 8.12 Searching an unsorted linked list

Let us now look at the searching algorithm which searches for a particular VAL in a sorted linked list. Figure 8.13 shows the algorithm to search a sorted linked list.

This algorithm is the same as the one above it with just the difference that if VAL is greater than PTR -> DATA, then we look for the next node's data. In case VAL is equal to PTR -> DATA, then the value of POS is set to contain the address of the node that has the value and the control jumps to the last statement of the algorithm.

```
Step 1: [INITIALIZE] SET PTR = START
Step 2: Repeat Steps 3 while PTR != NULL
Step 3:          IF PTR->DATA = VAL then
                    SET POS = PTR
                    Go to Step 5
                 ELSE IF PTR->DATA < VAL
                    SET PTR = PTR->NEXT
                 ELSE
                    Go To Step 4
                 [END OF IF]
        [END OF LOOP]
Step 4: SET POS = NULL
Step 5: EXIT
```

Figure 8.13 Algorithm to search a sorted linked list

Thus, we see that the worst case complexity of searching an unsorted or a sorted linked list is O(n). That is, the worst case complexity is proportional to the number of nodes in the linked list. The worst case of the searching algorithm would occur when either the VAL is not present in the linked list or it is equal to the DATA of the last node of the list. In the best case, the first node's data has a value equal to VAL, and in the average case, the middle node's data has VAL stored in it.

8.4.3 Inserting a New Node in a Linked List

In this section, we will see how a new node is added into an already existing linked list. To do the insertion, we will take five cases and then see how insertion is done in each case.

Case 1: The new node is inserted at the beginning.

Case 2: The new node is inserted at the end.

Case 3: The new node is inserted after a given node.

Case 4: The new node is inserted before a given node.

Case 5: The new node is inserted in a sorted linked list.

Before we start with the algorithms to do the insertions in all these five cases, let us first discuss an important term called `OVERFLOW`. Overflow is a condition that occurs when `AVAIL = NULL` or no free memory cell is present in the system. When this condition prevails, the programmer must give an appropriate message.

Insert at the beginning

Consider the linked list shown in Fig. 8.14. Suppose we want to add a new node with data 9 and add it as the first node of the list. Then, the following changes will be done in the linked list.

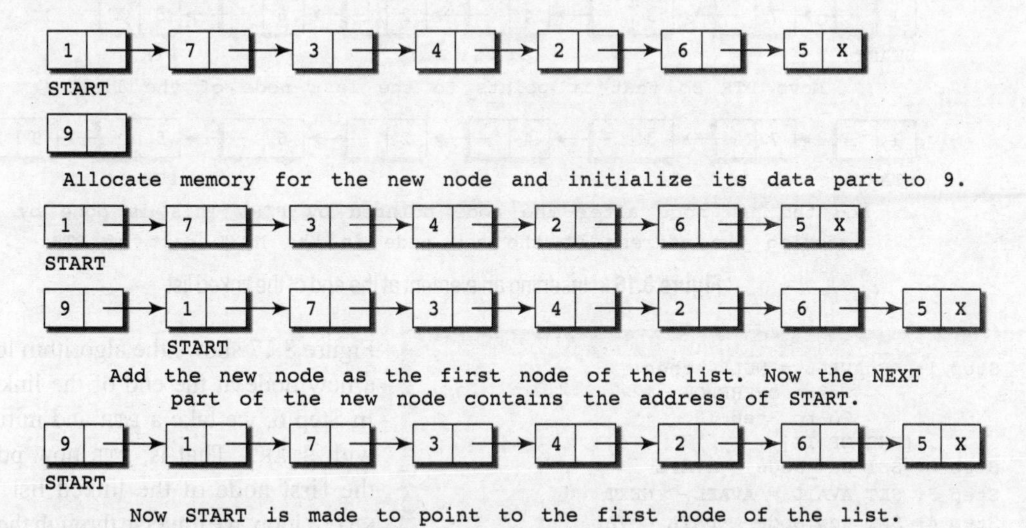

Figure 8.14 Inserting an element at the beginning of the linked list

```
Step 1: IF AVAIL = NULL, then
            Write OVERFLOW
            Go to Step 7
        [END OF IF]
Step 2: SET New_Node = AVAIL
Step 3: SET AVAIL = AVAIL -> NEXT
Step 4: SET New_Node -> DATA = VAL
Step 5: SET New_Node -> Next = START
Step 6: SET START = New_Node
Step 7: EXIT
```

Figure 8.15 Algorithm to insert a new node at the beginning of the linked list

Figure 8.15 shows the algorithm to insert a new node at the beginning of the linked list. In Step 1, we first check whether memory is available for the new node. If the free memory has exhausted, then an `OVERFLOW` message is printed. Otherwise, if a free memory cell is available, then we allocate space for the new node. Set its `data` part with the given `VAL` and the `next` part is initialized with the address of the first node of the list, which is stored in `START`. Now, since the new node is added as the first node of the list, it will now be known as the `START` node, that is, the `START` pointer variable will now hold the address of the `New_Node`. Note the following two steps.

```
Step 2: SET New_Node = AVAIL
Step 3: SET AVAIL = AVAIL->NEXT
```

These steps allocate memory for the new node. In C, there are functions like `malloc()`, `alloc`, and `calloc()` which do all this automatically on behalf of the user. We have already discussed these functions in Appendix B.

Insert at the end

Consider the linked list shown in Fig. 8.16. Suppose we want to add a new node with data 9 and add it as the last node of the list. Then, the following changes will be done in the linked list.

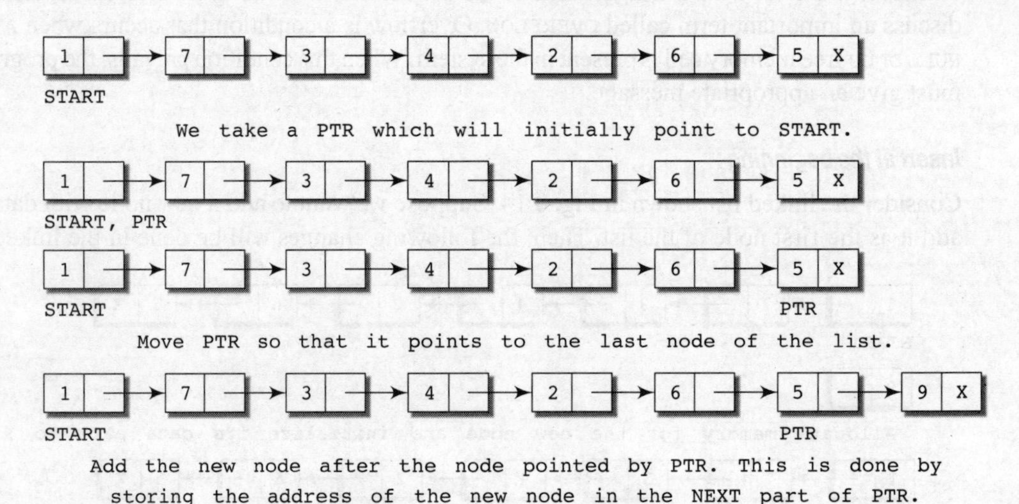

Figure 8.16 Inserting an element at the end of the linked list

```
Step 1: IF AVAIL = NULL, then
            Write OVERFLOW
            Go to Step 10
        [END OF IF]
Step 2: SET New_Node = AVAIL
Step 3: SET AVAIL = AVAIL->NEXT
Step 4: SET New_Node->DATA = VAL
Step 5: SET New_Node->Next = NULL
Step 6: SET PTR = START
Step 7: Repeat Step 8 while PTR->NEXT != NULL
Step 8:           SET PTR = PTR->NEXT
        [END OF LOOP]
Step 9: SET PTR->NEXT = New_Node
Step 10: EXIT
```

Figure 8.17 Algorithm to insert a new node at the end of the linked list

Figure 8.17 shows the algorithm to insert a new node at the end of the linked list. In Step 6, we take a `PTR` and initialize it with `START`. That is, `PTR` now points to the first node of the linked list. In the `while` loop, we traverse through the linked list to reach the last node. Once we reach the last node, in Step 9, we change the `NEXT` pointer of the last node to store the address of the new node. Remember that the `NEXT` field of the new node contains `NULL`, which signifies the end of the linked list.

Insert after a given node

Consider the linked list shown in Fig. 8.19. Suppose we want to add a new node with the value 9 and add it after the node containing data 3. Before discussing the changes that will be done in the linked list, let us first look at the algorithm shown in Fig. 8.18.

```
Step 1: IF AVAIL = NULL, then
            Write OVERFLOW
            Go to Step 12
        [END OF IF]
Step 2: SET New_Node = AVAIL
Step 3: SET AVAIL = AVAIL->NEXT
Step 4: SET New_Node->DATA = VAL
Step 5: SET PTR = START
Step 6: SET PREPTR = PTR
Step 7: Repeat Step 8 and 9 while PREPTR->DATA != NUM
Step 8:             SET PREPTR = PTR
Step 9:             SET PTR = PTR->NEXT
          [END OF LOOP]
Step 10: PREPTR->NEXT = New_Node
Step 11: SET New_Node->NEXT = PTR
Step 12: EXIT
```

Figure 8.18 Algorithm to insert a new node after a node that has value NUM

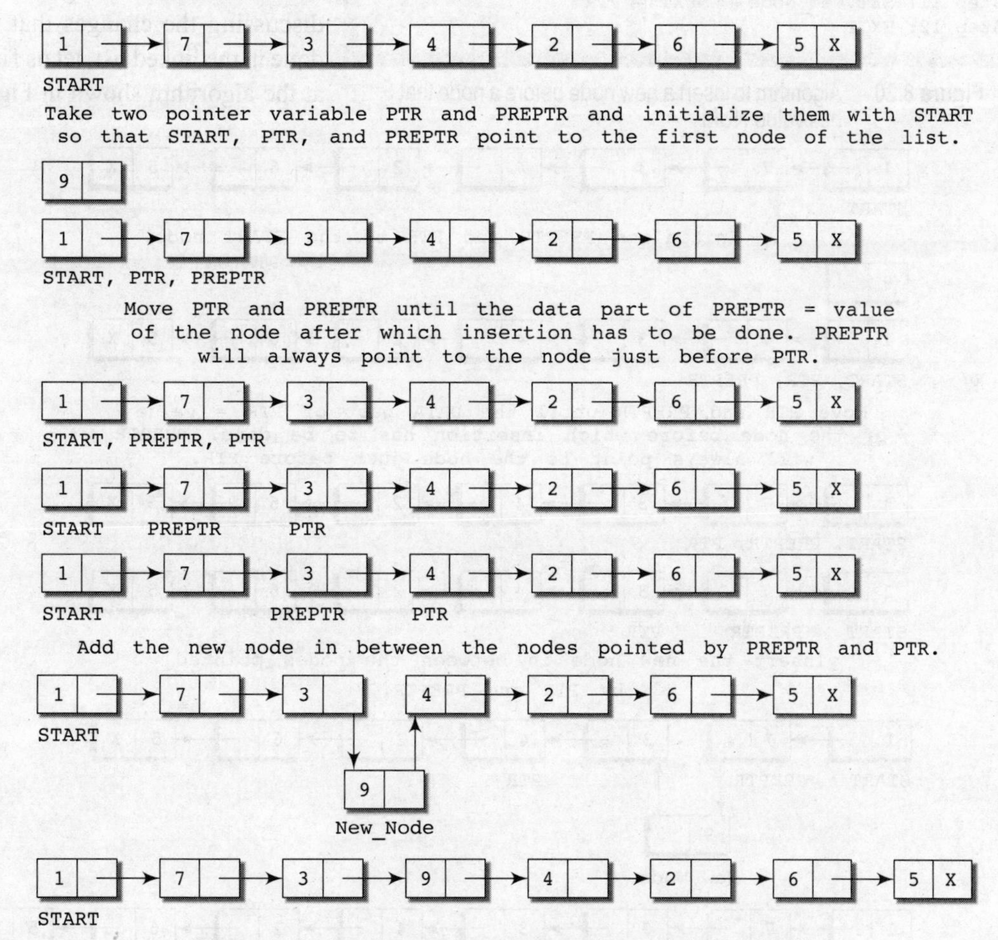

Figure 8.19 Inserting an element after a given node in the linked list

In Step 5, we take a PTR and initialize it with START. That is, PTR now points to the first node of the linked list. Then, we take another pointer variable PREPTR which will be used to store the address of the node preceding PTR. Initially, PREPTR is initialized to PTR. So now, PTR, PREPTR, and Start are all pointing to the first node of the linked list.

In the while loop, we traverse through the linked list to reach the node that has its value equal to NUM. We need to reach this node because the new node will be inserted after this node. Once we reach this node, in Step 10, we change the NEXT pointers in such a way that new node is inserted after the desired node.

```
Step 1: IF AVAIL = NULL, then
            Write OVERFLOW
            Go to Step 12
        [END OF IF]
Step 2: SET New_Node = AVAIL
Step 3: SET AVAIL = AVAIL->NEXT
Step 4: SET New_Node->DATA = VAL
Step 5: SET PTR = START
Step 6: SET PREPTR = PTR
Step 7: Repeat Step 8 and 9 while PTR->DATA != NUM
Step 8:           SET PREPTR = PTR
Step 9:           SET PTR = PTR->NEXT
        [END OF LOOP]
Step 10: PREPTR->NEXT = New_Node
Step 11: SET New_Node->NEXT = PTR
Step 12: EXIT
```

Figure 8.20 Algorithm to insert a new node before a node that has value NUM

Insert before a given node

Consider the linked list shown in Fig. 8.21. Suppose we want to add a new node with value 9 and add it before the node containing 3. Before discussing the changes that will be done in the linked list, let us first look at the algorithm shown in Fig. 8.20.

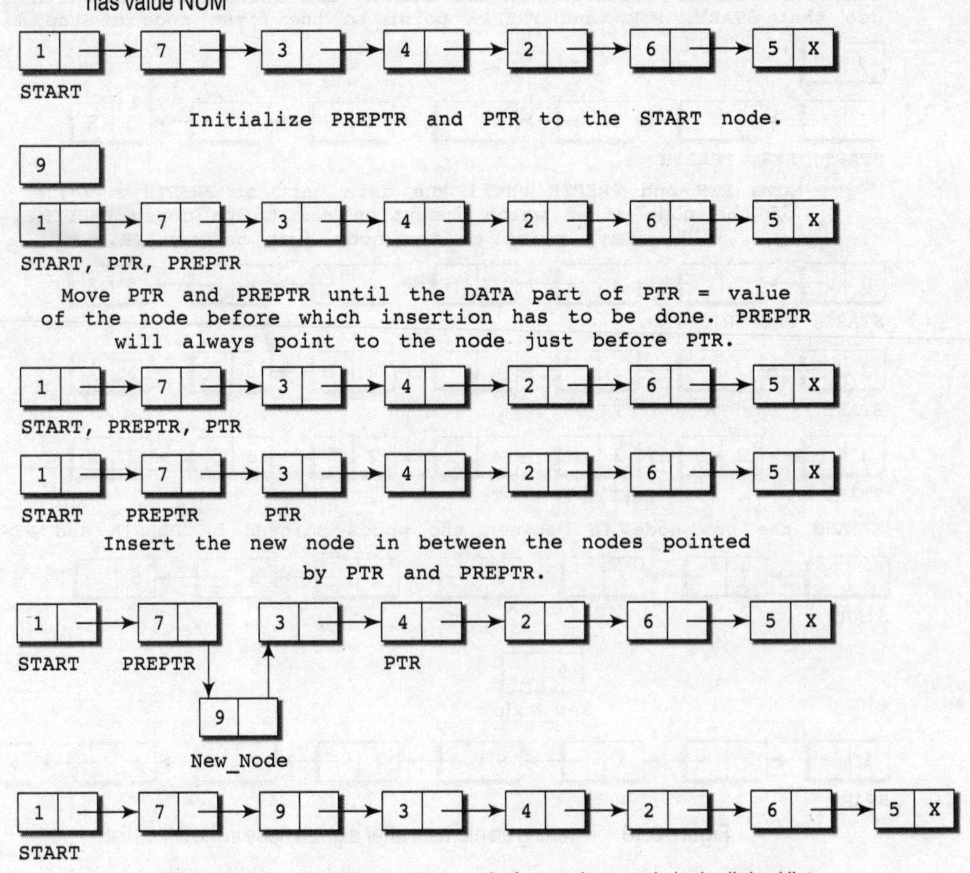

Figure 8.21 Inserting an element before a given node in the linked list

In Step 5, we take a pointer variable PTR, and initialize it with START. That is, PTR now points to the first node of the linked list. Then, we take another pointer variable PREPTR and initialize it with PTR. So now, PTR, PREPTR, and Start are all pointing to the first node of the linked list.

In the while loop, we traverse through the linked list to reach the node that has its value equal to NUM. We need to reach this node because the new node will be inserted before this node. Once we reach this node, in Step 10, we change the NEXT pointers in such a way that the new node is inserted after the desired node.

Insert in a sorted linked list

Consider the linked list shown in Fig. 8.22. Suppose we want to add a new node with the value 5, then the following changes will be made to the linked list.

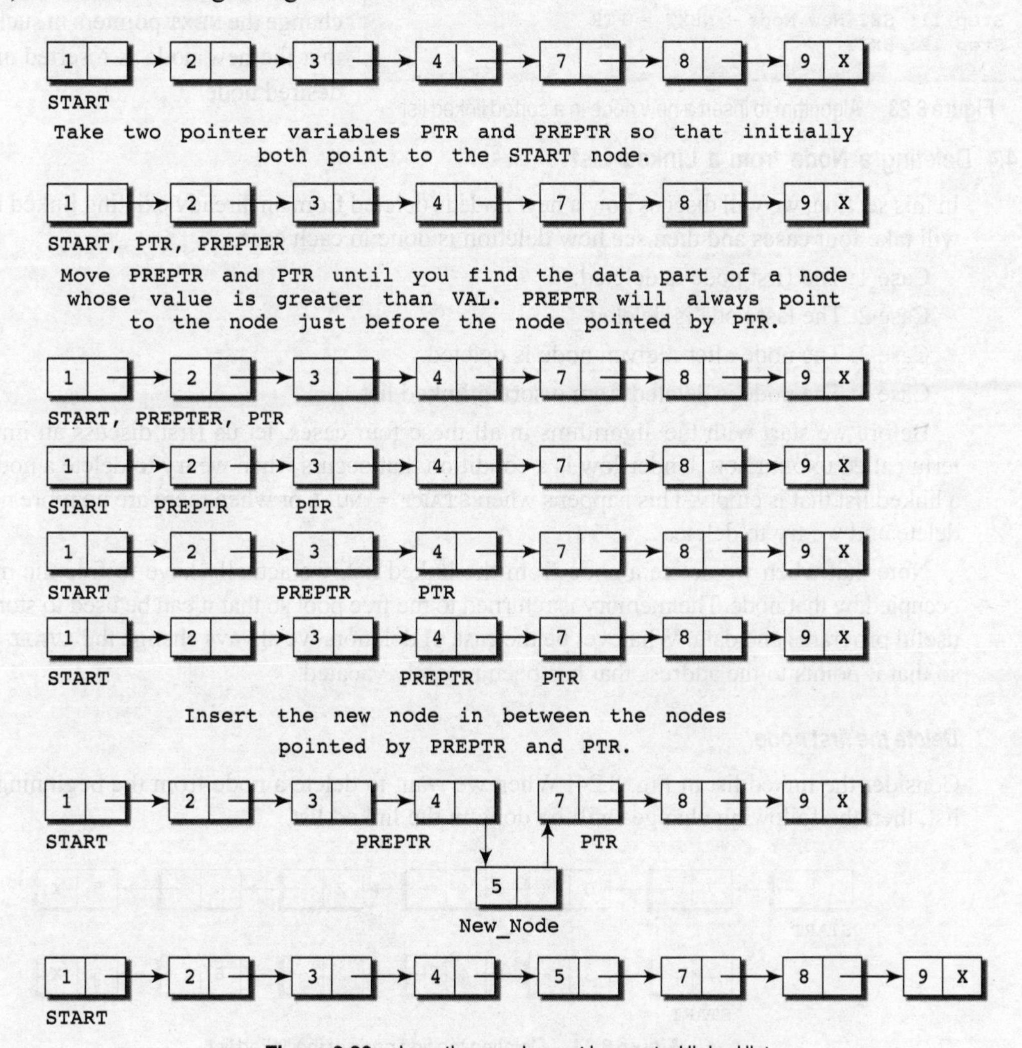

Figure 8.22 Inserting an element in a sorted linked list

Figure 8.23 shows the algorithm to insert a new node in a sorted linked list. In Step 5, we take a pointer variable PTR and initialize it with START. That is, PTR now points to the first node of the

```
Step 1: IF AVAIL = NULL, then
            Write OVERFLOW
            Go to Step 12
        [END OF IF]
Step 2: SET New_Node = AVAIL
Step 3: SET AVAIL = AVAIL -> NEXT
Step 4: SET New_Node -> DATA = VAL
Step 5: SET PTR = START
Step 6: SET PREPTR = PTR
Step 7: Repeat Step 8 and 9 while PTR -> DATA < VAL
Step 8:            SET PREPTR = PTR
Step 9:            SET PTR = PTR -> NEXT
        [END OF LOOP]
Step 10: SET PREPTR -> NEXT = New_Node
Step 11: SET New_Node -> NEXT = PTR
Step 12: EXIT
```

Figure 8.23 Algorithm to insert a new node in a sorted linked list

linked list. Then, we take another pointer variable `PREPTR` and initialize it with `PTR`. So now, `PREPTR`, `PTR`, and `Start` are all pointing to the first node of the linked list.

In the `while` loop, we traverse through the linked list to reach the node that has its value equal to `NUM`. We need to reach this node because the new node will be inserted before this node. Once we reach this node, in Step 10, we change the `NEXT` pointers in such a way that the new node is inserted after the desired node.

8.4.4 Deleting a Node from a Linked List

In this section, we will discuss how a new node is deleted from an already existing linked list. We will take four cases and then see how deletion is done in each case.

Case 1: The first node is deleted.

Case 2: The last node is deleted.

Case 3: The node after a given node is deleted.

Case 4: The node is deleted from a sorted linked list.

Before we start with the algorithms in all these four cases, let us first discuss an important term called `UNDERFLOW`. Underflow is a condition that occurs when we try to delete a node from a linked list that is empty. This happens when `START = NULL` or when there are no more nodes to delete and we try to delete.

Note that when we delete a node from the linked list, we actually have to free the memory occupied by that node. The memory is returned to the free pool so that it can be used to store other useful programs and data. Whatever be the case of deletion, we always change the `AVAIL` pointer so that it points to the address that has been recently vacated.

Delete the first node

Consider the linked list in Fig. 8.24. When we want to delete a node from the beginning of the list, then the following changes will be done in the linked list.

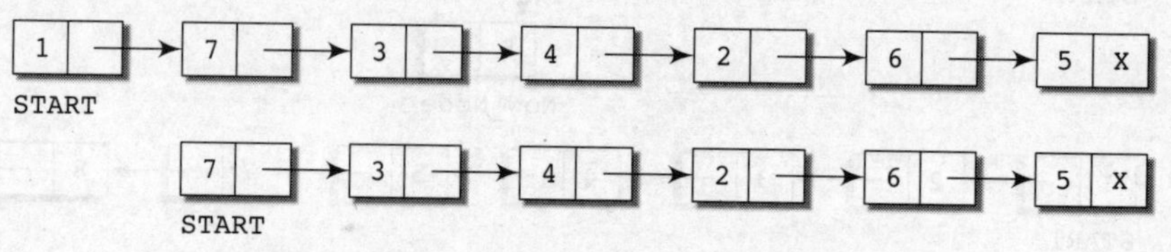

Figure 8.24 Deleting the first node of the linked list

Figure 8.25 shows the algorithm to delete the first node from a linked list. In Step 1, we check if the linked list exists or not. If `START = NULL`, then it signifies that there are no nodes in the list and the control is transferred to the last statement of the algorithm.

```
Step 1: IF START = NULL, then
            Write UNDERFLOW
            Go to Step 5
        [END OF IF]
Step 2: SET PTR = START
Step 3: SET START = START -> NEXT
Step 4: FREE PTR
Step 5: EXIT
```

Figure 8.25 Algorithm to delete the first node from a linked list

However, if there are nodes in the linked list, then we use a pointer variable that is set to point to the first node of the list. For this, we initiate `PTR` with `START` that stores the address of the first node of the list. In Step 3, `START` is made to point to the next node in sequence and finally the memory occupied by the node pointed by `PTR` (initially the first node of the list) is freed and returned to the free pool.

Delete the last node

Consider the linked list shown in Fig. 8.26. Suppose we want to delete the last node from the linked list, then the following changes will be done in the linked list.

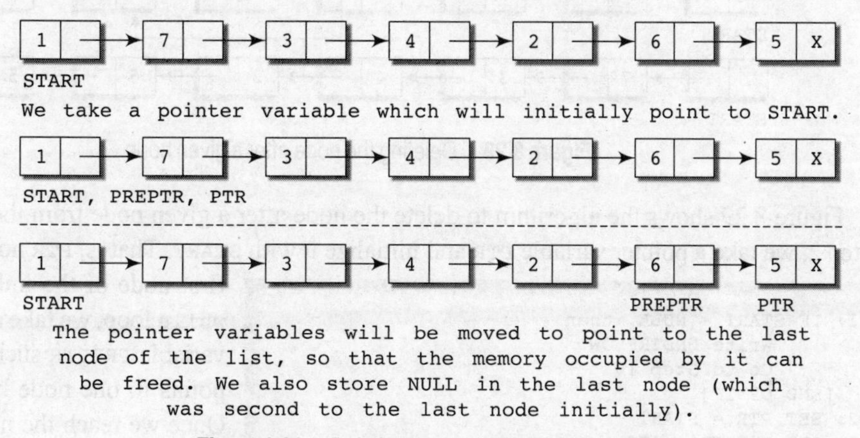

Figure 8.26 Deleting the last node of the linked list

Figure 8.27 shows the algorithm to delete the first node from the linked list. In Step 2, we take a pointer variable `PTR` and initialize it with `START`. That is, `PTR` now points to the first node of the linked list. In the `while` loop, we take another pointer variable `PREPTR` such that it always points to one node before the `PTR`. Once we reach the last node and the second last node, we set the next pointer of the second last node to `NULL`, so that it now becomes the (new) last node of the linked list. The memory of the previous last node is freed and returned back to the free pool.

```
Step 1: IF START = NULL, then
            Write UNDERFLOW
            Go to Step 8
        [END OF IF]
Step 2: SET PTR = START
Step 3: Repeat Step 4 and 5 while PTR -> NEXT != NULL
Step 4:             SET PREPTR = PTR
Step 5:             SET PTR = PTR -> NEXT
        [END OF LOOP]
Step 6: SET PREPTR -> NEXT = NULL
Step 7: FREE PTR
Step 8: EXIT
```

Figure 8.27 Algorithm to delete the last node from the linked list

Delete after a given node

Consider the linked list shown in Fig. 8.28. Suppose we want to delete the node that succeeds the node which contains the data value 4. Then, the following changes will be done in the linked list.

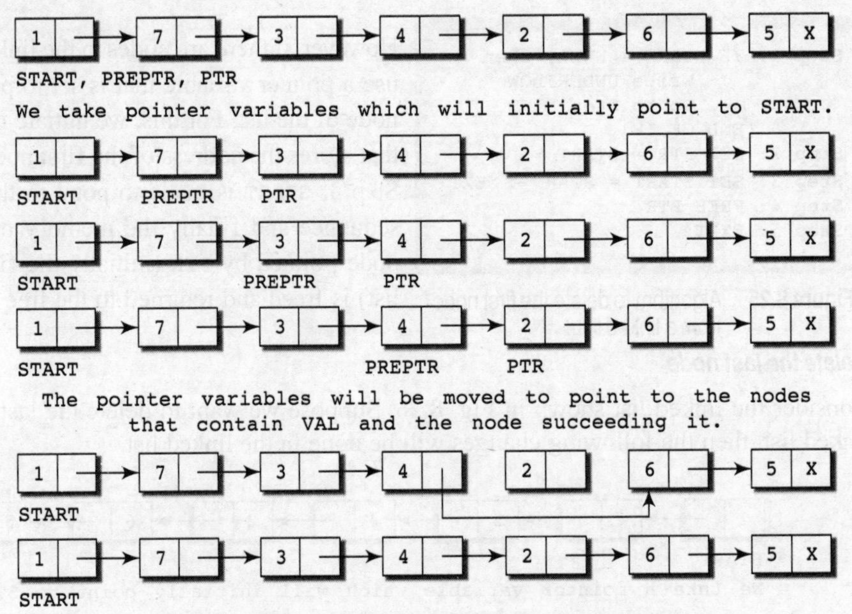

Figure 8.28 Deleting the node after a given node

Figure 8.29 shows the algorithm to delete the node after a given node from the linked list. In Step 2, we take a pointer variable PTR and initialize it with START. That is, PTR now points to the first node of the linked list. In the while loop, we take another pointer variable PREPTR such that it always points to one node before the PTR. Once we reach the node containing VAL and the node succeeding it, we set the next pointer of the node containing VAL to the address contained in next field of the node succeeding it. The memory of the node succeeding the given node is freed and returned back to the free pool.

```
Step 1: IF START = NULL, then
             Write UNDERFLOW
             Go to Step 10
        [END OF IF]
Step 2: SET PTR = START
Step 3: SET PREPTR = PTR
Step 4: Repeat Step 5 and 6 while PRETR->DATA != NUM
Step 5:             SET PREPTR = PTR
Step 6:             SET PTR = PTR->NEXT
        [END OF LOOP]
Step 7: SET TEMP = PTR->NEXT
Step 8: SET PREPTR->NEXT = TEMP->NEXT
Step 9: FREE TEMP
Step 10: EXIT
```

Figure 8.29 Algorithm to delete the node after a given node

Delete from a sorted linked list

Consider the linked list shown in Fig. 8.30. Suppose we want to delete a node with the value 4. Before discussing the changes that will be done in the linked list, let us first look at the algorithm.

Figure 8.31 shows the algorithm to delete the node from a sorted linked list. In Step 2, we take a pointer variable PTR and initialize it with START. That is, PTR now points to the first node of the linked list. In the while loop, we take another pointer variable PREPTR such that PREPTR always points to one node before the PTR. Once we reach the node containing VAL and the node preceding it, we set the next pointer of PREPTR so that it now stores the address of the node succeeding PTR. The memory of the node pointed by PTR is freed and returned to the free pool.

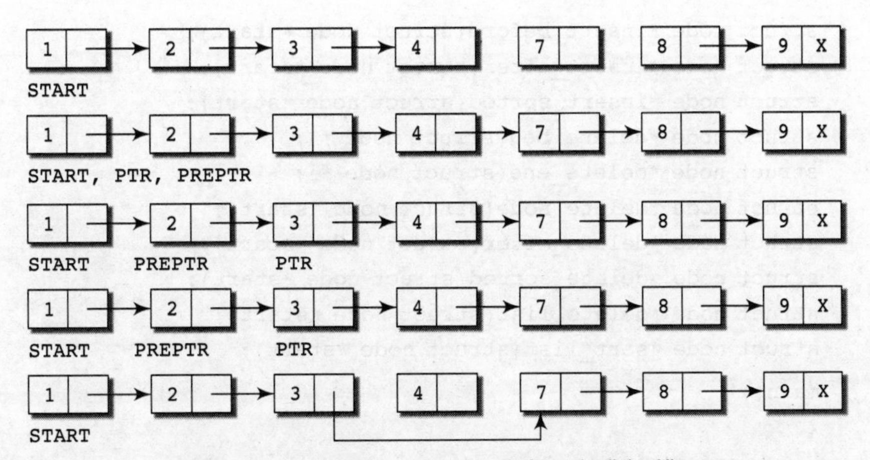

Figure 8.30 Deleting a given node from the linked list

```
Step 1: IF START = NULL, then
             Write UNDERFLOW
             Go to Step 9
        [END OF IF]
Step 2: SET PTR = START
Step 3: SET PREPTR = PTR
Step 4: Repeat Step 5 and 6 while PTR->DATA != NUM
Step 5:           SET PREPTR = PTR
Step 6:           SET PTR = PTR->NEXT
        [END OF LOOP]
Step 6: PREPTR->NEXT = PTR->NEXT
Step 7: FREE PTR
Step 8: EXIT
```

Figure 8.31 Algorithm to delete the node from a sorted linked list

PROGRAMMING EXAMPLE

1. Write a program to create a linked list and perform insertions and deletions of all cases. Write functions to sort and finally delete the entire list at once.

```
#include<stdio.h>
#include<conio.h>
struct node
{
        int data;
        struct node *next;
};
struct node *start = NULL;
struct node *create_ll(struct node *);
struct node *display(struct node *);
struct node *insert_beg(struct node *);
struct node *insert_end(struct node *);
```

```
struct node *insert_before(struct node *start);
struct node *insert_after(struct node *start);
struct node *insert_sorted(struct node *start);
struct node *delete_beg(struct node *);
struct node *delete_end(struct node *);
struct node *delete_node(struct node *start);
struct node *delete_after(struct node *start);
struct node *delete_sorted(struct node *start);
struct node *delete_list(struct node *start);
struct node *sort_list(struct node *start);
main()
{
      int option;
      clrscr();
      do
      {
            printf("\n\n *****MAIN MENU *****");
            printf("\n 1: Create a List");
            printf("\n 2: Display the list");
            printf("\n 3: Add a node in the beginning");
            printf("\n 4: Add a node at the end");
            printf("\n 5: Add a node before a given node");
            printf("\n 6: Add a node after a given node");
            printf("\n 7. Add a node in a sorted linked list");
            printf("\n 8: Delete a node from the beginning");
            printf("\n 9: Delete a node from the end");
            printf("\n 10: Delete a node a given node");
            printf("\n 11: Delete a node before a given node");
            printf("\n 12. Delete a node from a sorted linked list");
            printf("\n 13: Delete the entire list");
            printf("\n 14: Sort the list");
            printf("\n 15: EXIT");
            printf("\n ****************************");
            printf("\n\n ENter your option : ");
            scanf("%d", &option);
            switch(option)
            {
                  case 1:
                        start = create_ll(start);
                        printf("\n LINKED LIST CREATED");
                        break;
                  case 2:
                        start = display(start);
                        break;
```

```
                case 3:
                    start = insert_beg(start);
                    break;
                case 4:
                    start = insert_end(start);
                    break;
                case 5:
                    start = insert_before(start);
                    break;
                case 6:
                    start = insert_after(start);
                    break;
                case 7:
                    start = insert_sorted(start);
                    break;
                case 8:
                    start = delete_beg(start);
                    break;
                case 9:
                    start = delete_end(start);
                    break;
                case 10:
                    start = delete_node(start);
                    break;
                case 11:
                    start = delete_after(start);
                    break;
                case 12:
                    start = delete_sorted(start);
                    break;
                case 13:
                    start = delete_list(start);
                    printf("\n List is EMPTY");
                    break;
                case 14:
                    start = sort_list(start);
                    break;
            }
    }while(option !=15);
    getch();
    return 0;
}
struct node *create_ll(struct node *start)
{
    struct node *new_node;
    int num;
```

```
        printf("\n Enter -1 to end");
        printf("\n Enter the data : ");
        scanf("%d", &num);
        while(num!=-1)
        {
            new_node = (struct node*)malloc(sizeof(struct node*));
            new_node->data=num;
            if(start==NULL)
            {
                new_node->next = NULL;
                start = new_node;
            }
            else
            {
                new_node->next = start;
                start = new_node;
            }
            printf("\n Enter the data : ");
            scanf("%d", &num);
        }
        return start;
    }
    struct node *display(struct node *start)
    {
        struct node *ptr;
        ptr = start;
        printf("\n");
        while(ptr != NULL)
        {
            printf("\t %d", ptr->data);
            ptr = ptr->next;
        }
        return start;
    }
    struct node *insert_beg(struct node *start)
    {
        struct node *new_node;
        int num;
        printf("\n Enter the data : ");
        scanf("%d", &num);
        new_node = (struct node *)malloc(sizeof(struct node *));
        new_node->data = num;
        new_node->next = start;
        start = new_node;
        return start;
    }
```

```
struct node *insert_end(struct node *start)
{
      struct node *ptr, *new_node;
      int num;
      printf("\n Enter the data : ");
      scanf("%d", &num);
      new_node = (struct node *)malloc(sizeof(struct node *));
      new_node->data = num;
      ptr = start;
      while(ptr->next != NULL)
      ptr = ptr->next;
      ptr->next = new_node;
      new_node->next = NULL;
      return start;
}
struct node *insert_before(struct node *start)
{
      struct node *new_node, *ptr, *preptr;
      int num, val;
      printf("\n Enter the data : ");
      scanf("%d", &num);
      printf("\n Enter the value before which the data has to be inserted : ");
      scanf("%d", &val);
      new_node = (struct node *)malloc(sizeof(struct node *));
      new_node->data = num;
      ptr = start;
      while(ptr->data != val)
      {
            preptr = ptr;
            ptr = ptr->next;
      }
      new_node->next = ptr;
      preptr->next = new_node;
      return start;
}
struct node *insert_after(struct node *start)
{
      struct node *new_node, *ptr, *preptr;
      int num, val;
      printf("\n Enter the data : ");
      scanf("%d", &num);
      printf("\n Enter the value after which the data has to be inserted : ");
      scanf("%d", &val);
      new_node = (struct node *)malloc(sizeof(struct node *));
      new_node->data = num;
      ptr = start;
```

```
        while(preptr->data != val)
        {
            preptr = ptr;
            ptr = ptr->next;
        }
        new_node->next = ptr;
        preptr->next=new_node;
        return start;
    }
    struct node *insert_sorted(struct node *start)
    {
        struct node *new_node, *ptr, *preptr;
        int num;
        printf("\n Enter the data : ");
        scanf("%d", &num);
        new_node = (struct node *)malloc(sizeof(struct node *));
        new_node->data = num;
        ptr = start;
        while(ptr->data < num)
        {
            preptr = ptr;
            ptr = ptr->next;
            if(ptr == NULL)
                break;
        }
        if(ptr == NULL)
        {
            preptr->next = new_node;
            new_node->next = NULL;
        }
        else
        {
            new_node->next = ptr;
            preptr->next = new_node;
        }
        return start;
    }
    struct node *delete_beg(struct node *start)
    {
        struct node *ptr;
        ptr = start;
        start = start->next;
        free(ptr);
        return start;
    }
    struct node *delete_end(struct node *start)
```

```
{
        struct node *ptr, *preptr;
        ptr = start;
        while(ptr->next != NULL)
        {
                preptr = ptr;
                ptr = ptr->next;
        }
        preptr->next = NULL;
        free(ptr);
        return start;
}
struct node *delete_node(struct node *start)
{
        struct node *ptr, *preptr;
        int val;
        printf("\n Enter the value of the node which has to be deleted : ");
        scanf("%d", &val);
        ptr = start;
        if(ptr->data == val)
        {
                start = delete_beg(start);
                return start;
        }
        else
        {
                while(ptr->data != val)
                {
                        preptr = ptr;
                        ptr = ptr->next;
                }
                preptr->next = ptr->next;
                free(ptr);
                return start;
        }
}
struct node *delete_after(struct node *start)
{
        struct node *ptr, *preptr;
        int val;
        printf("\n Enter the value after which the node has to deleted : ");
        scanf("%d", &val);
        ptr = start;
        while(preptr->data != val)
        {
                preptr = ptr;
```

```
            ptr = ptr -> next;
        }
        preptr -> next=ptr -> next;
        free(ptr);
        return start;
    }
    struct node *delete_sorted(struct node *start)
    {
        struct node *ptr, *preptr;
        int val;
        printf("\n Enter the value of the node which has to be deleted : ");
        scanf("%d", &val);
        ptr = start;
        while(ptr -> data != val)
        {
            preptr = ptr;
            ptr = ptr -> next;
        }
        preptr -> next = ptr -> next;
        free(ptr);
        return start;
    }
    struct node *delete_list(struct node *start)
    {
        struct node *ptr;
        ptr=start;
        while(ptr -> next != NULL)
        {
            printf("\n %d is to be deleted next", ptr -> data);
            start = delete_beg(ptr);
            ptr = ptr -> next;
        }
        return start;
    }
    struct node *sort_list(struct node *start)
    {
        struct node *ptr1, *ptr2;
        int temp;
        ptr1 = start;
        while(ptr1 -> next != NULL)
        {
            ptr2 = ptr1 -> next;
            while(ptr2 != NULL)
            {
                if(ptr1 -> data > ptr2 -> data)
                {
```

```
                        temp = ptr1->data;
                        ptr1->data = ptr2->data;
                        ptr2->data = temp;
                    }
                    ptr2 = ptr2->next;
                }
                ptr1 = ptr1->next;
            }
            return start;
        }
```

8.5 POLYNOMIAL REPRESENTATION

Let us see how a polynomial is represented in the memory using linked list. A polynomial is given as $6x^3 + 9x^2 + 7x + 1$. Every individual term in a polynomial consists of two parts, a coefficient and a power. Here, 6, 9, 7, and 1 are the coefficients of the terms that have 3, 2, 1, and 0 as their powers respectively.

Every term of a polynomial can be represented as a node of the linked list. Figure 8.32 shows the linked representation of the terms of the above polynomial.

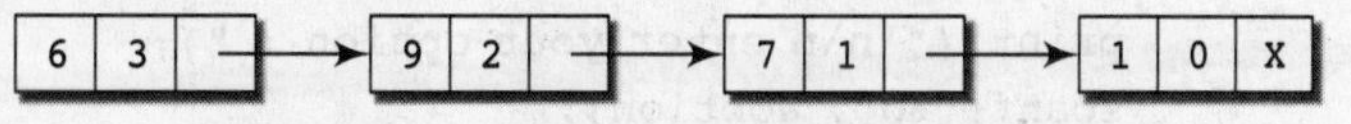

Figure 8.32 Linked representation of a polynomial

Now that we know how polynomials are represented using nodes of a linked list, let us write a program to perform operations on polynomials.

PROGRAMMING EXAMPLE

2. Write a program to store a polynomial using linked list. Also, perform addition and subtraction on two polynomials.

```
#include<stdio.h>
#include<conio.h>
struct node
{
        int num;
        int coeff;
        struct node *next;
};
struct node *start1 = NULL;
struct node *start2 = NULL;
struct node *start3 = NULL;
struct node *start4 = NULL;
struct node *last3 = NULL;
```

```
struct node *create_poly(struct node *);
struct node *display_poly(struct node *);
struct node *add_poly(struct node *, struct node *, struct node *);
struct node *sub_poly(struct node *, struct node *, struct node *);
struct node *add_node(struct node *, int, int);
main()
{
    int option;
    clrscr();
    do
    {
        printf("\n**** MAIN MENU *******");
        printf("\n 1. Enter the first polynomial");
        printf("\n 2. Display the first polynomial");
        printf("\n 3. Enter the second polynomial");
        printf("\n 4. Display the second polynomial");
        printf("\n 5. Add the polynomials");
        printf("\n 6. Display the result");
        printf("\n 7. Subtract the polynomials");
        printf("\n 8. Display the result");
        printf("\n 7. EXIT");
        printf("\n *******************************");
        printf("\n\n enter your option : ");
        scanf("%d", &option);
        switch(option)
        {
            case 1:
                start1 = create_poly(start1);
                break;
            case 2:
                start1 = display_poly(start1);
                break;
            case 3:
                start2 = create_poly(start2);
                break;
            case 4:
                start2 = display_poly(start2);
                break;
            case 5:
                start3 = add_poly(start1, start2, start3);
                break;
            case 6:
                start3 = display_poly(start3);
                break;
            case 7:
                start4 = sub_poly(start1, start2, start4);
                break;
```

```
                case 8:
                        start4 = display_poly(start4);
                        break;
                }
        }while(option!=9);
        getch();
        return 0;
}
struct node *create_poly(struct node *start)
{
        struct node *new_node, *ptr;
        int n, c;
        printf("\n Enter the number : ");
        scanf("%d", &n);
        printf("\n Enter its coefficient : ");
        scanf("%d", &c);
        while(n != -1)
        {
                if(start==NULL)
                {
                        start = (struct node *)malloc(sizeof(struct node *));
                        start -> num = n;
                        start -> coeff = c;
                        start -> next = NULL;
                }
                else
                {
                        ptr = start;
                        while(ptr -> next != NULL)
                        ptr = ptr -> next;
                        new_node = (struct node *)malloc(sizeof(struct node *));
                        new_node -> num = n;
                        new_node -> coeff = c;
                        new_node -> next = NULL;
                        ptr -> next = new_node;
                }
                printf("\n Enter the number : ");
                scanf("%d", &n);
                if(n == -1)
                        break;
                printf("\n Enter its coefficient : ");
                scanf("%d", &c);
        }
        return start;
}
struct node *display_poly(struct node *start)
```

```
{
        struct node *ptr;
        ptr = start;
        printf("\n");
        while(ptr != NULL)
        {
                printf("\n%d x%d\t", ptr->num, ptr->coeff);
                ptr = ptr->next;
        }
        return start;
}
struct node *add_poly(struct node *start1, struct node *start2, struct node
*start3)
{
        struct node *ptr1, *ptr2;
        int sum_num, c;
        ptr1 = start1, ptr2 = start2;
        while(ptr1->next != NULL || ptr2->next != NULL)
        {
                if(ptr1->coeff == ptr2->coeff)
                {
                        sum_num = ptr1->num + ptr2->num;
                        start3 = add_node(start3, sum_num, ptr1->coeff);
                        ptr1 = ptr1->next;
                        ptr2 = ptr2->next;
                }
                else if(ptr1->coeff > ptr2->coeff)
                {
                        start3 = add_node(start3, ptr1->num, ptr1->coeff);
                        ptr1 = ptr1->next;
                }
                else if(ptr1->coeff<ptr2->coeff)
                {
                        start3 = add_node(start3, ptr2->num, ptr2->coeff);
                        ptr2 = ptr2->next;
                }
        }
        if(ptr1->next == NULL)
        {
                while(ptr2->next != NULL)
                {
                        start3 = add_node(start3, ptr2->num, ptr2->coeff);
                        ptr2 = ptr2->next;
                }
        }
        if(ptr2->next == NULL)
        {
```

```
            while(ptr1->next != NULL)
            {
                  start3 = add_node(start3, ptr1->num, ptr1->coeff);
                  ptr1 = ptr1->next;
            }
      }
      last3->next = NULL;
      return start3;
}
struct node *sub_poly(struct node *start1, struct node *start2, struct node
*start4)
{
      struct node *ptr1, *ptr2;
      int sub_num, c;
      ptr1 = start1, ptr2 = start2;
      do
      {
            if(ptr1->coeff == ptr2->coeff)
            {
                  sub_num = ptr1->num - ptr2->num;
                  start4 = add_node(start4, sub_num, ptr1->coeff);
                  ptr1 = ptr1->next;
                  ptr2 = ptr2->next;
            }
            else if(ptr1->coeff > ptr2->coeff)
            {
                  start4 = add_node(start4, ptr1->num, ptr1->coeff);
                  ptr1 = ptr1->next;
            }
            else if(ptr1->coeff < ptr2->coeff)
            {
                  start4 = add_node(start4, ptr2->num, ptr2->coeff);
                  ptr2 = ptr2->next;
            }
      }while(ptr1 != NULL || ptr2 != NULL);
      if(ptr1 == NULL)
      {
            while(ptr2 != NULL)
            {
                  start4 = add_node(start4, ptr2->num, ptr2->coeff);
                  ptr2 = ptr2->next;
            }
      }
      if(ptr2 == NULL)
      {
            while(ptr1 != NULL)
```

```
                {
                        start4 = add_node(start4, ptr1->num, ptr1->coeff);
                        ptr1 = ptr1->next;
                }
        }
        return start4;
}
struct node *add_node(struct node *start, int n, int c)
{
        struct node *ptr, *new_node;
        if(start == NULL)
        {
                start = (struct node *)malloc(sizeof(struct node *));
                start->num = n;
                start->coeff = c;
                start->next = NULL;
        }
        else
        {
                ptr = start;
                while(ptr->next != NULL)
                ptr = ptr->next;
                new_node = (struct node *)malloc(sizeof(struct node *));
                new_node->num = n;
                new_node->coeff = c;
                new_node->next = NULL;
                ptr->next = new_node;
        }
        return start;
}
```

8.6 CIRCULAR LINKED LIST

In a circular linked list, the last node contains a pointer to the first node of the list. We can have a circular singly listed list as well as a circular doubly linked list. While traversing a circular linked list, we can begin at any node and traverse the list in any direction, forward or backward, until we reach the same node where we started. Thus, a circular linked list has no beginning and no ending. Figure 8.33 shows a circular linked list.

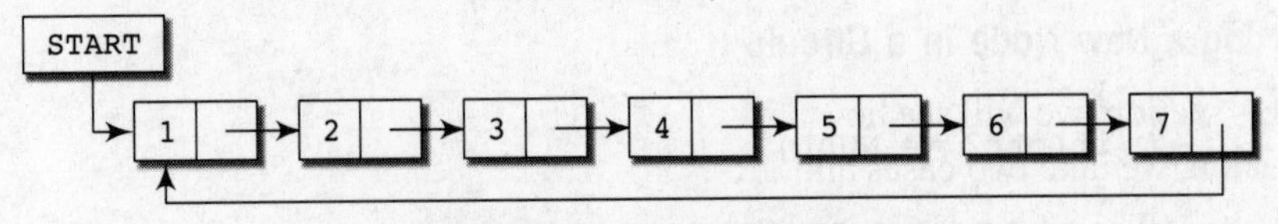

Figure 8.33 Circular linked list

The only downside of a circular linked list is the complexity of iteration. Note that there is no storing of NULL values in the list.

Circular linked lists are widely used in the operating systems for task maintenance. Take another example where a circular linked list is used. When we are surfing the Net, we can use the Back button and the Forward button to move to the previous pages that we have already visited. How is this done? The answer is simple. A circular linked list is used to maintain the sequence of the Web pages visited. Traversing this circular linked list either in forward or backward direction helps to revisit the pages again using Back and Forward buttons. Actually, this is done using either the circular stack or the circular queue. We will read more about it in the next chapter.

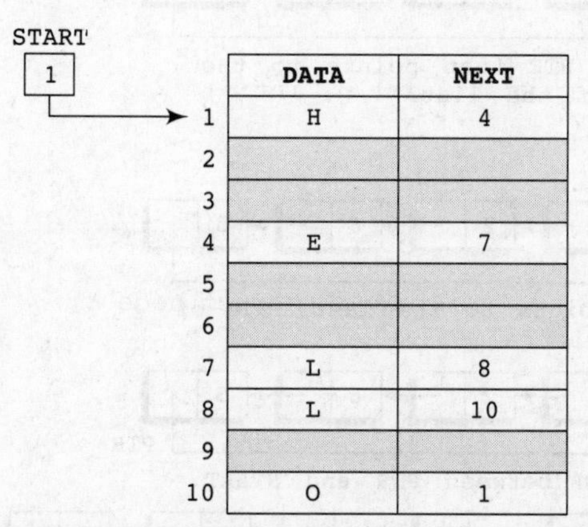

Figure 8.34 Memory representation of a circular linked list

Let us view how a linked list is maintained in the memory. In order to form a linked list, we need a structure called the node that has two fields, DATA and NEXT. DATA will store the information part and NEXT will store the address of the node in sequence. Consider Fig. 8.34.

We traverse the list until we find a the NEXT entry that contains the address of the first node of the list. This denotes the end of the linked list, that is, the node that contains the address of the first node is actually the last node of the list. When we traverse the DATA and NEXT in this manner, we will finally see that the linked list in the above example stores characters that when put together forms the word HELLO.

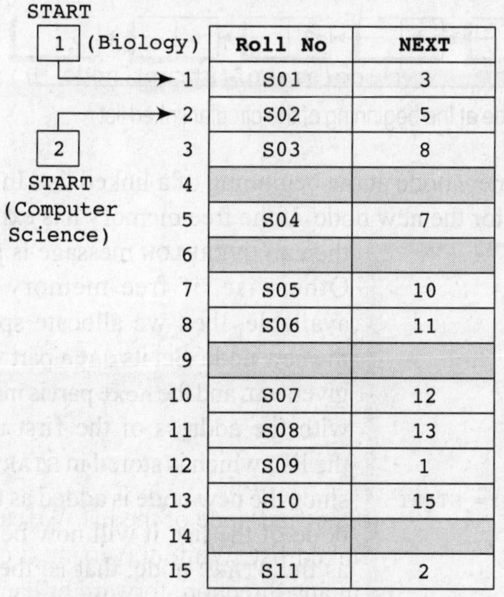

Figure 8.35 Memory representation of two circular linked lists stored in the memory simultaneously

Now, look at Fig. 8.35. Two different linked lists are simultaneously maintained in the memory. There is no ambiguity in traversing through the list because each list maintains a separate Start pointer which gives the address of the first node of the respective linked list. The remaining nodes are reached by looking at the value stored in the NEXT.

By looking at the figure, we can conclude that the roll numbers of the students who have opted for Biology are S01, S03, S06, S08, S10, and S11. Similarly, the roll numbers of the students who chose Computer Science are S02, S04, S05, S07, and S09.

8.6.1 Inserting a New Node in a Circular Linked List

In this section, we will see how a new node is added into an already existing linked list. To do the insertion, we take two cases and then see how insertion is done in each case.

Case 1: The new node is inserted at the beginning of the circular linked list.

Case 2: The new node is inserted at the end of the circular linked list.

Insert at the beginning

Consider the linked list shown in Fig. 8.36. Suppose we want to add a new node with data 9 and add it as the first node of the list. Then, the flowing changes will be done in the linked list.

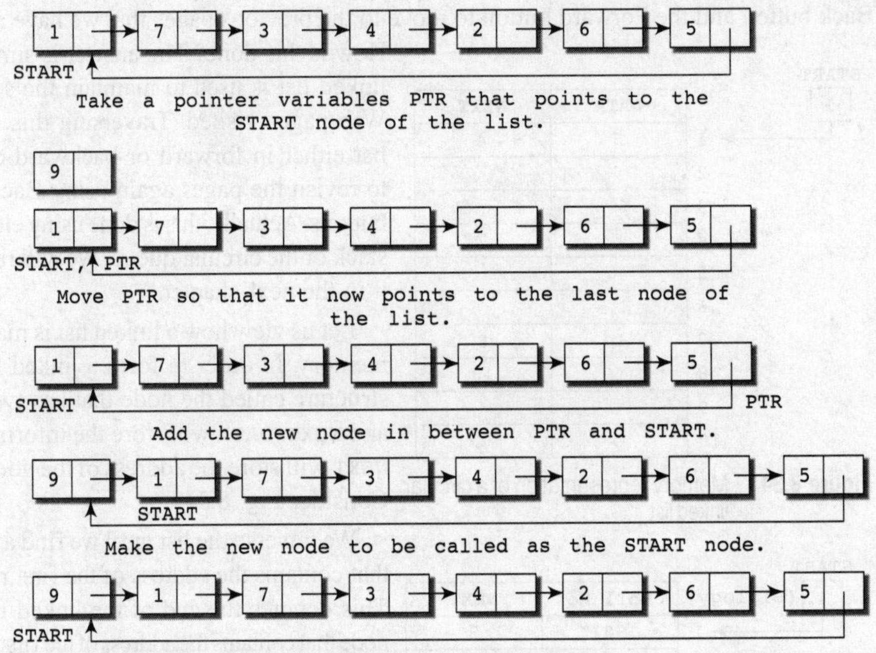

Figure 8.36 Inserting a new node at the beginning of the circular linked list

Figure 8.37 shows the algorithm to insert a new node at the beginning of a linked list. In Step 1, we first check whether memory is available for the new node. If the free memory has exhausted, then an `OVERFLOW` message is printed. Otherwise, if free memory cell is available, then we allocate space for the new node. Set its `data` part with the given `VAL` and the `next` part is initialized with the address of the first node of the list, which is stored in `START`. Now, since the new node is added as the first node of the list, it will now be known as the `START` node, that is, the `START` pointer variable will now hold the address of the `New_Node`.

```
Step 1: IF AVAIL = NULL, then
            Write OVERFLOW
            Go to Step 7
        [END OF IF]
Step 2: SET New_Node = AVAIL
Step 3: SET AVAIL = AVAIL->NEXT
Step 4: SET New_Node->DATA = VAL
Step 5: SET PTR = START
Step 6:     Repeat Step 7 while PTR->NEXT != START
Step 7:         PTR = PTR->NEXT
Step 8: SET New_Node->Next = START
Step 9: SET PTR->NEXT = New_Node
Step 10: SET START = New_Node
Step 11: EXIT
```

Figure 8.37 Algorithm to insert a new node at the beginning

While inserting a node in a circular linked list, we have to use a `while` loop to traverse to the last node of the list. Because the last node contains a pointer to the start, its next field should also be updated so that after insertion, it now points to the new node which will be now known as the `START`. Note the two steps.

```
Step 2: SET New_Node = AVAIL
Step 3: SET AVAIL = AVAIL->NEXT
```

These steps allocate memory for the new node. In C, there are functions like `malloc()`, `alloc`, and `calloc()` which do all this automatically on behalf of the user.

Insert at the end

Consider the linked list shown in Fig. 8.38. Suppose we want to add a new node with the data 9 as the last node of the list. Then, the following changes will be done in the linked list.

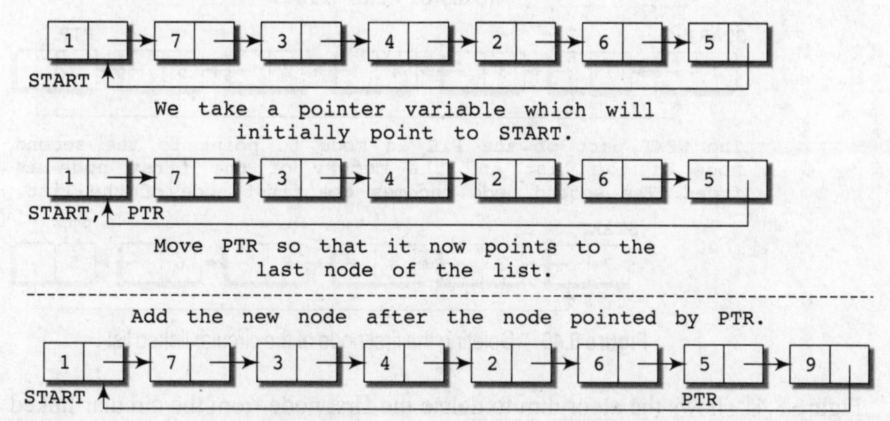

Figure 8.38 Inserting a new node at the end

```
Step 1: IF AVAIL = NULL, then
             Write OVERFLOW
             Go to Step 7
        [END OF IF]
Step 2: SET New_Node = AVAIL
Step 3: SET AVAIL = AVAIL -> NEXT
Step 4: SET New_Node -> DATA = VAL
Step 5: SET New_Node -> Next = START
Step 6: SET PTR = START
Step 7: Repeat Step 8 while PTR -> NEXT != START
Step 8:             SET PTR = PTR -> NEXT
        [END OF LOOP]
Step 9: SET PTR -> NEXT = New_Node
Step 10: EXIT
```

Figure 8.39 Algorithm to insert a new node at the end

Figure 8.39 shows the algorithm to insert a new node at the end of a circular linked list. In Step 6, we take a pointer variable `PTR` and initialize it with `START`. That is, `PTR` now points to the first node of the linked list. In the `while` loop, we traverse through the linked list to reach the last node. Once we reach the last node, in Step 9, we change the `NEXT` pointer of the last node to store the address of the new node. Remember that the `NEXT` field of the new node contains the address of the first node which is denoted by `START`.

8.6.2 Deleting a Node from a Circular Linked List

In this section, we will discuss how a new node is deleted from an already existing circular linked list. To do the deletion, we will take two cases and then see how deletion is done in each case. Rest of the cases of deletion are same as that given for singly linked list.

Case 1: The first node is deleted.

Case 2: The last node is deleted.

Delete the first node

Consider the circular linked list shown in Fig. 8.40. When we want to delete a node from the beginning of the list, then the flowing changes will be done in the linked list.

Take a variable PTR and make it point to the START node of the list.

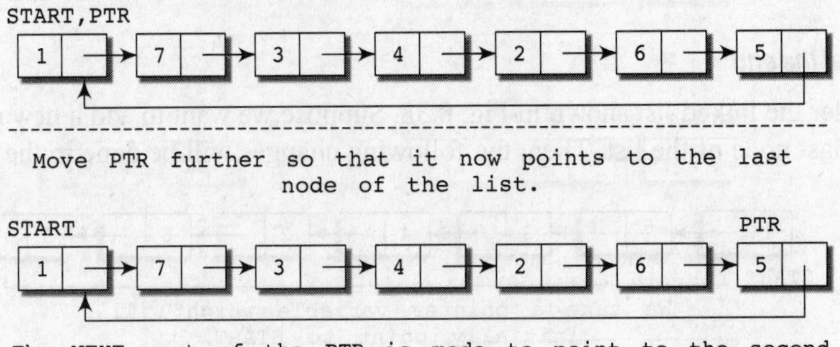

The NEXT part of the PTR is mode to point to the second node of the list and the memory of the first node is freed. The second node becomes the first node of the list.

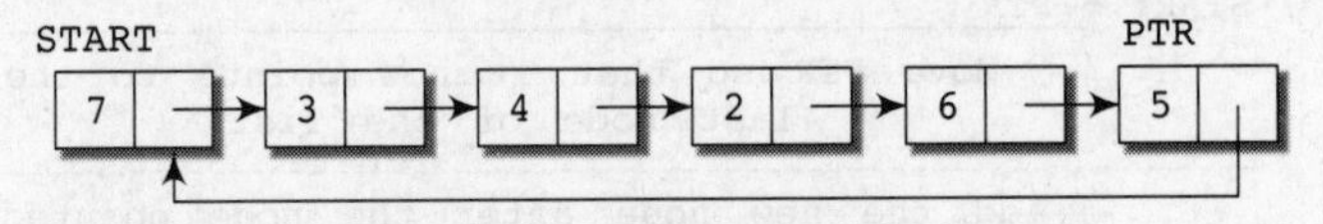

Figure 8.40 Deleting the first node of the circular linked list

Figure 8.41 shows the algorithm to delete the first node from the circular linked list. In Step 1 of the algorithm, we check if the linked list exists or not. If `START = NULL`, then it signifies that there are no nodes in the list and the control is transferred to the last statement of the algorithm.

```
Step 1: IF START = NULL, then
                Write UNDERFLOW
                Go to Step 8
        [END OF IF]
Step 2: SET PTR = START
Step 3: Repeat Step 4 while PTR->NEXT != START
Step 4:             SET PTR = PTR->NEXT
        [END OF IF]
Step 5: SET PTR->NEXT = START->NEXT
Step 6: FREE START
Step 7: SET START = PTR->NEXT
Step 8: EXIT
```

Figure 8.41 Algorithm to delete the first node from the circular linked list

However, if there are nodes in the linked list, then we use a pointer variable `PTR` which will be used to traverse the list to ultimately reach the last node. In Step 5, we change the next pointer of the last node to point to the second node of the circular linked list. In Step 6, the memory occupied by the first node is freed. Finally, in Step 7, the second node now becomes the first node of the list and its address is stored in the pointer variable `START`.

Delete the last node

Consider the circular linked list shown in Fig. 8.42. Suppose we want to delete the last node from the linked list, then the flowing changes will be done in the linked list.

Figure 8.43 shows the algorithm to delete the last node from the circular linked list. In Step 2, we take a pointer variable `PTR` and initialize it with `START`. That is, `PTR` now points to the first node of the linked list. In the `while` loop, we take another pointer variable `PREPTR` such that `PREPTR` always points to one node before `PTR`. Once we reach the last node and the second last node, we set the next pointer of the second last node to `START`, so that it now becomes the (new) last node of the linked list. The memory of the previous last node is freed and returned to the free pool.

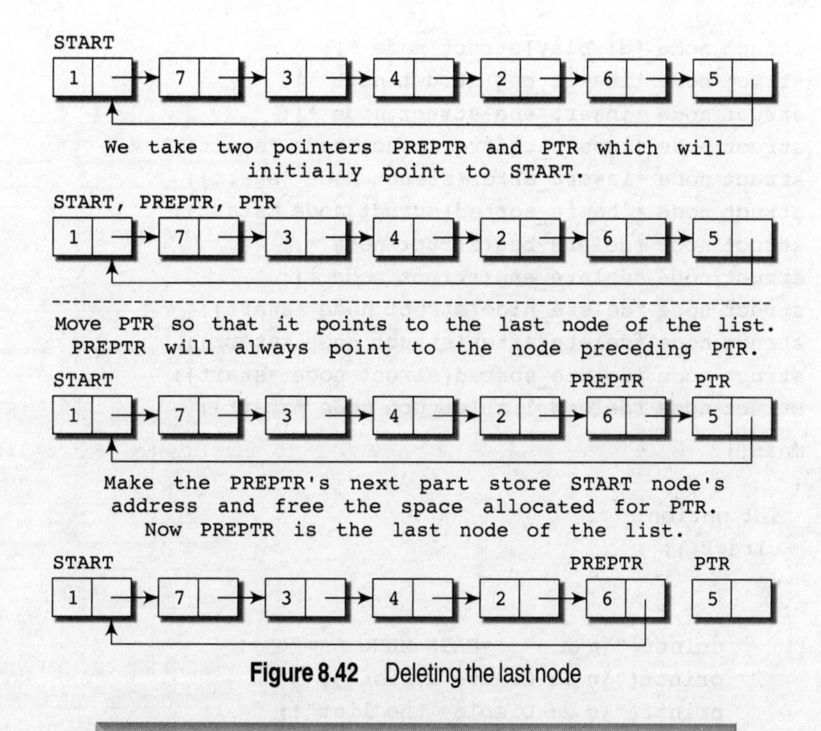

Figure 8.42 Deleting the last node

```
Step 1: IF START = NULL, then
              Write UNDERFLOW
              Go to Step 8
        [END OF IF]
Step 2: SET PTR = START
Step 3: Repeat Step 4  while PTR->NEXT != START
Step 4:                SET PREPTR = PTR
Step 5:                SET PTR = PTR->NEXT
        [END OF LOOP]
Step 6: SET PREPTR->NEXT = START
Step 7: FREE PTR
Step 8: EXIT
```

Figure 8.43 Algorithm to delete the last node

PROGRAMMING EXAMPLE

3. Write a program to create a circular linked list. Perform insertion and deletion in all cases.

```
#include<stdio.h>
#include<conio.h>
struct node
{
  int data;
  struct node *next;
};
struct node *start = NULL;
struct node *create_cll(struct node *);
```

```
struct node *display(struct node *);
struct node *insert_beg(struct node *);
struct node *insert_end(struct node *);
struct node *insert_before(struct node *start);
struct node *insert_after(struct node *start);
struct node *insert_sorted(struct node *start);
struct node *delete_beg(struct node *);
struct node *delete_end(struct node *);
struct node *delete_node(struct node *start);
struct node *delete_after(struct node *start);
struct node *delete_sorted(struct node *start);
struct node *delete_list(struct node *start);
main()
{
  int option;
  clrscr();
  do
  {
      printf("\n\n *****MAIN MENU *****");
      printf("\n 1: Create a List");
      printf("\n 2: Display the list");
      printf("\n 3: Add a node in the beginning");
      printf("\n 4: Add a node at the end");
      printf("\n 5: Add a node before a given node");
      printf("\n 6: Add a node after a given node");
      printf("\n 7. Add a node in a sorted linked list");
      printf("\n 8: Delete a node from the beginning");
      printf("\n 9: Delete a node from the end");
      printf("\n 10: Delete a given node");
      printf("\n 11: Delete a node after a given node");
      printf("\n 12. Delete a node from a sorted linked list");
      printf("\n 13: Delete the entire list");
      printf("\n 14: EXIT");
      printf("\n ***************************");
      printf("\n\n ENter your option : ");
      scanf("%d", &option);
      switch(option)
      {
              case 1:
                      start = create_cll(start);
                      printf("\n LINKED LIST CREATED");
                      break;
              case 2:
                      start = display(start);
                      break;
              case 3:
                      start = insert_beg(start);
                      break;
```

```
                case 4:
                        start = insert_end(start);
                        break;
                case 5:
                        start = insert_before(start);
                        break;
                case 6:
                        start = insert_after(start);
                        break;
                case 7:
                        start = insert_sorted(start);
                        break;
                case 8:
                        start = delete_beg(start);
                        break;
                case 9:
                        start = delete_end(start);
                        break;
                case 10:
                        start = delete_node(start);
                        break;
                case 11:
                        start = delete_after(start);
                        break;
                case 12:
                        start = delete_sorted(start);
                        break;
                case 13:
                        start = delete_list(start);
                        printf("\n List is EMPTY");
                        break;
        }
    }while(option !=14);
    getch();
    return 0;
}
struct node *create_cll(struct node *start)
{
        struct node *new_node, *ptr;
        int num;
        printf("\n Enter -1 to end");
        printf("\n Enter the data : ");
        scanf("%d", &num);
        while(num!=-1)
        {
                new_node = (struct node*)malloc(sizeof(struct node*));
                new_node->data = num;
                if(start == NULL)
```

```
        {
            new_node -> next = new_node;
            start = new_node;
        }
        else
        {   ptr = start;
            while(ptr -> next != start)
            ptr = ptr -> next;
            new_node -> next = start;
            ptr -> next = new_node;
        }
        printf("\n Enter the data : ");
        scanf("%d", &num);
    }
    return start;
}
struct node *display(struct node *start)
{
    struct node *ptr;
    ptr=start;
    printf("\n");
    while(ptr -> next != start)
    {
        printf("\t %d", ptr -> data);
        ptr = ptr -> next;
    }
    printf("\t %d", ptr -> data);
    return start;
}
struct node *insert_beg(struct node *start)
{
    struct node *new_node, *ptr;
    int num;
    printf("\n Enter the data : ");
    scanf("%d", &num);
    new_node = (struct node *)malloc(sizeof(struct node *));
    new_node -> data = num;
    ptr = start;
    while(ptr -> next != start)
        ptr = ptr -> next;
    ptr -> next = new_node;
    new_node -> next = start;
    start = new_node;
    return start;
}
struct node *insert_end(struct node *start)
{
```

```
        struct node *ptr, *new_node;
        int num;
        printf("\n Enter the data : ");
        scanf("%d", &num);
        new_node = (struct node *)malloc(sizeof(struct node *));
        new_node->data = num;
        ptr = start;
        while(ptr->next != start)
             ptr = ptr->next;
        ptr->next = new_node;
        new_node->next = start;
        return start;
}
struct node *insert_before(struct node *start)
{
        struct node *new_node, *ptr, *preptr;
        int num, val;
        printf("\n Enter the value before which the data has to be inserted : ");
        scanf("%d", &val);
        new_node = (struct node *)malloc(sizeof(struct node *));
        ptr=start;
        if(ptr->data == val)
        {
             printf("\n Enter the data : ");
             scanf("%d", &num);
             start = insert_beg(start);
        }
        else
        {
             new_node->data = num;
             while(ptr->data != val)
             {
                  preptr = ptr;
                  ptr = ptr->next;
             }
             new_node->next = ptr;
             preptr->next = new_node;
        }
        return start;
}
struct node *insert_after(struct node *start)
{
        struct node *new_node, *ptr, *preptr;
        int num, val;
        printf("\n Enter the data : ");
        scanf("%d", &num);
```

```
        printf("\n Enter the value after which the data has to be inserted : ");
        scanf("%d", &val);
        new_node = (struct node *)malloc(sizeof(struct node *));
        new_node -> data = num;
        ptr=start;
        while(preptr -> data != val)
        {
              preptr = ptr;
              ptr = ptr -> next;
        }
        new_node -> next = ptr;
        preptr -> next = new_node;
        return start;
}
struct node *insert_sorted(struct node *start)
{
        struct node *new_node, *ptr, *preptr;
        int num;
        printf("\n Enter the data : ");
        scanf("%d", &num);
        new_node = (struct node *)malloc(sizeof(struct node *));
        new_node -> data = num;
        ptr = start;
        while(ptr -> data<num)
        {
              preptr = ptr;
              ptr = ptr -> next;
              if(ptr == start)
                    break;
        }
        if(ptr == start)
        {
              preptr -> next = new_node;
              new_node -> next = start;
              start = new_node;
        }
        else
        {
              new_node -> next = ptr;
              preptr -> next = new_node;
        }
        return start;
}
struct node *delete_beg(struct node *start)
{
        struct node *ptr;
```

```
        ptr = start;
        while(ptr->next != start)
            ptr = ptr->next;
        ptr->next = start->next;
        free(start);
        start = ptr->next;
        return start;
}
struct node *delete_end(struct node *start)
{
        struct node *ptr, *preptr;
        ptr = start;
        while(ptr->next != start)
        {
            preptr = ptr;
            ptr = ptr->next;
        }
        preptr->next = ptr->next;
        free(ptr);
        return start;
}
struct node *delete_node(struct node *start)
{
        struct node *ptr, *preptr;
        int val;
        printf("\n Enter the value of the node which has to be deleted : ");
        scanf("%d", &val);
        ptr = start;
        if(ptr->data == val)
        {
            start = delete_beg(start);
            return start;
        }
        else
        {
            while(ptr->data!=val)
            {
                preptr = ptr;
                ptr = ptr->next;
            }
            preptr->next=ptr->next;
            free(ptr);
            return start;
        }
}
struct node *delete_after(struct node *start)
{
```

```
        struct node *ptr, *preptr;
        int val;
        printf("\n Enter the value after which the node has to deleted : ");
        scanf("%d", &val);
        ptr = start;
        while(preptr->data != val)
        {
              preptr = ptr;
              ptr = ptr->next;
        }
        preptr->next = ptr->next;
        if(ptr == start)
              start = preptr->next;
        free(ptr);
        return start;
}
struct node *delete_sorted(struct node *start)
{
        struct node *ptr, *preptr;
        int val;
        printf("\n Enter the value of the node which has to be deleted : ");
        scanf("%d", &val);
        ptr = start;
        if(ptr->data == val)
              start = delete_beg(start);
        else
        {
              while(ptr->data != val)
              {
                    preptr = ptr;
                    ptr = ptr->next;
              }
              preptr->next = ptr->next;
              free(ptr);
        }
        return start;
}
struct node *delete_list(struct node *start)
{
        struct node *ptr;
        ptr = start;
        while(ptr->next != start)
              start = delete_end(start);
        free(start);
        return start;
}
```

8.7 DOUBLY LINKED LIST

A doubly linked list or a two-way linked list is a more complex type of linked list which contains a pointer to the next as well as the previous node in the sequence. Therefore, it consists of three parts, and not just two. The three parts are data, a pointer to the next node, and a pointer to the previous node as shown in Fig. 8.44.

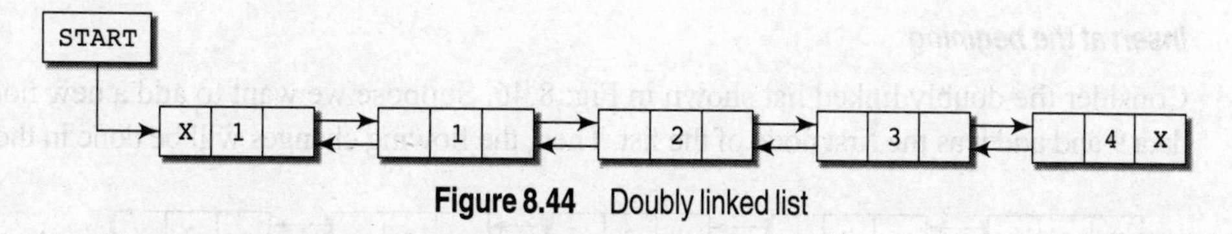

Figure 8.44 Doubly linked list

In C, the structure of a doubly linked list is given as,

```
struct node
{
    struct node *prev;
    int data;
    struct node *next;
};
```

The `prev` field of the first node and the `next` field of the last node will contain `NULL`. The `prev` field is used to store the address of the preceding node. This would enable to traverse the list in the backward direction as well.

START = 1

	DATA	PREV	Next
1	H	–1	3
2			
3	E	1	6
4			
5			
6	L	3	7
7	L	6	9
8			
9	O	7	–1

Figure 8.45 Memory representation of a doubly linked list

Thus, we see that a doubly linked list calls for more space per node and more expensive basic operations. However, a doubly linked list provides the ease to manipulate the elements of the list as it maintains pointers to nodes in both the directions (forward and backward). The main advantage of using a doubly linked list is that it makes searching twice as efficient. Let us view how a doubly linked list is maintained in the memory. Consider Fig. 8.45.

In the figure, we see that a variable `START` is used to store the address of the first node. Here in this example, `START = 1`, so the first data is stored at address 1, which is `H`. Since this is the first node, it has no previous node and hence stores `NULL` or –1 in the `prev` field. We will traverse the list until we reach a position where the `NEXT` entry contains –1 or `NULL`. This denotes the end of the linked list, that is, the node that contains the address of the first node is actually the last node of the list. When we traverse the `DATA` and `NEXT` in this manner, we will finally see that the linked list in the above example stores characters that when put together forms the word `HELLO`.

8.7.1 Inserting a New Node in a Doubly Linked List

In this section, we will discuss how a new node is added into an already existing doubly linked list. To do the insertion, we will take five cases and then see how insertion is done in each case.

Case 1: The new node is inserted at the beginning.

Case 2: The new node is inserted at the end.

Case 3: The new node is inserted after a given node.

Case 4: The new node is inserted before a given node.

Case 5: The new node has to be inserted in a sorted.

Insert at the begining

Consider the doubly linked list shown in Fig. 8.46. Suppose we want to add a new node with the data 9 and add it as the first node of the list. Then, the flowing changes will be done in the linked list.

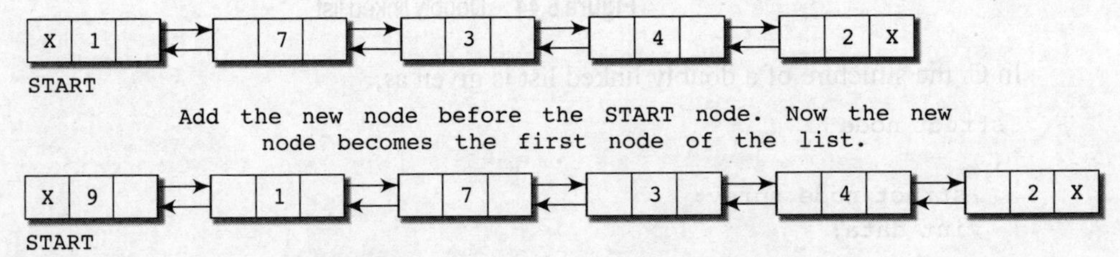

Figure 8.46 Inserting a new node at the beginning

Figure 8.47 shows the algorithm to insert a new node in the beginning of the doubly linked list. In Step 1, we first check whether memory is available for the new node. If the free memory has exhausted, then an `OVERFLOW` message is printed. Otherwise, if free memory cell is available, then we allocate space for the new node. Set its `data` part with the given `VAL` and the `next` part is initialized with the address of the first node of the list, which is stored in `START`. Now, since the new node is added as the first node of the list, it will now be known as the `START` node, that is, the `START` pointer variable will now hold the address of the `New_Node`.

```
Step 1: IF AVAIL = NULL, then
            Write OVERFLOW
            Go to Step 8
        [END OF IF]
Step 2: SET New_Node = AVAIL
Step 3: SET AVAIL = AVAIL->NEXT
Step 4: SET New_Node->DATA = VAL
Step 5: SET New_Node->PREV = NULL
Step 6: SET New_Node->Next = START
Step 7: SET START = New_Node
Step 8: EXIT
```

Figure 8.47 Algorithm to insert a new node at the beginning

Insert at the end

Consider the doubly linked list shown in Fig. 8.48. Suppose we want to add a new node with the data 9 and add it as the last node of the list. Then, the flowing changes will be done in the linked list.

Figure 8.49 shows the algorithm to insert a new node at the end of the doubly linked list. In Step 6, we take a pointer variable `PTR` and initialize it with `START`. In the `while` loop, we traverse through the linked list to reach the last node. Once we reach the last node, in Step 9, we change the `NEXT` pointer of the last node to store the address of the new node. Remember that the `NEXT` field of the new node contains `NULL` which signifies the end of the linked list. The `PREV` field of the `New_Node` will be set so that it points to the node pointed by `PTR` (now, the second last node of the list).

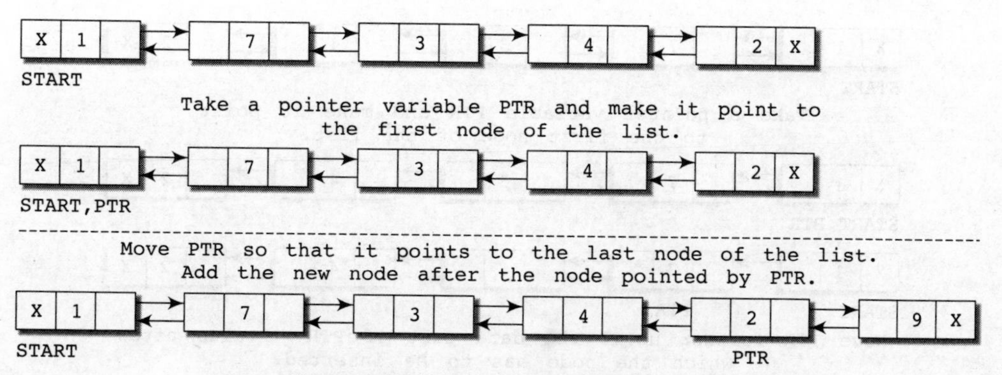

Figure 8.48 Inserting a new node at the end

```
Step 1: IF AVAIL = NULL, then
            Write OVERFLOW
            Go to Step 11
        [END OF IF]
Step 2: SET New_Node = AVAIL
Step 3: SET AVAIL = AVAIL->NEXT
Step 4: SET New_Node->DATA = VAL
Step 5: SET New_Node->Next = NULL
Step 6: SET PTR = START
Step 7: Repeat Step 8 while PTR->NEXT != NULL
Step 8:         SET PTR = PTR->NEXT
        [END OF LOOP]
Step 9: SET PTR->NEXT = New_Node
Step 10: New_Node->PREV = PTR
Step 11: EXIT
```

Figure 8.49 Algorithm to insert a new node at the end of the doubly linked list

Insert after a given node

Consider the doubly linked list shown in Fig. 8.51. Suppose we want to add a new node with the value 9 and add it after the node containing 3. Before discussing the changes that will be done in the linked list, let us first look at the algorithm as shown in Fig. 8.50.

```
Step 1: IF AVAIL = NULL, then
            Write OVERFLOW
            Go to Step 11
        [END OF IF]
Step 2: SET New_Node = AVAIL
Step 3: SET AVAIL = AVAIL->NEXT
Step 4: SET New_Node->DATA = VAL
Step 5: SET PTR = START
Step 6: Repeat Step 8 while PTR->DATA != NUM
Step 7:         SET PTR = PTR->NEXT
        [END OF LOOP]
Step 8: New_Node->NEXT = PTR->NEXT
Step 9: SET New_Node->PREV = PTR
Step 10: SET PTR->NEXT = New_Node
Step 11: EXIT
```

Figure 8.50 Algorithm to insert a new node after a given node

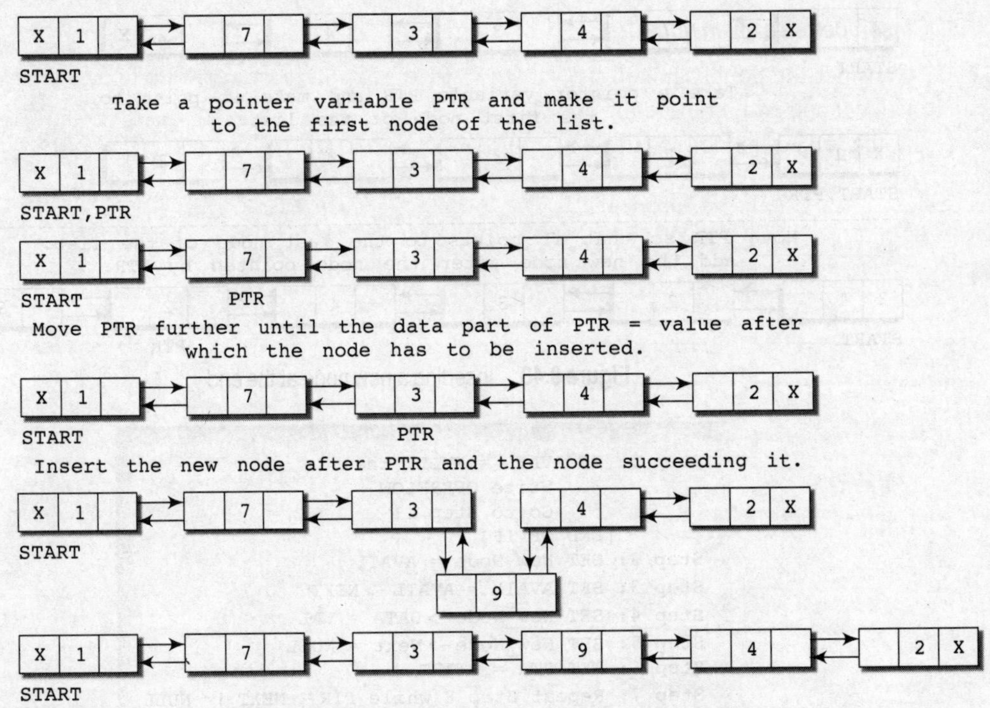

Figure 8.51 Inserting a new node after a given node

Figure 8.50 shows the algorithm to insert a new node after a given node in the doubly linked list. In Step 5, we take a pointer `P` and initialize it with `START`. That is, `PTR` now points to the first node of the linked list. In the `while` loop, we traverse through the linked list to reach the node that has its value equal to `NUM`. We need to reach this node because the new node will be inserted after this node. Once we reach this node, we change the `NEXT` and `PREV` field in such a way that the new node is inserted after the desired node.

```
Step 1: IF AVAIL = NULL, then
              Write OVERFLOW
              Go to Step 11
         [END OF IF]
Step 2: SET New_Node = AVAIL
Step 3: SET AVAIL = AVAIL -> NEXT
Step 4: SET New_Node -> DATA = VAL
Step 5: SET PTR = START
Step 6: Repeat Step 7 while PTR -> DATA != NUM
Step 7:                SET PTR = PTR -> NEXT
         [END OF LOOP]
Step 8: SET New_Node -> NEXT = PTR
Step 9: SET New_Node -> PREV = PTR -> PREV
Step 10: IF PTR -> PREV = NULL, then
              SET START = New_Node
         ELSE
              SET P -> PREV -> NEXT = New_Node
Step 11: EXIT
```

Figure 8.52 Algorithm to insert a new node before a given node

Insert before a given node

Consider the doubly linked list shown in Fig. 8.53. Suppose we want to add a new node with value 9 and add it before the node containing 3. Before discussing the changes that will be done in the linked list, let us first look at the algorithm as shown in Fig. 8.52.

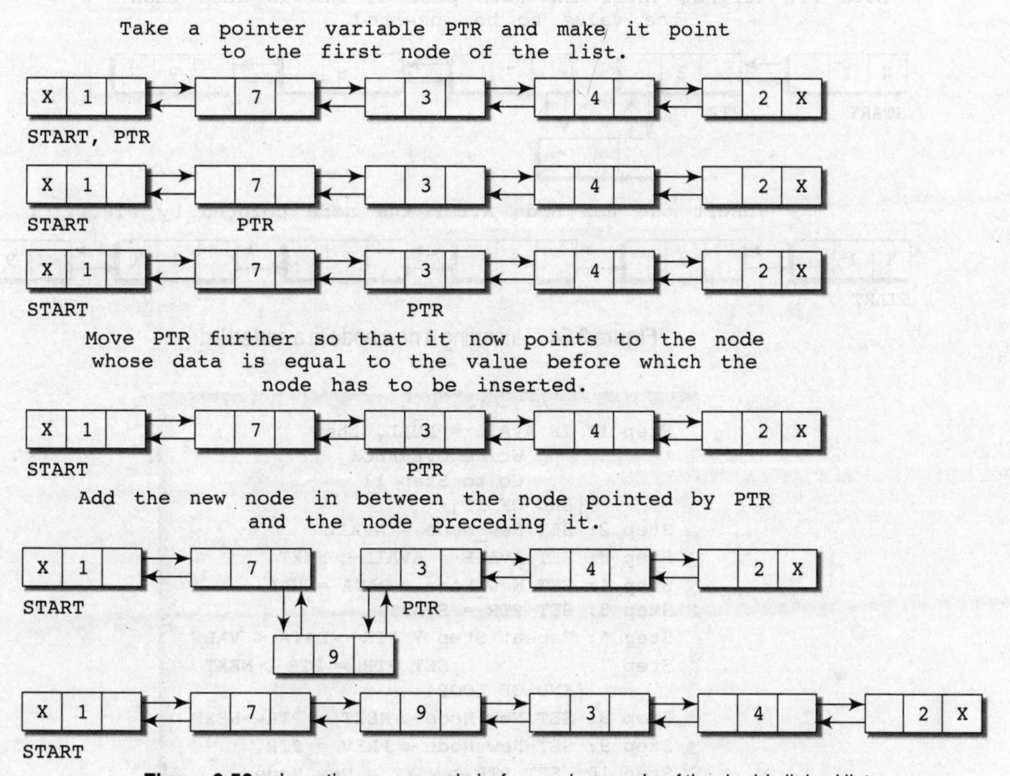

Figure 8.53 Inserting a new node before a given node of the doubly linked list

In Step 1, we first check whether memory is available for the new node. In Step 5, we take a pointer variable `PTR` and initialize it with `START`. That is, `PTR` now points to the first node of the linked list. In the `while` loop, we traverse through the linked list to reach the node that has its value equal to `NUM`. We need to reach this node because the new node will be inserted before this node. Once we reach this node, we change the `NEXT` and `PREV` fields in such a way that the new node is inserted after the desired node.

Insert in a sorted doubly linked list

Consider the doubly linked list shown in Fig. 8.54. Suppose we want to add a new node with the value 5. Before discussing the changes that will be done in the linked list, let us first look at the algorithm as shown in Fig. 8.55.

In Step 5, we take pointer variable `PTR` and initialize it with `START`. That is, `PTR` now points to the first node of the linked list. In the `while` loop, we traverse through the linked list to reach the node that has its value less than `VAL`. We need to reach this node because the new node will be inserted after this node. Once we reach this node, we change the `NEXT` and `PREV` fields in such a way that the new node is inserted after the desired node.

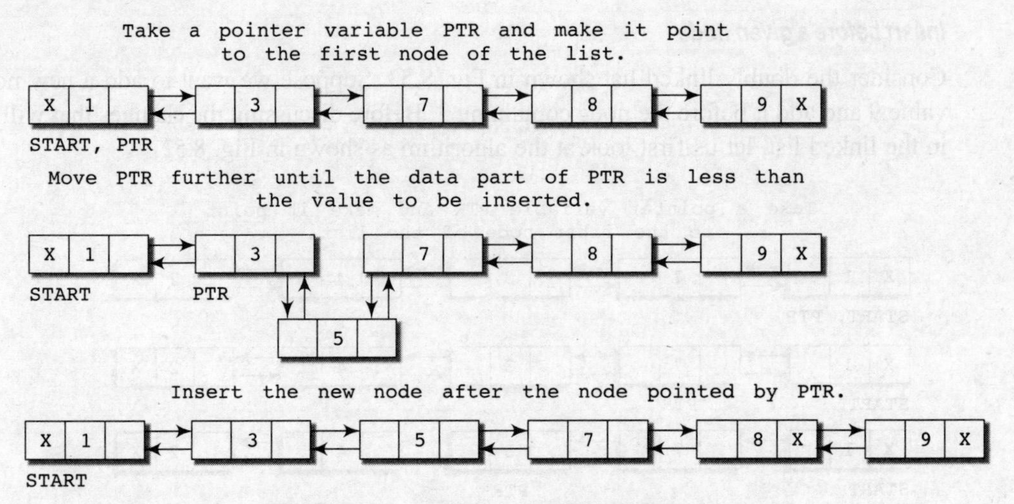

Figure 8.54 Inserting a new node in a sorted list

```
Step 1: IF AVAIL = NULL, then
             Write OVERFLOW
             Go to Step 11
        [END OF IF]
Step 2: SET New_Node = AVAIL
Step 3: SET AVAIL = AVAIL->NEXT
Step 4: SET New_Node->DATA = VAL
Step 5: SET PTR = START
Step 6: Repeat Step 7 PTR->DATA < VAL
Step 7:             SET PTR = PTR->NEXT
        [END OF LOOP]
Step 8: SET New_Node->NEXT = PTR->NEXT
Step 9: SET New_Node->PREV = PTR
Step 10: SET PTR->NEXT = New_Node
Step 11: EXIT
```

Figure 8.55 Algorithm to insert a new node in a sorted list

8.7.2 Deleting a Node from a Doubly Linked List

In this section, we will se how a new node is deleted from an already existing doubly linked list. To do the deletion, we will take five cases and then see how deletion is done in each case.

Case 1: The first node is deleted.

Case 2: The last node is deleted.

Case 3: The node after a given node is deleted.

Case 4: The node before a given node is deleted.

Case 5: The node is deleted from a sorted linked list.

Delete the first node

Consider the doubly linked list shown in Fig. 8.56. When we want to delete a node from the beginning of the list, then the following changes will be done in the linked list.

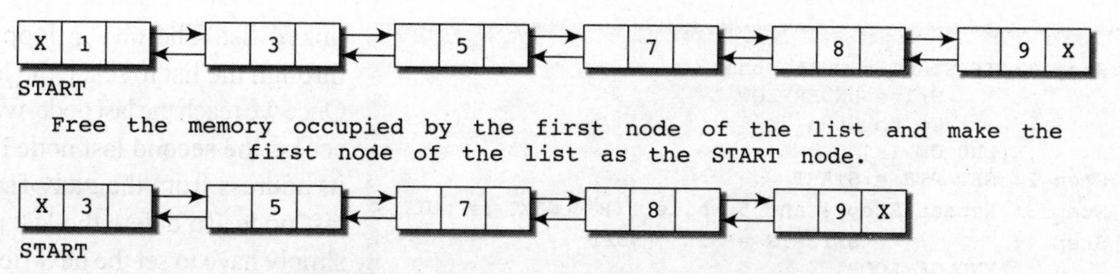

Figure 8.56 Deleting a node from a doubly linked list

Figure 8.57 shows the algorithm to delete the first node of a doubly linked list. In Step 1 of the algorithm, we check if the linked list exists or not. If `START = NULL`, then it signifies that there are no nodes in the list and the control is transferred to the last statement of the algorithm.

```
Step 1: IF START = NULL, then
            Write UNDERFLOW
            Go to Step 6
        [END OF IF]
Step 2: SET PTR = START
Step 3: SET START = START -> NEXT
Step 4: SET START -> PREV = NULL
Step 5: FREE PTR
Step 6: EXIT
```

Figure 8.57 Deleting the first element

However, if there are nodes in the linked list, then we use a temporary pointer variable that is set to point to the first node of the list. For this, we initiate `PTR` with `START` that stores the address of the first node of the list. In Step 3, `START` is made to point to the next node in sequence and finally the memory occupied by `PTR` (initially the first node of the list) is freed and returned to the free pool.

Delete the last node

Consider the doubly linked list shown in Fig. 8.58. Suppose we want to delete the last node from the linked list, then the following changes will be done in the linked list.

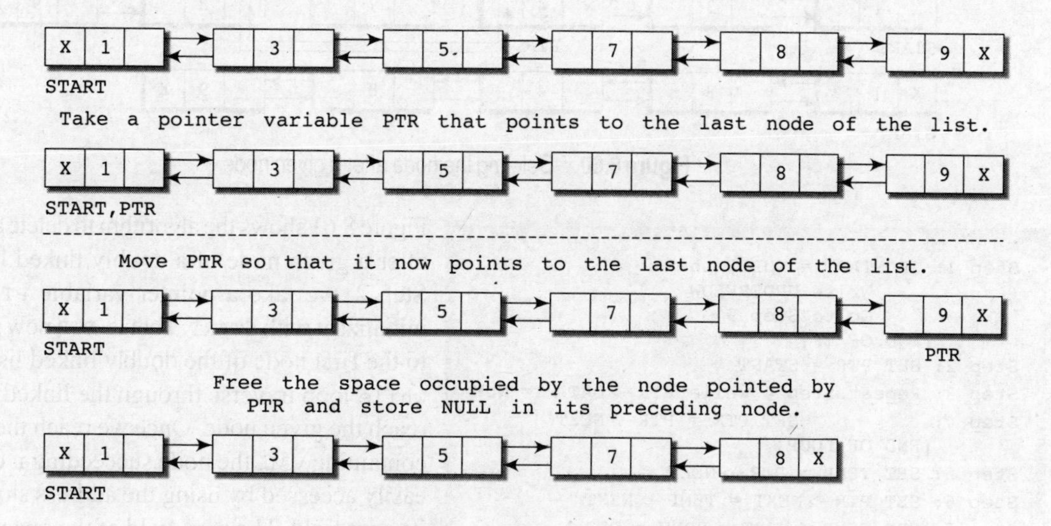

Figure 8.58 Deleting the last element

Figure 8.59 shows the algorithm to delete the last node of a doubly linked list. In Step 2, we take a pointer variable `PTR` and initialize it with `START`. That is, `PTR` now points to the first node of the

```
Step 1: IF START = NULL, then
            Write UNDERFLOW
            Go to Step 7
        [END OF IF]
Step 2: SET PTR = START
Step 3: Repeat Step 4 and 5 while PTR->NEXT != NULL
Step 4:         SET PTR = PTR->NEXT
        [END OF LOOP]
Step 5: SET PTR->PREV->NEXT = NULL
Step 6: FREE PTR
Step 7: EXIT
```

Figure 8.59 Algorithm to delete the last node

linked list. The `while` loop traverse through the list to reach the last node. Once we reach the last node, we can also access the second last node by taking its address from the `PREV` field of the last node. To delete the last node, we simply have to set the next field of that node to `NULL`, so that it now becomes the (new) last node of the linked list. The memory of the previous last node is freed and returned to the free pool.

Delete the node after a given node list

Consider the doubly linked list shown in Fig. 8.60. Suppose we want to delete the node that succeeds the node which contains the data value 4. Then, the following changes will be done in the linked list.

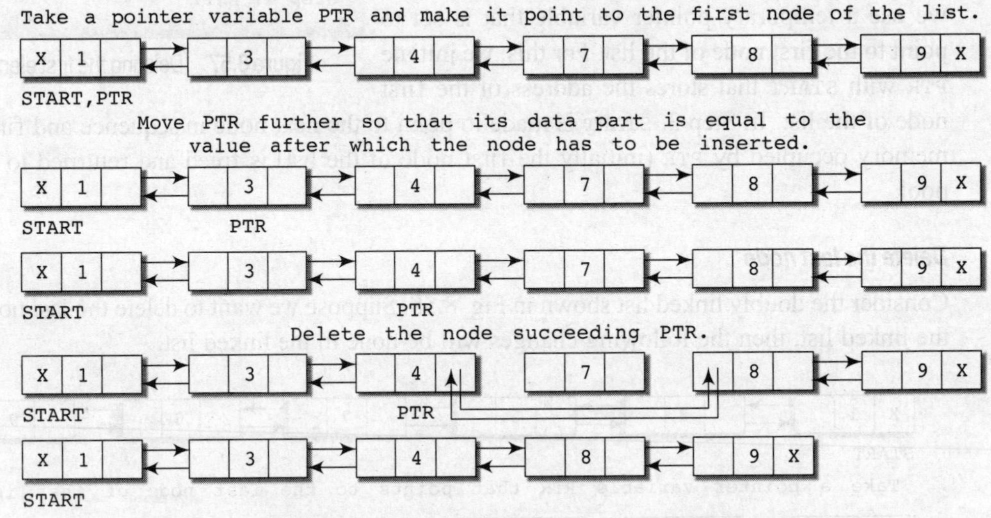

Figure 8.60 Deleting the node after a given node

```
Step 1: IF START = NULL, then
            Write UNDERFLOW
            Go to Step 9
        [END OF IF]
Step 2: SET PTR = START
Step 3: Repeat Step 4 while PTR->DATA != NUM
Step 4:         SET PTR = PTR->NEXT
        [END OF LOOP]
Step 5: SET TEMP = PTR->NEXT
Step 6: SET PTR->NEXT = TEMP->NEXT
Step 7: SET TEMP->NEXT->PREV = PTR
Step 8: FREE TEMP
Step 9: EXIT
```

Figure 8.61 Algorithm to delete a node after a given node

Figure 8.61 shows the algorithm to delete a node after a given node of a doubly linked list. In Step 2, we take a pointer variable `PTR` and initialize it with `START`. That is, `PTR` now points to the first node of the doubly linked list. The `while` loop traverse through the linked list to reach the given node. Once we reach the node containing `VAL`, the node succeeding it can be easily accessed by using the address stored in its next field. The next field of the given node is set to contain the contents in the next field of the succeeding node. Finally, the memory of the node succeeding the given node is freed and returned to the free pool.

Delete the node from a sorted list

Consider the doubly linked list shown in Fig. 8.62. Suppose we want to delete a node with the value 4. Before discussing the changes that will be done in the linked list, let us first look at the algorithm.

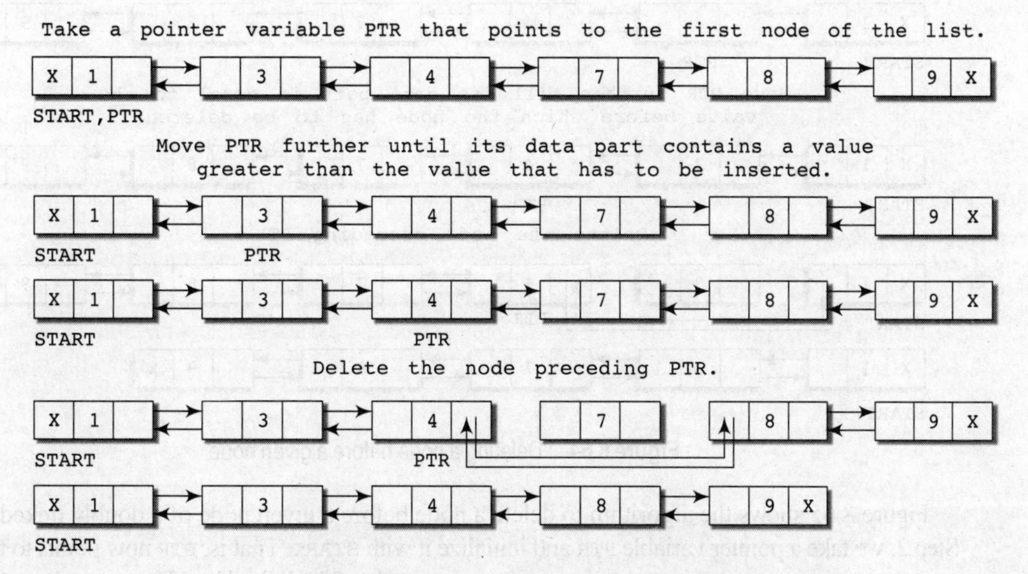

Figure 8.62 Deleting a node from a sorted list

Figure 8.63 shows the algorithm to delete a node from a sorted doubly linked list. In Step 2, we take a pointer variable `PTR` and initialize it with `START`. That is, `PTR` now points to the first node of the linked list. The `while` loop traverse through the linked list to reach the desired node. Once we reach the node containing `VAL`, the `NEXT` field of the preceding node is set to contain the address of the node succeeding `PTR`. The memory of the node is freed and returned to the free pool.

```
Step 1: IF START = NULL, then
            Write UNDERFLOW
            Go to Step 8
        [END OF IF]
Step 2: SET PTR = START
Step 3: Repeat Step 4 while PTR->DATA != NUM
Step 4:             SET PTR = PTR->NEXT
        [END OF LOOP]
Step 5: SET PTR->PREV->NEXT = PTR->NEXT
Step 6: SET PTR->NEXT->PREV = PTR->PREV
Step 7: FREE P
Step 8: EXIT
```

Figure 8.63 Algorithm to delete a node from a sorted list

Delete the node before a given node

Consider the doubly linked list shown in Fig. 8.64. Suppose we want to delete the node preceding the node with the value 4. Before discussing the changes that will be done in the linked list, let us first look at the algorithm.

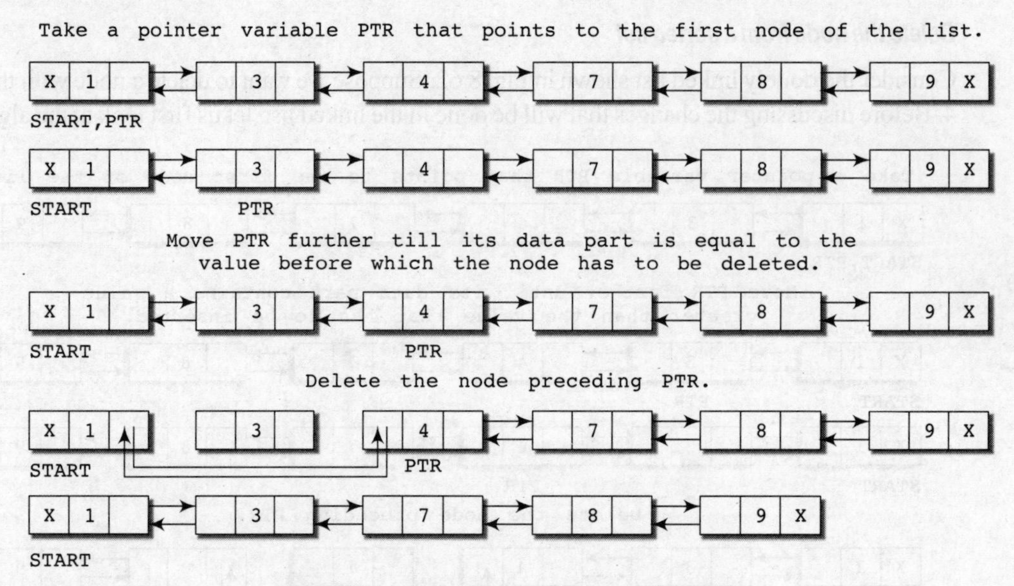

Figure 8.64 Deleting a node before a given node

Figure 8.65 shows the algorithm to delete a node before a given node of a doubly linked list. In Step 2, we take a pointer variable PTR and initialize it with START. That is, PTR now points to the first node of the linked list. The while loop traverse through the linked list to reach the desired node. Once we reach the node containing VAL, the PREV field of PTR is set to contain the address of the node preceding the node which comes before PTR. The memory of the node preceding PTR is freed and returned to the free pool.

```
Step 1: IF START = NULL, then
            Write UNDERFLOW
            Go to Step 9
        [END OF IF]
Step 2: SET PTR = START
Step 3: Repeat Step 4 while PTR->DATA != NUM
Step 4:          SET PTR = PTR->NEXT
        [END OF LOOP]
Step 5: SET TEMP = PTR->PREV
Step 6: SET TEMP->PREV->NEXT = PTR
Step 7: SET PTR->PREV = TEMP->PREV
Step 8: FREE TEMP
Step 9: EXIT
```

Figure 8.65 Algorithm to delete a node before a given node

Hence, we see that we can insert or delete a node in a constant number of operations given only that node's address. Note that this is not possible in the case of a singly linked list which requires the previous node's address also to perform the same operation.

PROGRAMMING EXAMPLE

4. Write a program to create a doubly linked list and perform insertions and deletions in all cases.

```
#include<stdio.h>
#include<conio.h>
struct node
{
        struct node *next;
        int data;
        struct node *prev;
```

```
};
struct node *start = NULL;
struct node *create_ll(struct node *);
struct node *display(struct node *);
struct node *insert_beg(struct node *);
struct node *insert_end(struct node *);
struct node *insert_before(struct node *start);
struct node *insert_after(struct node *start);
struct node *insert_sorted(struct node *start);
struct node *delete_beg(struct node *);
struct node *delete_end(struct node *);
struct node *delete_node(struct node *start);
struct node *delete_before(struct node *start);
struct node *delete_after(struct node *start);
struct node *delete_sorted(struct node *start);
struct node *delete_list(struct node *start);
main()
{
        int option;
        clrscr();
        do
        {
                printf("\n\n *****MAIN MENU *****");
                printf("\n 1: Create a List");
                printf("\n 2: Display the list");
                printf("\n 3: Add a node in the beginning");
                printf("\n 4: Add a node at the end");
                printf("\n 5: Add a node before a given node");
                printf("\n 6: Add a node after a given node");
                printf("\n 7. Add a node in a sorted linked list");
                printf("\n 8: Delete a node from the beginning");
                printf("\n 9: Delete a node from the end");
                printf("\n 10: Delete a node a given node");
                printf("\n 11: Delete a node before a given node");
                printf("\n 12: Delete a node after a given node");
                printf("\n 13. Delete a node from a sorted linked list");
                printf("\n 14: Delete the entire list");
                printf("\n 15: EXIT");
                printf("\n ***************************");
                printf("\n\n ENter your option : ");
                scanf("%d", &option);
                switch(option)
                {
                        case 1:
                                start = create_ll(start);
                                printf("\n LINKED LIST CREATED");
                                break;
```

```
                case 2:
                        start = display(start);
                        break;
                case 3:
                        start = insert_beg(start);
                        break;
                case 4:
                        start = insert_end(start);
                        break;
                case 5:
                        start = insert_before(start);
                        break;
                case 6:
                        start = insert_after(start);
                        break;
                case 7:
                        start = insert_sorted(start);
                        break;
                case 8:
                        start = delete_beg(start);
                        break;
                case 9:
                        start = delete_end(start);
                        break;
                case 10:
                        start = delete_node(start);
                        break;
                case 11:
                        start = delete_before(start);
                        break;
                case 12:
                        start = delete_after(start);
                        break;
                case 13:
                        start = delete_sorted(start);
                        break;
                case 14:
                        start = delete_list(start);
                        printf("\n List is EMPTY");
                        break;
        }
    }while(option !=15);
    getch();
    return 0;
}
struct node *create_ll(struct node *start)
{
```

```
        struct node *new_node;
        int num;
        printf("\n Enter -1 to end");
        printf("\n Enter the data : ");
        scanf("%d", &num);
        while(num != -1)
        {
                if(start == NULL)
                {
                        start = (struct node*)malloc(sizeof(struct node*));
                        start->prev = NULL;
                        start->data = num;
                        start->next = NULL;
                }
                else
                {
                        new_node = (struct node*)malloc(sizeof(struct node*));
                        new_node->prev = NULL;
                        new_node->data = num;
                        new_node->next = start;
                        start->prev = new_node;
                        start = new_node;
                }
                printf("\n Enter the data : ");
                scanf("%d", &num);
        }
        return start;
}
struct node *display(struct node *start)
{
        struct node *ptr;
        ptr=start;
        printf("\n");
        while(ptr!=NULL)
        {
                printf("\t %d", ptr->data);
                ptr = ptr->next;
        }
        return start;
}
struct node *insert_beg(struct node *start)
{
        struct node *new_node;
        int num;
        printf("\n Enter the data : ");
        scanf("%d", &num);
        new_node = (struct node *)malloc(sizeof(struct node *));
```

```
        start -> prev = new_node;
        new_node -> next = start;
        new_node -> prev = NULL;
        new_node -> data = num;
        start = new_node;
        return start;
}
struct node *insert_end(struct node *start)
{
        struct node *ptr, *new_node;
        int num;
        printf("\n Enter the data : ");
        scanf("%d", &num);
        new_node = (struct node *)malloc(sizeof(struct node *));
        new_node -> data = num;
        ptr=start;
        while(ptr -> next != NULL)
                ptr = ptr -> next;
        ptr -> next = new_node;
        new_node -> prev = ptr;
        new_node -> next = NULL;
        return start;
}
struct node *insert_before(struct node *start)
{
        struct node *new_node, *ptr;
        int num, val;
        printf("\n Enter the data : ");
        scanf("%d", &num);
        printf("\n Enter the value before which the data has to be inserted : ");
        scanf("%d", &val);
        new_node = (struct node *)malloc(sizeof(struct node *));
        new_node -> data = num;
        ptr = start;
        while(ptr -> data != val)
                ptr = ptr -> next;
        new_node -> next = ptr;
        ptr -> prev -> next = new_node;
        ptr -> prev = new_node;
        return start;
}
struct node *insert_after(struct node *start)
{
        struct node *new_node, *ptr;
        int num, val;
        printf("\n Enter the data : ");
        scanf("%d", &num);
```

```
        printf("\n Enter the value after which the data has to be inserted : ");
        scanf("%d", &val);
        new_node = (struct node *)malloc(sizeof(struct node *));
        new_node -> data = num;
        ptr = start;
        while(ptr -> data != val)
                ptr = ptr -> next;
        new_node -> prev = ptr;
        new_node -> next = ptr -> next;
        ptr -> next -> prev = new_node;
        ptr -> next = new_node;
        return start;
}
struct node *insert_sorted(struct node *start)
{
        struct node *new_node, *ptr;
        int num;
        printf("\n Enter the data : ");
        scanf("%d", &num);
        new_node = (struct node *)malloc(sizeof(struct node *));
        new_node -> data = num;
        ptr = start;
        while(ptr -> data<num)
        {
                ptr = ptr -> next;
                if(ptr -> next == NULL)
                        break;
        }
        if(ptr -> next == NULL)
        {
                ptr -> next = new_node;
                new_node -> prev = ptr;
                new_node -> next = NULL;
        }
        else
        {
                new_node -> next = ptr;
                new_node -> prev = ptr -> prev;
                ptr -> prev -> next = new_node;
                ptr -> prev = new_node;
        }
        return start;
}
struct node *delete_beg(struct node *start)
{
        struct node *ptr;
        ptr = start;
```

```
        start = start -> next;
        start -> prev = NULL;
        free(ptr);
        return start;
}
struct node *delete_end(struct node *start)
{
        struct node *ptr;
        ptr = start;
        while(ptr -> next != NULL)
        ptr = ptr -> next;
        ptr -> prev -> next = NULL;
        free(ptr);
        return start;
}
struct node *delete_node(struct node *start)
{
        struct node *ptr;
        int val;
        printf("\n Enter the value of the node which has to be deleted : ");
        scanf("%d", &val);
        ptr = start;
        if(ptr -> data == val)
        {
                start = delete_beg(start);
                return start;
        }
        else
        {
                while(ptr -> data!=val)
                        ptr = ptr -> next;
                ptr -> prev -> next = ptr -> next;
                ptr -> next -> prev = ptr -> prev;
                free(ptr);
                return start;
        }
}
struct node *delete_after(struct node *start)
{
        struct node *ptr, *temp;
        int val;
        printf("\n Enter the value after which the node has to deleted : ");
        scanf("%d", &val);
        ptr = start;
        while(ptr -> data != val)
                ptr = ptr -> next;
        temp = ptr -> next;
```

```
        ptr -> next = temp -> next;
        temp -> next -> prev = ptr;
        free(temp);
        return start;
}
struct node *delete_before(struct node *start)
{
        struct node *ptr, *temp;
        int val;
        printf("\n Enter the value before which the node has to deleted : ");
        scanf("%d", &val);
        ptr = start;
        while(ptr -> data != val)
        ptr = ptr -> next;
        temp = ptr -> prev;
        if(temp == start)
                start = delete_beg(start);
        else
        {
                ptr -> prev = temp -> prev;
                temp -> prev -> next = ptr;
        }
        free(temp);
        return start;
}
struct node *delete_sorted(struct node *start)
{
        struct node *ptr;
        int val;
        printf("\n Enter the value of the node which has to be deleted : ");
        scanf("%d", &val);
        ptr = start;
        while(ptr -> data != val)
                ptr = ptr -> next;
        ptr -> prev -> next = ptr -> next;
        free(ptr);
        return start;
}
struct node *delete_list(struct node *start)
{
        while(start != NULL)
                start = delete_beg(start);
        return start;
}
```

8.8 CIRCULAR DOUBLY LINKED LIST

A circular doubly linked list or a circular two-way linked list is a more complex type of linked list which contains a pointer to the next as well as the previous node in the sequence. The difference between a doubly linked and a circular doubly linked list is same as that exists between a singly linked list and a circular linked list. The circular doubly linked list does not contain `NULL` in the previous field of the first node and the next field of the last node. Rather, the next field of the last node stores the address of the first node of the list, i.e. `START`. Similarly, the previous field of the first field stores the address of the last node. A circular doubly linked list is shown in Fig. 8.66.

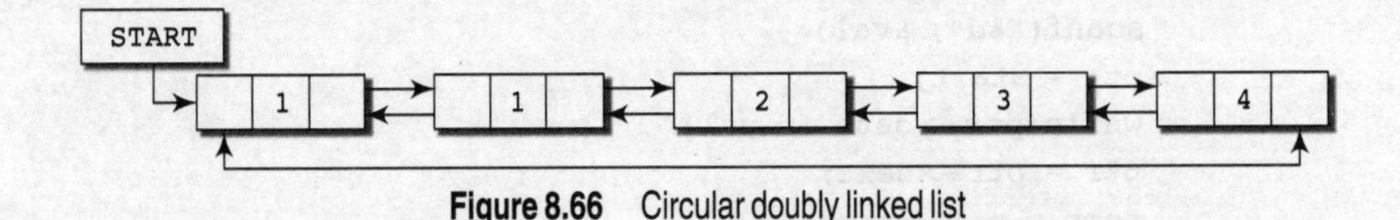

Figure 8.66 Circular doubly linked list

Since a circular doubly linked list contains three parts in its structure, it calls for more space per node and more expensive basic operations. However, a circular doubly linked list provides the ease to manipulate the elements of the list as it maintains pointers to nodes in both the directions (forward and backward). The main advantage of using a circular doubly linked list is that it makes searches twice as efficient.

Let us view how a circular doubly linked list is maintained in the memory. Consider Fig. 8.67.

START = 1

	DATA	PREV	Next
1	H	9	3
2			
3	E	1	6
4			
5			
6	L	3	7
7	L	6	9
8			
9	O	7	1

Figure 8.67 Memory representation of a circular doubly linked list

In the figure, we see that a variable `START` is used to store the address of the first node. Here in this example, `START = 1`, so the first data is stored at address 1, which is `H`. Since this is the first node, it stores the address of the last node of the list in its previous field. The corresponding `NEXT` stores the address of the next node, which is 3. So, we will look at address 3 to fetch the next data item. The previous field will contain the address of the first node. The second data element obtained from address 3 is `E`. We repeat this procedure until we reach a position where the `NEXT` entry stores the address of the first element of the list. This denotes the end of the linked list, that is, the node that contains the address of the first node is actually the last node of the list.

8.8.1 Inserting a New Node in a Circular Doubly Linked List

In this section, we will se how a new node is added into an already existing circular doubly linked list. To do the insertion, we will take two cases and then see how insertion is done in each case. Rest of the cases are similar to that given for doubly linked lists.

Case 1: The new node is inserted at the beginning.

Case 2: The new node is inserted at the end.

Insert at the beginning

Consider the circular doubly linked list shown in Fig. 8.68. Suppose we want to add a new node with the data 9 and add it as the first node of the list. Then, the following changes will be done in the linked list.

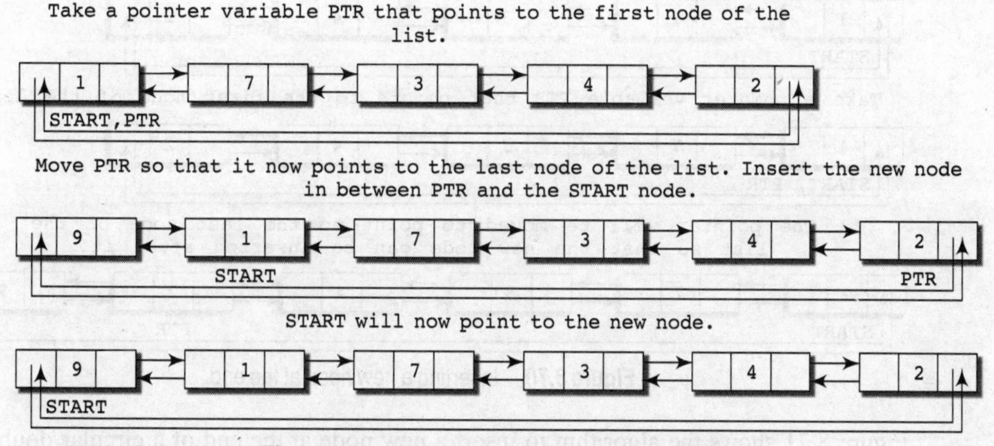

Figure 8.68 Inserting a new node at the beginning

Figure 8.69 shows the algorithm to insert a new node in the beginning of circular doubly linked list. In Step 1, we first check whether memory is available for the new node. If the free memory has exhausted, then an `OVERFLOW` message is printed. Otherwise, we allocate space for the new node. Set its data part with the given `VAL` and its next part is initialized with the address of the first node of the list, which is stored in `START`. Now since, the new node is added as the first node of the list, it will now be known as the `START` node, that is, the `START` pointer variable will now hold the address of the `New_Node`. The `while` loop traverse through the list to reach its last node. Since it is a circular doubly linked list, the `PREV` field of the `New_Node` is set to contain the address of the last node.

```
Step 1: IF AVAIL = NULL, then
             Write OVERFLOW
             Go to Step 12
        [END OF IF]
Step 2: SET New_Node = AVAIL
Step 3: SET AVAIL = AVAIL -> NEXT
Step 4: SET New_Node -> DATA = VAL
Step 5: SET PTR = START
Step 6: Repeat Step 7 while PTR -> NEXT != START
Step 7:      SET PTR = PTR -> NEXT
        [END OF LOOP}
Step 8: SET PTR -> NEXT = New_Node
Step 9: SET New_Node -> PREV = PTR
Step 10: SET New_Node -> NEXT = START
Step 11: SET START = New_Node
Step 12: EXIT
```

Figure 8.69 Algorithm to insert a new node at the beginning

Insert at the end

Consider the circular doubly linked list shown in Fig. 8.70. Suppose we want to add a new node with the data 9 and add it as the last node of the list, then the following changes will be done in the linked list.

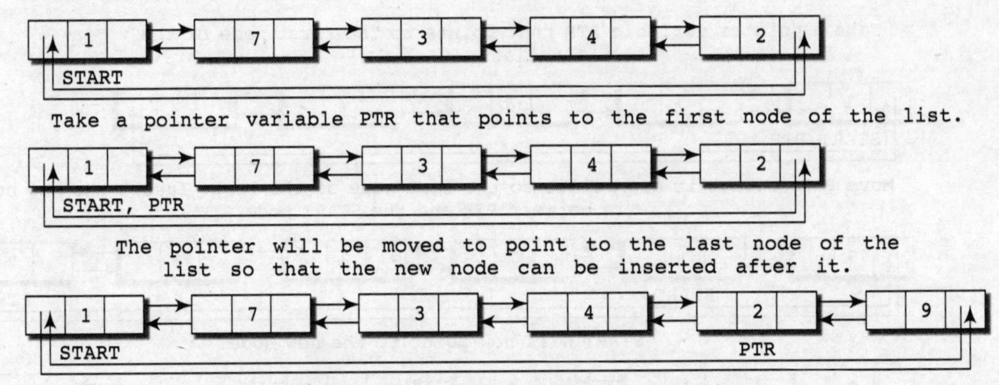

Figure 8.70 Inserting a new node at the end

Figure 8.71 shows the algorithm to insert a new node at the end of a circular doubly linked list. In Step 6, we take a pointer variable `PTR` and initialize it with `START`. That is, `PTR` now points to the first node of the linked list. In the `while` loop, we traverse through the linked list to reach the last node. Once we reach the last node, in Step 9, we change the `NEXT` pointer of the last node to store the address of the new node. The `PREV` field of the `New_Node` will be set so that it points to the node pointed by `PTR` (now, the second last node of the list).

```
Step 1: IF AVAIL = NULL, then
            Write OVERFLOW
            Go to Step 11
        [END OF IF]
Step 2: SET New_Node = AVAIL
Step 3: SET AVAIL = AVAIL->NEXT
Step 4: SET New_Node->DATA = VAL
Step 5: SET New_Node->Next = START
Step 6: SET PTR = START
Step 7: Repeat Step 8 while PTR->NEXT != START
Step 8:             SET PTR = PTR->NEXT
        [END OF LOOP]
Step 9: SET PTR->NEXT = New_Node
Step 10: New_Node->PREV = PTR
Step 11: EXIT
```

Figure 8.71 Algorithm to insert a new node at the end

8.8.2 Deleting a Node from a Circular Doubly Linked List

In this section, we will see how a new node is deleted from an already existing circular doubly linked list. To do the deletion, we will take two cases and then see how deletion is done in each case. Rest of the cases are same as that given for doubly linked lists.

Case 1: The first node is deleted.

Case 2: The last node is deleted.

Delete the first node

Consider the circular doubly linked list shown in Fig. 8.72. When we want to delete a node from the beginning of the list, then the following changes will be done in the linked list.

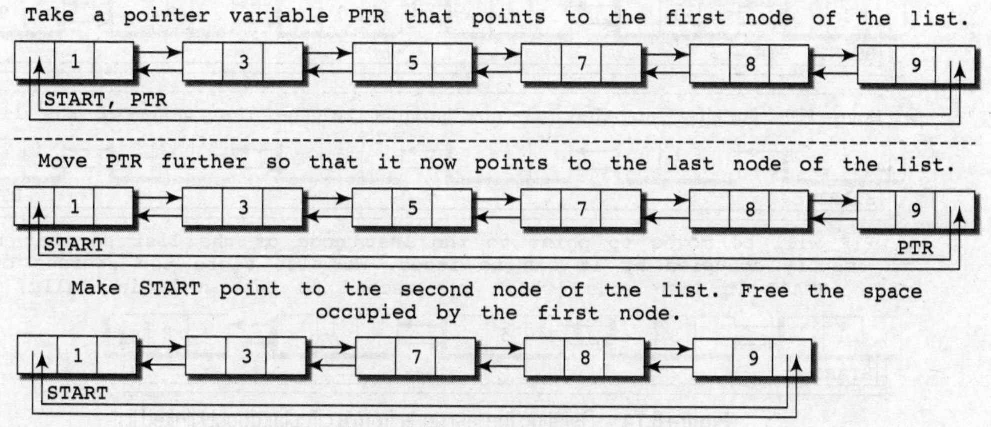

Figure 8.72 Deleting the first node

Figure 8.73 shows the algorithm to delete the first node if a circular doubly linked list. In Step 1 of the algorithm, we check if the linked list exists or not. If `START = NULL`, then it signifies that there are no nodes in the list and the control is transferred to the last statement of the algorithm.

```
Step 1: IF START = NULL, then
            Write UNDERFLOW
            Go to Step 8
        [END OF IF]
Step 2: SET PTR = START
Step 3: Repeat Step 4 while PTR->NEXT != START
Step 4:     SET PTR = PTR->NEXT
        [END OF LOOP]
Step 5: SET PTR->NEXT = START->NEXT
Step 6: SET START->NEXT->PREV = PTR
Step 6: FREE START
Step 7:  SET START = PTR->NEXT
```

Figure 8.73 Algorithm to delete the first node

However, if there are nodes in the linked list, then we use a pointer variable that is set to point to the first node of the list. For this, we initiate `PTR` with `START` that stores the address of the first node of the list. The `while` loop traverse through the list to reach the last node. Once we reach the last node, the `NEXT` pointer of `PTR` is set to contain the address of the node that succeeds `START`. Finally, `START` is made to point to the next node in the sequence and finally the memory occupied by initially the first node of the list is freed and returned to the free pool.

Delete the last node

Consider the circular doubly linked list shown in Fig. 8.74. Suppose we want to delete the last node from the linked list, then the following changes will be done in the linked list.

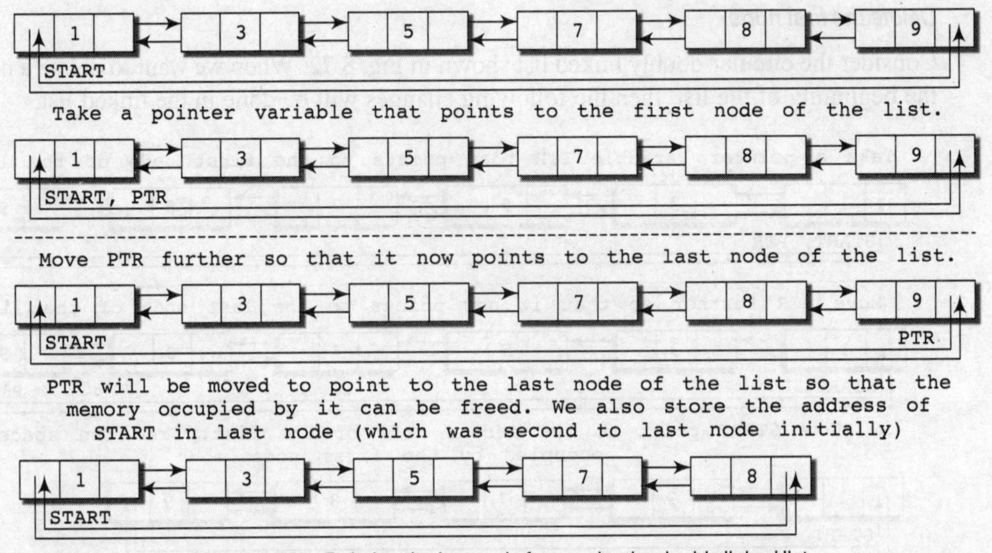

Figure 8.74 Deleting the last node from a circular doubly linked list

Figure 8.75 shows the algorithm to delete the last node if a circular doubly linked list. In Step 2, we take a pointer variable PTR and initialize it with START. That is, PTR now points to the first node of the linked list. The while loop traverse through the list to reach the last node. Once we reach the last node, we can also access the second last node by taking its address from the PREV field of the last node. To delete the last node, we simply have to set the next field of that node to contain the address of START, so that it now becomes the (new) last node of the linked list. The memory of the previous last node is freed and returned to the free pool.

```
Step 1: IF START = NULL, then
            Write UNDERFLOW
            Go to Step 8
        [END OF IF]
Step 2: SET PTR = START
Step 3: Repeat Step 4  while PTR->NEXT != START
Step 4:           SET PTR = PTR->NEXT
        [END OF LOOP]
Step 5: SET PTR->PREV->NEXT = START
Step 6: SET START->PREV = PTR->PREV
Step 7: FREE PTR
Step 8: EXIT
```

Figure 8.75 Algorithm to delete the last node

PROGRAMMING EXAMPLE

5. Write a program to create a circular doubly linked list. Also, perform insertions and deletions in all the cases.

```
#include<stdio.h>
#include<conio.h>
struct node
{
        struct node *next;
        int data;
        struct node *prev;
};
```

```
struct node *start = NULL;
struct node *create_ll(struct node *);
struct node *display(struct node *);
struct node *insert_beg(struct node *);
struct node *insert_end(struct node *);
struct node *insert_before(struct node *start);
struct node *insert_after(struct node *start);
struct node *insert_sorted(struct node *start);
struct node *delete_beg(struct node *);
struct node *delete_end(struct node *);
struct node *delete_node(struct node *start);
struct node *delete_before(struct node *start);
struct node *delete_after(struct node *start);
struct node *delete_sorted(struct node *start);
struct node *delete_list(struct node *start);
main()
{
      int option;
      clrscr();
      do
      {
            printf("\n\n *****MAIN MENU *****");
            printf("\n 1: Create a List");
            printf("\n 2: Display the list");
            printf("\n 3: Add a node in the beginning");
            printf("\n 4: Add a node at the end");
            printf("\n 5: Add a node before a given node");
            printf("\n 6: Add a node after a given node");
            printf("\n 7. Add a node in a sorted linked list");
            printf("\n 8: Delete a node from the beginning");
            printf("\n 9: Delete a node from the end");
            printf("\n 10: Delete a node a given node");
            printf("\n 11: Delete a node before a given node");
            printf("\n 12: Delete a node after a given node");
            printf("\n 13. Delete a node from a sorted linked list");
            printf("\n 14: Delete the entire list");
            printf("\n 15: EXIT");
            printf("\n **************************");
            printf("\n\n ENter your option : ");
            scanf("%d", &option);
            switch(option)
            {
                  case 1:
                        start = create_ll(start);
                        printf("\n LINKED LIST CREATED");
                        break;
```

```
                case 2:
                        start = display(start);
                        break;
                case 3:
                        start = insert_beg(start);
                        break;
                case 4:
                        start = insert_end(start);
                        break;
                case 5:
                        start = insert_before(start);
                        break;
                case 6:
                        start = insert_after(start);
                        break;
                case 7:
                        start = insert_sorted(start);
                        break;
                case 8:
                        start = delete_beg(start);
                        break;
                case 9:
                        start = delete_end(start);
                        break;
                case 10:
                        start = delete_node(start);
                        break;
                case 11:
                        start = delete_before(start);
                        break;
                case 12:
                        start = delete_after(start);
                        break;
                case 13:
                        start = delete_sorted(start);
                        break;
                case 14:
                        start = delete_list(start);
                        printf("\n List is EMPTY");
                        break;
        }
    }while(option != 15);
    getch();
    return 0;
}
struct node *create_ll(struct node *start)
```

```
{
        struct node *new_node, *ptr;
        int num;
        printf("\n Enter -1 to end");
        printf("\n Enter the data : ");
        scanf("%d", &num);
        while(num != -1)
        {
                if(start == NULL)
                {
                        start = (struct node*)malloc(sizeof(struct node*));
                        start -> prev = NULL;
                        start -> data = num;
                        start -> next = start;
                }
                else
                {
                        new_node = (struct node*)malloc(sizeof(struct node*));
                        new_node -> prev = NULL;
                        new_node -> data = num;
                        ptr = start;
                        while(ptr -> next != start)
                                ptr = ptr -> next;
                        new_node -> prev = ptr;
                        ptr -> next = new_node;
                        new_node -> next = start;
                        start -> prev = new_node;
                        start = new_node;
                }
                printf("\n Enter the data : ");
                scanf("%d", &num);
        }
        return start;
}
struct node *display(struct node *start)
{
        struct node *ptr;
        ptr = start;
        printf("\n");
        while(ptr -> next != start)
        {
                printf("\t %d", ptr -> data);
                ptr = ptr -> next;
        }
        printf("\t %d", ptr -> data);
```

```
        return start;
}
struct node *insert_beg(struct node *start)
{
        struct node *new_node, *ptr;
        int num;
        printf("\n Enter the data : ");
        scanf("%d", &num);
        new_node = (struct node *)malloc(sizeof(struct node *));
        ptr = start;
        while(ptr->next != start)
                ptr = ptr->next;
        new_node->prev = ptr;
        ptr->next = new_node;
        new_node->next = start;
        start->prev = new_node;
        start = new_node;
        return start;
}
struct node *insert_end(struct node *start)
{
        struct node *ptr, *new_node;
        int num;
        printf("\n Enter the data : ");
        scanf("%d", &num);
        new_node = (struct node *)malloc(sizeof(struct node *));
        new_node->data = num;
        ptr = start;
        while(ptr->next != start)
                ptr = ptr->next;
        ptr->next = new_node;
        new_node->prev = ptr;
        new_node->next = start;
        return start;
}
struct node *insert_before(struct node *start)
{
        struct node *new_node, *ptr;
        int num, val;
        printf("\n Enter the data : ");
        scanf("%d", &num);
        printf("\n Enter the value before which the data has to be inserted : ");
        scanf("%d", &val);
        new_node = (struct node *)malloc(sizeof(struct node *));
        new_node->data = num;
```

```
        ptr = start;
        if(ptr->data == val)
        {
                ptr = start;
                while(ptr->next != start)
                        ptr = ptr->next;
                new_node->prev = ptr;
                ptr->next = new_node;
                new_node->next = start;
                start->prev = new_node;
                start = new_node;
        }
        else
        {
                while(ptr->data!=val)
                        ptr = ptr->next;
                new_node->next = ptr;
                ptr->prev->next = new_node;
                ptr->prev = new_node;
        }
        return start;
}
struct node *insert_after(struct node *start)
{
        struct node *new_node, *ptr;
        int num, val;
        printf("\n Enter the data : ");
        scanf("%d", &num);
        printf("\n Enter the value after which the data has to be inserted : ");
        scanf("%d", &val);
        new_node = (struct node *)malloc(sizeof(struct node *));
        new_node->data = num;
        ptr = start;
        while(ptr->data != val)
                ptr = ptr->next;
        new_node->prev = ptr;
        new_node->next = ptr->next;
        ptr->next->prev = new_node;
        ptr->next = new_node;
        return start;
}
struct node *insert_sorted(struct node *start)
{
        struct node *new_node, *ptr;
        int num;
        printf("\n Enter the data : ");
```

```
        scanf("%d", &num);
        new_node = (struct node *)malloc(sizeof(struct node *));
        new_node->data = num;
        ptr = start;
        if(ptr->data > 0)
        {
        ptr = start;
        while(ptr->next != start)
            ptr = ptr->next;
        new_node->prev = ptr;
        ptr->next = new_node;
        new_node->next = start;
        start->prev = new_node;
        start = new_node;
        }
        else
        {
            while(ptr->data < num)
            {
                ptr = ptr->next;
                if(ptr->next == start)
                    break;
            }
            if(ptr->next == start)
            {
                ptr->next = new_node;
                new_node->prev = ptr;
                new_node->next = start;
            }
            else
            {
                new_node->next = ptr;
                new_node->prev = ptr->prev;
                ptr->prev->next = new_node;
                ptr->prev = new_node;
            }
        }
        return start;
    }
    struct node *delete_beg(struct node *start)
    {
        struct node *ptr;
        ptr = start;
        while(ptr->next != start)
            ptr = ptr->next;
        ptr->next = start->next;
```

```
		start -> next -> prev = ptr;
		free(start);
		start = start -> next;
		return start;
	}
	struct node *delete_end(struct node *start)
	{
		struct node *ptr;
		ptr=start;
		while(ptr -> next != start)
			ptr = ptr -> next;
		ptr -> prev -> next = start;
		start -> prev = ptr -> prev;
		free(ptr);
		return start;
	}
	struct node *delete_node(struct node *start)
	{
		struct node *ptr;
		int val;
		printf("\n Enter the value of the node which has to be deleted : ");
		scanf("%d", &val);
		ptr = start;
		if(ptr -> data == val)
		{
			start = delete_beg(start);
			return start;
		}
		else
		{
			while(ptr -> data != val)
			ptr = ptr -> next;
			ptr -> prev -> next = ptr -> next;
			ptr -> next -> prev = ptr -> prev;
			free(ptr);
			return start;
		}
	}
	struct node *delete_after(struct node *start)
	{
		struct node *ptr, *temp;
		int val;
		printf("\n Enter the value after which the node has to deleted : ");
		scanf("%d", &val);
		ptr = start;
		while(ptr -> data != val)
```

```
            ptr = ptr->next;
        temp = ptr->next;
        if(temp == start)
            start = delete_beg(start);
        else
        {
            ptr->next = temp->next;
            temp->next->prev = ptr;
            free(temp);
        }
        return start;
}
struct node *delete_before(struct node *start)
{
        struct node *ptr, *temp;
        int val;
        printf("\n Enter the value before which the node has to deleted : ");
        scanf("%d", &val);
        ptr = start;
        if(ptr->data == val)
            start = delete_end(start);
        else
        {
            while(ptr->data != val)
                ptr = ptr->next;
            temp = ptr->prev;
            if(temp == start)
                start = delete_beg(start);
            else
            {
                ptr->prev = temp->prev;
                temp->prev->next = ptr;
            }
            free(temp);
        }
        return start;
}
struct node *delete_sorted(struct node *start)
{
        struct node *ptr;
        int val;
        printf("\n Enter the value of the node which has to be deleted : ");
        scanf("%d", &val);
        ptr = start;
        if(ptr->data == val)
            start = delete_beg(start);
```

```
        else
        {
            while(ptr->data != val)
                ptr = ptr->next;
            ptr->prev->next = ptr->next;
            free(ptr);
        }
        return start;
    }
    struct node *delete_list(struct node *start)
    {
        struct node *ptr;
        ptr = start;
        while(ptr->next != start)
            start = delete_end(start);
        free(start);
        return start;
    }
```

8.9 HEADER LINKED LIST

A header linked list is a special type of linked list which contains a header node at the beginning of the list. So, in a header linked list, START will not point to the first node of the list but START will contain the address of the header node. The following are the two variants of a header linked list:

- *Grounded header linked list* which stores NULL in the next field of the last node.
- *Circular header linked list* which stores the address of the header node in the next field of the last node. Here, the header node will denote the end of the list.

Look at Fig. 8.76 which shows both the types of header linked lists.

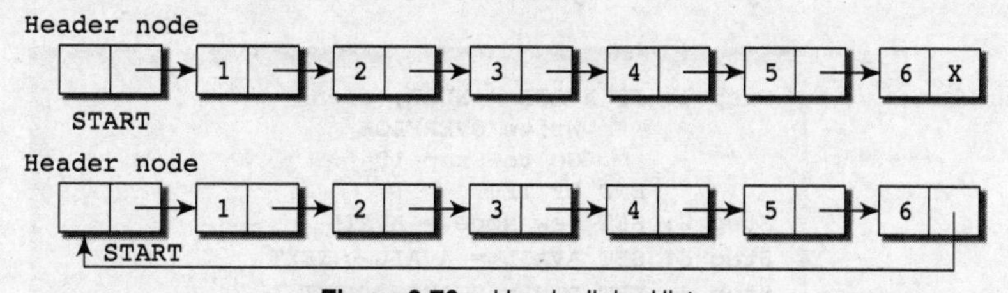

Figure 8.76 Header linked list

As in other linked lists, if START = NULL, then this denotes an empty header linked list. Let us see how a grounded header linked list is stored in the memory. In order to form a grounded header linked list, we need a structure called the node which has two fields, DATA and NEXT. The DATA will store the information part and the NEXT will store the address of the node in sequence. Consider Fig. 8.77.

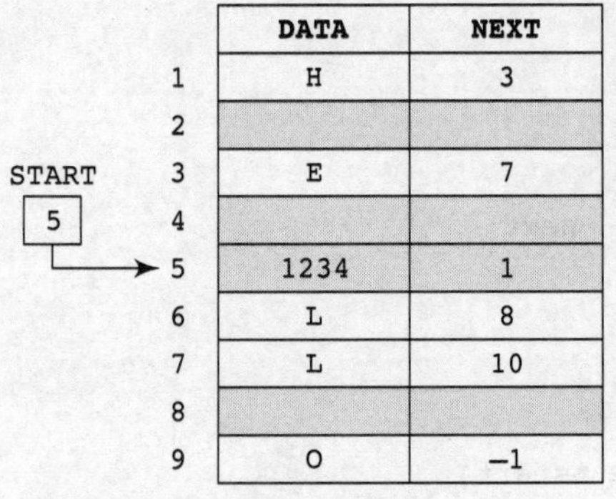

	DATA	NEXT
1	H	3
2		
3	E	7
4		
5	1234	1
6	L	8
7	L	10
8		
9	O	–1

START = 5

Figure 8.77 Memory representation of a header linked list

Note that START stores the address of the header node. The shaded row denotes a header node. The NEXT field of the header node stores the address of the first node of the list. This node stores H. The corresponding NEXT field stores the address of the next node, which is 3. So, we will look at address 3 to fetch the next data item.

Hence, we see that the first node can be accessed by writing first_node = START -> NEXT and not by writing START = first_node. This is because START points to the header node and the header node points to the first node of the header linked list.

Let us now see how a circular header linked list is stored in the memory. Look at Fig. 8.78.

	DATA	NEXT
1	H	3
2		
3	E	7
4		
5	1234	1
6	L	8
7	L	10
8		
9	O	5

START = 5

Figure 8.78 Memory representation of a circular header linked list

Note that the last node in this case stores the address of the header node (instead of –1).

Hence, we see that the first node can be accessed by writing first_node = START -> NEXT and not writing START = first_node. This is because START points to the header node and the header node points to the first node of the header linked list.

Let us quickly look at Figs 8.79, 8.80, and 8.81 that show the algorithms to traverse a circular header linked list, insert a new node in it, and delete an existing node from it.

```
Step 1: SET PTR = START -> NEXT
Step 2: Repeat Steps 3 and 4 while PTR != START
Step 3:           Apply PROCESS to PTR -> DATA
Step 4:           SET PTR = PTR -> NEXT
            [END OF LOOP]
Step 5: EXIT
```

Figure 8.79 Algorithm to traverse a circular header linked list

```
Step 1: IF AVAIL = NULL, then
                Write OVERFLOW
                Go to Step 10
            [END OF IF]
Step 2: SET New_Node = AVAIL
Step 3: SET AVAIL = AVAIL -> NEXT
Step 4: SET PTR = START -> NEXT
Step 5: SET New_Node -> DATA = VAL
Step 6: Repeat step 4 while PTR -> DATA != NUM
Step 7:         SET PTR = PTR -> NEXT
            [END OF LOOP]
Step 8: New_Node -> NEXT = PTR -> NEXT
Step 9: SET PTR -> NEXT = New_Node
Step 10: EXIT
```

Figure 8.80 Algorithm to insert a new node in a circular header linked list

```
Step 1: SET PTR = START->NEXT
Step 2: Repeat Steps 3 and 4 while PTR -> DATA != VAL
Step 3:      SET TEMP = PTR
Step 4:           SET PTR = PTR -> NEXT
          [END OF LOOP]
Step 5: SET TEMP -> NEXT = PTR -> NEXT
Step 6: FREE PTR
Step 7: EXIT
```

Figure 8.81 Algorithm to delete a node from a circular header linked list

After discussing linked lists in such detail, these algorithms are self-explanatory. There is actually just one small difference in these algorithms and the algorithms that we have discussed earlier. Like we have a header list and a circular header list, we also have a two-way (doubly) header list and a circular two-way (doubly) header list. The algorithms to perform all the basic operations will be exactly the same except that the first node will be accessed by writing `START -> NEXT` instead of `START`.

SUMMARY

- A linked list is a linear collection of data elements. These data elements are called nodes.
- Linked list is a data structure which, in turn, can be used to implement other data structures like stacks, queues, and their variations. Linked list contains a pointer variable `START` which stores the address of the first node in the list.
- `AVAIL` stores the address of the first free space.
- Before we insert a new node in the linked lists, we need to check for `OVERFLOW` condition, which occurs when `AVAIL = NULL` or no free memory cell is present in the system.
- Before we delete a node from a linked list, we must first check for `UNDERFLOW` condition which occurs when we try to delete a node from a linked list that is empty. This happens when `START = NULL` or when there are no more nodes to delete and even then we try to delete.
- When we delete a node from the linked list, we have to actually free the memory occupied by that node. The memory is returned back to the free pool so that it can be used to store other useful programs and data. Whatever be the case of deletion, we always change the `AVAIL` pointer so that it now points to the address that has been recently vacated.
- In a circular linked list, the last node contains a pointer to the first node of the list. While traversing a circular linked list, we can begin at any node and traverse the list in any direction forward or backward until we reach the same node where we had started.
- A doubly linked list or a two-way linked list is a more complex type of linked list which contains a pointer to the next as well as the previous node in the sequence. Therefore, it consists of three parts, and not just two. The three parts are data, a pointer to the next node, and a pointer to the previous node
- The `prev` field of the first node and the next field of the last node will contain `NULL`. It is used to store the address of the preceding node. This would enable to traverse the list in the backward direction as well.
- Thus, we see that a doubly linked list calls for more space per node and more expensive basic operations. However, a doubly linked list provides the ease to manipulate the elements of the list as it maintains pointers to nodes in both the directions (forward and backward). The main advantage of using a doubly linked list is that it makes searches twice as efficient.
- A circular doubly linked list or a circular two-way linked list is a more complex type of linked list which contains a pointer to the next as well as previous node in the sequence. The difference between a doubly linked and a circular doubly

linked list is the same as that exists between a singly linked list and a circular linked list. The circular doubly linked list does not contain NULL in the previous field of the first node and the next field of the last node. Rather, the next field of the last node stores the address of the first node of the list, i.e. START. Similarly, the previous field of the first field stores the address of the last node.

- A header linked list is a special type of linked list which contains a header node at the beginning of the list. So, in a header linked list START will not point to the first node of the list but START will contain the address of the header node.

GLOSSARY

Doubly linked list A type of a linked list in which each node has a link to the previous item as well as the next. This enables ease of accessing nodes backward as well as forward and deleting any node in constant time.

Linked list A list in which each node has a link to the next node.

Link A reference, pointer, or access handle to another part of the data structure. Usually, it stores the memory address of next node in the list.

Circular linked list A type of a linked list in which the last node is linked to the head. The nodes of the list may be accessed starting at any item and following links until one comes to the starting item again.

EXERCISES

Review Questions

1. Make a comparison between a linked list and a linear array. Which one will you prefer to use and when?
2. Why is a doubly linked list more useful than a singly linked list?
3. Give the advantages and uses of a circular linked list.
4. Specify the use of a header node in a header linked list.
5. Give the linked representation of the following polynomial.

   $$7x^3y^2 - 8x^2y + 3xy + 11x - 4$$

6. Explain the difference between a circular linked list and a singly linked list.
7. Form a linked list to store student's details.
8. Use the linked list of the above question to insert the record of a new student in the list.
9. Delete the record of a student with a specified roll number from the list maintained in Question 14.
10. Given a linked list that contains alphabets. The alphabets may be in upper case or in lower case. Create two linked lists- one which stores upper case alphabets and the other that stores lower case characters.
11. Create a linked list which stores names of the employees. Then, sort these names and re-display the contents of the linked list.
12. Why are linked lists more preferred than using arrays?

Programming Exercises

1. Write a program that removes all nodes that have duplicate information.
2. Write a program to print the total number of occurrences of a given item in the linked list.
3. Write a program to multiply every element of the linked list with 10.
4. Write a program to print the number of non-zero elements in the list.
5. Write a program that prints whether the given linked list is sorted (in ascending order) or not.
6. Write a program that copies a circular linked list.
7. Write a program to merge two linked lists.
8. Write a program to sort the values stored in a doubly circular linked list.
9. Write a program to merge two sorted linked lists. The resultant list must also be sorted.
10. Write a program to delete the first, last, and middle node of a header linked list.

11. Write a program to create a linked list from an already given list. The new linked list must contain every alternate element of the existing linked list.
12. Write a program to concatenate two doubly linked lists.
13. Write a program to delete the first element of a doubly linked list. Add this node as the last node of the list.
14. Write a program to-
    (a) Delete the first occurrence of a given character in a linked list
    (b) Delete the last occurrence of a given character
    (c) Delete all the occurrences of a given character
15. Write a program to reverse a linked list using recursion.
16. Write a program to input an *n* digit number. Now, break this number into its individual digits and then store every single digit in a separate node thereby forming a linked list. For example, if you enter 12345, now there will 5 nodes in the list containing nodes with values 1, 2, 3, 4, 5.
17. Write a program to sum the values of the nodes of a linked list and then calculate the mean.
18. Write a program that prints minimum and maximum value in a linked list that stores integer values.
19. Write a program to interchange the value of the first element with the last element, second element with second last element, so on and so forth of a doubly linked list.
20. Write a program to make the first element of singly linked list, the last element of the list.
21. Write a program to count the number of occurrences of a given value in a linked list.
22. Write a program that adds 10 to the values stored in the nodes of a doubly linked list.
23. Write a program to form a linked list of integer values. Calculate the sum of integers and then display the average of the numbers in the list.
24. Write a program to delete the kth node from a linked list.
25. Write a program to perform deletions in all the cases of a circular header linked list.
26. Write a program to multiply a polynomial with a given number.
27. Write a program to count the number of non-zero values in a circular linked list.
28. Write a program to create a linked list which stores the details of a student. Read and print the information stored in such a list.
29. Modify the program in Question 29 so that it displays the record of a given student only.
30. Using the structure created in Question 29, write a program to insert a new student's information in the list.
31. Using the structure created in Question 29, write a program to delete a student's information from the list.
32. Write a program to make a middle node of doubly link list to the top of the list.
33. Write a program to create a singly linked list and reverse the list by interchanging the links and not the data.
34. Write a program that prints the nth element from the end of a linked list in a single pass.
35. Write a program that creates a singly linked list. Use a function isSorted that returns 1 if the list is sorted and 0 otherwise.
36. Write a program to interchange the kth and the `(k+1)`th node of a circular doubly linked list.
37. Write a program to create a header linked list.
38. Write a program to delete a node from a circular header linked list.
39. Write a program to delete all nodes from a header linked list that has negative values in its data part.

Multiple Choice Questions

1. A linked list is a
    (a) Random access structure
    (b) Sequential access structure
    (c) Both
2. An array is a
    (a) Random access structure
    (b) Sequential access structure
    (c) Both
3. Linked list is used to implement data structures like
    (a) Stacks (b) Queues
    (c) Trees (d) Graphs
    (e) All of these

4. Which type of linked list contains a pointer to the next as well as previous node in the sequence?
   (a) Singly linked list
   (b) Circular linked list
   (c) Doubly linked list
   (d) All of these
5. Which type of linked list contains a pointer to the next as well as previous node in the sequence?
   (a) Singly linked list
   (b) Circular linked list
   (c) Doubly linked list
   (d) All of these
6. Which type of linked list does not store NULL in next field?
   (a) Singly linked list
   (b) Circular linked list
   (c) Doubly linked list
   (d) All of these
7. Which type of linked list stores the address of the header node in the next field of the last node?
   (a) Singly linked list
   (b) Circular linked list
   (c) Doubly linked list
   (d) Circular header linked list
8. The operation that involves accessing the nodes of the list in order to perform some processing on them
   (a) Inserting (b) Deleting
   (c) Searching (d) Traversing

True or False

1. A linked list is a linear collection of data elements.
2. A linked list can grow and shrink during run time.
3. A node in a linked list can point to only one node at a time.
4. A node in the singly linked list can reference the previous node.
5. A linked list can store only integer values.
6. Linked list is a random access structure.
7. Deleting a node from a doubly linked list is easier than deleting it from a singly linked list.
8. Every node in a linked list contains an integer part and a pointer.
9. START stores the address of the first node in the list.
10. Underflow is a condition that occurs when we try to delete a node from a linked list that is empty.

Fill in the blanks

1. ______ is used to store the address of the first free memory location.
2. The complexity to insert a node at the beginning of the linked list is ______.
3. The complexity to delete a node from the end of the linked list is ______.
4. Inserting a node at the beginning of the doubly linked list needs to modify ______ pointers.
5. Inserting a node in the middle of the singly linked list needs to modify ______ pointers.
6. Inserting a node at the end of the circular linked list needs to modify ______ pointers.
7. Inserting a node at the beginning of the circular doubly linked list needs to modify ______ pointers.
8. Deleting a node from the beginning of the singly linked list needs to modify ______ pointers
9. Deleting a node from the middle of the doubly linked list needs to modify ______ pointers
10. Deleting a node from the end of a circular linked list needs to modify ______ pointers
11. Each element in a linked list is known as a ______.
12. First node in the linked list is called the ______.
13. Data elements in a linked list is known as ______.
14. Overflow occurs when ______.
15. In a circular linked list, the last node contains a pointer to the ______ node of the list.

9 Stacks and Queues

Learning Objective

Stacks and queues are abstract data types. They can be implemented using either a linear array or a linked list. In this chapter, we will discuss what these data structures actually are, their applications, and the operations associated with them. We will also learn how to implement stacks and queues using arrays as well as linked lists. Apart from simple queues, we have different kinds of queues like priority queue, circular queue, dequeues, etc. This chapter deals with all these terms in detail.

9.1 STACKS

Stack is an important data structure which stores its elements in an ordered manner. Take an analogy; you must have seen a pile of plates where one plate is placed on top of another as shown in Fig. 9.1. Now, when you want to remove a plate, you remove the topmost plate first. Hence, you can add and remove an element (i.e. the plate) only at/from one position which is the topmost position.

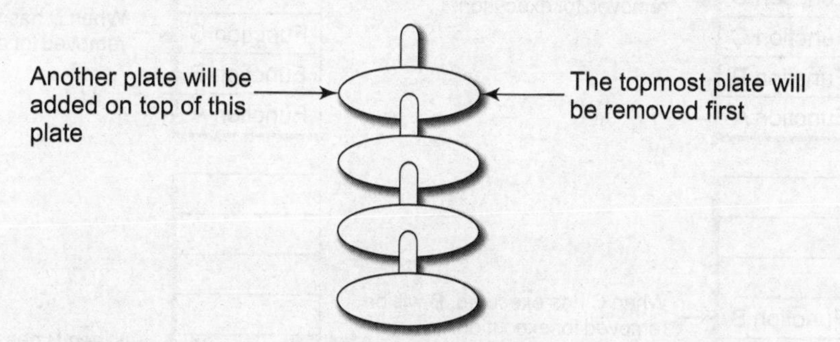

Figure 9.1 Stack of plates

Same is the case with a stack. A stack is a linear data structure which can be implemented by either using an array or a linked list. The elements in a stack are added and removed only from one end, which is called the `top`. Hence, a stack is called a LIFO (Last-In-First-Out) data structure, as the element that was inserted last is the first one to be taken out.

Now the big question is where do we need stacks in computer science? The answer is in function calls. Consider an example, where we are executing function A. In the course of its execution, function A calls another function B. Now, function B calls another function C, which in turn calls function D.

This scenario can be viewed in the form of a stack. Whenever a function calls another function, the calling function is pushed onto the top of the stack. This is because after the called function gets executed, the control is passed back to the calling function. Look at Fig. 9.2 which shows this concept.

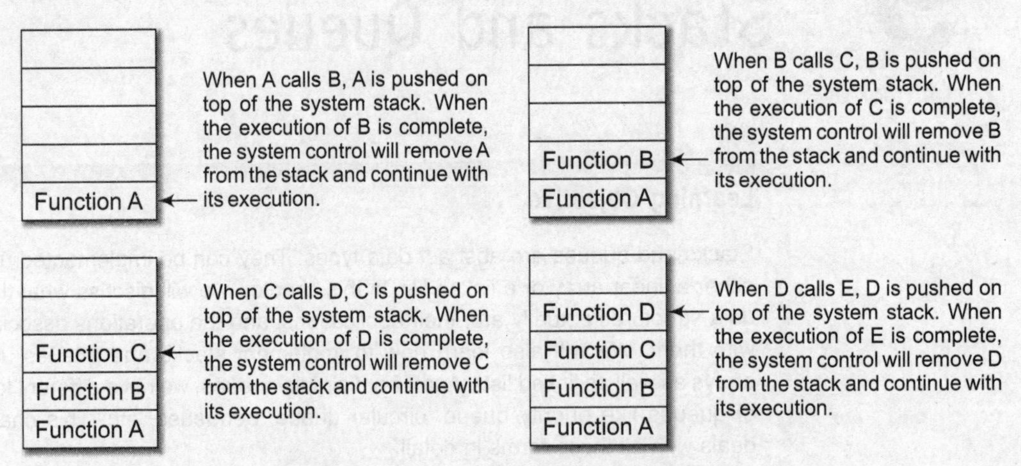

Figure 9.2 System stack in the case of function calls

Now when function `E` is executed, function `D` will be removed from the top of the stack and executed. Once function `D` gets completely executed, function `C` will be removed from the stack for execution. The whole procedure will be repeated until all the functions get executed. Let us look at the stack after each function is executed. This is shown in Fig. 9.3.

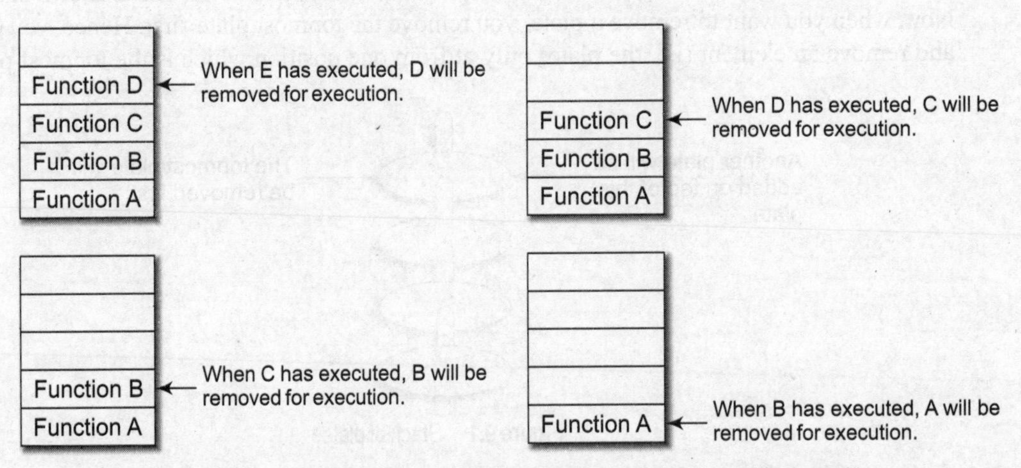

Figure 9.3 System stack when a called function returns to the calling function

Here, the stack ensures a proper execution order of functions. Therefore, stacks are frequently used in situations where the order of processing is very important, especially when the processing needs to be postponed until other conditions are fulfilled.

9.2 ARRAY REPRESENTATION OF STACKS

In the computer's memory, stacks can be represented as a linear array. Every stack has a variable called `TOP` associated with it. `TOP` is used to store the address of the topmost element of the stack.

It is in this position where the element will be added or deleted. There is another variable called MAX, which is used to store the maximum number of elements that the stack can hold.

If TOP = NULL, then it indicates that the stack is empty and if TOP = MAX–1, then the stack is full. (You must be wondering why we have written MAX–1. It is because array indices start from 0). Look at Fig. 9.4.

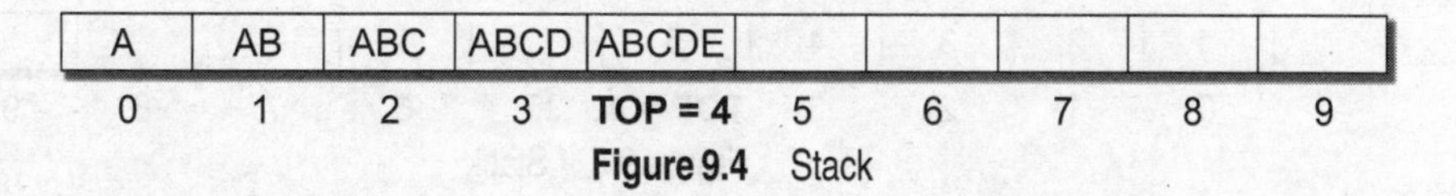

Figure 9.4 Stack

The stack in Fig. 9.4 shows that TOP = 4, so insertions and deletions will be done at this position. In the above stack, five more elements can still be stored.

9.3 OPERATIONS ON A STACK

A stack has three basic operations: push, pop, and peep. The push operation adds an element to the top of the stack and the pop operation removes the element from the top of the stack. The peep operation returns the value of the topmost element of the stack.

9.3.1 Push Operation

The push operation is used to insert an element into the stack. The new element is added at the topmost position of the stack. However, before inserting the value, we must first check if TOP=MAX–1, because if that is the case, then the stack is full and no more insertions can be further done. If an attempt is made to insert a value in a stack that is already full, an OVERFLOW message is printed. Consider the stack array given in Fig. 9.5.

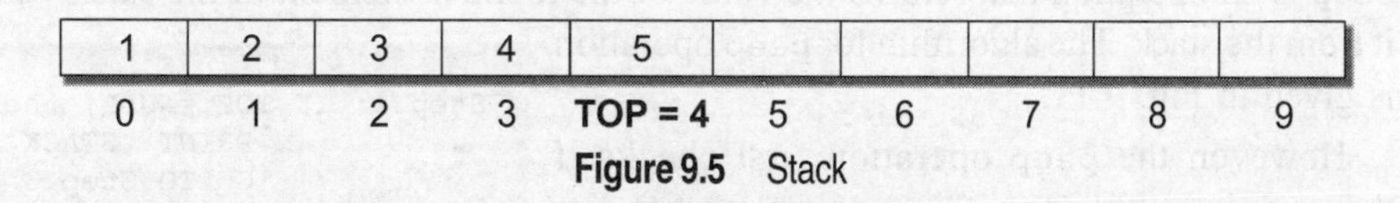

Figure 9.5 Stack

To insert an element with the value 6, we first check if TOP=MAX–1. If the condition is false, then we increment the value of TOP and store the new element at the position given by stack[TOP]. Thus, the updated stack becomes as shown in Fig. 9.6.

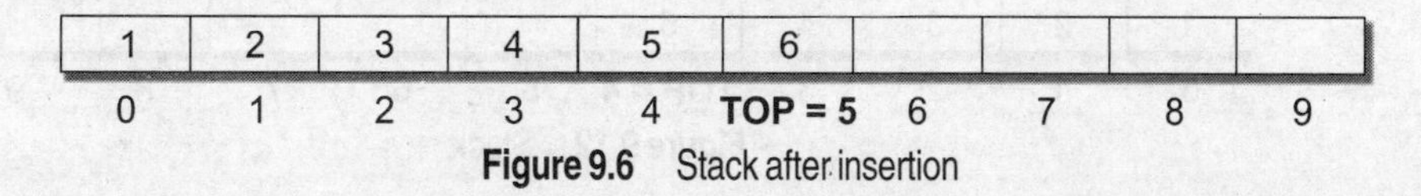

Figure 9.6 Stack after insertion

Figure 9.7 shows the algorithm to insert an element in the stack. In Step 1, we first check for the overflow condition. In Step 2, TOP is incremented so that it points to the next free location in the array. In Step 3, the value is stored in the stack array at the location pointed by the TOP.

```
Step 1: IF TOP = MAX-1, then
            PRINT "OVERFLOW"
        [END OF IF]
Step 2: SET TOP = TOP + 1
Step 3: SET STACK[TOP] = VALUE
Step 4: END
```

Figure 9.7 Algorithm to insert an element in the stack

9.3.2 Pop Operation

The pop operation is used to delete the topmost element from the stack. However, before deleting the value, we must first check if TOP=NULL because if that is the case, then it means the stack is empty and no more deletions can further be done. If an attempt is made to delete a value from a stack that is already empty, an UNDERFLOW message is printed. Consider the stack array given in Fig. 9.8.

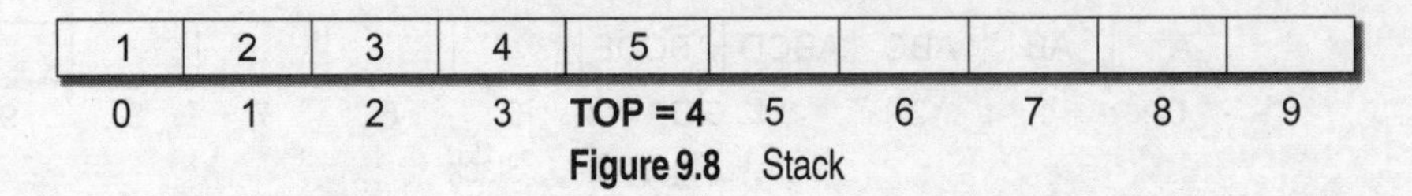

Figure 9.8 Stack

To delete the topmost element, we first check if TOP=NULL. If the condition is false, then we decrement the value of top. Thus, the updated stack becomes as shown in Fig. 9.9.

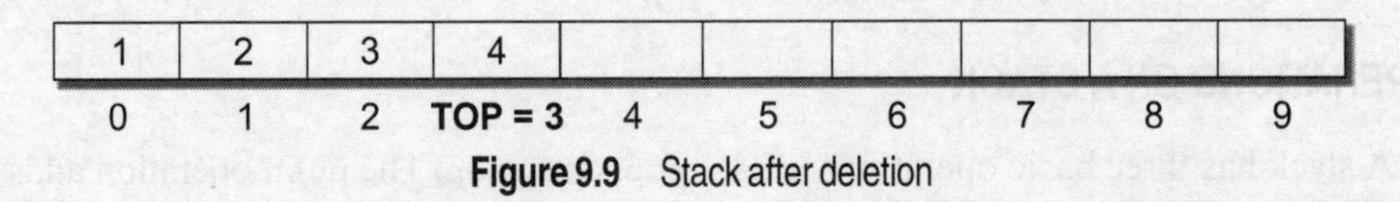

Figure 9.9 Stack after deletion

Figure 9.10 shows the algorithm to delete an element from the stack. In Step 1, we first check for the underflow condition. In Step 2, the value of the location in the stack array pointed by the TOP is stored in VAL. In Step 3, TOP is decremented.

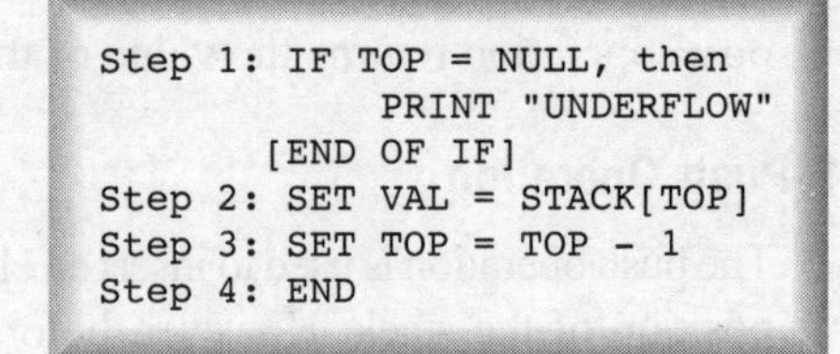

Figure 9.10 Algorithm to insert an element in the stack

9.3.3 Peep Operation

Peep is an operation that returns the value of the topmost element of the stack without deleting it from the stack. The algorithm for peep operation is given in Fig. 9.11.

However, the peep operation first checks if the stack is empty or contains some elements. For this, a condition is checked. If TOP = NULL, then an appropriate message is printed, else the value is returned. Consider the stack array given in Fig. 9.12.

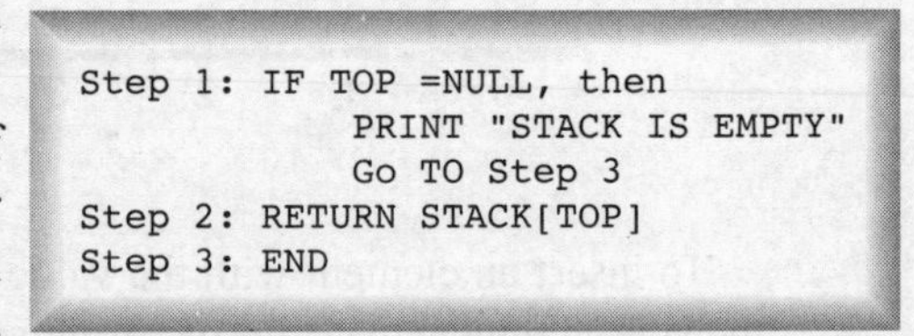

Figure 9.11 Algorithm for peep operation

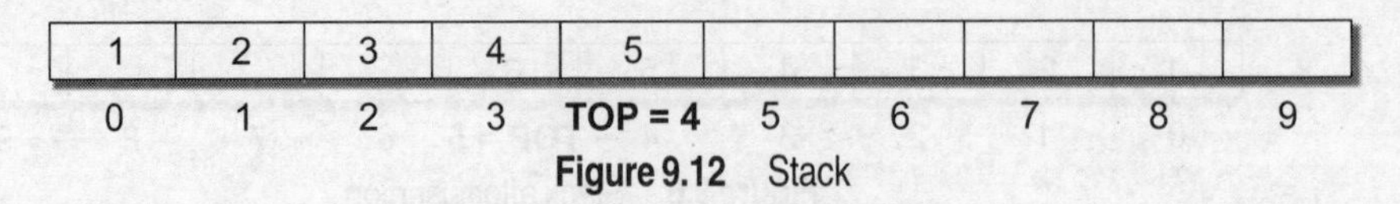

Figure 9.12 Stack

Here, the peep operation will return 5, as it is the value of the topmost element of the stack.

PROGRAMMING EXAMPLE

1. Write a program to perform Push, Pop, and Peep operations on a stack array.

```
#include<stdio.h>
#include<conio.h>
```

```
#define MAX 10
int st[MAX], top=-1;
void push(int st[], int val);
int pop(int st[]);
int peep(int st[]);
void display(int st[]);
main()
{
  int val, option;
  clrscr();
  do
  {
      printf("\n *****MAIN MENU*****");
      printf("\n 1. PUSH");
      printf("\n 2. POP");
      printf("\n 3. PEEK");
      printf("\n 4. DISPLAY");
      printf("\n *******************");
      printf("\n\n Enter your option : ");
      scanf("%d", &option);
      switch(option)
      {
            case 1:
                  printf("\n ENter the number to be pushed on to the stack
: ");
                  scanf("%d", &val);
                  push(st, val);
                  break;
            case 2:
                  val = pop(st);
                  printf("\n The value deleted from the stack is : %d", val);
                  break;
            case 3:
                  val = peek(st);
                  printf("\n The value stored at the top of the stack is :
%d", val);
                  break;
            case 4:
                  display(st);
                  break;
      }
  }while(option != 5);
  getch();
  return 0;
}
void push(int st[], int val)
{
      if(top == MAX-1)
```

```
        {
              printf("\n STACK OVERFLOW");
        }
        else
        {
              top++;
              st[top] = val;
        }
  }
  int pop(int st[])
  {
        int val;
        if(top == -1)
        {
              printf("\n STACK UNDERFLOW");
              return -1;
        }
        else
        {
              val = st[top];
              top--;
              return val;
        }
  }
  void display(int st[])
  {
        int i;
        if(top == -1)
              printf("\n STACK IS EMPTY");
        else
        {
              for(i=top;i>=0;i--)
              printf("\n%d",st[i]);
        }
  }
  int peep(int st[])
  {
        If(TOP == NULL)
        {
              printf("\n STACK IS EMPTY");
              return -1;
        }
        else
              return (st[top]);
  }
```

9.4 LINKED REPRESENTATION OF STACK

We have seen how an array is created using an array. Although this technique of creating a stack is easy, but the drawback is that the array must be declared to have some fixed size. In case the stack is a very small one or its maximum size is known in advance, then the array implementation of the stack gives an efficient implementation. But if the array size cannot be determined in advance, the other alternative, i.e. linked representation, is used.

The storage requirement of linked representation of the stack with n elements is `O(n)` and the typical time requirement for the operation is `O(1)`.

In a linked stack, every node has two parts– one that stores data and another that stores the address of the next node. The `START` pointer of the linked list is used as `TOP`. All insertions and deletions are done at the node pointed by the `TOP`. If `TOP = NULL`, then it indicates that the stack is empty.

9.5 OPERATIONS ON A LINKED STACK

A linked stack supports all the three stack operations, that is, `push`, `pop`, and `peep`. Consider the linked stack shown in Fig. 9.13.

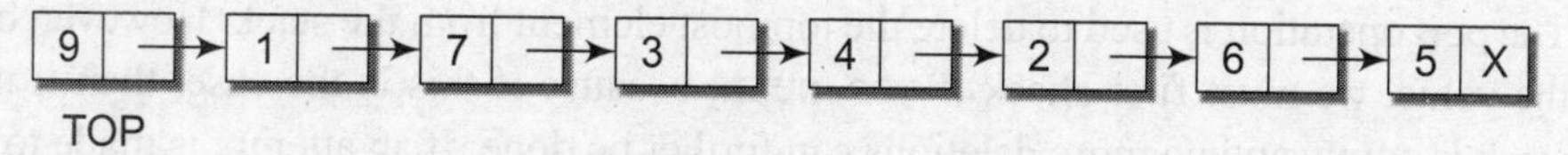

Figure 9.13 Linked stack

9.5.1 Push Operation

The push operation is used to insert an element into the stack. The new element is added at the topmost position of the stack. Consider the linked stack shown in Fig. 9.14.

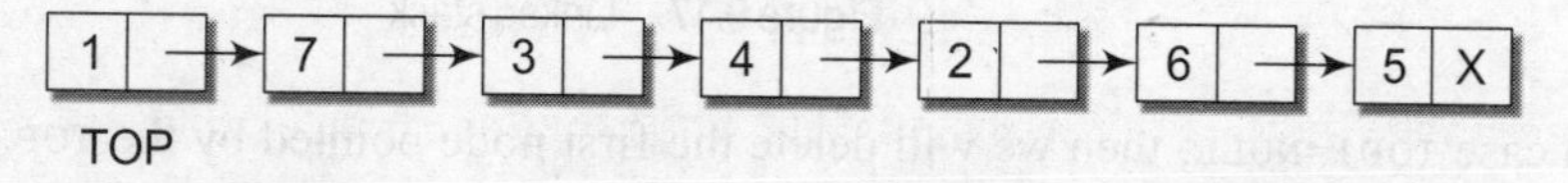

Figure 9.14 Linked stack

To insert an element with the value 9, we first check if `TOP=NULL`. If it is the case, then we allocate memory for a new node, store the value in its `data` part and `NULL` in its `next` part. The new node will then be called the `TOP`. However, if `TOP!=NULL`, then we insert the new node at the beginning of the linked stack and name this new node as `TOP`. Thus, the updated stack becomes as shown in Fig. 9.15.

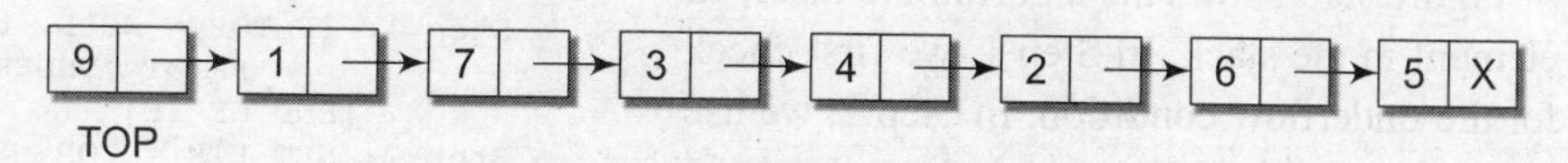

Figure 9.15 Linked stack after inserting a new node

Figure 9.16 shows the algorithm to push an element into a linked stack. In Step 1, the memory is allocated for the new node. In Step 2, the `DATA` part of the new node is initialized with the value

```
Step 1: Allocate memory for the new node and name it as New_Node
Step 2: SET New_Node -> DATA = VAL
Step 3: IF TOP = NULL, then
             SET New_Node -> NEXT = NULL
             SET TOP = New_Node
        ELSE
             SET New_node -> NEXT = TOP
             SET TOP = New_Node
        [END OF IF]
Step 4: END
```

Figure 9.16 Algorithm to insert an element in a linked stack

to be stored in the node. In Step 3, we check if the new node is the first node of the linked list. This is done by checking if `TOP = NULL`. In case the `if` statement evaluates to true, then `NULL` is stored in the `NEXT` part of the node and the new node is called the `TOP`. However, if the new node is not the first node in the list, then it is added before the first node of the list (that is, the `TOP` node) and termed as `TOP`.

9.5.2 Pop Operation

The pop operation is used to delete the topmost element from the stack. However, before deleting the value, we must first check if `TOP=NULL`, because if this is the case, then it means that the stack is empty and no more deletions can further be done. If an attempt is made to delete a value from a stack that is already empty, an `UNDERFLOW` message is printed. Consider the stack shown in Fig. 9.17.

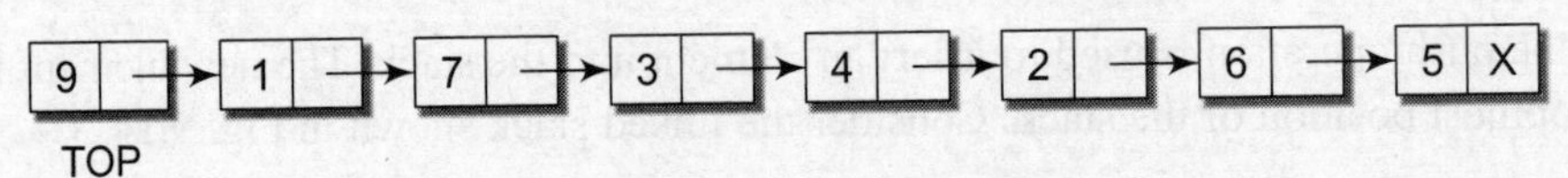

Figure 9.17 Linked stack

In case `TOP!=NULL`, then we will delete the first node pointed by the `TOP`. The `TOP` will now point to the second element of the linked stack. Thus, the updated stack becomes as shown in Fig. 9.18.

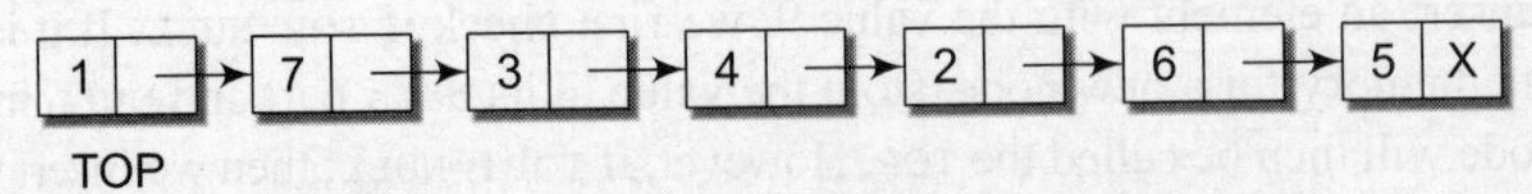

Figure 9.18 Linked stack after deletion of the topmost element

Figure 9.19 shows the algorithm to insert an element in the stack. In Step 1, we first check for the underflow condition. In Step 2, we use a `PTR` that points to the `TOP`. In Step 3, `TOP` is made to point to the next node in sequence. In Step 4, the memory occupied by `PTR` is given back to the free pool.

```
Step 1: IF TOP = NULL, then
             PRINT "UNDERFLOW"
        [END OF IF]
Step 2: SET PTR = TOP
Step 3: SET TOP = TOP -> NEXT
Step 4: FREE PTR
Step 5: END
```

Figure 9.19 Algorithm to delete an element from a stack

PROGRAMMING EXAMPLE

2. Write a program to implement a linked stack.

```
#include<stdio.h>
#include<conio.h>
struct stack
{
      int data;
      struct stack *next;
};
struct stack *top = NULL;
struct stack *push(struct stack *, int);
struct stack *display(struct stack *);
struct stack *pop(struct stack *);
int peek(struct stack *);
main()
{
      int val, option;
      clrscr();
      do
      {
            printf("\n *****MAIN MENU*****");
            printf("\n 1. PUSH");
            printf("\n 2. POP");
            printf("\n 3. PEEK");
            printf("\n 4. DISPLAY");
            printf("\n 5. EXIT");
            printf("\n *******************");
            printf("\n\n Enter your option : ");
            scanf("%d", &option);
            switch(option)
            {
                  case 1:
                        printf("\n ENter the number to be pushed on to the
stack : ");
                        scanf("%d", &val);
                        top = push(top, val);
                        break;
                  case 2:
                        top = pop(top);
                        break;
                  case 3:
                        val = peek(top);
                        printf("\n The value stored at the top of the stack
is : %d", val);
                        break;
```

```
                    case 4:
                            top = display(top);
                            break;
                }
        }while(option != 5);
        getch();
        return 0;
}
struct stack *push(struct stack *top, int val)
{
        struct stack *ptr;
        ptr = (struct stack*)malloc(sizeof(struct stack *));
        ptr->data = val;
        if(top == NULL)
        {
                top = ptr;
                top->next = NULL;
        }
        else
        {
                ptr->next = top;
                top = ptr;
        }
        return top;
}
struct stack *display(struct stack *top)
{
        struct stack *ptr;
        ptr = top;
        if(top == NULL)
                printf("\n STACK IS EMPTY");
        else
        {
                while(ptr != NULL)
                {
                        printf("\n%d", ptr->data);
                        ptr = ptr->next;
                }
        }
        return top;
}
struct stack *pop(struct stack *top)
{
        struct stack *ptr;
        ptr = top;
        if(top == NULL)
                printf("\n STACK UNDERFLOW");
        else
```

```
        {
                top = top->next;
                printf("\n The value being deleted is : %d", ptr->data);
                free(ptr);
        }
        return top;
}
int peek(struct stack *top)
{
        return top->data;
}
```

9.6 MULTIPLE STACKS

While implementing a stack array, we had seen that the size of the array must be known in advance. If the stack is allocated less space, then frequent `OVERFLOW` conditions will be encountered. To deal with this problem, the code will have to be modified to reallocate more space for the array.

In case we allocate a large amount of space for the stack, it will result in sheer waste of memory. Thus, there lies a trade-off between the frequency of overflows and the space allocated.

So, a better solution to deal with this problem is to have multiple stacks or to have more than one stack in the same array of sufficient size. Figure 9.20 illustrates this concept.

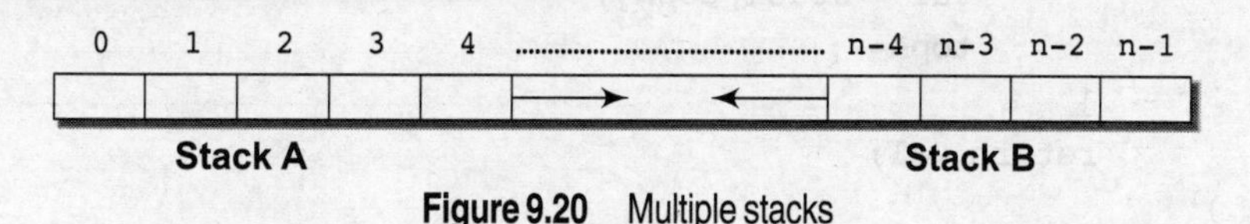

Figure 9.20 Multiple stacks

In Fig 9.20, an array `Stack[n]` is used to represent two stacks, `Stack A` and `Stack B`. The value of `n` is such that the combined size of both the stack will never exceed `n`. While operating on these stacks, it is important to note one thing. While `Stack A` will grow from left to right, `Stack B` will grow from right to left at the same time.

Extending the concept of multiple stacks, a `STACK` can also be used to represent `n` number of stacks in the same array. That is, if we have a `STACK[n]`, then each `stack I` will be allocated an equal amount of space bounded by indices `b[i]` and `e[i]`. This is shown in Fig. 9.21.

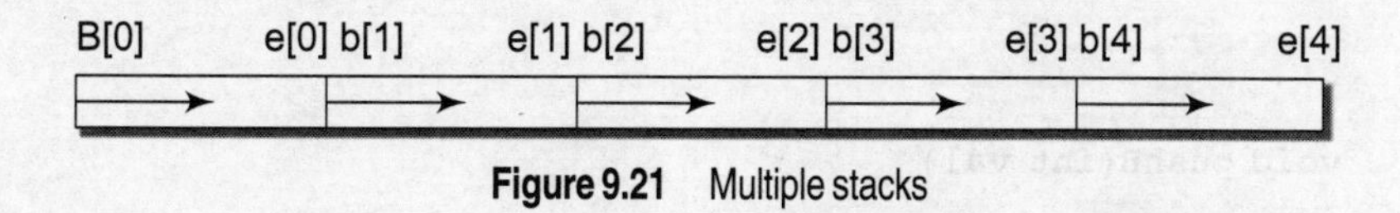

Figure 9.21 Multiple stacks

PROGRAMMING EXAMPLE

3. Write a program to implement multiple stacks.

```
#include<stdio.h>
#include<conio.h>
```

```
#define MAX 10
int stack[MAX],topA = -1,topB = MAX;
void pushA(int val)
{
        if(topA == topB)
                printf("\n OVERFLOW");
        else
        {
                topA += 1;
                stack[topA] = val;
        }
}
int popA()
{
        int val;
        if(topA == -1)
        {
                printf("\n UNDERFLOW");
                val = -999;
        }
        else
        {
                val = stack[topA];
                topA--;
        }
        return val;
}
void display_stackA()
{
        int i;
        if(topA == -1)
                printf("\n stack Empty");
        else
        {
                for(i=topA;i>=0;i-)
                        printf("\n %d",stack[i]);
        }
}
void pushB(int val)
{
        if(topB-1 == topA)
                printf("\n OVERFLOW");
        else
        {
                topB- = 1;
                stack[topB] = val;
```

```
        }
}
int popB()
{
        int val;
        if(topB == MAX)
        {
                printf("\n UNDERFLOW");
                val = -999;
        }
        else
        {
                val = stack[topB];
                topB--;
        }
}
void display_stackB()
{
        int i;
        if(topB == MAX)
                printf("\n stack Empty");
        else
        {
                for(i=topB;i<MAX;i++)
                        printf("\n %d",stack[i]);
        }
}
main()
{
        clrscr();
        int option, val;
        do
        {
                printf("\n *******MENU******");
                printf("\n 1. PUSH IN STACKA");
                printf("\n 2. PUSH IN STACKB");
                printf("\n 3. POP FROM STACKA");
                printf("\n 4. POP FROM STACKB");
                printf("\n 5. DISPLAY STACKA");
                printf("\n 6. DISPLAY STACKB");
                printf("\n 7. EXIT");
                printf("\nENter your CHoice");
                scanf("%d",&option);
                switch(option)
                {
```

```
                    case 1:printf("\n Enter the value to pushed on StackA :
    ");
                        scanf("%d",&val);
                        pushA(val);
                        break;
                    case 2: printf("\n Enter the value to be pushed in StackB
    : ");
                        scanf("%d",&val);
                        pushB(val);
                        break;
                    case 3: val=popA();
                        if(val!=-999)
                        printf("\n The value popped from StackA = %d",val);
                        break;
                    case 4 : val=popB();
                        if(val!=-999)
                        printf("\n The value popped from StackB = %d",val);
                        break;
                    case 5: printf("\n The contents of StackA are : \n");
                        showA();
                        break;
                    case 6: printf("\n The contents of StackB are : \n");
                        showB();
                        break;
                }
            }while(option!=7);
            getch();
    }
```

9.7 INFIX, POSTFIX, AND PREFIX NOTATION

Infix, postfix, and prefix notations are three different but equivalent notations of writing algebraic expressions. Let us first look at what an infix notation is. We are all familiar with the infix notation of writing algebraic expressions. While writing an arithmetic expression using the infix notation, the operator is placed in between the operands. For example, `A+B`; here, the '+' operator is placed between the two operands `A` and `B`. Although it is easy for us to write expressions using infix notation, but computers find it difficult to parse because they need a lot of information to evaluate the expression. Information is needed about operator precedence, associativity rules, and brackets which override these rules. So, computers work more efficiently with expressions written using prefix and postfix notations.

Postfix notation was given by Jan Lukasiewicz who was a Polish logician, mathematician, and philosopher. His aim was to develop a parenthesis-free prefix notation (also known as `Polish notation`) and a postfix notation which is better known as the `Reverse Polish Notation` or `RPN`.

In the postfix notation, as the name suggests, the operator is placed after the operands. For example, if an expression is written as A+B in infix notation, the same expression can be written as AB+ in postfix notation. The order of evaluation of a postfix expression is always from left to right. Even brackets cannot alter the order of evaluation.

Similarly, the expression (A + B) * C is written as

[AB+]*C

AB+C* in the postfix notation.

A postfix operation does not even follow the rules of operator precedence. The operator which occurs first in the expression is operated first on the operands. For example, given a postfix notation AB+C*, addition will be performed prior to multiplication.

EXAMPLE 9.1: Convert the following infix expressions into postfix expressions.

(i) (A-B) * (C+D)

[AB-] * [CD+]

AB-CD+*

(ii) (A + B) / (C + D) - (D * E)

[AB+] / [CD+] - [DE*]

[AB+CD+/] - [DE*]

AB+CD+/DE*-

Thus, we see that in the postfix notation, operators are applied to the operands that are immediately left to them. In the example, AB+C*, + is applied on A and B, then * is applied on the result of the addition and C.

Although a prefix expression is also evaluated from left to right, the operators in this case are placed before the operands. For example, if A+B is an expression in infix notation, then the corresponding expression in prefix notation is given by +AB.

While evaluating a prefix expression, the operators are applied to the operands that are present immediately on the right of the operator. As is the case with postfix expressions, prefix expressions also do not follow the rules of operator precedence and associativity, and even brackets cannot alter the order of evaluation.

EXAMPLE 9.2: Convert the following infix expressions into prefix expressions.

(i) (A + B) * C

(+AB)*C

*+ABC

(ii) (A-B) * (C+D)

[-AB] * [+CD]

*-AB+CD

(iii) (A + B) / (C + D) - (D * E)

[+AB] / [+CD] - [*DE]

[/ +AB+CD] - [*DE]

-/+AB+CD*DE

9.8 EVALUATION OF AN INFIX EXPRESSION

The ease of evaluation acts as the driving force for computers to translate an infix notation into a postfix notation. That is, given an algebraic expression written in infix notation, the computer first converts the expression into its equivalent postfix notation and then evaluates the postfix expression.

Both these tasks, converting the infix notation into postfix notation and evaluating the postfix expression, make extensive use of stacks as the primary tool.

Step 1: Convert the infix expression into its equivalent postfix expression

Let `I` be an algebraic expression written in infix notation. `I` may contain parentheses, operands, and operators. For simplicity of the algorithm, we only use `+`, `–`, `*`, `/`, `%` operators. The precedence of these operators can be given as:

Higher priority: `*`, `/`, `%`

Lower priority: `+`, `–`

The order of evaluation of these operators can be changed by using parentheses. For example, if we have an expression `A + B * C`, then `B * C` is evaluated first and the result is added to `A`. But, when the same expression is written as `(A + B) * C`, it will evaluate `A + B` first and then, the result will be multiplied with `C`.

The algorithm shown in Fig. 9.22 transforms an infix expression into postfix expression. The algorithm accepts an infix expression that may contain operators, operands, and parentheses. For simplicity, we assume that the infix operation contains only exponentiation (`^`), multiplication (`*`), division (`/`), addition (`+`), and subtraction (`-`) operators and that the operators with the same precedence are performed from left-to-right.

The algorithm uses a stack to temporarily hold operators. The postfix expression is obtained from left-to-right using the operands from the infix expression and the operators which are removed from the stack. The first step in this algorithm is to push a left parenthesis on the stack and adding a corresponding right parenthesis at the end of the infix expression. The algorithm is repeated until the stack is empty.

```
Step 1: Add ")" to the end of the infix expression
Step 2: Push "(" on to the stack
Step 3: Repeat until each character in the infix notation is scanned
        IF a "(" is encountered, push it on the stack
        IF an operand (whether a digit or an alphabet) is encountered, add it
        to the postfix expression.
        IF a ")" is encountered, then;
              a. Repeatedly pop from stack and add it to the postfix expression
                 until a "(" is encountered.
              b. Discard the "(". That is, remove the "(" from stack and do not
                 add it to the postfix Expression
        IF an operator O is encountered, then;
              a. Repeatedly pop from stack and add each operator (Popped from
                 the stack) to the postfix expression until it has the same
                 precedence or a Higher precedence than O
              b. Push the operator O to the stack
        [END OF IF]
Step 4: Repeatedly pop from the stack and add it to the postfix expression
        until the stack is empty
Step 5: EXIT
```

Figure 9.22 Algorithm to convert infix notation to postfix

EXAMPLE 9.3: Convert the following infix expression into postfix using the algorithm given in Fig. 9.22.

```
A - ( B / C + (D % E * F) / G )* H
A - ( B / C + (D % E * F) / G )* H )
```

Infix Character Scanned	STACK	Postfix Expression
	(	
A	(	A
-	(-	A
(	(- (	A
B	(- (	A B
/	(- (/	A B
C	(- (/	A B C
+	(- (+	A B C /
(	(- (+ (	A B C /
D	(- (+ (	A B C / D
%	(- (+ (%	A B C / D
E	(- (+ (%	A B C / D E
*	(- (+ (% *	A B C / D E
F	(- (+ (% *	A B C / D E F
)	(- (+	A B C / D E F * %
/	(- (+ /	A B C / D E F * %
G	(- (+ /	A B C / D E F * % G
)	(-	A B C / D E F * % G / +
*	(- *	A B C / D E F * % G / +
H	(- *	A B C / D E F * % G / + H
)		A B C / D E F * % G / + H * -

PROGRAMMING EXAMPLE

4. Write a program to convert an infix expression into its equivalent postfix notation.

```
#include<stdio.h>
#include<conio.h>
#define MAX 100

char st[MAX];
int top = -1;
void push(char st[], char);
char pop(char st[]);
void InfixtoPostfix(char source[], char target[]);
int getPriority(char);

main()
{
```

```
        char infix[100], postfix[100];
        clrscr();
        printf("\n Enter any infix expression : ");
        fflush(stdin);
        gets(infix);
        strcpy(postfix, "");
        InfixtoPostfix(infix, postfix);
        printf("\n The corresponding postfix expression is : ");
        puts(postfix);
        getch();
        return 0;
}
void InfixtoPostfix(char source[], char target[])
{
        int i=0, j=0;
        char temp;
        strcpy(target, "");
        while(source[i] != '\0')
        {
                if(source[i] == '(')
                {
                        push(st, source[i]);
                        i++;
                }
                else if(source[i] == ')')
                {
                        while((top != -1) && (st[top] != '('))
                        {
                                target[j] = pop(st);
                                j++;
                        }
                        if(top==-1)
                        {
                                printf("\n INCORRECT EXPRESSION");
                                exit(1);
                        }
                        temp = pop(st);//remove left paranthesis
                        i++;
                }
                else if(isdigit(source[i]) || isalpha(source[i]))
                {
                        target[j] = source[i];
                        j++;
                        i++;
                }
                else if( source[i] == '+' || source[i] == '-' || source[i] ==
'*' || source[i] == '/' || source[i] == '%')
```

```
        {
                while( (top != -1) && (st[top] != '(') &&
(getPriority(st[top]) > getPriority(source[i])))
                {
                        target[j] = pop(st);
                        j++;
                }
                push(st, source[i]);
                i++;
        }
        else
        {
                printf("\n INCORRECT ELEMENT IN EXPRESSION");
                exit(1);
        }
    }
    while((top != -1) && (st[top] != '('))
    {
        target[j] = pop(st);
        j++;
    }
    target[j] = '\0';
}
int getPriority( char op)
{
    if(op == '/' || op == '*' || op == '%')
        return 1;
    else if(op == '+' || op == '-')
        return 0;
}
void push(char st[], char val)
{
    if(top == MAX-1)
        printf("\n STACK OVERFLOW");
    else
    {
        top++;
        st[top] = val;
    }
}
char pop(char st[])
{
    char val = ' ';
    if(top == -1)
        printf("\n STACK UNDERFLOW");
    else
    {
```

```
            val = st[top];
            top--;
        }
        return val;
    }
```

Step 2: Evaluate the postfix expression

Using stacks, any postfix expression can be evaluated very easily. Every character of the postfix expression is scanned from left to right. If the character encountered is an operand, it is pushed onto the stack. However, if an operator is encountered, then the top two values are popped from the stack and the operator is applied on these values. The result is then pushed onto the stack. Let us look at Fig. 9.23 which shows the algorithm to evaluate a postfix expression.

```
Step 1: Add a ")" at the end of the postfix expression
Step 2: Scan every character of the postfix expression and repeat
        steps 3 and 4 until ")"is encountered
Step 3: IF an operand is encountered, push it on the stack
        IF an operator O is encountered, then
        a. pop the top two elements from the stack as A and B
        b. Evaluate B O A, where A was the topmost element and B
           was the element below A.
        c. Push the result of evaluation on the stack
        [END OF IF]
Step 4: SET RESULT equal to the topmost element of the stack
Step 5: EXIT
```

Figure 9.23 Algorithm to evaluate a postfix expression

Let us now take an example that makes use of this algorithm. Consider the infix expression given as `9 - (( 3 * 4) + 8) / 4`. Evaluate the expression.

The infix expression `9 - (( 3 * 4) + 8) / 4` can be written as `9 3 4 * 8 + 4 / -` using postfix notation. Look at Table 9.1 which shows the procedure.

Table 9.1 Evaluation of a postfix expression

Character scanned	Stack
9	9
3	9, 3
4	9, 3, 4
*	9, 12
8	9, 12, 8
+	9, 20
4	9, 20, 4
/	9, 5
-	4

PROGRAMMING EXAMPLE

5. Write a program to evaluate a postfix expression.

```
#include<stdio.h>
#include<conio.h>
#define MAX 100
float st[MAX];
int top = -1;
void push(float st[], float val);
float pop(float st[]);
float evaluatePostfixExp(char exp[]);
main()
{
        float val;
        char exp[100];
        clrscr();
        printf("\n Enter any postfix expression : ");
        fflush(stdin);
        gets(exp);
        val = evaluatePostfixExp(exp);
        printf("\n Value of the postfix expression = %.2f", val);
        getch();
        return 0;
}
float evaluatePostfixExp(char exp[])
{
        int i=0;
        float op1, op2, value;
        while(exp[i] != '\0')
        {
                if(isdigit(exp[i]))
                        push(st, (float)(exp[i]-'0'));
                else
                {
                        op2 = pop(st);
                        op1 = pop(st);
                        switch(exp[i])
                        {
                                case '+':
                                        value = op1 + op2;
                                        break;
                                case '-':
                                        value = op1 - op2;
                                        break;
```

```
                        case '/':
                                value = op1 / op2;
                                break;
                        case '*':
                                value = op1 * op2;
                                break;
                        case '%':
                                value = (int)op1 % (int)op2;
                                break;
                }
                push(st, value);
            }
            i++;
        }
        return(pop(st));
}
void push(float st[], float val)
{
        if(top == MAX-1)
            printf("\n STACK OVERFLOW");
        else
        {
            top++;
            st[top] = val;
        }
}
float pop(float st[])
{
        float val = -1;
        if(top == -1)
            printf("\n STACK UNDERFLOW");
        else
        {
            val = st[top];
            top--;
        }
        return val;
}
```

9.9 CONVERT INFIX EXPRESSION TO PREFIX EXPRESSION

There are two algorithms to convert an infix expression into its equivalent prefix expression. These two algorithms are given in Figs 9.24 and 9.25.

```
Step 1: Scan each character in the infix expression. For this, repeat Steps
        2-8 until the end of infix expression.
Step 2: Push the operator into the operator stack, operand into the operand
        stack, and ignore all the left parentheses until a right parenthesis
        is encountered.
Step 3: Pop operand 2 from operand stack.
Step 4: Pop operand 1 from operand stack.
Step 5: Pop operator from operator stack.
Step 6: Concatenate operator and operand 1.
Step 7: Concatenate result with operand 2.
Step 8: Push result into the operand stack.
```

Figure 9.24 First algorithm

The corresponding prefix expression is obtained in the operand stack.

```
Step 1: Reverse the infix string. Note that while reversing the string, you
        must interchange left and right parenthesis.
Step 2: Obtain the corresponding postfix expression of the infix expression
        obtained as a result of Step 1.
Step 3: Reverse the postfix expression to get the prefix expression
```

Figure 9.25 Second algorithm

Let us find out the postfix equivalent of the infix expression: `(A – B / C) * (A / K – L)`

Step 1: Reverse the infix string. Note that while reversing the string, you must interchange the left and right parenthesis.

`(L – K / A) * (C / B – A)`

Step 2: Obtain the corresponding postfix expression of the infix expression obtained as a result of Step 1.

The expression is: `(L – K / A) * (C / B – A)`

Therefore, `[L – (K A /)] * [(C B /) – A]`

`= [LKA/–] * [CB/A–]`

`= L K A / – C B / A – *`

Step 3: Reverse the postfix expression to get the prefix expression.

Therefore, the prefix expression is **`* – A / B C – / A K L`**

PROGRAMMING EXAMPLE

6. Write a program to convert an infix expression to its prefix equivalent.

```
#include<stdio.h>
#include<conio.h>
#include<string.h>
#define MAX 100
char st[MAX];
int top = -1;
```

```
void reverse(char str[]);
void push(char st[], char);
char pop(char st[]);
void InfixtoPostfix(char source[], char target[]);
int getPriority(char);
char infix[100], postfix[100], temp[100];
main()
{
        clrscr();
        printf("\\n Enter any infix expression : ");
        fflush(stdin);
        gets(infix);
        reverse(infix);
        puts(temp);
        strcpy(postfix, "");
        InfixtoPostfix(temp, postfix);
        printf("\n The corresponding postfix expression is : ");
        puts(postfix);
        strcpy(temp,"");
        reverse(postfix);
        printf("\n The prefix expression is : \n");
        puts(temp);
        getch();
        return 0;
}
void reverse(char str[])
{
        int len, i=0, j=0;;
        len=strlen(str);
        j=len-1;
        while(j >= 0)
        {
                if ( str[j] == '(')
                temp[i] = ')';
else if ( str[j] == ')')
temp[i] = '(';
                else
                        temp[i] = str[j];
                i++, j--;
        }
        temp[i] = '\0';
}
void InfixtoPostfix(char source[], char target[])
{
        int i=0, j=0;
```

```
        char temp;
        strcpy(target, "");
        while(source[i] != '\0')
        {
           if(source[i] == '(')
           {
                 push(st, source[i]);
                 i++;
           }
           else if(source[i] == ')')
           {
                 while((top != -1) && (st[top] != '('))
                 {
                       target[j] = pop(st);
                       j++;
                 }
                 if(top == -1)
                 {
                       printf("\n INCORRECT EXPRESSION");
                       exit(1);
                 }
                 temp = pop(st);                 //remove left anthesis
                 i++;
           }
           else if(isdigit(source[i]) || isalpha(source[i]))
           {
                 target[j] = source[i];
                 j++;
                 i++;
           }
           else if( source[i] == '+' || source[i] == '-' || source[i] == '*'
|| source[i] == '/' || source[i] == '%')
           {
                 while( (top!=-1) && (st[top]!= '(') &&
(getPriority(st[top]) > getPriority(source[i])))
                 {
                       target[j] = pop(st);
                       j++;
                 }
                       push(st, source[i]);
                       i++;
           }
           else
           {
              printf("\n INCORRECT ELEMENT IN EXPRESSION");
              exit(1);
```

```
            }
        }
        while((top!=-1) && (st[top]!='('))
        {
                target[j] = pop(st);
                j++;
        }
        target[j] = '\0';
}
int getPriority( char op)
{
        if(op=='/' || op == '*' || op=='%')
                return 1;
        else if(op=='+' || op=='-')
                return 0;
}
void push(char st[], char val)
{
        if(top == MAX-1)
                printf("\n STACK OVERFLOW");
        else
        {
                top++;
                st[top] = val;
        }
}
char pop(char st[])
{
        char val = ' ';
        if(top == -1)
                printf("\n STACK UNDERFLOW");
        else
        {
                val = st[top];
                top--;
        }
        return val;
}
```

9.10 APPLICATIONS OF STACK

Stacks are widely used to:

- Reverse the order of data (we have already seen its example in Recursion)
- Convert infix expression into postfix

- Convert postfix expression into infix
- Backtracking problem (as discussed in Appendix)
- System stack is used in every recursive function
- Converting a decimal number into its binary equivalent.

PROGRAMMING EXAMPLE

7. Write a program to convert a decimal number to its binary equivalent.

```
#include<stdio.h>
#include<conio.h>
#define MAX 10
int st[MAX], top = -1;
void push(int st[], int val);
int pop(int st[]);
main()
{
    int num, digit;
    clrscr();
    printf("\n ENter any decimal number : ");
    scanf("%d", &num);
    while(num > 0)
    {
        digit = num%2;
        push(st, digit);
        num /= 2;
    }
    printf("\n The binary equivalent is : ");
    while(top != -1)
        printf("%d", pop(st));
    getch();
    return 0;
}
void push(int st[], int val)
{
    if(top == MAX-1)
        printf("\n STACK OVERFLOW");
    else
    {
        top++;
        st[top] = val;
    }
}
int pop(int st[])
{
```

```
        int val;
        if(top == -1)
        {
                printf("\n STACK UNDERFLOW");
                return -1;
        }
        else
        {
                val = st[top];
                top--;
                return val;
        }
}
```

Note that we can use the same code to convert a decimal number into any base system by replacing 2 with the base.

9.11 QUEUES

Queue is an important data structure which stores its elements in an ordered manner. Take for example the analogies given below.

- People moving on an escalator. The people who got on the escalator first will be the first one to step out of it.
- People waiting for a bus. The first person standing in the line will be the first one to get into the bus.
- People standing outside the ticketing window of a cinema hall. The first person in the line will get the ticket first and thus will be the first one to move out of it.
- Luggage kept on conveyor belts. The bag which was placed first will be the first to come out at the other end.
- Cars lined for filling petrol. The car which came first will be filled first.
- Cars lined at a toll bridge. The first car to reach the bridge will be the first to leave.

Figure 9.26 Queue

In all these examples, we see that the element at the first position is served first (refer Fig. 9.26). Whenever a new car comes to get petrol filled, it is added at the end of the line (at one end of the queue). Similarly, when petrol is filled, the first car in the line leaves the queue and moves away (from the other end of the queue). Same is the case with queue data structure. A queue is a `FIFO` (First-In, First-Out) data structure in which the element that was inserted first is the first one to be taken out. The elements in a queue are added at one end called the `rear` and removed from the other end called the `front`.

Queues can be implemented by either using arrays or linked lists. In this section, we will see how queues are implemented using each of these data structures.

9.12 ARRAY REPRESENTATION OF QUEUES

Queues can be easily represented using linear arrays. As stated earlier, every queue has front and rear variables that point to the position from where deletions and insertions can be done respectively. Consider a queue shown in Fig. 9.27.

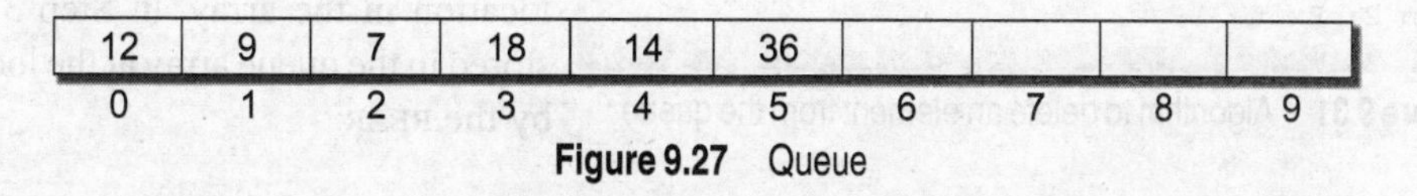

Figure 9.27 Queue

Here, `front` = `0` and `rear` = `5`. Suppose we want to add another element with the value 45, then the rear would be incremented by 1 and the value would be stored at the position pointed by the rear. The queue after addition would be as shown in Fig. 9.28.

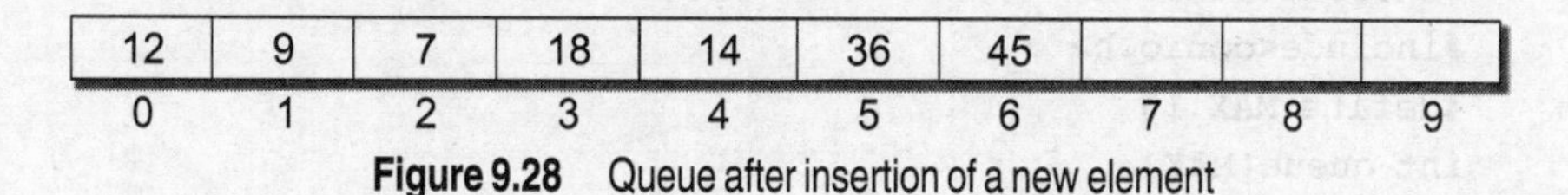

Figure 9.28 Queue after insertion of a new element

Here, `front` = `0` and `rear` = `6`. Every time a new element has to be added, we repeat the same procedure.

If we want to delete an element from the queue, then the value of the front will be incremented. Deletions are done from only this end of the queue. The queue after deletion will be as shown in Fig. 9.29.

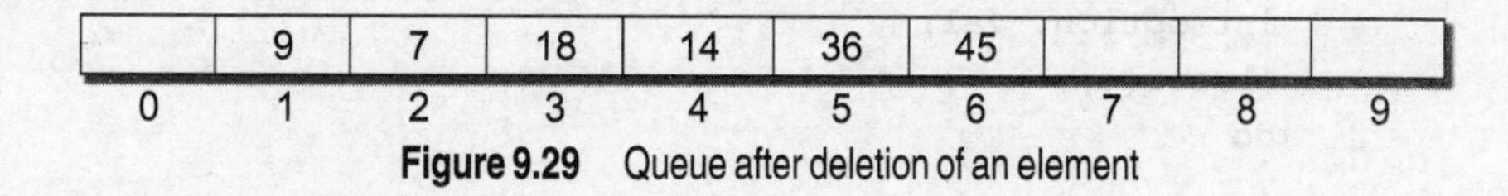

Figure 9.29 Queue after deletion of an element

Here, `front` = `1` and `rear` = `6`.

However, before inserting an element in the queue, we must check for `overflow` conditions. An `overflow` will occur when we try to insert an element into a queue that is already full. When `Rear` = `MAX - 1`, where `MAX` is the size of the queue, we have an overflow condition. Note that we have written `MAX - 1`, because the index starts from 0.

```
Step 1: IF REAR = MAX-1, then;
            Write OVERFLOW
        [END OF IF]
Step 2: IF FRONT == -1 and REAR = -1, then;
            SET FRONT = REAR = 0
        ELSE
            SET REAR = REAR + 1
        [END OF IF]
Step 3: SET QUEUE[REAR] = NUM
Step 4: Exit
```

Figure 9.30 Algorithm to insert an element in the queue

Similarly, before deleting an element from the queue, we must check for `underflow` conditions. An `underflow` condition occurs when we try to delete an element from a queue that is already empty. If `front` = `-1` and `rear` = `-1`, it means there is no element in the queue. Let us now look at Figs 9.30 and 9.31 which shows the algorithms to insert and delete an element from a queue.

Figure 9.31 shows the algorithm to insert an element in the queue. In Step 1, we first check for the `overflow` condition. In Step 2, we check if the queue is empty. In case the queue is

```
Step 1: IF FRONT = -1 OR FRONT > REAR, then;
            Write UNDERFLOW
        ELSE
            SET FRONT = FRONT + 1
            SET VAL = QUEUE[FRONT]
        [END OF IF]
Step 2: Exit
```

Figure 9.31 Algorithm to delete an element from the queue

initially empty, then the value of `front` and `rear` is set to zero, so that the new value can be stored at the 0th location. Otherwise, if the queue already has some values, then the `REAR` is incremented so that it points to the next free location in the array. In Step 3, the value is stored in the queue array at the location pointed by the `REAR`.

PROGRAMMING EXAMPLE

8. Write a program to implement a linear queue.

```
#include<stdio.h>
#include<conio.h>
#define MAX 10
int queue[MAX];
int front = -1, rear = -1;
void insert(void);
int delete_element(void);
int peek(void);
void display(void);
main()
{
        int option, val;
        clrscr();
        do
        {
                printf("\n\n ***** MAIN MENU *****");
                printf("\n 1. Insert an element");
                printf("\n 2. Delete an element");
                printf("\n 3. Peek");
                printf("\n 4. Display the queue");
                printf("\n 5. EXIT");
                printf("\n ************************");
                printf("\n\n Enter your option : ");
                scanf("%d", &option);
                switch(option)
                {
                    case 1:
                            insert();
                            break;
                    case 2:
                            val = delete_element();
                            printf("\n The number that was deleted is : %d", val);
                            break;
```

```
                        case 3:
                                val = peek();
                                printf("\n The first value in the queue is : %d", val);
                                break;
                        case 4:
                                display();
                                break;
                }
        }while(option != 5);
        getch();
        return 0;
}
void insert()
{
        int num;
        printf("\n Enter the number to be inserted in the queue : ");
        scanf("%d", &num);
        if(rear == MAX-1)
                printf("\n OVERFLOW");
        if(front == -1 && rear == -1)
                front = rear = 0;
        else
                rear++;
        queue[rear] = num;
}
int delete_element()
{
        int val;
        if(front == -1 || front>rear)
         {
                printf("\n UNDERFLOW");
                return -1;
         }
        else
        {
                front++;
                val = queue[front];
                return val;
        }
}
int peek()
{
        return queue[front];
}
void display()
{
```

```
        int i;
        printf("\n");
        for(i = front;i <= rear;i++)
                printf("\t %d", queue[i]);
}
```

9.13 CIRCULAR QUEUE

In linear lists, we have discussed so far that insertions can be done only at one end called the `rear` and deletion is always done from the other end called the `front`. Look at the queue shown in Fig. 9.32.

54	9	7	18	14	36	45	21	99	72
0	1	2	3	4	5	6	7	8	9

Figure 9.32 Queue

Here, `front = 0` and `rear = 9`.

Now, if you want to insert another value, it will not be possible because the queue is completely full. There is no empty space where the value can be inserted. Consider a scenario in which two successive deletions are made. The queue will then be given as shown in Fig. 9.33.

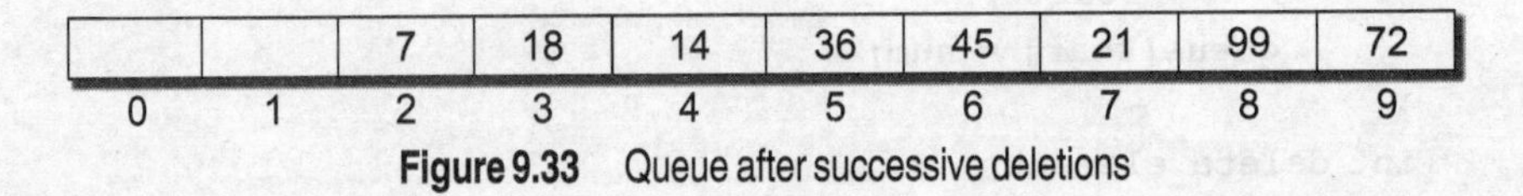

Figure 9.33 Queue after successive deletions

Here, `front = 2` and `rear = 9`.

Suppose we want to insert a new element in the queue shown in Fig. 9.33. Even though there is space available, the `OVERFLOW` condition still exists because the condition `rear = MAX - 1` still holds true. This is a major drawback of a linear queue.

To cater to this situation, we have two solutions. First, shift the elements to the left so that the vacant space can be occupied and utilized efficiently. But this can be very time-consuming, especially when the queue is quite large.

The second option is to use a circular queue. In the circular queue, the first index comes right after the last index. Conceptually, you can think of a circular queue as shown in Fig. 9.34.

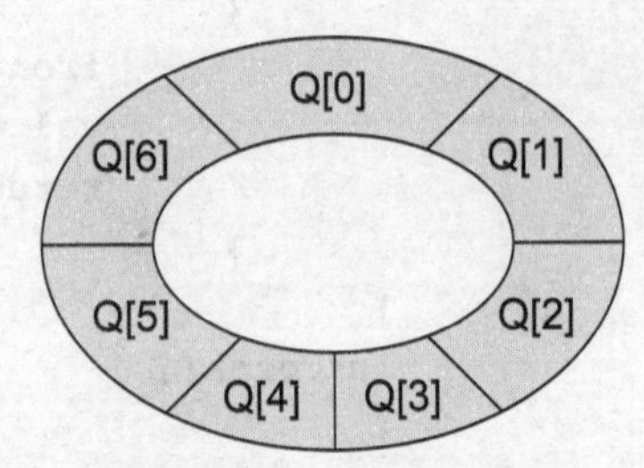

Figure 9.34 Circular queue

The circular queue will be full, only when `front = 0` and `rear = Max - 1`. A circular queue is implemented in the same manner as a linear queue is implemented. The only difference will be in the code that performs insertion and deletion operations. For insertion, we now have to check for the following three conditions:

- If `front = 0` and `rear = MAX - 1`, then print that the circular queue is full. Look at the queue given in Fig. 9.35 which illustrates this point.

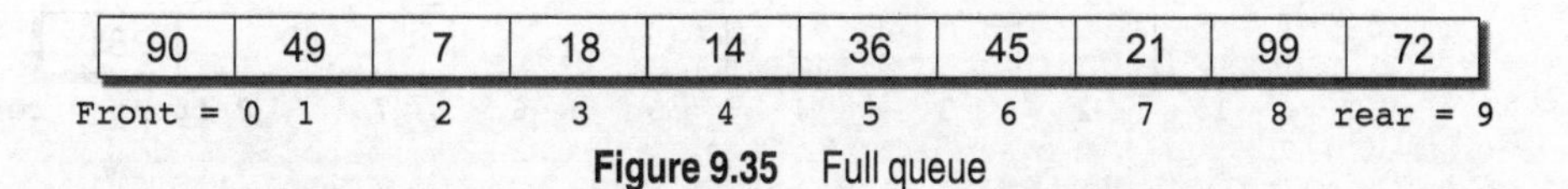

Figure 9.35 Full queue

- If `rear != MAX - 1`, then the value will be inserted and `rear` will be incremented as illustrated in Fig. 9.36.

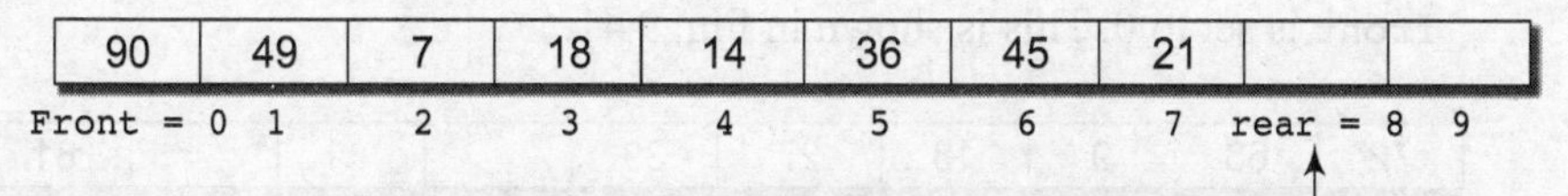

Figure 9.36 Queue with vacant locations

- If `front != 0` and `rear = MAX - 1`, then it means that the queue is not full. So, set `rear = 0` and insert the new element there, as shown in Fig. 9.37.

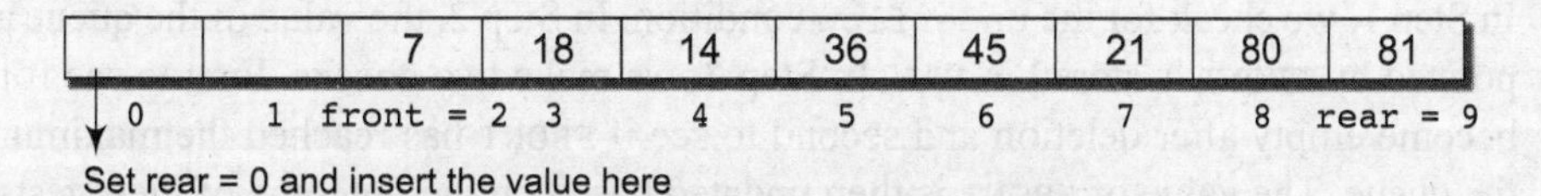

Figure 9.37 Inserting an element in a circular queue

Let us look at Fig. 9.38 which shows the algorithm to insert an element in a circular queue. In Step 1, we make three checks. First for the `overflow` condition, second to see if the queue is initially empty, and third to see if the `REAR` end has already reached the maximum capacity while there are certain free locations before the `FRONT` end. In Step 2, the value is stored in the queue array at the location pointed by the `REAR`.

```
Step 1: IF FRONT = 0 and Rear = MAX - 1, then
            Write "OVERFLOW"
        ELSE IF FRONT = -1 and REAR = -1, then;
            SET FRONT = REAR = 0
        ELSE IF REAR = MAX - 1 and FRONT != 0
            SET REAR = 0
        ELSE
            SET REAR = REAR + 1
        [END OF IF]
Step 2: SET QUEUE[REAR] = VAL
Step 3: Exit
```

Figure 9.38 Algorithm to insert an element in a circular queue

After seeing how a new element is added in a circular queue, let us now discuss how deletions are performed in this case. To delete an element, again we check for three conditions.

- Look at Fig. 9.39. If `front = -1`, then there are no elements in the queue. So, an `underflow` condition will be reported.

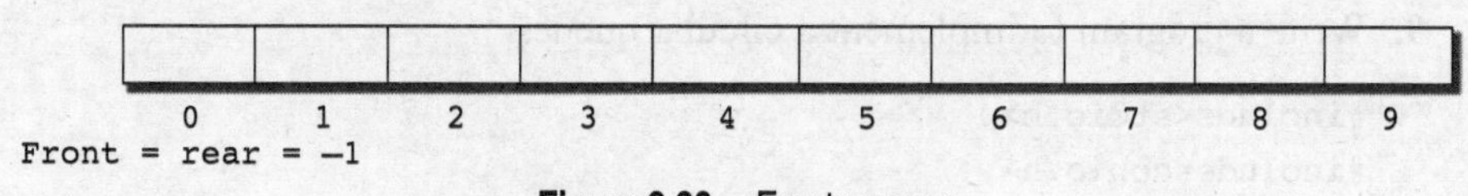

Figure 9.39 Empty queue

- If the queue is not empty and after returning the value on the `front`, `front = rear`, then the queue has now become empty and so, `front` and `rear` are set to `-1`. This is illustrated in Fig. 9.40.

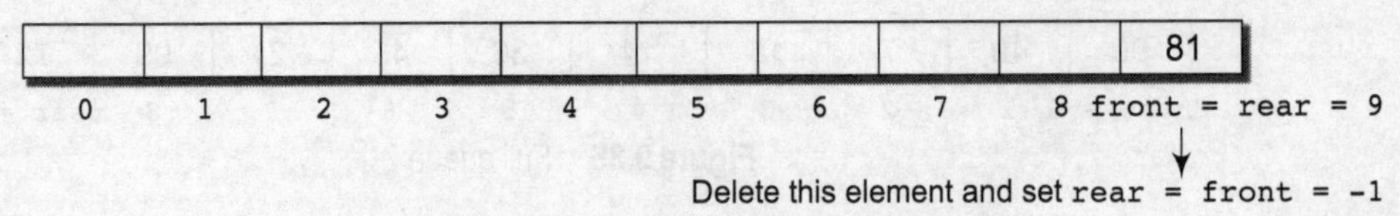

Figure 9.40 Queue with a single element

- If the queue is not empty and after returning the value on the `front`, `front = MAX-1`, then `front` is set to 0. This is shown in Fig. 9.41.

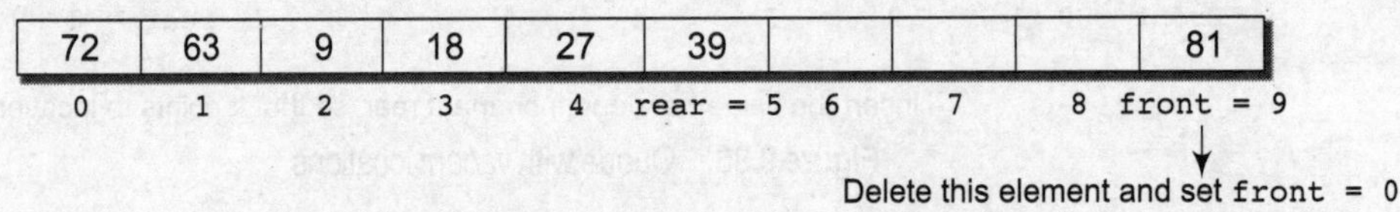

Figure 9.41 Queue where `front = MAX` before deletion

Let us look at Fig. 9.42 which shows the algorithm to delete an element from a circular queue. In Step 1, we check for the `underflow` condition. In Step 2, the value of the queue at the location pointed by `FRONT` is stored in `VAL`. In Step 3, we make two checks. First to see if the queue has become empty after deletion and second to see if `FRONT` has reached the maximum capacity of the queue. The value of `FRONT` is then updated based on the outcome of these tests.

```
Step 1: IF FRONT = -1, then
            Write "Underflow"
            SET VAL = -1
        [End of IF]
Step 2: SET VAL = QUEUE[FRONT]
Step 3: IF FRONT = REAR
            SET FRONT = REAR = -1
        ELSE
            IF FRONT = MAX -1
                   SET FRONT = 0
            ELSE
                   SET FRONT = FRONT + 1
            [End of IF]
        [END OF IF]
Step 4: Exit
```

Figure 9.42 Algorithm to delete an element from a circular queue

PROGRAMMING EXAMPLE

9. Write a program to implement a circular queue.

```
#include<stdio.h>
#include<conio.h>
#define MAX 10
int queue[MAX];
int front = -1, rear = -1;
void insert(void);
int delete_element(void);
```

```
int peek(void);
void display(void);
main()
{
        int option, val;
        clrscr();
        do
        {
                printf("\n\n ***** MAIN MENU *****");
                printf("\n 1. Insert an element");
                printf("\n 2. Delete an element");
                printf("\n 3. Peek");
                printf("\n 4. Display the queue");
                printf("\n 5. EXIT");
                printf("\n ************************");
                printf("\n\n Enter your option : ");
                scanf("%d", &option);
                switch(option)
                {
                    case 1:
                            insert();
                            break;
                    case 2:
                            val = delete_element();
                            printf("\n The number that was deleted is : %d", val);
                            break;
                    case 3:
                            val = peek();
                            printf("\n The first value in the queue is : %d", val);
                            break;
                    case 4:
                            display();
                            break;
                }
        }while(option != 5);
        getch();
        return 0;
}
void insert()
{
        int num;
        printf("\n Enter the number to be inserted in the queue : ");
        scanf("%d", &num);
        if(front == 0 && rear == MAX-1)
                printf("\n OVERFLOW");
        else if(front == -1 && rear == -1)
```

```
        {
                front = rear = 0;
                queue[rear] = num;
        }
        else if(rear == MAX-1 && front != 0)
        {
                rear = 0;
                queue[rear] = num;
        }
        else
        {
                rear++;
                queue[rear] = num;
        }
}
int delete_element()
{
        int val;
        if(front == -1)
        {
                printf("\n UNDERFLOW");
                return -1;
        }
        val = queue[front];
        if(front == rear)
                front = rear = -1;
        else
        {
                if(front == MAX-1)
                        front = 0;
                else
                        front++;
        }
        return val;
}
int peek()
{
        return queue[front];
}
void display()
{
        int i;
        printf("\n");
        if(front != -1 && rear != -1)
        {
                if(front<rear)
```

```
        {
            for(i = front;i <= rear;i++)
                printf("\t %d", queue[i]);
        }
        else
        {
            for(i = front;i<MAX;i++)
                printf("\t %d", queue[i]);
            for(i = 0;i<rear;i++)
                printf("\t %d", queue[i]);
        }
    }
}
```

9.14 LINKED REPRESENTATION OF A QUEUE

We have seen how a queue is created using an array. Although this technique of creating a queue is easy, but its drawback is that the array must be declared to have some fixed size. If we allocate space for 50 elements in the queue and it hardly uses 20–25 locations, then half of the space will be wasted. And in case we allocate less memory locations for a queue that might end up growing large and large, then a lot of re-allocations will have to be done, thereby creating a lot of overhead and consuming a lot of time.

In case the queue is a very small one or its maximum size is known in advance, then the array implementation of the stack gives an efficient implementation. But if the array size cannot be determined in advance, the other alternative, i.e. the linked representation is used.

The storage requirement of linked representation of a queue with `n` elements is `O(n)` and the typical time requirement for operations is `O(1)`.

In a linked queue, every element has two parts, one that stores the data and another that stores the address of the next element. The `START` pointer of the linked list is used as the `FRONT`. Here, we will also use another pointer called the `REAR`, which will store the address of the last element in the queue. All insertions will be done at the `rear` end and all the deletions are done at the `front` end. If `FRONT = REAR = NULL`, then it indicates that the queue is empty.

9.15 OPERATIONS ON A QUEUE

A queue has two basic operations: `insert` and `delete`. The insert operation adds an element to the end of the queue of the stack and the `delete` operation removes the element from the front or the start of the queue. Apart from this, there is another operation `Peek` which returns the value of the first element of the queue. Consider the linked queue shown in Fig. 9.43.

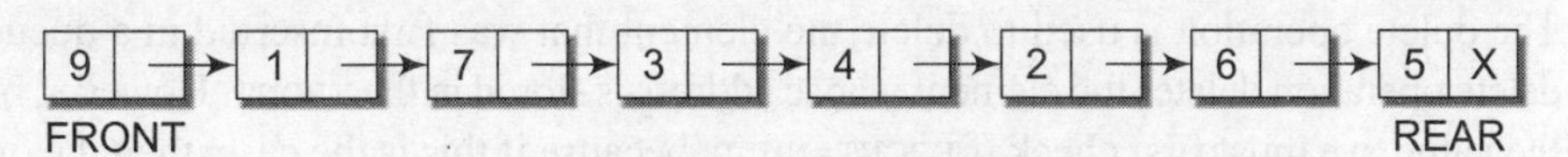

Figure 9.43 Linked queue

Insert operation

The insert operation is used to insert an element into the queue. The new element is added as the last element of the queue. Consider the linked queue shown in Fig. 9.44.

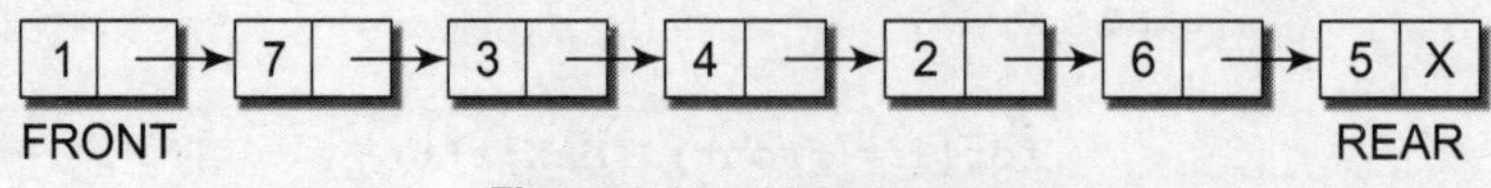

Figure 9.44 Linked queue

To insert an element with the value 9, we first check if `FRONT=NULL`. If the condition holds, then the queue is empty. So, we allocate memory for a new node, store the `value` in its `data` part and `NULL` in its `next` part. The new node will then be called the `FRONT`. However, if `FRONT != NULL`, then we will insert the new node at the beginning of the linked stack and name this new node as `TOP`. Thus, the updated stack becomes as shown in Fig. 9.45.

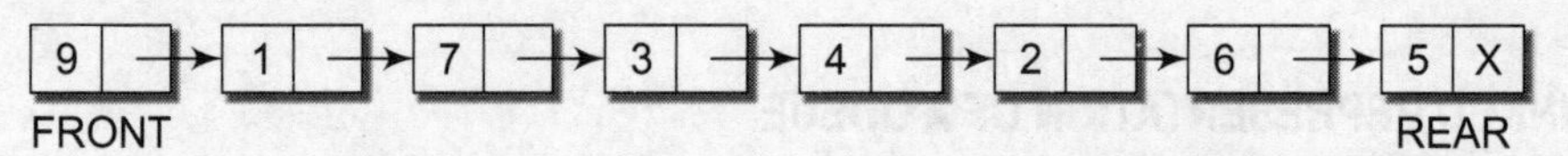

Figure 9.45 Linked queue after inserting a new node

Figure 9.46 shows the algorithm to insert an element in the circular queue. In Step 1, the memory is allocated for the new node. In Step 2, the `DATA` part of the new node is initialized with the value to be stored in the node. In Step 3, we check if the new node is the first node of the linked queue. This is done by checking if `FRONT = NULL`. In case the `if` statement evaluates to true, then the new node is tagged as `FRONT` as well as `REAR`. Also `NULL` is stored in the `NEXT` part of the node (which is also the `FRONT` and the `REAR` node). However, if the new node is not the first node in the list, then it is added at the `REAR` end of the linked queue (or the last node of the queue).

```
Step 1: Allocate memory for the new node and name it as PTR
Step 2: SET PTR -> DATA = VAL
Step 3: IF FRONT = NULL, then
            SET FRONT = REAR = PTR;
            SET FRONT -> NEXT = REAR -> NEXT = NULL
        ELSE
            SET REAR -> NEXT = PTR
            SET REAR = PTR
            SET REAR -> NEXT = NULL
        [END OF IF]
Step 4: END
```

Figure 9.46 Algorithm to insert an element in a linked queue

Delete operation

The delete operation is used to delete the element that was first inserted in a queue. That is, the delete operation deletes the element whose address is stored in the `FRONT`. However, before deleting the value, we must first check if `FRONT=NULL`, because if this is the case, then the queue is empty and no more deletions can be done. If an attempt is made to delete a value from a queue that is already empty, an `UNDERFLOW` message is printed. Consider the queue shown in Fig. 9.47.

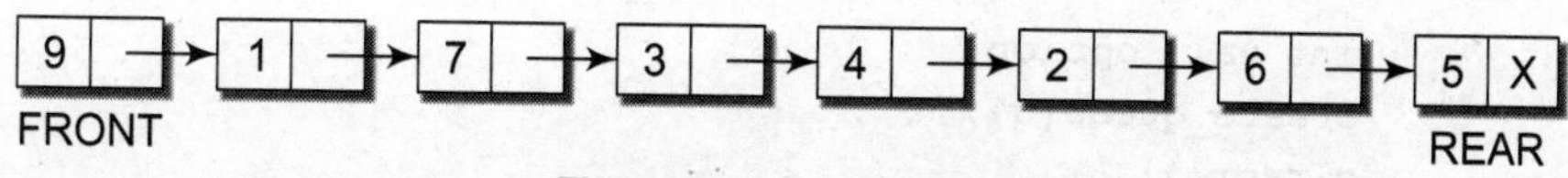

Figure 9.47 Linked queue

To delete an element, we first check if FRONT=NULL. If the condition is false, then we delete the first node pointed by FRONT. The FRONT will now point to the second element of the linked queue. Thus, the updated queue becomes as shown in Fig. 9.48.

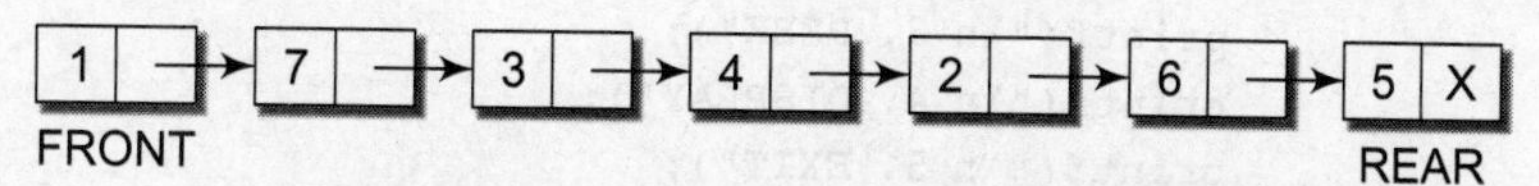

Figure 9.48 Linked queue after deletion of an element

Figure 9.49 shows the algorithm to delete an element from a linked queue. In Step 1, we first check for the underflow condition. If the condition is true, then an appropriate message is displayed, otherwise in Step 2, we use a PTR that points to FRONT. In Step 3, FRONT is made to point to the next node in sequence. In Step 4, the memory occupied by PTR is given back to the free pool.

```
Step 1: IF FRONT = NULL, then
             Write "Underflow"
             Go to Step 3
        [END OF IF]
Step 2: SET PTR = FRONT
Step 3: FRONT = FRONT -> NEXT
Step 4: FREE PTR
Step 5: END
```

Figure 9.49 Algorithm to delete an element from linked queue

PROGRAMMING EXAMPLE

10. Write a program to implement a linked queue.

```
#include<stdio.h>
#include<conio.h>
struct node
{
  int data;
  struct node *next;
};
struct queue
{
  struct node *front;
  struct node *rear;
};
struct queue *q;
void create_queue(struct queue *);
struct queue *insert(struct queue *,int);
struct queue *delete_element(struct queue *);
struct queue *display(struct queue *);
int peek(struct queue *);
main()
{
```

```
        int val, option;
        create_queue(q);
        clrscr();
        do
        {
                printf("\n *****MAIN MENU*****");
                printf("\n 1. INSERT");
                printf("\n 2. DELETE");
                printf("\n 3. PEEK");
                printf("\n 4. DISPLAY");
                printf("\n 5. EXIT");
                printf("\n *******************");
                printf("\n\n Enter your option : ");
                scanf("%d", &option);
                switch(option)
                {
                        case 1:
                                printf("\n ENter the number to be inserted in the
queue : ");
                                scanf("%d", &val);
                                q = insert(q,val);
                                break;
                        case 2:
                                q = delete_element(q);
                                break;
                        case 3:
                                val = peek(front);
                                printf("\n The value stored at the top of the stack
is : %d", val);
                                break;
                        case 4:
                                q = display(q);
                                break;
                }
        }while(option != 5);
        getch();
        return 0;
}
void create_queue(struct queue *q)
{
        q->rear = NULL;
        q->front = NULL;
}
struct queue *insert(struct queue *q,int val)
{
        struct node *ptr;
        ptr = (struct node*)malloc(sizeof(struct node *));
```

```
        ptr -> data = val;
        if(q -> front == NULL)
        {
                q -> front = ptr;
                q -> rear = ptr;
                q -> front -> next = q -> rear -> next = NULL;
        }
        else
        {
                q -> rear -> next = ptr;
                q -> rear = ptr;
                q -> rear -> next = NULL;
        }
        return q;
}
struct queue *display(struct queue *q)
{
        struct node *ptr;
        ptr = q -> front;
        if(ptr == NULL)
                printf("\n QUEUE IS EMPTY");
        else
        {
                printf("\n");
                while(ptr!=q -> rear)
                {
                        printf("%d\t", ptr -> data);
                        ptr = ptr -> next;
                }
                printf("%d\t", ptr -> data);
        }
        return q;
}
struct queue *delete_element(struct queue *q)
{
        struct node *ptr;
        ptr = q -> front;
        if(q -> front == NULL)
                printf("\n UNDERFLOW");
        else
        {
                q -> front = q -> front -> next;
                printf("\n The value being deleted is : %d", ptr -> data);
                free(ptr);
        }
        return q;
}
```

```
int peek(struct queue *q)
{
        return q->front->data;
}
```

9.16 DEQUES

A deque (pronounced as 'deck' or 'dequeue') is a list in which the elements can be inserted or deleted at either end. It is also known as a *head-tail linked list*, because elements can be added to or removed from either the front (head) or the back (tail) end.

However, no element can be added and deleted from the middle. In the computer's memory, a deque is implemented either using a circular array or a circular doubly linked list. In a deque, two pointers are maintained, `LEFT` and `RIGHT`, which point to either end of the deque. The elements in a deque stretch from the `LEFT` end to the `RIGHT` and since it is circular, `Dequeue[N-1]` is followed by `Dequeue[0]`.

Consider the deques shown in Fig. 9.50.

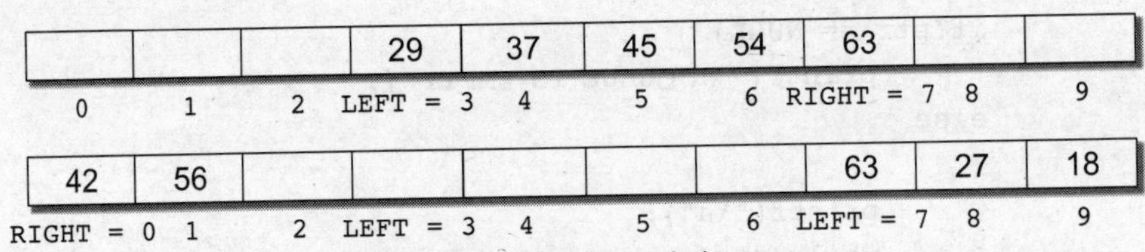

Figure 9.50 Double-ended queues

Basically, there are two variants of a double-ended queue. They include:

- *Input restricted deque:* In this dequeue, insertions can be done only at one of the dequeue, while deletions can be done from both ends.
- *Output restricted deque:* In this dequeue, deletions can be done only at one of the dequeue, while insertions can be done on both ends.

PROGRAMMING EXAMPLE

11. Write a program to implement input and output restricted deques.

```
# include<stdio.h>
# include<conio.h>
# define MAX 10
int deque[MAX];
int left = -1, right = -1;
void input_deque(void);
void output_deque(void);
void insert_left(void);
void insert_right(void);
void delete_left(void);
void delete_right(void);
```

```
void display(void);
main()
{
      int option;
      clrscr();
      printf("\n 1.Input restricted dequeue");
      printf("\n 2.Output restricted dequeue");
      printf("Enter your option : ");
      scanf("%d",&option);
      switch(option)
      {
            case 1:
                  input_que();
                  break;
            case 2:
                  output_que();
                  break;
      }
      return 0;
}
void input_deque()
{
      int option;
      do
      {
            printf("\n 1.Insert at right");
            printf("\n 2.Delete from left");
            printf("\n 3.Delete from right");
            printf("\n 4.Display");
            printf("\n 5.Quit");
            printf("\nEnter your option : ");
            scanf("%d",&option);
            switch(option)
            {
                  case 1:
                        insert_right();
                        break;
                  case 2:
                        delete_left();
                        break;
                  case 3:
                        delete_right();
                        break;
                  case 4:
                        display();
                        break;
```

```
            }
        }while(option != 5);
    }
    void output_deque()
    {
            int option;
            do
            {
                    printf("\n 1.Insert at right");
                    printf("\n 2.Insert at left");
                    printf("\n 3.Delete from left");
                    printf("\n 4.Display\n");
                    printf("\n 5.Quit");
                    printf("\n Enter your option : ");
                    scanf("%d",&option);
                    switch(option)
                    {
                            case 1:
                                    insert_right();
                                    break;
                            case 2:
                                    insert_left();
                                    break;
                            case 3:
                                    delete_left();
                                    break;
                            case 4:
                                    display();
                                    break;
                    }
        }while(option!=5);
    }
    void insert_right()
    {
        int val;
        printf("\n Enetr the value to be added:");
        scanf("%d", &val);
        if((left == 0 && right == MAX-1) || (left == right+1))
        {
            printf("\n OVERFLOW");
            return;
        }
        if (left == -1) /* if queue is initially empty */
        {
            left = 0;
```

```
            right = 0;
        }
        else
        {
            if(right == MAX-1)  /*right is at last position of queue */
                right = 0;
            else
                right = right+1;
        }
        deque[right] = val ;
}
insert_left()
{
        int val;
        printf("\n Enetr the value to be added:");
        scanf("%d", &val);
        if((left == 0 && right == MAX-1) || (left == right+1))
        {
            printf("Queue Overflow \n");
            return;
        }
        if (left == -1)            /*If queue is initially empty*/
        {
            left = 0;
            right = 0;
        }
        else
        if(left == 0)
            left = MAX-1;
        else
            left = left-1;
        deque[left] = val;
}
void delete_left()
{
        if (left == -1)
        {
            printf("\n UNDERFLOW");
            return ;
        }
        printf("\n The deleted element is : %d", deque[left]);
        if(left == right)        /*Queue has only one element */
        {
            left = -1;
            right = -1;
        }
```

```
          else
          {     if(left == MAX-1)
                     left = 0;
                else
                     left = left+1;
          }
     }
     void delete_right()
     {
          if (left == -1)
          {
                printf("\n UNDERFLOW");
                return ;
          }
          printf("\n The element deleted is : %d", deque[right]);
          if(left == right)       /*queue has only one element*/
          {
                left = -1;
                right = -1;
          }
          else
                if(right == 0)
                     right = MAX-1;
                else
                     right = right-1;
     }
     void display()
     {
          int front = left,rear = right;
          if(left == -1)
          {
                printf("\n QUEUE IS EMPTY");
                return;
          }
          printf("\n The elements of the queue are : ");
          if( front <= rear )
          {
                while(front <= rear)
                {
                     printf("%d",deque[front]);
                     front++;
                }
          }
          else
          {
                while(front <= MAX-1)
```

```
                {
                        printf("%d", deque[front]);
                        front++;
                }
                front = 0;
                while(front <= rear)
                {
                        printf("%d",deque[front]);
                        front++;
                }
        }
    }
    printf("\n");
    }
```

9.17 PRIORITY QUEUES

A priority queue is an abstract data type in which each element is assigned a priority. The priority of the element will be used to determine the order in which these elements will be processed. The general rule of processing the elements of a priority queue is:

- An element with higher priority is processed before an element with a lower priority.
- Two elements with the same priority are processed on a first-come-first-served (FCFS) basis.

A priority queue can be thought of as a modified `queue` in which when an element has to be taken off the queue, the one with the highest-priority is retrieved first. The priority of the element can be set based on various factors. Priority queues are widely used in operating systems to execute the highest priority process first. The priority of the process may be set based on the CPU time it requires to get executed completely. For example, if there are three processes, where the first process needs 5 nanoseconds to complete, the second process needs 4 nanoseconds, and the third process needs 7 nanoseconds, then the second process will have the highest priority and will thus be the first to be executed. However, CPU time is not the only factor that determines the priority, rather it is just one among several factors. Another factor is the importance of one process over another. In case we have to run two processes at the same time, where one process is concerned with online order booking and the second is printing the stock details, then obviously the online booking is more important and must be executed first.

9.17.1 Implementation of a Priority Queue

There are two ways to implement a priority queue. Either use a sorted list to store the elements so that when an element has to be taken out, the queue will not have to be searched for the element with the highest priority or use an unsorted list so that insertions are always done at the end of the list. Every time when an element has to be taken off the list, the element with the highest priority will be searched and removed. While a sorted list takes `O(n)` time to insert an element in the list, it takes only `O(1)` time to delete an element. On the contrary, an unsorted list will take `O(1)` time to insert an element and `O(n)` time delete an element from the list.

Practically, both these techniques are inefficient and usually a blend of these two approaches is adopted that takes roughly `O(log(n))` time or less.

9.17.2 Linked Representation of a Priority Queue

In the computer memory, a priority queue can be represented using arrays or linked lists. When a priority queue is implemented using a linked list, then every node of the list will have three parts: (a) the information or data part (b) the priority number of the element (c) the address of the next element. If we are using a sorted linked list, then the element with the higher priority will precede the element with the lower priority.

Note

Lower priority number means higher priority. For example, if there are two elements `A` and `B`, where `A` has a priority number 1 and `B` has a priority number 5, then `A` will be processed before `B` as it has higher priority than `B`.

Consider the priority queue shown in Fig. 9.51.

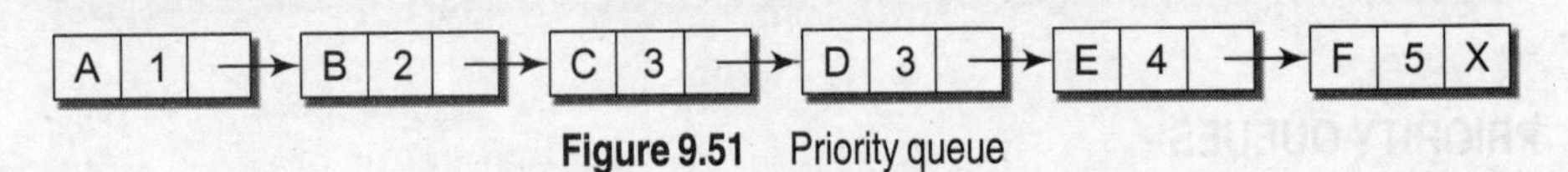

Figure 9.51 Priority queue

The priority queue in Fig. 9.51 is a sorted priority having six elements. From the queue, we cannot make out whether `A` was inserted before `E` or whether `E` joined the queue before `A` because the list is not sorted based on FCFS. Here, the element with a higher priority comes before the element with a lower priority. However, we can definitely say that `C` was inserted in the queue before `D` because, when two elements have the same priority, the elements are arranged and processed on FCFS principle.

Insertion Whenever a new element has to be inserted in a priority queue, we have to traverse the entire list until we find a node that has a priority lower than that of the new element. The new node is inserted before the node with a lower priority. However, if there exists an element that has the same priority as the new element, the new element is inserted after that element. For example, consider the priority queue shown in Fig. 9.52.

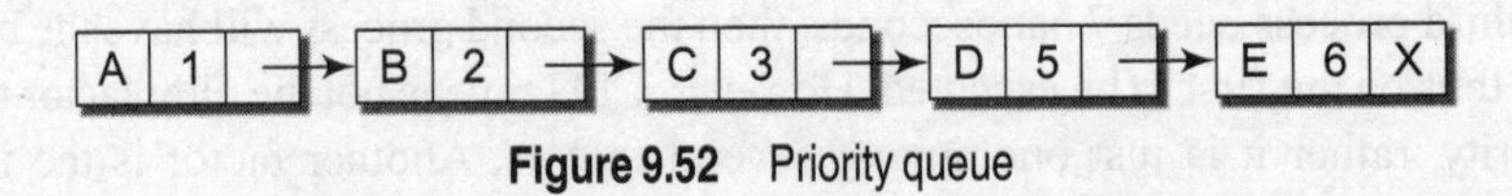

Figure 9.52 Priority queue

If we have to insert a new element with `data = F` and priority `number = 4`, then the element will be inserted before `D` that has the priority number 5, that is lower priority than that of the new element. So, the priority queue now becomes as shown in Fig. 9.53.

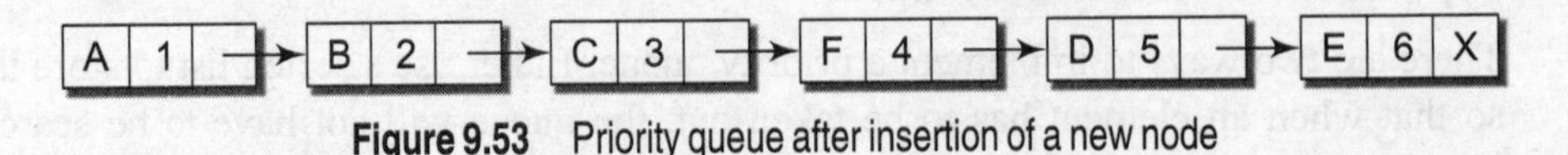

Figure 9.53 Priority queue after insertion of a new node

However, if we had a new element with `data = F` and `priority number = 2`, then the element will be inserted after `B`, as both these elements have the same priority but the insertions are done on FCFS basis as shown in Fig. 9.54.

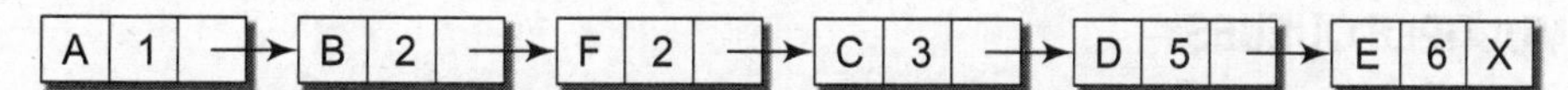

Figure 9.54 Priority queue after insertion of a new node

Deletion Deletion is a very simple process in this case. The first node of the list will be deleted and the data of that node will be processed first.

9.17.3 Array Representation of a Priority Queue

When arrays are used to implement a priority queue, then a separate queue for each priority number is maintained. Each of these queues will be implemented using circular arrays or circular queues. Every individual queue will have its own `FRONT` and `REAR` pointers.

We use a two-dimensional array for this purpose where each queue will be allocated the same amount of space. Look at the two-dimensional representation of a priority queue given below. Given the `front` and `rear` values of each queue, the two-dimensional matrix can be formed as shown in Fig. 9.55.

`FRONT[K]` and `REAR[K]` contain the front and rear values of row `K`, where `K` is the priority number. Note that here we are assuming that the row and column indices start from 1, not 0. Obviously, while programming, we will not take such assumptions.

FRONT	REAR
3	3
1	3
4	5
4	1

	1	2	3	4	5
1			A		
2	B	C	D		
3				E	F
4	I			G	H

Figure 9.55 Priority queue matrix

Insertion To insert a new element with priority `K` in the priority queue, add the element at the rear end of row `K`, where `K` is the row number as well as the priority number of that element. For example, if we have to insert an element `R` with priority number 3, then the priority queue will be given as shown in Fig. 9.56.

FRONT	REAR
3	3
1	3
4	1
4	1

	1	2	3	4	5
1			A		
2	B	C	D		
3	R			E	F
4	I			G	H

Figure 9.56 Priority queue matrix after insertion of a new element

Deletion To delete an element, we find the first non-empty queue and then process the front element of the first non-empty queue. In our priority queue, the first non-empty queue is the one with priority number 1 and the front element is `A`, so `A` will be deleted and processed first. In technical terms, find the element with the smallest `K`, such that `FRONT[K] != NULL`.

9.18 MULTIPLE QUEUES

When we had implemented a queue array, we have seen that the size of the array must be known in advance. If the queue is allocated less space, then frequent OVERFLOW conditions will be encountered. To deal with this problem, the code will have to be modified to reallocate more space for the array.

In case we allocate a large amount of space for the queue, it will result in sheer waste of the memory. Thus, there lies a tradeoff between the frequency of overflows and the space allocated.

So a better solution to deal with this problem is to have multiple queues or to have more than one queue in the same array of sufficient size. Figure 9.57 illustrates this concept.

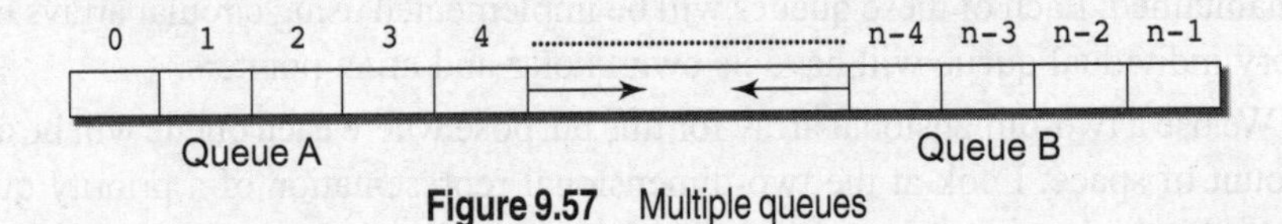

Figure 9.57 Multiple queues

In the figure, an array `Queue[n]` is used to represent two queues, `Queue A` and `Queue B`. The value of `n` is such that the combined size of both the queues will never exceed `n`. While operating on these queues, it is important to note one thing. While `queue A` will grow from left to right, `queue B` will grow from right to left at the same time.

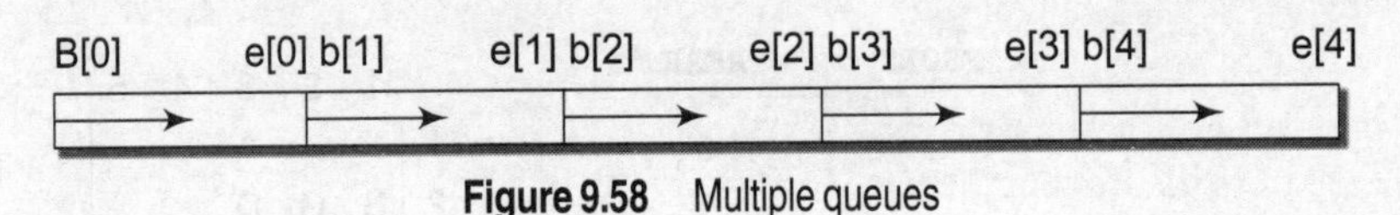

Figure 9.58 Multiple queues

Extending the concept of multiple queues, a `QUEUE` can also be used to represent `n` number of queues in the same array. That is, if we have a `QUEUE[n]`, then each queue `I` will be allocated an equal amount of space bounded by indices `b[i]` and `e[i]`. This is shown in Fig. 9.58.

SUMMARY

- A stack is a linear data structure in which elements are added and removed only from one end, which is called the top. Hence, a stack is called a LIFO (Last-In, First-Out) data structure, as the element that was inserted last is the first one to be taken out.
- In postfix notation, operators are placed after the operands; while in prefix notation, operators are placed before the operands.
- Postfix notations are evaluated using stacks. Every character of the postfix expression is scanned from left to right. If the character is an operand, it is pushed onto the stack. Else, if it is an operator, then the top two values are popped from the stack and the operator is applied on these values. The result is then pushed onto the stack.
- A queue is a FIFO data structure in which the element that was inserted first is the first one to be taken out. The elements in a queue are added at one end called the `rear` and removed from the other end called the `front`. In the circular queue, the first index comes right after the last index.
- The storage requirement of linked representation of stack/queue with `n` elements is `O(n)` and the typical time requirement for operations is `O(1)`.
- Multiple stacks/queues means to have more than one stack/queue in the same array of sufficient size.
- A deque is a list in which elements can be inserted or deleted at either end. It is also known as a head-tail linked list because elements can be added to or removed from the front (head) or back (tail).

However, no element can be added or deleted from the middle. In the computer's memory, a deque is implemented either using a circular array or a circular doubly linked list.

- In an input restricted deque, insertions can be done only at one end, while deletions can be done from both the ends. In an output restricted deque, deletions can be done only at one end, while insertions can be done at both the ends.
- A priority queue is an abstract data type in which each element is assigned a priority. The priority of the element will be used to determine the order in which these elements will be processed.
- When a priority queue is implemented using a linked list, then every node of the list will have three parts: (a) the information or data part (b) the priority number of the element and (c) the address of the next element.

GLOSSARY

Priority Queue An abstract data type that supports efficient search of the item with the highest priority across a series of operations.

Backtracking Backtracking is concerned with finding a solution by trying one of several choices. If the choice proves incorrect, computation backtracks or restarts at the point of choice and tries another choice. For this, choice points and alternate choices are maintained.

Deque A data structure in which values may be added to or deleted from the front or the rear.

Queue A collection of items in which only the first item added may be accessed first. It is also known as the first-in-first-out or FIFO data structure.

Stack A collection of items in which only the most recently added item may be removed. The latest added item is at the top. Basic operations are push and pop. It is also known as the last-in-first-out or LIFO data structure.

First-In-First-Out A policy in which items are processed in order of their arrival.

Last-In-Last-Out A policy in which the most recently arrived item is processed first.

EXERCISES

Review Questions

1. What is a priority queue? Discuss its applications.
2. Explain the concept of a circular queue? How is it better than a linear queue?
3. What do you understand by stack overflow? Write a program to implement a stack using a linear array that checks for stack overflow condition before inserting elements into it.
4. Differentiate between an array and a stack.
5. How does a linked stack differ from a linear stack?
6. Differentiate between `peek()` and `pop()` functions.
7. Why are parentheses not required in postfix/prefix expressions?
8. Explain how stacks are used in a non-recursive program?
9. Explain the terms overflow and underflow. Write a program to implement a stack that stores character data.
10. What do you understand by a multiple stack? How is it useful?
11. Explain the terms infix, prefix, and postfix expression. Convert the following infix expressions to their prefix equivalents.
    (a) A – B + C
    (b) A * B + C / D
    (c) (A – B) + C * D / E – C
    (d) (A * B) + (C / D) – (D + E)
    (e) ((A – B) + D / ((E + F) * G))
    (f) (A – 2 * (B + C) / D * E) + F
    (g) 14 / 7 * 3 – 4 + 9 / 2
12. Convert the following infix expressions to their postfix equivalents.
    (a) A – B + C
    (b) A * B + C / D
    (c) (A – B) + C * D / E - C

(d) $(A * B) + (C / D) - (D + E)$

(e) $((A - B) + D / ((E + F) * G))$

(f) $(A - 2 * (B + C) / D * E) + F$

(g) $14 / 7 * 3 - 4 + 9 / 2$

13. Find out the infix equivalents of the following postfix expressions.

(a) $A B + C * D -$ (b) $ABC * + D -$

14. Give the infix expression of the following prefix expressions

(a) $* - + A B C D$ (b) $+ - a * B C D$

15. Convert the expression given below into its corresponding postfix expression. Also, write a program to evaluate a postfix expression.

$10 + ((7 - 5) + 10) / 2$

16. Explain the concept of a circular queue. How is it better than an ordinary queue?

17. What do you understand by a priority queue? Discuss its applications. Also, explain its implementation details.

18. Why do we use multiple queues?

19. Write a function that accepts two stacks. Copy the contents of first stack in the second stack. Note that the order of elements must be preserved. (*Hint: use a temporary stack*)

20. Draw the stack structure in each case when the following operations are performed on an empty stack.

(a) Add A, B, C, D, E, F

(b) Delete two alphabets

(c) Add G

(d) Add H

(e) Delete four alphabets

(f) Add I

21. Draw the queue structure in each case when the following operations are performed on an empty queue.

(a) Add A, B, C, D, E, F

(b) Delete two alphabets

(c) Add G

(d) Add H

(e) Delete four alphabets

(f) Add I

22. Consider the queue given below which has `FRONT` = 1 and `REAR` = 5.

Now perform the following operations on the queue:

(a) Add F

(b) Delete two alphabets

(c) Add G

(d) Add H

(e) Delete four alphabets

(f) Add I

23. Consider the dequeue given below which has `LEFT` = 1 and `RIGHT` = 5.

Now perform the following operations on the queue:

(a) Add F on the left

(b) Add G on the right

(c) Add H on the right

(d) Delete two alphabets from left

(e) Add I on the right

(f) Add J on the left

(g) Delete two alphabets from right

Programming Exercises

1. Write a program to calculate the number of items in a queue.

2. Write a program to implement a stack using a linked list. How is a linked stack better than a linear stack? Do we have underflow and overflow situations in a linked stack?

3. Write a program to convert an infix expression into a postfix expression.

4. Write a program to convert an infix expression into a prefix expression.

5. Write a program to implement a simple queue.

6. Write a program to implement a dequeue with the help of a linear array.

7. Write a program to implement a dequeue with the help of a linked list.

8. Write a program to implement a stack that stores the names of students in a class.

9. Write a program to implement a circular queue.

10. Write a program for a input-restricted dequeue.

11. Write a program for a output-restricted dequeue.
12. Write a program to implement a priority queue.
13. Write a program to create a queue from a stack.
14. Write a program to create a stack from a queue.
15. Write a program to reverse the elements of a queue.
16. Write a program to input two queues and compare their contents.
17. Write a program to input two stacks and compare their contents.

Multiple Choice Questions

1. Stack is a
   (a) LIFO (b) FIFO
   (c) FILO (d) LILO
2. Which function places an element on the stack?
   (a) `Pop()` (b) `Push()`
   (c) `Peek()` (d) `isEmpty()`
3. Disks piled up one above the other represents a
   (a) Stack (b) Queue
   (c) Linked List (d) Array
4. A line in a grocery store represents a
   (a) Stack (b) Queue
   (c) Linked List (d) Array
5. In a queue, insertion is done at
   (a) Rear (b) Front
   (c) Back (d) Top
6. Reverse Polish notation is the other name of
   (a) Infix expression
   (b) Prefix expression
   (c) Postfix expression

True or False

1. A queue stores elements in a manner such that the first element is at the beginning of the list and the last element is at the end of the list.
2. Elements in a priority queue are processed sequentially.
3. `Pop()` is used to add an element on the top of the stack.
4. In a linked queue, a maximum of 100 elements can be added.
5. Conceptually a linked queue is same as that of a linear queue.
6. The size of a linked queue cannot change during run time.
7. In a priority queue, two elements with the same priority are processed on a FCFS basis.
8. Output-restricted deque allows deletions to be done only at one end of the dequeue, while insertions can be done at both the ends.
9. If `front=MAX - 1` and `rear= 0`, then the circular queue is full.
10. Postfix operation does not follow the rules of operator precedence.

Fill in the Blanks

1. New nodes are added at ______ of the queue.
2. ______ allows insertion of elements at either ends but not in the middle.
3. ______ is used to convert an infix expression into a postfix expression.
4. ______ is used in a non-recursive implementation of a recursive algorithm.
5. The storage requirement of a linked stack with n elements is ______.
6. The typical time requirement for operations in a linked queue is ______.
7. In ______, insertions can be done only at one end, while deletions can be done from both the ends.
8. Underflow takes below when ______.
9. Dequeue is implemented using ______.
10. The order of evaluation of a postfix expression is from ______.
11. ______ are appropriate data structures to process batch computer programs submitted to the computer center.
12. ______ are appropriate data structures to process a list of employees having a contract for a seniority system for hiring and firing.

Annexure C The Stack Abstract Data Type

STACK AND ADT DEFINITION

The stack definition lays a template to be used while using the stack ADT. Figure C.1 shows the standard template.

```
typedef struct node
{
        void *data;
        struct node *next;
}Stack_Node;

typedef struct
{
        Stack_Node *top;
}Stack;
```

Figure C.1 Stack definition

CREATE STACK

The `create_stack()` function allocates memory for the stack. It initializes the top pointer to NULL to indicate an empty stack and returns the address of the allocated memory to the caller. Hence, in the calling function, the return pointer value must be assigned to a stack pointer, by writing

```
stack = create_stack();
```

```
Stack *create_stack(void)
{
        Stack *stack;
        stack = (Stack*) malloc(sizeof (Stack));
        if(stack)
                stack -> top = NULL;
        return stack;
}
```

Figure C.2 Shows the standard `create_stack` function

PUSH

The first thing to be performed in a push operation is to find a place for the data by allocating memory using `malloc()`. Once the memory is allocated to the data, the data pointer is assigned to the node. Finally, the next pointer is set so that it now points to the top of the stack. Figure C.3 shows the code segment for push operation.

```
void push(Stack * stack, void *val)
{
    Stack_Node *new_node;
    new_node = (Stack_Node)malloc(sizeof(Stack_Node));
     if(new_node)
     {
          new_node -> data = val;
          new_node -> next = stack -> top;
          stack -> top = new_node;
     }
}
```

Figure C.3 Push operation

POP

The pop function returns the data in the node at the top of the stack and then deletes that node. If the stack is empty, a NULL is returned, else the node at the top of the stack is returned. Figure C.4 shows the code segment for pop operation.

```
void *pop(Stack * stack)
{
       void *val;
       Stack_Node *temp;
       if (stack -> top == NULL)
            return NULL;
       else
       {
              temp = stack -> top;
              val = stack -> top -> data;
              stack -> top = stack -> top -> next;
              free(temp);
                 return val;
       }
}
```

Figure C.4 Pop operation

10 Trees

Learning Objective

So far, we have discussed linear data structures such as strings, arrays, structures, stacks, and queues. In this chapter, we will learn about a non-linear data structure called trees. A tree is a structure which is mainly used to store data that is hierarchical in nature. In this chapter, we will first discuss the general binary trees. These binary trees are used to form binary search trees and heaps. They are widely used to manipulate arithmetic expression, construction of symbol tables, and for syntax analysis. Here, we will get an overview of binary trees.

10.1 BINARY TREES

A binary tree is a data structure which is defined as a collection of elements called *nodes*. Every node contains a `left` pointer, a `right` pointer, and a data element. Every binary tree has a root element pointed by a 'root' pointer. The root element is the topmost node in the tree. If `root` = `NULL`, then the tree is empty.

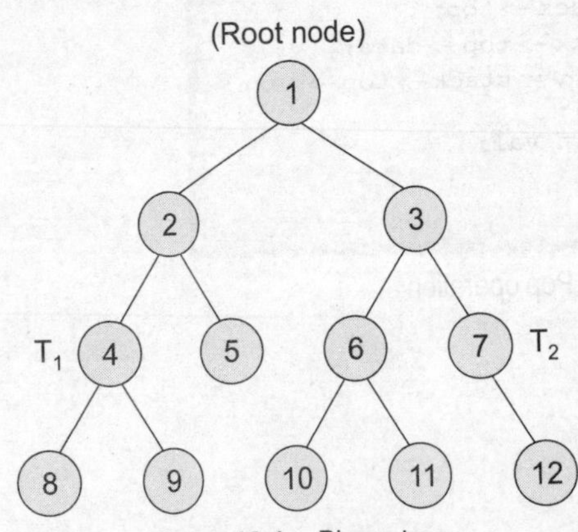

Figure 10.1 Binary tree

Figure 10.1 shows a binary tree. If the root node `R` is not `NULL`, then the two trees T_1 and T_2 are called the left and right sub-trees of `R`. If T_1 is non-empty, then T_1 is said to be the left successor of `R`. Likewise, if T_2 is non-empty, then it is called the right successor of `R`.

In Fig. 10.1, node 2 is the left successor and node 3 is the right successor of the root node 1. Note that the left sub-tree of the root node consists of the nodes 2, 4, 5, 8, and 9. Similarly, the right sub-tree of the root node consists of the nodes 3, 6, 7, 10, 11, and 12.

In a binary tree, every node has 0, 1, or at the most, 2 successors. A node that has no successors is called the *leaf node* or *terminal node*. In the tree, the root node R has two successors, 2 and 3. Node 2 has two successor nodes, 4 and 5. Node 4 has two successors, 8 and 9. Node 5 has no successor. Node 3 has two successor nodes, 6 and 7. Node 6 has two successors, 10 and 11. Finally, node 7 has only one successor, 12.

A binary tree is recursive by definition, as every node in the tree contains a left sub-tree and a right sub-tree. Even the terminal nodes contain an empty left sub-tree and an empty right sub-tree. In Fig. 10.1, nodes 5, 8, 9, 10, 11, and 12 have no or zero successors and thus are said to have empty sub-trees.

10.1.1 Key Terms

Sibling If N is a node in T that has a left successor S_1 and a right successor S_2, then N is called the *parent* of S_1 and S_2. Correspondingly, S_1 and S_2 are called the *left child* and *right child* of N. Also, S_1 and S_2 are said to be *siblings*. Every node, other than the root node, has a parent. In other words, all nodes that are at the same level and share the same parent are called siblings. For example, nodes 2 and 3, nodes 4 and 5, nodes 6 and 7, nodes 8 and 9, and nodes 10 and 11 are siblings.

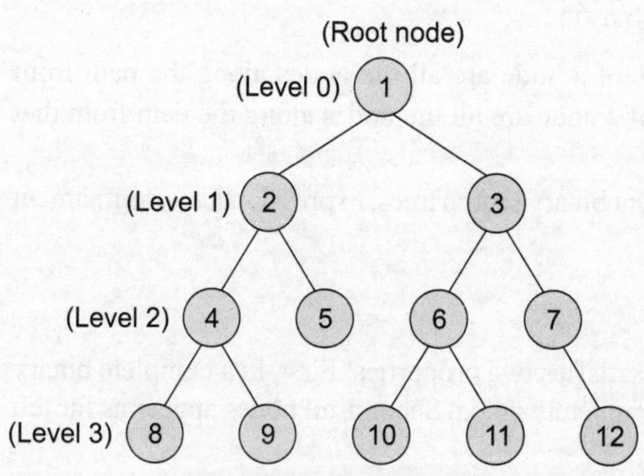

Figure 10.2 Binary Tree

Level number Every node in a binary tree is assigned a *level number* (refer Fig. 10.2). The root node is defined at level 0. The left and right child of the root node has a level number 1. Similarly, every node is a level higher than its parents. So, a child node's level number is defined as `parent's level number + 1`.

Degree The degree of a node is equal to the number of children that a node has. The degree of a leaf node is zero. For example, in Fig. 10.2, the degree of node 4 is 2, the degree of node 5 is zero, and the degree of node 7 is 1.

In-degree and out-degree In-degree of a node is the number of edges arriving at that node. The root node is the only node that has an in-degree equal to zero. Similarly, out-degree of a node is the number of edges leaving that node.

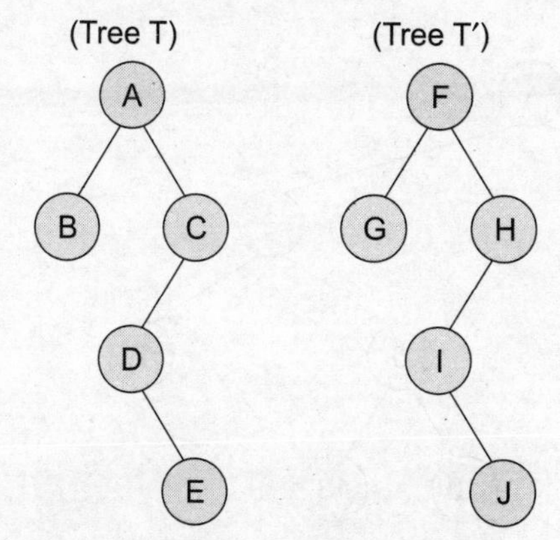

Figure 10.3 Similar binary trees

Leaf node A leaf node has no children. The leaf nodes in the tree are 8, 9, 10, 11, and 12.

Similar binary trees Two binary trees T and T′ are said to be similar if both these trees have the same structure. Figure 10.3 shows two similar binary trees.

Copies of binary trees Two binary trees T and T′ are said to be *copies* if they have the same structure and contents at the corresponding nodes. Figure 10.4 shows that T′ is a copy of T.

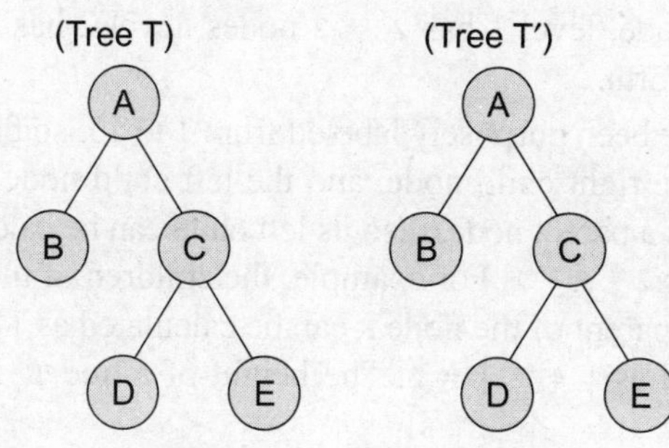

Figure 10.4 T′ is a copy of T

Directed edge The line drawn from a node N to any of its successor is called a directed edge. A binary tree of n nodes has exactly n - 1 edges (because every node except the root node is connected to its parent via an edge).

Path A sequence of consecutive edges is called a path. For example, in Fig. 10.4, the path from the root node to the node 8 is given as 1, 2, 4, and 8.

Depth The depth of a node N is given as the length of the path from the root R to the node N. The depth of the root node is zero. The height/depth of a tree is defined as the length of the path from the root node to the deepest node in the tree.

A tree with only a root node has a height of zero. A binary tree of height h has at least h nodes and at most $2^h - 1$ nodes. This is because every level will have at least one node and can have at most two nodes. So, if every level has two nodes, then a tree with height h will have at the most $2^h - 1$ nodes (as at level 0, there is only one element called the root). The height of a binary tree with n nodes is at least n and at the most $\log_2(n+1)$.

Ancestor and descendant nodes Ancestors of a node are all the nodes along the path from the root to that node. Similarly, descendants of a node are all the nodes along the path from that node to the leaf node.

Binary trees are commonly used to implement binary search trees, expression trees, tournament trees, and binary heaps.

10.1.2 Complete Binary Trees

A *complete binary tree* is a binary tree which satisfies two properties. First, in a complete binary tree, every level, except possibly the last, is completely filled. Second, all nodes appear as far left as possible.

In a complete binary tree T_n, there are exactly n nodes and level r of T can have at most 2^r nodes. Figure 10.5 shows a complete binary tree.

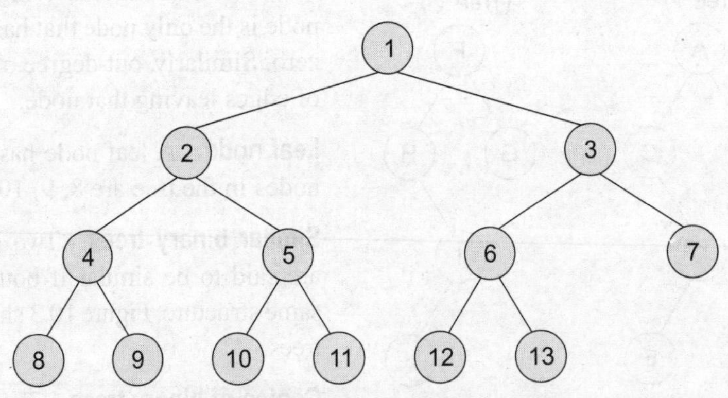

Figure 10.5 Complete binary tree

Note that in Fig. 10.5, level 0 has $2^0 = 1$ node, level 1 has $2^1 = 2$ nodes, level 2 has $2^2 = 4$ nodes, level 3 has $2^3 = 8$ nodes, so on and so forth.

The tree T_{13} has exactly 13 nodes. They have been purposely labeled from 1 to 13, so that it is easy for the reader to find the parent node, the right child node, and the left child node of the given node. The formula can be given as– if K is a parent node, then its left child can be calculated as 2 * K and its right child can be calculated as 2 * K + 1. For example, the children of the node 4 are 8 (2*4) and 9 (2* 4 + 1). Similarly, the parent of the node K can be calculated as | K/2 |. Given the node 4, its parent can be calculated as | 4/2 | = 2. The height of a tree T_n having exactly n nodes is given as:

$$H_n = | \log_2 n + 1 |$$

This means, if a tree T has 10,00,000 nodes, then its height is 21.

10.1.3 Extended Binary Trees

A binary tree T is said to be an extended binary tree (or a 2-tree) if each node in the tree has either no child or exactly two children. Figure 10.6 shows how an ordinary binary tree is converted into an extended binary tree.

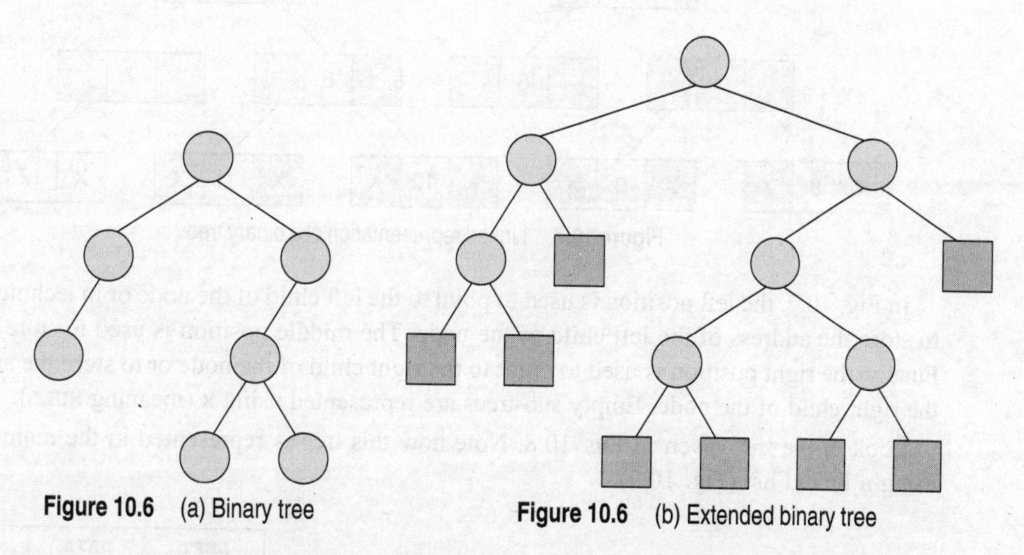

Figure 10.6 (a) Binary tree **Figure 10.6** (b) Extended binary tree

In an extended binary tree, nodes having two children are called *internal nodes* and nodes having no child are called *external nodes*. In Fig. 10.6, the internal nodes are represented using circles and the external nodes are represented using squares.

To convert a binary tree into an extended tree, every empty sub-tree is replaced by a new node. The original nodes in the tree are the internal nodes and the new nodes added are called the external nodes.

10.1.4 Representation of Binary Trees in the Memory

In the computer's memory, a binary tree can be maintained either by using a linked representation (as in case of a linked list) or by using a sequential representation (as in case of single arrays).

Linked representation of binary trees

In the linked representation of a binary tree, every node will have three parts: the data element, a pointer to the left node, and a pointer to the right node. So in C, the binary tree is built with a node type given below.

```
struct node {
    struct node *left;
    int data;
    struct node *right;
};
```

Every binary tree has a pointer ROOT, which points to the root element (topmost element) of the tree. If ROOT = NULL, then the tree is empty. Consider the binary tree given in Fig 10.1. The schematic diagram of the linked representation of the binary tree is shown in Fig. 10.7.

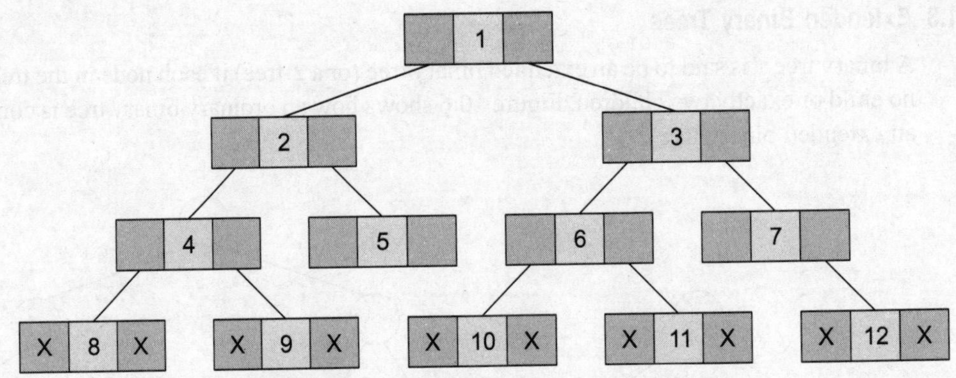

Figure 10.7 Linked representation of a binary tree

In Fig. 10.7, the left position is used to point to the left child of the node or in technical terms, to store the address of the left child of the node. The middle position is used to store the data. Finally, the right position is used to point to the right child of the node or to store the address of the right child of the node. Empty sub-trees are represented using `X` (meaning `NULL`).

Look at the tree given in Fig. 10.8. Note how this tree is represented in the main memory using a linked list (Fig. 10.9).

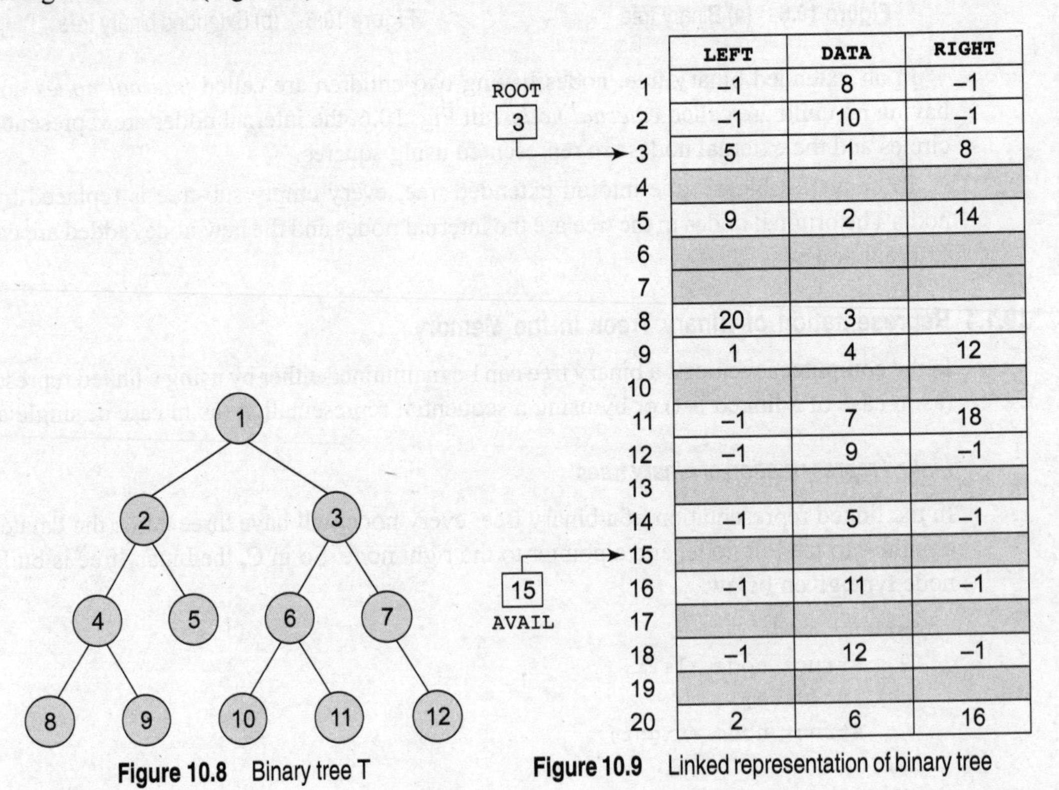

	LEFT	DATA	RIGHT
1	–1	8	–1
2	–1	10	–1
3	5	1	8
4			
5	9	2	14
6			
7			
8	20	3	
9	1	4	12
10			
11	–1	7	18
12	–1	9	–1
13			
14	–1	5	–1
15			
16	–1	11	–1
17			
18	–1	12	–1
19			
20	2	6	16

Figure 10.8 Binary tree T

Figure 10.9 Linked representation of binary tree

Example 10.1: Given the memory representation of a tree that stores the names of family members, construct the corresponding tree from the given data.

ROOT: 3

AVAIL: 7

	LEFT	NAMES	RIGHT
1	–1	Pallav	12
2			
3	9	Amar	
4			13
5			
6	17	Deepak	19
7			
8			
9	1	Janak	–1
10			
11	–1	Kuvam	–1
12	–1	Rudraksh	–1
13	6	Raj	20
14			
15	–1	Kunsh	–1
16			
17	–1	Tanush	–1
18			
19	–1	Ridhiman	–1
20	11	Sanjay	15

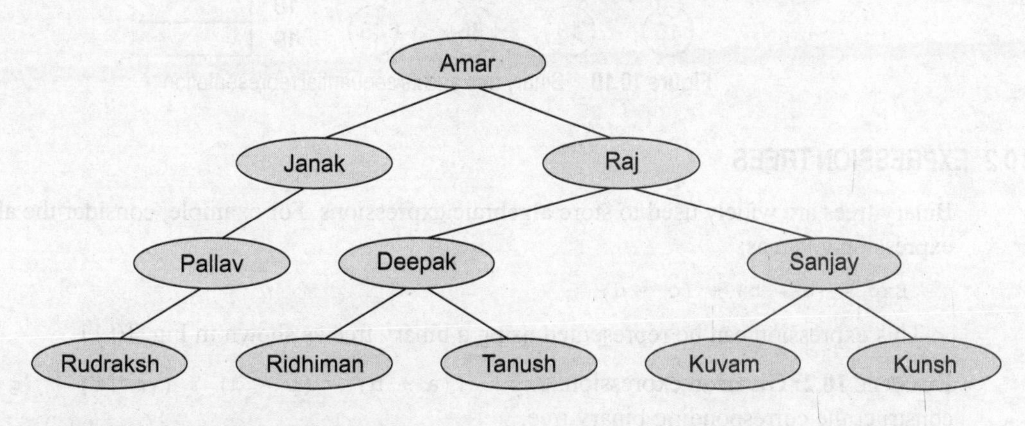

Sequential representation of binary trees

Sequential representation of trees is done using single or one-dimensional arrays. Though it is the simplest technique for memory representation, it is inefficient as it requires a lot of memory space. A sequential binary tree follows the following rules:

- A one-dimensional array, called the `TREE`, will be used.
- The root of the tree will be stored in the first location. That is, `TREE[0]` will store the data of the root element.
- The children of a node `K` will be stored in locations `(2*K)` and `(2*K+1)`.
- The maximum size of the array `TREE` is given as $(2^{d+1}-1)$, where `d` is the depth of the tree.
- An empty tree or sub-tree is specified using `NULL`. If `TREE[0] = NULL`, then the tree is empty.

Figure 10.10 shows a binary tree and its corresponding sequential representation. The tree has 11 nodes and its depth is 4.

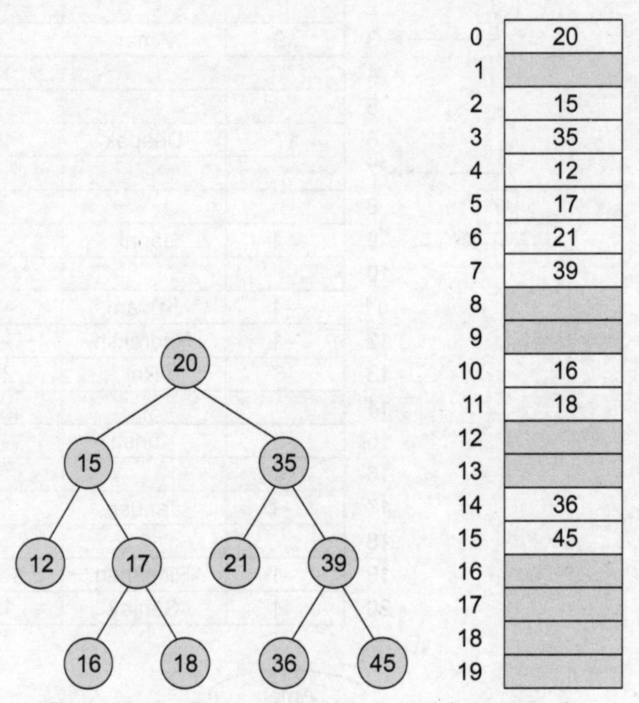

Figure 10.10 Binary tree and its sequential representation

10.2 EXPRESSION TREES

Binary trees are widely used to store algebraic expressions. For example, consider the algebraic expression given as:

```
Exp = (a - b) + (c * d)
```

This expression can be represented using a binary tree as shown in Fig. 10.11.

EXAMPLE 10.2: Given an expression, `Exp = ((a + b) - (c * d) % ((e ^f) / (g - h))`, construct the corresponding binary tree.

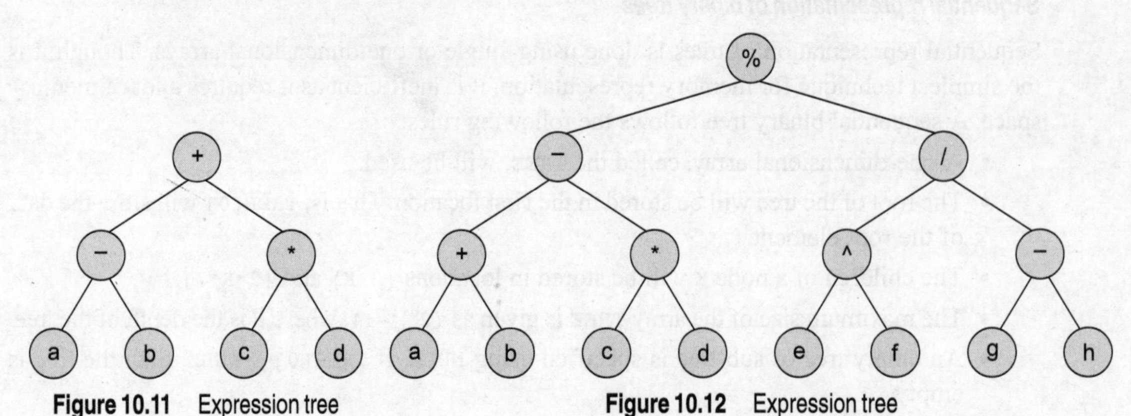

Figure 10.11 Expression tree

Figure 10.12 Expression tree

EXAMPLE 10.3: Given the binary tree, write down the expression that it represents.

`[{(a / b) + (c * d)} ^ {(f % g) / (h - i)}]`

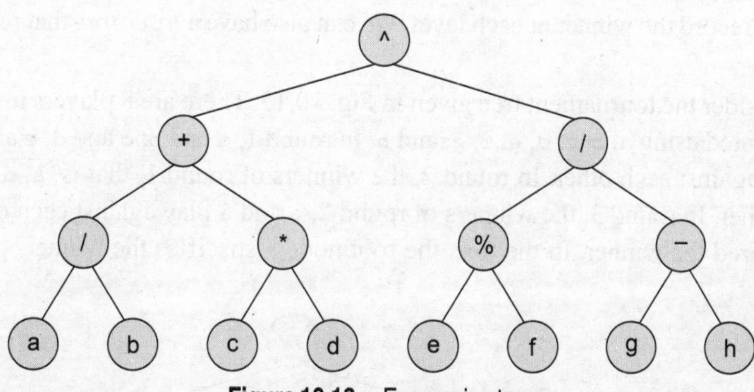

Figure 10.13 Expression tree

EXAMPLE 10.4: Given the expression, `Exp = a + b / c * d - e`, construct the corresponding binary tree.

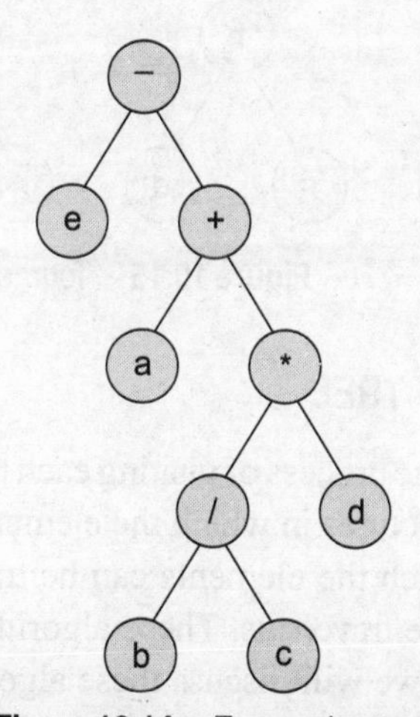

Figure 10.14 Expression tree

10.3 TOURNAMENT TREES

We all know that in a tournament, say of chess, n number of players participate. To declare the winner among all these players, a couple of matches are played and usually three rounds are played in the game.

In every match of round 1, a number of matches are played in which two players play the game against each other. The number of matches that will be played in round 1 will depend on the number of players. For example, if there are 8 players participating in a tournament, then 4 matches will be played in round 1. Every match of round 1 will be played between two players.

Then in round 2, the winners of round 1 play against each other. Similarly, in round 3, the winners of round 2 will play against each other and the person who wins round 3 is declared the winner. Tournament trees are used to represent this concept.

In a tournament tree (also called a *selection tree*), each external node represents a player and each internal node represents the winner of the match played between the players represented by its children nodes. These tournament trees are also called *winner trees* because they are being used to record the winner at each level. We can also have a *loser tree* that records the loser at each level.

Consider the tournament tree given in Fig. 10.15. There are 8 players in total whose names are represented using `a`, `b`, `c`, `d`, `e`, `f`, `g`, and `h`. In round 1, `a` and `b`; `c` and `d`; `e` and `f`; and finally `g` and `h` play against each other. In round 2, the winners of round 1, that is, `a`, `d`, `e`, and g play against each other. In round 3, the winners of round 2, `a` and `e` play against each other. Whosoever wins is declared the winner. In the tree, the root node `a` specifies the winner.

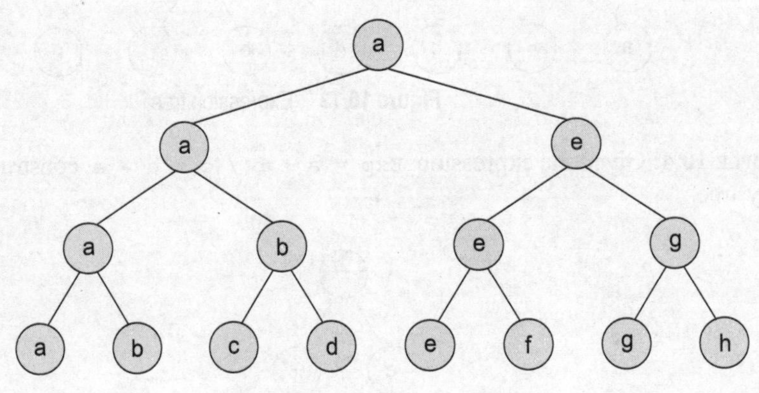

Figure 10.15 Tournament tree

10.4 TRAVERSING OF A BINARY TREE

Traversing a binary tree is the process of visiting each node in the tree exactly once, in a systematic way. Unlike linear data structures in which the elements are traversed sequentially, tree is a non-linear data structure in which the elements can be traversed in many different ways. There are different algorithms for tree traversals. These algorithms differ in the order in which the nodes are visited. In this section, we will discuss these algorithms.

10.4.1 Pre-order Algorithm

To traverse a non-empty binary tree in pre-order, the following operations are performed recursively at each node. The algorithm starts with the root node of the tree and continues by:

1. Visiting the root node,
2. Traversing the left sub-tree, and finally
3. Traversing the right sub-tree.

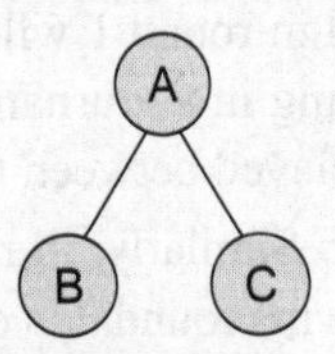

Figure 10.16 Binary tree

Consider the tree given in Fig. 10.16. The pre-order traversal of the tree is given as `A`, `B`, `C`. Root node first, the left sub-tree next, and then the right sub-tree. Pre-order traversal is also called as *depth-first traversal*. In this algorithm, the left sub-tree is always traversed before the right sub-tree. The word 'pre' in the pre-order specifies that the root node is accessed prior to any other nodes in the left and right sub-trees. Pre-order algorithm is also known as the NLR traversal algorithm (Node-Left-Right). The algorithm for pre-order travel is shown in Fig. 10.17.

```
Step 1: Repeat step 2 to 4 while TREE != NULL
Step 2:           Write "TREE -> DATA
Step 3:           PREORDER(TREE -> LEFT)
Step 4:           PREORDER(TREE -> RIGHT)
       [END OF WHILE]
Step 5: END
```

Figure 10.17 Algorithm for pre-order traversal

Pre-order traversal algorithms are used to extract a prefix notation from an expression tree. For example, consider the expression given below. When we traverse its elements using the pre-order traversal algorithm, the expression that we get is a prefix expression.

```
+ - a b * c d (from Fig. 10.11)
% - + a b * c d / ^ e f - g h (from Fig. 10.12)
^ + / a b * c d / % f g- h i (from Fig. 10.13)
- e + a * / b c d (from Fig. 10.14)
```

Example 10.5: In Figs 10.18 and 10.19, find out the sequence of nodes that will be visited using pre-order traversal algorithm.

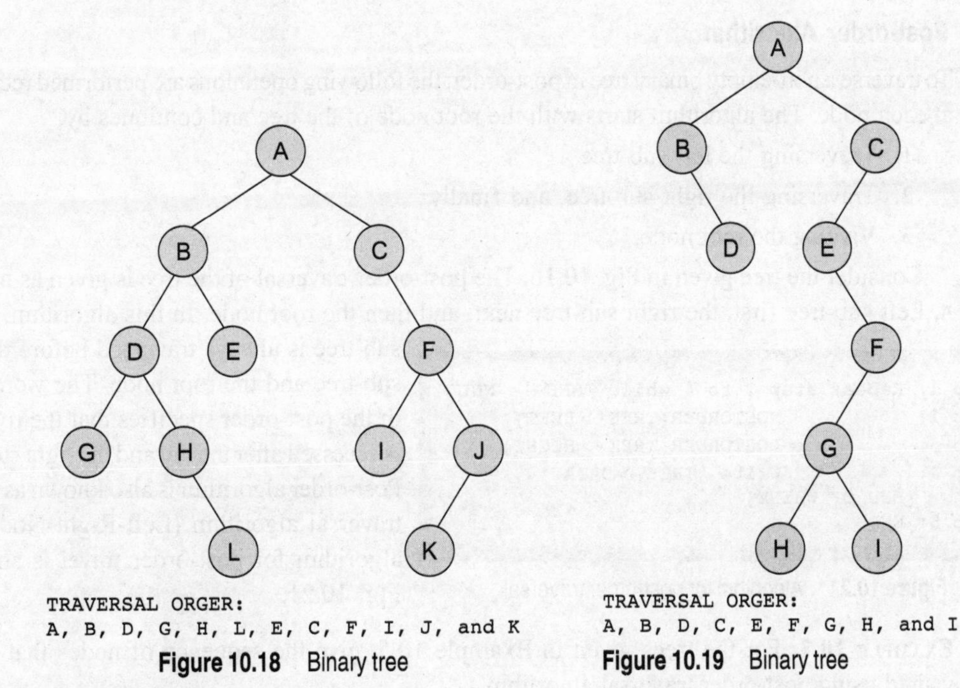

Figure 10.18 Binary tree

Figure 10.19 Binary tree

10.4.2 In-order Algorithm

To traverse a non-empty binary tree in in-order, the following operations are performed recursively at each node. The algorithm starts with the root node of the tree and continues by:

1. Traversing the left sub-tree,
2. Visiting the root node, and finally
3. Traversing the right sub-tree.

Consider the tree given in Fig 10.16. The in-order traversal of the tree is given as B, A, and C. Left sub-tree first, the root node next, and then the right sub-tree. Pre-order traversal is also called as *symmetric traversal*. In this algorithm, the left sub-tree is always traversed before the root node and the right sub-tree. The word 'in' in the in-order specifies that the root node is accessed in between the left and the right sub-trees. In-order algorithm is also known as the LNR traversal algorithm (Left-Node-Right). The algorithm for in-order travel is shown in Fig. 10.20.

```
Step 1: Repeat step 2 to 4 while TREE != NULL
Step 2:          INORDER(TREE -> LEFT)
Step 3:          Write "TREE -> DATA
Step 4:          INORDER(TREE -> RIGHT)
   [END OF WHILE]
Step 5: END
```

Figure 10.20 Algorithm for in-order traversal

In-order traversal algorithm is usually used to display the elements of a binary search tree. Here, all the elements with a lower value than a given value are accessed before the elements with a higher value. We will discuss binary search trees in detail in the next section.

Example 10.6: For the trees given in Example 10.5, find out the sequence of nodes that will be visited using in-order traversal algorithm.

```
TRAVERSAL ORDER: G, D, H, L, B, E, A, C, I, F, K, and J.
TRAVERSAL ORDER: B, D, A, E, H, G, I, F, AND C.
```

10.4.3 Post-order Algorithm

To traverse a non-empty binary tree in post-order, the following operations are performed recursively at each node. The algorithm starts with the root node of the tree and continues by:

1. Traversing the left sub-tree,
2. Traversing the right sub-tree, and finally
3. Visiting the root node.

Consider the tree given in Fig. 10.16. The post-order traversal of the tree is given as B, C, and A. Left sub-tree first, the right sub-tree next, and then the root node. In this algorithm, the left sub-tree is always traversed before the right sub-tree and the root node. The word 'post' in the post-order specifies that the root node is accessed after the left and the right sub-trees. Post-order algorithm is also known as the LRN traversal algorithm (Left-Right-Node). The algorithm for post-order travel is shown in Fig. 10.21.

```
Step 1: Repeat step 2 to 4 while TREE != NULL
Step 2:          POSTORDER(TREE -> LEFT)
Step 3:          POSTORDER(TREE -> RIGHT)
Step 4:          Write "TREE -> DATA
   [END OF WHILE]
Step 5: END
```

Figure 10.21 Algorithm for post-order traversal

Example 10.7: For the trees given in Example 10.5, give the sequence of nodes that will be visited using post-order traversal algorithm

```
TRAVERSAL ORDER: G, L, H, D, E, B, I, K, J, F, C, and A.
TRAVERSAL ORDER: D, B, H, I, G, F, E, C, and A.
```

10.4.4 Level-order Traversal

In level-order traversal, all the nodes at a level are accessed before going to the next level. This algorithm is also called as the *breadth-first traversal algorithm*. Consider the trees given in Fig. 10.22 and note the level order of these trees.

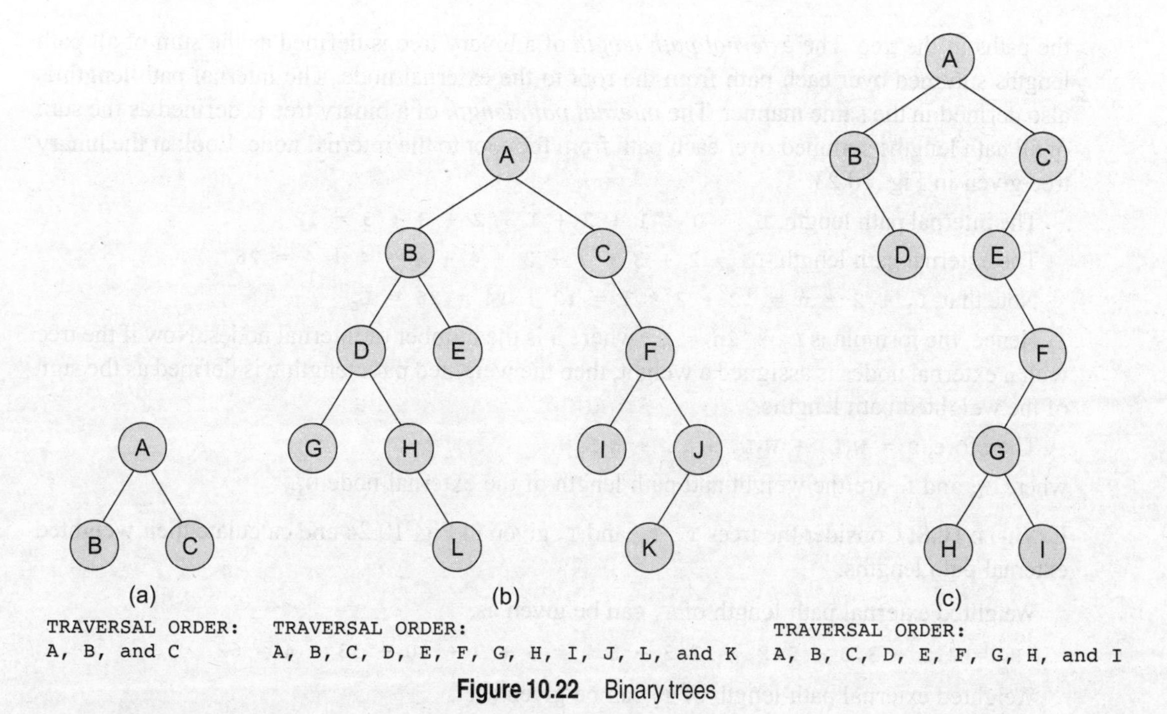

Figure 10.22 Binary trees

10.5 HUFFMAN'S TREE

Huffman coding is an entropy encoding algorithm developed by David A. Huffman that is widely used as a lossless data compression technique. The Huffman coding algorithm uses a variable-length code table to encode a source character where the variable-length code table is derived in a particular way based on the estimated probability of occurrence for each possible value of the source character.

The key idea behind Huffman algorithm is that it encodes the most common characters using shorter strings of bits than those used for less common source characters.

The algorithm works by creating a binary tree of nodes that are stored in a regular array. The size of this array depends on the number of nodes in the tree. A node can either be a leaf node or an internal node. Initially, all the nodes in the tree are at the leaf level and store the source character and its frequency of occurrence (also known as weight).

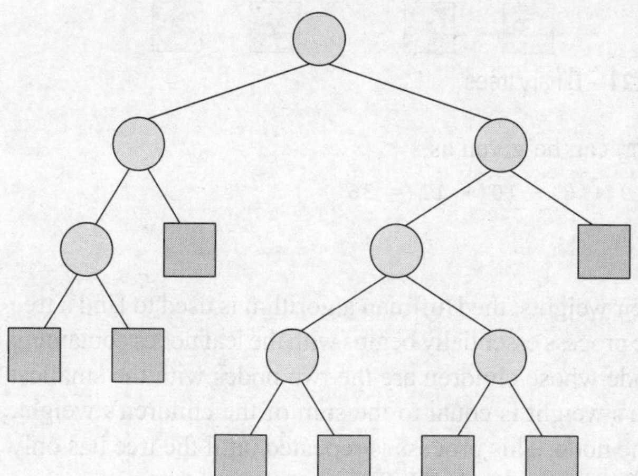

Figure 10.23 Binary tree

While the internal nodes are used to store the weight and contains links to its child nodes, the external node contains the actual character. Conventionally, a `'0'` represents following the left child and a `'1'` represents following the right child. A finished tree that has `n` leaf nodes will have `n - 1` internal nodes.

The running time of the algorithm depends on the length of the paths in the tree. So, before going into further details of the Huffman coding, let us first learn how to calculate the length of the paths in the tree. The *external path length* of a binary tree is defined as the sum of all path lengths summed over each path from the root to the external node. The internal path length is also defined in the same manner. The *internal path length* of a binary tree is defined as the sum of all path lengths summed over each path from the root to the internal node. Look at the binary tree given in Fig. 10.23.

The internal path length, $L_I = 0 + 1 + 2 + 1 + 2 + 3 + 3 = 12$

The external path length, $L_E = 2 + 3 + 3 + 2 + 4 + 4 + 4 + 4 = 26$

Note that, $L_I + 2 * n = 12 + 2 * 7 = 12 + 14 = 26 = L_E$

Hence, the formula is $L_I + 2n = L_E$, where n is the number of internal nodes. Now if the tree with n external nodes is assigned a weight, then the weighted path length P is defined as the sum of the weighted path lengths.

Therefore, $P = W_1L_1 + W_2L_2 + ... + W_nL_n$

where, W_i and L_i are the weight and path length of the external node N_i.

Example 10.8: Consider the trees T_1, T_2, and T_3 given in Fig. 10.24 and calculate their weighted external path lengths.

Weighted external path length of T_1 can be given as,

$P_1 = 2.3 + 3.3 + 5.2 + 11.3 + 2.2 = 6 + 9 + 10 + 33 + 4 = 62$

Weighted external path length of T_2 can be given as,

$P_2 = 5.2 + 7.2 + 3.3 + 4.3 + 2.2 = 10 + 14 + 9 + 12 + 4 = 49$

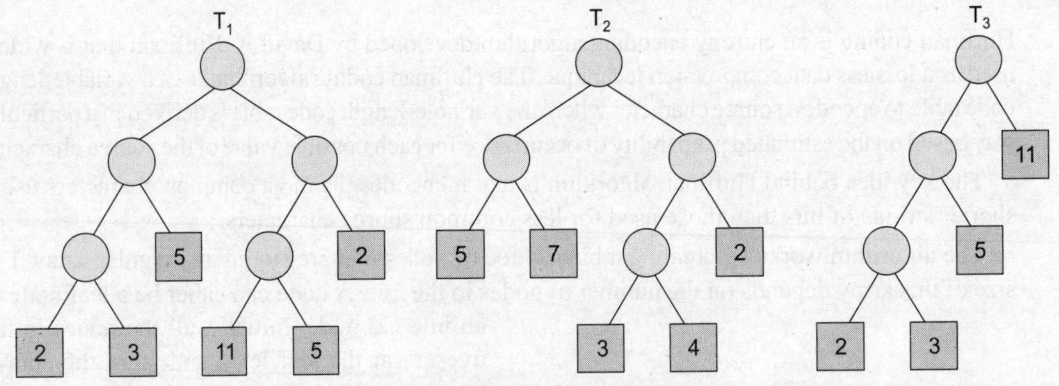

Figure 10.24 Binary trees

The weighted external path length of T_3 can be given as,

$P_3 = 3.3 + 2.3 + 5.2 + 11.1 = 9 + 6 + 10 + 11 = 36$

10.5.1 Technique

Given a tree with n external nodes and their weights, the Huffman algorithm is used to find a tree with a minimum-weighted path length. The process essentially begins with the leaf nodes containing the weights of the nodes. Then, a new node whose children are the two nodes with the smallest weight is created, such that the new node's weight is equal to the sum of the children's weight. That is, the two nodes are merged into one node. This process is repeated until the tree has only one node. Such a tree with only one node is known as the Huffman tree.

The Huffman algorithm can be implemented using a priority queue in which all the nodes are placed in such a way that the node with the lowest weight is given the highest priority. The algorithm is shown in Fig. 10.25.

```
Step 1: Create a leaf node for each character. Add the character and its
        weight or frequency of occurrence to the priority queue.
Step 2: Repeat Steps 3 to 5 while the total number of nodes in the queue is
        greater than 1.
Step 3: Remove two nodes that have the lowest weight (or highest priority).
Step 4: Create a new internal node by merging these two nodes as children
        and with weight equal to the sum of the two nodes' weights.
Step 5: Add the newly created node to the queue.
```

Figure 10.25 Huffman algorithm

Example 10.9: Create a Huffman tree with the following nodes arranged in a priority queue.

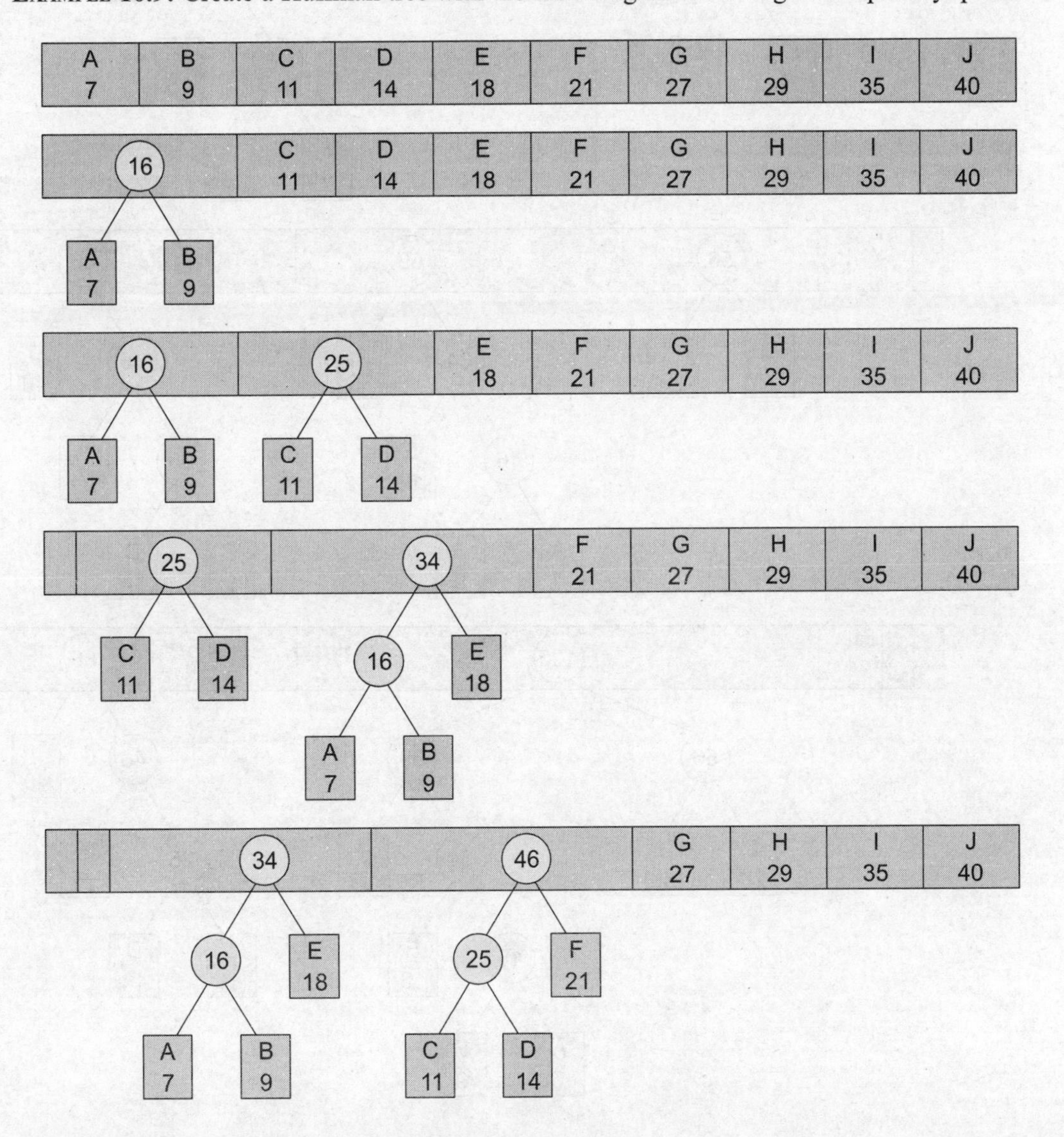

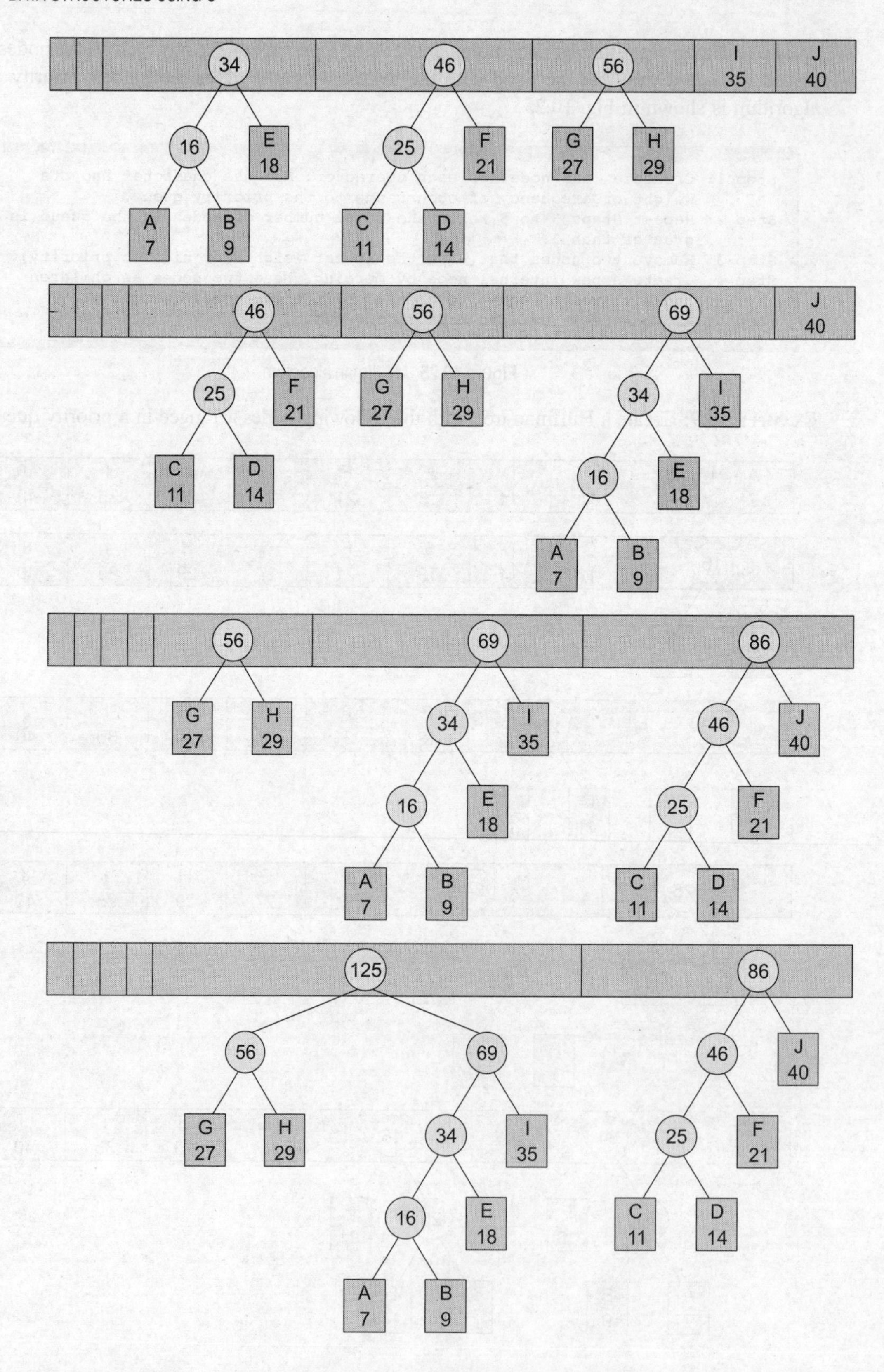

34
46
56
I 35
J 40
16
E 18
25
F 21
G 27
H 29
A 7
B 9
C 11
D 14
69
86
125

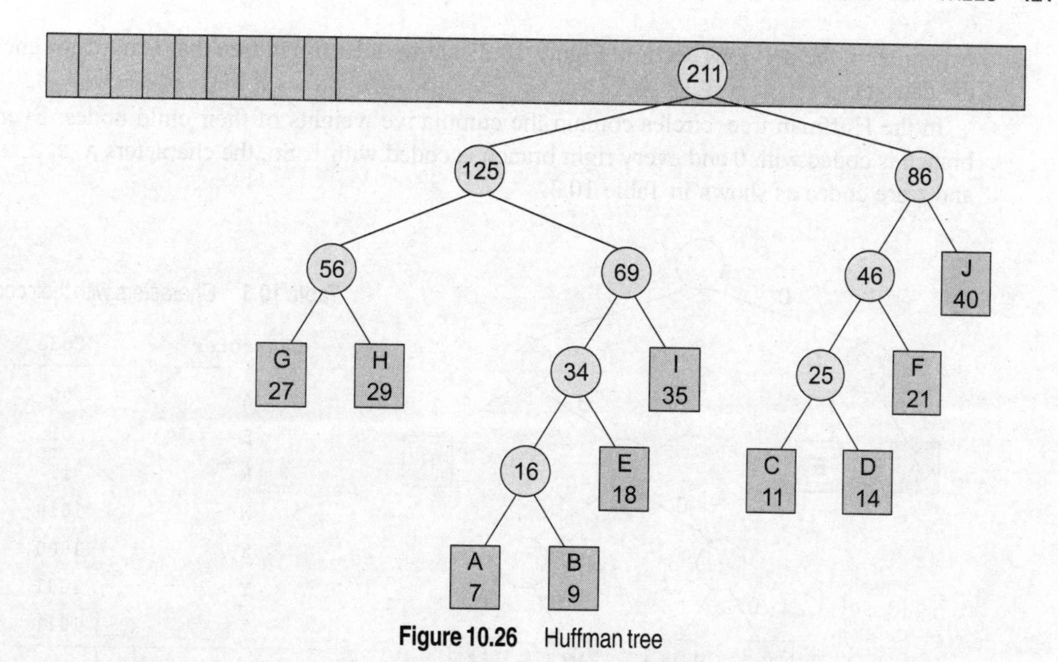

Figure 10.26 Huffman tree

10.5.2 Data Coding

When we want to code our data (character) using bits, then we use r bits to code 2^r characters. For example, if r=1, then two characters can be coded. If these two characters are A and B, then A can be coded as 0 and B can be coded as 1 and vice versa. Look at the following tables which show the range of characters that can be coded by using r=2 and r=3.

Table 10.1 Range of characters that can be coded using r=2

Code	Character
00	A
01	B
10	C
11	D

Table 10.2 Range of characters that can be coded using r=3

Code	Character
000	A
001	B
010	C
011	D
100	E
101	F
110	G
111	H

Now, if we have to code the data string ABBBBBBAAAACDEFGGGGH, then the corresponding code would be:

000001001001001001001000000000000010011100101110110110110111

This coding scheme has a fixed length code because every character is being coded using the same number of bits. Although this technique of coding is simple, but coding the data can be made more efficient by using a variable length code.

You might have observed that when we write a text in English, all the characters are not used frequently. For example, characters like a, e, i, and r are used more frequently than w, x, y, z and so on. So, the basic idea is to assign a shorter code to the frequently occurring characters and a longer code for less frequently occurring characters. Variable length coding is preferred over fixed length coding because it requires less number of bits to encode the same data.

For variable length encoding, we first build a Huffman tree. First, arrange all the characters in a priority queue in which the character with the highest frequency of occurrence has the lowest weight and thus the highest priority. Then, create a Huffman tree as explained in the previous section. Figure 10.27 shows a Huffman tree that is used for encoding the data set.

In the Huffman tree, circles contain the cumulative weights of their child nodes. Every left branch is coded with 0 and every right branch is coded with 1. So, the characters A, E, R, W, X, Y, and Z are coded as shown in Table 10.3.

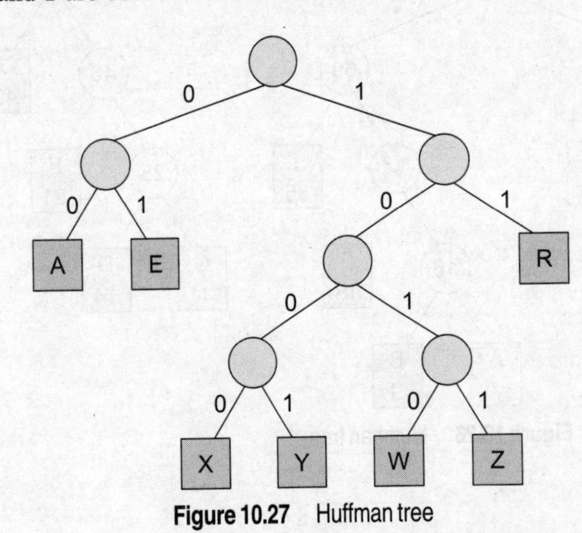

Figure 10.27 Huffman tree

Table 10.3 Characters with their codes

Character	Code
A	00
E	01
R	11
W	1010
X	1000
Y	1001
Z	1011

Thus, we see that frequent characters have a shorter code and infrequent characters have a longer code.

SUMMARY

- A tree is a structure which is mainly used to store hierarchical data. In a binary tree, every node has zero, one, or at the most two successors. A node that has no successors is called a leaf node or a terminal node. Every node other than the root node has a parent.
- The degree of a node is equal to the number of children that a node has. The degree of a leaf node is zero. All nodes that are at the same level and share the same parent are called siblings.
- Two binary trees having a similar structure are said to be copies if they have the same contents at the corresponding nodes.
- A binary tree of n nodes has exactly $n - 1$ edges. The depth of a node N is given as the length of the path from the root R to the node N. The depth of the root node is zero. A binary tree of height h has at least h nodes and at most $2^h - 1$ nodes.
- The height of a binary tree with n nodes is at least n and at most $\log_2(n+1)$. In-degree of a node is the number of edges arriving at that node. The root node is the only node that has an in-degree equal to zero. Similarly, out-degree of a node is the number of edges leaving that node.
- A binary tree T is said to be an extended binary tree (or a 2-tree) if each node in the tree has either no child or exactly two children.
- Pre-order traversal is also called as depth-first traversal. It is also known as the NLR traversal algorithm (Node-Left-Right) and is used to extract a prefix notation from an expression tree. In-order algorithm is known as the LNR traversal algorithm (Left-Node-Right). Similarly, post-order algorithm is known as the LRN traversal algorithm (Left-Right-Node).
- The Huffman coding algorithm uses a variable-length code table to encode a source character where the variable-length code table is derived in a particular way based on the estimated probability of occurrence for each possible value of the source character.

GLOSSARY

Binary tree A tree in which every node can have at most two children.

Complete binary tree A binary tree in which every level, except the last, is completely filled. At depth n, all nodes must be as far left as possible.

Depth-first search Search algorithm which considers the outgoing edges of a node before any of the node's siblings. That is, extremes are searched first.

Pre-order traversal The traversal technique in which all the nodes of a tree are processed by processing the root first and then recursively processing all sub-trees.

In-order traversal The traversal technique in which all the nodes of a tree are processed by recursively processing the left sub-tree first, then processing the root, and finally the right sub-tree.

Post-order traversal The traversal technique in which all the nodes of a tree are processed by recursively processing all sub-trees and then finally processing the root.

Level-order traversal The traversal technique in which all the nodes of a tree are processed by depth: first the root, then the children of the root, etc. It is equivalent to a breadth-first search from the root.

Tree A data structure that is accessed from the root node. Each node in the tree is either a leaf or an internal node. An internal node can have one or more child nodes and is called the parent of its child nodes. All children of the same node are siblings.

Descendant A child of a node in a tree.

Ascendant The parent of a node in a tree.

Height The maximum distance of any leaf node from the root node of a tree. If a tree has only one node (the root), the height is zero.

Depth Depth of a node is defined as the distance from the node to the root of the tree.

Degree Degree of a node is equal to the number of edges connected to it. That is, it is the number of child nodes it has.

Breadth-first search A search algorithm which considers the neighbours of a node, that is, outgoing edges of the vertex's predecessor in the search, before any outgoing edges of the node. That is, extremes are searched last.

Huffman coding A minimal variable-length character coding based on the frequency of each character. In this coding technique, each character becomes a trivial binary tree, with the character as the only node.

Full binary tree A binary tree in which every node has exactly zero or two children.

Leaf node A node in a tree that has zero or no children.

Root node The initial node of the tree that has no parents.

Internal node A node that is not a leaf is an internal node. That is, a node that has one or more child nodes.

Parent node It is a node that is conceptually above or closer to the root than the node and which has a link to the node.

Child node A node of a tree that is referred by a parent node. That is, every node in the tree except the root is the child of some parent.

Extended binary tree A binary tree with special nodes replacing every null sub-tree in such a way that every regular node has two children, and every special node has no children.

Recursive data structure A data structure that is partially composed of smaller or simpler instances of the same data structure. For example, a tree is composed of smaller trees (often known as sub-trees) and leaf nodes.

Sub-tree The tree which is the child of a node.

EXERCISES

Review Questions

1. Explain the concept of a tree. Discuss its applications.
2. How many binary trees are possible with four nodes?

3. Is it possible to implement binary trees using linear arrays? If yes, explain how?
4. What are the two ways of representing binary trees in the memory? Which one do you prefer and why?
5. List all possible non-similar binary trees having four nodes.
6. Draw the binary expression tree that represents the following postfix expression:
   A B + C * D –
7. Write short notes on:
   (a) Complete binary tree
   (b) Extended binary tree
   (c) Tournament tree
   (d) Expression tree
   (e) Huffman tree
8. Consider the given tree. Now, do the following:
   (a) Make a list of the leaf nodes
   (b) Name the leaf nodes
   (c) Name the non-leaf nodes
   (d) Name the ancestors of node E
   (e) Name the descendants of A
   (f) Name the siblings of C
   (g) Find the height of the tree
   (h) Find the height of sub-tree at E
   (i) Find the level of node E
   (j) Find out the in-order, pre-order, post-order, and level-order

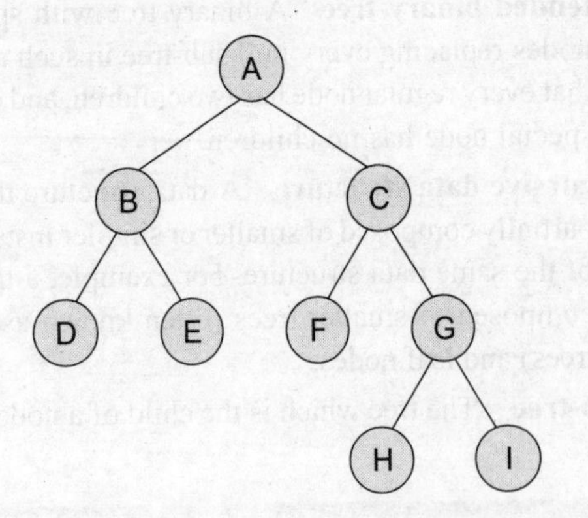

9. Use the expression tree given below, and do the following:
   (a) Extract the infix expression it represents
   (b) Find out the corresponding prefix and postfix expressions
   (c) Evaluate the infix expression, given a = 30, b = 10, c = 2, d = 30, e = 10

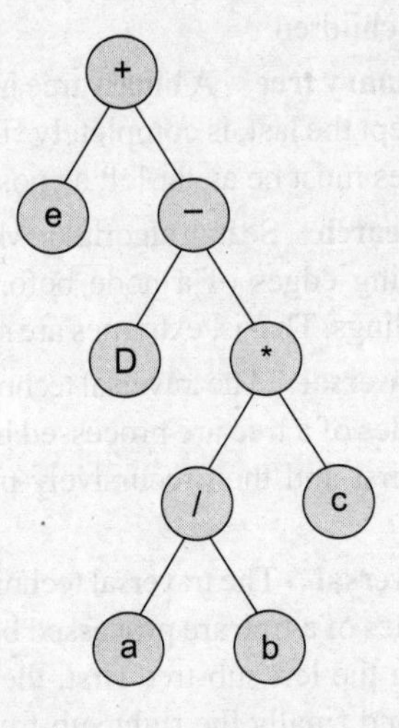

10. Convert the prefix expression `–/ab*+bcd` into infix expression and then draw the corresponding expression tree.
11. Consider the trees given below and state whether it is a complete binary tree or a full binary tree.

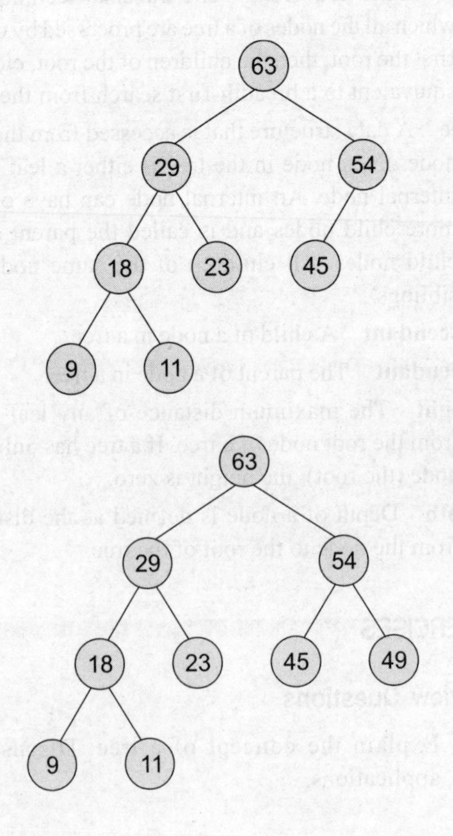

12. What is the maximum number of levels that a binary search tree with 100 nodes can have?

13. What is the maximum height of a tree with 32 nodes?

14. What is the maximum number of nodes that can be found in a binary node in level 3, 4, and 12?

15. Draw all possible non-similar binary trees having three nodes.

16. Draw the binary tree having the following memory representation:

ROOT = 3, AVAIL = 15

	LEFT	DATA	RIGHT
1	–1	8	–1
2	–1	10	–1
3	5	1	8
4			
5	9	2	14
6			
7			
8	20	3	11
9	1	4	12
10			
11	–1	7	18
12	–1	9	–1
13			
14	–1	5	–1
15			
16	–1	11	–1
17			
18	–1	12	–1
19			
20	2	6	16

17. Draw the memory representation of the binary tree given below.

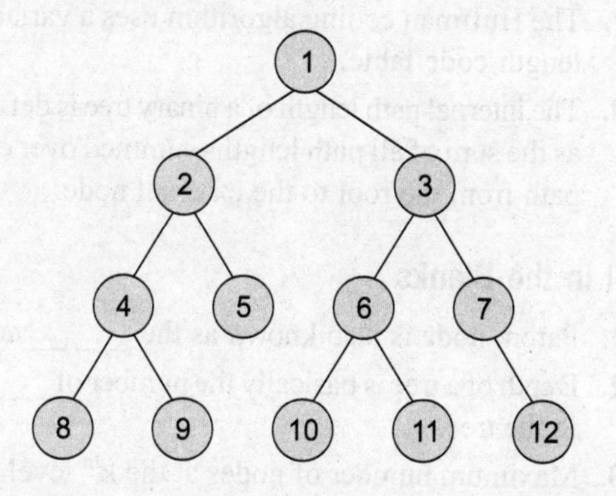

18. Consider the trees T_1, T_2, and T_3 given below and calculate their weighted path lengths.

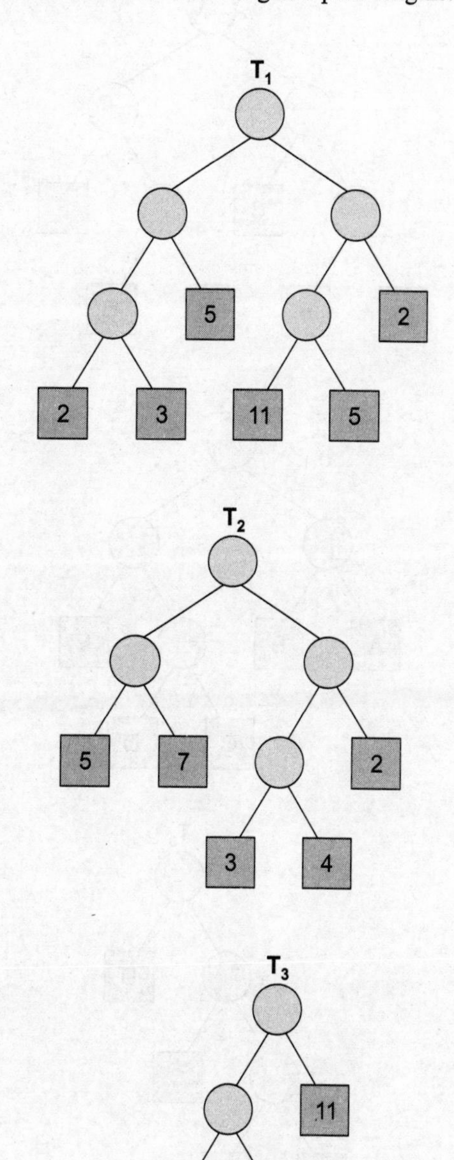

19. Consider the trees T_1, T_2, and T_3 given below and find the Huffman coding for the alphabets.

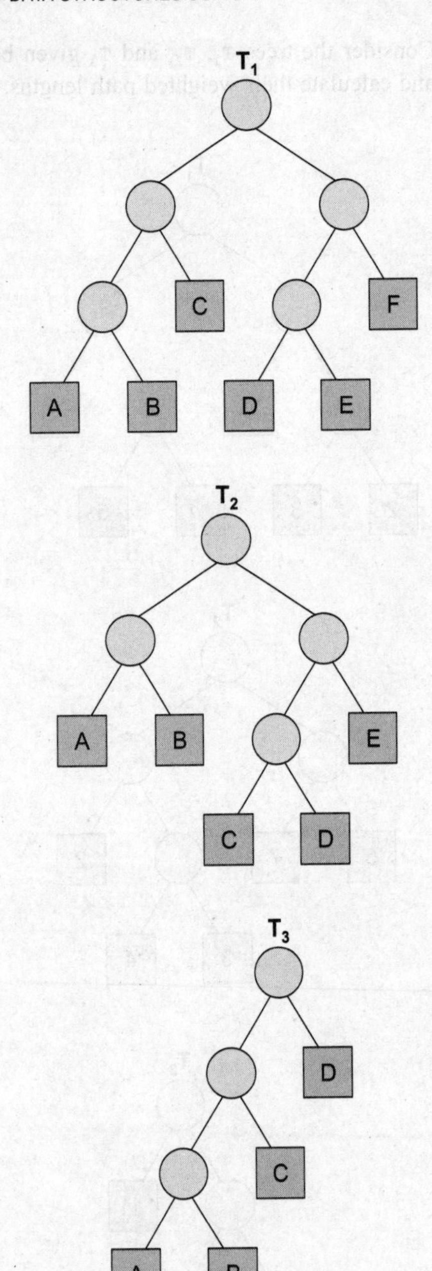

Programming Exercises

1. Write a program to traverse the nodes of a binary tree.
2. Write a program that traverses a binary tree using the following traversal techniques:
   (a) in-order (b) pre-order
   (c) post-order (d) level order
3. Write a program to delete all the nodes in a binary tree.

Multiple Choice Questions

1. Degree of a leaf node is ______.
   (a) 0 (b) 1 (c) 2 (d) 3
2. The depth of root node is ______.
   (a) 0 (b) 1 (c) 2 (d) 3
3. A binary tree of height h has at least h nodes and at most ______ nodes.
   (a) 2h (b) 2^h (c) 2^{h+1} (d) 2^{h-1}
4. Pre-order traversal is also called ______.
   (a) Depth first (b) Breadth first
   (c) Level order (d) In-order
5. The Huffman algorithm can be implemented using a ______.
   (a) Dequeue (b) Queue
   (c) Priority queue (d) None of these
6. Total number of nodes in the nth level of a binary tree can be given as
   (a) 2h (b) 2^h (c) 2^{h+1} (d) 2^{h-1}

True or False

1. Nodes that branch into child nodes are called parent nodes.
2. The size of a tree is equal to its total number of nodes.
3. A leaf node does not branch out further.
4. A node that has no successors is called the root node.
5. A binary tree of n nodes have exactly n – 1 edges.
6. Every node has a parent.
7. The Huffman coding algorithm uses a variable-length code table.
8. The internal path length of a binary tree is defined as the sum of all path lengths summed over each path from the root to the external node.

Fill in the Blanks

1. Parent node is also known as the ______ node.
2. Depth of a tree is basically the number of ______ in the tree.
3. Maximum number of nodes at the kth level of a binary tree is ______.

4. ______ is the best data structure to implement a priority queue.
5. In a binary tree, every node can have maximum ______ successors.
6. Nodes at the same level that share the same parent are called ______.
7. Two binary trees are said to be copies if they have similar ______ and ______.
8. The height of a binary tree with n nodes is at least ______ and at most ______.
9. A binary tree T is said to be an extended binary tree if ______.
10. ______ traversal algorithms are used to extract a prefix notation from an expression tree.
11. In a Huffman tree, the code of a character depends on ______.

11 Efficient Binary Trees

Learning Objective

In this chapter, we will discuss efficient binary trees such as binary search trees, AVL trees, M-way search trees, B-trees, B+ trees, threaded binary trees, trie, red-black trees, and splay trees. This chapter is an extension of binary trees.

11.1 BINARY SEARCH TREES

We have already discussed binary trees in the previous chapter. A binary search tree, also known as an ordered binary tree, is a variant of binary trees in which the nodes are arranged in an order. In a binary search tree, all the nodes in the left sub-tree have a value less than that of the root node. Correspondingly, all the nodes in the right sub-tree have a value either equal to or greater than the root node. The same rule is applicable to every sub-tree in the tree. (Note that a binary search tree may or may not contain duplicate values, depending on its implementation.)

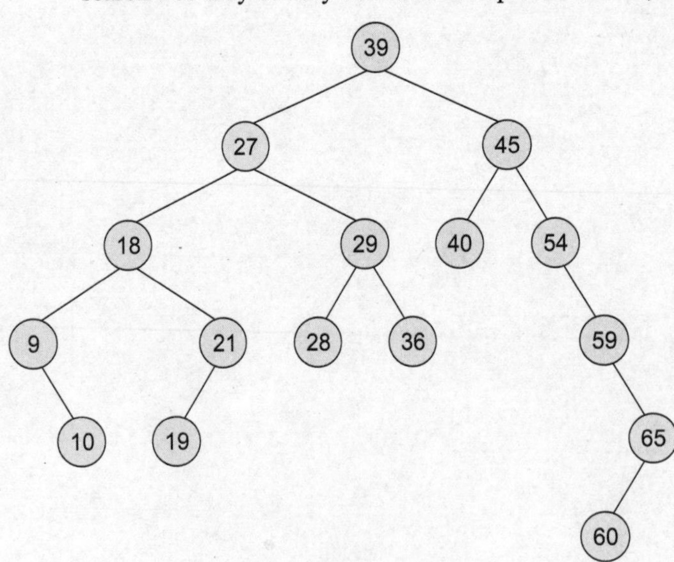

Figure 11.1 Binary search tree

Look at Fig. 11.1. The root node is 39. The left sub-tree of the root nodes consist of nodes 9, 10, 18, 21, 27, 28, 29, and 36. All these nodes have smaller values than the root node. The right sub-tree of the root node consists of nodes 40, 45, 54, 59, 60, and 65. Recursively, each of the sub-trees also obey the binary search tree constraint. For example, in the left sub-tree of the root node, 27 is the root and all elements in its left sub-tree (9, 10, 18, 21) are smaller than 27, while all nodes in its right sub-tree (28, 29, and 30) are greater than the root node's value.

Since the nodes in a binary search tree are ordered, the time needed to search an element from the tree is greatly reduced. Whenever we search for an element, we do not need to traverse the entire tree. At every node, we get a hint regarding which sub-tree to search in. For example, in the given tree, if we have to search for 29, then we know that we have to scan only the left sub-tree. If the value is present in the tree, it will only be in the left sub-tree, as 18 is smaller than 39 (the root node's value). The left sub-tree has a root node with the value 27. Since 29 is greater than 27, we will move to the right sub-tree, where we will find the element. Thus, the average running time of a search operation is $O(\log_2 n)$, as at every step, we eliminate half of the sub-tree from the search process. Due to its efficiency in searching elements, binary search trees are widely used in dictionary problems where the code always inserts and searches the elements that are indexed by some key value.

Binary search trees also speed up the insertion and deletion operations. The tree has a speed advantage when the data in the structure changes rapidly.

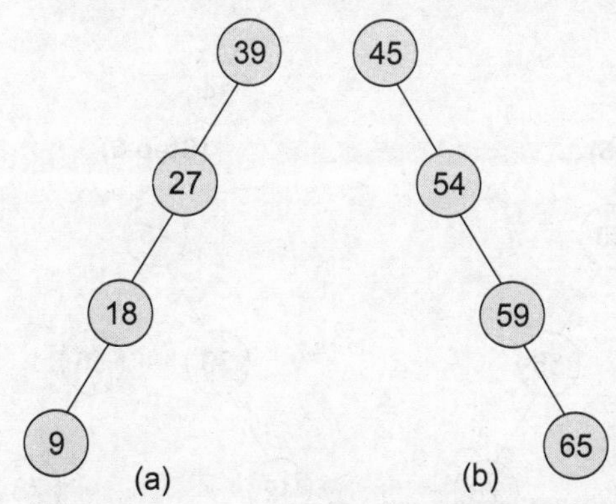

Figure 11.2 (a) Left skewed and (b) Right skewed binary search trees

Binary search trees are considered to be efficient data structures especially when compared with sorted linear arrays and linked lists. In a sorted array, searching can be done in $O(\log_2 n)$ time, but insertions and deletions are quite expensive. In contrast, inserting and deleting elements in a linked list is easier, but searching for an element is done in `O(n)` time.

However, in the worst case, a binary search tree will take `O(n)` time to search for an element from the tree. The worst case would occur when the tree is a linear chain of nodes as given in Fig. 11.2.

To summarize, a binary search tree is a binary tree with the following properties:

- The left sub-tree of a node `N` contains values that are less than `N`'s value.
- The right sub-tree of a node `N` contains values that are greater than `N`'s value.
- Both the left and the right binary trees also satisfies these properties and thus, are binary search trees.

EXAMPLE 11.1: State whether the binary trees in Fig. 11.3 are binary search trees or not.

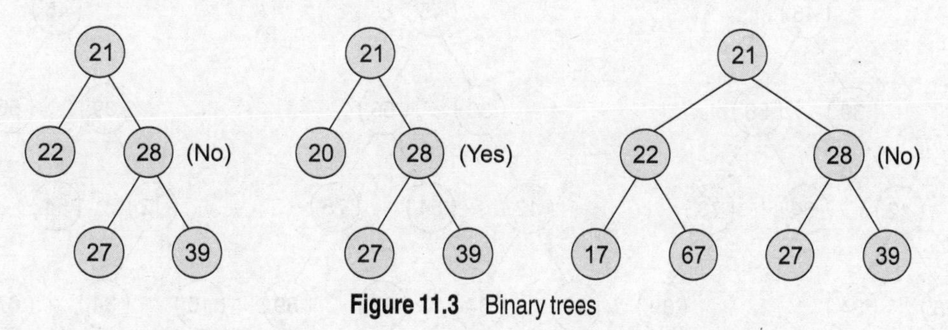

Figure 11.3 Binary trees

EXAMPLE 11.2: Create a binary search tree using the following data elements:

45, 39, 56, 12, 34, 78, 32, 10, 89, 54, 67, 81.

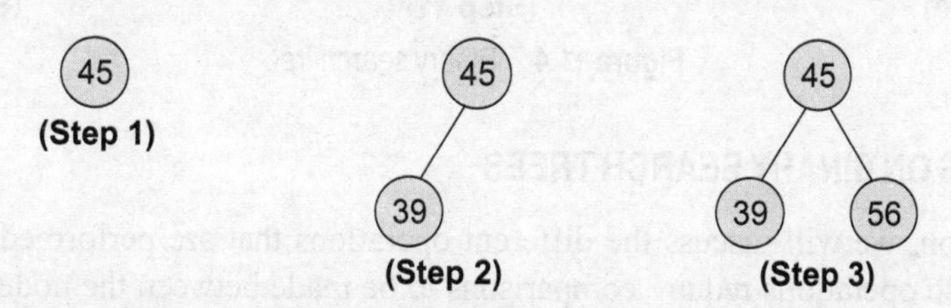

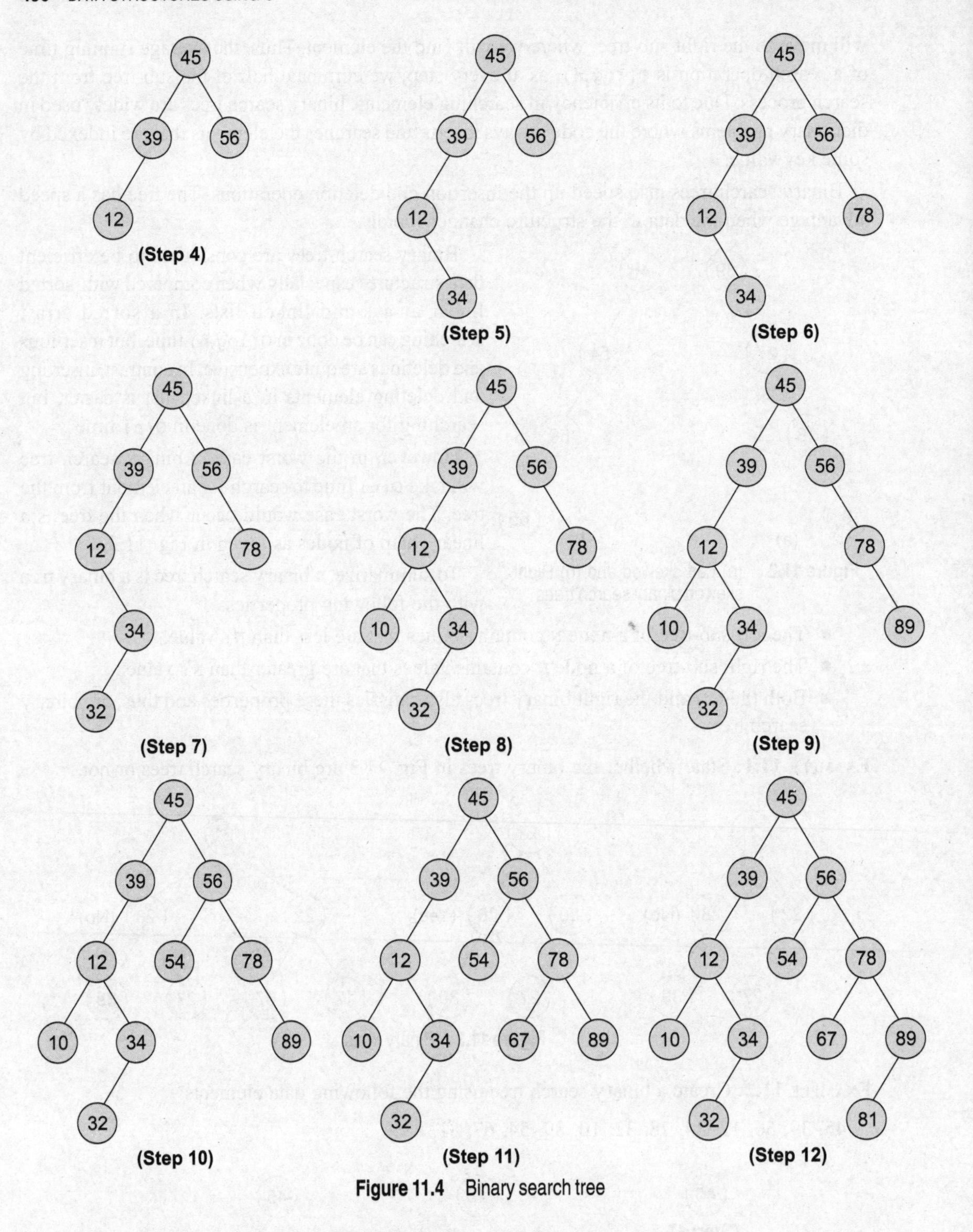

Figure 11.4 Binary search tree

11.2 OPERATIONS ON BINARY SEARCH TREES

In this section, we will discuss the different operations that are performed on a binary search tree. All these operations require comparisons to be made between the nodes.

11.2.1 Search

The search function is used to find whether a given value is present in the tree or not. The searching process begins at the root node. The function first checks if the binary search tree is empty. If it is empty, then the value we are searching for is not present in the tree. So, the search algorithm terminates by displaying an appropriate message. However, if there are nodes in the tree, then the search function checks to see if the key value of the current node is equal to the value to be searched. If not, it checks if the value to be searched for is less than the value of the node, in which case it should be recursively called on the left child node. In case the value is greater than the value of the node, it should be recursively called on the right child node.

Look at Fig. 11.5. The figure shows how a binary tree is searched to find a specific element. First, see how the tree will be traversed to find the node with the value 12.

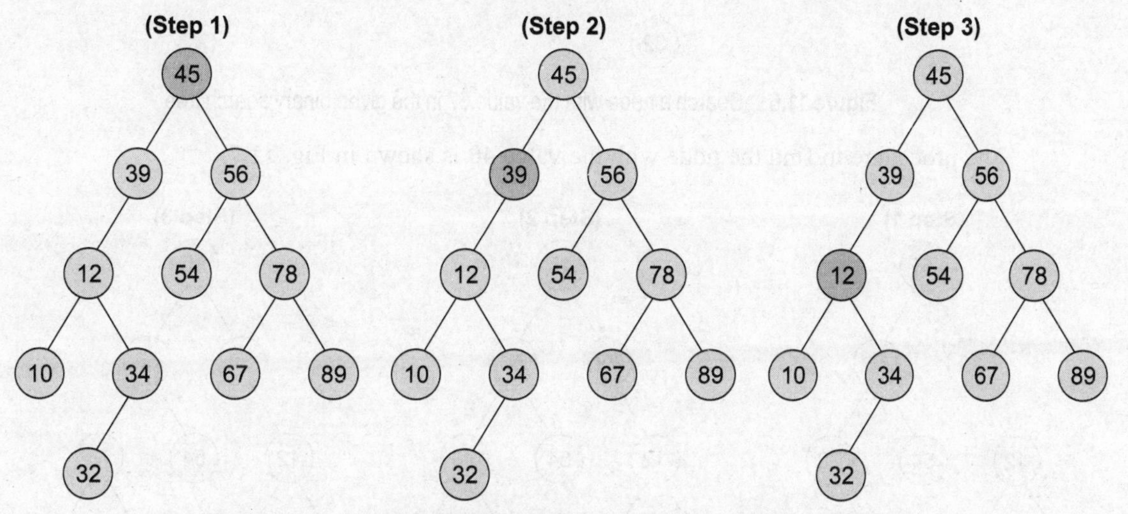

Figure 11.5 Search a node with the value 12 in the given binary search tree

The procedure to find the node with the value 67 is illustrated in Fig. 11.6.

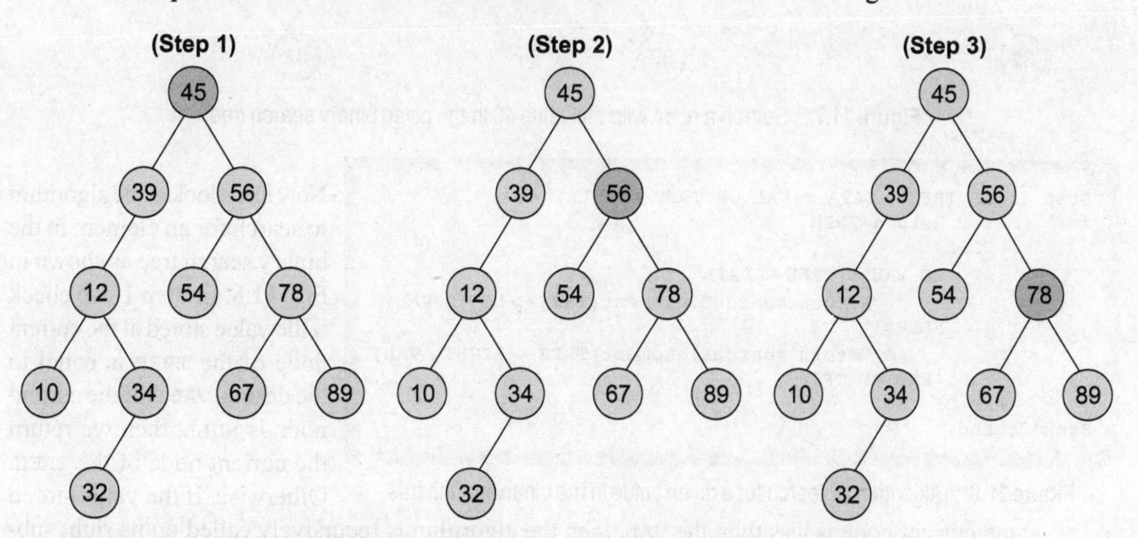

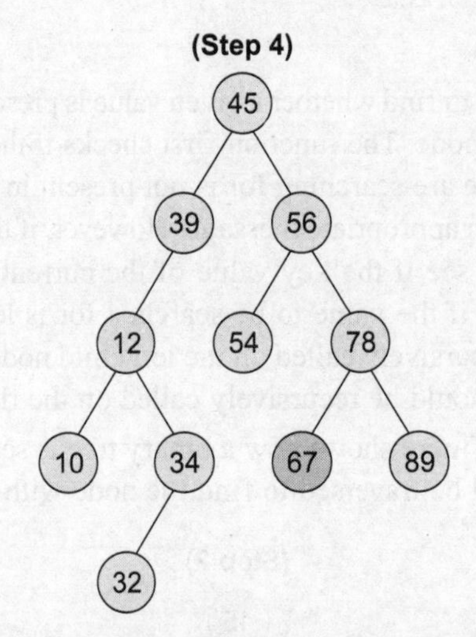

Figure 11.6 Search a node with the value 67 in the given binary search tree

The procedure to find the node with the value 40 is shown in Fig. 11.7.

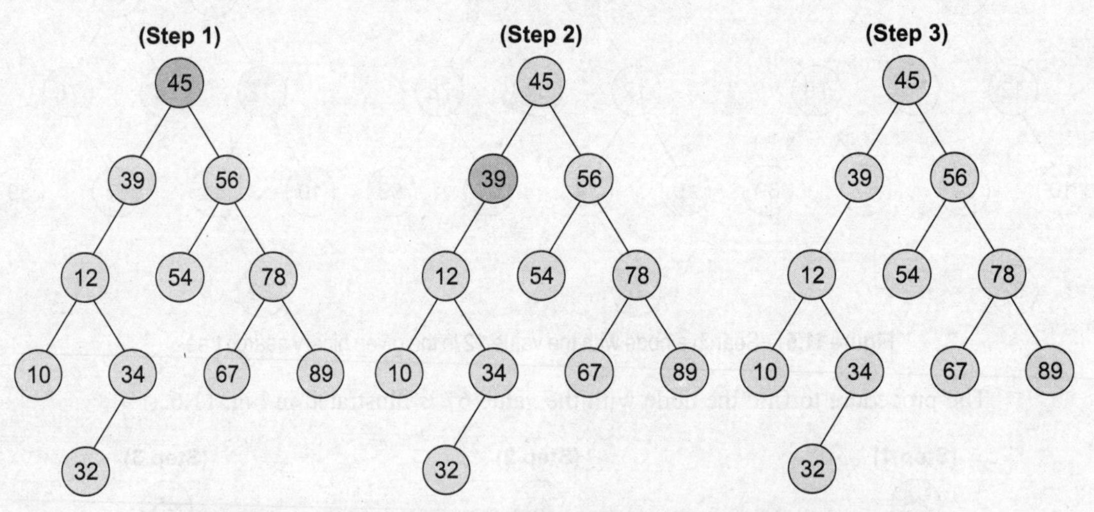

Figure 11.7 Search a node with the value 40 in the given binary search tree

```
Step 1: IF TREE->DATA = VAL OR TREE = NULL, then
            Return TREE
        ELSE
            IF VAL < TREE->DATA
                  Return searchElelement(TREE->LEFT, VAL)
            ELSE
                  Return searchElelement(TREE->RIGHT, VAL)
            [END OF IF]
      [END OF IF]
Step 2: End
```

Figure 11.8 Algorithm to search for a given value in the binary search tree

Now let us look at the algorithm to search for an element in the binary search tree as shown in Fig. 11.8. In Step 1, we check if the value stored at the current node of the `TREE` is equal to the desired `VAL` or if the current node is `NULL`, then we return the current node of the `TREE`. Otherwise, if the value stored at the current node is less than the `VAL`, then the algorithm is recursively called on its right sub-tree, else the algorithm is called on its left sub-tree.

11.2.2 Insertion

The insert function is used to add a new node with a given value at the correct position in the binary search. Adding the node at the correct position means that the new node should not violate the properties of the binary search tree. Figure 11.9 shows the algorithm to insert a given value in a binary search tree.

```
Step 1: IF TREE = NULL, then
              Allocate memory for TREE
              SET TREE -> DATA = VAL
              SET TREE -> LEFT = TREE -> RIGHT = NULL
        ELSE
              IF VAL < TREE -> DATA
                    Insert(TREE -> LEFT, VAL)
              ELSE
                    Insert(TREE -> RIGHT, VAL)
              [END OF IF]
        [END OF IF]
Step 2: End
```

Figure 11.9 Algorithm to insert a given value in the binary search tree

The initial code for the insert function is similar to the search function. This is because we first find the correct position where the insertion has to be done and then add the node at that position. The insertion function changes the structure of the tree. Therefore, when the insert function is called recursively, the function should return the new tree pointer for use to its caller.

In Step 1 of the algorithm, every call to the insert function checks if the current node of the `TREE` is `NULL`. If it is `NULL`, the algorithm simply adds the node, else it looks at the current node's value and then recurs down the left or right sub-tree.

However, if the current node's value is less than that of the new node, then the left sub-tree is traversed, else the right sub-tree is traversed. The insert function continues moving down the levels of a binary tree until it reaches a leaf node. The new node is added in place of the leaf node by following the rules of the binary search trees. That is, if the new node's value is greater than that of the parent node, the new node is inserted in the right sub-tree, else it is inserted in the left sub-tree. The insert function requires time proportional to the height of the tree in the worst case. It takes `O(log n)` time to execute in the average case and `O(n)` time in the worst case.

Look at Fig. 11.10 which shows insertion of values in a given tree. We will take up the case of inserting 12 and 55.

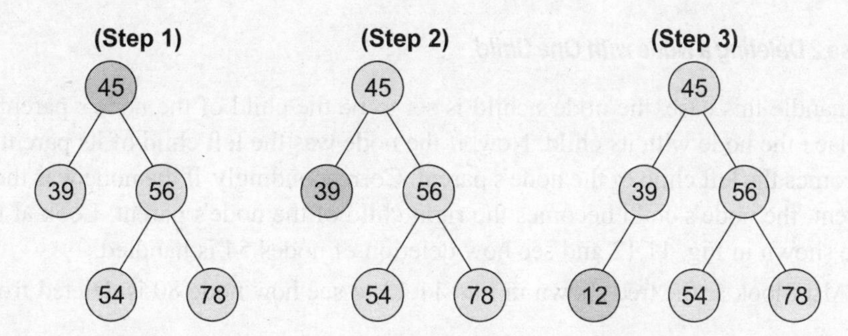

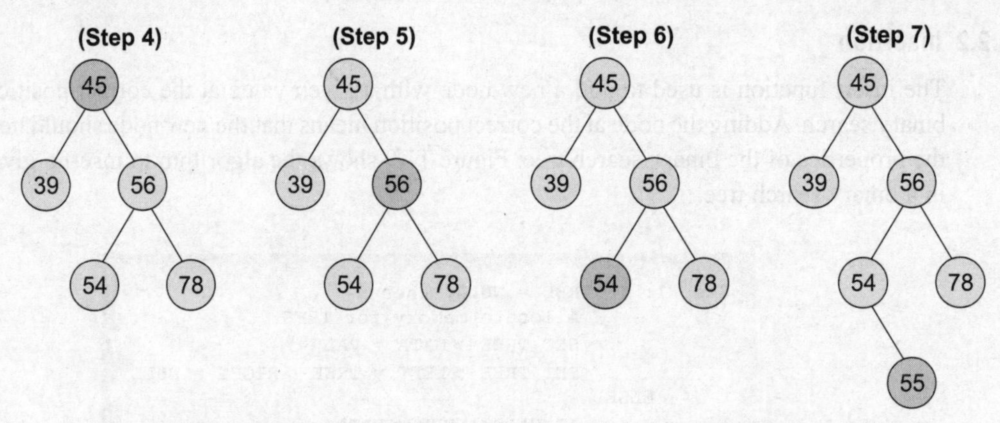

Figure 11.10 Inserting a node with the value 12 and 55 in the given binary search tree

11.2.3 Deletion

The delete function deletes a node from the binary search tree. However, utmost care should be taken that the properties of the binary search tree does not get violated and nodes are not lost in the process. We will take up three cases in this section and discuss how a node is deleted from a binary search tree. Each of the case has to be handled in its own way.

Case 1 Deleting a Node that has No Children

Look at the binary search tree given in Fig. 11.11. For example, if we have to delete node 78, we can simply remove this node without any issue. This is the simplest case in deletion.

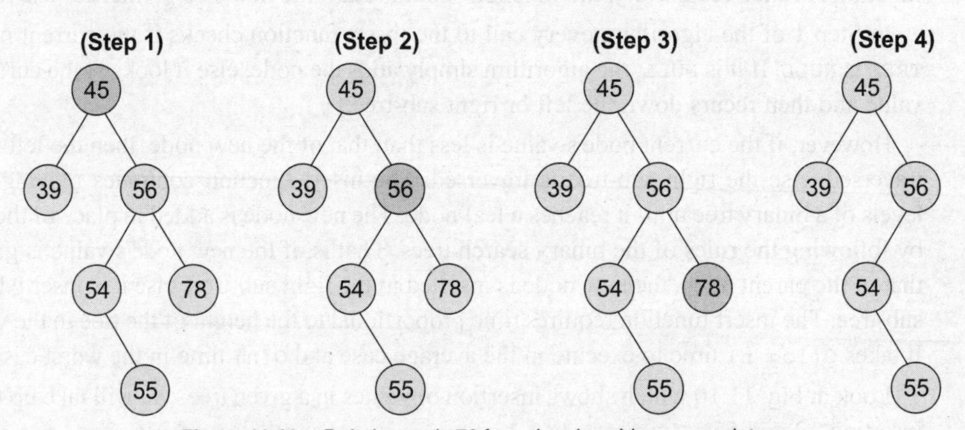

Figure 11.11 Deleting node 78 from the given binary search tree

Case 2 Deleting a Node with One Child

To handle this case, the node's child is set to be the child of the node's parent. In other words, replace the node with its child. Now, if the node was the left child of its parent, the node's child becomes the left child of the node's parent. Correspondingly, if the node was the right child of its parent, the node's child becomes the right child of the node's parent. Look at the binary search tree shown in Fig. 11.12 and see how deletion of nodes 54 is handled.

Also, look at the tree shown in Fig. 11.13 to see how node 80 is deleted from the tree.

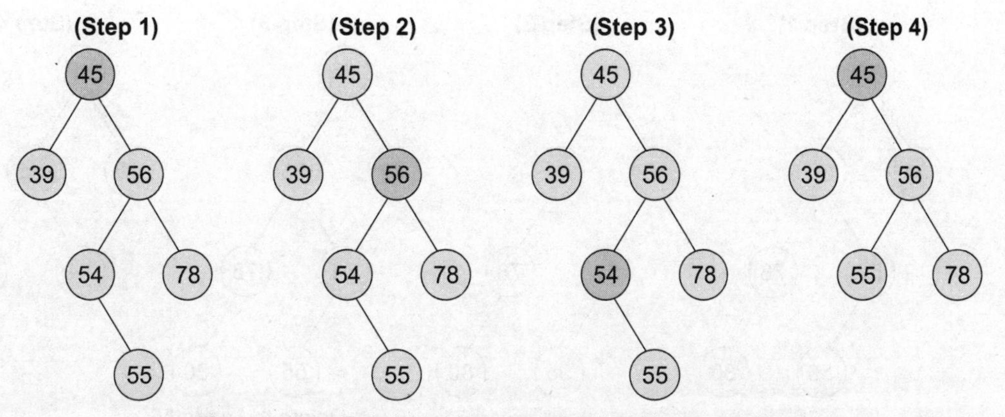

Figure 11.12 Deleting node 54 from the given binary search tree

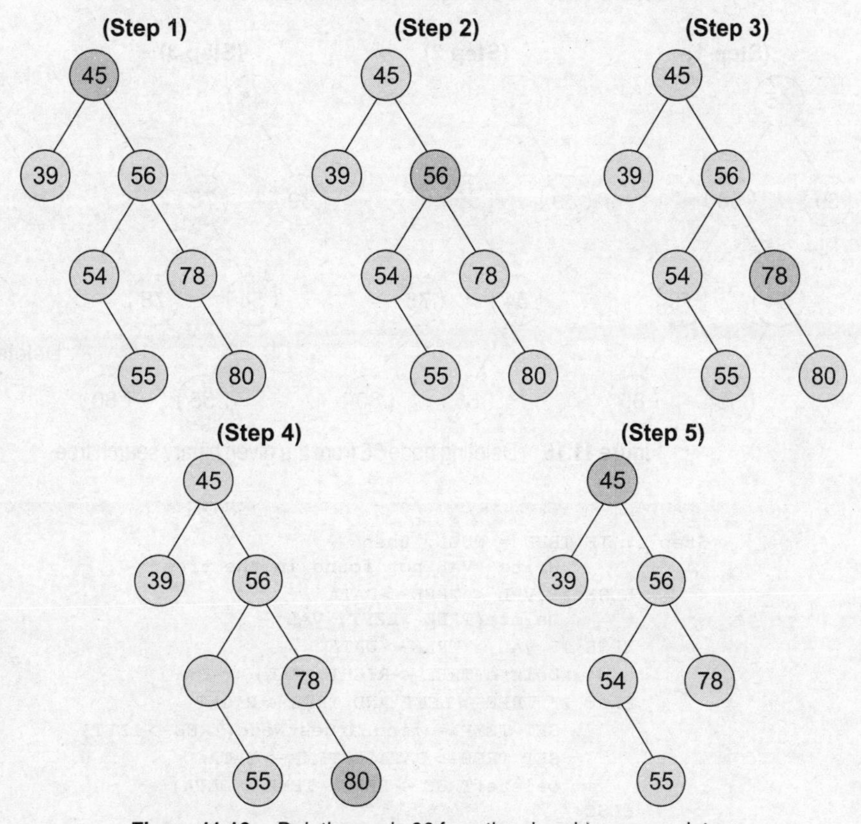

Figure 11.13 Deleting node 80 from the given binary search tree

Case 3 Deleting a Node with Two Children

To handle this case, replace the node's value with its *in-order predecessor* (right-most child of the left sub-tree) or *in-order successor* (left-most child of the right sub-tree). The in-order predecessor or the successor can then be deleted using any of the above cases. Look at the binary search tree given in Fig. 11.14 and see how deletion of node with the value 56 is handled.

This deletion could also be handled by replacing node 56 with its in-order successor, as shown in Fig. 11.15.

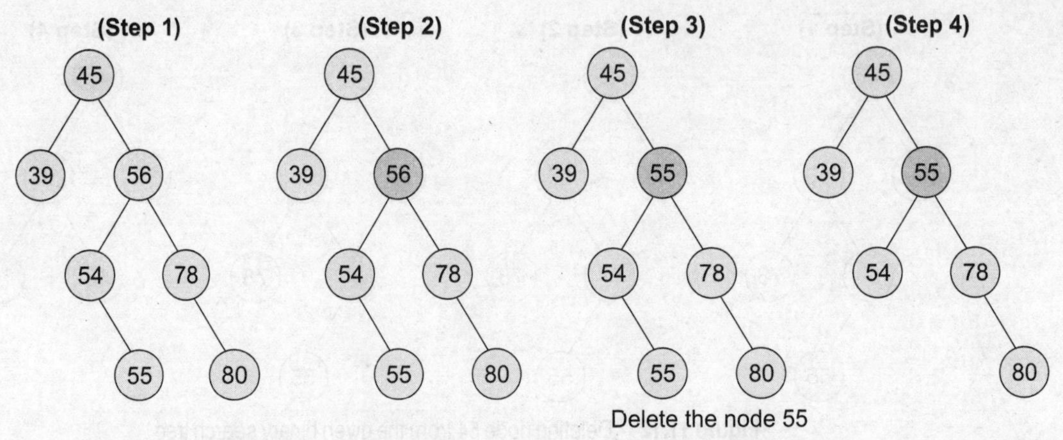

Figure 11.14 Deleting node 56 from the given binary search tree

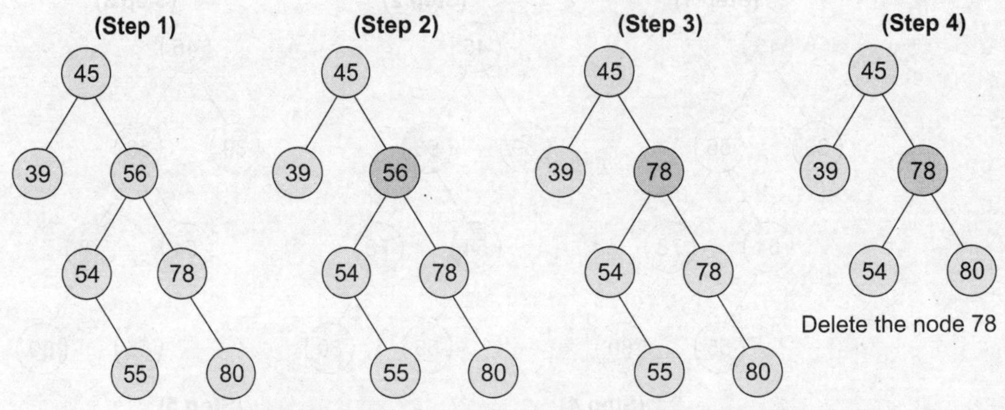

Figure 11.15 Deleting node 56 from the given binary search tree

```
Step 1: IF TREE = NULL, then
            Write "VAL not found in the tree"
      ELSE IF VAL < TREE->DATA
            Delete(TREE->LEFT, VAL)
      ELSE IF VAL > TREE->DATA
            Delete(TREE->RIGHT, VAL)
      ELSE IF TREE->LEFT AND TREE->RIGHT
            SET TEMP = findLargestNode(TREE->LEFT)
            SET TREE->DATA = TEMP->DATA
            Delete(TREE->LEFT, TEMP->DATA)
      ELSE
            SET TEMP = TREE
            IF TREE->LEFT = NULL AND TREE->RIGHT = NULL
                  SET TREE = NULL
            ELSE IF TREE->LEFT != NULL
                  SET TREE = TREE->LEFT
            ELSE
                  SET TREE = TREE->RIGHT
            FREE TEMP
      [END OF IF]
Step 2: End
```

Figure 11.16 Algorithm to delete a node from the binary search tree

Now, let us look at Fig. 11.16 which shows the algorithm to delete a node from the binary search tree.

In Step 1 of the algorithm, we first check if `TREE=NULL`, because if it is true, then the node to be deleted is not present in the tree. However, if that is not the case, then we check if the value to be deleted is less than the current node's data. In case the value is less, we call the algorithm recursively on the node's left sub-tree, otherwise the algorithm is called recursively on the node's right sub-tree.

Note that if we have found the node whose value is equal to `VAL`, then we check which case of deletion it is. If the node to be deleted has both left and right children, then we find the in-order predecessor of the node by calling `findLargestNode(TREE -> LEFT)` and replace the current node's value with that of its in-order predecessor. Then, we call `Delete(TREE -> LEFT, TEMP -> DATA)` to delete the initial node of the in-order predecessor. Thus, we reduce the case 3 of deletion into either case 1 or case 2 of deletion.

If the node to be deleted does not have any child, then we simply set the node to NULL. Last but not the least, if the node to be deleted has either left or right children but not both, then the current node is replaced by its child node and the initial child node is deleted from the tree.

The delete function requires time proportional to the height of the tree in the worst case. It takes `O(log n)` time to execute in the average case and `Ω(n)` time in the worst case.

11.2.4 Determining the Height of a Tree

In order to determine the height of a binary search tree, we calculate the height of the left sub-tree and the right sub-tree. Whichever height is greater, 1 is added to it. For example, if the height of the left sub-tree is greater than that of the right sub-tree, then 1 is added to the left sub-tree, else 1 is added to the right sub-tree. For example, the height of a given binary search tree can be calculated as:

height(left sub-tree) = 1   *height(right sub-tree) = 2*

Look at Fig. 11.17. Since the height of the right sub-tree is greater than the height of the left sub-tree, the height of the tree = height (right sub-tree) + 1= 2 + 1 = 3.

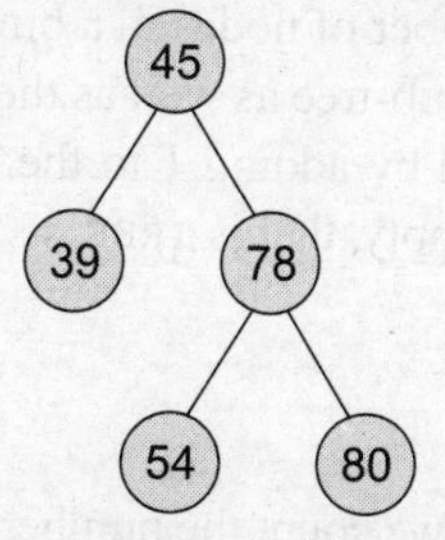

Figure 11.17 Binary search tree with height = 3

Figure 11.18 shows a recursive algorithm that determines the height of a binary search tree.

```
Step 1: IF TREE = NULL, then
            Return 0
        ELSE
            SET LeftHeight = Height(TREE -> LEFT)
            SET RightHeight = Height(TREE -> RIGHT)
            IF LeftHeight > RightHeight
                  Return LeftHeight + 1
            ELSE
                  Return RightHeight + 1
            [END OF IF]
        [END OF IF]
Step 2: End
```

Figure 11.18 Recursive algorithm to determine the height of binary search tree

In Step 1 of the algorithm, we first check if the `current node of the TREE = NULL`. If the condition is true, then 0 is returned to the calling code. Otherwise, for every node, we recursively call the algorithm to calculate the height of its left sub-tree as well as its right sub-tree. The height of the tree at that node is given by adding 1 to the height of the left sub-tree or the height of right sub-tree, whichever is greater.

11.2.5 Determining the Number of Nodes

Determining the number of nodes in a binary search tree is similar to determining its height. To calculate the total number of elements/nodes in the tree, we count the number of nodes in the left sub-tree and the right sub-tree.

```
Number of nodes = totalNodes(left sub-tree) +
                         totalNodes(right sub-tree) + 1
```

Consider the tree given in Fig. 11.19. The total number of nodes in the given tree can be calculated as:

```
Total nodes of left sub-tree = 1
Total nodes of left sub-tree = 5
Total nodes of tree = (1 + 5) + 1
                    = 7
```

Figure 11.19 Binary search tree

```
Step 1: IF TREE = NULL, then
              Return 0
        ELSE
              Return totalNodes(TREE -> LEFT) + totalNodes(TREE -> RIGHT) + 1
        [END OF IF]
Step 2: End
```

Figure 11.20 Recursive algorithm to calculate the number of nodes in a binary search tree

Figure 11.20 shows a recursive algorithm to calculate the number of nodes in a binary search tree. For every node, we recursively call the algorithm on its left sub-tree as well as the right sub-tree. The total number of nodes at a given node is then returned by adding 1 to the number of nodes in its left as well as right sub-tree. However if the tree is empty, that is `TREE = NULL`, then the number of nodes will be zero.

11.2.6 Number of Internal Nodes

To calculate the total number of internal nodes or non-leaf nodes, we count the number of internal nodes in the left sub-tree and the right sub-tree and add 1 to it (1 is added for the root node).

```
Number of internal nodes = totalInternalNodes(left
                                  sub-tree) +
                                  totalInternalNodes(right
                                  sub-tree) + 1
```

Consider the tree given in Fig. 11.21. The total number of internal nodes in the given tree can be calculated as:

```
Total internal nodes of left sub-tree = 0
Total internal nodes of right sub-tree = 3
Total internal nodes of tree = (0 + 3) + 1
                             = 4
```

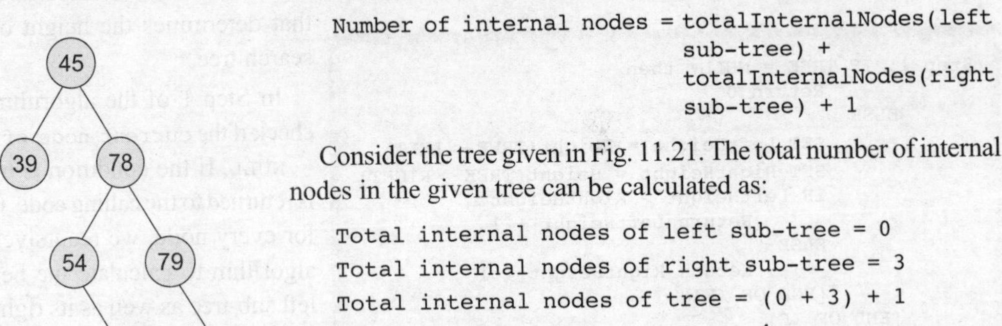

Figure 11.21 Binary search tree

Figure 11.22 shows a recursive algorithm to calculate the total number of internal nodes in a binary search tree. For every node, we recursively call the algorithm on its left sub-tree as well as the right sub-tree. The total

```
Step 1: IF TREE = NULL, then
        Return 0
    IF TREE -> LEFT = NULL AND TREE -> RIGHT = NULL, then
        Return 0
    ELSE
        Return totalInternalNodes(TREE -> LEFT) +
totalInternalNodes(TREE -> RIGHT) + 1
    [END OF IF]
Step 2: End
```

Figure 11.22 Recursive algorithm to calculate the total number of internal nodes in a binary search tree

number of internal nodes at a given node is then returned by adding internal nodes in its left as well as right sub-tree. However, if the tree is empty, that is `TREE = NULL`, then the number of internal nodes will be zero. Also if there is only one node in the tree, even the number of internal nodes will be zero.

11.2.7 Determining the Number of External Nodes

To calculate the total number of external nodes or leaf nodes, we add the number of external nodes in the left sub-tree and the right sub-tree. However if the tree is empty, that is `TREE = NULL`, then the number of external nodes will be zero. But if there is only one node in the tree, then the number of external nodes will be one.

```
Number of external nodes = totalExternalNodes(left
                           sub-tree) +
                           totalExternalNodes
                           (right sub-tree)
```

Consider the tree given in Fig. 11.23. The total number of external nodes in the given tree can be calculated as:

```
Total internal nodes of left sub-tree = 1
Total internal nodes of left sub-tree = 2
Total internal nodes of tree = 1 + 2
                             = 3
```

45, 39, 78, 54, 79, 55, 80

Figure 11.23 Binary search tree

Figure 11.24 shows a recursive algorithm to calculate the total number of external nodes in a binary search tree. For every node, we recursively call the algorithm on its left sub-tree as well as the right sub-tree. The total number of external nodes at a given node is then returned by adding the external nodes in its left as well as right sub-tree. However if the tree is empty, that is `TREE = NULL`, then the number of external nodes will be zero. Also if there is only one node in the tree, then there will be only one external node (that is the root node).

```
Step 1: IF TREE = NULL, then
        Return 0
    IF TREE -> LEFT = NULL AND TREE -> RIGHT = NULL, then
        Return 1
    ELSE
        Return totalExternalNodes(TREE -> LEFT) +
totalExternalNodes(TREE -> RIGHT)
    [END OF IF]
Step 2: End
```

Figure 11.24 Recursive algorithm to calculate the total number of external nodes in a binary search tree

11.2.8 Mirror Image

Mirror image of a binary search tree is obtained by interchanging the left sub-tree with the right sub-tree at every node of the tree. For example, given a tree `T`, the mirror image of `T` can be obtained as `T'`. Consider the tree T given in Fig. 11.25.

Figure 11.26 shows a recursive algorithm to obtain the mirror image of a binary search tree. In the algorithm, if `TREE != NULL`, that is if the current node in the tree has one or more nodes, then the algorithm is recursively called at every node in the tree to swap the nodes in its left and right sub-tree.

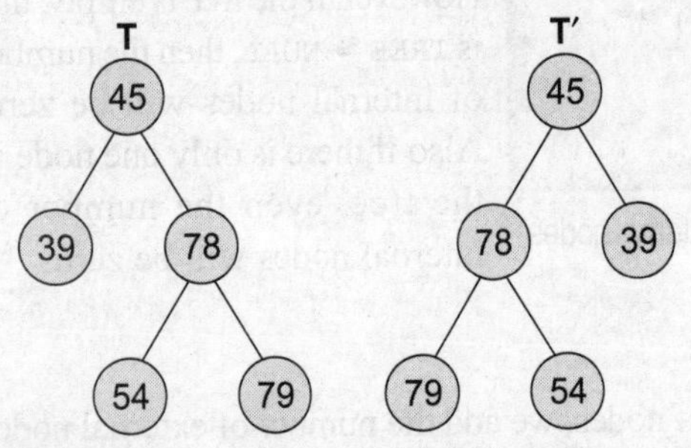

Figure 11.25 Binary search tree T and its mirror image T′

```
Step 1: IF TREE != NULL, then
            MirrorImage(TREE -> LEFT)
            MirrorImage(TREE -> RIGHT)
            SET TEMP = TREE -> LEFT
            SET TREE -> LEFT = TREE -> RIGHT
            SET TREE -> RIGHT = TEMP
        [END OF IF]
Step 2: End
```

Figure 11.26 Recursive algorithm to obtain the mirror image of a binary search tree

11.2.9 Removing the Tree

To delete/remove an entire binary search tree from the memory, we first delete the elements/nodes in the left sub-tree and then delete in the right sub-tree. The algorithm shown in Fig. 11.27 gives a recursive procedure to remove the binary search tree.

```
Step 1: IF TREE != NULL, then
            deleteTree (TREE -> LEFT)
            deleteTree (TREE -> RIGHT)
            Free (TREE)
        [END OF IF]
Step 2: End
```

Figure 11.27 Recursive procedure to remove a binary search tree

11.2.10 Finding the Smallest Node

The very basic property of the binary search tree states that the smaller value will occur in the left sub-tree. If the left sub-tree is `NULL`, then the value of the root node will be smallest as compared to the nodes in the right sub-tree. So, to find the node with the smallest value, we find the value of the leftmost node of the left sub-tree. However, if the left sub-tree is empty, then we find the value of the root node. The recursive algorithm to find the smallest node in a binary search tree is shown in Fig. 11.28.

```
Step 1: IF TREE = NULL OR TREE -> LEFT = NULL, then
            Returen TREE
        ELSE
            Return findSmallestElement(TREE -> LEFT)
        [END OF IF]
Step 2: End
```

Figure 11.28 Recursive algorithm to find the smallest node in a binary search tree

11.2.11 Finding the Largest Node

To find the node with the largest value, we find the value of the rightmost node of the right sub-tree. However, if the right sub-tree is empty, then we find the value of the root node. The recursive algorithm to find the largest node in a binary search tree is shown in Fig. 11.29.

```
Step 1: IF TREE = NULL OR TREE -> RIGHT = NULL, then
            Return TREE
        ELSE
            Return findLargestElement(TREE -> RIGHT)
        [END OF IF]
Step 2: End
```

Figure 11.29 Recursive algorithm to find the largest node in a binary search tree

Consider the tree given in Fig. 11.30. The smallest and the largest node can be given as:

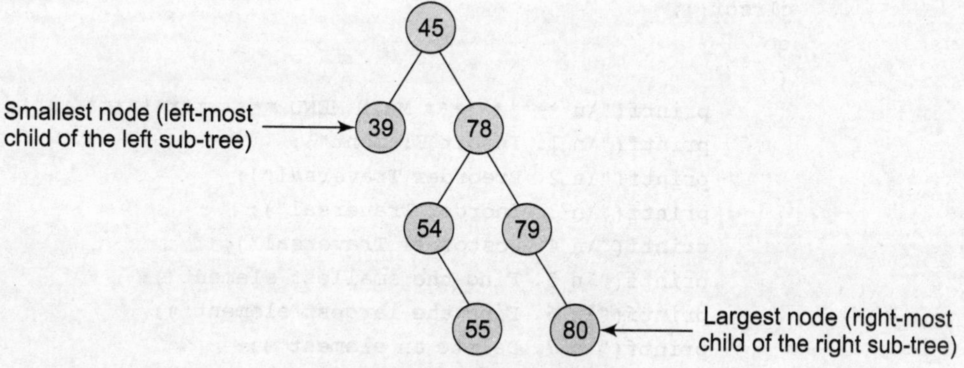

Figure 11.30 Binary search tree

PROGRAMMING EXAMPLE

1. Write a program to create a binary search tree and perform all the operations discussed above.

```
#include<stdio.h>
#include<conio.h>
struct node
{
    int data;
    struct node *left;
    struct node *right;
};
struct node *tree;
void create_tree(struct node *);
struct node *insertElement(struct node *, int);
void preorderTraversal(struct node *);
void inorderTraversal(struct node *);
void postorderTraversal(struct node *);
struct node *findSmallestElelemt(struct node *);
struct node *findLargestElelemt(struct node *);
struct node *deleteElement(struct node *, int);
struct node *mirrorImage(struct node *);
```

```
int totalNodes(struct node *);
int totalExternalNodes(struct node *);
int totalInternalNodes(struct node *);
int Height(struct node *);
struct node *deleteTree(struct node *);
main()
{
    int option, val;
    struct node *ptr;
    create_tree(tree);
    clrscr();
    do
    {
        printf("\n ********** MAIN MENU ************** \n");
        printf("\n 1. Insert Element");
        printf("\n 2. Preorder Traversal");
        printf("\n 3. Inorder Traversal");
        printf("\n 4. Postorder Traversal");
        printf("\n 5. Find the smallest element");
        printf("\n 6. Find the largest element");
        printf("\n 7. Delete an element");
        printf("\n 8. Count the total number of nodes");
        printf("\n 9. Count the total number of external nodes");
        printf("\n 10. Count the total number of internal nodes");
        printf("\n 11. Determine the height of the tree");
        printf("\n 12. Find the mirror image of the tree");
        printf("\n 13. Delete the tree");
        printf("\n 14. Exit");
    printf("\n\n*************************************************");
        printf("\n\n Enter your option : ");
        scanf("%d", &option);
        switch(option)
        {
            case 1:
                printf("\n Enter the value of the new node: ");
                scanf("%d", &val);
                tree = insertElement(tree, val);
                break;
            case 2:
                printf("\n The elements of the tree are : \n");
                preorderTraversal(tree);
                break;
            case 3:
                printf("\n The elements of the tree are : \n");
                inorderTraversal(tree);
                break;
```

```
                case 4:
                        printf("\n The elements of the tree are : \n");
                        postorderTraversal(tree);
                        break;
                case 5:
                        ptr = findSmallestElelemt(tree);
                        printf("\n The smallest element in the tree is :
%d", ptr->data);
                        break;
                case 6:
                        ptr = findLargestElelemt(tree);
                        printf("\n The smallest element in the tree is :
%d", ptr->data);
                        break;
                case 7:
                        printf("\n Enter the element to be deleted: ");
                        scanf("%d", &val);
                        tree = deleteElement(tree, val);
                        break;
                case 8:
                        printf("\n Total number of nodes in the tree is =
%d", totalNodes(tree));
                        break;
                case 9:
                        printf("\n Total number of external nodes in the
tree is = %d", totalExternalNodes(tree));
                        break;
                case 10:
                        printf("\n Total number of internal nodes in the
tree is = %d", totalInternalNodes(tree));
                        break;
                case 11:
                        printf("\n The height of the binary search tree is
= %d", Height(tree));
                        break;
                case 12:
                        tree = mirrorImage(tree);
                        break;
                case 13:
                        tree = deleteTree(tree);
                        break;
        }
    }while(option!=14);
    getch();
    return 0;
}
void create_tree(struct node *tree)
```

```
    {
        tree = NULL;
    }
    struct node *insertElement( struct node *tree, int val)
    {
        struct node *ptr, *nodeptr, *parentptr;
        ptr = (struct node*)malloc(sizeof(struct node*));
        ptr->data = val;
        ptr->left = NULL;
        ptr->right = NULL;
        if(tree==NULL)
        {
            tree=ptr;
            tree->left=NULL;
            tree->right=NULL;
        }
        else
        {
            parentptr=NULL;
            nodeptr=tree;
            while(nodeptr!=NULL)
            {
                parentptr=nodeptr;
                if(val<nodeptr->data)
                    nodeptr=nodeptr->left;
                else
                    nodeptr = nodeptr->right;
            }
            if(val<parentptr->data)
                parentptr->left = ptr;
            else
                parentptr->right = ptr;
        }
        return tree;
    }
    void preorderTraversal(struct node *tree)
    {
        if(tree != NULL)
        {
            printf("%d\t", tree->data);
            preorderTraversal(tree->left);
            preorderTraversal(tree->right);
        }
    }
    void inorderTraversal(struct node *tree)
    {
```

```
        if(tree != NULL)
        {
                inorderTraversal(tree->left);
                printf("%d\t", tree->data);
                inorderTraversal(tree->right);
        }
}
void postorderTraversal(struct node *tree)
{
        if(tree != NULL)
        {
                postorderTraversal(tree->left);
                postorderTraversal(tree->right);
                printf("%d\t", tree->data);
        }
}
struct node *findSmallestElelemt(struct node *tree)
{
        if( (tree == NULL) || (tree->left == NULL))
                return tree;
        else
                return findSmallestElelemt(tree->left);
}
struct node *findLargestElelemt(struct node *tree)
{
        if( (tree == NULL) || (tree->right == NULL))
                return tree;
        else
                return findLargestElelemt(tree->right);
}
struct node *deleteElement(struct node *tree, int val)
{
        struct node *ptr;
        if(tree==NULL)
                printf("\n %d is not present in the tree", val);
        else if(val<tree->data)
                deleteElement(tree->left, val);
        else if(val>tree->data)
                deleteElement(tree->right, val);
        else
        {
                if(tree->left && tree->right)
                {
                        ptr = findLargestElelemt(tree->left);
                        tree->data = ptr->data;
```

```
                deleteElement(tree->left, ptr->data);
            }
            else
            {
                ptr=tree;
                if(tree->left==NULL && tree->right==NULL)
                    tree=NULL;
                else if(tree->left!=NULL)
                    tree=tree->left;
                else
                    tree=tree->right;
                free(ptr);
            }
        }
        return tree;
}
int totalNodes(struct node *tree)
{
        if(tree==NULL)
            return 0;
        else
            return( totalNodes(tree->left) + totalNodes(tree->right) + 1);
}
int totalExternalNodes(struct node *tree)
{
        if(tree==NULL)
            return 0;
        else if((tree->left==NULL) && (tree->right==NULL))
            return 1;
        else
            return (totalExternalNodes(tree->left) +
totalExternalNodes(tree->right));
}
int totalInternalNodes(struct node *tree)
{
        if( (tree==NULL) || ((tree->left==NULL) && (tree->right==NULL)))
            return 0;
        else
            return (totalInternalNodes(tree->left) +
totalInternalNodes(tree->right) + 1);
}
int Height(struct node *tree)
{
        int leftheight, rightheight;
        if(tree==NULL)
            return 0;
```

```
    else
    {
        leftheight = Height(tree->left);
        rightheight = Height(tree->right);
        if(leftheight > rightheight)
            return (leftheight + 1);
        else
            return (rightheight + 1);
    }
}
struct node *mirrorImage(struct node *tree)
{
    struct node *ptr;
    if(tree!=NULL)
    {
        mirrorImage(tree->left);
        mirrorImage(tree->right);
        ptr=tree->left;
        ptr->left = ptr->right;
        tree->right = ptr;
    }
}
struct node *deleteTree(struct node *tree)
{
    if(tree!=NULL)
    {
        deleteTree(tree->left);
        deleteTree(tree->right);
        free(tree);
    }
}
```

11.3 THREADED BINARY TREE

A threaded binary tree is the same as that of a binary tree but with a difference in storing the `NULL` pointers. Consider the linked representation of a binary tree as given in Fig. 11.31.

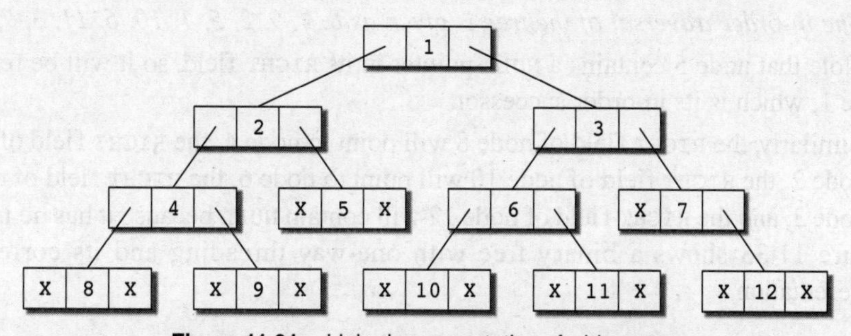

Figure 11.31 Linked representation of a binary tree

In the linked representation, a number of nodes contain a `NULL` pointer, either in their left or right fields or in both. This space that is wasted in storing a `NULL` pointer can be efficiently used to store some other useful peace of information. For example, the `NULL` entries can be replaced to store a pointer to the in-order predecessor, or the in-order successor of the node. These special pointers are called *threads* and binary trees containing threads are called *threaded trees*. In the linked representation of a threaded binary tree, threads will be denoted using dotted lines.

There are many ways of threading a binary tree and each type may vary according to the way the tree is traversed. In this book, we will discuss in-order traversal of the tree. Apart from this, a threaded binary tree may correspond to one-way threading or a two-way threading.

In one-way threading, a thread will appear either in the right field or the left field of the node. A one-way threaded tree is also called a single-threaded tree. If the thread appears in the left field, then the left field will be made to point to the in-order predecessor of the node. Such a one-way threaded tree is called a left-threaded binary tree. On the contrary, if the thread appears in the right field, then it will point to the in-order successor of the node. Such a one-way threaded tree is called a right-threaded binary tree.

In a two-way threaded tree, also called a double-threaded tree, threads will appear in both the left and the right field of the node. While the left field will point to the in-order predecessor of the node, the right field will point to its successor. A two-way threaded binary tree is also called a fully threaded binary tree. One-way threading and two-way threading of binary trees is explained below. Figure 11.32 shows a binary tree without threading and its corresponding linked representation.

Figure 11.32(a) Binary tree without threading

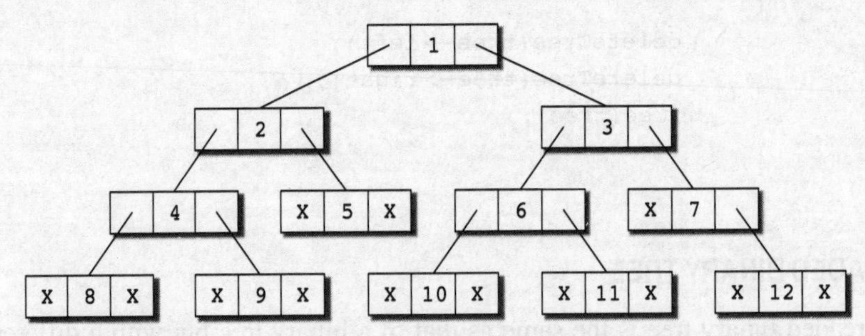

Figure 11.32(b) Linked representation of the binary tree (without threading)

The in-order traversal of the tree is given as 8, 4, 9, 2, 5, 1, 10, 6, 11, 3, 7, 12

Note that node 5 contains a `NULL` pointer in its `RIGHT` field, so it will be replaced to point to node 1, which is its in-order successor.

Similarly, the `RIGHT` field of node 8 will point to node 4, the `RIGHT` field of node 9 will point to node 2, the `RIGHT` field of node 10 will point to node 6, the `RIGHT` field of node 11 will point to node 3, and the `RIGHT` field of node 12 will contain `NULL` because it has no in-order successor. Figure 11.33 shows a binary tree with one-way threading and its corresponding linked representation.

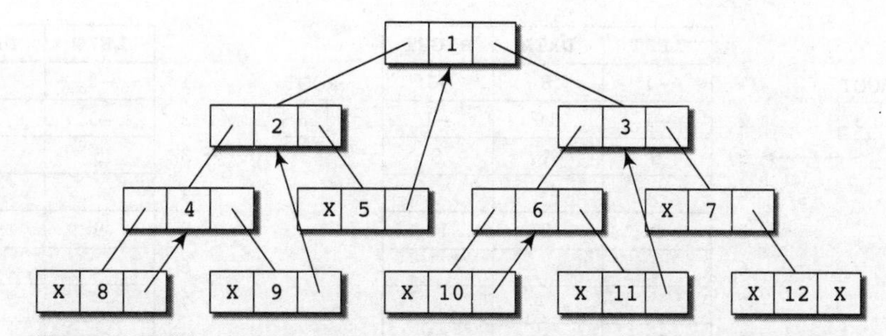

Figure 11.33(a) Linked representation of the binary tree with one-way threading

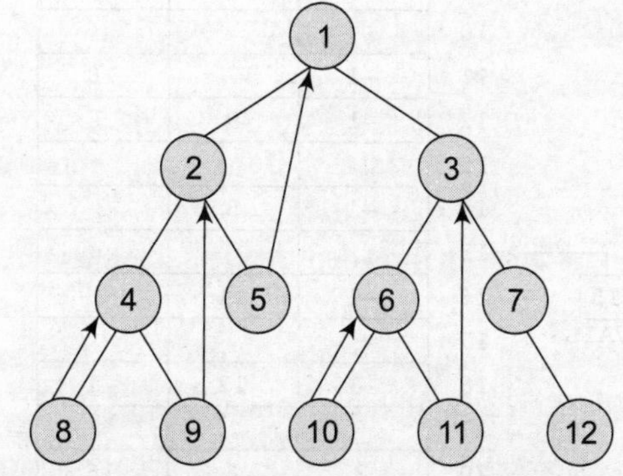

Figure 11.33(b) Binary tree with one-way threading

Note that node 5 contains a NULL pointer in its LEFT field, so it will be replaced to point to node 2, which is its in-order predecessor.

Similarly, the LEFT field of node 8 will contain NULL because it has no in-order successor, the LEFT field of node 7 will point to node 3, the LEFT field of node 9 will point to node 4, the LEFT field of node 10 will point to node 1, the LEFT field of node 11 will contain 6, and the LEFT field of node 12 will point to node 7. Figure 11.34 shows a binary tree with two-way threading and its corresponding linked representation.

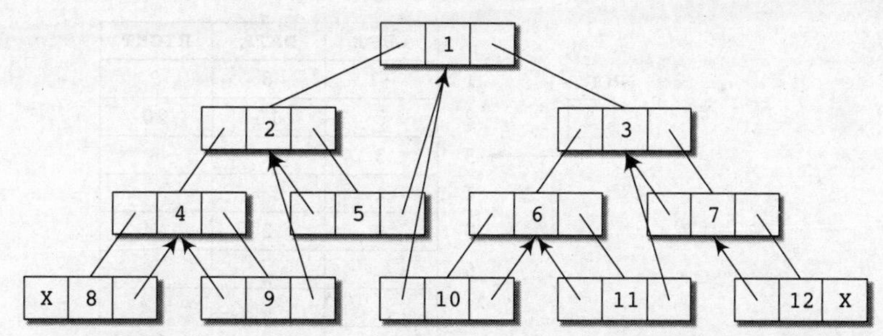

Figure 11.34(a) Linked representation of the binary tree with two-way threading

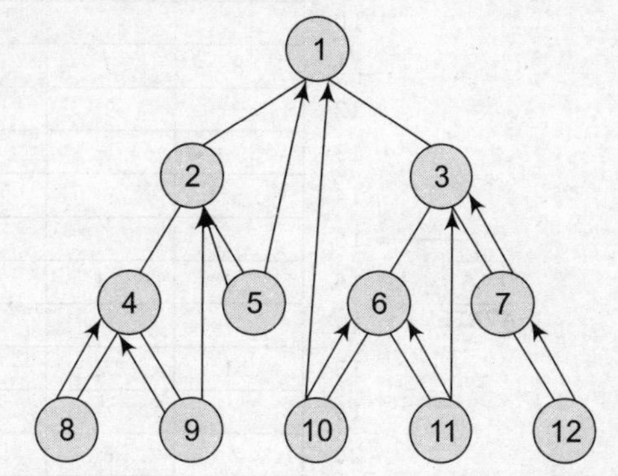

Figure 11.34(b) Binary tree with two-way threading

Now, let us look at the memory representation of a binary tree without threading, with one-way threading, and with two-way threading. This is illustrated in Fig. 11.35.

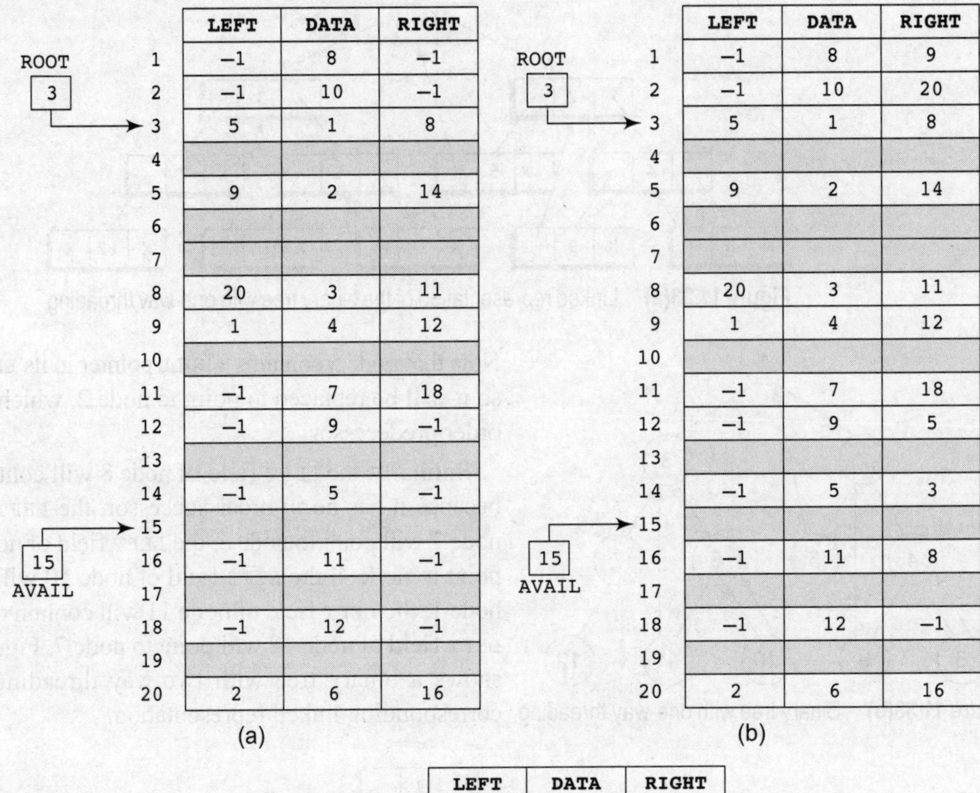

ROOT 3, AVAIL 15

	LEFT	DATA	RIGHT
1	–1	8	–1
2	–1	10	–1
3	5	1	8
4			
5	9	2	14
6			
7			
8	20	3	11
9	1	4	12
10			
11	–1	7	18
12	–1	9	–1
13			
14	–1	5	–1
15			
16	–1	11	–1
17			
18	–1	12	–1
19			
20	2	6	16

(a)

ROOT 3, AVAIL 15

	LEFT	DATA	RIGHT
1	–1	8	9
2	–1	10	20
3	5	1	8
4			
5	9	2	14
6			
7			
8	20	3	11
9	1	4	12
10			
11	–1	7	18
12	–1	9	5
13			
14	–1	5	3
15			
16	–1	11	8
17			
18	–1	12	–1
19			
20	2	6	16

(b)

ROOT 3, AVAIL 15

	LEFT	DATA	RIGHT
1	–1	8	9
2	3	10	20
3	5	1	8
4			
5	9	2	14
6			
7			
8	20	3	11
9	1	4	12
10			
11	8	7	18
12	9	9	5
13			
14	5	5	3
15			
16	20	11	8
17			
18	11	12	–1
19			
20	2	6	16

(c)

Figure 11.35 Memory representation of binary trees (a) without threading, (b) with one-way, and (c) two-way threading

Following are the advantages of having a threaded binary tree:

- It enables linear traversal of elements in the tree.
- Linear traversal eliminates the use of stacks which in turn consume a lot of memory space and computer time.
- It enables to find the parent of a given element without explicit use of parent pointers.
- Since nodes contain pointers to in-order predecessor and successor, the threaded tree enables forward and backward traversal of the nodes as given by in-order fashion.

Thus, we see the basic difference between a binary tree and a threaded binary tree is that, in binary trees, the nodes store a `NULL` pointer if it has no child and so there is no way to traverse back.

11.4 AVL TREES

AVL tree is a self-balancing binary search tree invented by G.M. Adelson-Velsky and E.M. Landis in 1962. The tree is named AVL in honour of its inventors. In an AVL tree, the heights of the two sub-trees of a node may differ by at most one. Due to this property, the AVL tree is also known as a height-balanced tree. The key advantage of using an AVL tree is that it takes `O(log n)` time to perform search, insert, and delete operations in an average case as well as the worst case (because the height of the tree is limited to `O(log n)`.

The structure of an AVL tree is the same as that of a binary search tree but with a little difference. In its structure, it stores an additional variable called the `BalanceFactor`. Thus, every node has a balance factor associated with it. The balance factor of a node is calculated by subtracting the height of its right sub-tree from the height of its left sub-tree. A binary search tree in which every node has a balance factor of –1, 0, or 1 is said to be height balanced. A node with any other balance factor is considered to be unbalanced and requires rebalancing of the tree.

Balance factor = Height (left sub-tree) – Height (right sub-tree)

- If the balance factor of a node is 1, then it means that the left sub-tree of the tree is one level higher than that of the right sub-tree. Such a tree is therefore called as a *left-heavy tree.*
- If the balance factor of a node is 0, then it means that the height of the left sub-tree (longest path in the left sub-tree) is equal to the height of the right sub-tree.
- If the balance factor of a node is –1, then it means that the left sub-tree of the tree is one level lower than that of the right sub-tree. Such a tree is therefore called as a *right-heavy tree.*

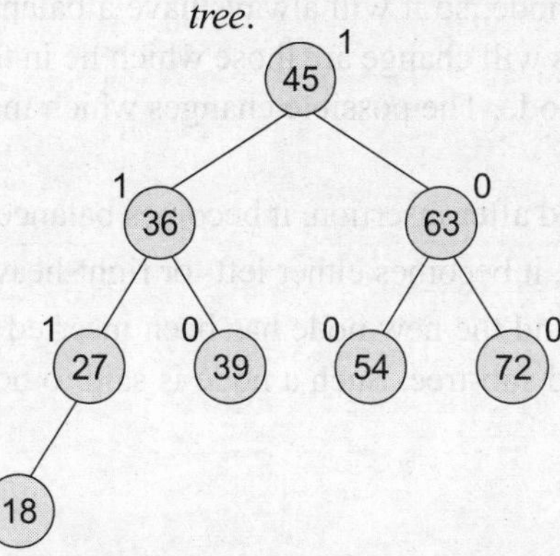

Figure 11.36 Left-heavy AVL tree

Look at Fig. 11.36. Note that the nodes 18, 39, 54, and 72 have no children, so their balance factor = 0. Node 27 has one left child and zero right child. So, the height of left sub-tree = 1, whereas the height of right sub-tree = 0. Thus, its balance factor = 1. Look at node 36, it has a left sub-tree with height = 2, whereas the height of right sub-tree = 1. Thus, its balance factor = 2 – 1 =1. Similarly, the balance factor of node 45 = 3 – 2 =1; and node 63 has a balance factor of 0 (1 – 1).

Now, look at Figs 11.37 and 11.38 which show a right-heavy AVL tree and a balanced AVL tree.

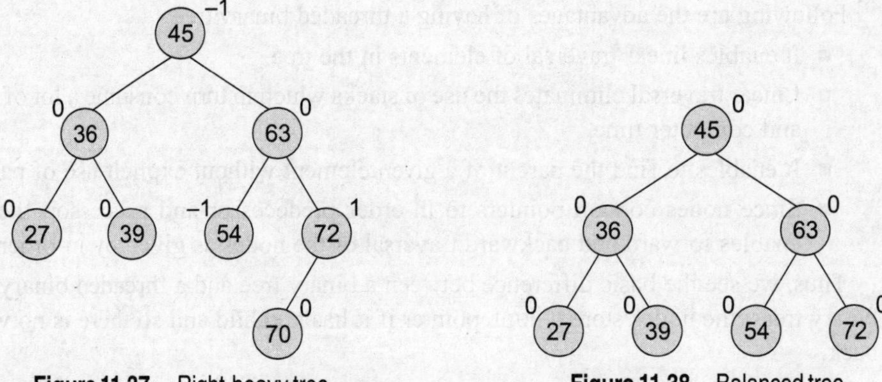

Figure 11.37 Right-heavy tree

Figure 11.38 Balanced tree

The trees given in Fig. 11.38 are typical candidates of AVL trees because the balancing factor of every node is either 1, 0, or –1. However, insertions and deletions from an AVL tree may disturb the balance factor of the nodes and thus, rebalancing of the tree may have to be done. The tree is rebalanced by performing rotation at the critical node. There are four types of rotations: LL rotation, RR rotation, LR rotation, and RL rotation. The type of rotation that has to be done will vary depending on the particular situation. In the following section, we will discuss insertion, deletion, searching, and rotations in AVL trees.

11.4.1 Search Operation

Searching in an AVL tree is performed exactly the same way as it is performed in a binary search tree. Due to the height-balancing of the tree, the search operation takes `O(log n)` time to complete. Since the operation does not modify the structure of the tree, no special provisions are required.

11.4.2 Insertion Operation

Insertion in an AVL tree is also done in the same way as it is done in a binary search tree. In the AVL tree, the new node is always inserted as the leaf node. But the step of insertion is usually followed by an additional step of rotation. Rotation is done to restore the balance of the tree. However, if insertion of the new node does not disturb the balance factor, that is, if the balance factor of every node is still –1, 0, or 1, then rotations are not required.

During insertion, the new node is inserted as the leaf node, so it will always have a balance factor equal to zero. The only nodes whose balance factors will change are those which lie in the path between the root of the tree and the newly inserted node. The possible changes which may take place in any node on the path are as follows:

- Initially, the node was either left- or right-heavy and after insertion, it becomes balanced.
- Initially, the node was balanced and after insertion, it becomes either left- or right-heavy.
- Initially, the node was heavy (either left or right) and the new node has been inserted in the heavy sub-tree, thereby creating an unbalanced sub-tree. Such a node is said to be a *critical node*.

Consider the AVL tree given in Fig. 11.39.

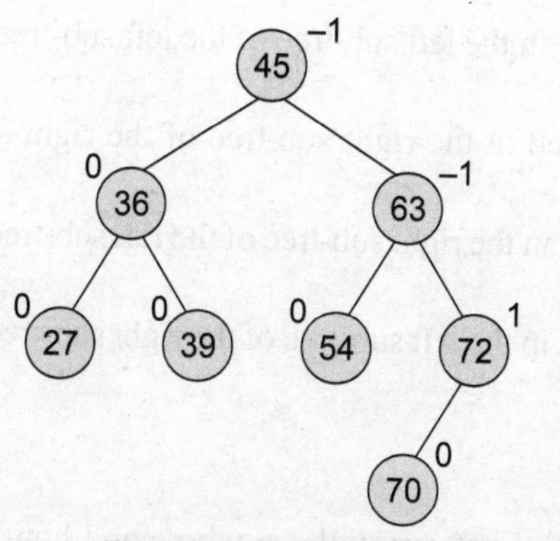

Figure 11.39 AVL tree

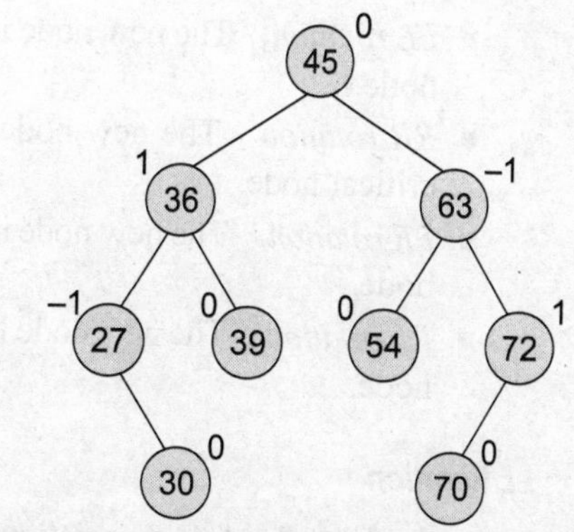

Figure 11.40 AVL tree after inserting a node with the value 30

If we insert a new node with the value 30, then the new tree will still be balanced and no rotations will be required in this case. Look at the tree given in Fig. 11.40 which shows the tree after inserting node 30.

Let us take another example to see how insertion can disturb the balance factors of the nodes and how rotations are done to restore the AVL property of a tree. Look at the tree given in Fig. 11.41.

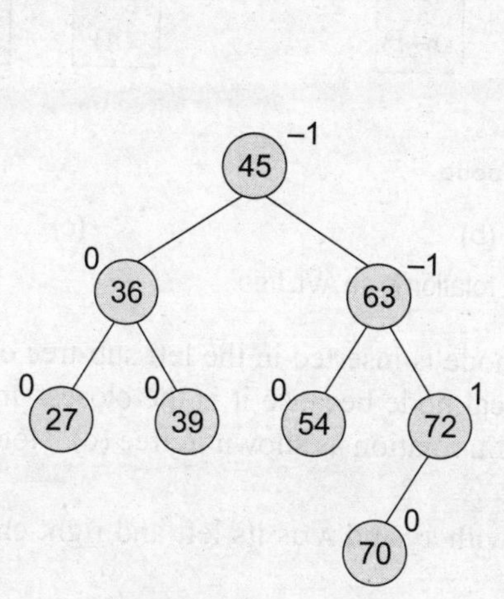

Figure 11.41 AVL tree

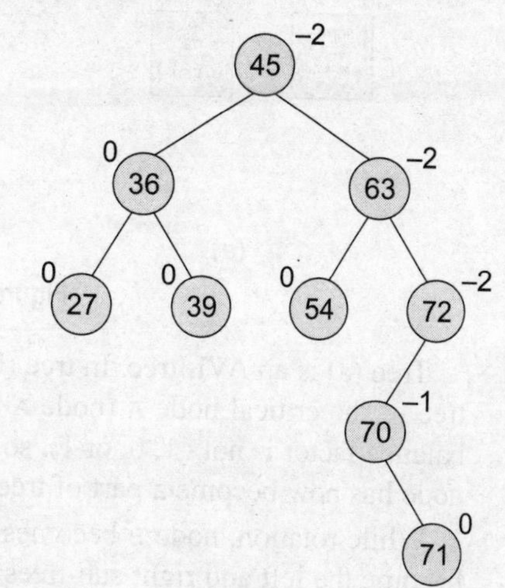

Figure 11.42 AVL tree after inserting a node with the value 71

After inserting a new node with the value 71, the new tree will be as shown in Fig. 11.42.

Note that there are three nodes in the tree that have their balance factors 2, –2, and –2, thereby disturbing the *AVLness* of the tree. So, here comes the need to perform rotation. To perform rotation, our first task is to find the critical node. Critical node is the nearest ancestor node on the path from the root to the inserted node whose balance factor is neither –1, 0, nor 1 (In the tree given above, the critical node is 72). The second task in rebalancing the tree is to determine which type of rotation has to be done. There are four types of rebalancing rotations and application of these rotations depends on the position of the inserted node with reference to the critical node. The four categories of rotations are:

- *LL rotation* The new node is inserted in the left sub-tree of the left sub-tree of the critical node.
- *RR rotation* The new node is inserted in the right sub-tree of the right sub-tree of the critical node.
- *LR rotation* The new node is inserted in the right sub-tree of the left sub-tree of the critical node.
- *RL rotation* The new node is inserted in the left sub-tree of the right sub-tree of the critical node.

LL Rotation

Let us study each of these rotations in detail. First, we will see where and how LL rotation is applied. Consider the tree given in Fig. 11.43 which shows an AVL tree.

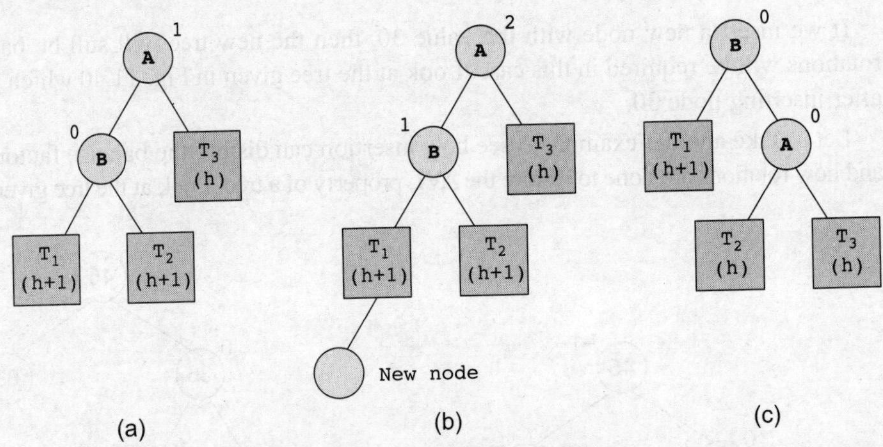

Figure 11.43 LL rotation in an AVL tree

Tree (a) is an AVL tree. In tree (b), a new node is inserted in the left sub-tree of the left sub-tree of the critical node `A` (node `A` is the critical node because it is the closest ancestor whose balance factor is not –1, 0, or 1), so we apply `LL` rotation as shown in tree (c). Note that the new node has now become a part of tree T_1.

While rotation, node `B` becomes the root, with T_1 and `A` as its left and right child. T_2 and T_3 become the left and right sub-trees of `A`.

Example 11.3: Consider the AVL tree given below and insert 9 into it.

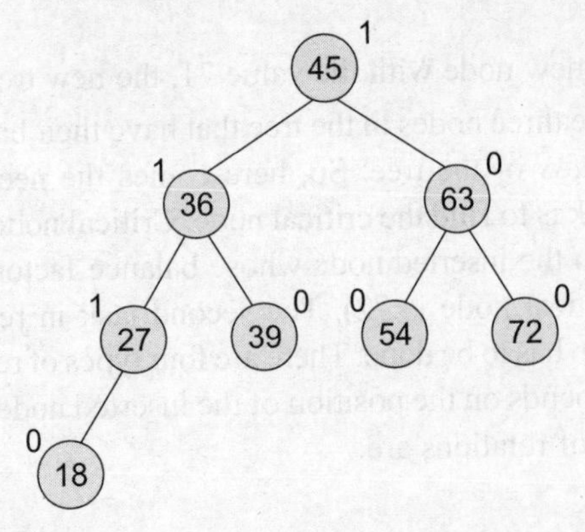

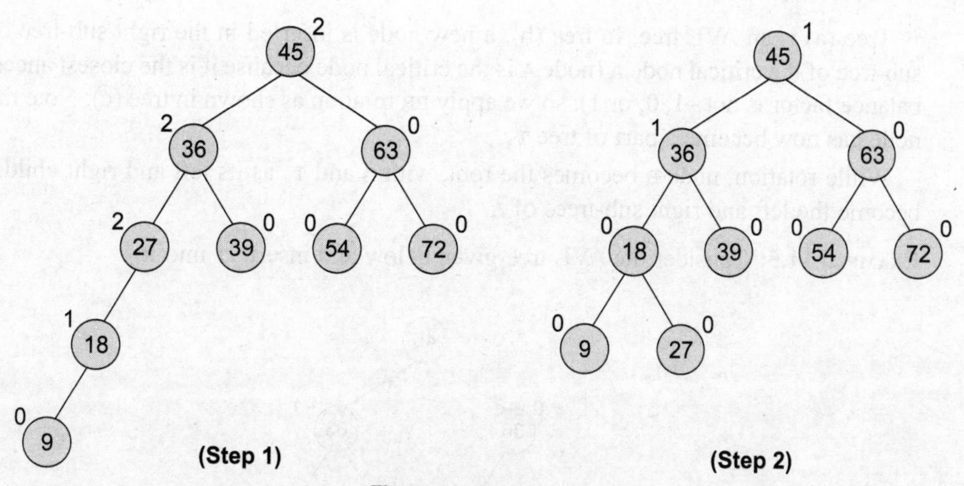

Figure 11.44 AVL tree

EXAMPLE 11.4: Consider the AVL tree given below and insert 18 into it.

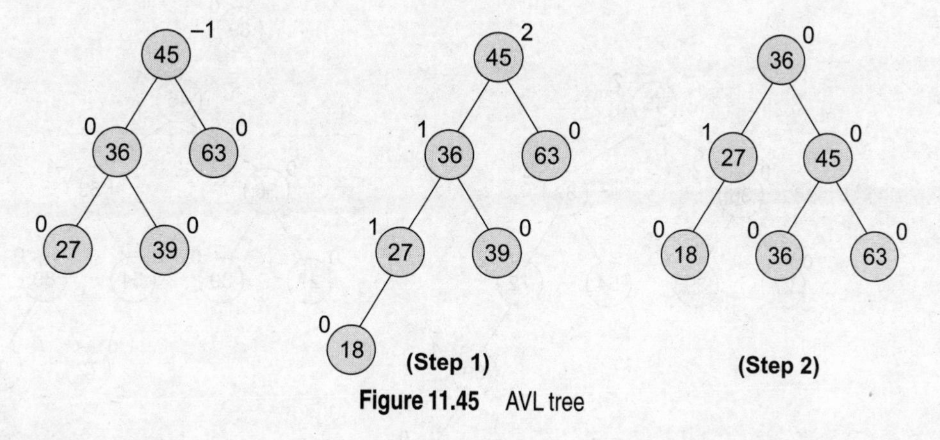

Figure 11.45 AVL tree

RR Rotation

Let us now discuss where and how RR rotation is applied. Consider the tree given in Fig. 11.46 which shows an AVL tree.

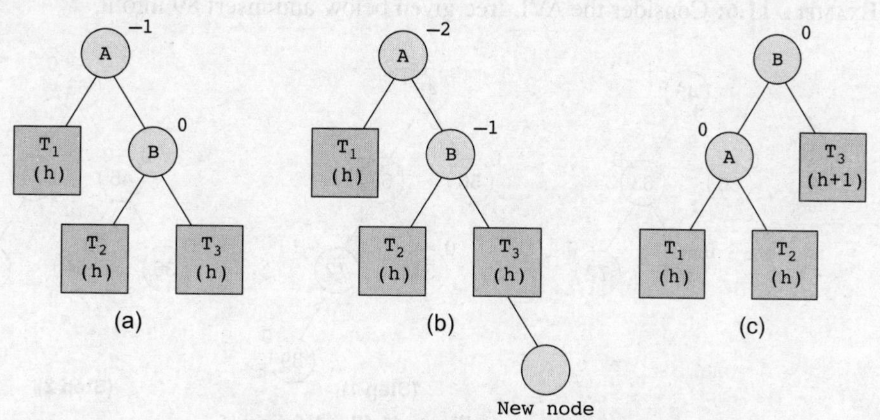

Figure 11.46 RR rotation in an AVL tree

Tree (a) is an AVL tree. In tree (b), a new node is inserted in the right sub-tree of the right sub-tree of the critical node A (node A is the critical node because it is the closest ancestor whose balance factor is not –1, 0, or 1), so we apply RR rotation as shown in tree (c). Note that the new node has now become a part of tree T_3.

While rotation, node B becomes the root, with A and T_3 as its left and right child. T_1 and T_2 become the left and right sub-trees of A.

EXAMPLE 11.5: Consider the AVL tree given below and insert 91 into it.

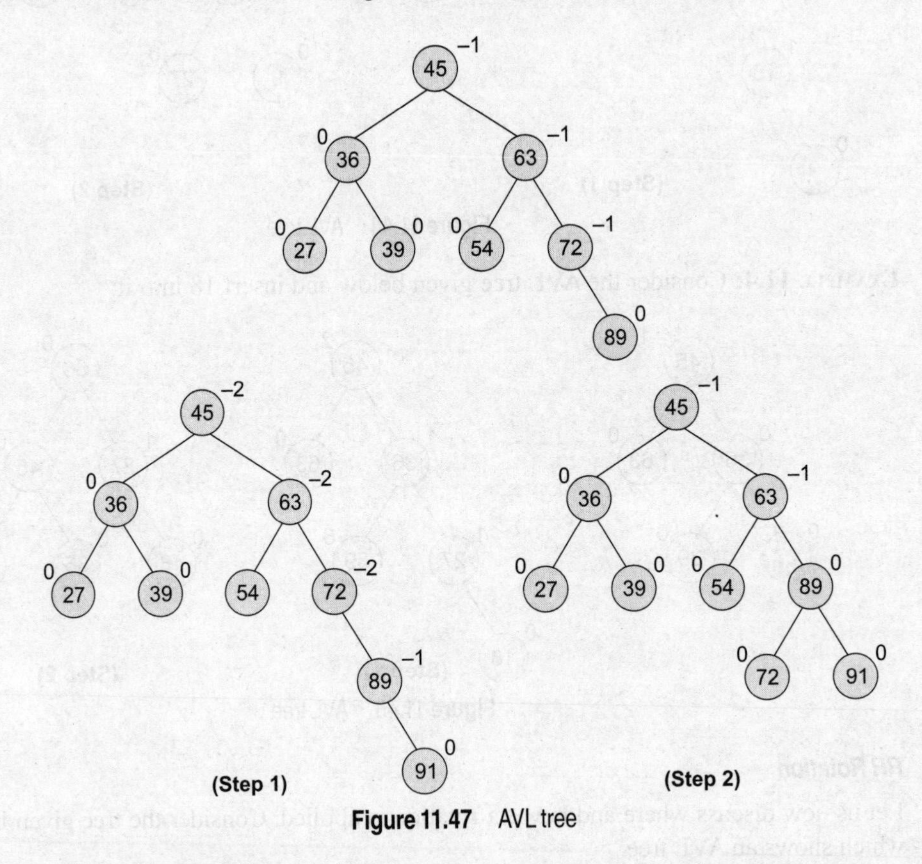

Figure 11.47 AVL tree

EXAMPLE 11.6: Consider the AVL tree given below and insert 89 into it.

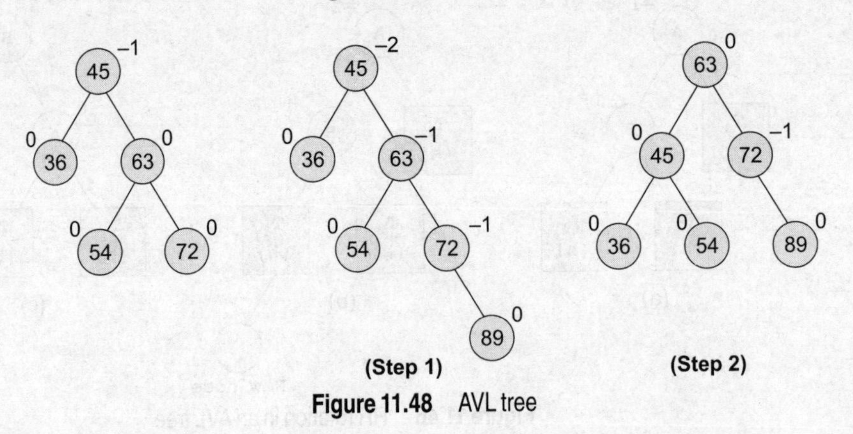

Figure 11.48 AVL tree

LR and RL Rotations

Consider the AVL tree given in Fig. 11.49 and see how LR rotation is done to rebalance the tree.

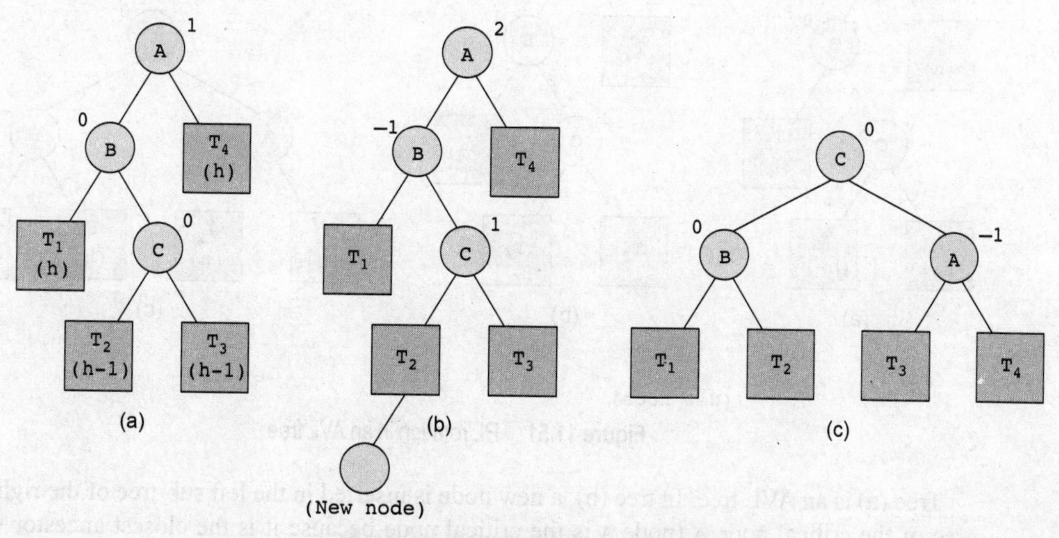

Figure 11.49 LR rotation in an AVL tree

Tree (a) is an AVL tree. In tree (b), a new node is inserted in the right sub-tree of the left sub-tree of the critical node A (node A is the critical node because it is the closest ancestor whose balance factor is not –1, 0 or 1), so we apply LR rotation as shown in tree (c). Note that the new node has now become a part of tree T_2.

While rotation, node C becomes the root, with B and A as its left and right child. Node B has T_1 and T_2 as its left and right sub-trees and T_3 and T_4 become the left and right sub-trees of node A.

Example 11.7: Consider the AVL tree given below and insert 37 into it.

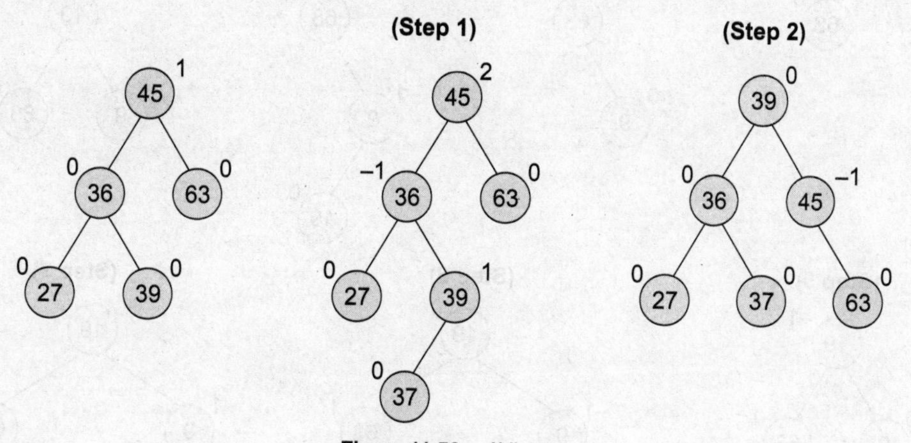

Figure 11.50 AVL tree

Now, consider the AVL tree given in Fig. 11.51 and see how RL rotation is done to rebalance the tree.

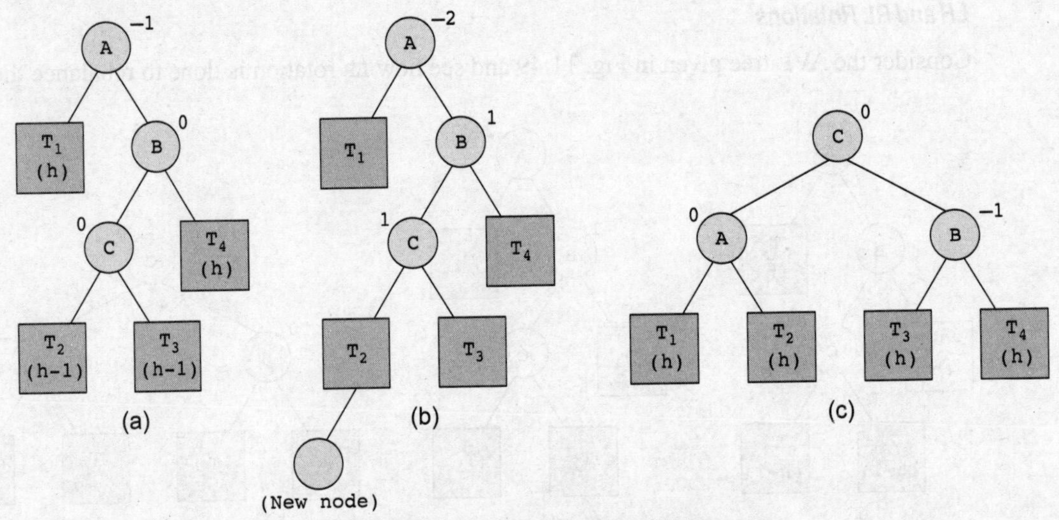

Figure 11.51 RL rotation in an AVL tree

Tree (a) is an AVL tree. In tree (b), a new node is inserted in the left sub-tree of the right sub-tree of the critical node A (node A is the critical node because it is the closest ancestor whose balance factor is not –1, 0, or 1), so we apply RL rotation as shown in tree (c). Note that the new node has now become a part of tree T_2.

While rotation, node C becomes the root, with A and B as its left and right child. Node A has T_1 and T_2 as its left and right sub-trees and T_3 and T_4 become the left and right sub-trees of node B.

Example 11.8: Construct an AVL tree by inserting the following elements in the given order. 63, 9, 19, 27, 18, 108, 99, 81.

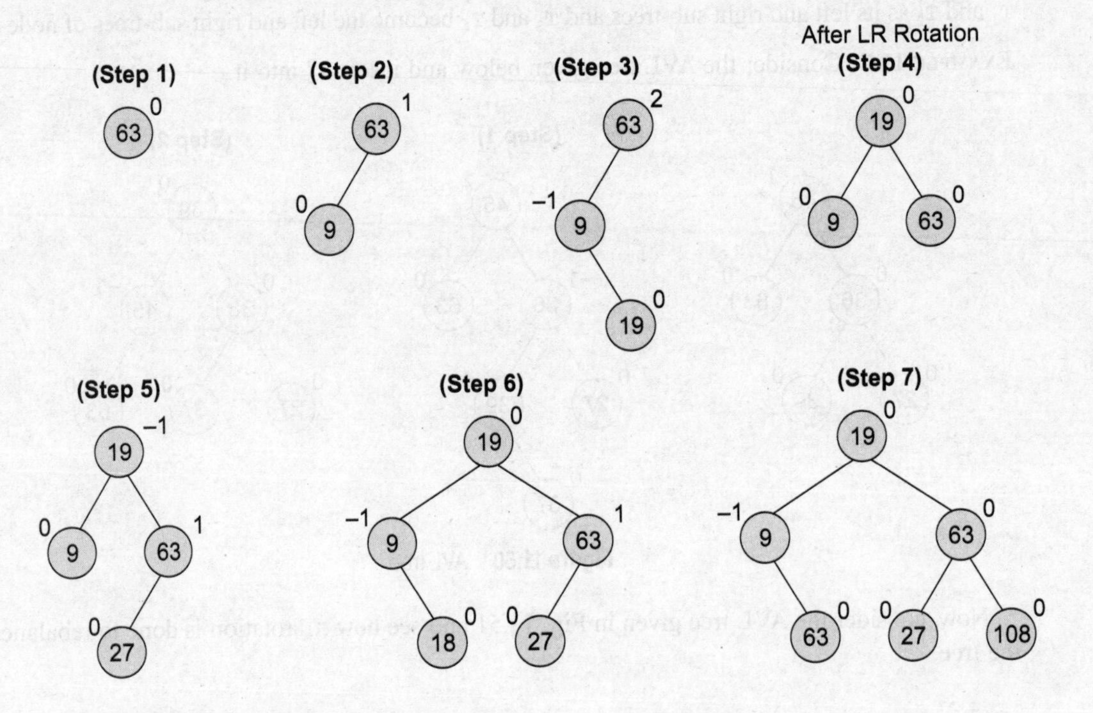

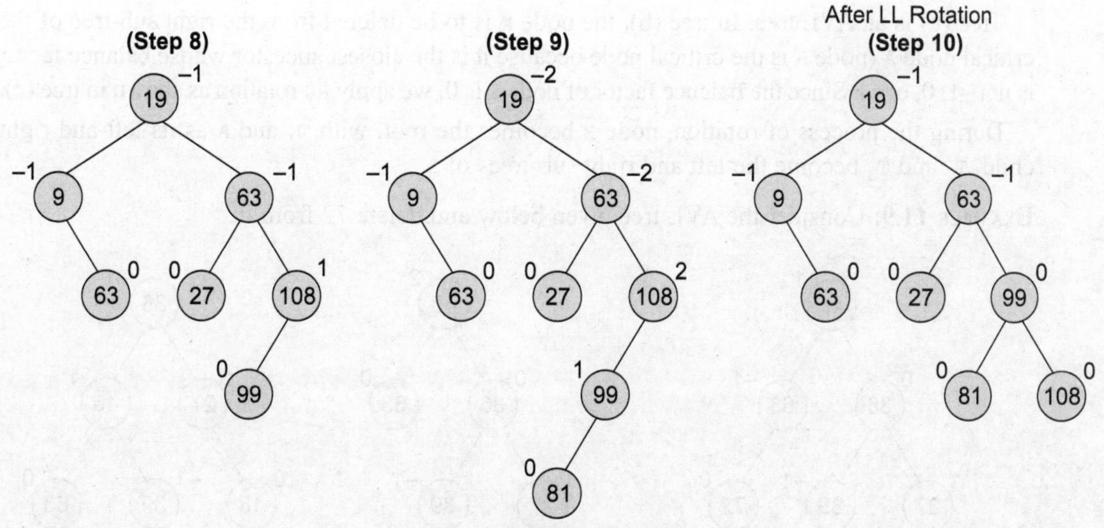

Figure 11.52 AVL tree

11.4.3 Deletion

Deletion of a node in the AVL tree is similar to that of binary search trees. But it goes one step ahead. Deletion may disturb the AVLness of the tree, so to rebalance the AVL tree, we need to perform rotations. There are two classes of rotation that can be performed on an AVL tree after deleting a given node. These rotations are R rotation and L rotation.

On deletion of node X from the AVL tree, if node A becomes the critical node (closest ancestor node on the path from the root node to X that does not have its balance factor as 1, 0, or 1), then the type of rotation depends on whether X is on the left sub-tree of A or on its right sub-tree. If the node to be deleted is present in the left sub-tree of A, then L rotation is applied, else if X is on the right sub-tree, R rotation is performed.

Further, there are three categories of L and R rotations. The variations of L rotation are L-1, L0, and L1 rotation. Correspondingly for R rotation, there are R0, R-1, and R1 rotations. In this section, we will discuss only R rotation. L rotations are the mirror images of R rotations.

R0 Rotation

Let B be the root of the left or right sub-tree of A (critical node). R0 rotation is applied if the balance factor of B is 0. This is illustrated in Fig. 11.53.

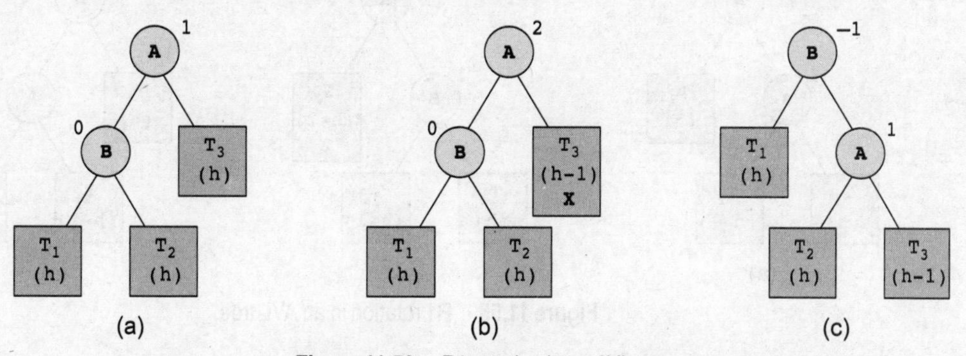

Figure 11.53 R0 rotation in an AVL tree

Tree (a) is an AVL tree. In tree (b), the node X is to be deleted from the right sub-tree of the critical node A (node A is the critical node because it is the closest ancestor whose balance factor is not –1, 0, or 1). Since the balance factor of node B is 0, we apply R0 rotation as shown in tree (c).

During the process of rotation, node B becomes the root, with T_1 and A as its left and right child. T_2 and T_3 become the left and right sub-trees of A.

EXAMPLE 11.9: Consider the AVL tree given below and delete 72 from it.

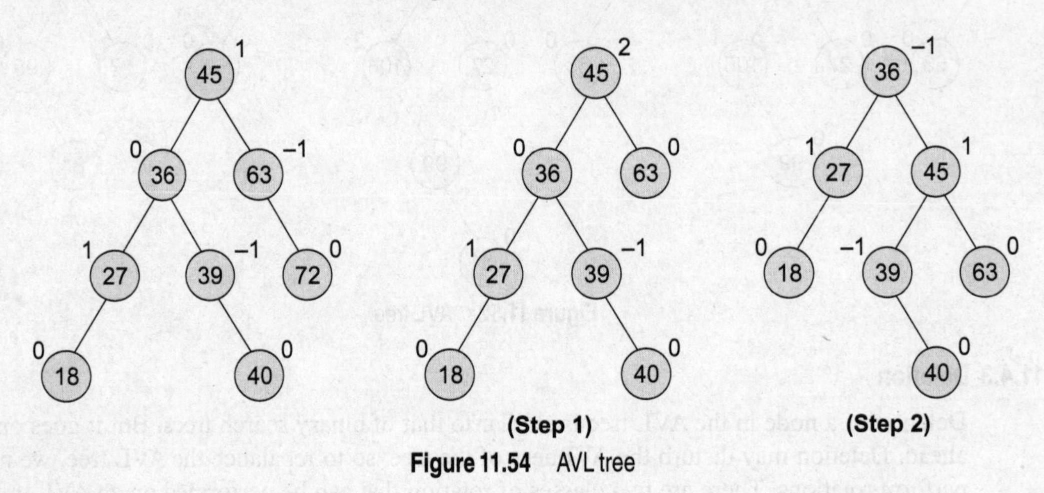

Figure 11.54 AVL tree

R1 Rotation

Let B be the root of the left or right sub-tree of A (critical node). R1 rotation is applied if the balance factor of B is 1. Observe that R0 and R1 rotations are similar to LL rotations; the only difference is that R0 and R1 rotations yield different balance factors. This is illustrated in Fig. 11.55.

Tree (a) is an AVL tree. In tree (b), the node X is to be deleted from the right sub-tree of the critical node A (node A is the critical node because it is the closest ancestor whose balance factor is not –1, 0, or 1). Since the balance factor of node B is 1, we apply R1 rotation as shown in tree (c).

During the process of rotation, node B becomes the root, with T_1 and A as its left and right child. T_2 and T_3 become the left and right sub-trees of A.

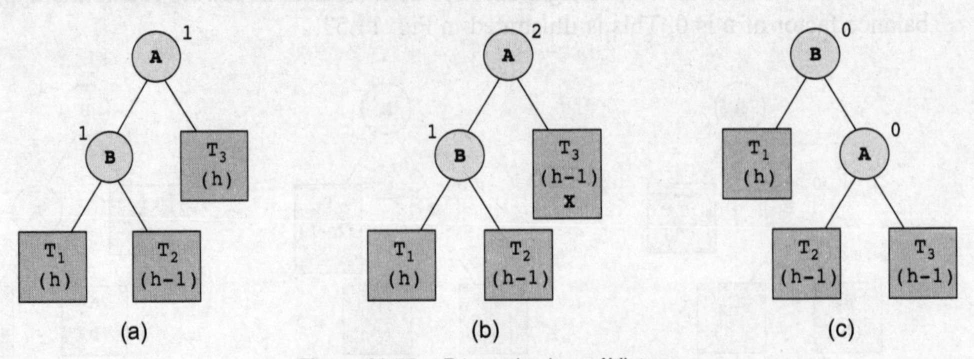

Figure 11.55 R1 rotation in an AVL tree

EXAMPLE 11.10: Consider the AVL tree given below and delete 72 from it.

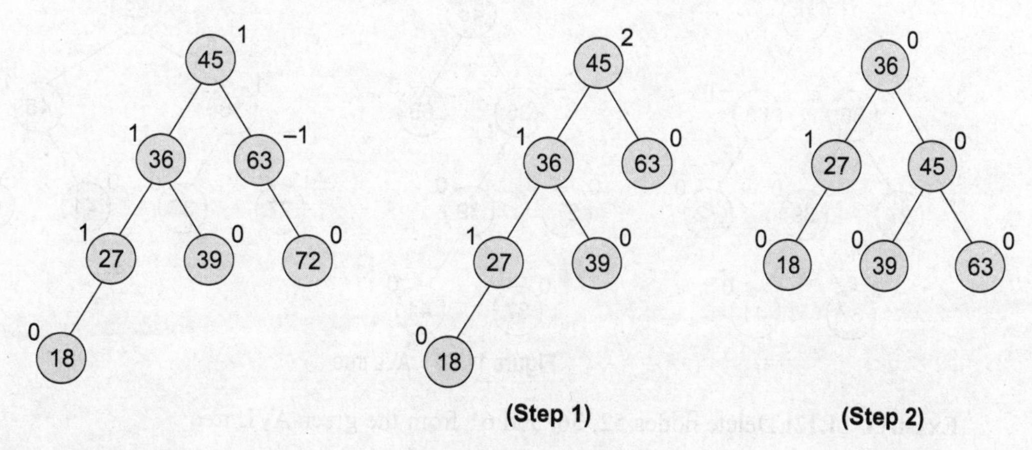

Figure 11.56 AVL tree

R–1 Rotation

Let B be the root of the left or right sub-tree of A (critical node). R-1 rotation is applied if the balance factor of B is –1. Observe that R-1 rotation is similar to LR rotation. This is illustrated in Fig. 11.57.

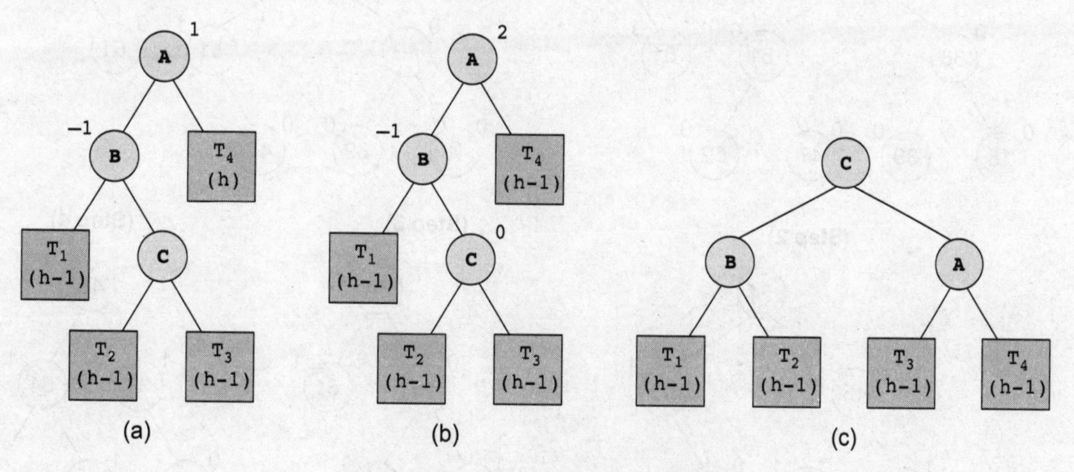

Figure 11.57 R–1 Rotation in an AVL tree

Tree (a) is an AVL tree. In tree (b), the node X is to be deleted from the right sub-tree of the critical node A (node A is the critical node because it is the closest ancestor whose balance factor is not –1, 0 or 1). Since the balance factor of node B is –1, we apply R–1 rotation as shown in tree (c).

While rotation, node B becomes the root, with T_1 and A as its left and right child. T_2 and T_3 become the left and right sub-trees of A.

EXAMPLE 11.11: Consider the following AVL tree and delete 72 from it.

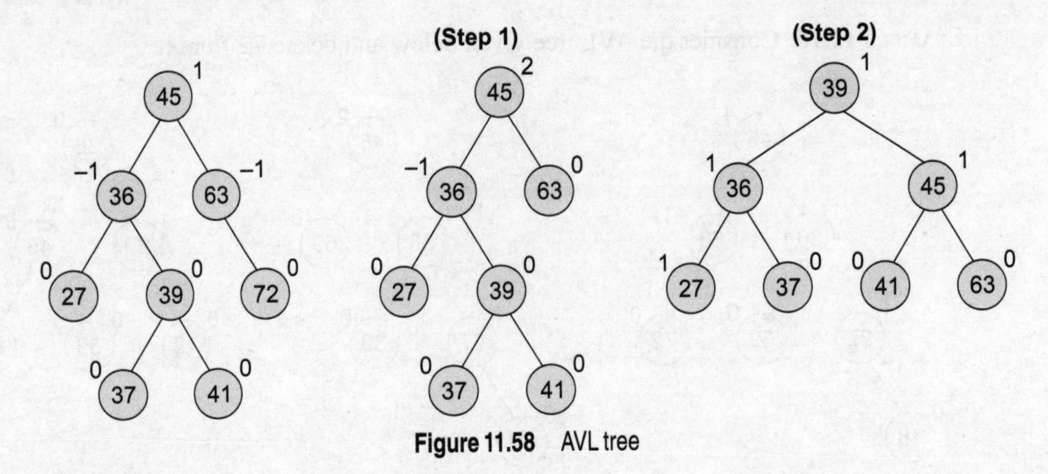

Figure 11.58 AVL tree

EXAMPLE 11.12: Delete nodes 52, 36, and 61 from the given AVL tree.

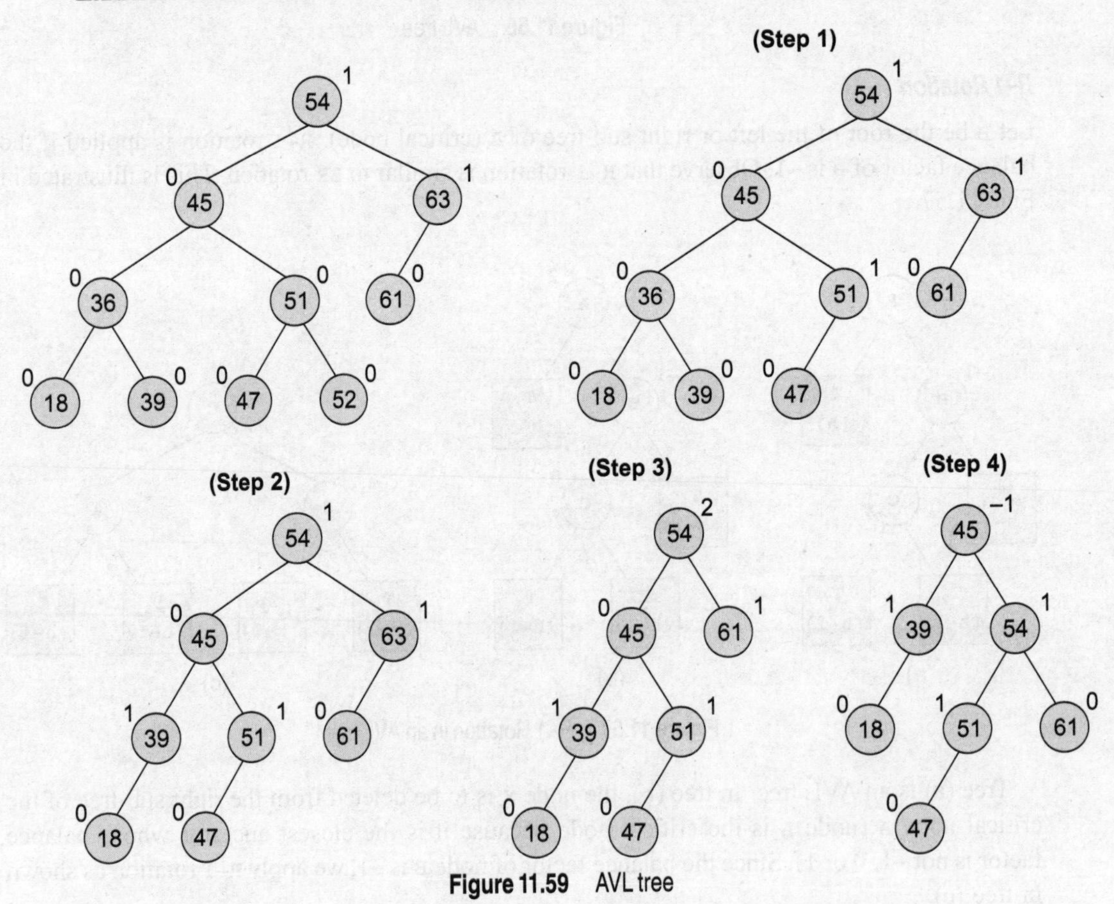

Figure 11.59 AVL tree

11.5 M-WAY SEARCH TREES

We have discussed that every node in a binary search tree contains one value and two pointers, *left* and *right*, which point to the node's left and right sub-trees respectively. The structure of a binary search tree node is shown in Fig. 11.60.

Pointer to left sub-tree	Value or Key of the node	Pointer to right sub-tree

Figure 11.60 Structure of a binary search tree node

The same concept is used in an *M-way* search tree which has `(M - 1)` values per node and `M` sub-trees. In such a tree, `M` is called the degree of the tree. Note that in a binary search tree `M = 2`, so it has one value and two sub-trees. In other words, every internal node of an M-way search tree consists of pointers to `M` sub-trees and contains `(M - 1)` keys, where `M > 2`.

The structure of an M-way search tree is shown in Fig. 11.61.

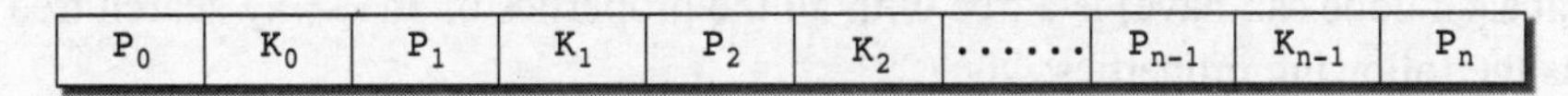

Figure 11.61 Structure of an M-way search tree node

In the structure, $P_0, P_1, P_2, ..., P_n$ are pointers to the node's sub-trees and $K_0, K_1, K_2, ..., K_{n-1}$ are the key values of the node. All the key values are stored in ascending order. That is, $(K_i < K_{i+1})$ for $(0 \le i \le n-2)$.

In an M-way search tree, it is not compulsory that every node has exactly `(M-1)` values and `M` sub-trees. Rather, the node can have anywhere from 1 to `(M-1)` values, and the number of sub-trees may vary from 0 (for a leaf node) to `1 + i`, where `i` is the number of key values in the node. `M` is thus a fixed upper limit that defines how many key values can be stored in the node.

Consider an M-way search tree as shown in Fig. 11.62. Here `M = 3`. So a node can store a maximum of two key values and contain pointers to three sub-trees.

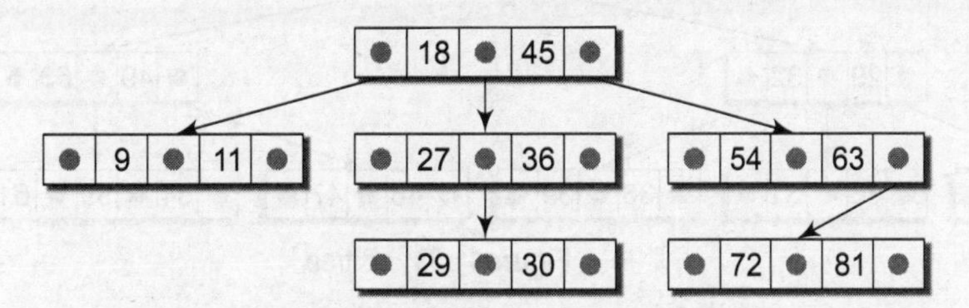

Figure 11.62 M-way search tree

In our example, we have taken a very small value of `M` so that the concept becomes easier for the reader, but in practice, `M` is usually very large. Using a 3-way search tree, let us lay down some of the basic properties of an M-way search tree.

- Note that the key values in the sub-tree pointed by P_0 are less than the key value K_0. Similarly, all the key values in the sub-tree pointed by P_1 are less than K_1, so on and so forth. Thus, the generalized rule is that all the key values in the sub-tree pointed by P_i are less than K_i, where $(0 \le i \le n-1)$.
- Note that the key values in the sub-tree pointed by P_1 are greater than the key value K_0. Similarly, all the key values in the sub-tree pointed by P_2 are greater than K_1, so on and so forth. Thus, the generalized rule is that all the key values in the sub-tree pointed by P_i are greater than K_{i-1}, where $(0 \le i \le n-1)$.

In an M-way search tree, every sub-tree is also an M-way search tree and follows the same rules.

11.6 B-TREES

A B-tree is a specialized M-way tree created by Rudolf Bayer and Ed McCreight that is widely used for disk access. A B-tree of order `m` can have a maximum of `m-1` keys and `m` pointers to its sub-trees. A B-tree may contain a large number of key values and pointers to sub-trees. Storing a large number of keys in a single node keeps the height of the tree relatively small.

A B-tree is designed to store sorted data and allows search, insert, and delete operations to be performed in logarithmic amortized time. A B-tree of order `m` (the maximum number of children that each node can have) is a tree with all the properties of an `m-way` search tree and in addition has the following properties:

1. Every node in the B-tree has at most (maximum) `m` children.
2. Every node in the B-tree except the root node and leaf nodes has at least (minimum) $m/2$ children. This condition helps to keep the tree bushy so that the path from the root node to the leaf is very short, even in a tree that stores a lot of data.
3. The root node has at least two children if it is not a terminal (leaf) node.
4. All leaf nodes are at the same level.

An internal node in the B-tree can have `n` number of children, where `(0 ≤ n ≤ m)`. It is not necessary that every node has the same number of children, but the only restriction is that the node should have at least `m/2` children. A sample B-tree is given in Fig. 11.63.

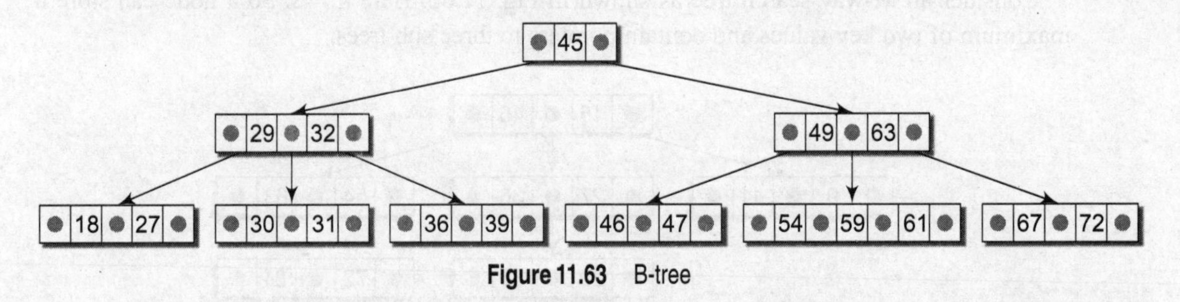

Figure 11.63 B-tree

While performing insert and delete operations in a B-tree, the number of child nodes may change. So, in order to maintain a minimum number of children, the internal nodes may be joined or split. We will discuss search, insert, and delete operations in this section.

11.6.1 Search Operation

Searching for an element in a B-tree is similar to that of binary search trees. Consider the B-tree given in Fig. 11.64. To search for 59, we begin at the root node. The root node has a value 45 which is less than 59. So, we traverse in the right sub-tree. The right sub-tree of the root node has two key values, 49 and 63. Since 49 ≤ 59 ≤ 63, we traverse the right sub-tree of 49, that is, the left sub-tree of 63. This sub-tree has three values, 54, 59, and 61. On finding the value 59, the search is successful.

Take another example. If you want to search for 9, then we traverse the left sub-tree of the root node. The left sub-tree has two key values, 29 and 32. Again, we traverse the left sub-tree of 29. We find that it has two key values, 18 and 27. There is no left sub-tree of 18, hence the value 9 is not stored in the tree.

Since the running time of the search operation depends upon the height of the tree, the algorithm to search for an element in a B-tree takes $O(\log_t n)$ time to execute.

11.6.2 Insert Operation

In a B-tree, all insertions are done at the leaf node level. A new value is inserted in the B- tree using the algorithm given below.

1. Search the B-tree to find the leaf node where the new key value should be inserted.
2. If the leaf node is not full, that is, it contains less than `m-1` key values, then insert the new element in the node keeping the node's elements ordered.
3. If the leaf node is full, that is, the leaf node already contains `m-1` key values, then
   (a) insert the new value in order into the existing set of keys,
   (b) split the node at its median into two nodes (note that the split nodes are half full),
   (c) push the median element up to its parent's node. If the parent's node is already full, then split the parent node by following the same steps.

Example 11.13: Look at the B-tree of order 5 given below and insert 8, 9, 39, and 4 into it.

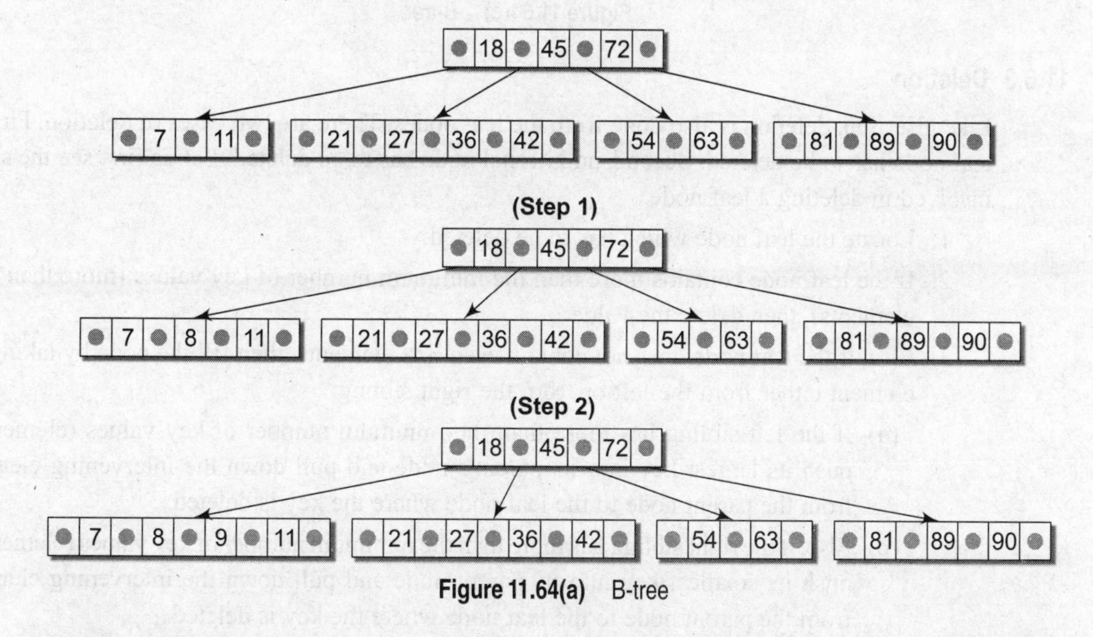

Figure 11.64(a) B-tree

Till now, we have easily inserted 8 and 9 in the tree because the leaf nodes were not full. But now, the node where 39 should be inserted is already full as it contains four values. Here we split the nodes to form two separate nodes. But before splitting, arrange the key values in order (including the new value). The ordered set of values is given as 21, 27, 36, 39, and 42. The median value is 36, so push 36 into its parent's node and split the leaf nodes.

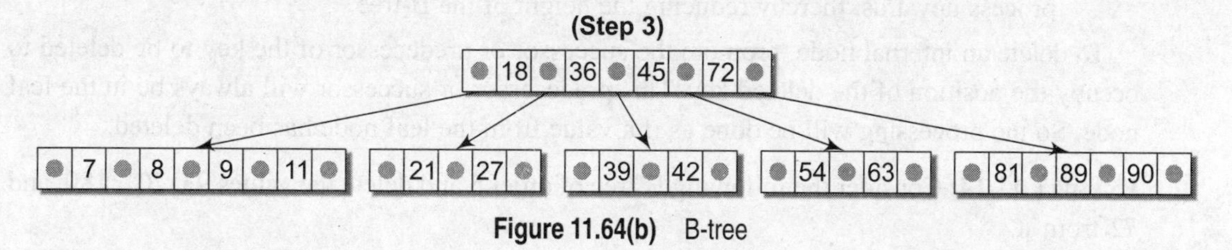

Figure 11.64(b) B-tree

Now the node where 4 should be inserted is already full as it contains four key values. Here we split the nodes to form two separate nodes. But before splitting, we arrange the key values in order (including the new value). The ordered set of values is given as 4, 7, 8, 9, and 11. The median value is 8, so push 8 into its parent's node and split the leaf nodes. But again, we see that the parent's node is already full, so we split the parent node using the same procedure.

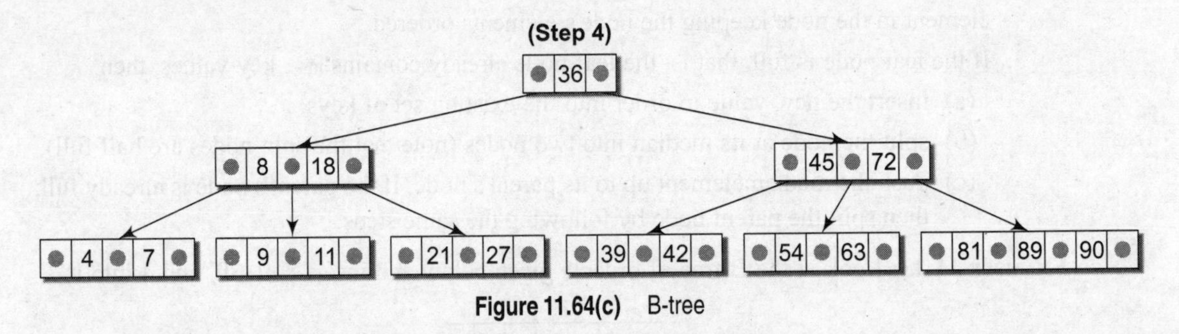

Figure 11.64(c) B-tree

11.6.3 Deletion

Like insertion, deletion is also done from the leaf nodes. There are two cases of deletion. First, a leaf node has to be deleted. Second, an internal node has to be deleted. Let us first see the steps involved in deleting a leaf node.

1. Locate the leaf node which has to be deleted.
2. If the leaf node contains more than the minimum number of key values (more than m/2 elements), then delete the value.
3. Else, if the leaf node does not contain even `m/2` elements, then fill the node by taking an element either from the left or from the right sibling.
   (a) If the left sibling has more than the minimum number of key values (elements), push its largest key into its parent's node and pull down the intervening element from the parent node to the leaf node where the key is deleted.
   (b) Else, if the right sibling has more than the minimum number of key values (elements), push its smallest key into its parent node and pull down the intervening element from the parent node to the leaf node where the key is deleted.
4. Else, if both left and right siblings contain only the minimum number of elements, then create a new leaf node by combining the two leaf nodes and the intervening element of the parent node (ensuring that the number of elements do not exceed the maximum number of elements a node can have, that is, `m`). If pulling the intervening element from the parent node leaves it with less than the minimum number of keys in the node, then propagate the process upwards, thereby reducing the height of the B-tree.

To delete an internal node, promote the successor or predecessor of the key to be deleted to occupy the position of the deleted key. This predecessor or successor will always be in the leaf node. So the processing will be done as if a value from the leaf node has been deleted.

EXAMPLE 11.14: Consider the following B-tree of order 5 and delete the values 93, 201, 180, and 72 from it.

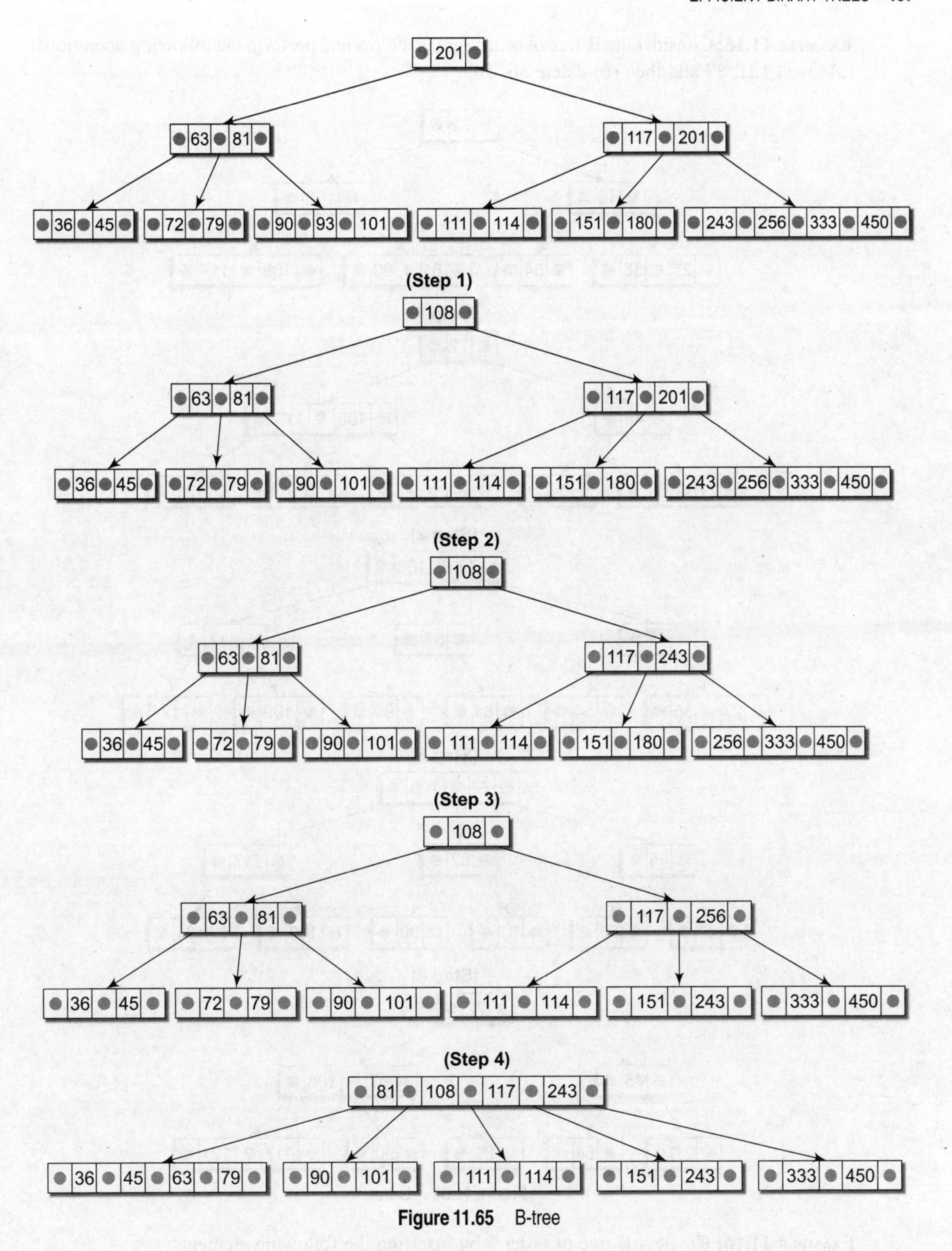

Figure 11.65 B-tree

Example 11.15: Consider the B-tree of order 3 given below and perform the following operations: (a) insert 121, 87 and then (b) delete 36, 109.

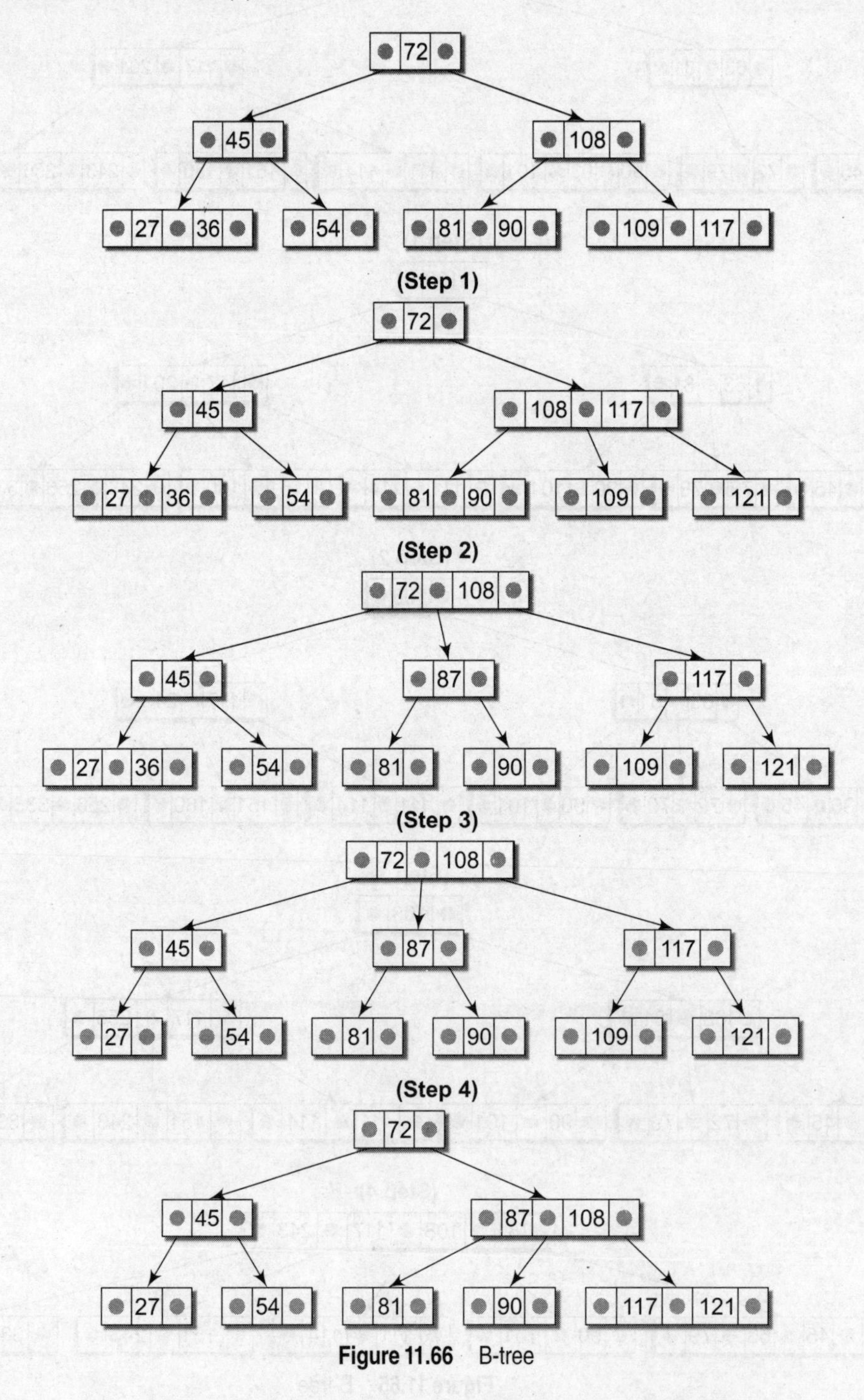

Figure 11.66 B-tree

Example 11.16: Create a B-tree of order 5 by inserting the following elements:

3, 14, 7, 1, 8, 5, 11, 17, 13, 6, 23, 12, 20, 26, 4, 16, 18, 24, 25, and 19.

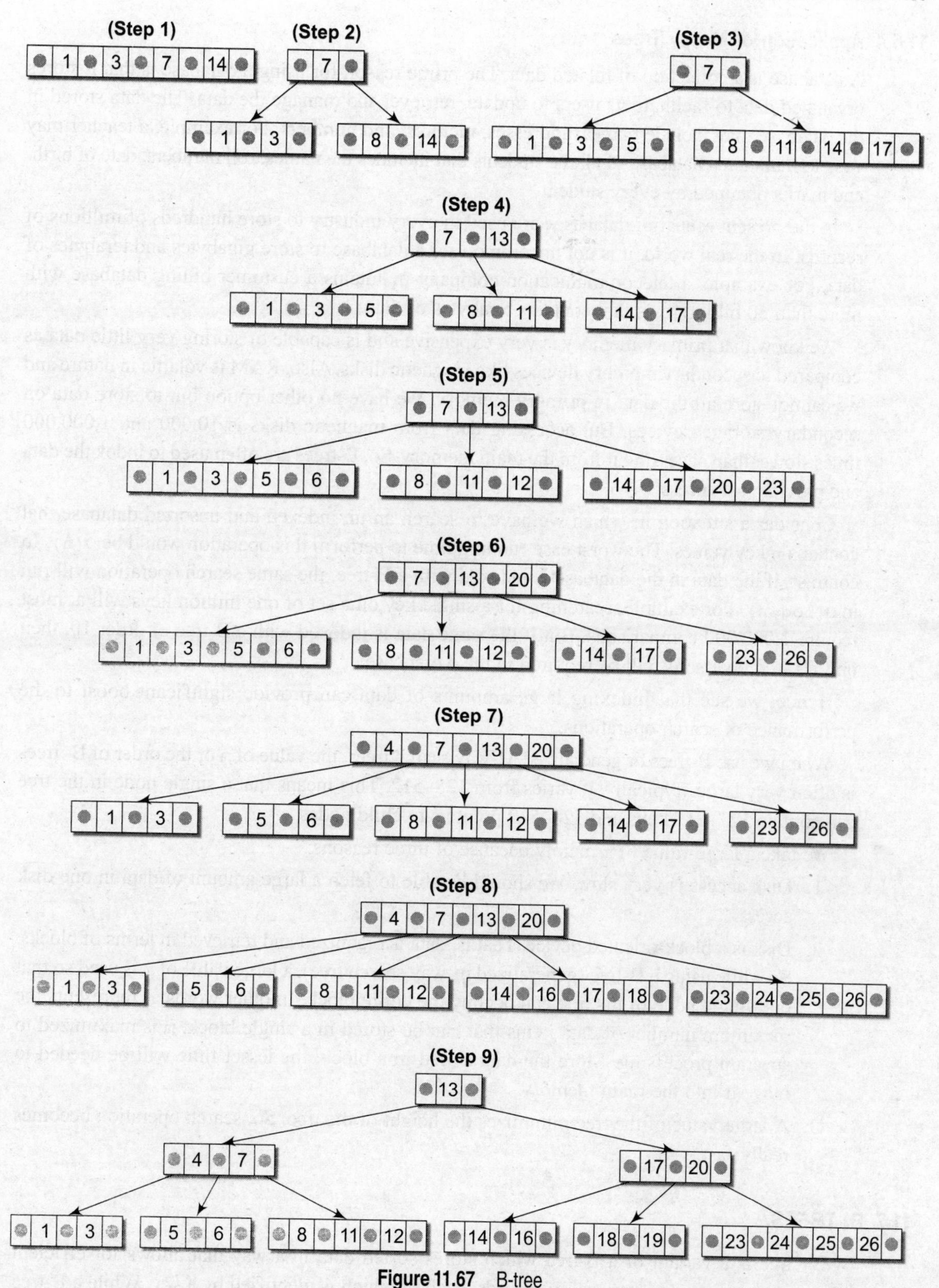

Figure 11.67 B-tree

11.6.4 Application of M-way Trees

A database is a collection of related data. The prime reason for using a database is that it stores organized data to facilitate its users to update, retrieve, and manage the data. The data stored in the database may include names, addresses, pictures, and numbers. For example, a teacher may wish to maintain a database of all the students that includes the names, roll numbers, date of birth, and marks obtained by every student.

In the present scenario, databases are used in every industry to store hundreds of millions of records. In the real world, it is not uncommon for a database to store gigabytes and terabytes of data. For example, a telecommunication company maintains a customer billing database with more than 50 billion rows that contains terabytes of data.

We know that primary memory is very expensive and is capable of storing very little data as compared to secondary memory devices like magnetic disks. Also, RAM is volatile in nature and we cannot store all the data in primary memory. We have no other option but to store data on secondary storage devices. But accessing data from magnetic disks is 10,000 and 1,000,000 times slower than accessing it from the main memory. So, B-trees are often used to index the data and provide fast access.

Consider a situation in which we have to search an un-indexed and unsorted database that contains `n` key values. The worst case running time to perform this operation would be `O(n)`. In contrast, if the data in the database is indexed with a B-tree, the same search operation will run in `O(log n)`. For example, searching for a single key on a set of one million keys will at most require 1,000,000 comparisons. But if the same data is indexed with a B-tree of order 10, then only 114 comparisons will be required in the worst case.

Hence, we see that indexing large amounts of data can provide significant boost to the performance of search operations.

When we use B-trees or generalized M-way search trees, the value of `m` or the order of B- trees is often very large. Typically, it varies from 128–512. This means that a single node in the tree can contain 127–511 keys and 128–512 pointers to child nodes.

We take a large value of `m` mainly because of three reasons:

1. Disk access is very slow. We should be able to fetch a large amount of data in one disk access.
2. Disk is a block-oriented device. That is, data is organized and retrieved in terms of blocks. So while using a B-tree (generalized m-way search tree), a large value of `m` is used so that one single node of the tree can occupy the entire block. In other words, `m` represents the maximum number of data items that can be stored in a single block. `m` is maximized to speedup processing. More the data stored in a block, the lesser time will be needed to move it into the main memory.
3. A large value of the tree minimizes the height of the tree. So, search operation becomes really fast.

11.7 B+ TREES

A B+ tree is a variant of a B-tree which stores sorted data in a way that allows for efficient insertion, retrieval, and removal of records, each of which is identified by a *key*. While a B-tree can store both keys and records in its interior nodes, a B+ tree, in contrast, stores all the records at the leaf level of the tree; only keys are stored in the interior nodes.

The leaf nodes of a B+ tree are often linked to one another in a linked list. This has an added advantage of making the queries simpler and more efficient.

Typically, B+ trees are used to store large amounts of data that cannot be stored in the main memory. With B+ trees, the secondary storage (magnetic disk) is used to store the leaf nodes of the tree and the internal nodes of the tree are stored in the main memory.

B+trees store data only in the leaf nodes. All other nodes (internal nodes) are called *index nodes* or *i-nodes* and store index values which allow us to traverse the tree from the root down to the leaf node that stores the desired data item. Figure 11.68 shows a B+ tree.

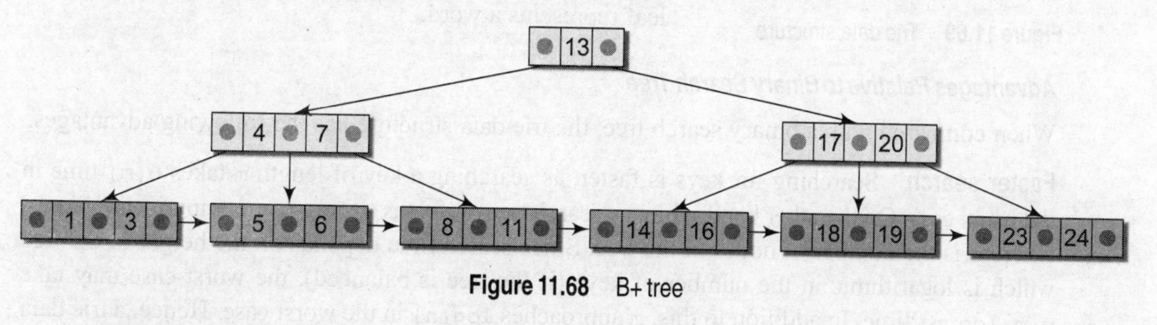

Figure 11.68 B+ tree

Insertion A new element is simply added in the leaf node if there is space for it. But if the data node in the tree where insertion has to be done is full, then that node is split into two nodes. This calls for adding a new index value in the parent index node so that future queries can arbitrate between the two new nodes.

However, adding the new index value in the parent node may cause it, in turn, to split. In fact, all the nodes on the path from a leaf to the root may split when a new value is added to a leaf node. If the root node splits, a new leaf node is created and the tree grows by one level.

Deletion As in B trees, deletion is always done from a leaf node. If deleting a data element may leave that node empty, then the neighbouring nodes are examined and merged with the *underfull* node.

This process calls for the deletion of an index value from the parent index node which, in turn, may cause it to become empty. Similar to the insertion process, deletion may cause a merge-delete wave to run from a leaf node all the way up to the root. This leads to shrinking of the tree by one level.

Hence, insertion and deletion operations are recursive in nature and can cascade up or down the B+tree, thereby affecting its shape dramatically.

11.8 TRIE

The term *trie* has been taken from the word 'ret**rie**val'. A trie is an ordered tree data structure introduced in the 1960s by Fredkin. Trie stores keys that are usually strings. It is basically a *k-ary* position tree.

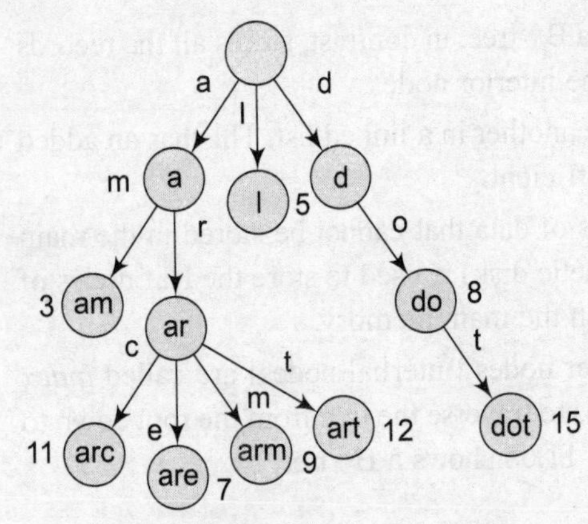

Figure 11.69 Trie data structure

In contrast to binary search trees, nodes in a trie do not store the keys associated with them. Rather, a node's position in the tree represents the key associated with that node. All the descendants of a node have a common prefix of the string associated with that node, and the root is associated with the empty string. Figure 11.69 shows a trie.

In the given tree, keys are listed in the nodes and the values below them. Note that each complete English word is assigned an arbitrary integer value. We can find each of these words by traversing the various branches in the tree until a leaf node is encountered. Any path from the root to a leaf represents a word.

Advantages Relative to Binary Search Tree

When compared with a binary search tree, the trie data structure has the following advantages.

Faster search Searching for keys is faster, as searching a key of length `m` takes `O(m)` time in the worst case. On the other hand, a binary search tree performs `O(log(n))` comparisons of keys, where `n` is the number of nodes in the tree. Since search time depends on the height of the tree which is logarithmic in the number of keys (if the tree is balanced), the worst case may take `O(m log n)` time. In addition to this, `m` approaches `log(n)` in the worst case. Hence, a trie data structure provides a faster search mechanism.

Less space Trie occupies less space, especially when it contains a large number of short strings. Since keys are not stored explicitly and nodes are shared between the keys with common initial subsequences, a trie calls for less space as compared to a binary search tree.

Longest prefix-matching Trie facilitates the longest-prefix matching which enables us to find the key sharing the longest possible prefix of all unique characters.

Since trie provides more advantages, it can be thought of as a good replacement for binary search trees.

Advantages Relative to Hash Table

Trie can also be used to replace a *hash table*, as it provides the following advantages:

- Searching for data in a trie is faster in the worst case, `O(m)` time, compared to an imperfect hash table which may have numerous key collisions. Trie is free from collision of keys problem.
- Unlike a hash table, there is no need to choose a hash function or to change it when more keys are added to a trie.
- A trie can sort the keys using a predetermined alphabetical ordering.

Disadvantages

The disadvantages of having a trie are listed below:

- In some cases, tries can be slower than hash tables while searching data. This is true in cases when the data is directly accessed on a hard disk drive or some other secondary storage device that has high random access time as compared to the main memory.

- All the key values cannot be easily represented as strings. For example, the same floating point number can be represented as a string in multiple ways (1 is equivalent to 1.0, 1.00, +1.0, etc.).

Applications

Tries are commonly used to store a dictionary (for example, on a mobile telephone). These applications take advantage of a trie's ability to quickly search, insert, and delete the entries. Tries are also used to implement approximate matching algorithms, including those used in spell-checking software.

11.9 RED-BLACK TREE

A red-black tree is a self-balancing binary search tree that was invented in 1972 by Rudolf Bayer who called it the 'symmetric binary B-tree'. Although a red-black tree is complex, but it has good worst-case running time for its operations and is efficient to use as searching, insertion, and deletion can all be done in `O(log n)` time, where `n` is the number of nodes in the tree. Practically, a red-black tree is a binary search tree which inserts and removes intelligently, to keep the tree reasonably balanced. A special point to note about the red-black tree is that in these trees, no data is stored in the leaf nodes.

11.9.1 Properties of a Red-Black Tree

A red-black tree is a binary search tree in which every node has a color which is either red or black. Apart from the other restrictions of a binary search tree, the red-black tree has the following additional requirements:

1. The color of a node is either red or black.
2. The color of the root node is always black.
3. All leaf nodes are black.
4. Every red node has both the children coloured in black.
5. Every simple path from a given node to any of its leaf nodes has an equal number of black nodes.

Look at Fig. 11.70 which shows a red-black tree.

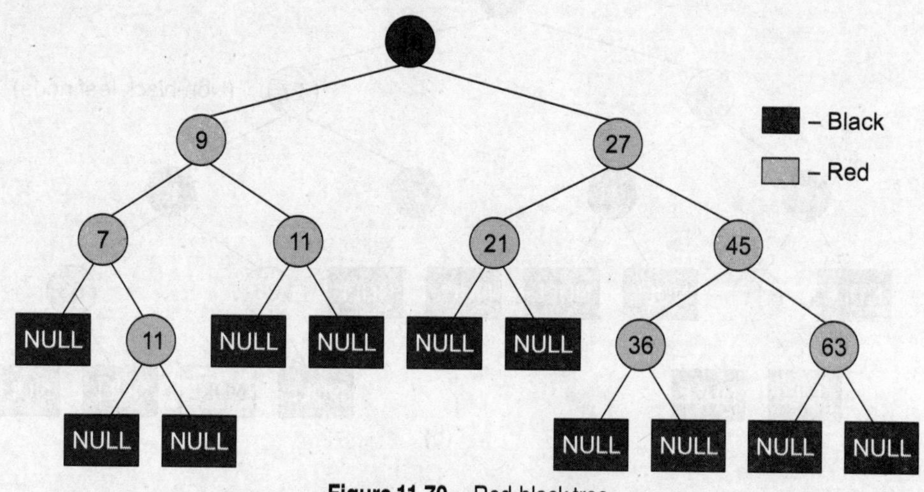

Figure 11.70 Red-black tree

These constraints enforce a critical property of red-black trees. *The longest path from the root node to any leaf node is no more than twice as long as the shortest path from the root to any other leaf in that tree.*

This results in a roughly balanced tree. Since operations such as insertion, deletion, and searching require worst-case times proportional to the height of the tree, this theoretical upper bound on the height allows red-black trees to be efficient in the worst case, unlike ordinary binary search trees.

To understand the importance of these properties, it suffices to note that according to property 4, no path can have two red nodes in a row. The shortest possible path will have all black nodes, and the longest possible path would alternately have a red and a black node. Since all maximal paths have the same number of black nodes (property 5), *no path is more than twice as long as any other path.*

Figure 11.71 shows some binary search trees that are not red-black trees.

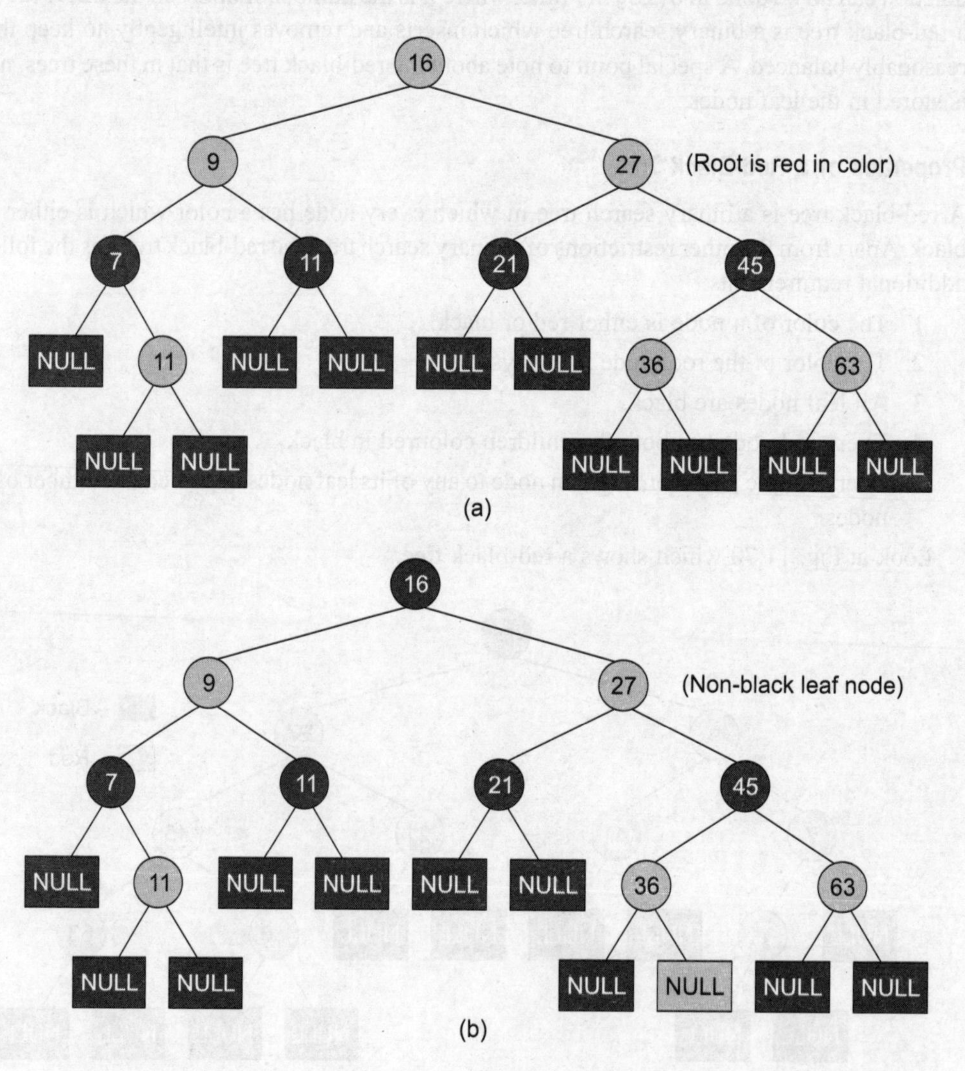

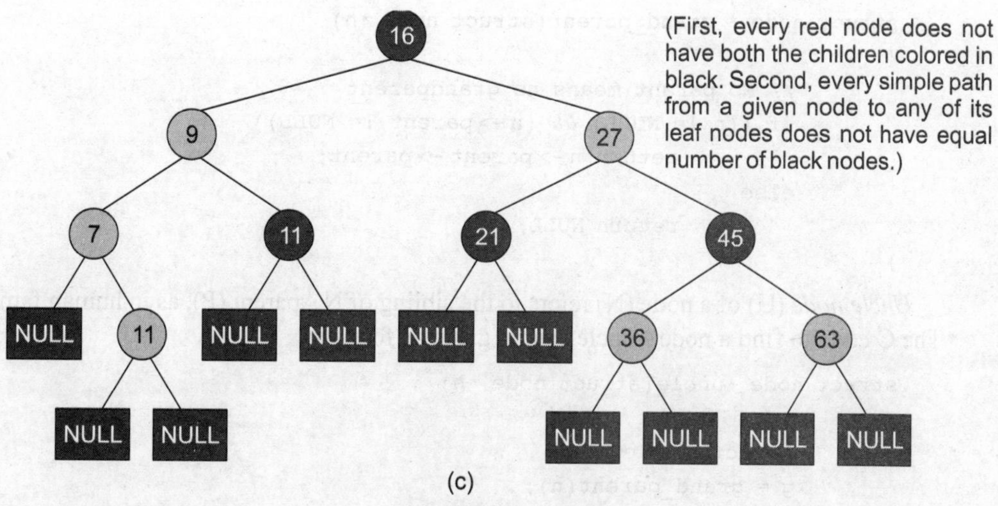

(c)

Figure 11.71 Trees

11.9.2 Applications

Red-black trees are efficient binary search trees, as they offer worst case time guarantee for insertion, deletion, and search operations. Red-black trees are not only valuable in time-sensitive applications such as real-time applications, but are also preferred to be used as a building block in other data structures which provide worst-case guarantee.

AVL trees also support `O(log n)` search, insertion, and deletion operations, but they are more rigidly balanced than red-black trees, thereby resulting in slower insertion and removal but faster retrieval of data.

11.9.3 Operations

Preforming a read-only operation (like traversing the nodes in a tree) on a red-black tree requires no modification from those used for binary search trees. Remember that every red-black tree is a special case of a binary search tree. However, insertion and deletion operations may violate the properties of a red-black tree. Therefore, these operations may create a need to restore the red-black properties that may require a small number `(O(log n)` or amortized `O(1))` of color changes.

Insertion

The insertion operation starts in the same way as we add a new node in the binary search tree. However, in a binary search tree, we always add the new node as a leaf, while in a red-black tree, leaf nodes contain no data. So instead of adding the new node as a leaf node, we add a red interior node that has two black leaf nodes. Note that the color of the new node is red and its leaf nodes are colored in black.

Once a new node is added, it may violate some properties of the red-black tree. So in order to restore their property, we check for certain cases and restore the property depending on the case that turns up after insertion. But before learning these cases in detail, first let us discuss certain important terms that will be used.

Grandparent node (G) of a node (N) refers to the parent of N's parent (P), as in human family trees. The C code to find a node's grandparent can be given as follows:

```
struct node * grand_parent(struct node *n)
{
        // No parent means no grandparent
        if ((n != NULL) && (n->parent != NULL))
                return n->parent->parent;
        else
                return NULL;
}
```

Uncle node (U) of a node (N) refers to the sibling of N's parent (P), as in human family trees. The C code to find a node's uncle can be given as follows:

```
struct node *uncle(struct node *n)
{
        struct node *g;
        g = grand_parent(n);
        //With no grandparent, there cannot be any uncle
        if (g == NULL)
                return NULL;
        if (n->parent == g->left)
                return g->right;
        else
                return g->left;
}
```

When we insert a new node in a red-black tree, note the following:

- All leaf nodes are always black. So property 3 always holds true.
- Property 4 (both children of every red node are black) is threatened only by adding a red node, repainting a black node red, or a rotation.
- Property 5 (all paths from any given node to its leaf nodes has equal number of black nodes) is threatened only by adding a black node, repainting a red node black, or a rotation.

Case 1: The New Node N is Added as the Root of the Tree

In this case, N is repainted black, as the root of the tree is always black. Since N adds one black node to every path at once, Property 5 is not violated. The C code for case 1 can be given as follows:

```
void case1(struct node *n)
{
        if (n->parent == NULL) // Root node
                n->color = BLACK;
        else
                case2(n);
}
```

Case 2: The New Node's Parent P is Black

In this case, both children of every red node are black, so Property 4 is not invalidated. Property 5 is also not threatened. This is because the new node N has two black leaf children, but because N is red, the paths through each of its children have the same number of black nodes. The C code to check for case 2 can be given as follows:

```
void case2(struct node *n)
{
        if (n->parent->color == BLACK)
                return; /* Red black tree property is not violated*/
        else
                case3(n);
}
```

In the following cases, it is assumed that N *has a grandparent node* G*, because its parent* P *is red, and if it were the root, it would be black. Thus,* N *also has an uncle node* U *(irrespective of whether* U *is a leaf node or an internal node).*

Case 3: If Both the Parent (P) and the Uncle (U) are Red

In this case, Property 5 which says all paths from any given node to its leaf nodes have an equal number of black nodes is violated. Insertion in the third case is illustrated in Fig. 11.72.

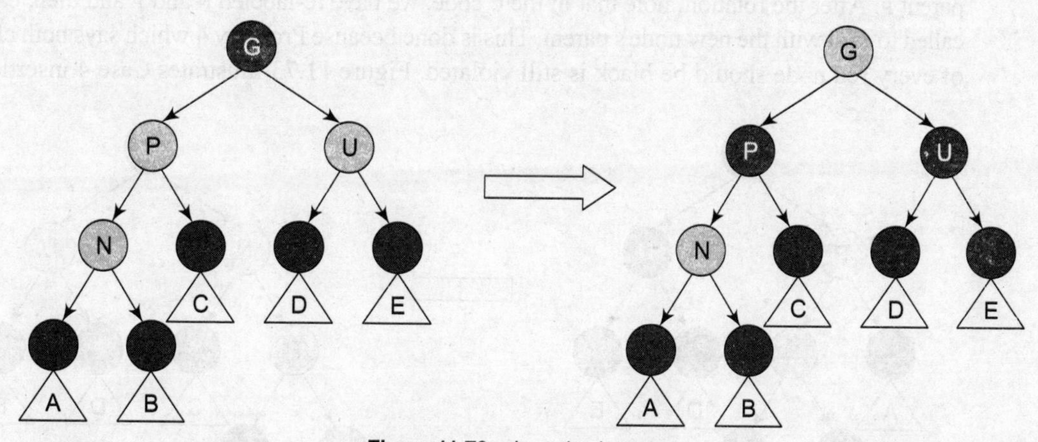

Figure 11.72 Insertion in case 3

In order to restore Property 5, both nodes (P and U) are repainted black and the grandparent G is repainted red. Now, the new red node N has a black parent. Since any path through the parent or uncle must pass through the grandparent, the number of black nodes on these paths has not changed.

However, the grandparent G may now violate Property 2 which says that the root node is always black or Property 4 which states that both children of every red node are black. Property 4 will be violated when G has a red parent. In order to fix this problem, this entire procedure is recursively performed on G from Case 1. The C code to deal with case 3 insertion is as follows:

```
void case3(struct node *n)
{
        struct node *u, *g;
        u = uncle (n);
        g = grand_parent(n);
```

```
            if ((u != NULL) && (u->color == RED)) {
                      n->parent->color = BLACK;
                      u->color = BLACK;
                      g->color = RED;
                      case1(g);
            }
            else {
                      insert_case4(n);
            }
   }
```

In the remaining cases, we assume that the parent node P is the left child of its parent. If it is the right child, then interchange *left* and *right* in cases 4 and 5.

Case 4: The Parent P is Red but the Uncle U is Black and N is the Right Child of P and P is the Left Child of G

In order to fix this problem, a left rotation is done to switch the roles of the new node N and its parent P. After the rotation, note that in the C code, we have re-labeled N and P and then, case 5 is called to deal with the new node's parent. This is done because Property 4 which says both children of every red node should be black is still violated. Figure 11.73 illustrates Case 4 insertion.

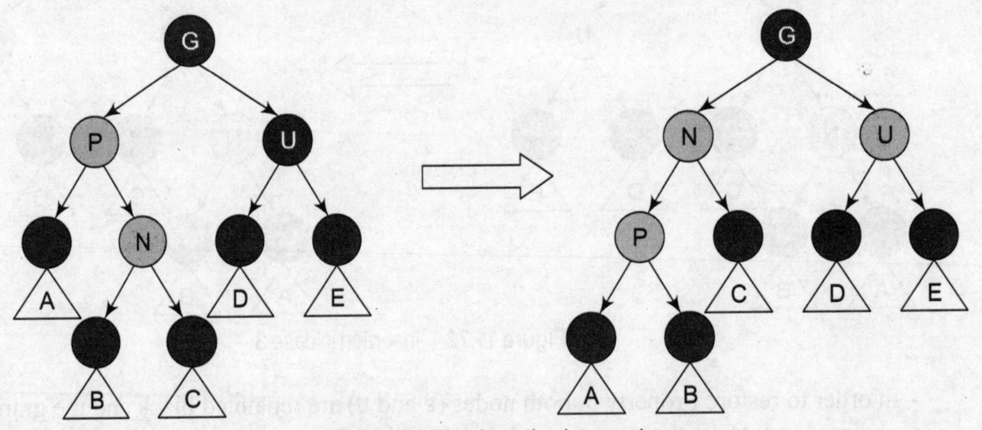

Figure 11.73 Insertion in case 4

Note that in case N is the left child of P and P is the right child of G, we have to perform a right rotation. In the C code that handles case 4, we check for P and N and then, perform either a left or a right rotation.

```
void case4(struct node *n)
{
          struct node *g = grand_parent(n);
          if ((n == n->parent->right) && (n->parent == g->left))
          {
                    rotate_left(n->parent);
                    n = n->left;
          }
```

```
        else if ((n == n->parent->left) && (n->parent == g->right))
        {
                rotate_right(n->parent);
                n = n->right;
        }
        case5(n);
}
```

Case 5: The Parent P is Red but the Uncle U is Black and the New Node N is the Left Child of P, and P is the Left Child of its Parent G.

In order to fix this problem, a right rotation on G (the grandparent of N) is performed. After this rotation, the former parent P is now the parent of both the new node N and the former grandparent G.

We know that the color of G is black (because otherwise its former child P could not have been red), so now switch the colors of P and G so that the resulting tree satisfies Property 4 which states that both children of a red node are black. Case 5 insertion is illustrated in Fig. 11.74.

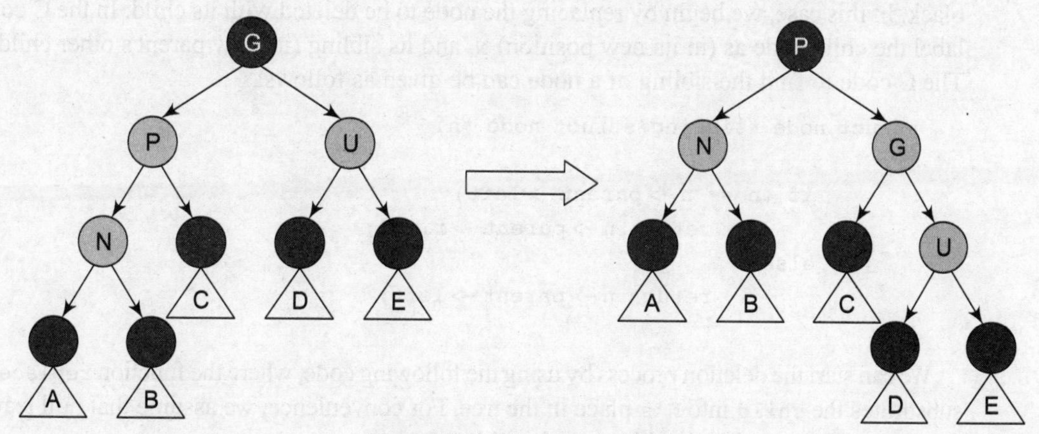

Figure 11.74 Insertion in case 5

Note that in case N is the right child of P and P is the right child of G, we perform a left rotation. In the C code that handles case 5, we check for P and N and then, perform either a left or a right rotation.

```
void case5(struct node *n)
{
        struct node *g;
        g = grandparent(n);
        if ((n == n->parent->left) && (n->parent == g->left))
                rotate_right(g);
        else if ((n == n->parent->right) && (n->parent == g->right))
                rotate_left(g);
        n->parent->color = BLACK;
        g->color = RED;
}
```

Deletion

We start deleting a node from a red-black tree in the same way as we do in case of a binary search tree. In a binary search tree, when we delete a node with two non-leaf children, we find either the maximum element in its left sub-tree of the node or the minimum element in its right sub-tree, and move its value into the node being deleted. After that, we delete the node from which we had copied the value. Note that this node must have less than two non-leaf children. Therefore, merely copying a value does not violate any red-black properties, but it just reduces the problem of deleting to the problem of deleting a node with at most one non-leaf child.

In this section, we will assume that we are deleting a node with at most one non-leaf child, which we will call its child. In case this node has both leaf children, then let one of them be its child.

While deleting a node, if its color is red, then we can simply replace it with its child, which must be black. All paths through the deleted node will simply pass through one less red node, and both the deleted node's parent and child must be black, so none of the properties will be violated.

Another simple case is when we delete a black node that has a red child. In this case, property 4 and property 5 could be violated, so to restore them, just repaint the deleted node's child with black.

However, a complex situation arises when both the node to be deleted as well as its child is black. In this case, we begin by replacing the node to be deleted with its child. In the C code, we label the child node as (in its new position) `N`, and its sibling (its new parent's other child) as `S`. The C code to find the sibling of a node can be given as follows:

```
struct node *sibling(struct node *n)
{
        if (n == n->parent->left)
                return n->parent->right;
        else
                return n->parent->left;
}
```

We can start the deletion process by using the following code, where the function `replace_node` substitutes the `child` into `N's` place in the tree. For convenience, we assume that null leaves are represented by actual node objects, rather than `NULL`.

```
void delete_child(struct node *n)
{
        /* If N has at most one non-null child */
        struct node *child;
        if (is_leaf(n->right))
                child = n->left;
        else
                child = n->right;
        replace_node(n, child);
        if (n->color == BLACK) {
                if (child->color == RED)
                        child->color = BLACK;
                else
                        del_case1(child);
        }
```

```
        free(n);
}
```

When both N and its parent P are black, then deleting P will cause paths which precede through N to have one fewer black node than the other paths. This will violate Property 5. Therefore, the tree needs to be rebalanced. There are several cases to consider, which are discussed below.

Case 1: N is the New Root

In this case, we have removed one black node from every path, and the new root is black, so none of the properties are violated.

```
void del_case1(struct node *n)
{
        if (n->parent != NULL)
                del_case2(n);
}
```

In cases 2, 5, and 6, we assume N is the left child of its parent P. If it is the right child, left and right should be interchanged throughout these three cases.

Case 2: Sibling S is Red

In this case, interchange the colors of P and S, and then rotate left at P. In the resultant tree, S will become N's grandparent. Figure 11.75 illustrates Case 2 deletion.

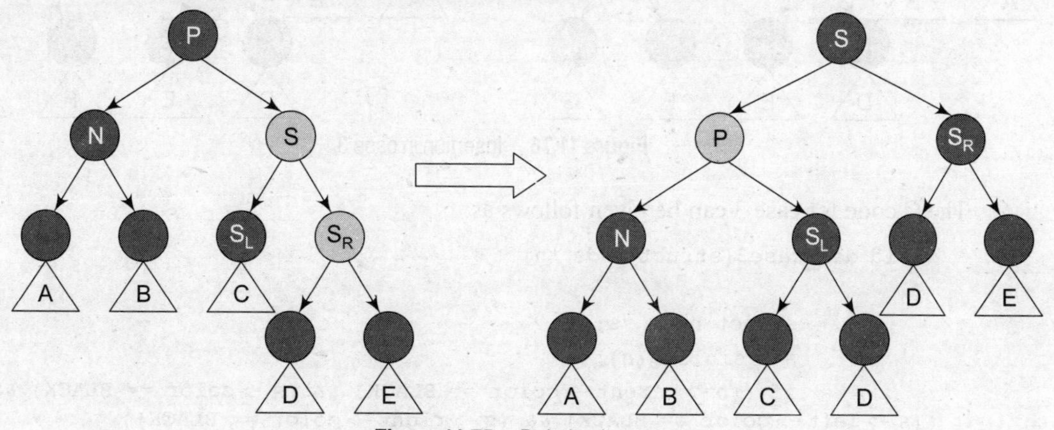

Figure 11.75 Deletion in case 2

The C code that handles case 2 deletion can be given as follows:

```
void del_case2(struct node *n)
{
        struct node *s;
        s = sibling(n);
        if (s->color == RED)
        {
                if (n == n->parent->left)
                        rotate_left(n->parent);
```

```
                else
                        rotate_right(n->parent);
                n->parent->color = RED;
                s->color = BLACK;
            }
        del_case3(n);
}
```

Case 3: P, S, and S's Children are Black

In this case, simply repaint S with red. In the resultant tree, all the paths passing through S will have one less black node. Therefore, all the paths that pass through P now have one fewer black node than the paths that do not pass through P, so Property 5 is still violated. To fix this problem, we perform the rebalancing procedure on P, starting at Case 1. Case 3 is illustrated in Fig. 11.76.

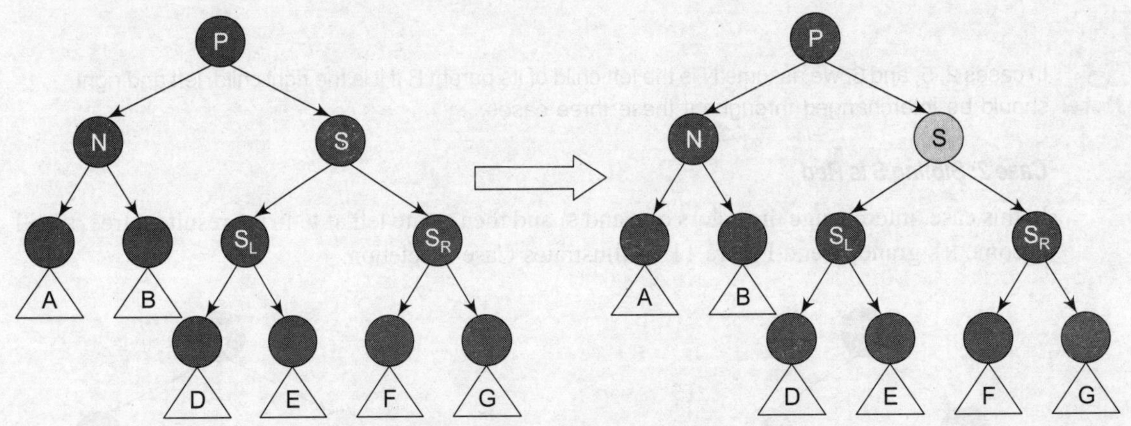

Figure 11.76 Insertion in case 3

The C code for case 3 can be given follows as:

```
void del_case3(struct node *n)
{
        struct node *s;
        s = sibling(n);
        if ((n->parent->color == BLACK) && (s->color == BLACK) &&
(s->left->color == BLACK) && (s->right->color == BLACK))
        {
                s->color = RED;
                del_case1(n->parent);
        } else
                del_case4(n);
}
```

Case 4: S and S's Children are Black, but P is Red

In this case, we interchange the colors of S and P. Although this will not affect the number of black nodes on the paths going through S, it will add one black node to the paths going through N, making up for the deleted black node on those paths. Figure 11.77 illustrates this case.

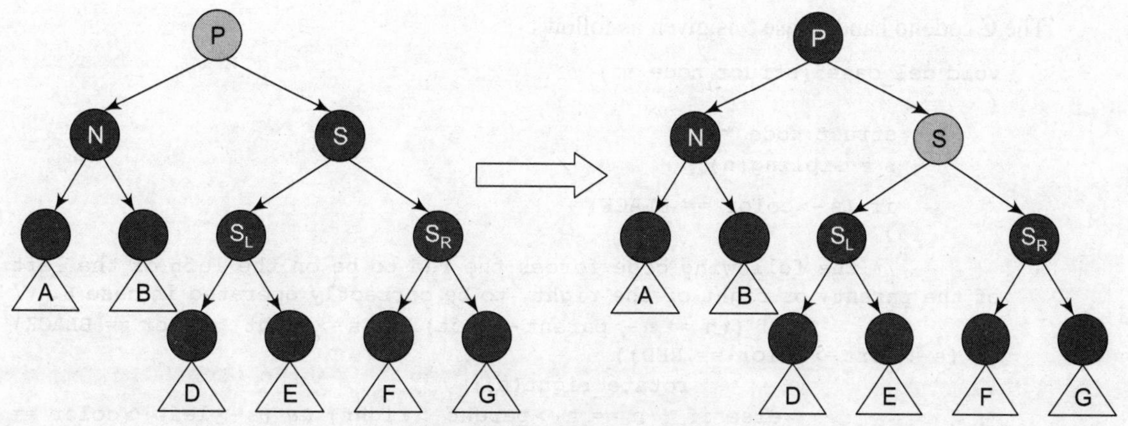

Figure 11.77 Insertion in case 4

The C code to handle Case 4 is as follows:

```
void del_case4(struct node *n)
{
        struct node *s;
        s = sibling(n);
        if ((n->parent->color == RED) && (s->color == BLACK) &&
(s->left->color == BLACK) && (s->right->color == BLACK))
        {
                s->color = RED;
                n->parent->color = BLACK;
        } else
                del_case5(n);
}
```

Case 5: N is the Left Child of P and S is Black, S's Left Child is Red, S's Right Child is Black.

In this case, perform a right rotation at S. After the rotation, S's left child becomes S's parent and N's new sibling. Also, interchange the colors of S and its new parent.

Note that now all paths still have equal number of black nodes, but N has a black sibling whose right child is red, so we fall into Case 6. Refer Fig. 11.78.

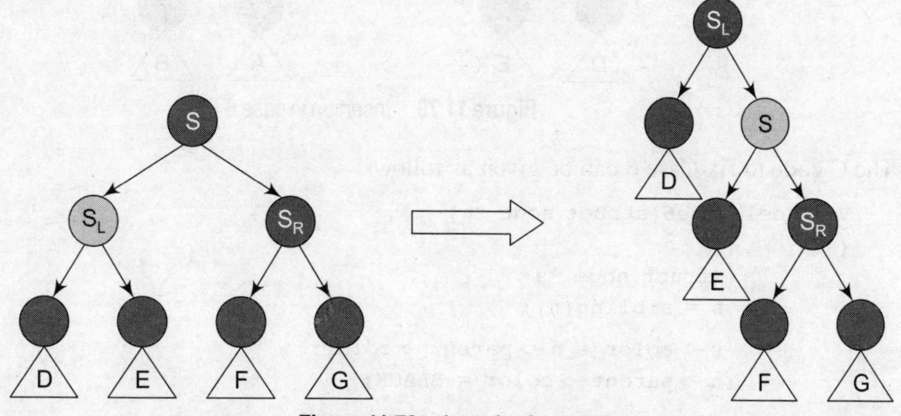

Figure 11.78 Insertion in case 5

The C code to handle case 5 is given as follows:

```
void del_case5(struct node *n)
{
        struct node *s;
        s = sibling(n);
        if (s -> color == BLACK)
        {
        /* the following code forces the red to be on the left of the left
of the parent, or right of the right, to be correctly operated in case 6. */
                if ((n == n -> parent -> left) && (s -> right -> color == BLACK)
&& (s -> left -> color == RED))
                        rotate_right(s);
                else if ((n == n -> parent -> right) && (s -> left -> color ==
BLACK) && (s -> right -> color == RED))
                        rotate_left(s);
                s -> color = RED;
                s -> right -> color = BLACK;
        }
        del_case6(n);
}
```

Case 6: S is Black, S's Right Child is Red, and N is the Left Child of its Parent P

In this case, a left rotation is done at `P` to make `S` the parent of `P` and `S`'s right child. After the rotation, the colors of `P` and `S` are interchanged and `S`'s right child is colored black. Once these steps are followed, you will observe that property 4 and property 5 remain valid. Case 6 is explained in Fig. 11.79.

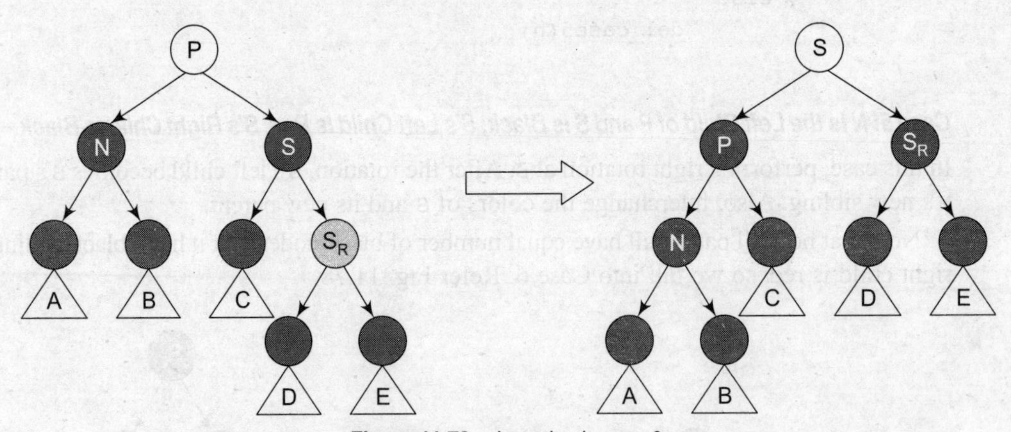

Figure 11.79 Insertion in case 6

The C code to fix Case 6 can be given as follows:

```
Void del_case6(struct node *n)
{
        struct node *s;
        s = sibling(n);
        s -> color = n -> parent -> color;
        n -> parent -> color = BLACK;
        if (n == n -> parent -> left) {
```

```
                        s->right->color = BLACK;
                        rotate_left(n->parent);
                } else {
                        s->left->color = BLACK;
                        rotate_right(n->parent);
                }
}
```

11.10 SPLAY TREE

Splay trees were invented by Daniel Sleator and Robert Tarjan. A splay tree is a self-balancing binary search tree with an additional property that recently accessed elements can be re-accessed fast. It is said to be an efficient binary tree because it performs basic operations such as insertion, search, and deletion operations in `O(log(n))` amortized time. For many non-uniform sequences of operations, splay trees perform better than other search trees, even when the specific pattern of the sequence is unknown.

Splay tree consists of a binary tree, with no additional fields. When a node in a splay tree is accessed, it is rotated or 'splayed' to the root, thereby changing the structure of the tree. Since the most frequently accessed node is always moved closer to the starting point of the search (or the root node), these nodes are therefore located faster. A simple idea behind it is that if an element is accessed, it is likely that it will be accessed again.

In a splay tree, operations such as insert, search, and delete are combined with one basic operation called *splaying*. Splaying the tree for a particular node rearranges the tree to place that node at the root. A technique to do this is to first perform a standard binary tree search for that node and then use rotations in a specific fashion to bring the node on top.

11.10.1 Advantages and Disadvantages

The advantages of using a splay tree are:

- A splay tree gives good performance for search, insert, and delete operations. This advantage centers on the fact that the splay tree is self-balancing, and a self- optimizing data structure in which the frequently accessed nodes are moved closer to the root so that they can be accessed quickly. This advantage is particularly useful for implementing caches and garbage collection algorithms.
- Splay trees are considerably simpler to implement than the other self-balancing binary search trees, such as red-black trees or AVL trees, while their average-case performance is just as efficient.
- Splay trees minimize memory requirements, as they do not store any book-keeping data.
- Unlike other types of self-balancing trees, splay trees provide good performance (amortized `O(log n)`) with nodes containing identical keys.

However, the negative side of a splay tree includes:

- While sequentially accessing all the nodes of a tree in a sorted order, the resultant tree becomes completely unbalanced. This takes n accesses of the tree in which each access takes `O(log n)` time. For example, re-accessing the first node triggers an operation that in turn takes `O(n)` operations to rebalance the tree before returning the first node. Although this creates a significant delay for that final operation, the amortized performance over the entire sequence is still `O(log n)`.

- For uniform access, the performance of a splay tree will be considerably worse than a somewhat balanced simple binary search tree. For uniform access, unlike splay trees, these other data structures provide worst-case time guarantees, and can be more efficient to use.

11.10.2 Operations on a Splay Tree

In this section, we will discuss the four main operations that are performed on a splay tree. These include splaying, insert, search, and delete operations.

Splaying

When we access a node N, splaying is performed on `N` to move it to the root. To perform a splay operation, certain *splay steps* are performed where each step moves `N` closer to the root. Splaying a particular node of interest after every access ensures that the recently accessed nodes are kept closer to the root and the tree remains roughly balanced, so that the desired amortized time bounds can be achieved.

Each splay step depends on three factors:

- Whether `N` is the left or right child of its parent `P`,
- Whether `P` is the root or not, and if not,
- Whether `P` is the left or right child of its parent, `G` (`N`'s grandparent).

Depending on these three factors, we have one splay step based on each factor.

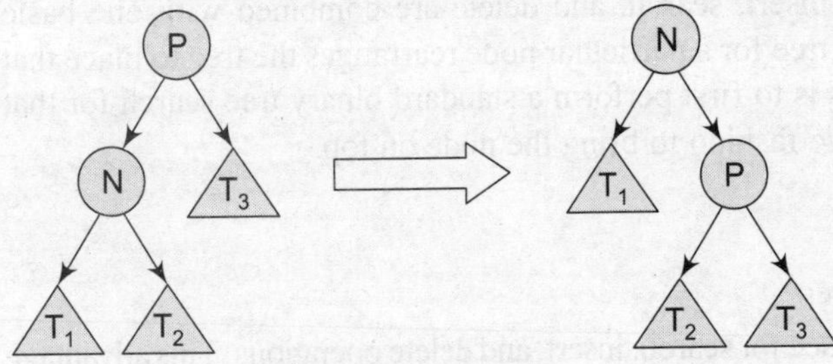

Figure 11.80 Zig step

Zig step The `zig` operation is done when `P` (the parent of `N`) is the root of the splay tree. In the `zig` step, the tree is rotated on the edge between `N` and `P`. `Zig` step is usually performed as the last step in a splay operation and only when `N` has an odd depth at the beginning of the operation. Refer Fig. 11.80.

Zig-zig step The `zig-zig` operation is performed when `P` is not the root. In addition to this, `N` and `P` are either both right or left children of their parents. Figure 11.81 shows the case where `N` and `P` are the left children. During the `zig-zig` step, first the tree is rotated on the edge joining `P` and its parent `G`, and then again rotated on the edge joining `N` and `P`.

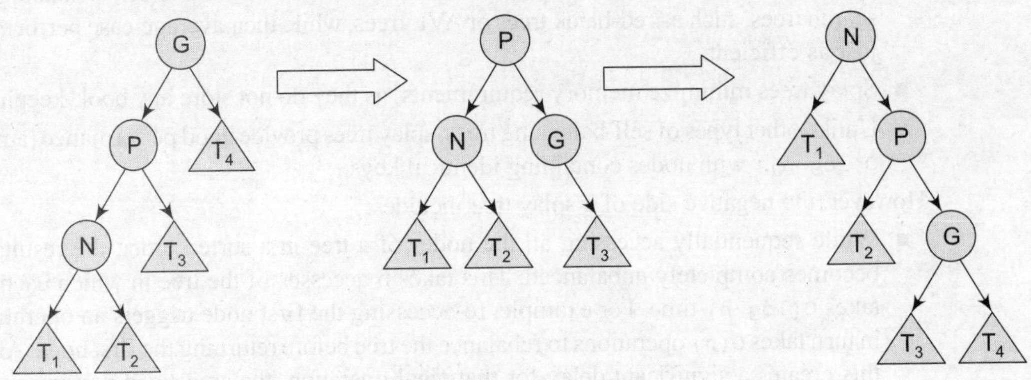

Figure 11.81 Zig-zig step

Zig-zag step The `zig-zag` operation is performed when `P` is not the root. In addition to this, `N` is a right child of `P` and `P` is a left child of `G` or vice versa. In `zig-zag` step, the tree is first rotated on the edge between `N` and `P`, and then rotated on the edge between `P` and `G`. Refer Fig. 11.82.

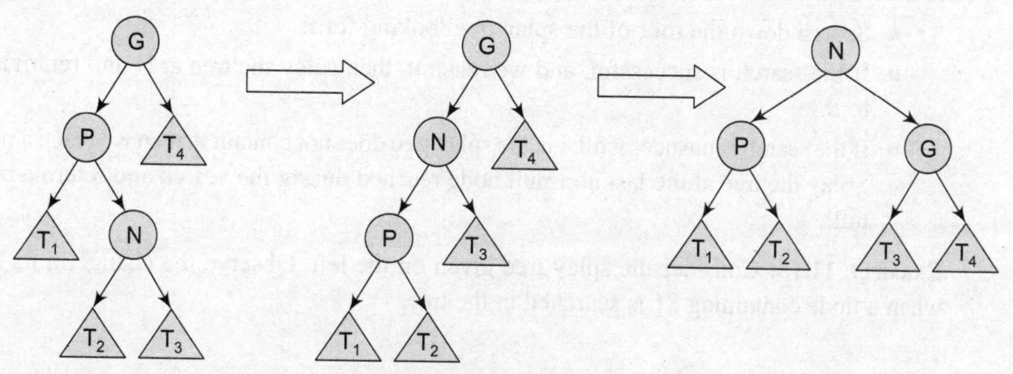

Figure 11.82 Zig-zag step

Insertion

Although the process of inserting a new node `N` into a splay tree begins in the same way as we insert a node in a binary search tree, but after the insertion, `N` is made the new root of the splay tree. The steps performed to insert a new node `N` in a splay tree can be given as follows:

Step 1 Search `N` in the splay tree. If the search is successful, splay at the node `N`.

Step 2 If the search is unsuccessful, add the new node `N` in such a way that it replaces the `NULL` pointer reached during the search by a pointer to a new node `N`. Splay the tree at `N`.

EXAMPLE 11.17: Consider the splay tree given on the left. Observe the change in its structure when 81 is added to it.

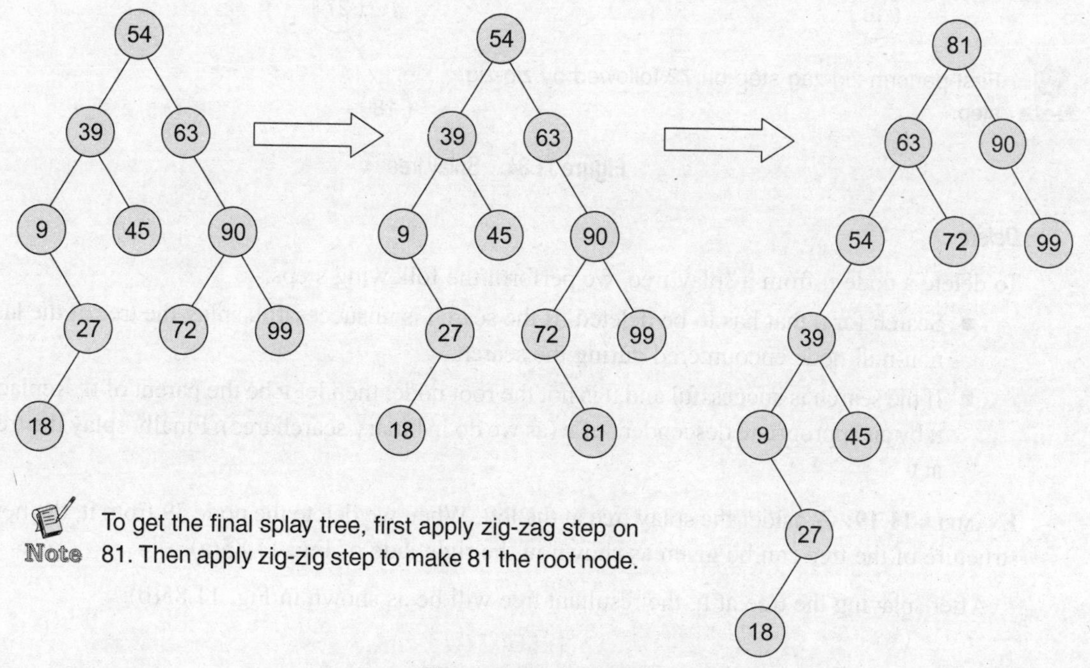

Note: To get the final splay tree, first apply zig-zag step on 81. Then apply zig-zig step to make 81 the root node.

Figure 11.83 Splay tree

Search

If a particular node N is present in the splay tree, then a pointer to N is returned; otherwise a pointer to the null node is returned. The steps performed to search a node N in a splay tree include:

- Search down the root of the splay tree looking for N.
- If the search is successful, and we reach N, then splay the tree at N and return a pointer to N.
- If the search is unsuccessful, i.e. the splay tree does not contain N, then we reach a null node. Splay the tree at the last non-null node reached during the search and return a pointer to null.

EXAMPLE 11.18: Consider the splay tree given on the left. Observe the change in its structure when a node containing 81 is searched in the tree.

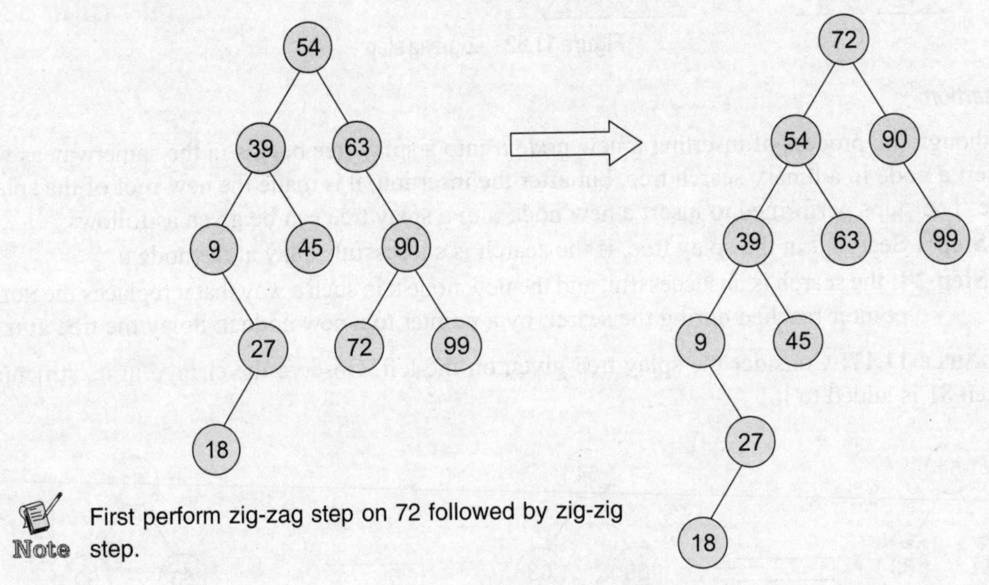

Figure 11.84 Splay tree

Delete

To delete a node N from a splay tree, we perform the following steps:

- Search for N that has to be deleted. If the search is unsuccessful, splay the tree at the last non-null node encountered during the search.
- If the search is successful and N is not the root node, then let P be the parent of N. Replace N by an appropriate descendent of P (as we do in binary search tree). Finally splay the tree at P.

EXAMPLE 11.19: Consider the splay tree at the left. When we delete the node 39 from it, the new structure of the tree can be given as shown in the right side of Fig. 11.85(a).

After splaying the tree at P, the resultant tree will be as shown in Fig. 11.85(b):

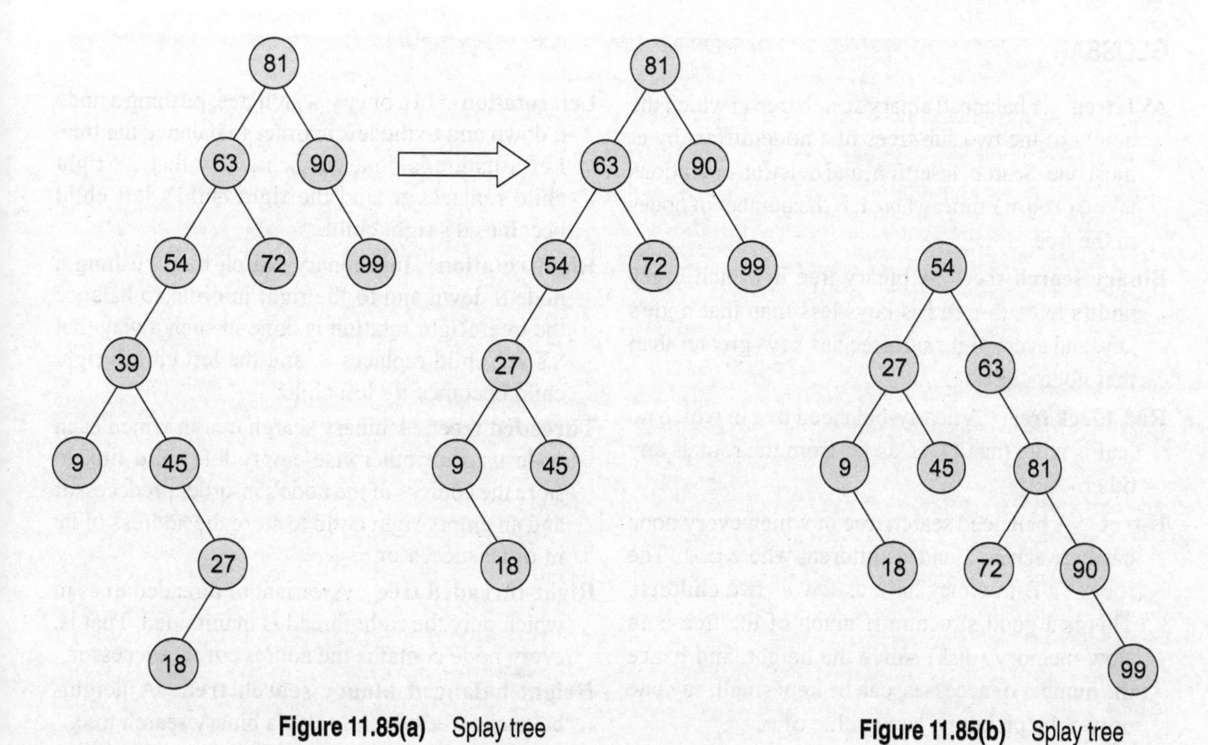

Figure 11.85(a) Splay tree

Figure 11.85(b) Splay tree

SUMMARY

- A binary search tree, also known as an ordered binary tree, is a variant of binary tree in which the nodes are arranged in order.
- The average running time of a search operation is `O(log₂n)`. However, in the worst case, a binary search tree will take `O(n)` time to search an element from the tree.
- Mirror image of a binary search tree is obtained by interchanging the left sub-tree with the right sub-tree at every node of the tree.
- A one-way threaded tree is also called a single threaded tree. In a two-way threaded tree, also called a double threaded tree, threads will appear in both the left and the right field of the node.
- An M-way search tree has `(M - 1)` values per node and `M` sub-trees. In such a tree, `M` is called the degree of the tree. M-way search tree consists of pointers to `M` sub-trees and contains `(M - 1)` keys, where `M` > 2.
- A B-tree of order `m` can have a maximum of `(m-1)` keys and `m` pointers to its sub-trees. A B-tree may contain a large number of key values and pointers to its sub-trees.
- A B+ tree is a variant of B-tree which stores sorted data in a way that allows for efficient insertion, retrieval, and removal of records, each of which is identified by a key. B+ tree record data at the leaf level of the tree; only keys are stored in interior nodes.
- A red-black tree is a self-balancing binary search tree that was invented in 1972 by Rudolf Bayer who called it as 'symmetric binary B-trees'. Although a red-black tree is complex, but it has good worst case running time for its operations and is efficient to use, as searching, insertion, and deletion can all be done in `O(log n)` time.
- Splay trees were invented by Daniel Sleator and Robert Tarjan. A splay tree is a self-balancing binary search tree with an additional property that recently accessed elements can be re-accessed fast.

GLOSSARY

AVL tree A balanced binary search tree in which the height of the two sub-trees of a node differs by at most one. Search, insertion, and deletion operations take `O(log n)` time, where `n` is the number of nodes in the tree.

Binary search tree A binary tree in which every node's left sub-tree has keys less than that node's key, and every right sub-tree has keys greater than that node's key.

Red-black tree A nearly-balanced tree in which no leaf is more than twice as far from the root as any other.

B-tree A balanced search tree in which every node has between `m/2` and `m` children, where `m>1`. The root of a B-tree may have as few as two children. This is a good structure if much of the tree is in slow memory (disk), since the height, and hence the number of accesses, can be kept small, say one or two, by picking a large value of `m`.

Multi-way tree A tree that has any number of children for each node.

B+ tree A variant of B-tree in which keys are stored in the leaves.

Left rotation In a binary search tree, pushing a node `N` down and to the left in order to balance the tree. Left rotation is done in such a way that `N`'s right child replaces `N`, and the right child's left child becomes `N`'s right child.

Right rotation In a binary search tree, pushing a node `N` down and to the right in order to balance the tree. Right rotation is done in such a way that `N`'s left child replaces `N`, and the left child's right child becomes `N`'s left child.

Threaded tree A binary search tree in which each node uses an otherwise-empty left child link to store the address of the node's in-order predecessor and an empty right child to store the address of its in-order successor.

Right-threaded tree A variant of threaded trees in which only the right thread is maintained. That is, every node contains the address of its successor.

Height-balanced binary search tree A height-balanced tree which is also a binary search tree.

Splay tree A binary search tree in which operations that access the nodes alter the structure of the tree.

Trie A tree for storing strings in which there is one node for every common prefix. In tries, strings are stored in extra leaf nodes.

EXERCISES

Review Questions

1. Explain the concept of binary search trees.
2. Explain the operations on binary search trees.
3. How does the height of a binary search tree effect its performance?
4. How many nodes will a complete binary tree with 27 nodes have in the last level? What will be the height of the tree?
5. Write a short note on threaded binary trees.
6. Why are threaded binary trees called efficient binary trees? Give the merits of using a threaded binary tree.
7. Discuss the advantages of an AVL tree.
8. How is an AVL tree better than a binary search tree?
9. Why is a large value of `m` needed in a B-tree?
10. Compare B-trees with B+ trees.
11. In what conditions will you prefer a B+ tree over a B-tree?
12. Write a short note on trie data structure.
13. Compare binary search trees with trie. Also, list the merits and demerits of using the trie data structure.
14. Compare hash tables with trie.
15. How does a red-black tree perform better than a binary search tree?
16. List the merits and demerits of a splay tree.
17. Create a binary search tree with the input given below:

    98, 2, 48, 12, 56, 32, 4, 67, 23, 87, 23, 55, 46

(a) Insert 21, 39, 45, 54, and 63 into the tree
(b) Delete values 23, 56, 2 and 45 from the tree

18. Consider the binary search tree given below.

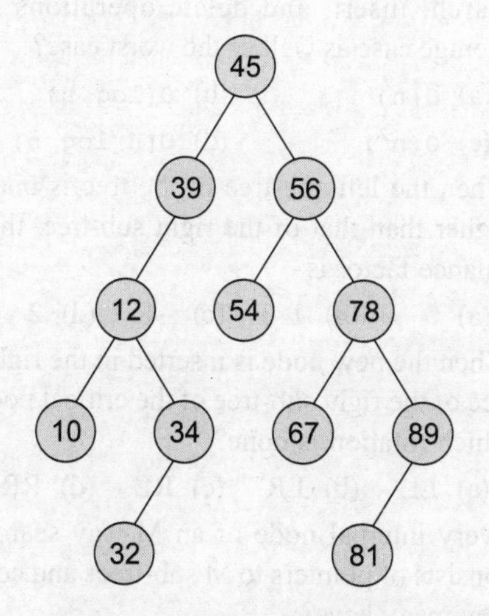

Now do the following operations:

- Find out the result of in-order, pre-order, and post-order traversals.
- Show the deletion of the root node
- Insert 11, 22, 33, 44, 55, 66, and 77 in the tree

19. Consider the AVL tree given below and insert 18, 81, 29, 15, 19, 25, 26, and 1 in it.
Delete nodes 39, 63, 15 and 1 from the AVL tree formed after solving the above question.

20. Give a brief summary of m-way search trees.

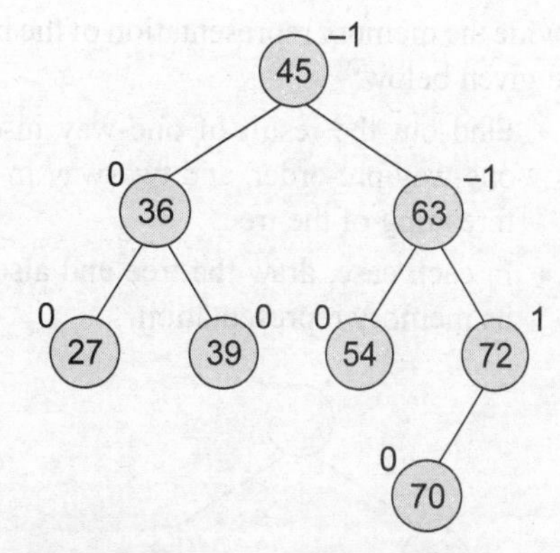

21. Consider the B-tree given below

(a) Insert 1, 5, 7, 11, 13, 15, 17, and 19 in the tree.
(b) Delete 30, 59 and 67 from the tree.

22. Write an essay on B+ trees.

23. Create a B+ tree of order 5 for the following data arriving in sequence:
90, 27, 7, 9, 18, 21, 3, 4, 16, 11, 21, 72

24. List down the applications of B-trees.

25. What do you understand by a trie data structure? How is it useful?

26. Discuss the properties of a red-black tree. Explain the insertion cases.

27. Explain splay trees in detail with relevant examples.

28. B-trees of order 2 are full binary trees. Justify this statement.

29. Consider the 3-way search tree given below. Insert 23, 45, 67, 87, 54, 32, and 11 in the tree. Then, delete 9, 36, and 54 from it.

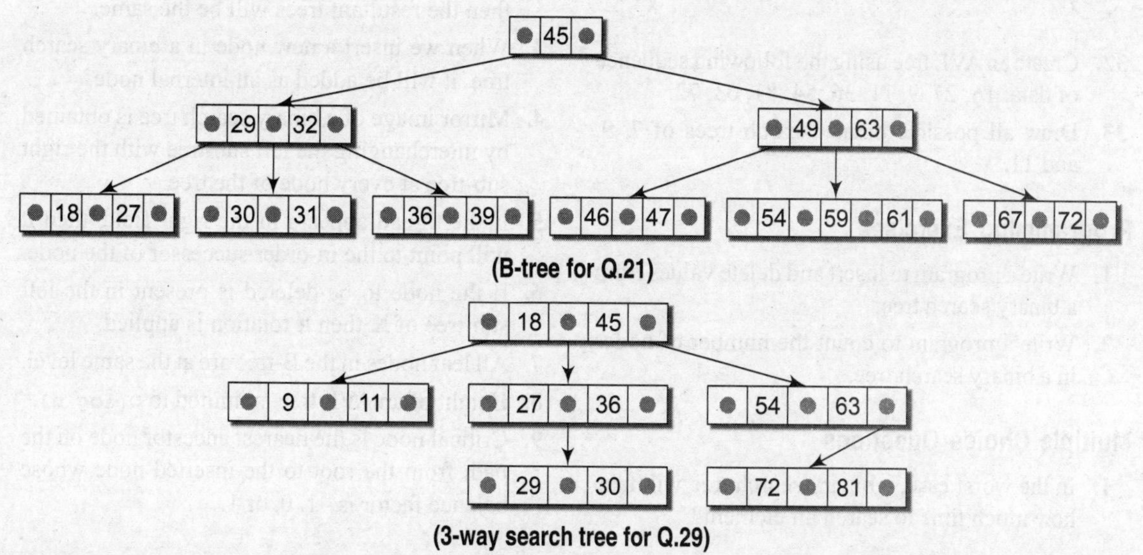

(B-tree for Q.21)

(3-way search tree for Q.29)

30. Provide the memory representation of the binary tree given below:
- Find out the result of one-way in-order, one-way pre-order, and two-way in-order threading of the tree.
- In each case, draw the tree and also give its memory representation.

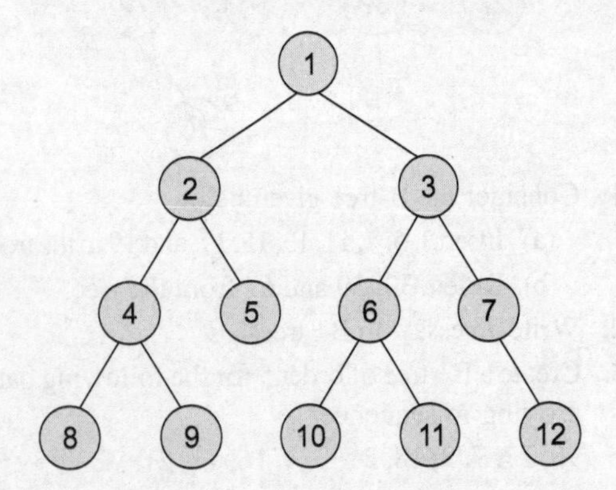

31. Balance the AVL trees given below.

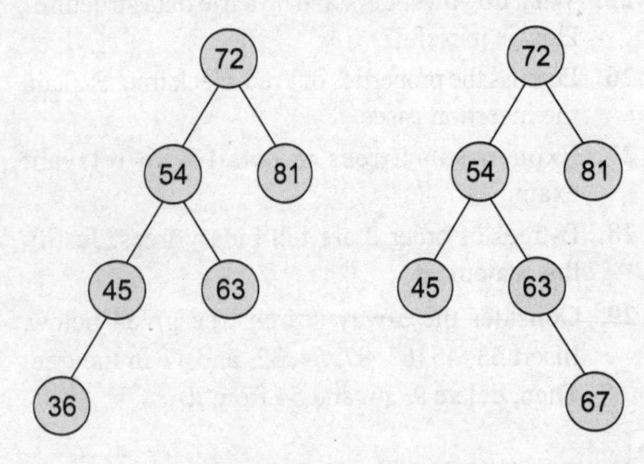

32. Create an AVL tree using the following sequence of data: 16, 27, 9, 11, 36, 54, 81, 63, 72.

33. Draw all possible binary search trees of 7, 9, and 11.

Programming Exercises

1. Write a program to insert and delete values from a binary search tree.
2. Write a program to count the number of nodes in a binary search tree.

Multiple Choice Questions

1. In the worst case, a binary search tree will take how much time to search an element?

   (a) `O(n)` (b) `O(log n)`
   (c) $O(n^2)$ (d) `O(n log n)`
2. How much time does an AVL tree take to perform search, insert, and delete operations in the average case as well as the worst case?

   (a) `O(n)` (b) `O(log n)`
   (c) $O(n^2)$ (d) `O(n log n)`
3. When the left sub-tree of the tree is one level higher than that of the right sub-tree, then the balance factor is

   (a) 0 (b) 1 (c) –1 (d) 2
4. When the new node is inserted in the right sub-tree of the right sub-tree of the critical noe, then which rotation is done?

   (a) LL (b) LR (c) RL (d) RR
5. Every internal node of an M-way search tree consists of pointers to M sub-trees and contains how many keys?

   (a) M (b) M–1 (c) 2 (d) M+1
6. Every node in a B-tree has at most ______ children.

   (a) M (b) M–1 (c) 2 (d) M+1

True or False

1. In a binary search tree, all the nodes in the left sub-tree have a value less than that of the root node.
2. If we take two empty binary search trees and insert the same elements but in a different order, then the resultant trees will be the same.
3. When we insert a new node in a binary search tree, it will be added as an internal node.
4. Mirror image of a binary search tree is obtained by interchanging the left sub-tree with the right sub-tree at every node of the tree.
5. If the thread appears in the right field, then it will point to the in-order successor of the node.
6. If the node to be deleted is present in the left sub-tree of `A`, then `R` rotation is applied.
7. All leaf nodes in the B-tree are at the same level.
8. Height of an AVL tree is limited to `O(log n)`.
9. Critical node is the nearest ancestor node on the path from the root to the inserted node whose balance factor is –1, 0, or 1.

10. RL rotation is done when the new node is inserted in the right sub-tree of the right sub-tree of the critical node.
11. All leaf nodes in a B-tree are at the same level.
12. A B+ tree stores data only in the i-nodes.

Fill in the Blanks

1. ______ is also called a fully threaded binary tree.
2. To find the node with the largest value, we will find the value of the rightmost node of the ______.
3. If the thread appears in the right field, then it will point to the ______ of the node.
4. The balance factor of a node is calculated by ______.
5. Balance factor –1 means ______.
6. An M-way search tree consists of pointers to ______ sub-trees and contains ______ keys.
7. Searching an AVL tree takes ______ time.
8. ______ rotation is done when the new node is inserted in the left sub-tree of the left sub-tree of the critical node.
9. In a red-black tree, the color of the root node is ______ and the color of leaf node is ______.
10. The zig operation is done when ______.

Annexure D
Program for Simple Right In-threaded Binary Tree

The following program shows how to create and traverse a simple right in-threaded binary tree.

```
#include<stdio.h>
#include<conio.h>
struct tree
{
    int val;
    struct tree *right;
    struct tree *left;
    int thread;
};
typedef struct tree *node;
node insert_node(node root, node ptr, node rt)
{
    if(root == NULL)
    {
        root = ptr;
        if(rt != NULL)
        {
            root->right = rt;
            root->thread = 1;
        }
    }
    else if(ptr->val < root->val)
        root->left=insert_node(root->left, ptr, root);
    else
        if(root->thread == 1)
        {
            root->right = insert_node(NULL, ptr, rt);
            root->thread=0;
        }
        else
            root->right=insert_node(root->right, ptr, rt);
    return root;
}
node create_threaded_tree()
{
    node root = NULL, ptr;
    int num;
```

```
    printf("\n Enter the elements, press -1 to terminate");
    scanf("%d", &num);
    while(num != -1)
    {
        ptr = (node)malloc(sizeof(struct node));
        ptr -> val = num;
        ptr -> left = ptr -> right = NULL;
        ptr -> thread = 0;
        root = insert_node(root, ptr, NULL);
        printf(" \n Enter the next element");
        scanf("%d", &num);
    }
    return root;
}
void inorder(node root)
{
    node ptr = root, prev;
    do
    {
        while(ptr != NULL)
        {
            prev = ptr;
            ptr = ptr -> left;
        }
        if(prev != NULL)
        {
            printf("% d", prev -> info);
            ptr = prev -> right;
            while(prev != NULL && prev -> thread)
            {
                printf("% d", ptr -> val);
                prev = ptr;
                ptr = ptr -> right;
            }
        }
    }while(ptr != NULL);
}
void main()
{
    node root;
    clrscr();
    root = create_threaded_tree();
    printf(" \n The in-order traversal of the tree can be given as");
    inorder(root);
    getch();
}
```

The following program shows how to perform insertion operation in an AVL tree.

```
#include <stdio.h>
typedef enum { FALSE ,TRUE } bool;
struct node
{
        int val;
        int balance;
        struct node *left_child;
        struct node *right_child;
};
struct node* search(struct node *ptr, int data)
{
       if(ptr!=NULL)
           if(data < ptr->val)
               ptr = search(ptr->left_child,data);
           else if( data > ptr->val)
               ptr = search(ptr->right_child, data);
       return(ptr);
}
struct node *insert (int data, struct node *ptr, int *ht_inc)
{
       struct node *aptr;
       struct node *bptr;
       if(ptr==NULL)
        {
           ptr = (struct node *) malloc(sizeof(struct node));
           ptr->val = data;
           ptr->left_child = NULL;
           ptr->right_child = NULL;
           ptr->balance = 0;
           *ht_inc = TRUE;
           return (ptr);
        }
        if(data < ptr->val)
        {
           ptr->left_child = insert(data, ptr->left_child, ht_inc);
           if(*ht_inc==TRUE)
           {
               switch(ptr->balance)
               {
                   case -1: /* Right heavy */
                          ptr->balance = 0;
                         *ht_inc = FALSE;
                          break;
```

```
                case 0: /* Balanced */
                        ptr->balance = 1;
                        break;
                case 1: /* Left heavy */
                        aptr = ptr->left_child;
                        if(aptr->balance == 1)
                        {
                              printf("Left to Left Rotation\n");
                              ptr->left_child= aptr->right_child;
                              aptr->right_child = ptr;
                              ptr->balance = 0;
                              aptr->balance=0;
                              ptr = aptr;
                        }
                        else
                        {
                              printf("Left to right rotation\n");
                              bptr = aptr->right_child;
                              aptr->right_child = bptr->left_child;
                              bptr->left_child = aptr;
                              ptr->left_child = bptr->right_child;
                              bptr->right_child = ptr;
                              if(bptr->balance == 1 )
                                    ptr->balance = -1;
                              else
                                    ptr->balance = 0;
                              if(bptr->balance == -1)
                                    aptr->balance = 1;
                              else
                                    aptr->balance = 0;
                              bptr->balance=0;
                              ptr = bptr;
                        }
                        *ht_inc = FALSE;
            }
        }
    }
    if(data > ptr->val)
    {
        ptr->right_child = insert(info, ptr->right_child, ht_inc);
        if(*ht_inc==TRUE)
        {
            switch(ptr->balance)
            {
              case 1: /* Left heavy */
                        ptr->balance = 0;
```

```
                    *ht_inc = FALSE;
                    break;
                case 0: /* Balanced */
                    ptr -> balance = -1;
                    break;
                case -1: /* Right heavy */
                    aptr = ptr -> right_child;
                    if(aptr -> balance == -1)
                    {
                        printf("Right to Right Rotation\n");
                        ptr -> right_child= aptr -> left_child;
                        aptr -> left_child = ptr;
                        ptr -> balance = 0;
                        aptr -> balance=0;
                        ptr = aptr;
                    }
                    else
                    {
                        printf("Right to Left Rotation\n");
                        bptr = aptr -> left_child;
                        aptr -> left_child = bptr -> right_child;
                        bptr -> right_child = aptr;
                        ptr -> right_child = bptr -> left_child;
                        bptr -> left_child = pptr;
                        if(bptr -> balance == -1)
                            ptr -> balance = 1;
                        else
                            ptr -> balance = 0;
                        if(bptr -> balance == 1)
                            aptr -> balance = -1;
                        else
                            aptr -> balance = 0;
                        bptr -> balance=0;
                        ptr = bptr;
                    }/*End of else*/
                    *ht_inc = FALSE;
            }
        }
    }
    return(ptr);
}
void display(struct node *ptr, int level)
{
    int i;
    if ( ptr!=NULL )
```

```
        {
            display(ptr->right_child, level+1);
            printf("\n");
            for (i = 0; i < level; i++)
                printf(" ");
            printf("%d", ptr->val);
            display(ptr->left_child, level+1);
        }
}
void inorder(struct node *ptr)
{
        if(ptr!=NULL)
        {
          inorder(ptr->left_child);
          printf("%d ",ptr->val);
          inorder(ptr->right_child);
        }
}
main()
{
   bool ht_inc;
   int data ;
   int option;
   struct node *root = (struct node *)malloc(sizeof(struct node));
   root = NULL;
   while(1)
   {
        printf("1.Insert\n");
        printf("2.Display\n");
        printf("3.Quit\n");
        printf("Enter your option : ");
        scanf("%d",&option);
        switch(choice)
            {
            case 1:
                    printf("Enter the value to be inserted : ");
                    scanf("%d", &data);
                          if( search(root,data) == NULL )
                                root = insert(data, root, &ht_inc);
                                else
                                printf("Duplicate value ignored\n");
                            break;
            case 2:
                    if(root==NULL)
                     {
```

```
                        printf("Tree is empty\n");
                        continue;
                    }
                        printf("Tree is :\n");
                        display(root, 1);
                        printf("\n\n");
                        printf("Inorder Traversal is: ");
                        inorder(root);
                        printf("\n");
                        break;
            case 3:
                    exit(1);
                    default:
                        printf("Wrong option\n");
                    }
            }
    }
```

12 Heaps

Learning Objective

A heap is a specialized tree-based data structure. There are several variants of heaps which are the prototypical implementations of priority queues. We have already discussed priority queues in Chapter 9. Heaps are also crucial in several efficient graph algorithms.

In this chapter, we will discuss three types of heaps. They include binary heap, binomial heap, and Fibonacci heap.

12.1 BINARY HEAPS

A *binary heap* is defined as a complete binary tree with elements at every node being either less than or equal to the element at its left and the right child. Binary heaps were first introduced by Williams in 1964. It has the following properties:

- Since a heap is defined as a complete binary tree, all its elements can be stored sequentially in an array. It follows the same rules as that of a complete binary tree. That is, if an element is at position `i` in the array, then its left child is stored at position `2i` and its right child at position `(2i + 1)`. Conversely, an element at position `i` have its position stored at position i/2.
- Being a complete binary tree, all the levels of the tree except the last level are completely filled.
- The height of a binary tree is given as $\log_2 n$, where `n` is the number of elements.
- Heaps (also known as partially ordered trees) are a very popular data structure for implementing priority queues.

A binary heap is a useful data structure in which elements can be added randomly but only that element with the highest value is removed in case of max heap and lowest (value in case of mean heap). A binary tree is better, but a binary heap is more space efficient and simpler.

There are two types of binary heaps: *min heap* and *max heap*. Figure 12.1 shows a binary min heap and a binary max heap.

- In *min heap*, the elements at every node will either be less than or equal to the element at its left and right child.
- In *max heap*, the elements at every node will either be greater than or equal to the element at its left and right child.

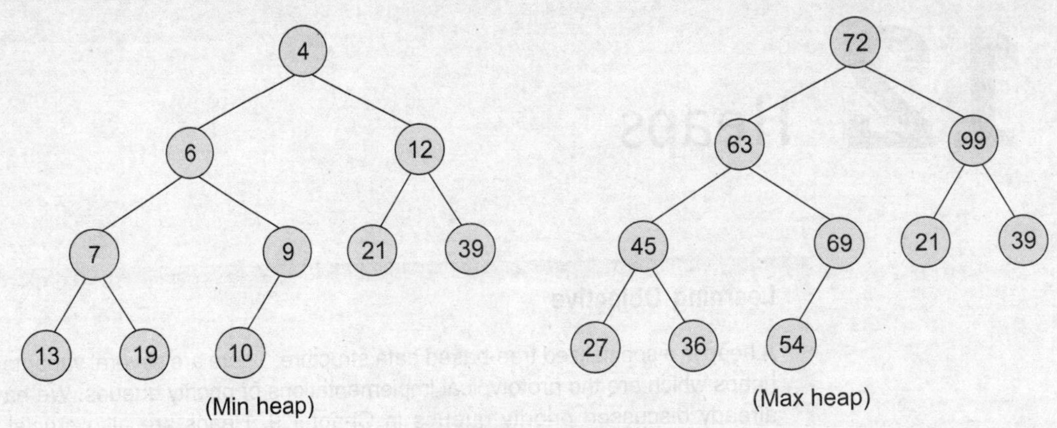

Figure 12.1 Binary heaps

12.1.1 Insertion in a Binary Heap

Consider a max heap H with n elements. Inserting a new value into the heap is done in the following two steps:

1. Add the new value at the bottom of H in such a way that H is still a complete binary tree but not necessarily a heap.
2. Let the new value rise to its appropriate place in H so that H now becomes a heap as well.

To do this, compare the new value with its parent to check if they are in the correct order. If they are, then the procedure halts, else the new value and its parent's value are swapped and Step 2 is repeated.

Example 12.1: Consider the heap given in Fig. 12.2 and insert 99 in it.

The first step says, insert the element in the heap so that the heap is a complete binary tree. So, insert the new value as the right child of node 27 in the heap. This is illustrated in Fig. 12.3.

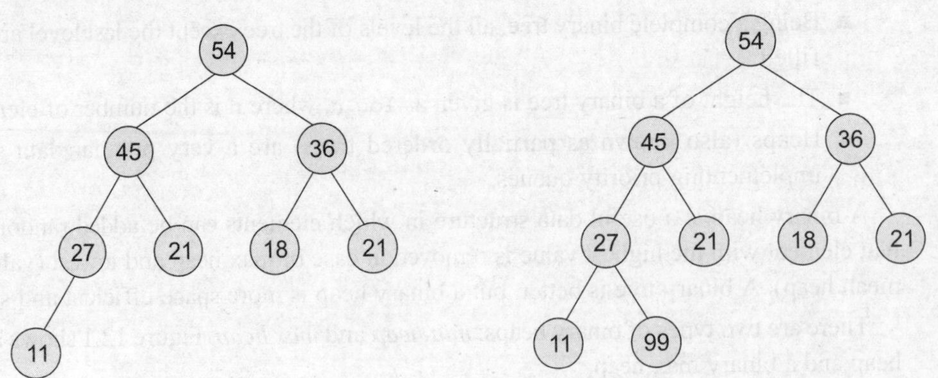

Figure 12.2 Binary heap

Figure 12.3 Binary heap after insertion of 27

Now, as per the second step, let the new value rise to its appropriate place in H so that H becomes a heap as well. Compare 99 with its parent node value. If it is less than its parent's value, then the new node is in its appropriate place and H has become a heap. If the new value is greater than that of its parent's node, then replace the two values. Repeat the whole process until H becomes a heap. This is illustrated in Fig. 12.4.

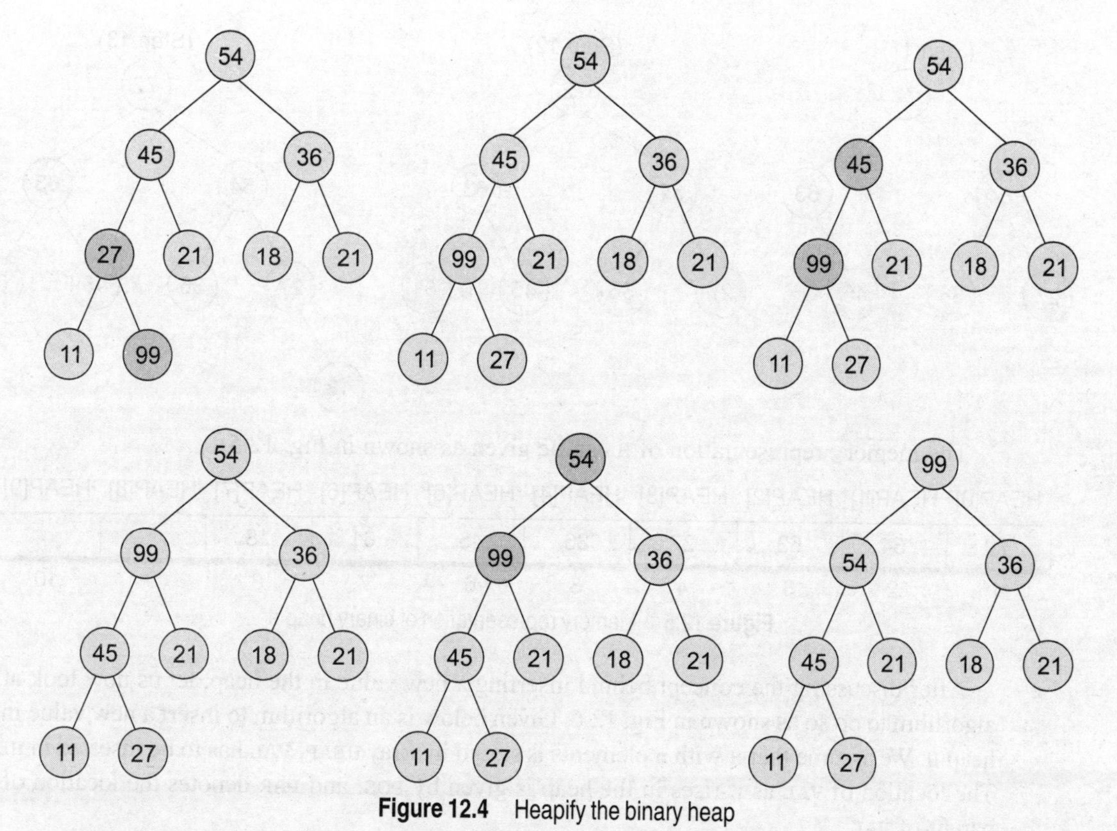

Figure 12.4 Heapify the binary heap

EXAMPLE 12.2: Build a heap H from the given set of numbers: 45, 36, 54, 27, 63, 72, 61, and 18. Also, draw the memory representation of the heap.

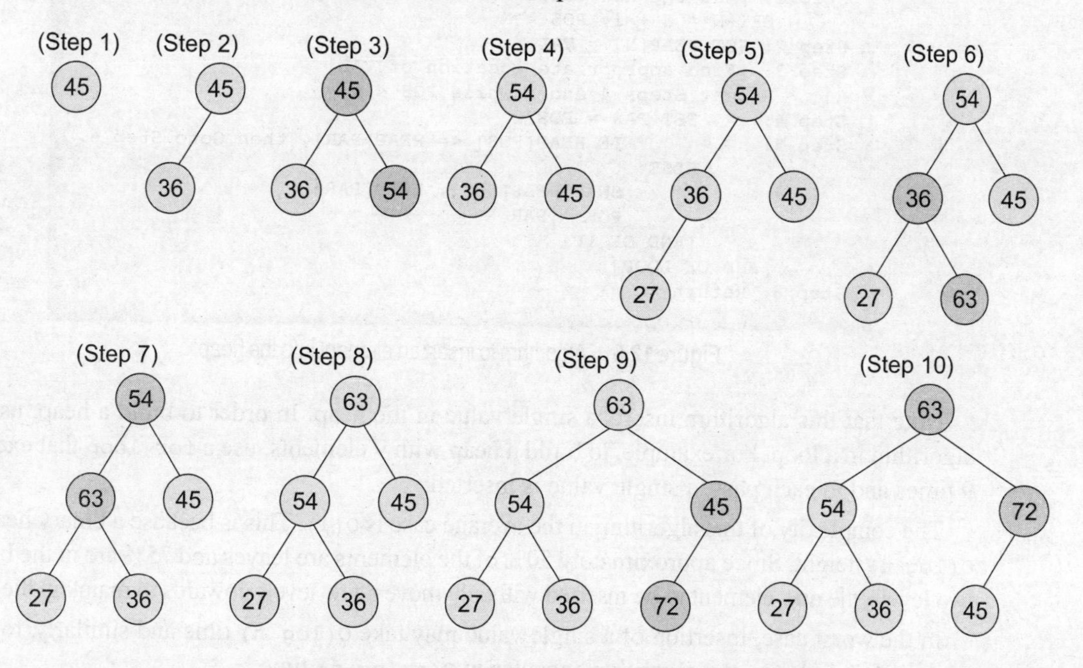

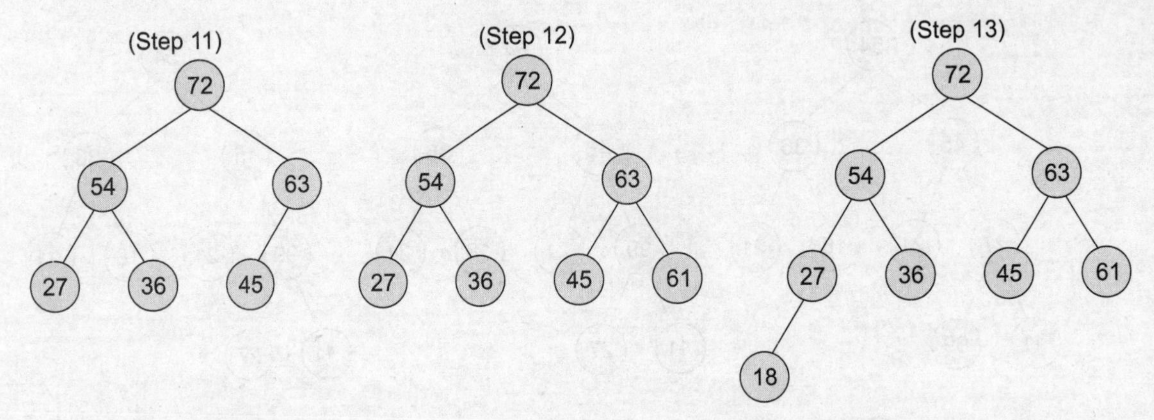

The memory representation of H can be given as shown in Fig. 12.5.

HEAP[0]	HEAP[1]	HEAP[2]	HEAP[3]	HEAP[4]	HEAP[5]	HEAP[6]	HEAP[7]	HEAP[8]	HEAP[9]
72	54	63	27	36	45	61	18		
1	2	3	4	5	6	7	8	9	10

Figure 12.5 Memory representation of binary heap H

After discussing the concept behind inserting a new value in the heap, let us now look at the algorithm to do so as shown in Fig. 12.6. Given below is an algorithm to insert a new value in the heap H. We assume that H with n elements is stored in array HEAP. VAL has to be inserted in HEAP. The location of VAL as it rises in the heap is given by POS, and PAR denotes the location of the parent of VAL.

```
Step 1: [Add the new value and set its POS]
        SET N = N + 1, POS = N
Step 2: SET HEAP[N] = VAL
Step 3: [Find appropriate location of VAL]
        Repeat Steps 4 and 5 while POS < 0
Step 4:         SET PAR = POS/2
Step 5:                 IF HEAP[POS] <= HEAP[PAR], then Goto Step 6.
                ELSE
                        SWAP HEAP[POS], HEAP[PAR]
                        POS = PAR
                [END OF IF]
        [END OF LOOP]
Step 6: Return
```

Figure 12.6 Algorithm to insert an element into the heap

Note that this algorithm inserts a single value in the heap. In order to build a heap, use this algorithm in a loop. For example, to build a heap with 9 elements, use a `for loop` that executes 9 times and in each pass, a single value is inserted.

The complexity of this algorithm in the average case is `O(1)`. This is because a binary heap has `O(log n)` height. Since approximately 50% of the elements are leaves and 75% are in the bottom two levels, the new element to be inserted will only move a few levels upwards to maintain the heap.

In the worst case, insertion of a single value may take `O(log n)` time and similarly, to build a heap of n elements, the algorithm executes in `O(n log n)` time.

12.1.2 Deleting an Element from a Binary Heap

Consider a max heap H having n elements. An element is always deleted from the root of the heap. So, deleting an element from the heap is done in the following two steps:

1. Replace the root node's value with the last node's value so that H is still a complete binary tree but not necessarily a heap.
2. Delete the last node.
3. Sink down the new root node's value so that H satisfies the heap property. In this step, interchange the root node's value with its child node's value (whichever is largest among its children).

Example 12.3: Consider the given heap H and delete the root node's value.

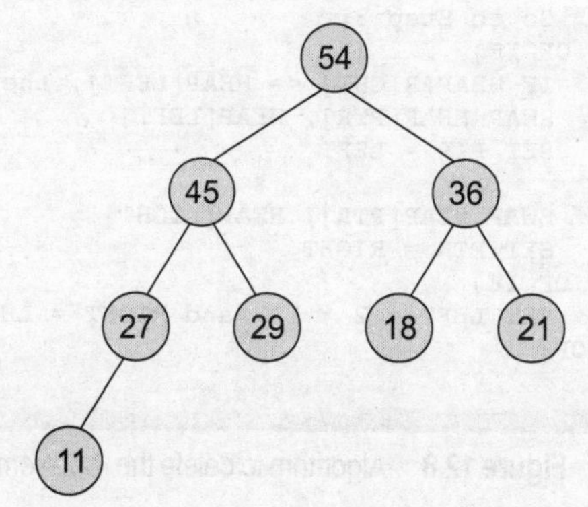

Here, the value of root node = 54 and the value of the last node = 11.
So, replace 45 with 11 and delete the last node.

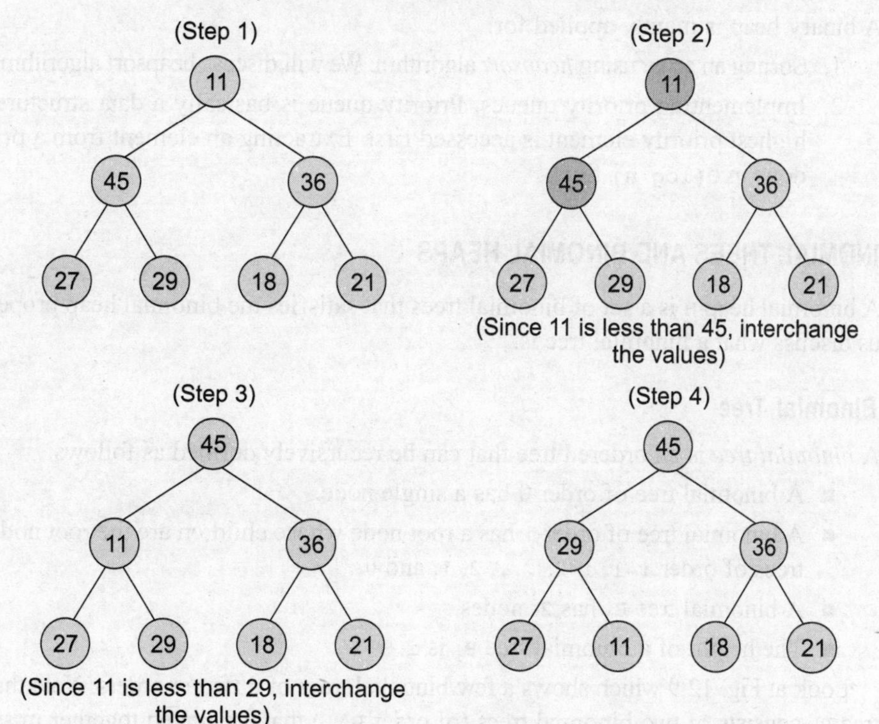

Figure 12.7 Binary heap

After discussing the concept behind deleting the root element from the heap, let us look at the algorithm given in Fig. 12.8. We assume that heap `H` with `n` elements is stored using a sequential array called `HEAP`. `LAST` is the position of the last element in the heap and `PTR`, `LEFT`, and `RIGHT` denote the position of `LAST` and its left and right children respectively as it moves down the heap.

```
Step 1: [ Remove the last node from the heap]
      SET LAST = HEAP[N], SET N = N - 1
Step 2: [Initialization]
      SET PTR = 0, LEFT = 1, RIGHT = 2
Step 3: SET HEAP[PTR] = LAST
Step 4: Repeat Steps 5 to 7 while LEFT <= N
Step 5:            IF HEAP{PTR] >= HEAP[LEFT] AND HEAP[PTR] >= HEAP[RIGHT], then
                   Go to Step 8
             [END OF IF]
Step 6:            IF HEAP[RIGHT] <= HEAP[LEFT], then
                   SWAP HEAP[PTR], HEAP[LEFT]
                   SET PTR = LEFT
             ELSE
                   SWAP HEAP[PTR], HEAP[RIGHT]
                   SET PTR = RIGHT
             [END OF IF]
Step 7:            SET LEFT = 2 * PTR and RIGHT = LEFT + 1
        [END OF LOOP]
Step 8: Return
```

Figure 12.8 Algorithm to delete the root element from heap

12.1.3 Applications of Binary Heap

A binary heap is mainly applied for:

1. Sorting an array using *heapsort* algorithm. We will discuss heapsort algorithm in Chapter 12.
2. Implementing priority queues. Priority queue is basically a data structure in which the highest priority element is accessed first. Extracting an element from a priority queue is done in `O(log n)` time.

12.2 BINOMIAL TREES AND BINOMIAL HEAPS

A binomial heap `H` is a set of binomial trees that satisfies the binomial heap properties. First, let us discuss what a binomial tree is.

12.2.1 Binomial Tree

A *binomial tree* is an ordered tree that can be recursively defined as follows:

- A binomial tree of order 0 has a single node.
- A binomial tree of order `i` has a root node whose children are the root nodes of binomial trees of order `i-1, i-2, ..., 2, 1,` and `0`.
- A binomial tree B_i has 2^i nodes.
- The height of a binomial tree B_i is i.

Look at Fig. 12.9 which shows a few binomial trees of different orders. Note that the binomial tree B_i consists of two binomial trees (of order B_{i-1}) that are linked together in such a way that the root of one is the leftmost child of the root of another.

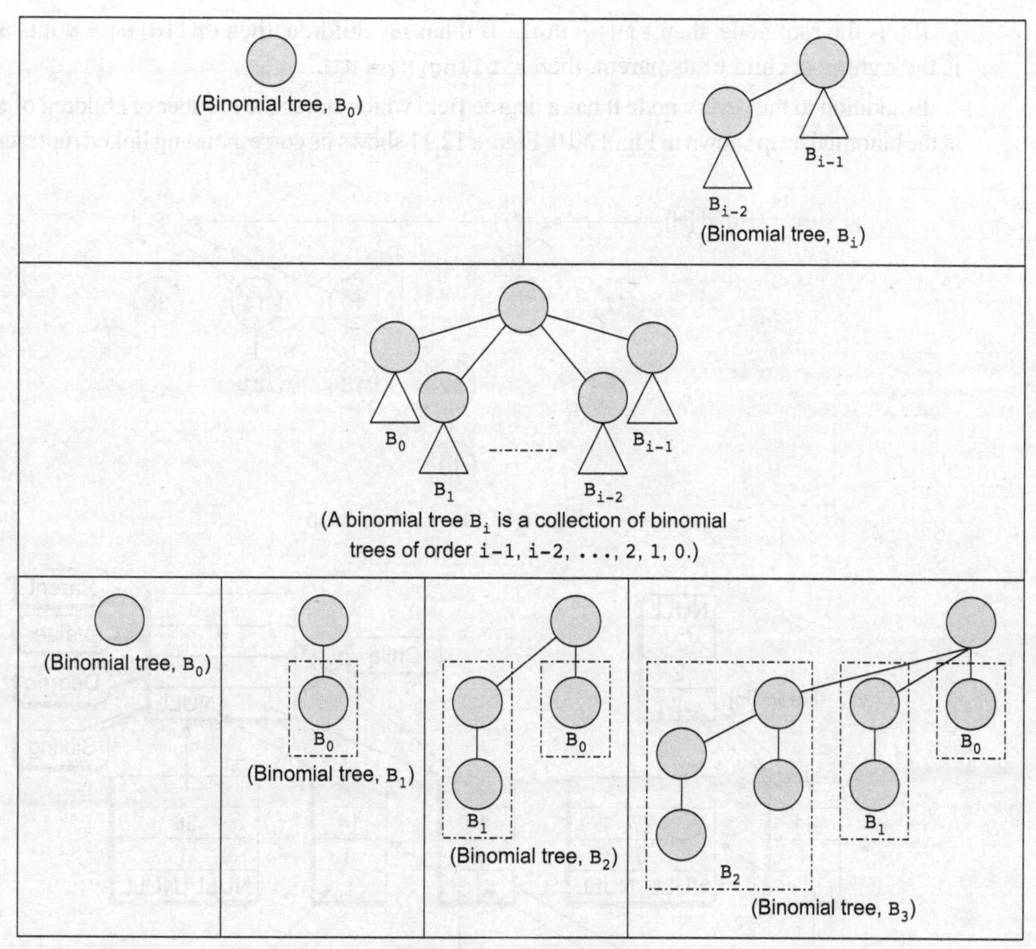

Figure 12.9 Binomial trees

12.2.2 Binomial Heap

A *binomial heap* `H` is a collection of binomial trees that satisfies the following properties:

- Every binomial tree in `H` satisfies the minimum heap property (according to which the key of a node is either greater than or equal to the key of its parent).
- There can be one or zero binomial trees for each order including zero order.

According to the first property, the root of a heap-ordered tree contains the smallest key in the tree. The second property, on the other hand, implies that a binomial heap H having N nodes contains at most `log (N + 1)` binomial trees.

Linked Representation of Binomial Heaps in Memory

Each node in a binomial heap `H` has a `val` field that stores its value. In addition, each node `N` has certain pointers such as:

- `P[N]` that points to the parent of `N`,
- `Child[N]` that points to the leftmost child, and
- `Sibling[N]` that points to the sibling of `N` which is immediately to its right.

If `N` is the root node, then `P[N] = NULL`. If `N` has no children, then `Child[N] = NULL`, and if `N` is the rightmost child of its parent, then `sibling[N] = NIL`.

In addition to this, every node `N` has a degree field which stores the number of children of `N`. Look at the binomial heap shown in Fig. 12.10. Figure 12.11 shows its corresponding linked representation.

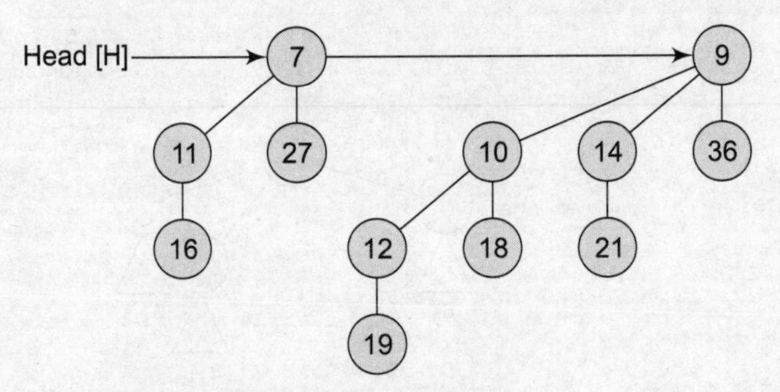

Figure 12.10 Binomial heap

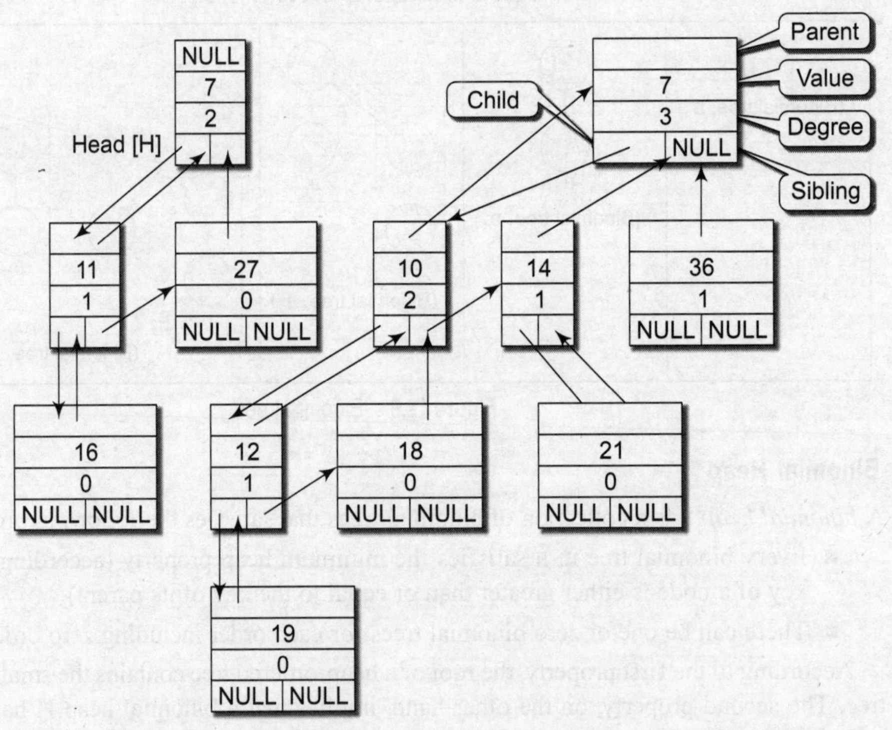

Figure 12.11 Linked representation of the binomial tree shown in Fig. 12.9

12.2.3 Operations on Binomial Heaps

In this section, we will discuss the different operations that are performed on binomial heaps.

Create a New Binomial Heap

The procedure `Create_Binomial-Heap()` allocates and returns an object `H`, where `Head[H]` is set to `NULL`. The running time of this procedure can be given as `O(1)`.

Finding the Minimum Key

The procedure `Min_Binomial-Heap()` returns a pointer to the node which has the minimum value in the binomial heap `H`. The algorithm for `Min_Binomial-Heap()` is shown in Fig. 12.12.

We have already discussed that a binomial heap is heap-ordered, therefore, the node with the minimum value in a particular binomial tree will appear as the root node in the binomial heap. Thus, the `Min_ Binomial-Heap()` procedure checks all roots. Since there are at most `lg (n + 1)` roots to check, the running time of this procedure is `O(lg n)`.

```
Min_Binomial-Heap()

Step 1: INITIALIZE SET y = NULL, X = Head[H] and Min = ∞
Step 2: REPEAT Steps 3 and 4 While X ≠ NULL
Step 3:       IF Val[X] < Min, then
                    SET Min = Val[X]
                    SET Y = X
              [END OF IF]
Step 4:       SET y = Sibling[X]
Step 5: RETURN y
```

Figure 12.12 Algorithm to find the node with minimum value

Example 12.4: Consider the binomial heap given below and see how the procedure works in this case.

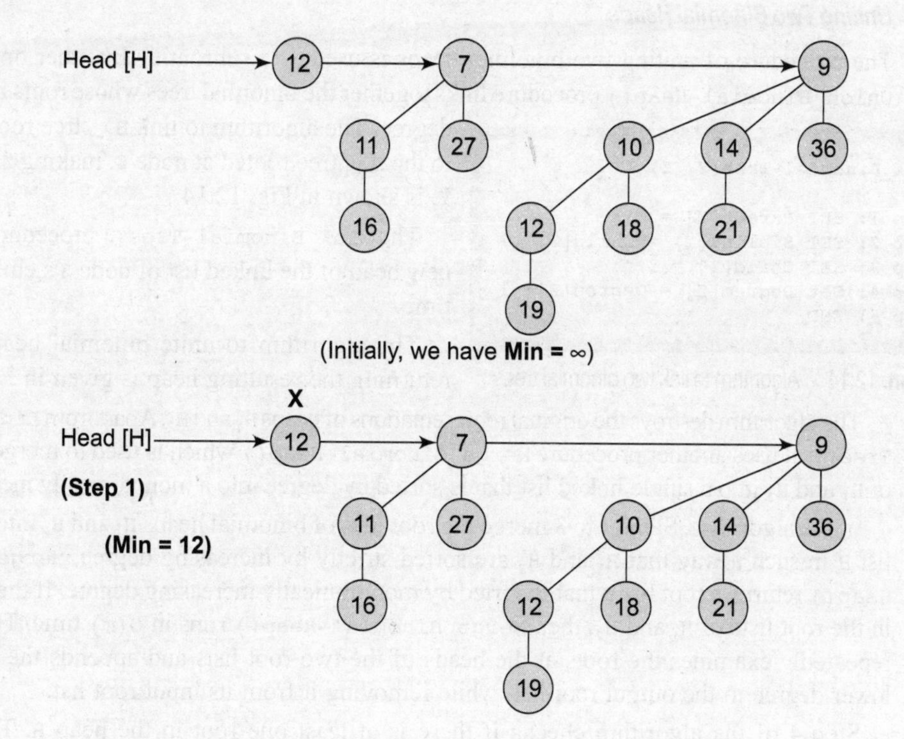

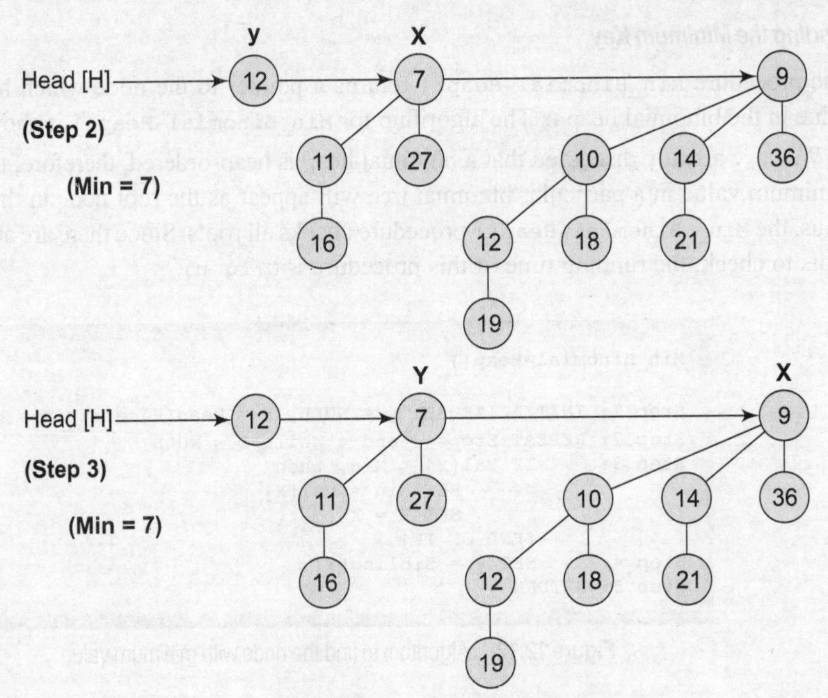

Figure 12.13 Binomial heap

Uniting Two Binomial Heaps

The procedure of uniting two binomial heaps is used as a subroutine by other operations. The `Union_Binomial-Heap()` procedure links together the binomial trees whose roots have the same degree. The algorithm to link B_{i-1} tree rooted at node `Z` to the B_{i-1} tree rooted at node `Z`, making `Z` the parent of `Y`, is shown in Fig. 12.14.

```
Link_Binomial-Tree(Y, Z)

Step 1: SET Parent[Y] = Z
Step 2: SET Sibling[Y} = Child[Z]
Step 3: SET Child[Z] = Y
Step 4: Set Degree[Z] = Degree[Z]+ 1
Step 5: END
```

Figure 12.14 Algorithm to link two binomial trees

The `Link_Binomial-Tree()` procedure makes `Y` the new head of the linked list of node `Z`'s children in `O(1)` time.

The algorithm to unite binomial heaps H_1 and H_2 returning the resulting heap is given in Fig. 12.15.

The algorithm destroys the original representations of heaps H_1 and H_2. Apart from `Link_Binomial-Tree()`, it uses another procedure `Merge_Binomial-Heap()` which is used to merge the root lists of H_1 and H_2 into a single linked list that is sorted by degree into a monotonically increasing order.

In the algorithm, Steps 1 to 3 merge the root lists of binomial heaps H_1 and H_2 into a single root list `H` in such a way that H_1 and H_2 are sorted strictly by increasing degree. `Merge_Binomial-Heap()` returns a root list `H` that is sorted by monotonically increasing degree. If there are `m` roots in the root lists of H_1 and H_2, then `Merge_Binomial-Heap()` runs in `O(m)` time. This procedure repeatedly examines the roots at the heads of the two root lists and appends the root with the lower degree to the output root list, while removing it from its input root list.

Step 4 of the algorithm checks if there is at least one root in the heap `H`. The algorithm proceeds only if `H` has at least one root. In Step 5, we initialize three pointers: `PTR` which points to the root that is currently being examined, `PREV` which points to the root preceding `x` on the root list, and `NEXT` which points to the root following `x` on the root list.

```
Union_Binomial-Heap(H1, H2)

Step 1: SET H = Create_Binomial-Heap()
Step 2: SET Head[H] = Merge_Binomial-Heap()
Step 3: Free the memory occupied by H1 and H2
Step 4: If Head[H] = NULL, then RETURN
Step 5: SET PREV = NULL, PTR = Head[H] and NEXT =
        Sibling[PTR]
Step 6: Repeat Step 7 while NEXT ≠ NULL
Step 7:     IF Degree[PTR] ≠ Degree[NEXT] OR
            (Sibling[NEXT] ≠ NULL AND
            Degree[Sibling[NEXT] = Degree[PTR]), then
                  SET PREV = PTR, PTR = NEXT
            ELSE IF Val[PTR] ≤ Val[NEXT], then
                  SET Sibling[PTR] = Sibling[NEXT]
                  Link_Binomial-Tree(NEXT, PTR)
                  Else
                        IF PREV = NULL, then
                           Head[H] = NEXT
                        ELSE
                           Sibling[PREV] = NEXT
                           Link_Binomial-Tree(PTR, NEXT)
                           SET PTR = NEXT
            SET NEXT = Sibling[PTR]
Step 8: RETURN H
```

Figure 12.15 Algorithm to unite two binomial heaps

In Step 6, we have a while loop in which at each iteration, we decide whether to link PTR to NEXT or NEXT to PTR depending on their degrees and possibly the degree of sibling[NEXT].

If Step 7, we check for two conditions. First, if degree[PTR] ≠ degree[NEXT], that is, when PTR is the root of a B_i tree and NEXT is the root of a B_j tree for some $(j > i)$, then PTR and NEXT are not linked to each other, but we move the pointers one position further down the list. Second, we check if PTR is the first of three roots of equal degree, that is,

degree[PTR] = degree[NEXT] = degree[sibling[NEXT]]

In this case also, we just move the pointers one position further down the list by writing PREV = PTR, PTR = NEXT.

However, if the above IF conditions do not satisfy, then the case that pops up is that PTR is the first of two roots of equal degree, that is,

degree[PTR] = degree[NEXT] ≠ degree[sibling[NEXT]]

In this case, we link either PTR with NEXT or NEXT with PTR depending on whichever has the smaller key. Of course, the node with the smaller key will be the root after the two nodes are linked.

The running time of Union_Binomial-Heap() can be given as $O(\lg n)$, where n is the total number of nodes in binomial heaps H_1 and H_2. If H_1 contains n_1 nodes and H_2 contains n_2 nodes, then H_1 contains at most $\lg (n_1 + 1)$ roots and H_2 contains at most $\lg (n_2 + 1)$ roots, so H contains at most $(\lg n_2 + \lg n_1 + 2) \le (2 \lg n + 2) = O(\lg n)$ roots when we call Merge_Binomial-Heap(). Since, $n = n_1 + n_2$, the Merge_Binomial-Heap() takes $O(\lg n)$ to execute. Each iteration of the while loop takes $O(1)$ time, and because there are at most $(\lg n_1 + \lg n_2 + 2)$ iterations, the total time is thus $O(\lg n)$.

Example 12.5: Unite the binomial heaps given below.

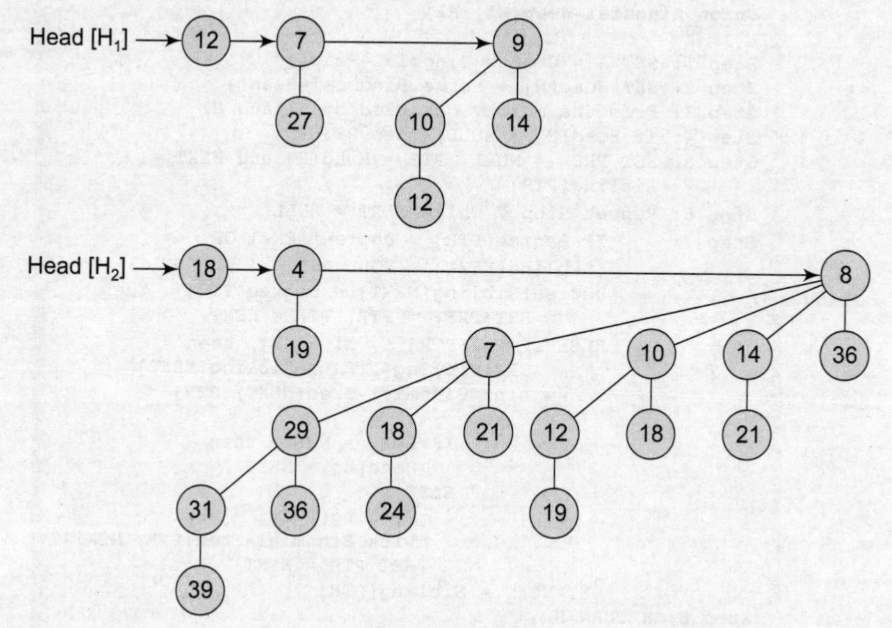

After `Merge_Binomial-Heap()`, the resultant heap can be given as follows:

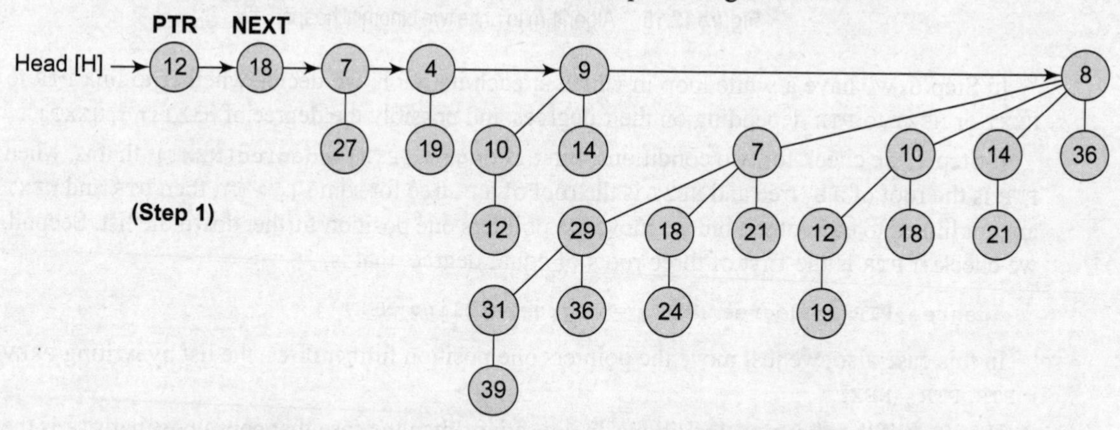

Link `NEXT` to `PTR`, making `PTR` the parent of the node pointed by `NEXT`.

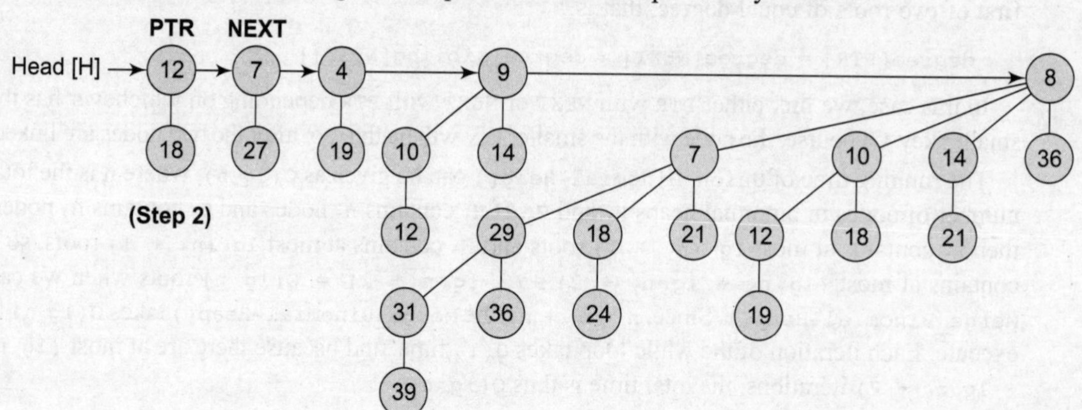

Now PTR is the first of the three roots of equal degree, that is, `degree[PTR] = degree[NEXT] = degree[sibling[NEXT]]`. Therefore, move the pointers one position further down the list by writing `PREV = PTR`, `PTR = NEXT`, and `NEXT = sibling[PTR]`.

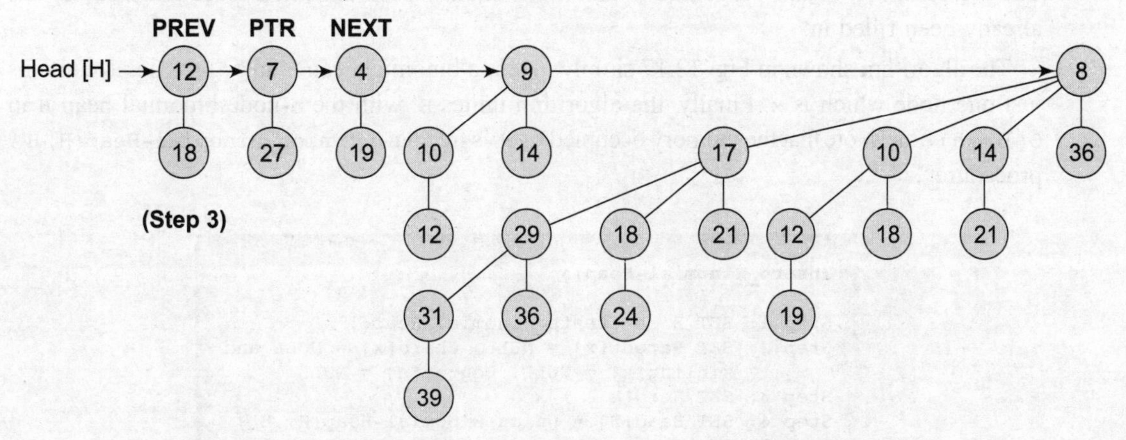

Link PTR to NEXT, making NEXT the parent of the node pointed by PTR.

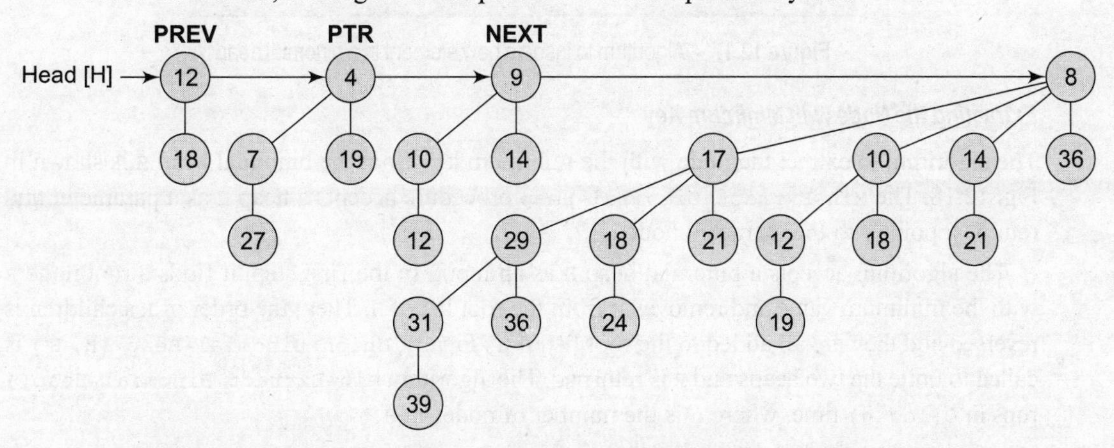

Link NEXT to PTR, making PTR the parent of the node pointed by NEXT.

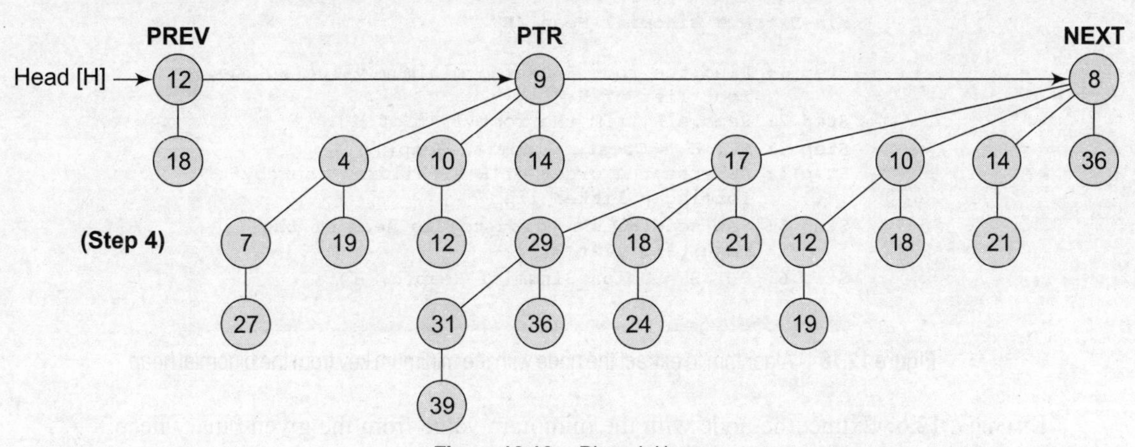

Figure 12.16 Binomial heap

Inserting a New Node

The `Insert_Binomial-Heap()` procedure is used to insert a node `x` into the binomial heap H. The pre-condition of this procedure is that `x` has already been allocated space and `val[x]` has already been filled in.

The algorithm shown in Fig. 12.17 simply makes a binomial heap `H′` in `O(1)` time. `H′` contains just one node which is `x`. Finally, the algorithm unites `H′` with the `n`-node binomial heap `H` in `O(log n)` time. Note that the memory occupied by `H′` is freed in the `Union_Binomial-Heap(H, H′)` procedure.

```
Insert_Binomial-Heap()

Step 1: SET H' = Create_Binomial-Heap()
Step 2: SET Parent[x] = NULL, Child[x] = NULL and
        sibling[x] = NULL, Degree[x] = NULL
Step 3: SET Head[H'] = x
Step 4: SET Head[H] = Union_Binomial-Heap(H, H')
Step 5: END
```

Figure 12.17 Algorithm to insert a new element in a binomial heap

Extracting the Node with Minimum Key

The algorithm to extract the node with the minimum key from the binomial heap `H` is shown in Fig. 12.18. The `Min-Extract_Binomial-Heap` procedure accepts a heap `H` as a parameter and returns a pointer to the extracted node.

The algorithm accepts a binomial heap `H` as an input. In the first step, it finds a root node `R` with the minimum value and removes it from the root list of `H`. Then, the order of `R`'s children is reversed and they are all added to the root list of `H′`. Finally, `Union_Binomial-Heap (H, H′)` is called to unite the two heaps and `R` is returned. The algorithm `Min-Extract_Binomial-Heap()` runs in `O(log n)` time, where `n` is the number of nodes in `H`.

```
Min-Extract_Binomial Heap (H)

Step 1: Find the root R having minimum value in the
        root list of H
Step 2: Remove R from the root list of H
Step 3: SET H′ = Create_Binomial-Heap()
Step 4: Reverse the order of R's children thereby
        forming a linked list
Step 5: Set head[H′] to point to the head of the
        resulting list
Step 6: SET H = Union_Binomial-Heap(H, H′)
```

Figure 12.18 Algorithm to extract the node with the minimum key from the binomial heap

EXAMPLE 12.6: Extract the node with the minimum value from the given binary heap.

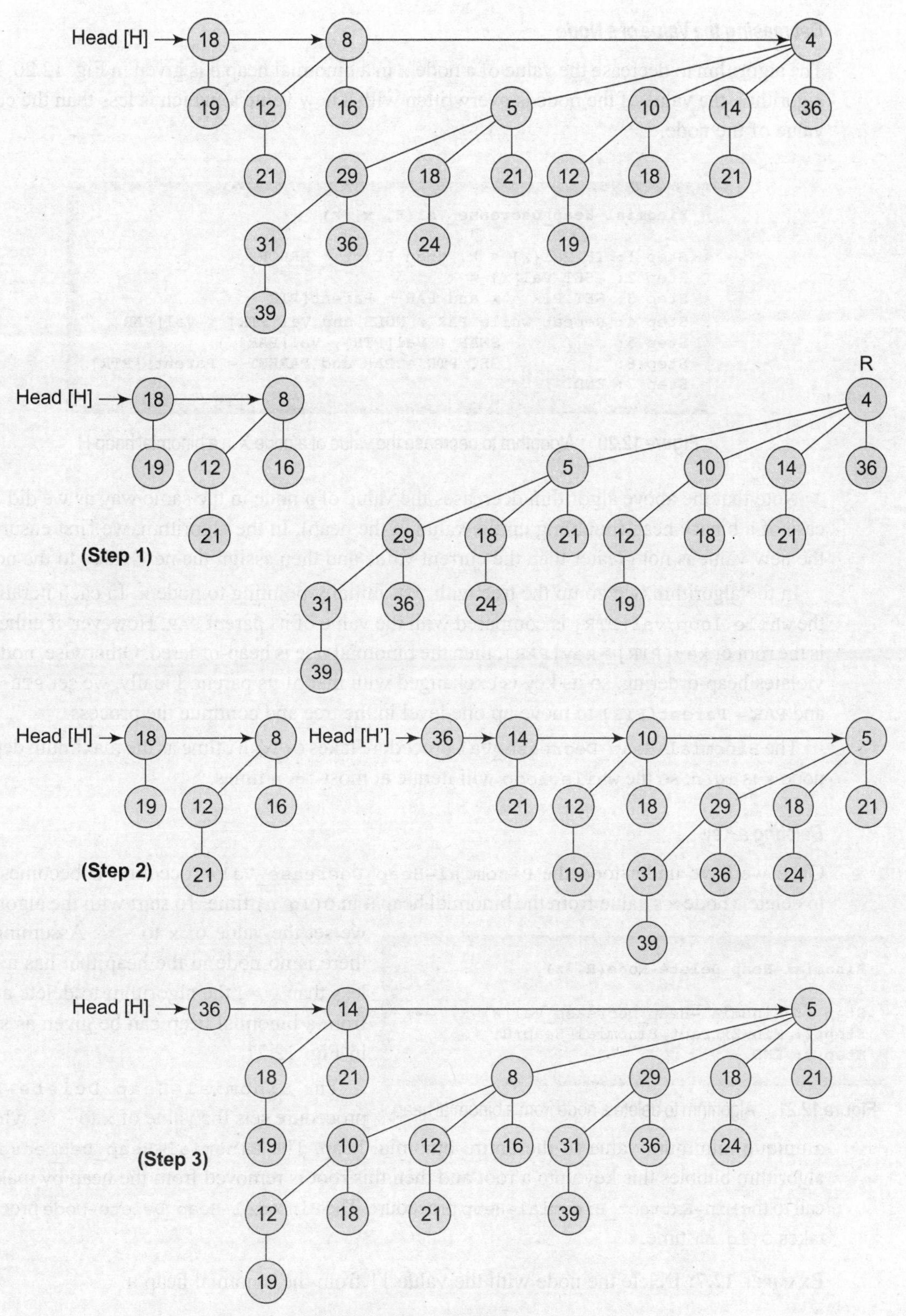

Figure 12.19 Binomial heap

Decreasing the Value of a Node

The algorithm to decrease the value of a node x in a binomial heap H is given in Fig. 12.20. In the algorithm, the value of the node is overwritten with a new value k, which is less than the current value of the node.

```
Binomial-Heap_Decrease_Val(H, x, k)

Step 1: IF Val{x] < k, then Print " ERROR"
Step 2:  SET Val[x] = k
Step 3: SET PTR = x and PAR = Parent[PTR]
Step 4: Repeat while PAR ≠ NULL and Val[PTR] < Val[PAR]
Step 5:           SWAP ( Val[PTR}, Val[PAR] )
Step 6:           SET PTR = PAR and PARENT = Parent [PTR]
Step 7: END
```

Figure 12.20 Algorithm to decrease the value of a node X in a binomial heap H

Note that the above algorithm decreases the value of a node in the same way as we did in the case of a binary heap (bubbling up the value in the heap). In the algorithm, we first ensure that the new value is not greater than the current value and then assign the new value to the node.

In the algorithm, we go up the tree with PTR initially pointing to node x. In each iteration of the while loop, val[PTR] is compared with the value of its parent PAR. However, if either PTR is the root or key[PTR] ≥ key[PAR], then the binomial tree is heap-ordered. Otherwise, node PTR violates heap-ordering, so its key is exchanged with that of its parent. Finally, we set PTR = PAR and PAR = Parent[PTR] to move up one level in the tree and continue the process.

The Binomial-Heap_Decrease_Val procedure takes O(lg n) time as the maximum depth of node x is lg n, so the while loop will iterate at most lg n times.

Deleting a Key

Once we have understood the Binomial-Heap_Decrease_Val procedure, it becomes easy to delete a node x's value from the binomial heap H in O(lg n) time. To start with the algorithm, we set the value of x to – ∞. Assuming that there is no node in the heap that has a value less than – ∞, the algorithm to delete a node from a binomial heap can be given as shown in Fig. 12.21.

```
Binomial-Heap_Delete-Node(H, x)

Step 1: Binomial-Heap_Decrease_Val(H, x, -∞)
Step 2: Min-Extract_Binomial-Heap(H)
Step 3: END
```

Figure 12.21 Algorithm to delete a node from a binomial heap

The Binomial-Heap_Delete-Node procedure sets the value of x to – ∞, which is a unique minimum value in the entire binomial heap. The Binomial-Heap_Decrease_Val algorithm bubbles this key upto a root and then this root is removed from the heap by making a call to the Min-Extract_Binomial-Heap procedure. The Binomial-Heap_Delete-Node procedure takes O(lg n) time.

EXAMPLE 12.7: Delete the node with the value 11 from the binomial heap H.

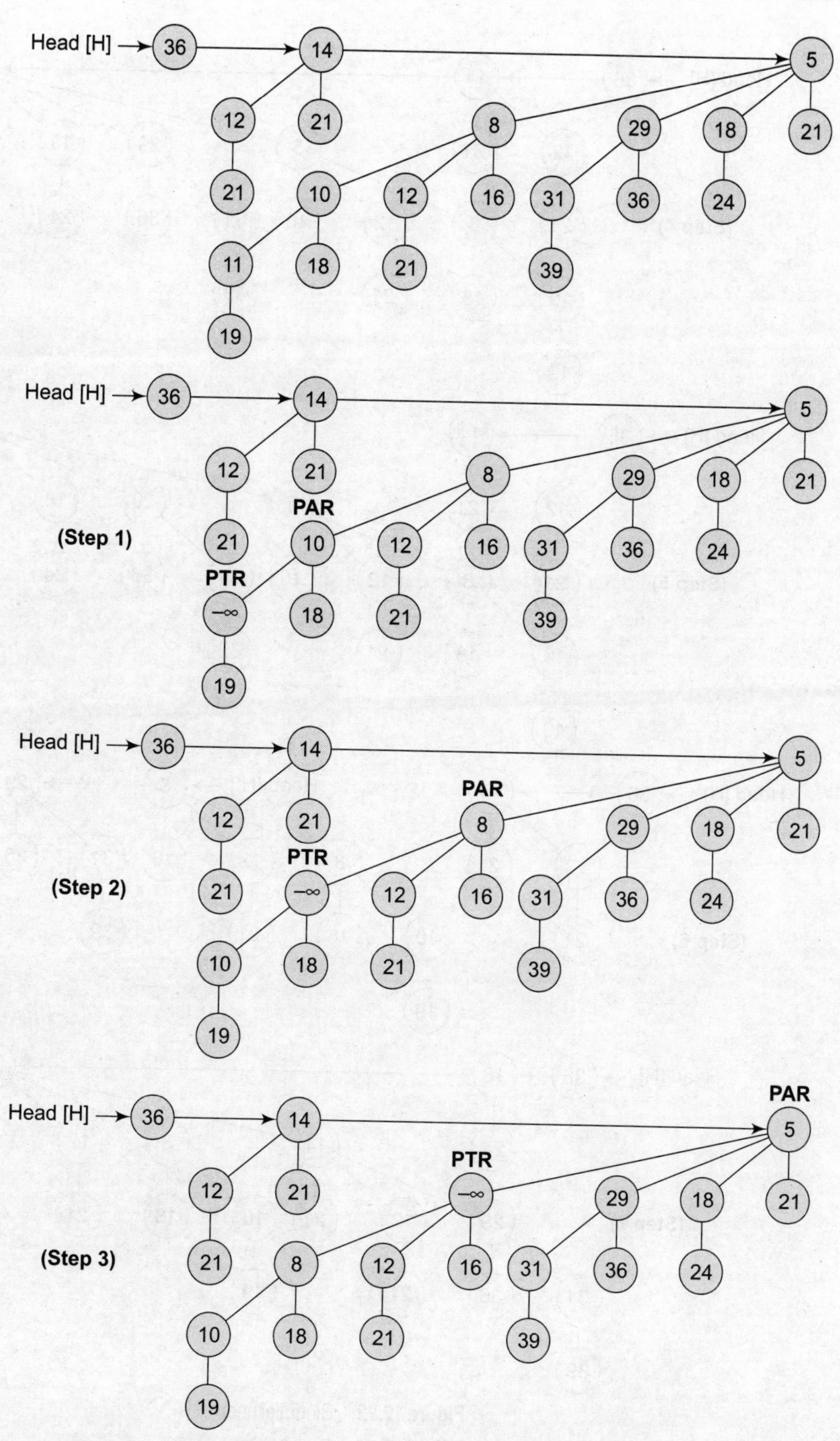
Head [H]
36
14
5
12
21
8
29
18
21
21
10
12
16
31
36
24
11
18
21
39
19
Head [H]
PAR
(Step 1)
PTR
−∞
Head [H]
PAR
PTR
(Step 2)
−∞
Head [H]
PAR
PTR
(Step 3)
−∞

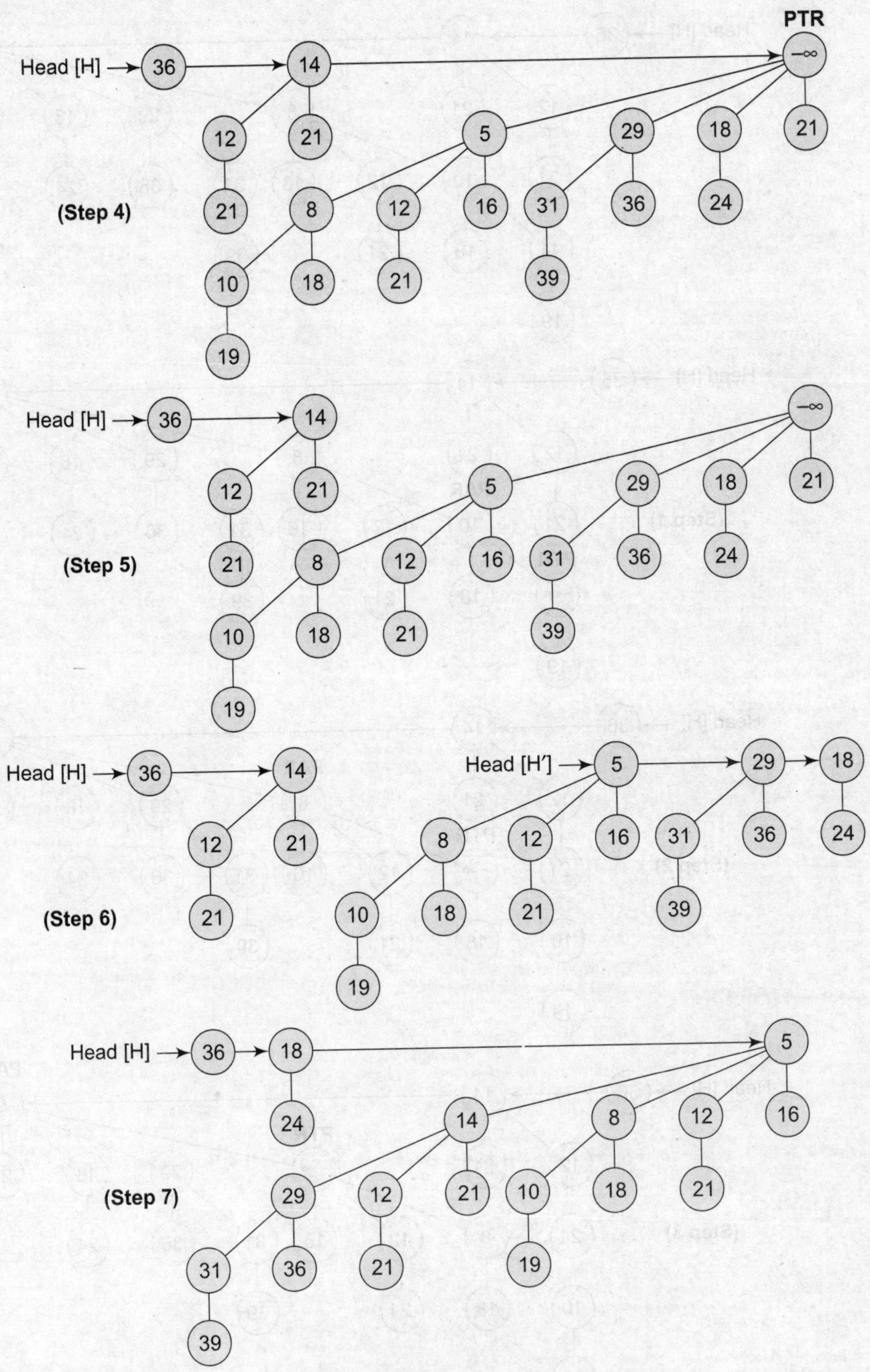

Figure 12.22 Binomial heap

12.3 FIBONACCI HEAPS

In the last section, we have seen that binomial heaps support operations such as insert, extract-minimum, decrease-value, delete, and union in `O(log n)` worst-case time. In this section, we will discuss Fibonacci heaps which support the same operations but have the advantage that operations that do not involve deleting an element run in `O(1)` amortized time. So, theoretically, Fibonacci heaps are especially desirable when the number of extract-minimum and delete operations are small relative to the number of other operations performed. This situation arises in many applications, for example, some algorithms for graph problems may call the decrease-value once per edge. However, the programming complexity of Fibonacci heaps make them less desirable to use.

A Fibonacci heap is a collection of trees. It is loosely based on binomial heaps. If neither the decrease-value nor the delete operation is performed, each tree in the heap is like a binomial tree. Fibonacci heaps differ from binomial heaps, as they have a more relaxed structure, allowing improved asymptotic time bounds.

12.3.1 Structure of Fibonacci Heaps

Although a Fibonacci heap is a collection of heap-ordered trees, the trees in a Fibonacci heap are not constrained to be binomial trees. That is, while the trees in a binomial heap are ordered, those within Fibonacci heaps are rooted but unordered.

Look at the Fibonacci heap given in Fig. 12.23. The figure shows that each node in the Fibonacci heap contains the following pointers:

- a pointer to its parent, and
- a pointer to any one of its children.

Note that the children of each node are linked together in a circular doubly linked list which is known as the child list of that node. Each child `x` in a child list contains pointers to its left and right sibling. If node `x` is the only child of its parent, then `left[x] = right[x] = x` (Refer Fig. 12.24).

Circular doubly linked lists provide an added advantage, as they allow a node to be removed in `O(1)` time. Also, given two circular doubly linked lists, the lists can be concatenated to form one list in `O(1)` time.

Apart from this information, every node will store two other fields. First, the number of children in the child list of node `x` is stored in `degree[x]`. Second, a boolean value `mark[x]` indicates whether node `x` has lost a child since the last time `x` was made the child of another node. Of course, the newly created nodes are unmarked. Also, when the node `x` is made the child of another node, it becomes unmarked.

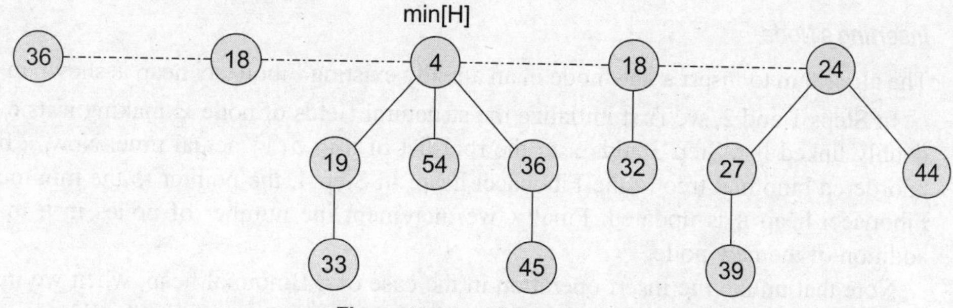

Figure 12.23 Fibonacci heap

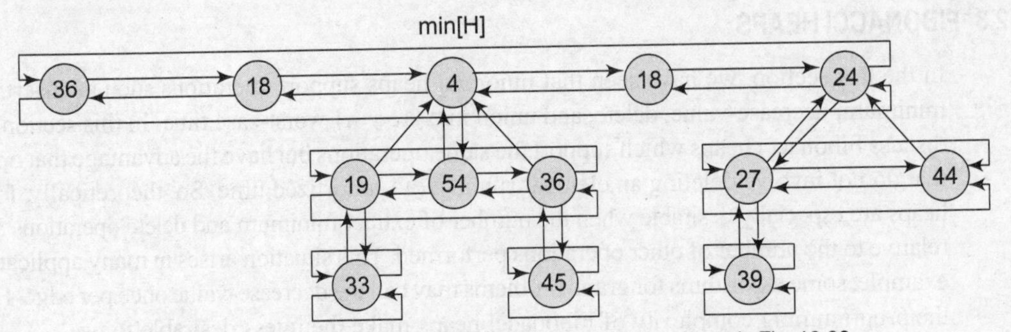

Figure 12.24 Linked representation of fibonacci heap shown in Fig. 12.23

Fibonacci heap `H` is generally accessed by a pointer called `min[H]` which points to the root that has a minimum value. If the Fibonacci heap `H` his empty, then `min[H] = NULL`.

As seen in Fig. 12.24, roots of all the trees in a Fibonacci heap are linked together using their left and right pointers into a circular doubly linked list called the *root list* of the Fibonacci heap. Also note that the order of the trees within a root list is arbitrary.

In a Fibonacci heap `H`, the number of nodes in `H` are stored in `n[H]` and the degree of nodes are stored in `D(n)`.

12.3.2 Fibonacci Heap Operations

In this section, we will discuss the operations as implemented in Fibonacci heaps. If we perform operations such as create-heap, insert, minimum, extract-minimum, and union, then each Fibonacci heap is simply a collection of unordered binomial trees. An *unordered binomial tree* is like a binomial tree which is defined recursively. The unordered binomial tree U_0 consists of a single node, and an unordered binomial tree U_i consists of two unordered binomial trees U_{i-1} for which the root of one is made into any child of the root of another. All the properties of a binomial tree also holds for unordered binomial trees but for an unordered binomial tree U_i, the root has degree `i`, which is greater than that of any other node. The children of the root are roots of sub-trees $U_0, U_1, \ldots, U_{i-1}$ in some order. Thus, if an n-node Fibonacci in a heap is a collection of unordered binomial trees, then `D(n) = lg n`. The main principle of operations on Fibonacci heaps is to delay the work as long as possible.

Creating a New Fibonacci Heap

To create an empty Fibonacci heap, the `Create_Fib-Heap` procedure is used to allocate and return the Fibonacci heap object `H`, where `n[H] = 0` and `min[H] = NULL`. The amortized cost of `Create_Fib-Heap` is equal to its `O(1)` actual cost.

Inserting a Node

The algorithm to insert a new node in an already existing Fibonacci heap is shown in Fig. 12.25.

In Steps 1 and 2, we first initialize the structural fields of node `x`, making it its own circular doubly linked list. Step 3 adds `x` to the root list of `H` in `O(1)` actual time. Now, `x` becomes an unordered binomial tree in the Fibonacci heap. In Step 4, the pointer to the minimum node of Fibonacci heap `H` is updated. Finally, we increment the number of nodes in `H` to reflect the addition of the new node.

Note that unlike the insert operation in the case of a Binomial heap, when we insert a node in a Fibonacci heap, no attempt is made to consolidate the trees within the Fibonacci heap. So,

```
Insert_Fib-Heap(H, x)

Step 1: [Initialize] SET Degree[x] = 0, Parent[x] = Null,
        Child[x] = NULL, mark[x] = False
Step 2: SET Left[x] = x and Right[x] = x
Step 3: Concatenate the root list containing x with the
        root list of H
Step 4: IF min[H] = NULL OR Val[x] < Val[min[H]], then
             SET min[H] = x
        [END OF IF]
Step 5: SET n[H] = n[h]+ 1
Step 6: END
```

Figure 12.25 Algorithm to insert a new node in a fibonacci heap

even if `k` consecutive insert operations are performed, then `k` single-node trees are added to the root list.

EXAMPLE 12.8: Insert node 16 in the Fibonacci heap given below.

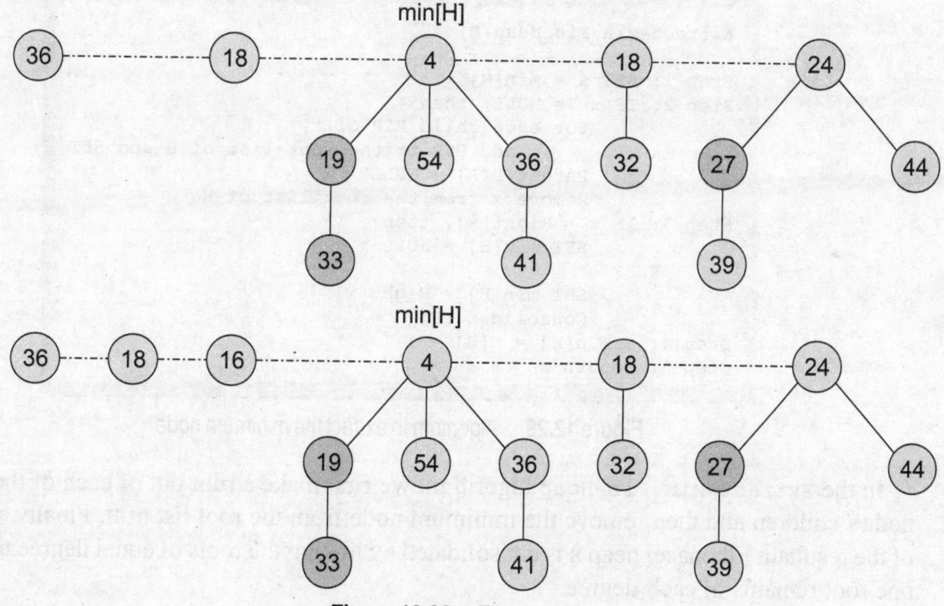

Figure 12.26 Fibonacci heap

Finding the Minimum Node

Fibonacci heaps maintain a pointer `min[H]` that points to the root having the minimum value. Therefore, finding the minimum node is a straightforward task that can be performed in just `O(1)` time.

Uniting Two Fibonacci Heaps

The algorithm given in Fig. 12.27 unites two Fibonacci heaps H_1 and H_2.

In the algorithm, we first concatenate the root lists of H_1 and H_2 into a new root list `H`. Then, the minimum node of `H` is set and the total number of nodes in `H` is updated. Finally, the memory occupied by H_1 and H_2 is freed and the resultant heap `H` is returned.

```
Union_Fib-Heap(H1, H2)

Step 1: H = Create_Fib-Heap()
Step 2: SET min[H] = min[H1]
Step 3: Concatenate root list of H2 with that of H.
Step 4: IF (min[H1] = NULL) OR (min[H2] != NULL and min[H2] < min[H1]), then
              SET min[H] = min[H2]
Step 5: SET n[H] = n[H1] + n[H2]
Step 6: Free H1 and H2
Step 7: Return H
```

Figure 12.27 Algorithm to unite two fibonacci heaps

Extracting the Minimum Node

The process of extracting the minimum node from a Fibonacci heap is the most complicated operation of all the operations that we have discussed so far. Till now, we had been delaying the work of consolidating the trees, but in this operation, the work of consolidating the tree is finally done. The algorithm to extract the minimum node is given in Fig. 12.28.

```
Extract-Min_Fib_Heap(H)

Step 1: SET x = min[H]
Step 2: IF x != NULL, then
              for each child PTR of x,
                     add PTR to the root list of H and SET
              Parent[PTR] = NULL
              Remove x from the root list of H
Step 3: IF x = Right[x], then
              SET min[H] = NULL
        ELSE
              SET min[H] = Right[x]
              Consolidate(H)
Step 4: SET n[H] = n[H] - 1
Step 5: Return x
```

Figure 12.28 Algorithm to extract the minimum node

In the `Extract-Min_Fib-Heap` algorithm, we first make a root out of each of the minimum node's children and then remove the minimum node from the root list of `H`. Finally, the root list of the resultant Fibonacci heap `H` is consolidated by linking the roots of equal degree until at most one root remains of each degree.

Note that in Step 1, we save a pointer `x` to the minimum node; this pointer is returned at the end. However, if `x = NULL`, then the heap is already empty and we are done. Otherwise, the node `x` is deleted from `H` by making all its children the roots of `H` and then removing `x` from the root list (as done in Step 3).

If `x = right[x]`, then `x` is the only node on the root list and it has no children, so now `H` is empty. However, if `x != Right[x]`, then we set the pointer `min[H]` to the node whose address is stored in the right field of `x`.

Consolidate the Heap

The Fibonacci heap is consolidated to reduce the number of trees in the heap. While consolidating the root list of H, the following steps are repeatedly executed until every root in the root list has a distinct *degree* value.

- Find two roots x and y in the root list that has the same degree and where Val[x] ≤ Val[y].
- Link y to x. That is, remove y from the root list of H and make it a child of x. This operation is actually done in the Link_Fib-Heap procedure. Finally, the degree[x] is incremented and the mark on y, if any, is cleared.

In the consolidate algorithm shown in Fig. 12.29, we have used an auxiliary array A[0 ... D(n[H])], such that if A[i] = x, then x is currently a node in the root list of H and degree[x] = i.

```
Consolidate(H)

Step 1: Repeat for i=0 to D(n[H]), SET A[i] = NULL
Step 2: Repeat Steps 3 to 12 for each node x in the
        root list of H
Step 3:     SET PTR = x
Step 4:     SET deg = Degree[PTR]
Step 5:     Repeat Steps 6 to 10 while A[deg] != NULL
Step 6:           SET TEMP = A[deg]
Step 7:           IF Val[PTR] > Val[TEMP], then
Step 8:                 EXCHANGE PTR and TEMP
Step 9:           Link_Fib-Heap(H, TEMP, PTR)
Step 10:          SET A[deg] = NULL
Step 11:    SET deg = deg + 1
Step 12:    SET A[deg] = PTR
Step 13: SET min[H] = NULL
Step 14: Repeat for i = 0 to D(n(H))
Step 15:     IF A[i] != NULL, then
Step 16:          Add A[i] to the root list of H
Step 17:          IF min[H] = NULL OR Val[A[i]]<
                  Val[min[H]], then
Step 18:          SET min[H] = A[i]
Step 19: END
```

Figure 12.29 Consolidate algorithm

In Step 1, we set every entry in the array A to NULL. When Step 1 is over, we end up in a tree that is rooted at some node x. Initially, the array entry A[degree[x]] is set to point to x. In the for loop, each root node in H is examined. In each iteration of the while loop, A[d] points to some root TEMP because d = degree[PTR] = degree[TEMP], so these two nodes must be linked with each other. Of course, the node with the smaller key becomes the parent of the other as a result of the link operation and so if need arises, we exchange the pointers to PTR and TEMP.

```
Link_Fib-Heap (H, x, y)

Step 1: Remove node y from the root list of H
Step 2: Make x the parent of y
Step 3: Increment the degree of x
Step 4: SET mark[y] = FALSE
Step 5: END
```

Figure 12.30 Algorithm to link two fibonacci heaps

Next, we link TEMP to PTR using the Link_Fib-Heap procedure. The Link_Fib-Heap procedure (Fig. 12.30) increments the degree of x but leaves the degree of y unchanged. Since, node y is no longer a root, the pointer to it in array A is removed in Step 10. Note that the value of degree of x is incremented in the Link_Fib-Heap procedure, so Step 13 restores the value of d = degree[x]. The while loop is repeated until A[d] = NULL, that is until no other root with the same degree as x exists in the root list of H.

EXAMPLE 12.9: Remove the minimum node from the Fibonacci heap given below.

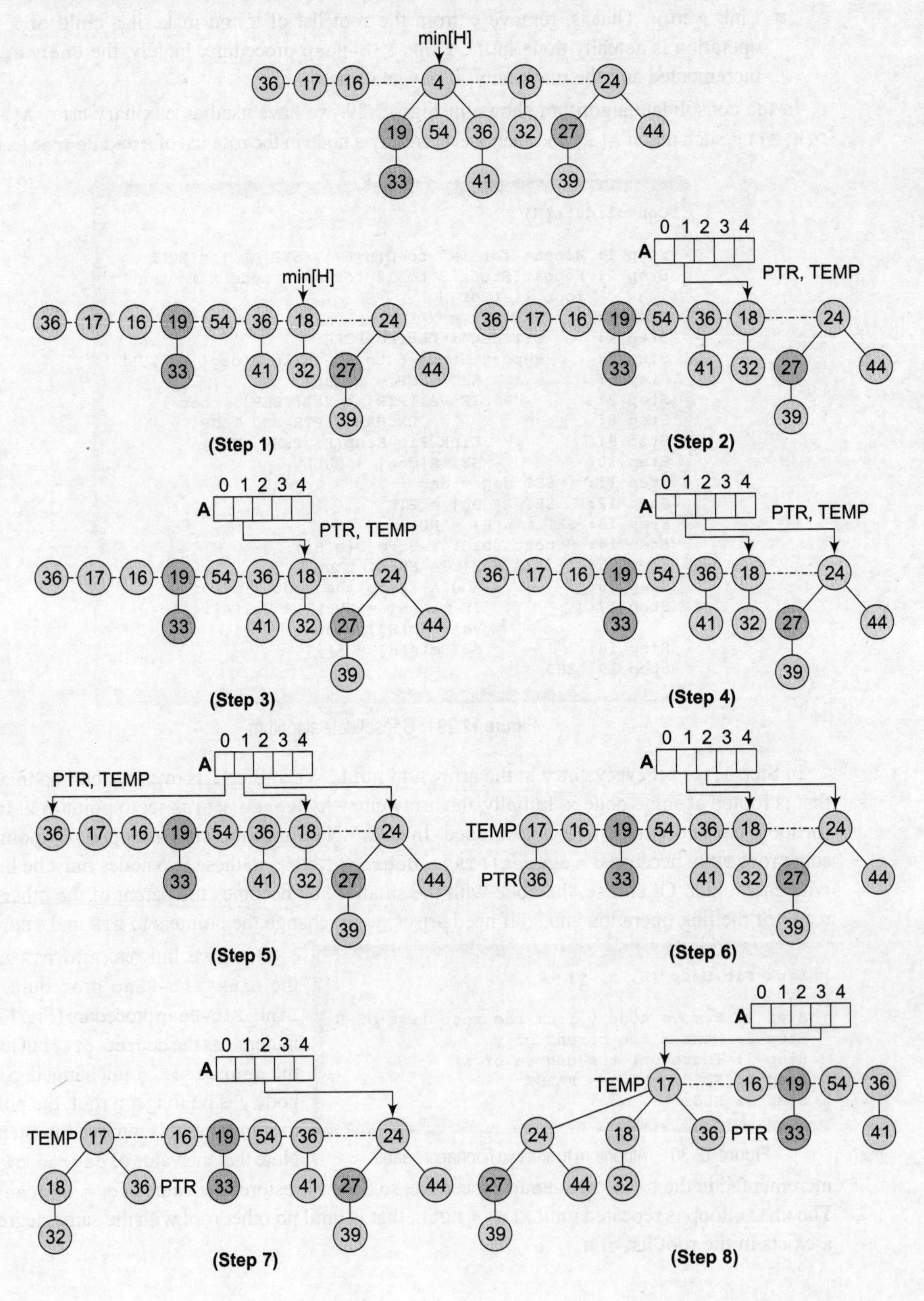

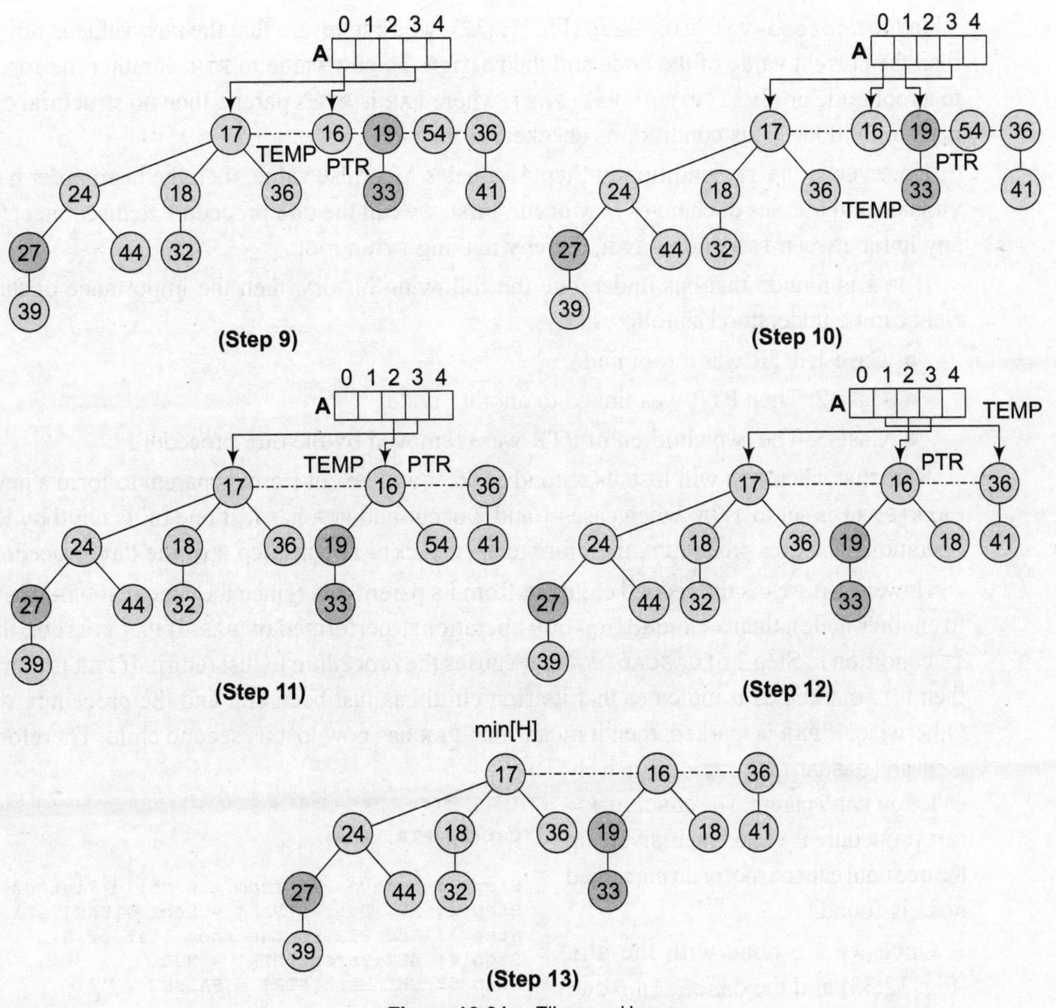

Figure 12.31 Fibonacci heap

Decreasing a Key

The algorithm to decrease the value of a node in O(1) amortized time is given in Fig. 12.32.

```
Decrease-Val_Fib-Heap (H, PTR, v)

Step 1: IF v > Val[PTR], then
             PRINT "ERROR"
Step 2: SET Val[PTR] = v
Step 3: SET PAR = Parent[PTR]
Step 4: IF PTR != NULL and Val[PTR] < Val[PAR], then
             Cut (H, PTR, PAR)
             Cascading-Cut(H, PAR)
Step 5: IF Val[PTR] < Val[min[H]], then
             SET min[H] = PTR
Step 6: END
```

Figure 12.32 Algorithm to decrease the value of a node

In the Decrese-Val_Fib-Heap (Fig. 12.32), we first ensure that the new value is not greater than the current value of the node and then assign the new value to PTR. If either the PTR points to a root node or if Val[PTR] = Val[PAR], where PAR is PTR's parent, then no structural changes need to be done. This condition is checked in Step 4.

However, if the IF condition in Step 4 valuates to a false value, then the heap order has been violated and a series of changes may occur. First, we call the Cut procedure to disconnect (or cut) any link between PTR and its PAR, thereby making PTR a root.

If PTR is a node that has undergone the following history, then the importance of the mark field can be understood as follows:

- Case 1: PTR was a root node.
- Case 2: Then PTR was linked to another node.
- Case 3: The two children of PTR were removed by the cut procedure.

Note that when PTR will lose its second child, it will be cut from its parent to form a new root. mark[PTR] is set to TRUE when cases 1 and 2 occur and PTR has lost one of its child by the Cut operation. The Cut procedure, therefore, clears mark[PTR] in Step 4 of the Cut procedure.

However, if PTR is the second child cut from its parent PAR (since the time that PAR was linked to another node), then a cascading-cut operation is performed on PAR. If PAR is a root, then the IF condition in Step 2 of CASCADING-CUT causes the procedure to just return. If PAR is unmarked, then it is marked as it indicates that its first child has just been cut, and the procedure returns. Otherwise, if PAR is marked, then it means that PAR has now lost its second child. Therefore, PTR is cut and CASCADING-CUT is recursively called on PAR's parent. The CASCADING-CUT procedure is called recursively up the tree until either a root or an unmarked node is found.

```
Cut(H, PTR, PAR)

Step 1: Remove PTR from the child list of PAR
Step 2: SET Degree[PAR] = Degree[PAR] - 1
Step 3: Add PTR to the root list of H
Step 4: SET Parent[PTR] = NULL
Step 5: SET Mark[PTR] = FALSE
Step 6: END
```

Figure 12.33 Algorithm to perform cut procedure

Once we are done with the Cut (Fig. 12.33) and the Cascading-Cut (Fig. 12.34) operations, Step 5 of the Decrease-Val_Fib-Heap finish up by updating min[H].

Note that the amortized cost of DECREASE-VAL_FIB-HEAP is O(1). The actual cost of DECREASE-VAL_FIB-HEAP is O(1) time plus the time required to perform the cascading cuts. If CASCADING-CUT procedure is recursively called c times, then each call of CASCADING-CUT takes O(1) time exclusive of recursive calls. Therefore, the actual cost of DECREASE-VAL_FIB-HEAP including all recursive calls is O(c).

```
Cascading-Cut (H, PTR)

Step 1: SET PAR = Parent[PTR]
Step 2: IF PAR != NULL, then
              IF mark[PTR] = FALSE, then
                    SET mark[PTR] = TRUE
              ELSE
                    Cut (H, PTR, PAR)
                    Cascading-Cut(H, PAR)
Step 3: END
```

Figure 12.34 Algorithm to perform cascade

EXAMPLE 12.10: Decrease the value of node 44 to 19 in the Fibonacci heap given below.

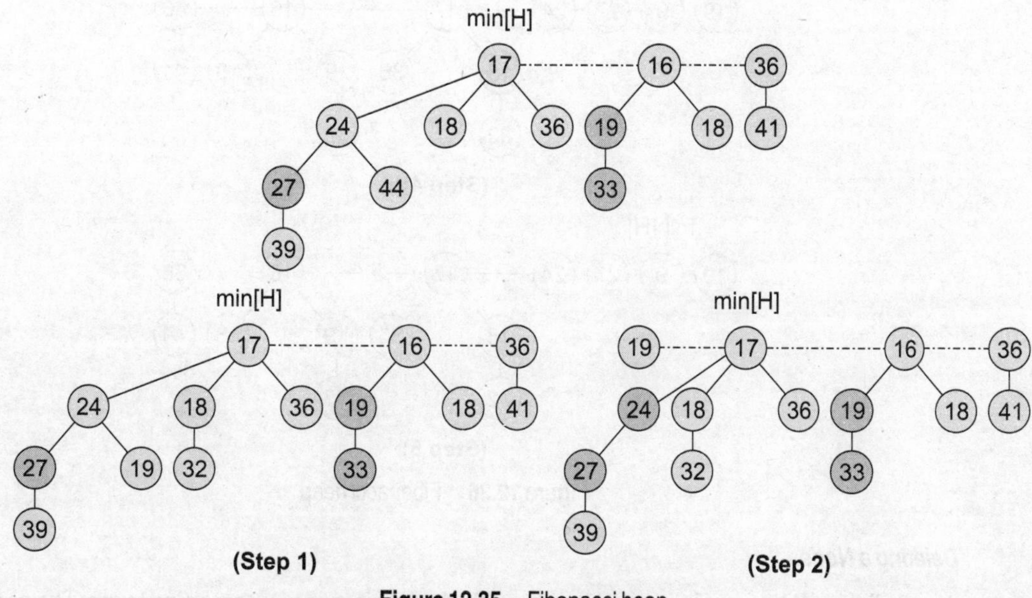

Figure 12.35 Fibonacci heap

EXAMPLE 12.11: Decrease the value of node 39 to 9 in the Fibonacci heap given below.

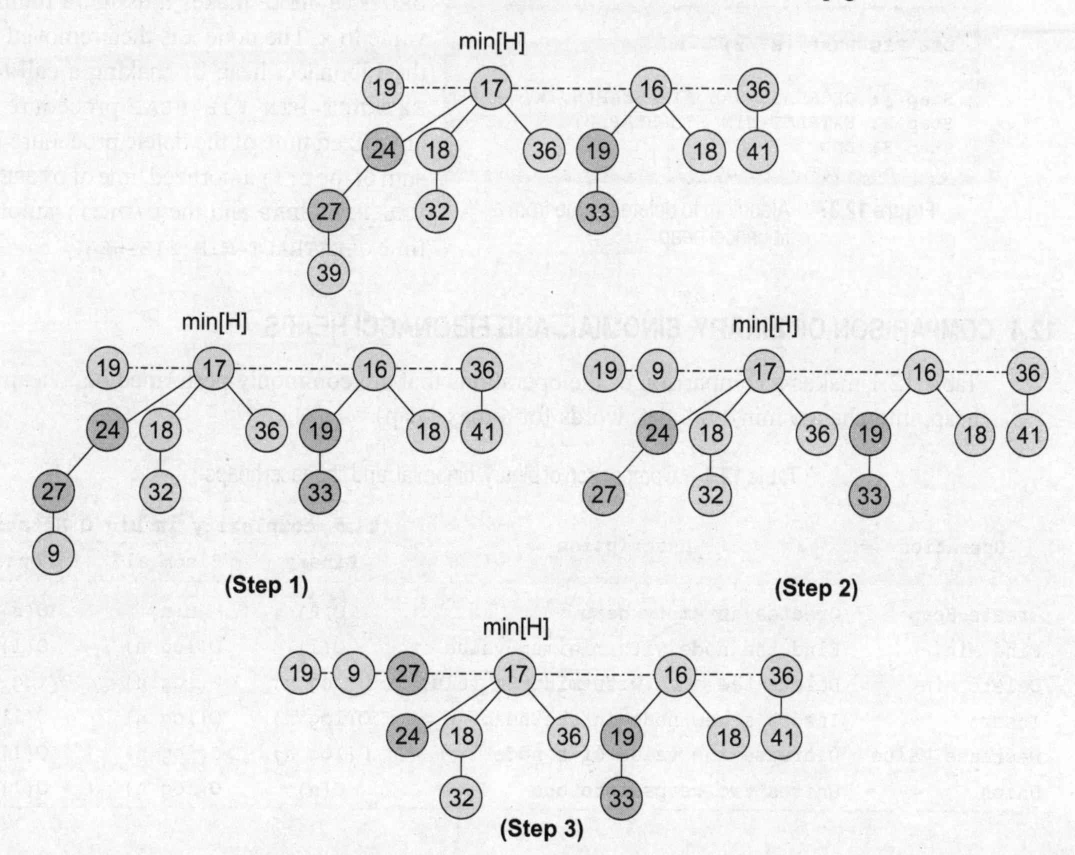

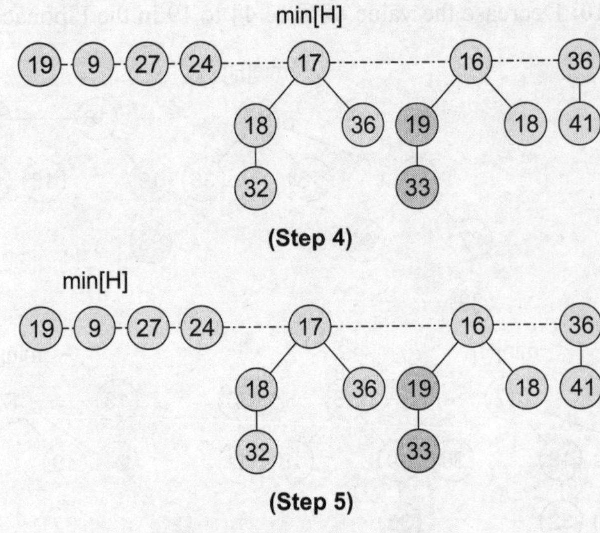

Figure 12.36 Fibonacci heap

Deleting a Node

A node from a Fibonacci heap can be very easily deleted in `O(D(n))` amortized time. The procedure to delete a node is given in Fig. 12.37.

```
DEL_FIB-HEAP (H, x)

Step 1: DECREASE-VAL_FIB-HEAP(H, x, -∞)
Step 2: EXTRACT-MIN_FIB-HEAP(H)
Step 3: END
```

Figure 12.37 Algorithm to delete a node from a fibonacci heap

`DEL_FIB-HEAP` makes `x` assign a minimum value to `x`. The node `x` is then removed from the Fibonacci heap by making a call to the `EXTRACT-MIN_FIB-HEAP` procedure. The amortized time of the delete procedure is the sum of the `O(1)` amortized time of `DECREASE-VAL_FIB-HEAP` and the `O(D(n))` amortized time of `EXTRACT-MIN_FIB-HEAP`.

12.4 COMPARISON OF BINARY, BINOMIAL, AND FIBONACCI HEAPS

Table 12.1 makes a comparison of the operations that are commonly performed on a heap (min heap, interchange min, and max words for a max-heap).

Table 12.1 Comparison of binary, binomial, and fibonacci heaps

Operation	Description	Time complexity in Big O Notation		
		Binary	Binomial	Fibonacci
Create Heap	Creates an empty heap	O(n)	O(n)	O(n)
Find Min	Find the node with minimum value	O(1)	O(log n)	O(1)
Delete Min	Delete the node with minimum value	O(log n)	O(log n)	O(log n)
Insert	Insert a new node in the heap	O(log n)	O(log n)	O(1)
Decrease Value	Decrease the value of a node	O(log n)	O(log n)	O(1)
Union	Unites two heaps into one	O(n)	O(log n)	O(1)

12.5 APPLICATIONS OF HEAPS

Heaps are preferred for applications that include:

- ***Heap sort*** It is one of the best sorting methods that has no quadratic worst-case scenarios.
- ***Selection algorithms*** These algorithms are used to find the minimum and maximum values in linear or sub-linear time.
- ***Graph algorithms*** Heaps can be used as internal traversal data structures. This guarantees that runtime is reduced by an order of polynomial. Heaps are therefore used for implementing Prim's minimal spanning tree algorithm and Dijkstra's shortest path problem.

SUMMARY

- A binary heap is defined as a complete binary tree with elements at every node being either less than or equal to the element at its left and the right child.
- In min heap, elements at every node will either be less than or equal to the element at its left and the right child. Similarly, in max heap, the elements at every node will either be greater than the element at its left and the right child.
- A binomial tree of order i has a root node whose children are the root nodes of binomial trees of order `i-1, i-2, ..., 2, 1, 0`.
- A binomial tree B_i of height `i` has 2^i nodes.
- A binomial heap `H` is a collection of binomial trees that satisfies the following properties:
  - Every binomial tree in `H` satisfies the minimum heap property.
  - There can be one or zero binomial trees for each order including zero order.
- A Fibonacci heap is a collection of trees. Fibonacci heaps differ from binomial heaps, as they have a more relaxed structure, allowing improved asymptotic time bounds.

GLOSSARY

Binary heap A complete binary tree in which every node has a key more extreme (greater or less) than or equal to the key of its parent.

Build heap The process of converting an array into a heap by executing heapify progressively closer to the root.

Binomial heap A heap made of a forest of binomial trees satisfying the heap property numbered `k=0, 1, 2, ..., n`, each containing either 0 or 2^k nodes. Each binomial tree in the heap is formed by linking two of its predecessors, that is, by joining one at the root of the other.

Heapify The process to rearrange a heap in order to maintain the heap property, that is, the key of the root node is more extreme (greater or less) than or equal to the keys of its children. If the root node's key is not more extreme, then swap it with the most extreme child key, and then recursively heapify that child's sub-tree.

Fibonacci heap A heap made of a forest of trees. While the amortized cost of the operations, create, insert, decrease, find minimum, and merge, is a constant $\Theta(1)$, the delete operation takes $O(\log n)$ time.

Binomial tree An ordered tree of order `k`, where $k \geq 0$. A binomial tree B_k has a root with `k` children where the ith child is a binomial tree of order `k-i`.

Heap property Each node of a heap has a key which is more extreme (greater or less) than or equal to the key of its parent.

Max-heap property The property by which each node in the heap has a key which is less than or equal to the key of its parent.

Min-heap property The property by which each node in the heap has a key which is greater than or equal to the key of its parent.

EXERCISES

Review Questions

1. Define a binary heap.
2. Differentiate between a min-heap and a max-heap.
3. Explain the insertion process in a binary heap with suitable examples.
4. Illustrate the deletion operation in a binary heap with examples.
5. Compare binary trees with binary heaps.
6. Explain the steps involved in inserting of a new value in a binary heap with the help of a suitable example.
7. Explain the steps involved in deleting a value from a binary heap with the help of a suitable example.
8. Discuss the applications of a binary heap.
9. Form a binary max-heap and a min-heap from the following sequence of data:
   50, 40, 35, 25, 20, 27, 33.
10. Heaps are excellent data structures to implement priority queues. Justify this statement.
11. Define a binomial heap. Draw its structure.
12. Differentiate among binary heap, binomial heap, and Fibonacci heap.
13. Explain the operations performed on a Fibonacci heap.
14. Why are Fibonacci heaps preferred over binary and binomial heaps?
15. Analyse the complexity of the algorithm to unite two binomial heaps.
16. The running time of the algorithm to find the minimum key in a binomial heap is `O(logn)`. Comment.
17. Discuss the process of inserting a new node in a binomial heap. Explain with the help of an example.
18. The algorithm `Min-Extract_Binomial-Heap()` runs in `O(lg n)` time where n is the number of nodes in H. Justify this statement.
19. Explain how an existing node is deleted from a binomial heap with the help of a relevant example.
20. Explain the process of inserting a new node in a Fibonacci heap.
21. Write down the algorithm to unite two Fibonacci heaps.
22. What is the procedure to extract the node with the minimum value from a Fibonacci heap? Give the algorithm and analyse its complexity.
23. Make a comparison of binary, binomial, and Fibonacci heaps.
24. Consider the figure given below and state whether it is a heap or not.

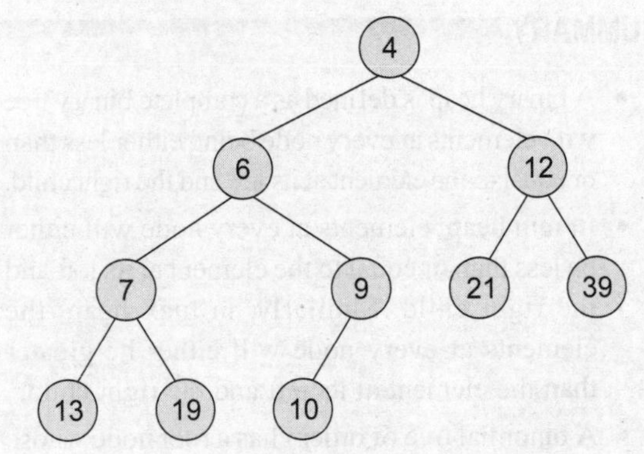

25. Reheap the following structure to make it a heap.

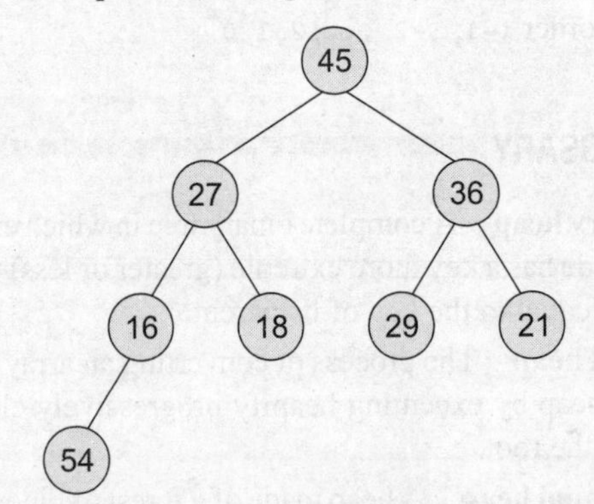

26. Show the array implementation of the following heap.

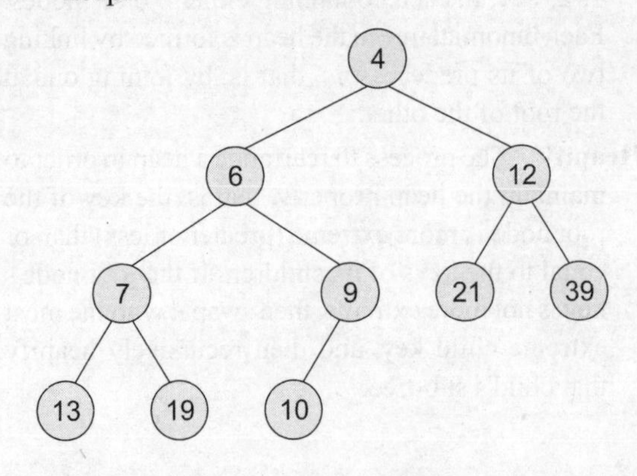

27. Given the following array structure, draw the heap.

45	27	36	18	16	21	23	10

Also, find out

(a) the parent of node 10, 21 and 23, and

(b) tndex of left and right child of node 23

28. Which of the following sequences represent a binary heap?

(a) 40, 33, 35, 22, 12, 16, 5, 7

(b) 44, 37, 20, 22, 16, 32, 12

(c) 15, 15, 15, 15, 15, 15

29. A heap sequence is given as: 52, 32, 42, 22, 12, 27, 37, 12, 7. Which element will be deleted when the delete algorithm is called thrice?

30. Show the resulting heap when 35, 24, and 10 are added to the heap of the above question.

31. Draw a heap that is also a binary search tree.

32. Analyse the complexity of heapify algorithm.

33. Consider the Fibonacci heap given below and then decrease the value of node 33 to 9. Insert a new node with the value 5 and finally delete the node 19 from it.

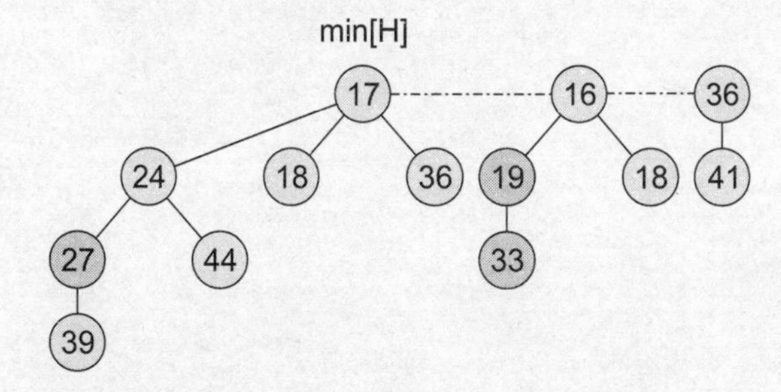

Multiple Choice Questions

1. The height of a binary heap with `n` nodes is equal to

(a) `O(n)` (b) `O(log n)`
(c) `O(n log n)` (d) $O(n^2)$

2. An element at position `i` in the array has its left child stored at position

(a) `2i` (b) `2i + 1`
(c) `i/2` (d) `i/2 + 1`

3. In the worst case, it takes how much time to build a binary heap of `n` elements?

(a) `O(n)` (b) `O(log n)`
(c) `O(n log n)` (d) $O(n^2)$

4. The height of a binomial tree B_i is

(a) `2i` (b) `2i + 1`
(c) `i/2` (d) `i`

5. How many nodes does a binomial tree of order 0 have?

(a) 0 (b) 1 (c) 2 (d) 3

6. The running time of `Link_Binomial-Tree()` procedure is

(a) `O(n)` (b) `O(log n)`
(c) `O(n log n)` (d) `O(1)`

7. In a Fibonacci heap, how much time does it take to find the minimum node?

(a) `O(n)` (b) `O(log n)`
(c) `O(n log n)` (d) `O(1)`

True or False

1. A binary heap is a complete binary tree.

2. In a min heap, the root node has the highest key value in the heap.

3. An element at position `i` has its position stored at position `i/2`.

4. All levels of a tree except the last level are completely filled.

5. In a min-heap, elements at every node will be greater than its left and the right child.

6. A binomial tree B_i has `2i` nodes.

7. Binomial heaps are ordered.

8. Fibonacci heaps are rooted and ordered.

9. The running time of `Min_Binomial-Heap()` procedure is `O(lg n)`.

10. If there are `m` roots in the root lists of H_1 and H_2, then `Merge_Binomial-Heap()` runs in `O(m log m)` time.

11. Fibonacci heaps are preferred over binomial heaps.

Fill in the Blanks

1. An element at position `i` in the array has its right child stored at position ______.

2. Heaps are used to implement ______.

3. Heaps are also known as ______.
4. In ______, elements at every node will either be less than or equal to the element at its left and the right child.
5. An element is always deleted from the ______.
6. The height of a binomial tree B_i is ______.
7. A boolean value `mark[x]` indicates ______.
8. A binomial heap is defined as ______.
9. A binomial tree B_i has ______ nodes.
10. A binomial heap is created in ______ time.
11. A Fibonacci heap is a ______.
12. In a Fibonacci heap, `mark[x]` indicates ______.

13 Graphs

Learning Objective

In this chapter, we will discuss another abstract type data structure called graphs. We will discuss the representation of graphs in the memory as also the different operations that can be performed on them. Last but not the least, we will discuss some of the real-world applications of graphs.

13.1 INTRODUCTION

A graph is an abstract data structure that is used to implement the graph concept from mathematics. It is basically a collection of vertices (also called nodes) and edges that connect these vertices. A graph is often viewed as a generalization of the tree structure, where instead of having a purely parent-to-child relationship between the tree nodes, any kind of complex relationship can be represented.

13.1.1 Why are Graphs Useful?

Graphs are widely used to model any situation where entities or things are related to each other in pairs. For example, the following information can be represented by graphs:

- *Family trees* in which the member nodes have an edge from parent to each of their children.
- *Transportation networks* in which nodes are airports, intersections, ports, etc. The edges can be airline flights, one-way roads, shipping routes, etc.

13.1.2 Definition

A graph `G` is defined as an ordered set `(V, E)`, where `V(G)` represents the set of vertices and `E(G)` represents the edges that connect the vertices.

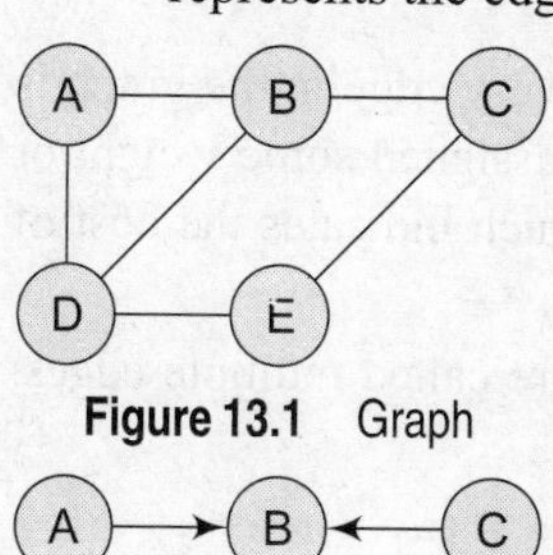

Figure 13.1 Graph

Figure 13.2 Directed graph

Figure 13.1 shows a graph with `V(G) = {A, B, C, D and E}` and `E(G) = {(A, B), (B, C), (A, D), (B, D), (D, E), (C, E)}`. Note that there are five vertices or nodes and six edges in the graph.

A graph can be directed or undirected. In an undirected graph, the edges do not have any direction associated with them. That is, if an edge is drawn between nodes `A` and `B`, then the nodes can be traversed from `A` to `B` as well as from `B` to `A`. Figure 13.1 shows an undirected graph because it does not give any information about the direction of the edges.

Look at Fig. 13.2 which shows a directed graph. In a directed graph, edges form an ordered pair. If there is an edge from `A` to `B`, then there is a path from `A` to `B` but not from `B` to `A`. The edge `(A, B)` is said to initiate from node `A` (also known as initial node) and terminate at node `B` (terminal node).

13.1.3 Graph Terminology

Adjacent nodes or neighbours For every edge, `e = (u, v)` that connects nodes `u` and `v`, the nodes `u` and `v` are the end-points and are said to be the adjacent nodes or neighbours.

Degree of a node Degree of a node `u`, `deg(u)`, is the total number of edges containing the node `u`. If `deg(u) = 0`, it means that `u` does not belong to any edge and such a node is known as an isolated node.

Regular graph It is a graph where each vertex has the same number of neighbours. That is, every node has the same degree. A regular graph with vertices of degree `k` is called a `k-regular graph` or a regular graph of degree `k`. Figure 13.3 shows regular graphs.

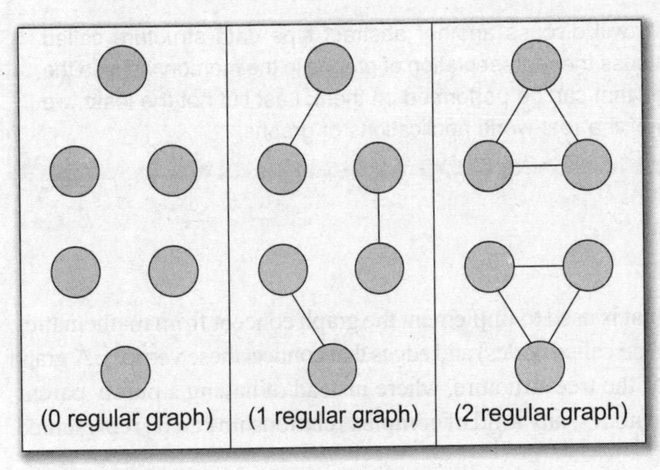

Figure 13.3 Regular graphs

Path A path `P` written as `P = {`v_0`,` v_1`,` v_2`, ...,` v_n`)`, of length `n` from a node `u` to `v` is defined as a sequence of `(n+1)` nodes. Here, `u =` v_0, `v =` v_n and v_{i-1} is adjacent to v_i for `i = 1, 2, 3, ..., n`.

Closed path A path `P` is known as a closed path if the edge has the same end-points. That is, if $v_0 = v_n$.

Simple path A path `P` is known as a simple path if all the nodes in the path are distinct with an exception that v_0 may be equal to v_n. If $v_0 = v_n$, then the path is called a closed simple path.

Cycle A closed simple path with length 3 or more is known as a cycle. A cycle of length `k` is called a `k-cycle`.

Connected graph A graph in which there exists a path between any two of its nodes is called a connected graph. That is to say that there are no isolated nodes in a connected graph. A connected graph that does not have any cycle is called a tree. Therefore, a tree is treated as a special graph (Refer Fig.13.4 (b)).

Complete graph A graph `G` is said to be complete, if all its nodes are fully connected. That is, there is a path from one node to every other node in the graph. A complete graph has `n(n-1)/2` edges, where `n` is the number of nodes in `G`.

Labeled graph or weighted graph A graph is said to be labeled if every edge in the graph is assigned some data. In a weighted graph, the edges of the graph are assigned some weight or length. The weight of an edge denoted by `w(e)` is a positive value which indicates the cost of traversing the edge. Figure 13.4 (c) shows a weighted graph.

Multiple edges Distinct edges which connect the same end-points are called multiple edges. That is, `e = (u, v)` and `e' = (u, v)` are known as multiple edges of `G`.

Loop An edge that has identical end-points is called a loop. That is, `e = (u, u)`.

Multi-graph A graph with multiple edges and/or a loop is called a multi-graph. Figure 13.4(a) shows a weighted graph.

Size of a graph The size of a graph is the total number of edges in it.

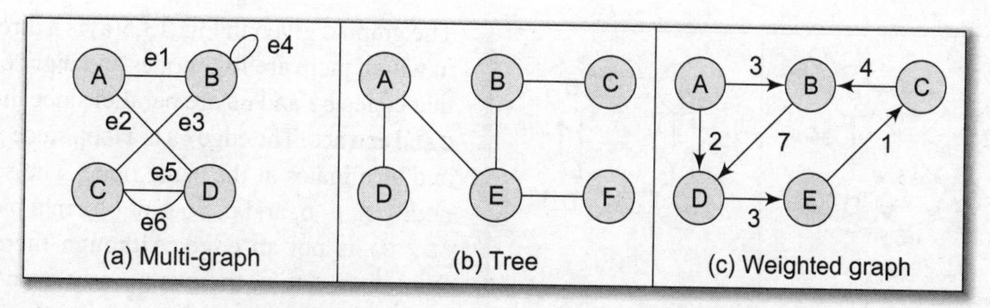

Figure 13.4 Multi-graph, tree, and weighted graph

13.1.4 Directed Graph

A directed graph `G`, also known as a *digraph*, is a graph in which every edge has a direction assigned to it. An edge of a directed graph is given as an ordered pair `(u, v)` of nodes in `G`. For an edge `(u, v)`,

- The edge begins at `u` and terminates at `v`.
- `u` is known as the origin or initial point of `e`. Correspondingly, `v` is known as the destination or terminal point of `e`.
- `u` is the predecessor of `v`. Correspondingly, `v` is the successor of `u`.
- Nodes `u` and `v` are adjacent to each other.

Terminology of a Directed Graph

Out-degree of a node The out-degree of a node `u`, written as `outdeg(u)`, is the number of edges that originate at `u`.

In-degree of a node The in-degree of a node `u`, written as `indeg(u)`, is the number of edges that terminate at `u`.

Degree of a node The degree of a node, written as `deg(u)` is equal to the sum of in-degree and out-degree of that node. Therefore, `deg(u) = indeg(u) + outdeg(u)`.

Source A node `u` is known as a source if it has a positive out-degree but a zero in-degree.

Sink A node `u` is known as a sink if it has a positive in-degree but a zero out-degree.

Reachability A node `v` is said to be reachable from node `u`, if and only if there exists a (directed) path from node `u` to node `v`. For example, if you consider the directed graph given in Fig. 13.5(a), you will observe that node `D` is reachable from node `A`.

Strongly connected directed graph A digraph is said to be strongly connected if and only if there exists a path from every pair of nodes in `G`. That is, if there is a path from node `u` to `v`, then there must be a path from node `v` to `u`.

Unilaterally connected graph A digraph is said to be unilaterally connected if there exists a path from any pair of nodes `u` or `v` in `G` such that there is a path from `u` to `v` or a path from `v` to `u`, but not both.

Parallel/Multiple edges Distinct edges which connect the same end-points are called multiple edges. That is, `e = (u, v)` and `e' = (u, v)` are known as multiple edges of `G`. In Fig. 13.5(a), `e3` and `e5` are multiple-edged connecting nodes `C` and `D`.

Simple directed graph A directed graph `G` is said to be a simple directed graph if and only if it has no parallel edges. However, a simple directed graph may contain cycles with an exception that it cannot have more than one loop at a given node.

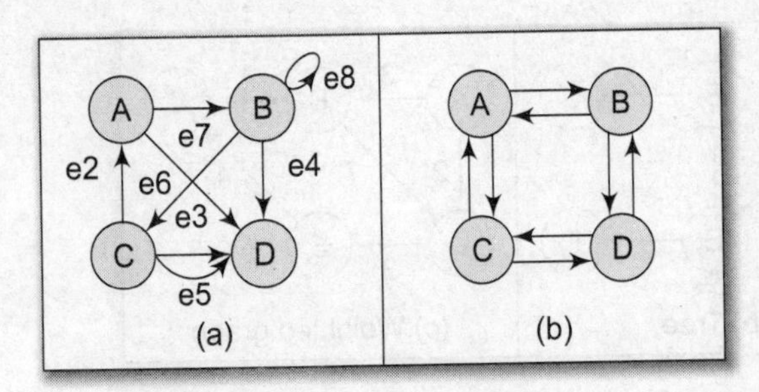

Figure 13.5 (a) Directed acyclic graph and (b) strongly connected directed acyclic graph

The graph G given in Fig. 13.5(a) is a directed graph in which there are four nodes and eight edges. Note that edges e3 and e5 are parallel since they begin at C and end at D. The edge e8 is a loop since it originates and terminates at the same node. The sequence of nodes, A, B, D, and C, does not form a path because (D, C) is not an edge. Although there is a path from node C to D, there is no way from D to C.

In the graph, we see that there is no path from node D to any other node in G, so the graph is not strongly connected. However, G is said to be unilaterally connected. We also observe that node D is a sink since it has a positive out-degree but a zero in-degree.

13.2 TRANSITIVE CLOSURE OF A DIRECTED GRAPH

A transitive closure of a graph is constructed to answer reachability questions. That is, is there a path from a node A to node E in one or more hops? A binary relation indicates only whether the node A is connected to node B, and that node B is connected to node C, etc. But once the transitive closure is constructed as shown in Fig. 13.6, we can easily determine in O(1) time whether node E is reachable from node A or not. Like the adjacency list, transitive closure is also stored as a matrix T, so if T[1][5] = 1; then the case is that node 1 can reach node 5 in one or more hops.

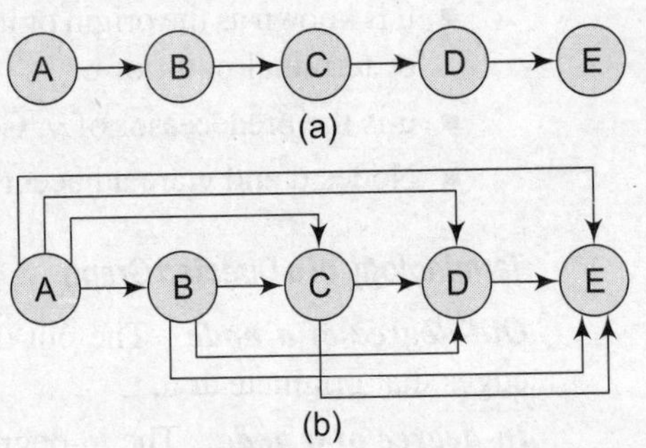

Figure 13.6 (a) A graph G and its (b) transitive closure G*

Definition

For a directed graph G = (V,E), where V is the set of vertices and E is the set of edges, the transitive closure of G is a graph G* = (V,E*). In G*, for every vertex v, w in V there is an edge (v, w) in E* if and only if there is a valid path from v to w in G.

Where and Why is it Needed?

Finding the transitive closure of a directed graph is an important problem in the following computational tasks:

- Transitive closure is used to find the reachability analysis of transition networks representing distributed and parallel systems.
- It is used in the construction of parsing automata in compiler construction.
- Recently, transitive closure computation is being used to evaluate recursive database queries (because almost all practical recursive queries are transitive in nature).

Algorithm

The algorithm to find the transitive enclosure of a graph G is given in Fig. 13.8. In order to determine the transitive closure of a graph, we define a matrix t where $t^k_{ij} = 1$, (for i, j, k = 1, 2, 3, ... n)

When k = 0	$T^0_{ij} = \begin{cases} 0 \text{ if (i, j) is not in E} \\ 1 \text{ if (I, j) is in E} \end{cases}$
When K ≥ 1	$T^k_{ij} = T^{k-1}_{ij} \vee (T^{k-1}_{ik} \wedge T^{k-1}_{kj})$

Figure 13.7 Relation between k and T^k_{ij}

if there exists a path in G from the vertex i to vertex j with intermediate vertices in the set (1, 2, 3, ..., k) and 0 otherwise. That is, G* is constructed by adding an edge (i, j) into E* if and only if $t^k_{ij} = 1$. Look at Fig. 13.7 which shows the relation between k and T^k_{ij}.

```
Transitive_Closure(A, t, n)

Step 1: SET i=1, j=1, k=1
Step 2: Repeat Steps 3 and 4 while i<=n
Step 3:     Repeat Step 4 while j<=n
Step 4:          IF (A[i][j] = 1), then
                    SET t[i][j] = 1
                 ELSE
                    SET t[i][j] = 0
              INCREMENT j
           INCREMENT i
Step 5: Repeat Steps 6 to 11 while k<=n
Step 6:    Repeat Step 7 to 10 while i<=n
Step 7:        Repeat Step 8 and 9 while j<=n
Step 8:           SET t[i,j] = t[i][j] V (t[i][k] Λ t[k][j])
Step 9:           INCREMENT j
Step 10:    INCREMENT i
Step 11:  INCREMENT k
Step 12: END
```

Figure 13.8 Algorithm to find the transitive enclosure of a graph G

13.3 BI-CONNECTED GRAPHS

A vertex v of G is called an articulation point, if removing v along with the edges incident to v, results in a graph that has at least two connected components.

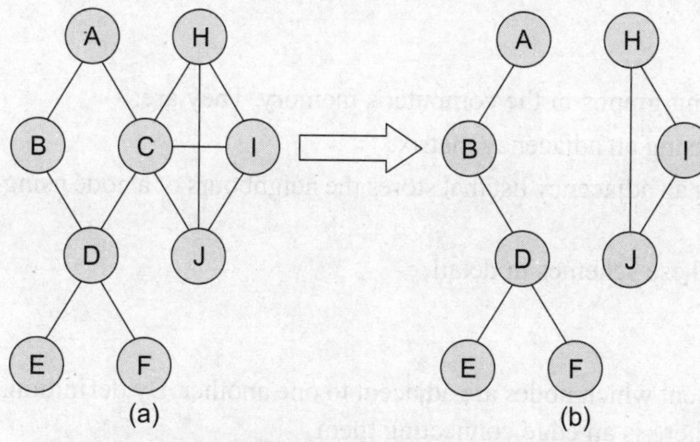

Figure 13.9 Non bi-connected graph

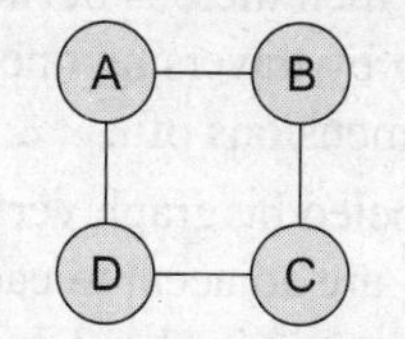

Figure 13.10 Bi-connected graph

A bi-connected graph (shown in Fig. 13.10) is defined as a connected graph that has no articulation vertices. That is, a bi-connected graph is connected and non-separable in the sense that even if we remove any vertex from the graph, the resultant graph is still connected. By definition,

- A bi-connected undirected graph is a connected graph that is not broken into disconnected pieces by deleting any single vertex.
- In a bi-connected directed graph for any two vertices v and w, there are two directed paths from v to w which have no vertices in common other than v and w.

Note that the graph shown in Fig. 13.9 (a) is not a bi-connected graph, as deleting vertex C from the graph results in two disconnected components of the original graph (Fig. 13.9 (b)).

As in graphs, there is a related concept for edges. An edge in a graph is called a *bridge* if removing that edge results in a disconnected graph. Also, an edge on the graph that does not lie on a cycle is a bridge. This means that a bridge has at least one articulation point at its end, although it is not necessary that the articulation point is linked in a bridge. Look at the graph shown in Fig. 13.11.

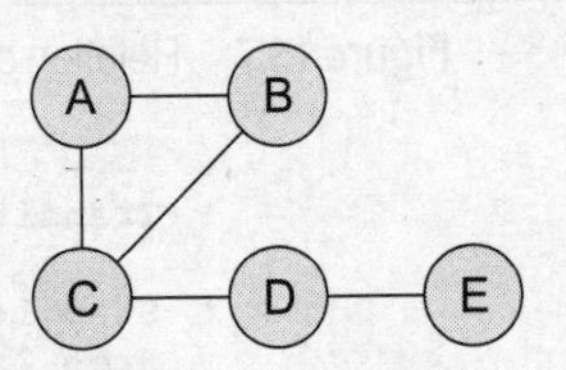

Figure 13.11 Graph with bridges

In the graph, CD and DE are bridges. Consider some more examples shown in Fig. 13.12.

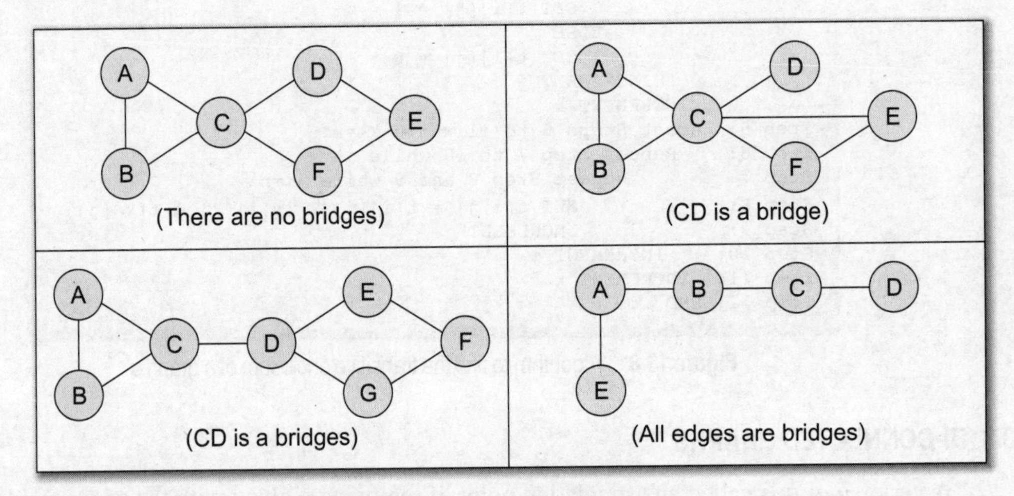

Figure 13.12 Graph with bridges

13.4 REPRESENTATION OF GRAPHS

There are two common ways of storing graphs in the computer's memory. They are:

- *Sequential representation* by using an adjacency matrix.
- *Linked representation* by using an adjacency list that stores the neighbours of a node using a linked list.

In this section, we will discuss both these schemes in detail.

13.4.1 Adjacency Matrix Representation

An adjacency matrix is used to represent which nodes are adjacent to one another. By definition, two nodes are said to be adjacent if there is an edge connecting them.

In a directed graph G, if node v is adjacent to node u, then there is definitely an edge from u to v. That is, if v is adjacent to u, we can get from u to v by traversing one edge. For any graph G having n nodes, the adjacency matrix will have the dimensions of $n \times n$.

In an adjacency matrix, the rows and columns are labeled by graph vertices. An entry a_{ij} in the adjacency matrix will contain 1, if vertices v_i and v_j are adjacent to each other. However, if the nodes are not adjacent, a_{ij} will be set to zero. It is summarized in Fig. 13.13.

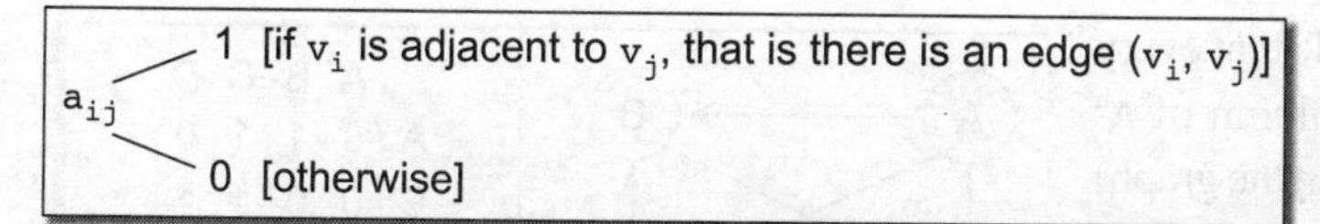

Figure 13.13 Adjacency matrix entry

Since an adjacency matrix contains only 0s and 1s, it is called a *bit matrix* or a *Boolean matrix*. The entries in the matrix depend on the ordering of the nodes in G. Therefore, a change in the order of nodes will result in a different adjacency matrix. Figure 13.14 shows some graphs and their corresponding adjacency matrices.

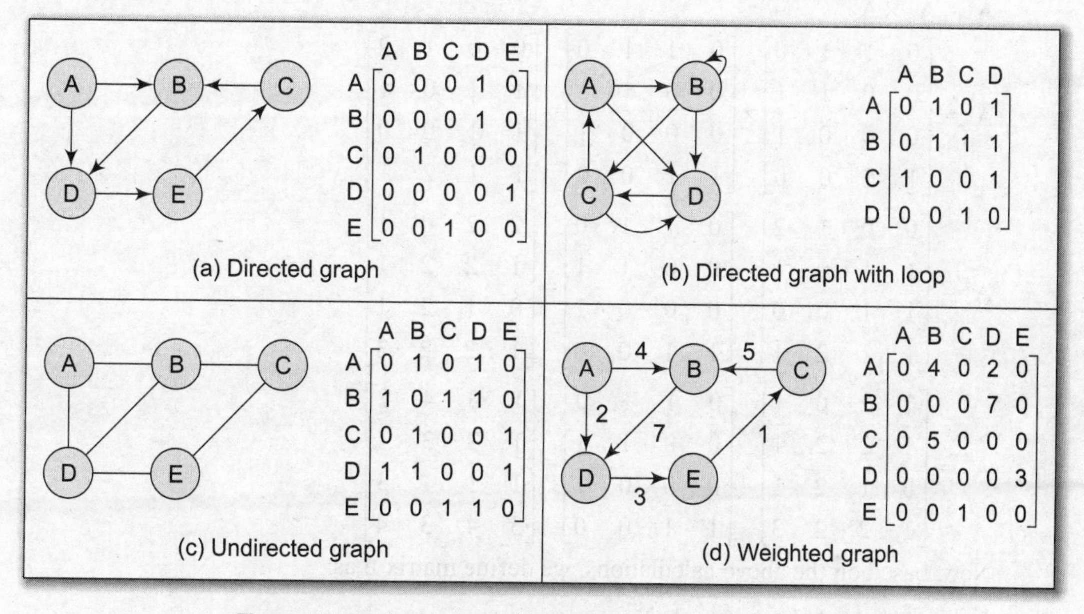

Figure 13.14 Graphs and their corresponding adjacency matrices

From the above examples, we can draw the following conclusions:

- For a simple graph (that has no loops), the adjacency matrix has 0s on the diagonal.
- The adjacency matrix of an undirected graph is symmetric.
- The memory use of an adjacency matrix is $O(n^2)$, where n is the number of nodes in the graph.
- Number of 1s (or non-zero entries) in an adjacency matrix is equal to the number of edges in the graph.
- The adjacency matrix for a weighted graph contains the weights of the edges connecting the nodes.

Now let us discuss the powers of an adjacency matrix. From adjacency matrix A^1, we can conclude that an entry 1 in the ith row and j^{th} column means that there exists a path of length 1 from v_i to v_j. Now consider, A^2, A^3, and A^4.

$$(a_{ij})^2 = \Sigma\, a_{ik}\, a_{kj}$$

Any entry $a_{ij} = 1$ if $a_{ik} = a_{kj} = 1$. That is, if there is an edge (v_i, v_k) and (v_k, v_j), then there is a path from v_i to v_j of length 2.

Similarly, every entry in the ith row and j^{th} column of A^3 gives the number of paths of length 3 from node v_i to v_j.

In general terms, we can conclude that every entry in the ith row and j^{th} column of A^n (where n is the number of nodes in the graph) gives the number of paths of length n from node v_i to v_j. Consider a directed graph given in Fig. 13.15. Given its adjacency matrix A, let us calculate A^2, A^3, and A^4.

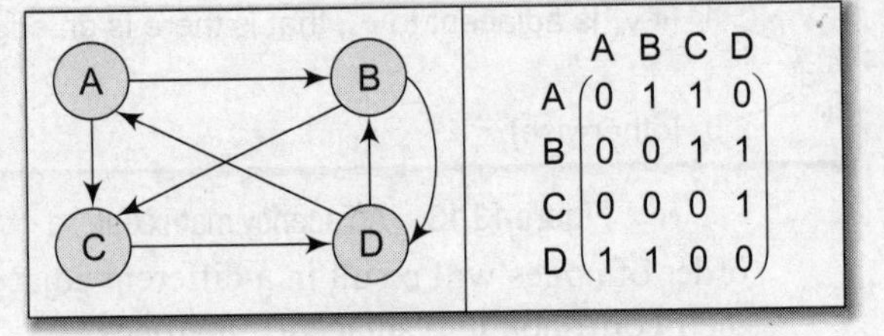

Figure 13.15 Directed graph with its adjacency matrix

$$A^2 = A^1 \times A^1$$

$$A^2 = \begin{bmatrix} 0 & 1 & 1 & 0 \\ 0 & 0 & 1 & 1 \\ 0 & 0 & 0 & 1 \\ 1 & 1 & 0 & 0 \end{bmatrix} \times \begin{bmatrix} 0 & 1 & 1 & 0 \\ 0 & 0 & 1 & 1 \\ 0 & 0 & 0 & 1 \\ 1 & 1 & 0 & 0 \end{bmatrix} = \begin{bmatrix} 0 & 0 & 1 & 2 \\ 1 & 1 & 0 & 1 \\ 1 & 1 & 0 & 0 \\ 1 & 1 & 2 & 1 \end{bmatrix}$$

$$A^3 = \begin{bmatrix} 0 & 0 & 1 & 2 \\ 1 & 1 & 0 & 1 \\ 1 & 1 & 0 & 0 \\ 1 & 1 & 2 & 1 \end{bmatrix} \times \begin{bmatrix} 0 & 1 & 1 & 0 \\ 0 & 0 & 1 & 1 \\ 0 & 0 & 0 & 1 \\ 1 & 1 & 0 & 0 \end{bmatrix} = \begin{bmatrix} 2 & 2 & 0 & 1 \\ 1 & 2 & 2 & 1 \\ 0 & 1 & 2 & 1 \\ 1 & 2 & 2 & 3 \end{bmatrix}$$

$$A^4 = \begin{bmatrix} 2 & 2 & 0 & 1 \\ 1 & 2 & 2 & 1 \\ 0 & 1 & 2 & 1 \\ 1 & 2 & 2 & 3 \end{bmatrix} \times \begin{bmatrix} 0 & 1 & 1 & 0 \\ 0 & 0 & 1 & 1 \\ 0 & 0 & 0 & 1 \\ 1 & 1 & 0 & 0 \end{bmatrix} = \begin{bmatrix} 1 & 3 & 4 & 2 \\ 1 & 2 & 3 & 4 \\ 1 & 1 & 1 & 3 \\ 3 & 4 & 3 & 4 \end{bmatrix}$$

Now, based on the above calculations, we define matrix B as:

$$B^r = A^1 + A^2 + A^3 + \ldots + A^r$$

An entry in the ith row and j^{th} column of matrix B^r gives the number of paths of length r or less than r from vertex v_i to v_j. The main goal to define matrix B is to obtain the path matrix P. The path matrix P can be calculated from B by setting an entry $P_{ij} = 1$, if B_{ij} is non-zero; and $P_{ij} = 0$, otherwise. The path matrix is used to show whether there exists a simple path from node v_i to v_j or not. This is shown in Fig. 13.16.

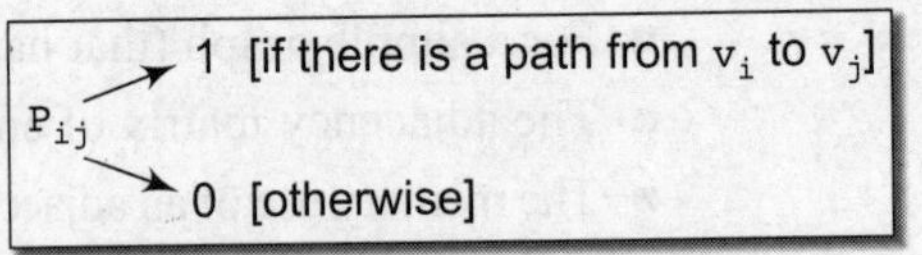

Figure 13.16 Path matrix entry

Let us now calculate matrix B and matrix P using the above discussion.

$$B = \begin{bmatrix} 0 & 1 & 1 & 0 \\ 0 & 0 & 1 & 1 \\ 0 & 0 & 0 & 1 \\ 1 & 1 & 0 & 0 \end{bmatrix} + \begin{bmatrix} 0 & 1 & 1 & 2 \\ 1 & 1 & 0 & 1 \\ 1 & 1 & 0 & 0 \\ 1 & 1 & 2 & 1 \end{bmatrix} + \begin{bmatrix} 2 & 2 & 0 & 1 \\ 1 & 2 & 2 & 1 \\ 0 & 1 & 2 & 1 \\ 1 & 2 & 2 & 3 \end{bmatrix} + \begin{bmatrix} 1 & 3 & 4 & 2 \\ 1 & 2 & 3 & 4 \\ 1 & 1 & 1 & 3 \\ 3 & 4 & 3 & 4 \end{bmatrix} = \begin{bmatrix} 3 & 7 & 6 & 5 \\ 3 & 5 & 6 & 7 \\ 2 & 3 & 3 & 5 \\ 6 & 6 & 7 & 8 \end{bmatrix}$$

Now the path matrix P can be given as:

$$P = \begin{bmatrix} 1 & 1 & 1 & 1 \\ 1 & 1 & 1 & 1 \\ 1 & 1 & 1 & 1 \\ 1 & 1 & 1 & 1 \end{bmatrix}$$

13.4.2 Adjacency List

The adjacency list is another way in which graphs can be represented in the computer's memory. This structure consists of a list of all nodes in G. Furthermore, every node is in turn linked to its own list that contains the names of all other nodes that are adjacent to it.

The key advantage of using an adjacency list includes:

- It is easy to follow and clearly shows the adjacent nodes of a particular node.
- It is often used for storing graphs that have a small-to-moderate number of edges. That is, an adjacency list is preferred for representing sparse graphs in the computer's memory; otherwise, an adjacency matrix is a good choice.
- Adding new nodes in G is easy and straightforward when G is represented using an adjacency list. Adding new nodes in an adjacency matrix is a difficult task, as the size of the matrix needs to be changed and existing nodes may have to be reordered.

Consider the graph given in Fig. 13.17 and see how its adjacency list is stored in the memory.

For a directed graph, the sum of the lengths of all adjacency lists is equal to the number of edges in G. However, for an undirected graph, the sum of the lengths of all adjacency lists is equal to twice the number of edges in G because an edge (u, v) means an edge from node u to v as well as an edge from v to u. The adjacency list can also be modified to store weighted graphs. Let us now see an adjacency list for an undirected graph as well as a weighted graph. This is shown in Fig. 13.18.

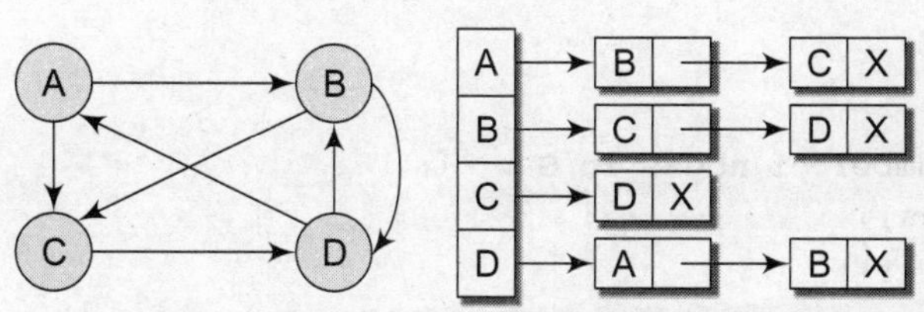

Figure 13.17 Graph G and its adjacency list

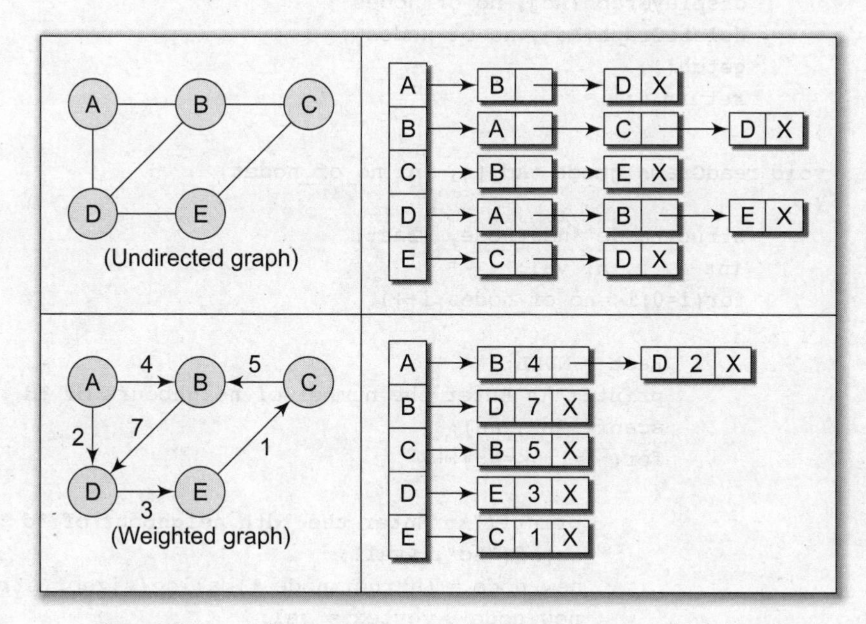

Figure 13.18 Adjacency list for an undirected graph and a weighted graph

PROGRAMMING EXAMPLE

1. Write a program to create a graph of *n* vertices using an adjacency list. Also, write the code to read and print its information and finally to delete a desired node.

```
#include<stdio.h>
#include<conio.h>
#include<alloc.h>
typedef struct node
{
        char vertex;
        struct node *next;
};
void displayGraph(struct node *adj[], int no_of_nodes);
void deleteGraph(struct node *adj[], int no_of_nodes);
void readGraph(struct node *adj[], int no_of_nodes);
main()
{
        struct node *Adj[10];
        int no_of_nodes, i;
        clrscr();
        printf("\n Enter the number of nodes in G : ");
        scanf("%d", &no_of_nodes);
        for(i=0;i<=no_of_nodes;i++)
                Adj[i] = NULL;
        readGraph(Adj, no_of_nodes);
        printf("\n The graph is : ");
        displayGraph(Adj, no_of_nodes);
        deleteGraph(Adj, no_of_nodes);
        getch();
        return 0;
}
void readGraph(gnode *Adj[], int no_of_nodes)
{
        struct node *new_node, *last;
        int i, j, n, val;
        for(i=0;i<=no_of_nodes;i++)
        {
                Last = NULL;
                printf("\n Enter the number of neighbours of %d : ", i);
                scanf("%d", &n);
                for( j=1;j<=n;j++)
                {
                        printf("\n Enter the %dth neighbour of %d : j, i);
                        scanf("%d", &val);
                        new_node = (struct node *)malloc(sizeof(struct node));
                        new_node->vertex = val;
                        new_node->next = NULL;
                        if (Adj[i] == NULL)
```

```
                    Adj[i] = new_node;
                else
                    last->next = new_node;
                last = new_node;
            }
        }
    }
    void printGraph ( struct node *Adj[], int no_of_nodes)
    {
        struct node *ptr;
        int i;
        for(i=0;i<=no_of_nodes;i++)
        {
            ptr = Adj[i];
            printf("\n The neighbours of node %d are : ", i);
            while(ptr != NULL)
            {
                printf("\t%d", ptr->vertex);
                ptr = ptr->next;
            }
        }
    }
    void deleteGraph ( struct node *Adj[], int no_of_nodes)
        {
            int i;
            struct node *temp, *ptr;
            for(i=0;i<= no_of_nodes;i++)
            {
                ptr = Adj[i];
                while(ptr!=NULL)
                {    temp = ptr;
                     ptr = ptr->next;
                     free(temp);
                }
                Adj[i] = NULL;
            }
    }
```

Note Note that, had it been a weighted graph, then the structure of the node would have been:

```
typedef struct node
{
        int vertex;
        int weight;
        struct node *next;
};
```

13.5 GRAPH TRAVERSAL ALGORITHMS

In this section, we will discuss how to traverse graph. By traversing a graph, we mean the method of examining the nodes and edges of the graph. There are two standard methods of graph traversal which we will discuss in this section. These two methods are:

(a) Breadth-first search

(b) Depth-first search

While breadth-first search uses a queue as an auxiliary data structure to store nodes for further processing, the depth-first search scheme uses a stack. But both these algorithms make use of a variable STATUS. During the execution of the algorithm, every node in the graph will have the variable STATUS set to 1 or 2, depending on its current state. Table 13.1 shows the value of STATUS and its significance.

Table 13.1 Value of status and its significance

Status	State of the node	Description
1	Ready	The initial state of the node N
2	Waiting	Node N is placed on the queue or stack and waiting to be processed
3	Processed	Node N has been completely processed

13.5.1 Breadth-First Search

Breadth-first search (BFS) is a graph search algorithm that begins at the root node and explores all the neighbouring nodes. Then for each of those nearest nodes, the algorithm (Fig. 13.19) explores their unexplored neighbour nodes, and so on, until it finds the goal.

That is, we start examining the node A and then all the neighbours of A are examined. In the next step, we examine the neighbours of neighbours of A, so on and so forth. This means that we need to track the neighbours of the node and guarantee that every node in the graph is processed and no node is processed more than once. This is accomplished by using a queue that will hold the nodes that are waiting for further processing and a variable STATUS to represent the current state of the node.

```
Step 1: SET STATUS = 1 (ready state) for each node in G.
Step 2: Enqueue the starting node A and set its STATUS = 2
        (waiting state)
Step 3: Repeat Steps 4 and 5 until QUEUE is empty
Step 4: Dequeue a node N. Process it and set its STATUS = 3
        (processed state).
Step 5: Enqueue all the neighbors of N that are in the ready
        state (whose STATUS = 1) and set their STATUS = 2
        (waiting state)
          [END OF LOOP]
Step 6: EXIT
```

Figure 13.19 Algorithm for breadth-first search

Example 13.1: Consider the graph G given in Fig. 13.20. The adjacency list of G is also given. Assume that G represents the daily flights between different cities and we want to fly from city A to H with minimum stops. That is, find the minimum path P from A to H given that every edge has a length of 1.

The minimum path P can be found by applying the breadth-first search algorithm that begins at city A and ends when H is encountered. During the execution of the algorithm, we use two arrays

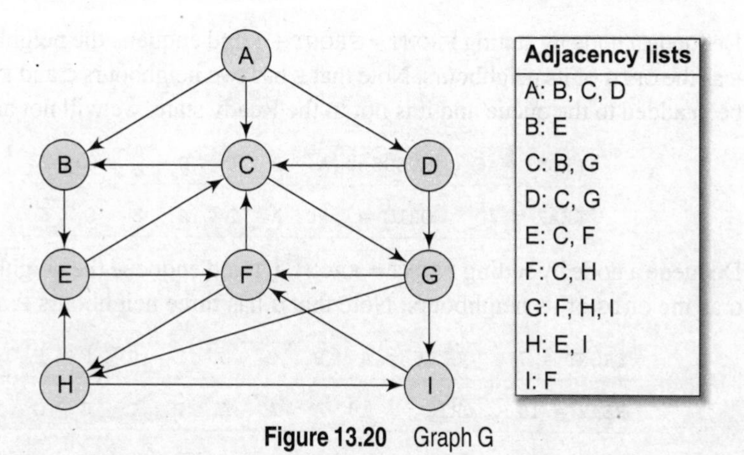

Figure 13.20 Graph G

QUEUE and ORIG. While QUEUE is used to hold the nodes that have to be processed; ORIG is used to keep track of the origin of each edge. The algorithm for this is as follows:

(a) Initially, add A to QUEUE and add NULL to ORIG.

```
FRONT = 1    QUEUE = A
REAR = 1     ORIG =  \0
```

(b) Dequeue a node by setting FRONT = FRONT + 1 (remove the FRONT element of QUEUE) and enqueue the neighbours of A. Also, add A as the ORIG of its neighbours.

```
FRONT = 2    QUEUE = A  B  C  D
REAR = 4     ORIG =  \0 A  A  A
```

(c) Dequeue a node by setting FRONT = FRONT + 1 and enqueue the neighbours of B. Also, add B as the ORIG of its neighbours.

```
FRONT = 3    QUEUE = A  B  C  D  E
REAR = 5     ORIG =  \0 A  A  A  B
```

(d) Dequeue a node by setting FRONT = FRONT + 1 and enqueue the neighbours of C. Also, add C as the ORIG of its neighbours. Note that C has two neighbours B and G. Since B has already been added to the queue and it is not in the Ready state, we will not add B and only add G.

```
FRONT = 4    QUEUE = A  B  C  D  E  G
REAR = 6     ORIG =  \0 A  A  A  B  C
```

(e) Dequeue a node by setting FRONT = FRONT + 1 and enqueue the neighbours of D. Also, add D as the ORIG of its neighbours. Note that D has two neighbours C and G. Since both of them have already been added to the queue and they are not in the Ready state, we will not add them again.

```
FRONT = 5    QUEUE = A  B  C  D  E  G
REAR = 6     ORIG =  \0 A  A  A  B  C
```

(f) Dequeue a node by setting FRONT = FRONT + 1 and enqueue the neighbours of E. Also, add E as the ORIG of its neighbours. Note that E has two neighbours C and F. Since C has already been added to the queue and it is not in the Ready state, we will not add C and add only F.

```
FRONT = 6    QUEUE = A  B  C  D  E  G  F
REAR = 7     ORIG =  \0 A  A  A  B  C  E
```

(g) Dequeue a node by setting FRONT = FRONT + 1 and enqueue the neighbours of G. Also, add G as the ORIG of its neighbours. Note that G has three neighbours F, H, and I.

```
FRONT = 7    QUEUE = A  B  C  D  E  G  F  H  I
REAR = 10    ORIG =  \0 A  A  A  B  C  G  G  G
```

Since I is our final destination, we stop the execution of this algorithm as soon as it is encountered and added to the QUEUE. Now, backtrack from I using ORIG to find the minimum path P. Thus, we have P as A -> C -> G -> I.

Features of Breadth-First Search Algorithm

Space complexity In the breadth-first search algorithm, all the nodes at a particular level must be saved until their child nodes in the next level have been generated. The space complexity is therefore proportional to the number of nodes at the deepest level of the graph. Given a graph with branching factor b (number of children at each node) and depth d, the asymptotic space complexity is the number of nodes at the deepest level $O(b^d)$.

If the number of vertices and edges in the graph are known ahead of time, the space complexity can also be expressed as $O(|E| + |V|)$, where $|E|$ is the total number of edges in G and $|V|$ is the number of nodes or vertices.

Time complexity In the worst case, breadth-first search has to traverse through all paths to all possible nodes, the time complexity of this algorithm asymptotically approaches $O(b^d)$. However, the time complexity can also be expressed as $O(|E| + |V|)$, since every vertex and every edge will be explored in the worst case.

Completeness Breadth-first search is said to be a complete algorithm because if there is a solution, breadth-first search will find it regardless of the kind of graph. But in case of an infinite graph where there is no possible solution, it will diverge.

Optimality Breadth-first search is optimal for a graph that has edges of equal length, since it always returns the result with the fewest edges between the start node and the goal node. But generally, in real-world applications, we have weighted graphs that have costs associated with each edge, so the goal next to the start does not have to be the cheapest goal available.

Applications of Breadth-First Search

Breadth-first search can be used to solve many problems such as:

- Finding all connected components in a graph G.
- Finding all nodes within an individual connected component.
- Finding the shortest path between two nodes, u and v, of an unweighted graph.
- Finding the shortest path between two nodes, u and v, of a weighted graph.

PROGRAMMING EXAMPLE

2. Write a program to implement the breadth-first search algorithm.

```
include <stdio.h>
#define MAX 10
void breadth_first_search(int adj[][MAX],int visited[],int start)
{
        int queue[MAX],rear=-1,front=-1,i;
        queue[++rear]=start;
        visited[start]=1;
        while(rear != front)
        {
                start = queue[++front];
                if(start==4)
                        printf("5\t");
                else
                        printf("%c \t",start+65);
                for(i=0;i<MAX;i++)
                {
                        if(adj[start][i]==1 && visited[i]==0)
                        {
                                queue[++rear]=i;
                                visited[i]=1;
                        }
                }
        }
}
int main()
{
        int visited[MAX]={0};
        int adj[MAX][MAX], i, j;
        printf("\n Enter the adjacency matrix : ");
        for(i=0, i<MAX;i++)
          for(j=0;j<MAX;j++)
                scanf("%d", &adj[i][j]);
        breadth_first_search(adj,visited,0);
        return 0;
}
```

13.5.2 Depth-first Search Algorithm

The depth-first search algorithm (Fig. 13.21) progresses by expanding the starting node of G and thus going deeper and deeper until a goal node is found, or until a node that has no children is encountered. When a dead-end is reached, the algorithm backtracks, returning to the most recent node that has not been completely explored.

In other words, depth-first search begins at a starting node A which becomes the current node. Then, it examines each node N along a path P which begins at A. That is, we process a neighbour of A, then a neighbour of neighbour of A, and so on. During the execution of the algorithm, if we reach a path that has a node N that has already been processed, then we backtrack to the current node. Otherwise, the unvisited (unprocessed) node becomes the current node.

The algorithm proceeds like this until we reach a dead-end (end of path P). On reaching the dead-end, we backtrack to find another path P′. The algorithm terminates when backtracking leads back to the starting node A. In this algorithm, edges that lead to a new vertex are called *discovery edges* and edges that lead to an already visited vertex are called *back edges*.

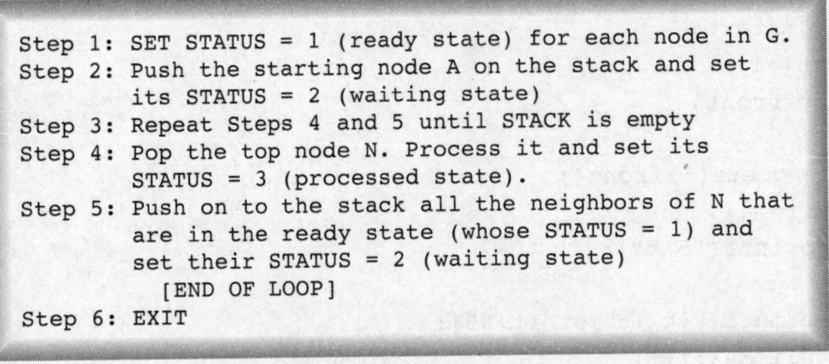

```
Step 1: SET STATUS = 1 (ready state) for each node in G.
Step 2: Push the starting node A on the stack and set
        its STATUS = 2 (waiting state)
Step 3: Repeat Steps 4 and 5 until STACK is empty
Step 4: Pop the top node N. Process it and set its
        STATUS = 3 (processed state).
Step 5: Push on to the stack all the neighbors of N that
        are in the ready state (whose STATUS = 1) and
        set their STATUS = 2 (waiting state)
          [END OF LOOP]
Step 6: EXIT
```

Figure 13.21 Algorithm for depth-first search

Observe that this algorithm is similar to the in-order traversal of a binary tree. Its implementation is similar to that of the breadth-first search algorithm but here we use a stack instead of a queue. Again, we use a variable STATUS to represent the current state of the node.

Example 13.2: Consider the graph G given in Fig. 13.22. The adjacency list of G is also given. Suppose we want to print all the nodes that can be reached from the node H (including H itself). One alternative is to use a depth-first search of G starting at node H. The procedure can be explained here.

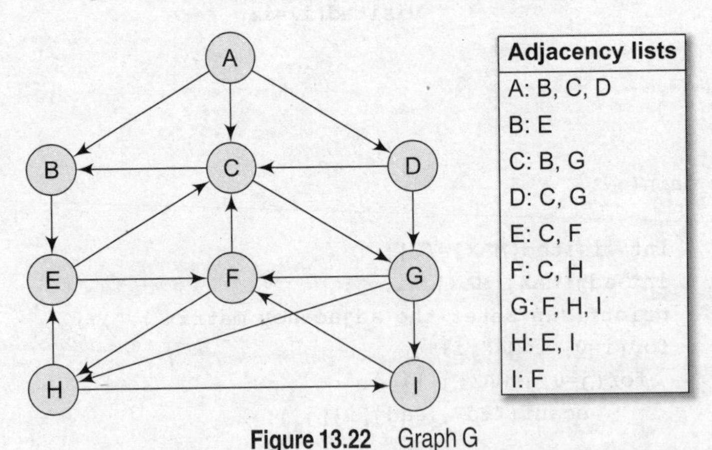

Figure 13.22 Graph G

(a) Push H onto the stack.

STACK: H

(b) Pop and print the top element of the STACK, that is, H. Push all the neighbours of H onto the stack that are in the ready state. The STACK now becomes

PRINT: H STACK: E, I

(c) Pop and print the top element of the STACK, that is, I. Push all the neighbours of I onto the stack that are in the ready state. The STACK now becomes

PRINT: I STACK: E, F

(d) Pop and print the top element of the `STACK`, that is, `F`. Push all the neighbours of `F` onto the stack that are in the ready state. (Note F has two neighbours, `C` and `H`. But only `C` will be added, as `H` is not in the ready state). The `STACK` now becomes

PRINT: F | STACK: E, C

(e) Pop and print the top element of the `STACK`, that is, `C`. Push all the neighbours of `C` onto the stack that are in the ready state. The `STACK` now becomes

PRINT: C | STACK: E, B, G

(f) Pop and print the top element of the `STACK`, that is, `G`. Push all the neighbours of `G` onto the stack that are in the ready state. Since there are no neighbours of `G` that are in the ready state, no push operation is performed. The `STACK` now becomes

PRINT: G | STACK: E, B

(g) Pop and print the top element of the `STACK`, that is, `B`. Push all the neighbours of `B` onto the stack that are in the ready state. Since there are no neighbours of `B` that are in the ready state, no push operation is performed. The `STACK` now becomes

PRINT: B | STACK: E

(h) Pop and print the top element of the `STACK`, that is, `E`. Push all the neighbours of `E` onto the stack that are in the ready state. Since there are no neighbours of `E` that are in the ready state, no push operation is performed. The `STACK` now becomes empty

PRINT: E | STACK:

Since the `STACK` is now empty, the depth-first search of `G` starting at node `H` is complete and the nodes which were printed are:

```
H, I, F, C, G, B, E.
```

These are the nodes which are reachable from the node `H`.

Features of Depth-First Search Algorithm

Space complexity The space complexity of a depth-first search is lower than that of a breadth-first search.

Time complexity The time complexity of a depth-first search is proportional to the number of vertices plus the number of edges in the graphs that are traversed. The time complexity can be given as `(O(|V| + |E|))`.

Completeness Breadth-first search is said to be a complete algorithm. If there is a solution, breadth-first search will find it regardless of the kind of graph. But in case of an infinite graph, where there is no possible solution, it will diverge.

Applications of Depth-First Search Algorithm

Depth-first search is useful for:

- Finding a path between two specified nodes, `u` and `v`, of an unweighted graph.
- Finding a path between two specified nodes, `u` and `v`, of an weighted graph.
- Finding whether a graph is connected or not.
- Computing the spanning tree of a connected graph.

PROGRAMMING EXAMPLE

3. Write a program to implement the depth-first search algorithm.

```
#include<stdio.h>
#define MAX 5
void depth_first_search(int adj[][MAX],int visited[],int start)
{
     int stack[MAX];
     int top=-1,i;
     printf("%c-",start+65);
     visited[start]=1;
     stack[++top]=start;
     while(top!=-1)
     {
          start=stack[top];
          for(i=0;i<MAX;i++)
          {
               if(adj[start][i]&&visited[i]==0)
               {
                    stack[++top]=i;
                    printf("%c-",i+65);
                    visited[i]=1;
                    break;
               }
          }
          if(i==MAX)
               top--;
     }
}
int main()
{
     int adj[MAX][MAX];
     int visited[MAX]={0}, i, j;
     printf("\n Enter the adjacency matrix : ");
     for(i=0, i<MAX;i++)
          for(j=0;j<MAX;j++)
               scanf("%d", &adj[i][j]);
     printf("DFS Traversal : ");
     depth_first_search(adj,visited,0);
     printf("\n");
     return 0;
}
```

13.6 TOPOLOGICAL SORTING

Topological sort of a directed acyclic graph (DAG) G is defined as a linear ordering of its nodes in which each node comes before all nodes to which it has outbound edges. Every DAG has one or more number of topological sorts.

That is, a topological sort of a DAG G is an ordering of the vertices of G such that if G contains an edge (u, v), then u appears before v in the ordering. Note that topological sort is possible only on directed acyclic graphs that do not have any cycle. For a DAG that contains cycle, no linear ordering of its vertices is possible.

In simple words, a topological ordering of a DAG G is an ordering of its vertices such that any directed path in G traverses the vertices in increasing order.

Topological sorting is widely used in scheduling applications, jobs, or tasks. The jobs that have to be completed are represented by nodes, and there is an edge from node u to v if job u must be completed before job v can be started. A topological sort of such a graph gives an order in which the given jobs must be performed.

EXAMPLE 13.3: Consider three DAGs shown in Fig. 13.23 and their possible topological sorts.

More the number of edges in a DAG, fewer the number of topological orders it has. This is because each edge (u, v) forces node u to occur before v, which restricts the number of valid permutations of the nodes.

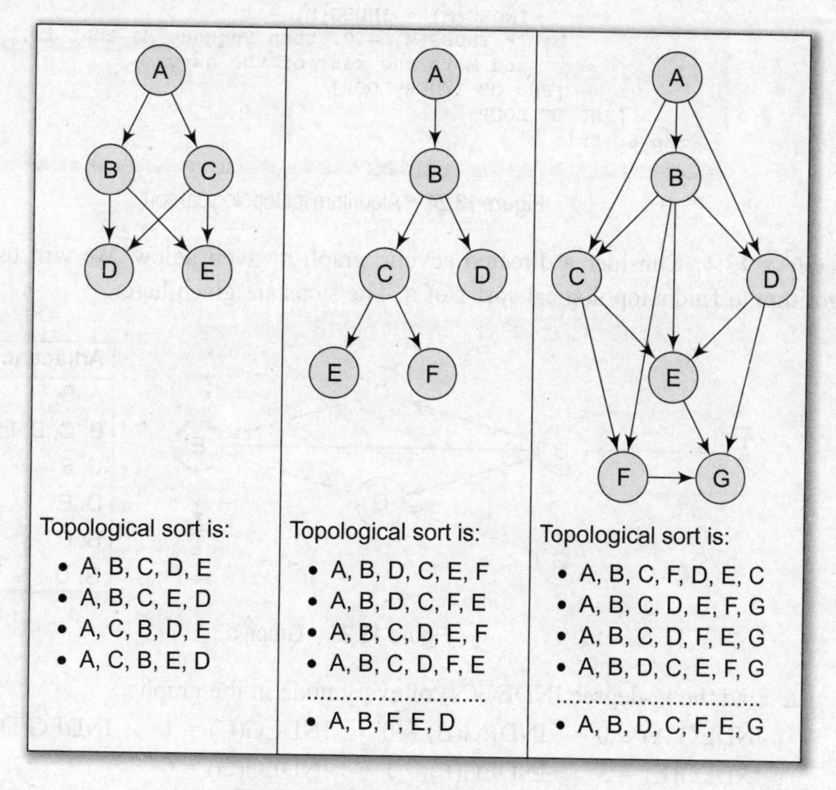

Figure 13.23 Topological sort

Algorithm

The algorithm for the topological sort of a graph (Fig. 13.24) that has no cycles focuses on selecting a node N with zero in-degree, that is a node that has no predecessor. The two main steps involved in the topological sort algorithm include:

- Selecting a node with zero in-degree, and
- Deleting N from the graph along with its edges.

We will use a QUEUE to hold the nodes with zero in-degree. The order in which the nodes will be deleted from the graph will depend on the sequence in which the nodes are inserted in the QUEUE. Then, we use a variable INDEG, where INDEG(N) will represent the in-degree of node N. Note that the in-degree can be calculated in two ways, either by counting the incoming edges from the graph or traversing through the adjacency list.

Note that the running time of the algorithm for topological sorting can be given as linear in the number of nodes plus the number of edges O(|V|+|E|).

```
Step 1: Find the in-degree INDEG(N) of every node in the
        graph
Step 2: Enqueue all the nodes with a zero in-degree
Step 3: Repeat Steps 4 and 5 until the QUEUE is empty
Step 4:     Remove the front node N of the QUEUE by setting
            FRONT = FRONT + 1
Step 5:     Repeat for each neighbor M of node N:
              a) Delete the edge from N to M by setting
                 INDEG(M) = INDEG(M) - 1
              b) IF INDEG(M) = 0, then Enqueue M, that is,
                 add M to the rear of the queue
            [END OF INNER LOOP]
        [END OF LOOP]
Step 6: Exit
```

Figure 13.24 Algorithm for topological sort

Example 13.4: Consider a directed acyclic graph G given below. We will use the discussed algorithm to find a topological sort T of G. The steps are given here:

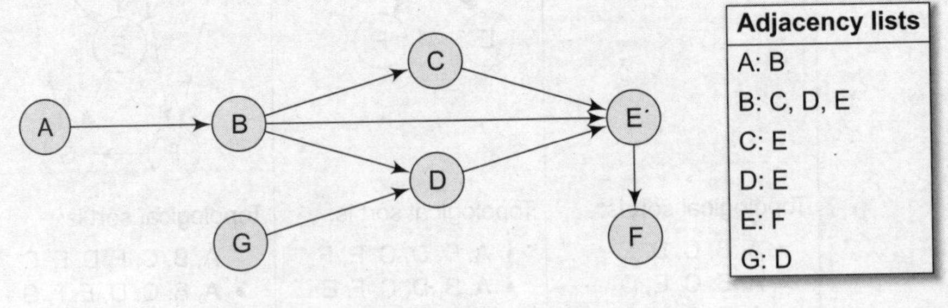

Figure 13.25 Graph G

Step 1: Find the in-degree INDEG(N) of every node in the graph

INDEG(A) = 0 INDEG(B) = 0 INDEG(C) = 1 INDEG(D) = 2

INDEG(E) = 3 INDEG(F) = 1 INDEG(G) = 0

Step 2: Enqueue all the nodes with a zero in-degree

FRONT = 1 REAR = 3 QUEUE = A, B, G

Step 3: Remove the front element A from the queue by setting `FRONT = FRONT + 1`, so

FRONT = 2 REAR = 3 QUEUE = A, B, G

Step 4: Set INDEG(B) = INDEG(B) – 1, since B is the neighbour of A. Note that `INDEG(B)` is 0 and it is already on the queue. So, delete the edge from A to B. The graph now becomes,

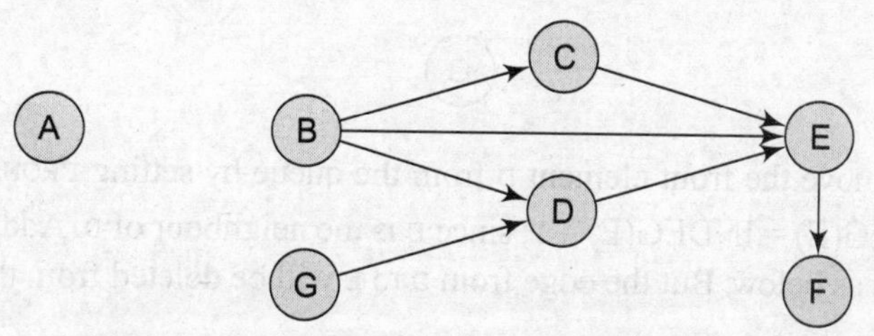

Step 5: Remove the front element B from the queue by setting `FRONT = FRONT + 1`, so

FRONT = 3 REAR = 3 QUEUE = A, B, G

Step 6: Set INDEG(C) = INDEG(C) – 1, INDEG(D) = INDEG(D) – 1, INDEG(E) = INDEG(E) – 1, since C, D, and E are the neighbours of B. Now,

INDEG(C) = 0, INDEG(D) = 1 and INDEG(E) = 2

Since in-degree of node C is zero, add C at the rear of the queue. The queue and the corresponding graph can be given as below:

FRONT = 3 REAR = 4 QUEUE = A, B, G, C

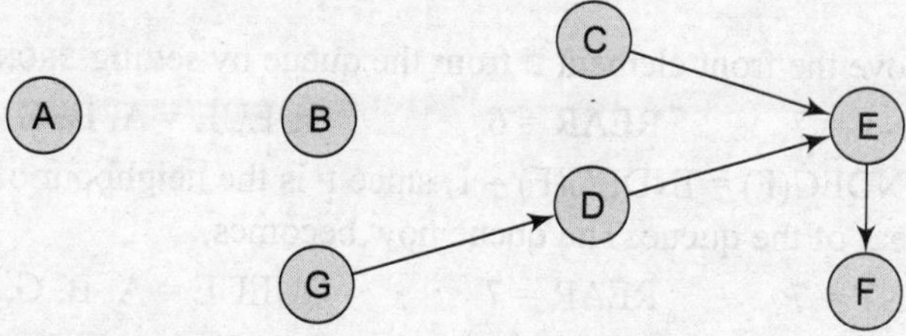

Step 7: Remove the front element G from the queue by setting `FRONT = FRONT + 1`, so

FRONT = 4 REAR = 4 QUEUE = A, B, G, C

Step 8: Set INDEG(D) = INDEG(D) – 1, since D is the neighbour of G. Now, INDEG(D) = 0

Since the in-degree of node D is zero, add D at the rear of the queue. The queue and the corresponding graph can be given as below:

FRONT = 4 REAR = 5 QUEUE = A, B, G, C, D

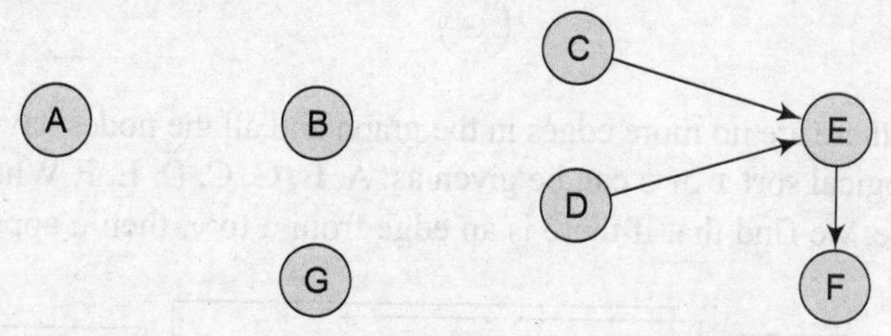

Step 9: Remove the front element C from the queue by setting `FRONT = FRONT + 1`, so

FRONT = 5 REAR = 5 QUEUE = A, B, G, C, D

Step 10: Set INDEG(E) = INDEG(E) – 1, since E is the neighbour of C. Now, INDEG(E) = 1

The queue and the corresponding graph can be given as below:

FRONT = 5 REAR = 5 QUEUE = A, B, G, C, D

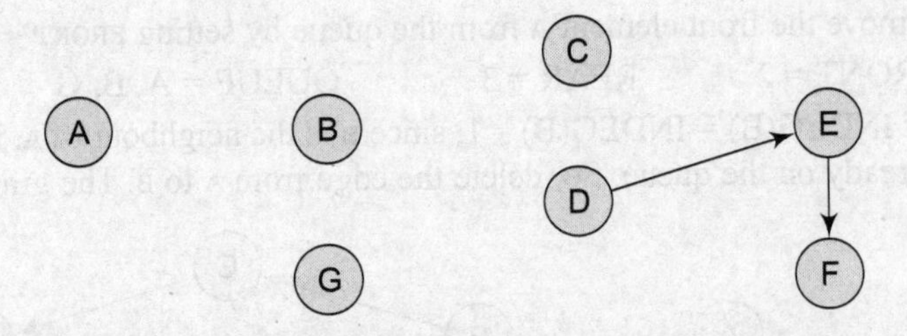

Step 11: Remove the front element D from the queue by setting `FRONT = FRONT + 1`, so

Set INDEG(E) = INDEG(E) – 1, since E is the neighbour of D. Add, E in the queue. The queue can be given as below. But the edge from D to E will be deleted from the graph, so the graph now becomes,

FRONT = 6 REAR = 6 QUEUE = A, B, G, C, D, E

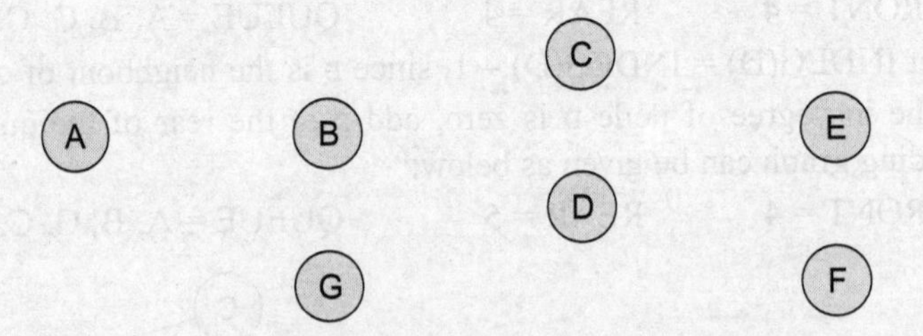

Step 13: Remove the front element E from the queue by setting `FRONT = FRONT + 1`, so

FRONT = 7 REAR = 6 QUEUE = A, B, G, C, D, E

Step 14: Set INDEG(F) = INDEG(F) – 1, since F is the neighbour of E. Now `INDEG(F) = 0`, so add F to the rear of the queue. The queue now becomes,

FRONT = 7 REAR = 7 QUEUE = A, B, G, C, D, E, F

The edge from D to E will be deleted from the graph, so the graph now becomes,

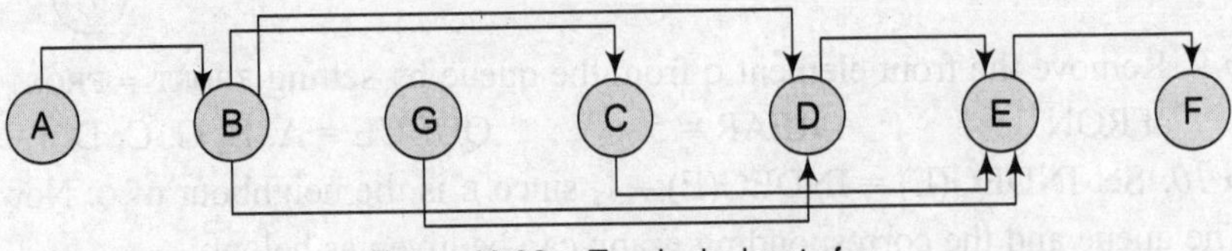

Note that there are no more edges in the graph and all the nodes have been added to the queue, so the topological sort T of G can be given as: A, B, G, C, D, E, F. When we arrange these nodes in a sequence, we find that if there is an edge from u to v, then u appears before v.

Figure 13.26 Topological sort of G

PROGRAMMING EXAMPLE

4. Write a program to implement topological sorting.

```
#include<stdio.h>
#include<conio.h>
#define MAX 10
int n,adj[MAX][MAX];
int front=-1,rear=-1,queue[MAX];
void create_graph(void);
void display();
void insert_queue(int);
int delete_queue(void);
int find_indegree(int);
main()
{
    int node,j=0,del_node, node;
    int topsort[MAX],indeg[MAX];
    create_graph();
    printf("\n The adjacency matrix is :");
    display();
    /*Find the indegree of each node*/
    for(node=1;node<=n;node++)
    {
        indeg[node]= find_indegree(node);
        if( indeg[node]==0 )
            insert_queue(node);
    }
    while(front<=rear) /*Continue loop until queue is empty */
    {
        del_node = delete_queue();
        topsort[j]= del_node; /*Add the deleted node to topsort*/
        j++;
        /*Delete the del_node edges */
        for(node=1;node<=n;node++)
        {
            if( adj[del_node][node]==1 )
            {
                adj[del_node][node]=0;
                indeg[node]=indeg[node]-1;
                if(indeg[node]==0)
                    insert_queue(node);
            }
        }
    }
    printf("The topological sorting can be given as :\n");
    for(node=0;i<j;node++)
```

```
                printf( "%d ",topsort[node] );
}
void create_graph()
{
        int i,max_edges,org,dest;
        printf("\n Enter number of vertices : ");
        scanf("%d",&n);
        max_edges=n*(n-1);
        for(i=1;i<=max_edges;i++)
        {
              printf("\n Enter edge %d(0 to quit): ",i);
              scanf("%d %d",&org,&dest);
              if((org==0) && (dest==0))
                      break;
              if( org > n || dest > n || org<=0 || dest<=0)
              {
                      printf("\n Invalid edge");
                      i--;
              }
              else
                      adj[org][dest]=1;
        }
}
void display()
{
        int i,j;
        for(i=1;i<=n;i++)
        {
              printf("\n");
              for(j=1;j<=n;j++)
                    printf("%3d",adj[i][j]);
        }
}
void insert_queue(int node)
{
        if (rear==MAX-1)
              printf("\n OVERFLOW ");
        else
        {
              if (front==-1) /*If queue is initially empty */
                      front=0;
              queue[++rear] = node ;
        }
}
int delete_queue()
{
        int del_node;
        if (front == -1 || front > rear)
```

```
        {
            printf("\n UNDERFLOW ");
            return ;
        }
        else
        {
            del_node=queue[front++];
            return del_node;
        }
    }
int find_indegree(int node)
{
        int i,in_deg=0;
        for(i=1;i<=n;i++)
            if( adj[i][node] == 1 )
                in_deg++;
        return in_deg;
}
```

13.7 SHORTEST PATH ALGORITHMS

In this section, we will discuss three different algorithms to calculate the shortest path between the vertices of a graph G. These algorithms include:

- Minimum spanning tree
- Dijkstra's algorithm
- Warshall's algorithm

While the first two use an adjacency list to find the shortest path, Warshall's algorithm uses an adjacency matrix to do the same.

13.7.1 Minimum Spanning Tree

A spanning tree of a connected, undirected graph G is a sub-graph of G which is a tree that connects all the vertices together. A graph G can have many different spanning trees. We can assign *weights* to each edge (which is a number that represents how unfavorable the edge is), and use it to assign a weight to a spanning tree by calculating the sum of the weights of the edges in that spanning tree. A *minimum spanning tree* (MST) is defined as a spanning tree with weight less than or equal to the weight of every other spanning tree. In other words, a minimum spanning tree is a spanning tree that has weights associated with its edges, and the total weight of the tree (the sum of the weights of its edges) is at a minimum.

An Analogy

Take an analogy of a cable TV company laying cable to a new neighbourhood. If it is restricted to bury the cable only along particular paths, then we can make a graph that represents the points that are connected by those paths. Some paths may be more expensive (due to their length or the depth at which the cable should be buried) than the others, we can represent these paths by edges with larger weights.

Therefore, a spanning tree for such a graph would be a subset of those paths that has no cycles but still connects to every house. Many distinct spanning trees can be obtained from this graph, but a minimum spanning tree would be the one with the lowest total cost.

Properties

Possible multiplicity There can be multiple minimum spanning trees of the same weight. Particularly, if all the weights are the same, then every spanning tree will be minimum.

Uniqueness When each edge in the graph is assigned a different weight, then there will only be one, unique minimum spanning tree.

Minimum-cost subgraph If the edges of a graph are assigned *non-negative* weights, then a minimum spanning tree is in fact the minimum-cost subgraph or a tree that connects all vertices.

Cycle property If there exists a cycle C in the graph G that has a weight larger than that of other edges of C, then this edge cannot belong to an MST.

Usefulness Minimum spanning trees can be computed quickly and easily to provide optimal solutions. These trees create a sparse subgraph that reflects a lot about the original graph.

Simplicity The minimum spanning tree of a weighted graph is nothing but a spanning tree of the graph which comprises of n-1 edges of minimum total weight. Note that for an unweighted graph, any spanning tree is a minimum spanning tree.

EXAMPLE 13.5: Consider an unweighted graph G given below. From G, we can draw many distinct spanning trees. Eight of them are given here. For an unweighted graph, every spanning tree is a minimum spanning tree.

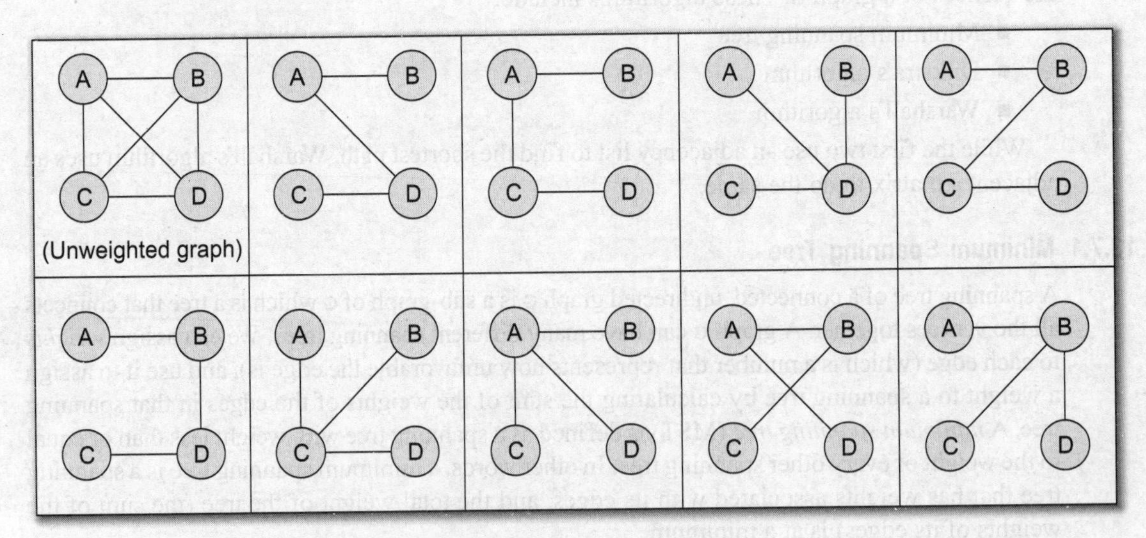

Figure 13.27 Spanning tree

EXAMPLE 13.6: Consider a weighted graph G given below. From G, we can draw three distinct spanning trees. But only a single minimum spanning t ree can be obtained, that is, the one that has the minimum weight (cost) associated with it.

Of all the spanning trees given inside Fig. 13.28, the one that is highlighted is called the minimum spanning tree, as it has the lowest cost associated with it.

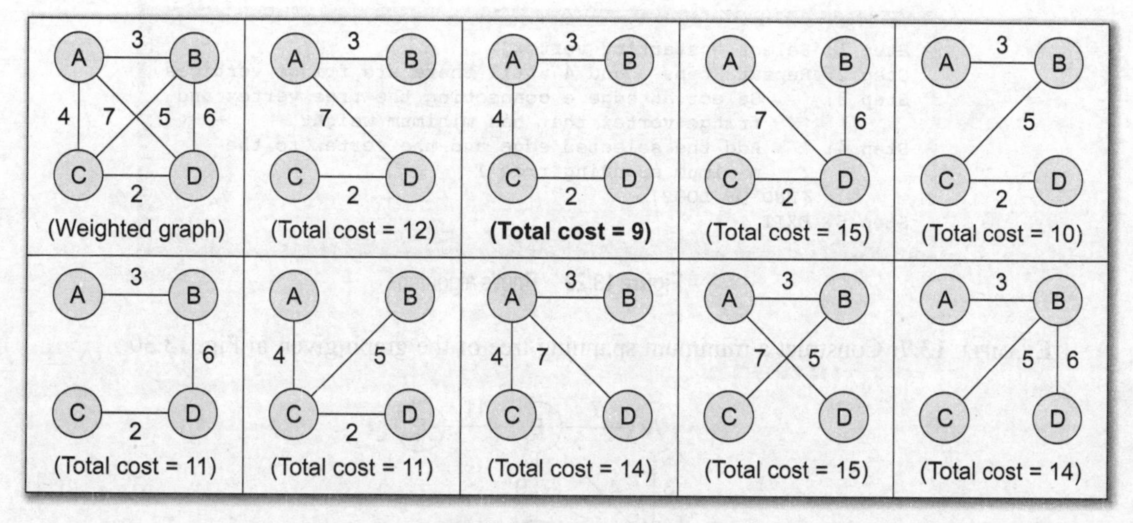

Figure 13.28 Spanning tree

Applications of MST

1. It is widely used for designing networks. For instance, people separated by varying distances wish to be connected together through a telephone network. A minimum spanning tree is used to determine the least costly paths with no cycles in this network, thereby providing a connection that has the minimum cost involved.
2. MSTs are used to find airline routes. While the vertices in the graph denote cities, edges represent the routes between these cities. No doubt, more the distance between the cities, higher will be the amount charged. Therefore, MSTs are used to optimize airline routes by finding the least costly path with no cycles.
3. It is also used to find the cheapest way to connect terminals, such as cities, electronic components or computers via roads, airlines, railways, wires or telephone lines.
4. MSTs are applied in routing algorithms for finding the most efficient path.

13.7.2 Prim's Algorithm

Let us first discuss the three kinds of vertices that we will be using during the execution of Prim's algorithm.

- ***Tree vertices*** Vertices that are a part of the minimum spanning tree `T`.
- ***Fringe vertices*** Vertices that are currently not a part of `T`, but are adjacent to some tree vertex.
- ***Unseen vertices*** Vertices that are neither tree vertices nor fringe vertices fall under this category.

The steps involved in the Prim's algorithm are shown in Fig. 13.29.

- Choose a starting vertex.
- Branch out from the starting vertex and during each iteration, select a new vertex and an edge. Basically, during each iteration of the algorithm, we have to select a vertex from the fringe vertices in such a way that the edge connecting the tree vertex and the new vertex has the minimum weight assigned to it.

The running time of Prim's algorithm can be given as `O(E log V)` where `E` is the number of edges and `V` is the number of vertices in the graph.

```
Step 1: Select a starting vertex
Step 2: Repeat Steps 3 and 4 until there are fringe vertices
Step 3:        Select an edge e connecting the tree vertex and
               fringe vertex that has minimum weight
Step 4:        Add the selected edge and the vertex to the
               minimum spanning tree T
          [END OF LOOP]
Step 5: EXIT
```

Figure 13.29 Prim's algorithm

EXAMPLE 13.7: Construct a minimum spanning tree of the graph given in Fig. 13.30.

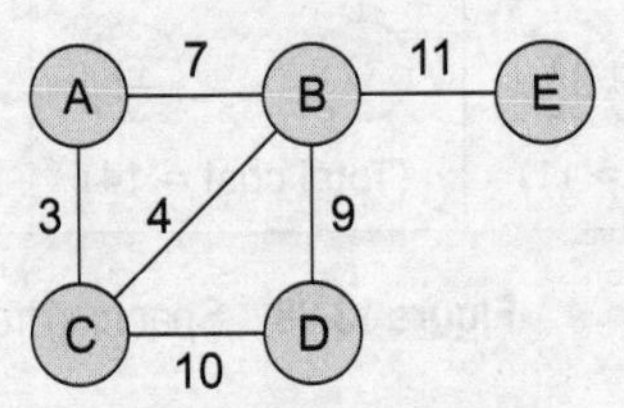

Figure 13.30 Graph G

Step 1: Choose a starting vertex A.

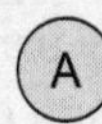

Step 2: Add the fringe vertices (that are adjacent to A). The edges connecting the vertex and fringe vertices are shown with dotted lines.

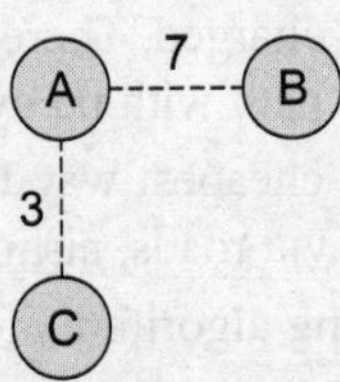

Step 3: Select an edge connecting the tree vertex and the fringe vertex that has the minimum weight and add the selected edge and the vertex to the minimum spanning tree T. Since the edge connecting A and C has less weight, add C to the tree. Now C is not a fringe vertex but a tree vertex.

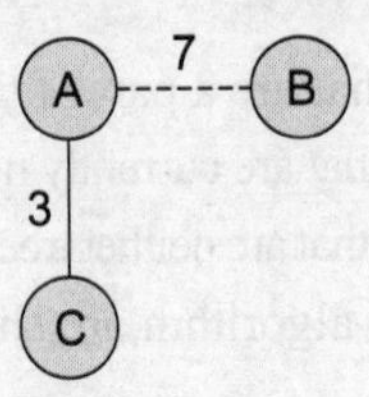

Step 4: Add the fringe vertices (that are adjacent to C).

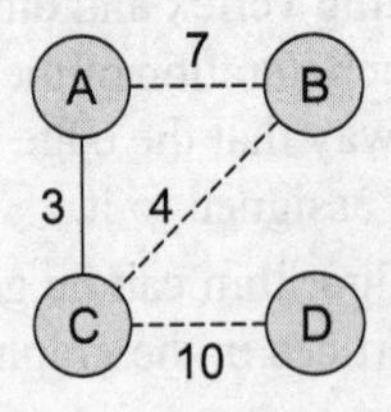

Step 5: Select an edge connecting the tree vertex and the fringe vertex that has the minimum weight and add the selected edge and the vertex to the minimum spanning tree T. Since the edge connecting C and B has less weight, add B to the tree. Now B is not a fringe vertex but a tree vertex.

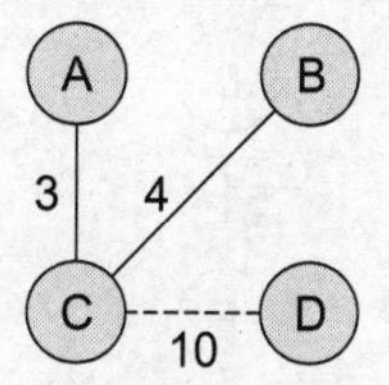

Step 6: Add the fringe vertices (that are adjacent to B).

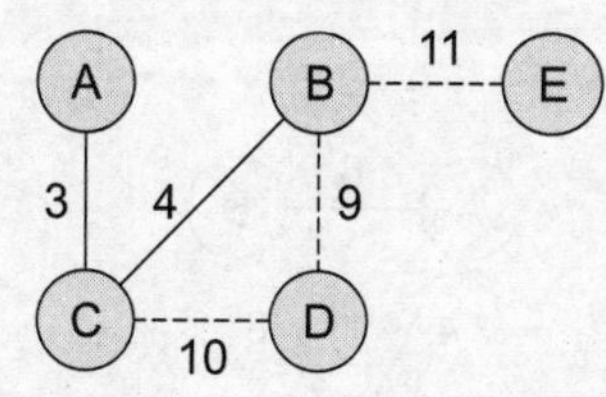

Step 7: Select an edge connecting the tree vertex and the fringe vertex that has the minimum weight and add the selected edge and the vertex to the minimum spanning tree T. Since the edge connecting B and D has less weight, add D to the tree. Now D is not a fringe vertex but a tree vertex.

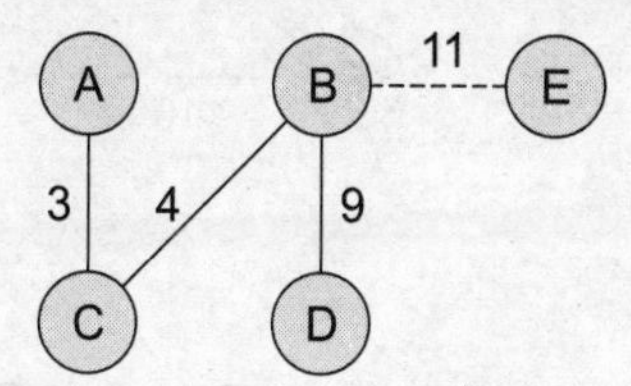

Step 8: Note, now node E is not connected, so we will add it in the tree because a minimum spanning tree is one in which all the n nodes are connected with n-1 edges that have minimum weight. So, the minimum spanning tree can now be given as,

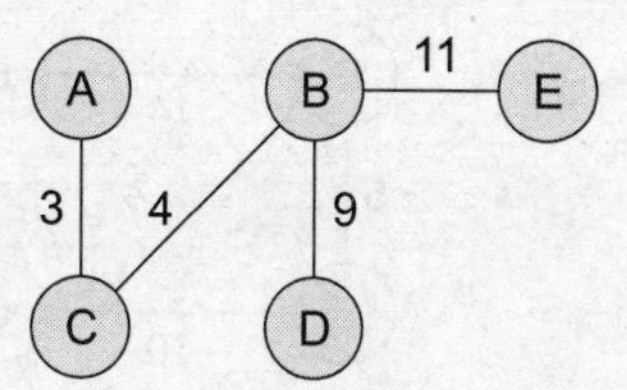

EXAMPLE 13.8: Construct a minimum spanning tree of the graph given in Fig.13.31. Start the Prim algorithm from vertex D.

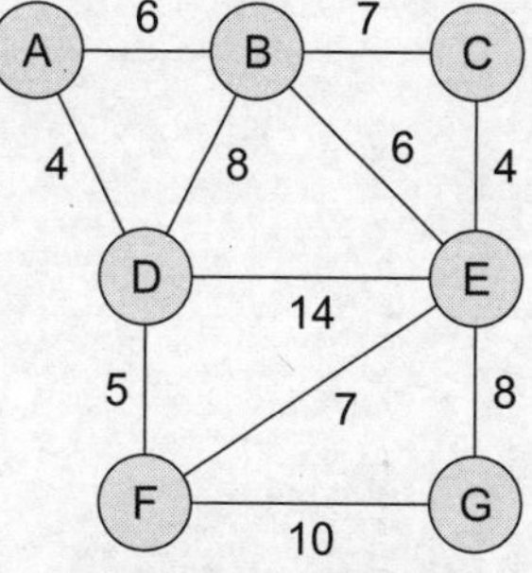

Figure 13.31 Graph G

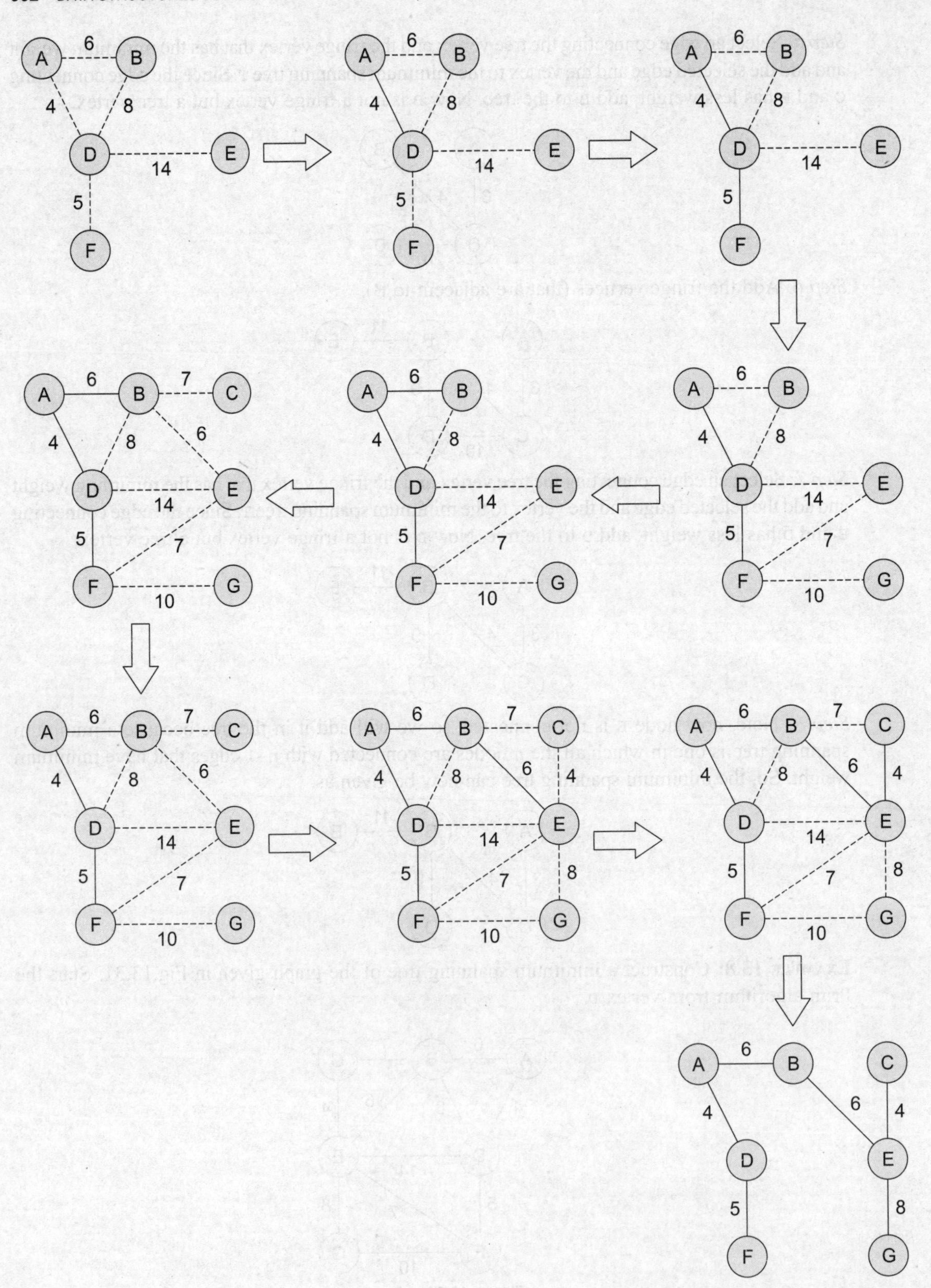

13.7.3 Kruskal's Algorithm

Kruskal's algorithm is used to find the minimum spanning tree for a connected weighted graph. The algorithm aims to find a subset of the edges that forms a tree that includes every vertex. The total weight of all the edges in the tree is minimized. However, if the graph is not connected, then it finds a *minimum spanning forest*. Note that a forest is a collection of trees. Similarly, a minimum spanning forest is a collection of minimum spanning trees.

Kruskal's algorithm is an example of a greedy algorithm, as it makes the locally optimal choice at each stage with the hope of finding the global optimum. The algorithm is shown in Fig. 13.32.

```
Step 1: Create a forest in such a way that each graph is a separate
        tree.
Step 2: Create a priority queue Q that contains all the edges of the
        graph.
Step 3: Repeat Steps 4 and 5 while Q is NOT EMPTY
Step 4:      Remove an edge from Q
Step 5: IF the edge obtained in Step 4 connects two different trees,
        then Add it to the forest (for combining two trees into one
        tree).
        ELSE
             Discard the edge
Step 6: END
```

Figure 13.32 Kruskal's algorithm

In the algorithm, we use a priority queue `Q` in which edges that have minimum weight takes a priority over any other edge in the graph. When the Kruskal's algorithm terminates, the forest has only one component and forms a minimum spanning tree of the graph. The running time of Kruskal's algorithm can be given as `O(E log V)`, where `E` is the number of edges and `V` is the number of vertices in the graph.

EXAMPLE 13.9: Apply Kruskal's algorithm on the graph given in Fig. 13.33.

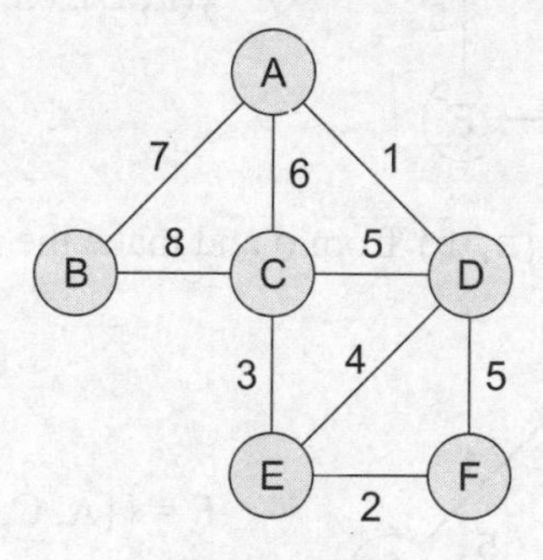

Figure 13.33

Initially, we have F = {{A}, {B}, {C}, {D}, {E}, {F}}

MST = {}

Q = {(A, D), (E, F), (C, E), (E, D), (C, D), (D, F), (A, C), (A, B), (B, C)}

Step 1: Remove the edge `(A, D)` from `Q` and make the following changes:

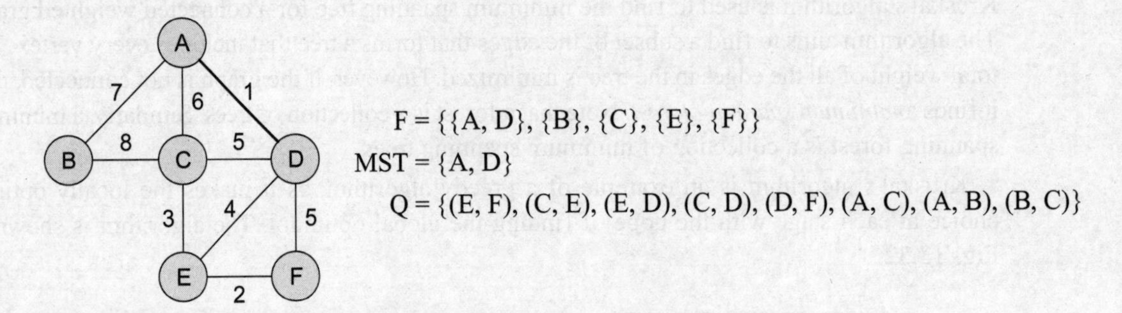

Step 2: Remove the edge `(E, F)` from `Q` and make the following changes:

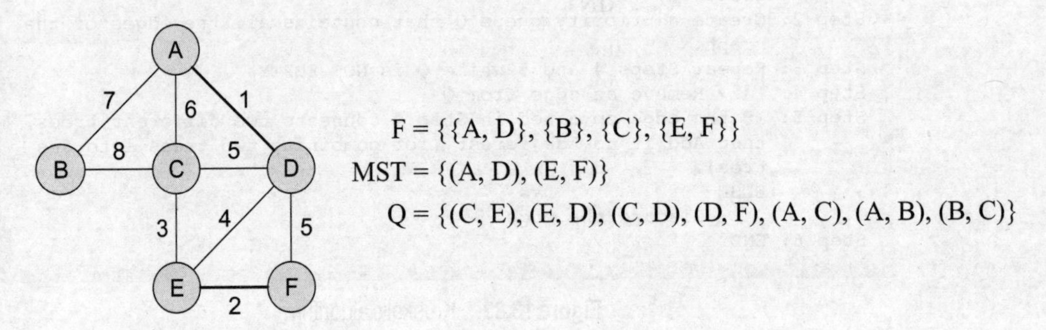

Step 3: Remove the edge `(C, E)` from `Q` and make the following changes:

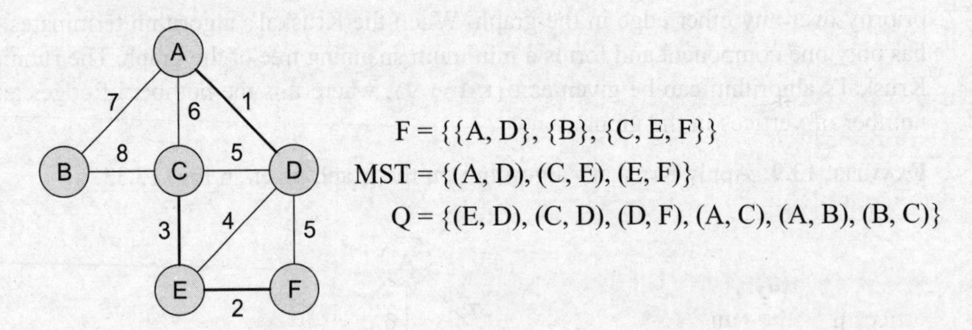

Step 4: Remove the edge `(E, D)` from `Q` and make the following changes:

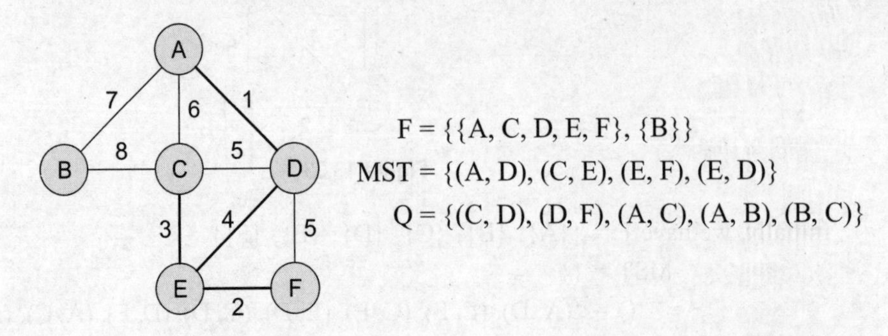

Step 5: Remove the edge `(C, D)` from `Q`. Note that this edge does not connect different trees, so simply discard this edge. Only an edge connecting `(A, D, C, E, F)` to `B` will be added to the MST. Therefore,

F = {{A, C, D, E, F}, {B}}
MST = {(A, D), (C, E), (E, F), (E, D)}
Q = {(D, F), (A, C), (A, B), (B, C)}

Step 6: Remove the edge `(D, F)` from `Q`. Note that this edge does not connect different trees, so simply discard this edge. Only an edge connecting `(A, D, C, E, F)` to `B` will be added to the MST.

F = {{A, C, D, E, F}, {B}}
MST = {(A, D), (C, E), (E, F), (E, D)}
Q = {(A, C), (A, B), (B, C)}

Step 7: Remove the edge `(A, C)` from `Q`. Note that this edge does not connect different trees, so simply discard this edge. Only an edge connecting `(A, D, C, E, F)` to `B` will be added to the MST.

F = {{A, C, D, E, F}, {B}}
MST = {(A, D), (C, E), (E, F), (E, D)}
Q = {(A, B), (B, C)}

Step 8: Remove the edge `(A, B)` from `Q` and make the following changes:

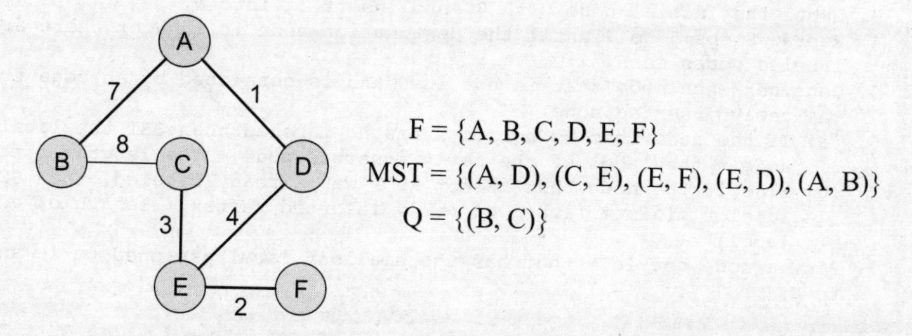

F = {A, B, C, D, E, F}
MST = {(A, D), (C, E), (E, F), (E, D), (A, B)}
Q = {(B, C)}

Step 9: The algorithm continues until `Q` is empty. Since the entire forest has become one tree, all the remaining edges will simply be discarded. The resultant MST can be given as shown below.

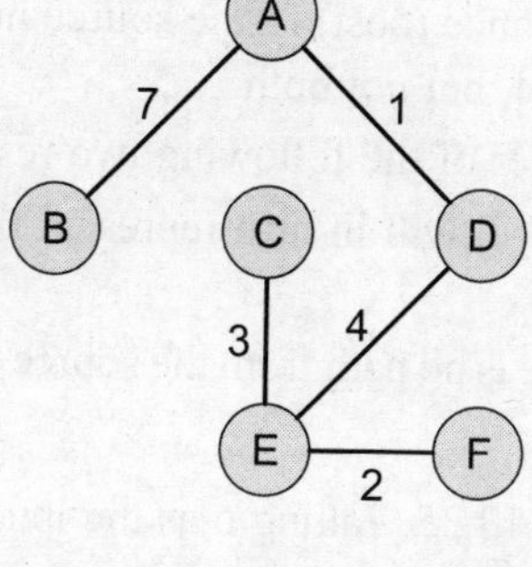

F = {A, B, C, D, E, F}
MST = {(A, D), (C, E), (E, F), (E, D), (A, B)}
Q = {}

13.7.4 Dijkstra's Algorithm

Dijkstra's algorithm, given by a Dutch scientist Edsger Dijkstra in 1959, is used to find the shortest path tree. This algorithm is widely used in network routing protocols, most notably IS-IS and OSPF (Open Shortest Path First).

Given a graph G and a source node A, the algorithm is used to find the shortest path (one having the lowest cost) between A (source node) and every other node. Moreover, Dijkstra's algorithm is also used for finding the costs of the shortest paths from a source node to a destination node.

For example, if we draw a graph in which nodes represent the cities and weighted edges represent the driving distances between pairs of cities connected by a direct road, then Dijkstra's algorithm when applied gives the shortest route between one city and all other cities.

Algorithm

Dijkstra's algorithm is used to find the length of an *optimal* path between two nodes in a graph. The term *optimal* can mean anything, shortest, cheapest, or fastest. If we start the algorithm with an initial node, then the distance of a node Y can be given as the distance from the initial node to that node. Figure 13.34 explains the Dijkstra's algorithm.

```
1. Select the source node also called the initial node
2. Define an empty set N that will be used to hold nodes to which a shortest
   path has been found.
3. Label the initial node with 0, and insert it into N.
4. Repeat Steps 5 to 7 until the destination node is in N or there are no more
   labeled nodes in N.
5. Consider each node that is not in N and is connected by an edge from
   the newly inserted node.
6. (a) If the node that is not in N has no labeled then SET the label of the
       node = the label of the newly inserted node + the length of the edge.
   (b) Else if the node that is not in N was already labeled, then SET its new
       label = minimum (label of newly inserted vertex + length of edge, old
       label)
7. Pick a node not in N that has the smallest label assigned to it and add it
   to N
```

Figure 13.34 Dijkstra's algorithm

Dijkstra's algorithm labels every node in the graph where the labels represent the distance (cost) from the source node to that node. There are two kinds of labels: *temporary* and *permanent*. Temporary labels are assigned to nodes that have not been reached, while permanent labels are given to nodes that have been reached and their distance (cost) to the source node is known. A node must be a permanent label or a temporary label, but not both.

The execution of this algorithm will produce either of the following two results:

(a) If the destination node is labeled, then the label will in turn represent the distance from the source node to the destination node.

(b) If the destination node is not labeled, then there is no path from the source to the destination node.

Example 13.10: Consider the graph G given in Fig. 13.35. Taking D as the initial node, execute the Dijkstra's algorithm on it.

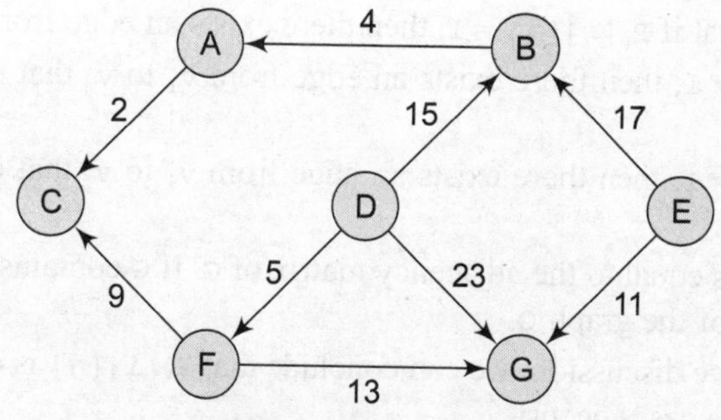

Figure 13.35 Graph G

Step 1: Set the label of `D = 0` and `N = {D}`.

Step 2: Label of `D = 0, B = 15, G = 23,` and `F = 5`. Therefore, `N = {D, F}`.

Step 3: Label of `D = 0, B = 15, G` has been re-labeled 18 because minimum `(5 + 13, 23) = 18,` C has been re-labeled 14 (5 + 9). Therefore, `N = {D, F, C}`.

Step 4: Label of `D = 0, B = 15, G = 18`. Therefore, `N = {D, F, C, B}`.

Step 5: Label of `D = 0, B = 15, G = 18` and `A = 19 (15 + 4)`. Therefore, `N = {D, F, C, B, G}`.

Step 6: Label of `D = 0` and `A = 19`. Therefore, `N = {D, F, C, B, G, A}`

Note that we have no labels for node `E`; this means that E is not reachable from `D`. Only the nodes that are in `N` are reachable from `B`.

The running time of Dijkstra's algorithm can be given as $O(|V|^2+|E|)=O(|V|^2)$ where `V` is the set of vertices and `E` in the graph.

Difference Between Dijkstra's Algorithm and Minimum Spanning Tree

Minimum spanning tree algorithm is used to traverse a graph in the most efficient manner, but Dijkstra's algorithm calculates the distance from a given vertex to every other vertex in the graph.

Dijkstra's algorithm is very similar to Prim's algorithm. Both the algorithms begin at a specific node and extend outward within the graph, until all other nodes in the graph have been reached. The point where these algorithms differ is that while Prim's algorithm stores a minimum cost edge, Dijkstra's algorithm stores the total cost from a source node to the current node. Moreover, Dijkstra's algorithm is used to store the summation of minimum cost edges, while Prim's algorithm stores at most one minimum cost edge.

13.7.5 Warshall's Algorithm

If a graph `G` is given as `G=(V, E)`, where `V` is the set of vertices and `E` is the set of edges, the path matrix of `G` can be found as, $P = A + A^2 + A^3 + \ldots + A^n$. This is a lengthy process, so Warshall has given a very efficient algorithm to calculate the shortest path between two vertices. Warshall's algorithm defines matrices $P_0, P_1, P_2, \ldots, P_n$ as given in Fig. 13.36.

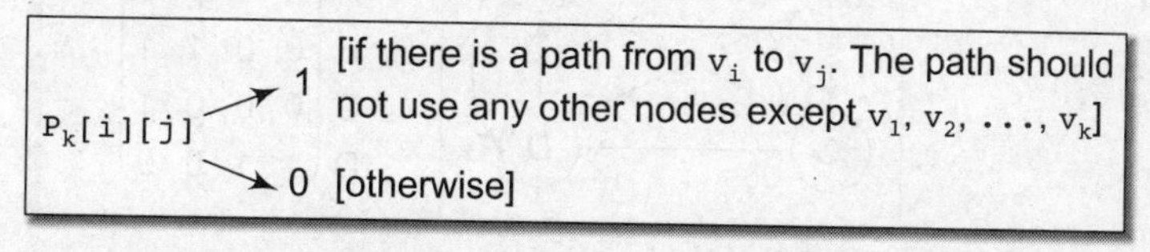

Figure 13.36 Path matrix entry

This means that if $P_0[i][j] = 1$, then there exists an edge from node v_i to v_j.

If $P_1[i][j] = 1$, then there exists an edge from v_i to v_j that does not use any other vertex except v_1.

If $P_2[i][j] = 1$, then there exists an edge from v_i to v_j that does not use any other vertex except v_1 and v_2.

Note that P_0 is equal to the adjacency matrix of G. If G contains n nodes, then $P_n = P$ which is the path matrix of the graph G.

From the above discussion, we can conclude that $P_k[i][j]$ is equal to 1 only when either of the two following cases occur:

- There is a path from v_i to v_j that does not use any other node except $v_1, v_2, ..., v_{k-1}$. Therefore, $P_{k-1}[i][j] = 1$.
- There is a path from v_i to v_k and a path from v_k to v_j where all the nodes use $v_1, v_2, ..., v_{k-1}$. Therefore,

$$P_{k-1}[i][k] = 1 \text{ AND } P_{k-1}[k][j] = 1$$

Hence, the path matrix P_n can be calculated with the formula given as:

$$P_k[i][j] = P_{k-1}[i][j] \vee (P_{k-1}[i][k] \wedge P_{k-1}[k][j])$$

where $\vee$ indicates logical OR operation and $\wedge$ indicates logical AND operation.

Figure 13.37 shows the Warshall's algorithm to find the path matrix P using the adjacency matrix A.

```
Step 1: [Initialize the Path Matrix] Repeat Step 2 for I = 0 to n-1,
        where n is the number of nodes in the graph
Step 2:         Repeat Step 3 for J = 0 to n-1
Step 3:                 IF A[I][J] = 0, then SET P[I][J] = 0
                        ELSE P[I][J] = 1
                [END OF LOOP]
        [END OF LOOP]
Step 4: [Calculate the path matrix P] Repeat Step 5 for K = 0 to n-1
Step 5:         Repeat Step 6 for I = 0 to n-1
Step 6:                 Repeat Step 7 for J=0 to n-1
Step 7:                         SET PK[I][J] = PK-1[I][J] V (PK-1[I][K]
                                Λ PK-1[K][J])
Step 8: EXIT
```

Figure 13.37 Warshall's algorithm

EXAMPLE 13.11: Consider the graph in Fig. 13.38 and its adjacency matrix A. We can straightaway calculate the path matrix P using the Warshall's algorithm.

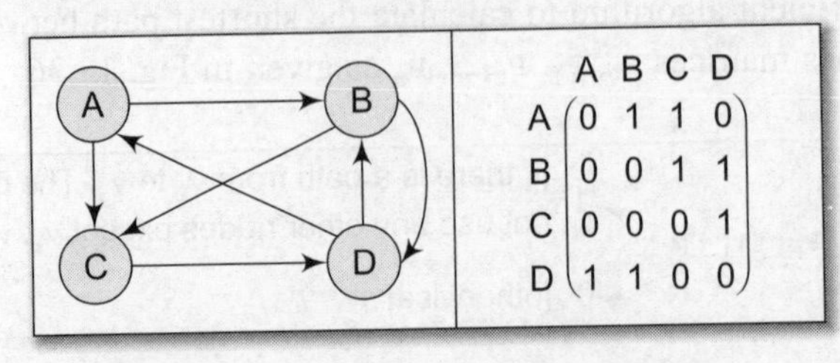

Figure 13.38 Graph G and its path matrix P

The path matrix P can be given in a single step as:

$$P = \begin{array}{c} \\ A \\ B \\ C \\ D \end{array} \begin{array}{c} \begin{array}{cccc} A & B & C & D \end{array} \\ \begin{pmatrix} 1 & 1 & 1 & 1 \\ 1 & 1 & 1 & 1 \\ 1 & 1 & 1 & 1 \\ 1 & 1 & 1 & 1 \end{pmatrix} \end{array}$$

Thus, we see that calculating A, A2, A3, A4, . . ., A5 to calculate P is a very slow and inefficient technique as compared to the Warshall's technique.

PROGRAMMING EXAMPLE

5. Write a program to implement Warshall's algorithm to find the path matrix.

```
#include<stdio.h>
#include<conio.h>
void read (int mat[5][5], int n);
void display (int mat[5][5], int n);
void mul(int mat[5][5], int n);
main()
{
        int adj[5][5], P[5][5], n, i, j, k;
        clrscr();
        printf("\n Enter the number of nodes in the graph : ");
        scanf("%d", &n);
        printf("\n Enter the adjacency matrix : ");
        read(adj, n);
        clrscr();
        printf("\n The adjacency matrix is : ");
        display(adj, n);
        for(i=0;i<n;i++)
        {
                for(j=0;j<n;j++)
                {
                        if(adj[i][j] == 0)
                                P[i][j] = 0;
                        else
                                P[i][j] = 1;
                }
        }
        for(k=0; k<n;k++)
        {
                for(i=0;i<n;i++)
                {
                        for(j=0;j<n;j++)
```

```
                    P[i][j] = P[i][j] | ( P[i][k] & P[k][j]);
        }
    }
    pintf("\n The Path Matrix is :");
    display (P, n);
    getch();
    return 0;
}
void read(int mat[5][5], int n)
{
    int i, j;
    for(i=0;i<n;i++)
    {
        for(j=0;j<n;j++)
        {
            printf("\n mat[%d][%d] = ", i, j);
            scanf("%d", &mat[i][j]);
        }
    }
}
void display(int mat[5][5], int n)
{
    int i, j;
    for(i=0;i<n;i++)
    {printf("\n");
        for(j=0;j<n;j++)
            printf("%d\t", mat[i][j]);
    }
}
```

13.7.6 Modified Warshall's Algorithm

The modified Warshall's algorithm is used to obtain a matrix that gives the shortest paths between the nodes in a graph G. As an input to the algorithm, we take the adjacency matrix A of G and replace all the values of A which are zero by infinity (∞). Infinity (∞) denotes a very large number and indicates that there no path between the vertices. In Warshall's modified algorithm, we obtain a set of matrices $Q_0, Q_1, Q_2, \ldots, Q_m$ using the formula given below.

$$Q_k[i][j] = \text{Minimum}(\ M_{k-1}[i][j],\ M_{k-1}[i][k] + M_{k-1}[k][j])$$

Q_0 is exactly the same as A with a little difference that every element having a zero value in A is replaced by (∞) in Q_0. Using the given formula, the matrix Q_n will give the path matrix that has the shortest path between the vertices of the graph. Warshall's modified algorithm is shown in Fig. 13.39.

```
Step 1: [Initialize the Shortest Path Matrix, Q] Repeat Step 2 for I = 0
        to n-1, where n is the number of nodes in the graph
Step 2:     Repeat Step 3 for J = 0 to n-1
Step 3:           IF A[I][J] = 0, then SET Q[I][J] = Infinity (or 9999)
                  ELSE Q[I][J] = A[I][j]
            [END OF LOOP]
      [END OF LOOP]
Step 4: [Calculate the shortest path matrix Q] Repeat Step 5 for K = 0
        to n-1
Step 5:     Repeat Step 6 for I = 0 to n-1
Step 6:           Repeat Step 7 for J=0 to n-1
Step 7:                       IF Q[I][J] <= Q[I][K] + Q[K][J]
                              SET Q[I][J] = Q[I][J]
                        ELSE SET Q[I][J] = Q[I][K] + Q[K][J]
                        [END OF IF]
            [END OF LOOP]
      [END OF LOOP]
      [END OF LOOP]
Step 8: EXIT
```

Figure 13.39 Modified Warshall's algorithm

Example 13.12: Consider the unweighted graph G given in Fig. 13.40 and apply Warshall's algorithm to it.

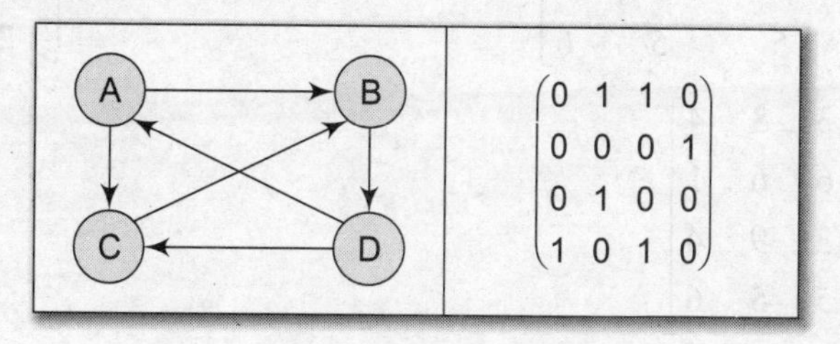

Figure 13.40 Graph G

$$Q0 = \begin{bmatrix} 9999 & 1 & 1 & 9999 \\ 9999 & 9999 & 9999 & 1 \\ 9999 & 1 & 9999 & 9999 \\ 1 & 9999 & 1 & 9999 \end{bmatrix} \quad Q1 = \begin{bmatrix} 9999 & 1 & 1 & 9999 \\ 9999 & 9999 & 9999 & 1 \\ 9999 & 1 & 9999 & 9999 \\ 1 & 2 & 1 & 9999 \end{bmatrix}$$

$$Q2 = \begin{bmatrix} 9999 & 1 & 1 & 2 \\ 9999 & 9999 & 9999 & 1 \\ 9999 & 1 & 9999 & 2 \\ 1 & 2 & 9999 & 3 \end{bmatrix} \quad Q3 = \begin{bmatrix} 9999 & 1 & 1 & 2 \\ 9999 & 9999 & 9999 & 1 \\ 9999 & 1 & 9999 & 2 \\ 1 & 2 & 9999 & 3 \end{bmatrix}$$

$$Q4 = Q \begin{bmatrix} 3 & 1 & 1 & 2 \\ 2 & 3 & 2 & 1 \\ 3 & 1 & 3 & 2 \\ 1 & 2 & 1 & 3 \end{bmatrix}$$

EXAMPLE 13.13: Consider a weighted graph G given in Fig. 13.41 and apply Warshall's shortest path algorithm to it.

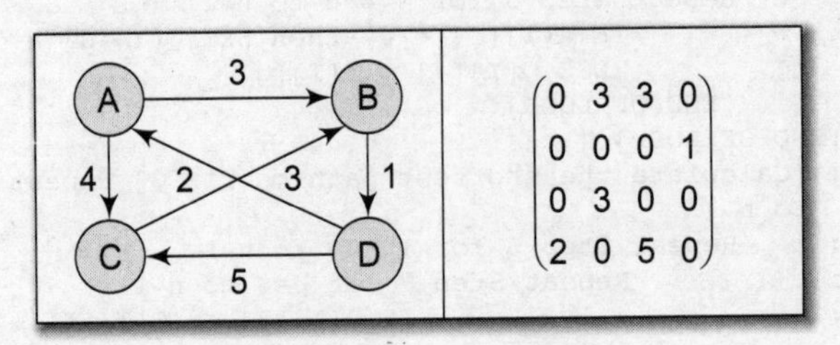

Figure 13.41 Graph G

$$Q0 = \begin{bmatrix} 9999 & 3 & 3 & 9999 \\ 9999 & 9999 & 9999 & 1 \\ 9999 & 3 & 9999 & 9999 \\ 2 & 9999 & 5 & 9999 \end{bmatrix} \quad Q1 = \begin{bmatrix} 9999 & 3 & 3 & 9999 \\ 9999 & 9999 & 9999 & 1 \\ 9999 & 3 & 9999 & 9999 \\ 2 & 5 & 5 & 9999 \end{bmatrix}$$

$$Q2 = \begin{bmatrix} 9999 & 3 & 3 & 4 \\ 9999 & 9999 & 9999 & 1 \\ 9999 & 3 & 9999 & 4 \\ 2 & 5 & 5 & 6 \end{bmatrix} \quad Q3 = \begin{bmatrix} 9999 & 3 & 3 & 4 \\ 9999 & 9999 & 9999 & 1 \\ 9999 & 3 & 9999 & 6 \\ 2 & 5 & 5 & 6 \end{bmatrix}$$

$$Q4 = Q \begin{bmatrix} 6 & 3 & 3 & 4 \\ 3 & 6 & 6 & 1 \\ 6 & 3 & 9 & 4 \\ 1 & 5 & 5 & 6 \end{bmatrix}$$

PROGRAMMING EXAMPLE

6. Write a program to implement Warshall's modified algorithm to find the shortest path.

```
#include<stdio.h>
#include<conio.h>
#define INFINITY 9999
void read (int mat[5][5], int n);
void display(int mat[5][5], int n);
main()
{
        int adj[5][5], Q[5][5], n, i, j, k;
        clrscr();
        printf("\n Enter the number of nodes in the graph : ");
        scanf("%d", &n);
        printf("\n Enter the adjacency matrix : ");
        read(adj, n);
        clrscr();
```

```
        printf("\n The adjacency matrix is : ");
        display(adj, n);
        for(i=0;i<n;i++)
        {
                for(j=0;j<n;j++)
                {
                        if(adj[i][j] == 0)
                                Q[i][j] = INFINITY;
                        else
                                Q[i][j] = adj[i][j];
                }
        }
        for(k=0; k<n;k++)
        {
                for(i=0;i<n;i++)
                {
                        for(j=0;j<n;j++)
                        {
                                if(Q[i][j] <= Q[i][k] + Q[k][j])
                                        Q[i][j] = Q[i][j];
                                else
                                        Q[i][j] = Q[i][k] + Q[k][j];
                        }
                }
                printf("\n\n");
                display(Q, n);
        }
        getch();
        return 0;
}
void read(int mat[5][5], int n)
{
        int i, j;
        for(i=0;i<n;i++)
        {
                for(j=0;j<n;j++)
                {
                        printf("\n mat[%d][%d] = ", i, j);
                        scanf("%d", &mat[i][j]);
                }
        }
}
void display(int mat[5][5], int n)
{
        int i, j;
        for(i=0;i<n;i++)
        {printf("\n");
```

```
            for(j=0;j<n;j++)
                printf("%d\t", mat[i][j]);
        }
    }
```

13.8 APPLICATIONS OF GRAPHS

Graphs are constructed for various types of applications such as:

- In circuit networks where points of connection are drawn as vertices and component wires become the edges of the graph.
- In transport networks where stations are drawn as vertices and routes become the edges of the graph.
- In maps that draw cities/states/regions as vertices and adjacency relations as edges.
- In program flow analysis where procedures or modules are treated as vertices and calls to these procedures are drawn as edges of the graph.
- Once we have a graph of a particular concept, they can be easily used for finding shortest paths, project planning, etc.

SUMMARY

- A graph is basically a collection of vertices (also called nodes) and edges that connect these vertices.
- Degree of a node u is the total number of edges containing the node u. When the degree of a node is zero, it is also called an isolated node. A path P is known as a closed path if the edge has the same end-points. A closed simple path with length 3 or more is known as a cycle.
- A graph in which there exists a path between any two of its nodes is called a connected graph. An edge that has identical end-points is called a loop. The size of a graph is the total number of edges in it.
- The out-degree of a node is the number of edges that originate at u.
- The in-degree of a node is the number of edges that terminate at u. A node u is known as a sink if it has a positive in-degree but a zero out-degree.
- A transitive closure of a graph is constructed to answer reachability questions.
- Since an adjacency matrix contains only 0s and 1s, it is called a bit matrix or a Boolean matrix. The memory use of an adjacency matrix is $O(n^2)$, where n is the number of nodes in the graph.
- Topological sort of a directed acyclic graph G is defined as a linear ordering of its nodes in which each node comes before all the nodes to which it has outbound edges. Every DAG has one or more number of topological sorts.
- A vertex v of G is called an articulation point if removing v along with the edges incident to v results in a graph that has at least two connected components.
- A biconnected graph is defined as a connected graph that has no articulation vertices.
- Breadth-first search is a graph search algorithm that begins at the root node and explores all the neighbouring nodes. Then for each of those nearest nodes, the algorithm explores their unexplored neighbour nodes, and so on, until it finds the goal.
- The depth-first search algorithm progresses by expanding the starting node of G and thus going deeper and deeper until a goal node is found, or until a node that has no children is encountered.
- A spanning tree of a connected, undirected graph G is a sub-graph of G which is a tree that connects all the vertices together.
- Kruskal's algorithm is an example of a greedy algorithm, as it makes the locally optimal choice

at each stage with the hope of finding the global optimum.

- Dijkstra's algorithm is used to find the length of an optimal path between two nodes in a graph.

GLOSSARY

Directed acyclic graph A directed graph that has no path starting and ending at the same node.

Acyclic graph A graph (un-directed) that has no path starting and ending at the same node.

Directed graph A graph that has edges which are ordered pairs of nodes. In a directed graph, each edge can be followed from one vertex to another vertex.

Weighted directed graph A directed graph in which each edge is associated with a weight (or numeric value).

Weighted graph A graph in which each edge has a weight associated with it.

Strongly connected graph A directed graph that has a path from a node to every other node in the graph.

Connected graph An undirected graph in which there exists a path between every pair of nodes.

Complete graph An undirected graph that has an edge between every pair of nodes.

Undirected graph A graph that has edges that are unordered pairs of nodes. That is, each edge in the graph connects two nodes.

Multi-graph A graph whose edges are unordered pairs of nodes and the same pair of nodes can be connected by multiple edges.

Graph A set of nodes connected by edges. That is, a graph is a set of nodes and a binary relation between nodes (adjacency).

Edge A connection between two nodes of a graph. In a weighted graph, each edge has a weight associated with it. In a directed graph, an edge goes from one node (source) to another node (destination), thereby making a connection in only one direction.

In-degree In-degree of a node is equal to the number of in-coming edges of that node in a directed graph.

Out-degree The number of out-going edges of a node in a directed graph.

Sink A node with zero out-degree. That is, a node of a directed graph with no outgoing edges.

Source A node of a directed graph with no incoming edges, that is, a node with zero in-degree.

Adjacency list A representation of a directed graph with `n` nodes using an array of `n` lists of nodes. List `i` contains vertex `j` if there is an edge from vertex `i` to vertex `j`. However, a weighted graph may be represented with a list of node/weight pairs.

Adjacency matrix A representation of a directed graph using an $n \times n$ matrix where `n` is the number of nodes. An entry at `(i,j)` is 1 if there is an edge from vertex `i` to vertex `j`; otherwise the entry is 0. A weighted graph contains the weight as the entry. However, an undirected graph may be represented using the same entry in both `(i,j)` and `(j,i)`.

Biconnected graph A connected graph that cannot be broken into disconnected pieces on deleting any single node.

Path A list of nodes of a graph where each node has an edge from it to the next node.

Dijkstra's algorithm An algorithm used to find the shortest path from a single source node to all other nodes in a weighted, directed graph in which all weights must be non-negative.

Bridge An edge of a connected graph which when removed would make the graph unconnected.

Connected components Connected components comprises of the set of maximally connected components of an undirected graph.

Cycle A path that is starting and ending at the same node.

Kruskal's algorithm An algorithm that is used to compute a minimum spanning tree. It maintains a set of partial minimum spanning trees, and repeatedly adds the shortest edge in the graph whose nodes are in different partial minimum spanning trees.

Minimum spanning tree A minimum weight tree in a weighted graph which contains all of the nodes of the graph.

Spanning tree A minimum weight tree in a weighted graph which contains all of the graph's vertices.

Labeled graph A graph that has labels associated with each edge or node.

Node A unit of reference in data structure. In the graphs and trees terminology, a node is also called a vertex. Node is basically a collection of information which must be kept at a single memory location.

Shortest path The problem of finding the shortest path in a graph from one node to another. The word 'shortest' may denote least number of edges, least total weight, etc.

Sibling A node in a tree that has the same parent as another node.

Sub-graph A graph whose nodes and edges are subsets or a part of another graph.

Topological order A numbering of the nodes of a directed acyclic graph in such a way that every edge from a node numbered `i` to a node numbered `j` satisfies `i<j`.

Topological sort To arrange items according to a partial order. This is basically done when some pairs of items have no comparison.

Transitive closure An extension or superset of a binary relation in such a way that whenever `(a,b)` and `(b,c)` are in the extension, `(a,c)` is also in the extension.

Depth-first search A traversal algorithm that marks all nodes in a directed graph in the order they are discovered and finished, partitioning the graph into a forest. This is typically implemented using a stack.

EXERCISES

Review Questions

1. Explain the relationship between a linked list structure and a digraph?
2. What is a graph? Explain its key terms.
3. How are graphs represented inside a computer's memory? Which method do you prefer and why?
4. Consider the graph given below.
   (a) Write the adjacency matrix of G.
   (b) Write the path matrix of G.
   (c) Is the graph biconnected?
   (d) Is the graph complete?
   (e) Find the shortest path matrix using Warshall's algorithm.
   (f) Find the shortest path matrix using Warshall's modified algorithm.

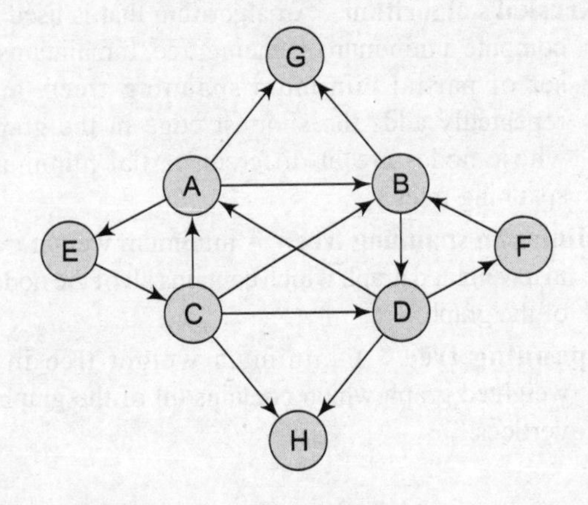

5. Explain the graph traversal algorithms in detail with example.
6. Draw a complete undirected graph having five nodes.
7. Consider the graph given below and find out the degree of each node.

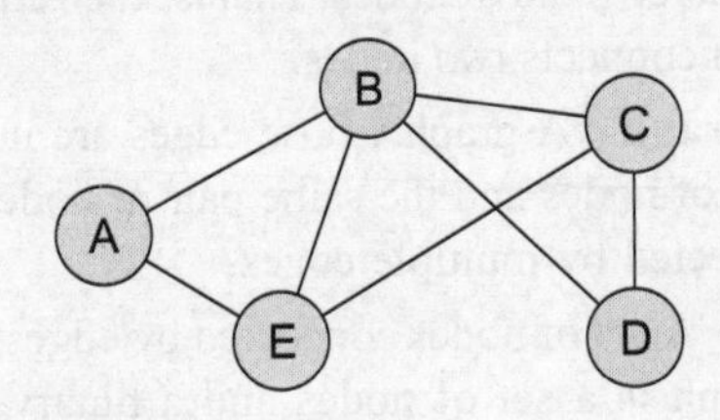

8. Consider the graph given below. State all the simple paths from A to D, B to D, and C to D. Also, find out the in-degree and out-degree of each node. Is there any source or sink in the graph?

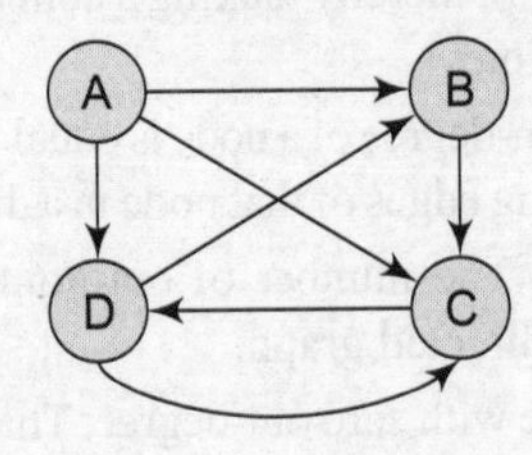

9. Consider the graph given below. Find out its depth-first and breadth-first traversal scheme.

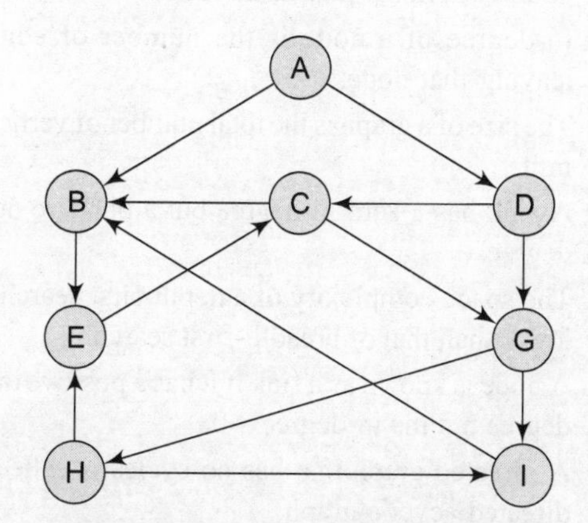

10. Differentiate between depth-first search and breadth-first search traversal of a graph.
11. Explain the topological sorting of a graph `G`.
12. Define spanning tree.
13. When is a spanning tree called a minimum spanning tree? Take a weighted graph of your choice and find out its minimum spanning tree.
14. Explain Prim's algorithm.
15. Write a brief note on Kruskal's algorithm.
16. Write a short note on Dijkstra's algorithm.
17. Differentiate between Dijkastra's algorithm and minimum spanning tree algorithm.
18. Consider the graph given below. Find the minimum spanning tree of this graph using (a) Prim's algorithm, (b) Kruskal's algorithm, and (c) Dijkstra's algorithm.

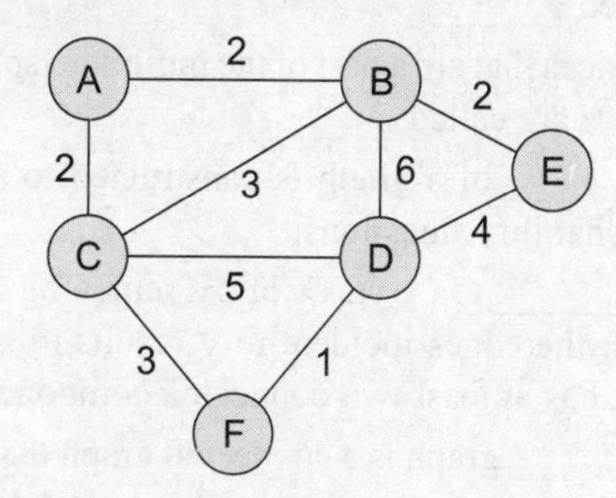

19. Briefly discuss Warshall's algorithm. Also, discuss its modified version.
20. Show the working of Floyd-Warshall's algorithm to find the shortest paths between all pair of nodes in the following graph.

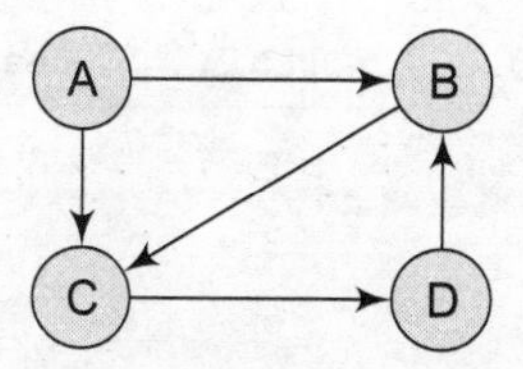

21. Write a short note on transitive closure of a graph `G`.
22. Given the adjacency matrix of a graph, write a program to calculate the degree of a node `N` in the graph.
23. Given the adjacency matrix of a graph, write a program to calculate the in-degree and the out-degree of a node `N` in the graph.
24. Given the adjacency matrix of a graph, write a function `isFullConnectedGraph` which returns 1 if the graph is fully connected and 0 otherwise.
25. In which kind of graph do we use topological sorting?
26. Consider the graph given below and show its adjacency list in the memory.

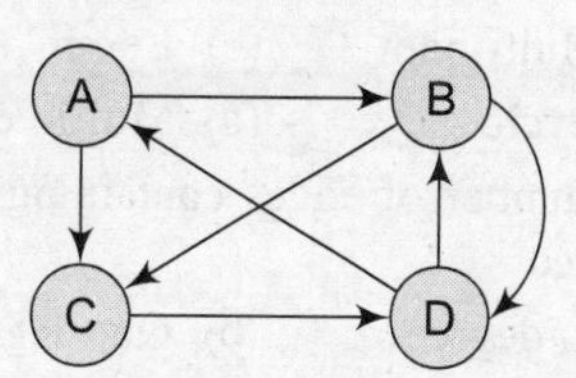

27. Consider the graph given in Question 26 and show the changes in the graph as well as its adjacency list when node `E` and edges `(A, E)` and `(C, E)` are added to it. Also, delete edge `(B, D)` from the graph.
28. Given the following adjacency matrix, draw the weighted graph.

$$\begin{pmatrix} 0 & 4 & 0 & 2 & 0 \\ 0 & 0 & 0 & 7 & 0 \\ 0 & 5 & 0 & 0 & 0 \\ 0 & 0 & 0 & 0 & 3 \\ 0 & 0 & 1 & 0 & 0 \end{pmatrix}$$

29. Given five cities: (1) New Delhi, (2) Mumbai, (3) Chennai, (4) Bangalore and (5) Kolkata, and a list of flights that connect these cities as shown in the following table. Use the given information to construct a graph.

Flight No.	Origin	Destination
101	2	3
102	3	2
103	5	3
104	3	4
105	2	5
106	5	2
107	5	1
108	1	4
109	5	4
110	4	5

Programming Exercises

1. Write a program to create and print a graph.
2. Write a program to determine whether there is at least one path from the source to the destination.

Multiple Choice Questions

1. An edge that has identical end-points is called a
   (a) Multi-path (b) Loop
   (c) Cycle (d) Multi-edge
2. Total number of edges containing the node u are called
   (a) In-degree (b) Out-degree
   (c) Degree (d) None of these
3. A graph in which there exists a path between any two of its nodes is called
   (a) Complete graph (b) Connected graph
   (c) Digraph (d) In-directed graph
4. The number of edges that originate at u are called
   (a) In-degree (b) Out-degree
   (c) Degree (d) source
5. The memory use of an adjacency matrix is
   (a) `O(n)` (b) $O(n^2)$
   (c) $O(n^3)$ (d) `O(log n)`
6. The term optimal can mean
   (a) Shortest (b) Cheapest
   (c) Fastest (d) All of these
7. A biconnected graph contains how many articulation vertices?
   (a) 0 (b) 1 (c) 2 (d) 3

True or False

1. Graph is a linear data structure.
2. In-degree of a node is the number of edges leaving that node.
3. The size of a graph is the total number of vertices in it.
4. A sink has a zero in-degree but a positive out-degree.
5. The space complexity of a depth-first search is lower than that of breadth-first search.
6. A node is known as a sink if it has a positive out-degree but the in-degree = 0.
7. A directed graph that has no cycles is called a directed acyclic graph.
8. A graph G can have many different spanning trees.
9. Fringe vertices are not a part of T, but are adjacent to some tree vertex.
10. Kruskal's algorithm is an example of a greedy algorithm.

Fill in the Blanks

1. ______ has a zero degree.
2. In-degree of a node is the number of edges that ______ at u. Adjacency matrix is also known as a ______.
3. A path P is known as a ______ path if the edge has the same end-points.
4. A closed simple path with length 3 or more is known as a ______.
5. A graph with multiple edges and/or a loop is called a ______.
6. Vertices that are a part of the minimum spanning tree T are called ______.
7. A ______ of a graph is constructed to answer reachability questions.
8. An ______ is a vertex v of G if removing v along with the edges incident to v results in a graph that has at least two connected components.
9. A ______ graph is a connected graph that is not broken into disconnected pieces by deleting any single vertex.
10. An edge is called a ______ if removing that edge results in a disconnected graph.

Program for Finding Minimum Spanning Tree

The following program shows how to find the cost of a minimum spanning tree.

```
include<stdio.h>
#define MAX 10
int adj[MAX][MAX], tree[MAX][MAX], n;
void readmatrix()
{
        int i, j;
        printf("\n Enter the number of nodes in the Graph : ");
        scanf("%d", &n);
        printf("\n Enter the adjacency matrix of the Graph");
        for (i = 1; i <= n; i++)
                for (j = 1; j <= n; j++)
                        scanf("%d", &adj[i][j]);
}
int spanningtree(int src)
{
        int visited[MAX], d[MAX], parent[MAX];
        int i, j, k, min, u, v, cost;
        for (i = 1; i <= n; i++)
        {
                d[i] = adj[src][i];
                visited[i] = 0;
                parent[i] = src;
        }
        visited[src] = 1;
        cost = 0;
        k = 1;
        for (i = 1; i < n; i++)
        {
                min = 9999;
                for (j = 1; j <= n; j++)
                {
                        if (visited[j]==0 && d[j] < min)
                        {
                                min = d[j];
                                u = j;
                        }
```

```
            }
            visited[u] = 1;
            cost = cost + d[u];
            tree[k][1] = parent[u];
            tree[k++][2] = u;
            for (v = 1; v <= n; v++)
                if (visited[v]==0 && (adj[u][v] < d[v]))
                {
                    d[v] = adj[u][v];
                    parent[v] = u;
                }
        }
        return (cost);
}
void display(int cost)
{
        int i;
        printf("\n The Edges of the Mininum Spanning Tree are");
        for (i = 1; i < n; i++)
            printf(" %d %d \n", tree[i][1], tree[i][2];
        printf("\n The Total cost of the Minimum Spanning Tree is : ", cost);
}
main()
{
        int source, treecost;
        readmatrix();
        printf("\n Enter the Source : ");
        scanf("%d", &source);
        treecost = spanningtree(source);
        display(treecost);
        return 0;
}
```

14 Sorting

Learning Objective

In this chapter, we will discuss the different techniques of sorting an array of numbers or characters. After discussing the sorting algorithms in detail, we will make a comparison between these algorithms to know which algorithm performs better under what condition.

14.1 INTRODUCTION

Sorting means arranging the elements of an array so that they are placed in some relevant order which may either be ascending or descending. That is, if `A` is an array, then the elements of `A` are arranged in a sorted order (ascending order) in such a way that `A[0] < A[1] < A[2] < ...... < A[N]`.

For example, if we have an array that is declared and initialized as

```
int A[] = {21, 34, 11, 9, 1, 0, 22};
```

then the sorted array (ascending order) can be given as:

```
A[] = {0, 1, 9, 11, 21, 22, 34}
```

A sorting algorithm is defined as an algorithm that puts the elements of a list in a certain order (that can either be numerical order, lexicographical order, or any user-defined order). Efficient sorting algorithms are widely used to optimize the use of other algorithms like search and merge algorithms which require sorted lists to work correctly. There are two types of sorting:

- ***Internal sorting*** which deals with sorting the data stored in the computer's memory
- ***External sorting*** which deals with sorting the data stored in files. External sorting is applied when there is voluminous data that cannot be stored in the memory.

In this chapter, we will only discuss internal sorting. External sorting will be discussed in Chapter 16.

14.2 BUBBLE SORT

Bubble sort is a very simple method that sorts the array elements by repeatedly moving the largest element to the highest index position of the array (in case of arranging elements in ascending order). In *bubble sorting*, consecutive adjacent pairs of elements in the array are compared with each other. If the element at the lower index is greater than the element at the higher index, the two elements are interchanged so that the smaller element is placed before the bigger one. This process will continue till the list of unsorted elements exhausts.

This procedure of sorting is called bubble sorting because the smaller elements 'bubble' to the top of the list. Note that at the end of the first pass, the largest element in the list will be placed at its proper position (i.e. at the end of the list).

If the elements are to be sorted in descending order, then with each pass the smallest element is moved to the lowest index of the array.

Bubble Sort Example

To discuss bubble sort in detail, let us consider an array that has the following elements:

A[] = {30, 52, 29, 87, 63, 27, 18, 54}

Pass 1:

(a) Compare 30 and 52. Since 30 < 52, no swapping is done.

(b) Compare 52 and 29. Since 52 > 29, swapping is done.

30, **29, 52**, 87, 63, 27, 19, 54

(c) Compare 52 and 87. Since 52 < 87, no swapping is done.

(d) Compare, 87 and 63. Since 87 > 83, swapping is done.

30, 29, 52, **63, 87**, 27, 19, 54

(e) Compare87 and 27. Since 87 > 27, swapping is done.

30, 29, 52, 63, **27, 87**, 19, 54

(f) Compare 87 and 19. Since 87 > 19, swapping is done.

30, 29, 52, 63, 27, **19, 87**, 54

(g) Compare 87 and 54. Since 87 > 54, swapping is done.

30, 29, 52, 63, 27, 19, **54, 87**

Observe that after the end of the first pass, the largest element is placed at the highest index of the array. All the other elements are still unsorted.

Pass 2:

(a) Compare 30 and 29. Since 30 > 29, swapping is done.

29, 30, 52, 63, 27, 19, 54, 87

(b) Compare 30 and 52. Since 30 < 52, no swapping is done.

(c) Compare 52 and 63. Since 52 < 63, no swapping is done.

(d) Compare 63 and 27. Since 63 > 27, swapping is done.

29, 30, 52, **27, 63**, 19, 54, 87

(e) Compare 63 and 19. Since 63 > 19, swapping is done.

29, 30, 52, 27, **19, 63**, 54, 87

(f) Compare 63 and 54. Since 63 > 54, swapping is done.

29, 30, 52, 27, 19, **54, 63**, 87

Observe that after the end of the second pass, the second largest element is placed at the second highest index of the array. All the other elements are still unsorted.

Pass 3:

(a) Compare 29 and 30. Since 29 < 30, no swapping is done.

(b) Compare 30 and 52. Since 30 < 52, no swapping is done.

(c) Compare 52 and 27. Since 52 > 27, swapping is done.

29, 30, **27, 52**, 19, 54, 63, 87

(d) Compare 52 and 19. Since 52 > 19, swapping is done.

```
29, 30, 27, 19, 52, 54, 63, 87
(e) Compare 52 and 54. Since 52 < 54, no swapping is done.
```

Observe that after the end of the third pass, the third largest element is placed at the third highest index of the array. All the other elements are still unsorted.

```
Pass 4:
(a) Compare 29 and 30. Since 29 < 30, no swapping is done.
(b) Compare 30 and 27. Since 30 > 27, swapping is done.
    29, 27, 30, 19, 52, 54, 63, 87
(c) Compare 30 and 19. Since 30 > 19, swapping is done.
    29, 27, 19, 30, 52, 54, 63, 87
(d) Compare 30 and 52. Since 30 < 52, no swapping is done.
```

Observe that after the end of the fourth pass, the fourth largest element is placed at the fourth highest index of the array. All the other elements are still unsorted.

```
Pass 5:
(a) Compare 29 and 27. Since 29 > 27, swapping is done.
    27, 29, 19, 30, 52, 54, 63, 87
(b) Compare 29 and 19. Since 29 > 19, swapping is done.
    27, 19, 29, 30, 52, 54, 63, 87
(c) Compare 29 and 30. Since 29 < 30, no swapping is done.
```

Observe that after the end of the fifth pass, the fifth largest element is placed at the fifth highest index of the array. All the other elements are still unsorted.

```
Pass 6:
(a) Compare 27 and 19. Since 27 > 19, swapping is done.
    19, 27, 29, 30, 52, 54, 63, 87
(b) Compare 27 and 29. Since 27 < 29, no swapping is done.
```

Observe that after the end of the sixth pass, the sixth largest element is placed at the sixth largest index of the array. All the other elements are still unsorted.

```
Pass 7:
(a) Compare 19 and 27. Since 19 < 27, no swapping is done.
```

Observe that the entire list is sorted now.

Algorithm for Bubble Sort

If we look at the basic methodology of the working of bubble sort, we can generalize its working as follows:

(a) In Pass 1, `A[1]` and `A[2]` are compared, then `A[2]` is compared with `A[3]`, `A[3]` is compared with `A[4]`, and so on. Finally, `A[N-1]` is compared with `A[N]`. Pass 1 involves `n-1` comparisons and places the biggest element at the highest index of the array.

(b) In Pass 2, `A[1]` and `A[2]` are compared, then `A[2]` is compared with `A[3]`, `A[3]` is compared with `A[4]`, and so on. Finally, `A[N-2]` is compared with `A[N-1]`. Pass 2 involves `n-2` comparisons and places the second biggest element at the second highest index of the array.

(c) In Pass 3, `A[1]` and `A[2]` are compared, then `A[2]` is compared with `A[3]`, `A[3]` is compared with `A[4]`, and so on. Finally, `A[N-3]` is compared with `A[N-2]`. Pass 3 involves `n-3` comparisons and places the third biggest element at the second highest index of the array.

(d) In Pass `n-1`, `A[1]` and `A[2]` are compared so that `A[1]<A[2]`. After this step, all the elements of the array are arranged in ascending order.

Figure 14.1 shows the algorithm for bubble sort. In this algorithm, the outer loop is for the total number of passes which is `N-1`. The inner loop will be executed for every pass. However, the frequency of the inner loop will decrease with every pass because after every pass, one element will be in its correct position. Therefore, for every pass, the inner loop will be executed `N-I` times, where `N` is the number of elements in the array and `I` is the count of the pass.

```
BUBBLE_SORT(A, N)

Step 1: Repeat steps 2 For I = 0 to N-1
Step 2:       Repeat For J = 0 to N - I
Step 3:                 If A[J] > A[J + 1], then
                        SWAP A[J] and A[J+1]
             [End of Inner Loop]
         [End of Outer Loop]
Step 4: EXIT
```

Figure 14.1 Algorithm for bubble sort

Complexity of Bubble Sort

The complexity of any sorting algorithm depends upon the number of comparisons. In bubble sort, we have seen that there are `N-1` passes in total. In the first pass, `N-1` comparisons are made to place the highest element in its correct position. Then, in Pass 2, there are `N-2` comparisons and the second highest element is placed in its position. Therefore, to compute the complexity of bubble sort, we need to calculate the total number of comparisons. It can be given as:

$$f(n) = (n - 1) + (n - 2) + (n - 3) + \dots + 3 + 2 + 1$$
$$= n(n - 1)/2$$
$$= n^2/2 + O(n) = O(n^2)$$

Therefore, the complexity of bubble sort algorithm is $O(n^2)$. It means the time required to execute bubble sort is proportional to n^2, where n is the total number of elements in the array.

PROGRAMMING EXAMPLES

1. Write a program to enter *n* numbers in an array. Redisplay the array with elements being sorted in ascending order.

```
#include<stdio.h>
#include<conio.h>
int main()
{
        int i, n, temp, j, arr [10];
        clrscr();
        printf("\n Enter the number of elements in the array : ");
        scanf("%d", &n);
        for(i=0;i<n;i++)
        {
                printf("\n Arr[%d] = ",i);
                scanf("%d", &arr [i]);
        for(i=0;i<n;i++)
        {
                for(j= 0;j<n-i;j++)
```

```
		{
			if(arr [j] > arr [j+1])
			{
				temp = arr [j];
				arr [j] = arr [j+1];
				arr [j+1] = temp;
			}
		}
	}
	printf("\n The array sorted in ascending order is :");
	for(i=0;i<n;i++)
		printf("\n Arr[%d] = %d", i, arr[i]);
	getch();
	return 0;
}
```

2. Write a program to enter *n* numbers in an array. Redisplay the array with elements being sorted in descending order.

```
#include<stdio.h>
#include<conio.h>
int main()
{
	int i, n, j, temp, arr[10];
	clrscr();
	printf("\n Enter the number of elements in the array : ");
	scanf("%d", &n);
	for(i=0;i<n;i++)
	{
		printf("\n Arr[%d] = ", i);
		scanf("%d", &arr[i]);
	}
	for(i=0;i<n;i++)
	{
		for(j= 0;j<n-i;j++)
		{
			if(arr[j] > arr[j+1])
			{
				temp = arr[j];
				arr[j] = arr[j+1];
				arr[j+1] = temp;
			}
		}
	}
	printf("\n The array sorted in descending order is :");
	for(i=0;i<n;i++)
		printf("\n Arr[%d] = %d", i, arr[i]);
```

```
        getch();
        return 0;
}
```

14.3 INSERTION SORT

Insertion sort is a very simple sorting algorithm in which the sorted array (or list) is built one element at a time. We all are familiar with this technique of sorting, as we usually use it for ordering a deck of cards while playing bridge.

The main idea behind insertion sort is that it inserts each item into its proper place in the final list. To save memory, most implementations of the insertion sort algorithm works by moving the current data element past the already sorted values and repeatedly interchanging it with the preceding value until it is in its correct place.

Insertion sort is less efficient as compared to other more advanced algorithms such as quick sort, heap sort, and merge sort.

Technique

Insertion sort works as follows:

- The array of values to be sorted is divided into two sets. One that stores sorted values and another that contains unsorted values.
- The sorting algorithm will proceed until there are elements in the unsorted set.
- Suppose there are `n` elements in the array. Initially, the element with index 0 (assuming `LB, Lower Bound = 0`) is in the sorted set Rest of the elements are in the unsorted set.
- The first element of the unsorted partition has array index 1 (if `LB = 0`).
- During each iteration of the algorithm, the first element in the unsorted set is picked up and inserted into the correct position in the sorted set.

EXAMPLE 14.1: Consider an array of integers given below. Sort the values in the array using insertion sort.

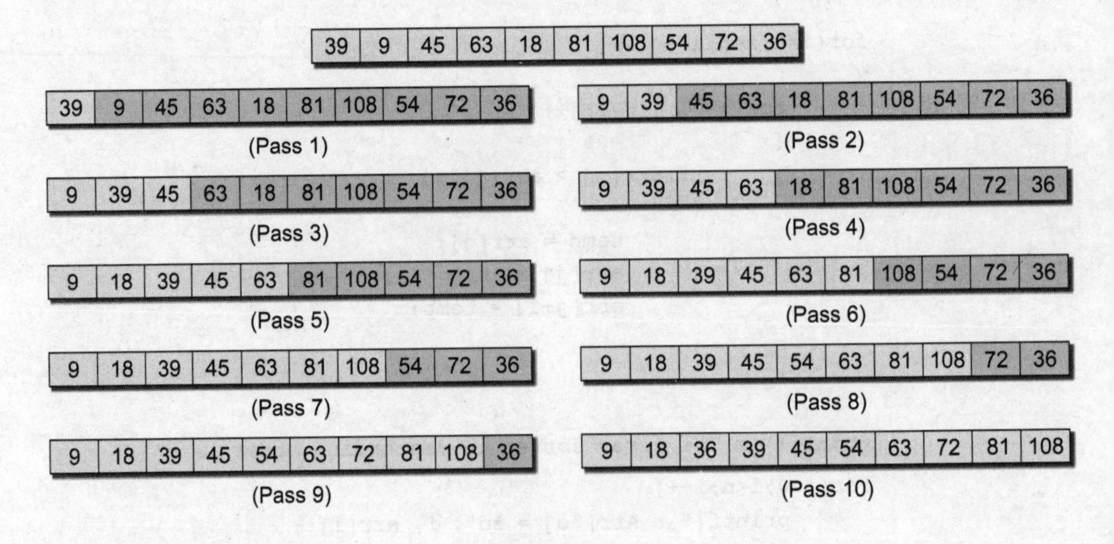

In Pass 1, A[0] is the only element in the sorted set. In Pass 2, A[1] will be placed either before or after A[0], so that the array A is sorted. In Pass 3, A[2] will be placed either before A[0], in-between A[0] and A[1], or after A[1]. In Pass 4, A[4] will be placed in its proper place. In Pass N, A[N-1] will be placed in its proper place to keep the array sorted.

To insert an element A[K] in a sorted list A[0], A[1], ..., A[K-1], we need to compare A[K] with A[K-1], then with A[K-2], A[K-3], and so on until we meet an element A[J] such that A[J] <= A[K]. In order to insert A[K] in its correct position, we need to move each element A[K-1], A[K-2], ..., A[J] by one position and then A[K] is inserted at the (J+1)th location. The algorithm for insertion sort is given in Fig. 14.2.

```
Insertion sort (ARR, N)

Step 1: Repeat Steps 2 to 5 for K = 1 to N
Step 2:        SET TEMP = ARR[K]
Step 3:        SET J = K - 1
Step 4:        Repeat while TEMP <= ARR[J]
                      SET ARR[J + 1] = ARR[J]
                      SET J = J - 1
               [END OF INNER LOOP]
Step 5:        SET ARR[J + 1] = TEMP
    [END OF LOOP]
Step 6: EXIT
```

Figure 14.2 Algorithm for insertion sort

In the algorithm, Step 1 executes a for loop which will be repeated for each element in the array. In Step 2, we store the value of the Kth element in TEMP. In Step 3, we set the Jth index in the array. In Step 4, a for loop is executed that will create space for the new element from the unsorted list to be stored in the list of sorted elements. Finally, in Step 5, the element is stored at the Jth location.

Advantages of Insertion Sort

The advantages of this sorting algorithm are as follows:

- It is easy to implement and efficient to use on small sets of data.
- It can be efficiently implemented on data sets that are already substantially sorted.
- It performs better than algorithms like selection sort and bubble sort. Insertion sort algorithm is simpler than shell sort, with only a small trade-off in efficiency. It is over twice as fast as the bubble sort and almost 40 percent faster than the selection sort.
- It requires less memory space (only O(1) of additional memory space).
- It is said to be online, as it can sort a list as and when it receives new elements.

Complexity of Insertion Sort

For insertion sort, the best case occurs when the array is already sorted. In this case, the running time of the algorithm has a linear running time (i.e. O(n)). This is because, during each iteration, the first element from the unsorted set is compared only with the last element of the sorted set of the array.

Similarly, the worst case of the insertion sort algorithm occurs when the array is sorted in the reverse order. In the worst case, the first element of the unsorted set has to be compared with almost every element in the sorted set. Furthermore, every iteration of the inner loop will have to shift the elements of the sorted set of the array before inserting the next element. Therefore, in the worst case, insertion sort has a quadratic running time (i.e. $O(n^2)$).

Even in the average case, the insertion sort algorithm will have to make at least (K-1)/2 comparisons. Thus, the average case also has a quadratic running time.

PROGRAMMING EXAMPLE

3. Write a program to sort an array using insertion sort algorithm.

```
#include<stdio.h>
#include<conio.h>
void insertion_sort( int arr[], int n);
void main()
{
    int arr[10], i, n, j,k;
    clrscr();
    printf("\n ENter the number of elements in the array : ");
    scanf("%d", &n);
    printf("\n Enter the elements of the array ");
    for(i=0;i<n;i++)
    {
        printf("\n arr[%d] = ", i);
        scanf("%d", &arr[i]);
    }
    insertion_sort(arr, n);
    printf("\n The sorted array is : \n");
    for(i=0;i<10;i++)
        printf("%d\t", arr[i]);
    getch();
}
void insertion_sort( int arr[], int n)
{
    int k, j, temp;
    for(k=1;i<n;i++)
    {
        temp = arr[k];
        j = k-1;
        while((temp < arr[j]) && (j>=0))
        {
            arr[j+1] = arr[j];
            j—;
        }
        arr[j+1] = temp;
    }
}
```

14.4 SELECTION SORT

Selection sort is a sorting algorithm that has a quadratic running time complexity of $O(n^2)$, thereby making it inefficient to be used on large lists. Although selection sort performs worse than insertion sort algorithm, it is noted for its simplicity and also has performance advantages over more complicated algorithms in certain situations. Selection sort is generally used for sorting files with very large objects (records) and small keys.

Technique

Consider an array ARR with N elements. The selection sort takes N-1 passes to sort the entire array and works as follows:

First find the smallest value in the array and place it in the first position. Then, find the second smallest value in the array and place it in the second position. Repeat this procedure until the entire array is sorted. Therefore,

- In Pass 1, find the position POS of the smallest value in the array and then swap ARR[POS] and ARR[0]. Thus, ARR[0] is sorted.
- In Pass 2, find the position POS of the smallest value in sub-array of N-1 elements. Swap ARR[POS] with ARR[1]. Now, A[0] and A[1] is sorted.
- In Pass N-1, find the position POS of the smaller of the elements ARR[N-2] and ARR[N-1]. Swap ARR[POS] and ARR[N-2] so that ARR[0], ARR[1], ..., ARR[N-1] is sorted.

Example 14.2: Sort the array given below using selection sort.

39	9	81	45	90	27	72	18

PASS	LOC	ARR[0]	ARR[1]	ARR[2]	ARR[3]	ARR[4]	ARR[5]	ARR[6]	ARR[7]
1	1	9	39	81	45	90	27	72	18
2	7	9	18	81	45	90	27	72	39
3	5	9	18	27	45	90	81	72	39
4	7	9	18	27	39	90	81	72	45
5	7	9	18	27	39	45	81	72	90
6	6	9	18	27	39	45	72	81	90

The algorithm for selection sort is shown in Fig. 14.3. In the algorithm, during the Kth pass, we need to find the position POS of the smallest elements from ARR[K], ARR[K+1], ..., ARR[N]. To find the smallest element, we use a variable SMALL to hold the smallest value in the sub-array ranging from ARR[K] to ARR[N]. Then, swap ARR[K] with ARR[POS]. This procedure is repeated until all the elements in the array are sorted.

```
SMALLEST (ARR, K, N, POS)

Step 1: [Initialize] SET SMALL = ARR[K]
Step 2: [Initialize] SET POS = K
Step 3: Repeat for J = K+1 to N
          IF SMALL > ARR[J], then
                SET SMALL = ARR[J]
                SET POS = J
          [END OF IF]
     [END OF LOOP]
Step 4: Exit
```

```
SELECTION SORT(ARR, N)

Step 1: Repeat Steps 2 and 3 for K = 1
          to N-1
Step 2:      CALL SMALLEST(ARR, K, N, POS)
Step 3:      SWAP A[K] with ARR[POS]
          [END OF LOOP]
Step 4: Exit
```

Figure 14.3 Algorithm for selection sort

Complexity of Selection Sort

Selection sort is a sorting algorithm that is independent of the original order of elements in the array. In Pass 1, selecting the element with the smallest value calls for scanning of all n elements; thus, n-1 comparisons are required in the first pass. Then, the smallest value is swapped with the element in the first position. In Pass 2, selecting the second smallest value requires scanning the remaining (n - 1) elements and so on. Therefore,

$$(n - 1) + (n - 2) + \dots + 2 + 1$$
$$= n(n - 1) / 2 = \cup(n^2) \text{ comparisons}$$

Advantages of Selection Sort

- It is simple and easy to implement.
- It can be used for small data sets.
- It is 60 percent more efficient than bubble sort.

However, in case of large data sets, the efficiency of selection sort drops as compared to insertion sort.

PROGRAMMING EXAMPLE

4. Write a program to sort an array using selection sort algorithm.

```
#include<stdio.h>
#include<conio.h>
int smallest(int arr[], int k, int n);
void selection_sort(int arr[], int n);
void main()
{
        int arr[10], i, n, j,k;
        clrscr();
        printf("\n ENter the number of elements in the array : ");
        scanf("%d", &n);
        printf("\n Enter the elements of the array ");
        for(i=0;i<n;i++)
        {
                printf("\n arr[%d] = ", i);
                scanf("%d", &arr[i]);
        }
        selection_sort(arr, n);
        printf("\n The sorted array is : \n");
        for(i=0;i<10;i++)
                printf("%d\t", arr[i]);
        getch();
}
int smallest(int arr[], int k, int n)
{
        int pos = k, small=arr[k], i;
```

```
        for(i=k+1;i<n;i++)
        {
            if(arr[i]< small)
            {
                small = arr[i];
                pos = i;
            }
        }
        return pos;
}
void selection_sort(int arr[],int n)
{
        int k, pos, temp;
        for(k=0;k<n;k++)
        {
            pos = smallest(arr, k, n);
            temp = arr[k];
            arr[k] = arr[pos];
            arr[pos] = temp;
        }
}
```

14.5 MERGE SORT

Merge sort is a sorting algorithm that uses the divide, conquer, and combine algorithmic paradigm.

- ***Divide*** means partitioning the n-element array to be sorted into two sub-arrays of n/2 elements in each sub-array. If A is an array containing zero or one element, then it is already sorted. However, if there are more elements in the array, divide A into two sub-arrays, A_1 and A_2, each containing about half of the elements of A.
- ***Conquer*** means sorting the two sub-arrays recursively using merge sort.
- ***Combine*** means merging the two sorted sub-arrays of size n/2 each to produce the sorted array of n elements.

Merge sort algorithms focuses on two main concepts to improve its performance (running time):

- A smaller list takes fewer steps and thus less time to sort than a large list.
- Less steps, thus less time is needed to create a sorted list from two sorted lists rather than creating it using two unsorted lists.

The basic steps of a merge sort algorithm are as follows:

- If the array is of length 0 or 1, then it is already sorted.
- Otherwise, divide the unsorted array into two sub-arrays of about half the size.
- Use merge sort algorithm recursively to sort each sub-array.
- Merge the two sub-arrays to form a single sorted list.

EXAMPLE 14.3: Sort the array given below using merge sort.

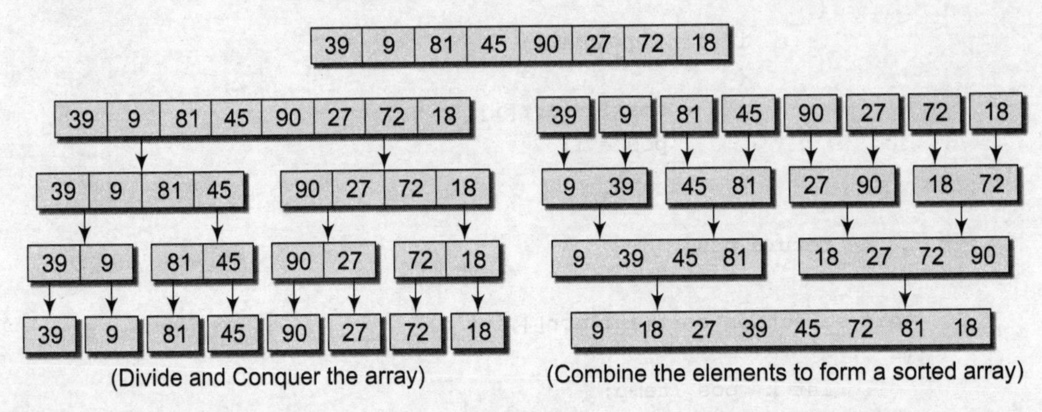

(Divide and Conquer the array) (Combine the elements to form a sorted array)

The merge sort algorithm (Fig. 14.4) uses a function `merge` which combines the sub-arrays to form a sorted array. While the merge sort algorithm recursively divides the list into smaller lists, the merge algorithm conquers the list to sort the elements in individual lists. Finally, the smaller lists are merged to form one list.

To understand the merge algorithm, consider Fig. 14.4 which shows how we merge two lists to form one list. For ease of understanding, we have taken two sub-lists each containing four elements. The same concept can be utilized to merge four sub-lists containing two elements, and eight sub-lists having just one element.

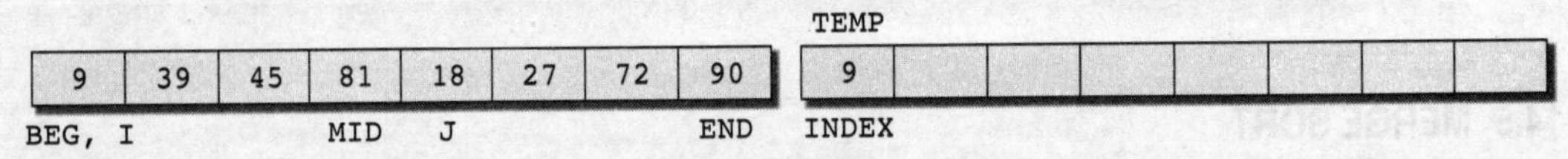

Compare `ARR[I]` and `ARR[J]`, the smaller of the two is placed in `TEMP` at the location specified by `INDEX` and subsequently the value `I` or `J` is incremented.

TEMP

9	39	45	81	18	27	72	90		9	18						
BEG	I		MID	J			END			INDEX						

9	39	45	81	18	27	72	90		9	18	27					
BEG	I		MID		J		END				INDEX					

9	39	45	81	18	27	72	90		9	18	27	39				
BEG	I		MID			J	END					INDEX				

9	39	45	81	18	27	72	90		9	18	27	39	45			
BEG		I	MID			J	END						INDEX			

9	39	45	81	18	27	72	90		9	18	27	39	45	72		
BEG			I, MID			J	END							INDEX		

When `I` is greater than `MID`, copy the remaining elements of the right sub-array in `TEMP`.

9	39	45	81	18	27	72	90		9	18	27	39	45	72	81	
BEG			MID	I			J END								INDEX	

9	39	45	81	18	27	72	90		9	18	27	39	45	72	81	90
BEG			MID	I			J END									INDEX

```
MERGE (ARR, BEG, MID, END)

Step 1: [Initialize] SET I = BEG, J = MID + 1, INDEX = 0
Step 2: Repeat while (I <= MID) AND (J<=END)
            IF ARR[I] < ARR[J], then
                  SET TEMP[INDEX] = ARR[I]
                  SET I = I + 1
            ELSE
                  SET TEMP[INDEX] = ARR[J]
                  SET J = J + 1
            [END OF IF]
      SET INDEX = INDEX + 1
      [END OF LOOP]
Step 3: [Copy the remaining elements of right sub-array, if any]
            IF I > MID, then
               Repeat while J <= END
                  SET TEMP[INDEX] = ARR[J]
                  SET INDEX = INDEX + 1, SET J = J + 1
            [END OF LOOP]
      [Copy the remaining elements of left sub-array, if any]
            Else
            Repeat while I <= MID
                  SET TEMP[INDEX] = ARR[I]
                  SET INDEX = INDEX + 1, SET I = I + 1
            [END OF LOOP]
      [END OF IF]
Step 4: [Copy the contents of TEMP back to ARR] SET K=0
Step 5: Repeat while K < INDEX
            a. SET ARR[K] = TEMP[K]
            b. SET K = K + 1
      [END OF LOOP]
Step 6: END
```

```
MERGE_SORT(ARR, BEG, END)

Step 1: IF BEG < END, then
            SET MID = (BEG + END)/2
            CALL MERGE_SORT (ARR, BEG, MID)
            CALL MERGE_SORT (ARR, MID + 1, END)
            MERGE (ARR, BEG, MID, END)
      [END OF IF]
Step 2: END
```

Figure 14.4 Algorithm for merge sort

Complexity of Merge Sort

The running time of merge sort in the average case and the worst case can be given as `O(n logn)`. Although merge sort has an optimal time complexity, it needs an additional space of `O(n)` for the temporary array `TEMP`.

PROGRAMMING EXAMPLE

5. Write a program to implement merge sort.

```
#include<stdio.h>
#include<conio.h>
```

```
void merge(int a[], int, int, int);
void merge_sort(int a[],int, int);
void main()
{
        int arr[10], i, n, j,k;
        clrscr();
        printf("\n ENter the number of elements in the array : ");
        scanf("%d", &n);
        printf("\n Enter the elements of the array ");
        for(i=0;i<n;i++)
        {
                printf("\n arr[%d] = ", i);
                scanf("%d", &arr[i]);
        }
        merge_sort(arr, 0, n-1);
        printf("\n The sorted array is : \n");
        for(i=0;i<n;i++)
                printf("%d\t", arr[i]);
        getch();
}
void merge(int arr[], int beg, int mid, int end)
{
        int i=beg, j=mid+1, index=beg, temp[10], k;
        while((i<=mid) && (j<=end))
        {
                if(arr[i] < arr[j])
                {
                        temp[index] = arr[i];
                        i++;
                }
                else
                {
                        temp[index] = arr[j];
                        j++;
                }
                index++;
        }
        if(i>mid)
        {
                while(j<=end)
                {
                        temp[index] = arr[j];
                        j++;
                        index++;

                }
        }
        else
        {
```

```
            while(i<=mid)
            {
                  temp[index] = arr[i];
                  i++;
                  index++;
            }
      }
      for(k=beg;k<index;k++)
            arr[k] = temp[k];
}
void merge_sort(int arr[], int beg, int end)
{
      int mid;
      if(beg<end)
      {
            mid = (beg+end)/2;
            merge_sort(arr, beg, mid);
            merge_sort(arr, mid+1, end);
            merge(arr, beg, mid, end);
      }
}
```

14.6 QUICK SORT

Quick sort is a widely used sorting algorithm developed by C. A. R. Hoare that makes O(n log n) comparisons in the average case to sort an array of n elements. However, in the worst case, it has a quadratic running time given as $O(n^2)$. Basically, the quick sort algorithm is faster than other O(n log n) algorithms, because its efficient implementation can minimize the probability of requiring quadratic time. Quick sort is also known as partition exchange sort.

Like merge sort, this algorithm works by using a divide and conquer strategy to divide a single unsorted array into two smaller sub-arrays.

The quick sort algorithm works as follows:

1. Select an element pivot from the array elements.
2. Rearrange the elements in the array in such a way that all elements that are less than the pivot appear before the pivot and all elements greater than the pivot element come after it (equal values can go either way). After such a partitioning, the pivot is placed in its final position. This is called the *partition* operation.
3. Recursively, sort the two sub-arrays thus obtained (One with sub-list of lesser values than that of the pivot element and the other having higher value elements).

Like merge sort, the *base case* of the recursion occurs when the array has zero or one element because in that case, the array is already sorted. After each iteration, one element (pivot) is always in its final position. Hence, with every iteration, there is one less element to be sorted in the array.

Thus, the main task is to find the pivot element, which will partition the array into two halves. To understand how we find the pivot element, follow the steps given below (We take the first element in the array as pivot).

1. Set the index of the first element in the array to `loc` and `left` variables. Also, set the index of the last element of the array to the `right` variable.

   That is, `loc = 0`, `left = 0`, and `right = n-1` (where n in the number of elements in the array)

2. Start from the element pointed by right and scan the array from right to left, comparing each element on the way with the element pointed by the variable `loc`.

   That is, `a[loc]` should be less than `a[right]`.

   (a) If that is the case, then simply continue comparing until `right` becomes equal to `loc`. Once `right = loc`, it means the pivot has been placed in its correct position.

   (b) However, if at any point, we have `a[loc] > a[right]`, then interchange the two values and jump to Step 3.

   (c) Set `loc = right`

3. Start from the element pointed by `left` and scan the array from left to right, comparing each element on the way with the element pointed by `loc`.

   That is, `a[loc]` should be greater than `a[left]`.

   (a) If that is the case, then simply continue comparing until `left` becomes equal to `loc`. Once `left = loc`, it means the pivot has been placed in its correct position.

   (b) However, if at any point, we have `a[loc] < a[left]`, then interchange the two values and jump to Step 2.

   (c) Set `loc = left`.

EXAMPLE 14.4: Sort the elements given in the following array using quick sort algorithm.

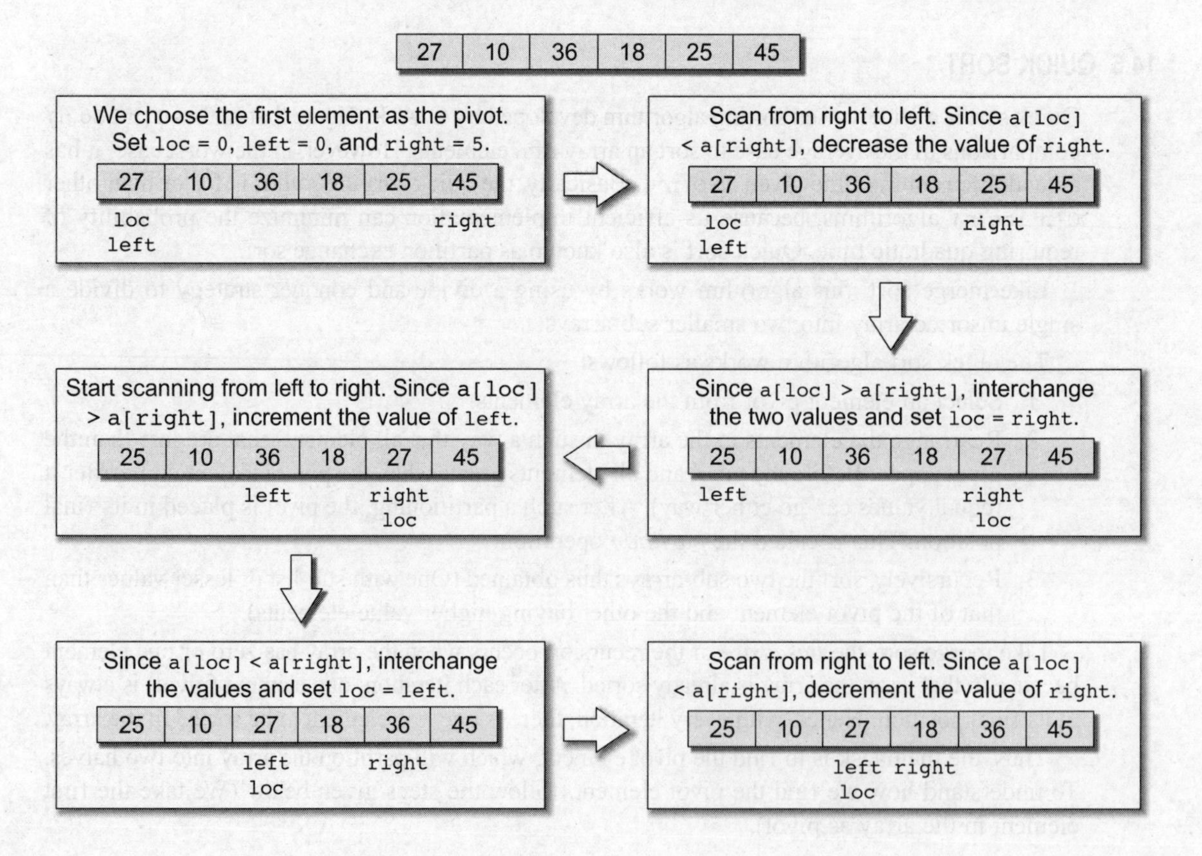

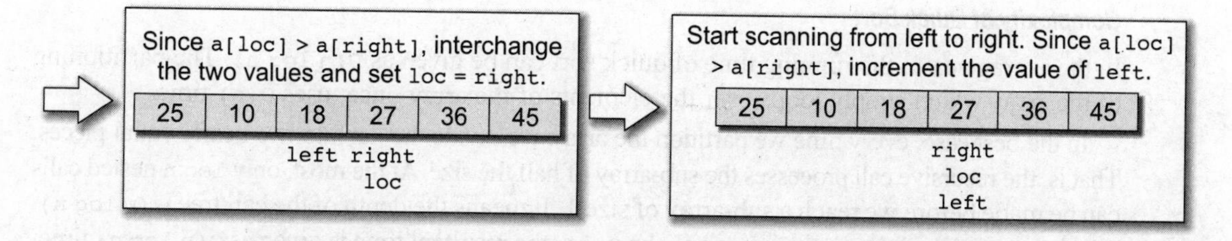

Now `left = loc`, so the procedure terminates, as the pivot element (the first element of the array, that is, 27) is placed in its correct position. All the elements smaller than 25 are placed before it and those the greater than 27 are placed after it.

The left sub-array containing 25, 10, 18 and right sub-array containing 36 and 45 are sorted in the same manner.

The quick sort algorithm (Fig. 14.5) makes use of a function `Partition` to divide the array into two sub-arrays.

```
PARTITION (ARR, BEG, END, LOC)

Step 1: [Initialize] SET LEFT = BEG, RIGHT = END, LOC = BEG, FLAG = 0
Step 2: Repeat Steps 3 to while FLAG = 0
Step 3: Repeat while ARR[LOC] <= ARR[RIGHT] AND LOC != RIGHT
                  SET RIGHT = RIGHT - 1
        [END OF LOOP]
Step 4: IF LOC == RIGHT, then
                  SET FLAG = 1
        ELSE IF ARR[LOC] > ARR[RIGHT], then
                  SWAP ARR[LOC] with  ARR[RIGHT]
                  SET LOC = RIGHT
        [END OF IF]
Step 5: IF FLAG = 0, then
                  Repeat while ARR[LOC] >= ARR[LEFT] AND LOC != LEFT
                  SET LEFT = LEFT + 1
                  [END OF LOOP]
Step 6:                 IF LOC == LEFT, then
                        SET FLAG = 1
                  ELSE IF ARR[LOC] < ARR[LEFT], then
                  SWAP ARR[LOC] with  ARR[LEFT]
                  SET LOC = LEFT
                  [END OF IF]
        [END OF IF]
Step 7: [END OF LOOP]
Step 8: END
```

```
QUICK_SORT (ARR, BEG, END)

Step 1: IF (BEG < END), then
                CALL PARTITION (ARR, BEG, END, LOC)
                CALL QUICKSORT(ARR, BEG, LOC - 1)
                CALL QUICKSORT(ARR, LOC + 1, END)
        [END OF IF]
Step 2: END
```

Figure 14.5 Algorithm for quick sort

Complexity of Quick Sort

In the average case, the running time of quick sort can be given as `O(n log n)`. The partitioning of the array which simply loops over the elements of the array once, uses `O(n)` time.

In the best case, every time we partition the array, we divide the list into two nearly equal pieces. That is, the recursive call processes the sub-array of half the size. At the most, only `log n` nested calls can be made before we reach a sub-array of size 1. It means the depth of the call tree is `O(log n)`. And because at each level, there can only be `O(n)`, the resultant time is given as `O(n log n)` time.

Practically, the efficiency of quick sort depends on the element which is chosen as the pivot. Its worst-case efficiency is given as $O(n^2)$. The worst case occurs when the array is already sorted (either in ascending or descending order) and the left-most element is chosen as the pivot.

However, many implementations randomly choose the pivot element. The randomized version of the quick sort algorithm always has an algorithmic complexity of `O(n log n)`.

Pros and Cons

It is faster than other algorithms such as bubble sort, selection sort, and insertion sort. Quick sort can be used to sort arrays of small size, medium size, or large size. On the flip side, quick sort is complex and massively recursive.

PROGRAMMING EXAMPLE

6. Write a program to implement quick sort algorithm.

```
#include<stdio.h>
#include<conio.h>
int partition(int a[], int beg, int end);
void quick_sort(int a[], int beg, int end);
void main()
{
        int arr[10], i, n, j,k;
        clrscr();
        printf("\n Enter the number of elements in the array : ");
        scanf("%d", &n);
        printf("\n Enter the elements of the array ");
        for(i=0;i<n;i++)
        {
                printf("\n arr[%d] = ", i);
                scanf("%d", &arr[i]);
        }
        quick_sort(arr, 0, n-1);
        printf("\n The sorted array is : \n");
        for(i=0;i<n;i++)
                printf("%d\t", arr[i]);
        getch();
}
int partition( int a[], int beg, int end)
{
        int left, right, temp, loc, flag;
```

```
            loc = left = beg;
            right = end;
            flag = 0;
            while(flag != 1)
            {
                while((a[loc] <= a[right]) && (loc!=right))
                    right--;
                if(loc==right)
                    flag =1;
                else if(a[loc]>a[right])
                {
                    temp = a[loc];
                    a[loc] = a[right];
                    a[right] = temp;
                    loc = right;
                }
                if(flag!=1)
                {
                    while((a[loc] >= a[left]) && (loc!=left))
                        left++;
                    if(loc==left)
                        flag =1;
                    else if(a[loc] <a[left])
                    {
                        temp = a[loc];
                        a[loc] = a[left];
                        a[left] = temp;
                        loc = left;
                    }
                }
            }
            return loc;
        }
        void quick_sort(int a[], int beg, int end)
        {
            int loc;
            if(beg<end)
            {
                loc = partition(a, beg, end);
                quick_sort(a, beg, loc-1);
                quick_sort(a, loc+1, end);
            }
        }
```

14.7 RADIX SORT

Radix sort is a linear sorting algorithm for integers that uses the concept of sorting names in alphabetical order. When we have a list of sorted names, the *radix* is 26 (or 26 buckets) because there are 26 letters of the alphabet. So radix sort is also known as bucket sort. Observe that words are first sorted according to the first letter of the name. That is, 26 classes are used to arrange the names, where the first class stores the names that begin with A, the second class contains the names with B, and so on.

During the second pass, names are grouped according to the second letter. After the second pass, names are sorted on the first two letters. This process is continued till the nth pass, where n is the length of the name with maximum number of letters.

After every pass, all the names are collected in order of buckets. That is, first pick up the names in the first bucket that contains the names beginning with A. In the second pass, collect the names from the second bucket, and so on.

When radix sort is used on integers, sorting is done on each of the digits in the number. The sorting procedure proceeds by sorting the least significant to the most significant digit. While sorting the numbers, we have ten buckets, each for one digit (0, 1, 2..., 9) and the number of passes will depend on the length of the number having maximum number of digits.

```
Algorithm for RadixSort (ARR, N)

Step 1: Find the largest number in ARR as LARGE
Step 2: [Initialize] SET NOP = Number of digits in LARGE
Step 3: SET PASS = 0
Step 4: Repeat Step 5 while PASS <= NOP-1
Step 5:             SET I = 0 AND Initialize buckets
Step 6:             Repeat Step 7 to Step 9 while I<N-1
Step 7:                   SET DIGIT  = digit at PASSth place in A[I]
Step 8:                   Add A[I] to the bucket numbered DIGIT
Step 9:                   INCEREMENT bucket count for bucket numbered DIGIT
              [END OF LOOP]
Step 10:            Collect the numbers in the bucket
         [END OF LOOP]
Step 11: END
```

Figure 14.6 Algorithm for radix sort

EXAMPLE 14.5: Sort the numbers given below using radix sort.

345, 654, 924, 123, 567, 472, 555, 808, 911

In the first pass, the numbers are sorted according to the digit at ones place. The buckets are pictured upside down as shown below.

Number	0	1	2	3	4	5	6	7	8	9
345						345				
654					654					
924					924					
123				123						
567								567		
472			472							
555						555				
808									808	
911		911								

After this pass, the numbers are collected bucket by bucket. The new list thus formed is used as an input for the next pass. In the second pass, the numbers are sorted according to the digit at the tens place. The buckets are pictured upside down.

Number	0	1	2	3	4	5	6	7	8	9
911		911								
472								472		
123			123							
654						654				
924			924							
345					345					
555						555				
567							567			
808	808									

In the third pass, the numbers are sorted according to the digit at the hundreds place. The buckets are pictured upside down.

Number	0	1	2	3	4	5	6	7	8	9
808									808	
911										911
123		123								
924										924
345				345						
654							654			
555						555				
567						567				
472					472					

The numbers are collected bucket by bucket. The new list thus formed is the final sorted result. After the third pass, the list can be given as

123, 345, 472, 555, 567, 654, 808, 911, 924.

Complexity of Radix Sort

To calculate the complexity of radix sort algorithm, assume that there are `n` numbers that have to be sorted and `k` is the number of digits in the largest number. In this case, the radix sort algorithm is called a total of `k` times. The inner loop is executed `n` times. Hence, the entire radix sort algorithm takes `O(kn)` time to execute. When radix sort is applied on a data set of finite size (very small set of numbers), then the algorithm runs in `O(n)` asymptotic time.

Pros and Cons

Radix sort is a very simple algorithm. When programmed properly, radix sort is one of the fastest sorting algorithms for numbers or strings of letters.

But there are certain trade-offs for radix sort that can make it less preferable as compared to other sorting algorithms. Radix sort takes more space than other sorting algorithms. Besides the array of numbers, we need 10 buckets to sort numbers, 26 buckets to sort strings containing only alphabets, and at least 40 buckets to sort a string containing alphanumeric characters.

Another drawback of radix sort is that the algorithm is dependant on the digits or letters. This feature compromises with the flexibility to sort input of any data type. For every different data type, the algorithm has to be rewritten. Even if the sorting order changes, the algorithm has to be rewritten. Thus, radix sort takes more time to write and writing a general purpose radix sort algorithm that can handle all kinds of data is not a trivial task.

Radix sort is a good choice for many programs that need a fast sort, but there are faster sorting algorithms available. This is the main reason why radix sort is not as widely used as other sorting algorithms.

PROGRAMMING EXAMPLE

7. Write a program to implement radix sort algorithm.

```
#include<stdio.h>
#include<conio.h>
int largest(int arr[], int n);
void radix_sort(int arr[], int n);
void main()
{
        int arr[10], i, n, j,k;
        clrscr();
        printf("\n ENter the number of elements in the array : ");
        scanf("%d", &n);
        printf("\n Enter the elements of the array ");
        for(i=0;i<n;i++)
        {
                printf("\n arr[%d] = ", i);
                scanf("%d", &arr[i]);
        }
        radix_sort(arr, n);
        printf("\n The sorted array is : \n");
        for(i=0;i<n;i++)
                printf("%d\t", arr[i]);
        getch();
}
int largest(int arr[], int n)
{
        int large=arr[0], i;
        for(i=1;i<n;i++)
        {
                if(arr[i]>large)
                        large = arr[i];
        }
```

```
        return large;
    }
    void radix_sort(int arr[], int n)
    {
        int bucket[10][10], bucket_count[10];
        int i, j, k, remainder, NOP=0, divisor=1, large, pass;
        large = largest(arr, n);
        while(large>0)
        {
            NOP++;
            large/=10;
        }
        for(pass=0;pass<NOP;pass++)    // Initialize the buckets
        {
            for(i=0;i<10;i++)
                bucket_count[i]=0;
            for(i=0;i<n;i++)
            {
            // sort the numbers according to the digit at passth place
                remainder = (arr[i]/divisor)%10;
                bucket[remainder][bucket_count[remainder]] = arr[i];
                bucket_count[remainder] += 1;
            }
            // collect the numbers after PASS pass
            i=0;
            for(k=0;k<10;k++)
            {
                for(j=0;j<bucket_count[k];j++)
                {
                    arr[i] = bucket[k][j];
                    i++;
                }
            }
            divisor *= 10;
        }
    }
```

14.8 HEAP SORT

We have discussed binary heaps in Chapter 9. Therefore, we already know how to build a heap H from an array, how to insert a new element in the already existing heap, and how to delete an element from H. Now, using these basic concepts, we will discuss the application of heaps to write an efficient algorithm of heap sort (also known as tournament sort) that has a running time complexity of O(n log n).

Given an array ARR with n elements, the heap sort algorithm can be used to sort ARR in two phases:

- In phase 1, build a heap H using the elements of ARR.
- In phase 2, repeatedly delete the root element of the heap formed in phase 1.

In a max heap, we know that the largest value in H is always present at the root node. So in phase B, when the root element is deleted, we are actually collecting the elements of ARR in decreasing order. Let us take a look at the algorithm of heap sort shown in Fig. 14.7.

```
HEAPSORT(ARR, N)

Step 1: [Build Heap H]
     Repeat for I = 0 to N-1
          CALL Insert_Heap(ARR, N, ARR[I])
     [END OF LOOP]
Step 2: [Repeatedly delete the root element)
     Repeat while N>0
          CALL Delete_Heap(ARR, N, VAL)
          SET N = N + 1
     [END OF LOOP]
Step 3: END
```

Figure 14.7 Algorithm for heap sort

Complexity of Heap Sort

Heap sort uses two heap operations: *insertion* and *root deletion*. Each element extracted from the root is placed in the last empty location of the array.

In phase 1, when we build a heap, the number of comparisons to find the right location of the new element in H cannot exceed the depth of H. Since H is a complete tree, its depth cannot exceed m, where m is the number of elements in the heap H.

Thus, the total number of comparisons g(n) to insert n elements of ARR in H is bounded as:

g(n) <= n log n

Hence, the running time of the first phase of the heap sort algorithm is O(n log n).

In phase 2, we have H which is a complete tree with m elements having left and right sub-trees as heaps. Assuming L to be the root of the tree, *reheaping* the tree would need 4 comparisons to move L one step down the tree H. Since the depth of H cannot exceed O(log m), reheaping the tree will require a maximum of 4 log m comparisons to find the right location of L in H.

Since n elements will be deleted from the heap H, reheaping will be done n times. Therefore, the number of comparisons to delete n elements is bounded as:

h(n) <= 4n log n

Hence, the running time of the second phase of the heap sort algorithm is O(n log n).

Each phase requires time proportional to O(n log n). Therefore, the running time to sort an array of n elements in the worst case is proportional to O(n log n).

Therefore, we can conclude that heap sort is simple, fast, and a stable sorting algorithm that can be used to sort large sets of data efficiently.

PROGRAMMING EXAMPLE

8. Write a program to implement heap sort algorithm.

```
#include<stdio.h>
#include<conio.h>
#define MAX 10
void RestoreHeapUp(int *,int);
```

```
void RestoreHeapDown(int*,int,int);
void main()
{
        int Heap[MAX],n,i,j,k;
        clrscr();
        printf("\n Enter the number of elements : ");
        scanf("%d",&n);
          printf("\n Enter the elements : ");
          for(i=1;i<=n;i++)
          {
                  scanf("%d",&Heap[i]);
                  RestoreHeapUp(Heap, i); // Heapify
          }
          // Delete the root element and heapify the heap
          j=n;
          for(i=1;i<=j;i++)
          {
                  int temp;
                  temp=Heap[1];
                  Heap[1]=[n];
                  Heap[n]=temp;
                  n = n-1;  // The element Heap[n] is supposed to be deleted
                  RestoreHeapDown(Heap,1,n); // Heapify
          }
          n=j;
          printf("\n The sorted elements are: ");
          for(i=1;i<=n;i++)
                  printf("%4d",Heap[i]);
          return 0;
}
void RestoreHeapUp(int *Heap,int index)
{
          int val = Heap[index];
          while( (index>1) && (Heap[index/2] < val) ) // Check parent's value
          {
                  Heap[index]=Heap[index/2];
                  index /= 2;
          }
          Heap[index]=val;
}
void RestoreHeapDown(int *Heap,int index,int n)
{
          int val = Heap[index];
          int j=index*2;
```

```
        while(j<=n)
        {
                if( (j<n) && (Heap[j] < Heap[j+1]) )
        // Check sibling's value
                        j++;
                if(Heap[j] < Heap[j/2])  // Check parent's value
                break;
                Heap[j/2]=Heap[j];
                j=j*2;
        }
        Heap[j/2]=v;
    }
```

14.9 SHELL SORT

Shell sort, invented by Donald Shell in 1959, is a sorting algorithm that is a generalization of insertion sort. While discussing insertion sort, we have observed two things:

- First, insertion sort works well when the input data is 'almost sorted'.
- Second, insertion sort is quite inefficient to use, as it moves the values just one position at a time.

Shell sort is considered an improvement over insertion sort, as it compares elements separated by a gap of several positions. This enables the element to take bigger steps towards its expected position. In Shell sort, the elements are sorted in multiple passes and in each pass, data are taken with smaller and smaller gap sizes. However, the last step of Shell sort is a plain insertion sort. But by the time we reach the last step, the elements are already 'almost sorted', and hence it provides good performance.

If we take a scenario in which the smallest element is stored in the other end of the array, then sorting such an array with either bubble sort or insertion sort will execute in $O(n^2)$ time and take roughly n comparisons and exchanges to move this value all the way to its correct position. On the other hand, Shell sort first moves small values using giant step sizes, so a small value will move a long way towards its final position, with just a few comparisons and exchanges.

Technique

To visualize the way in which Shell sort works, perform the following steps:

- *Step 1:* Arrange the elements of the array in the form of a table and sort the columns (using insertion sort).
- *Step 2:* Repeat Step 1, each time with smaller number of longer columns in such a way that at the end, there is only one column of data to be sorted.

Note that we are only visualizing the elements being arranged in a table, the algorithm does its sorting in-place.

EXAMPLE 14.6: Sort the elements given below using Shell sort.

63, 19, 7, 90, 81, 36, 54, 45, 72, 27, 22, 9, 41, 59, 33

Arrange the elements of the array in the form of a table and sort the columns.

63	19	7	90	81	36	54	45
72	27	22	9	41	59	33	

Result:

63	19	7	9	41	36	33	45
72	27	22	90	81	59	54	

The elements of the array can be given as:

63, 19, 7, 9, 41, 36, 33, 45, 72, 27, 22, 90, 81, 59, 54.

Repeat Step 1 with smaller number of long columns.

63	19	7	9	41
36	33	45	72	27
22	90	81	59	54

Result:

22	19	7	9	27
36	33	45	59	41
63	90	81	72	54

The elements of the array can be given as:

22, 19, 7, 9, 27, 36, 33, 45, 59, 41, 63, 90, 81, 72, 54.

Repeat Step 1 with smaller number of long columns.

22	19	7
9	27	36
33	45	59
41	63	90
81	72	54

Result:

9	19	7
22	27	36
33	45	54
41	63	59
81	72	90

The elements of the array can be given as:

9, 19, 7, 22, 27, 36, 33, 45, 54, 41, 63, 59, 81, 72, 90.

Finally, arrange the elements of the array in a single column and sort the column.

7
19
9
22
27
36
33
41
54
45
63
59
81
72
90

The elements of the array can be given as:

7, 19, 9, 22, 27, 36, 33, 41, 54, 45, 63, 59, 81, 72, 90

Algorithm

The algorithm to sort an array of elements using Shell sort is shown in Fig. 14.8. In the algorithm, we sort the elements of the array `Arr` in multiple passes. In each pass, we reduce the `gap_size` (visualize it as the number of columns) by a factor of half as done in Step 4. In each iteration of the `for loop` in Step 5, we compare the values of the array and interchange them if we have a larger value preceding the smaller one.

```
Shell_Sort(Arr, n)

Step 1: SET Flag = 1, gap_size = n
Step 2: Repeat Steps 3 to 6 while Flag = 1 OR gap_size > 1
Step 3:       SET Flag = 0
Step 4:      SET gap_size = (gap_size + 1) / 2
Step 5:      Repeat Step 6 FOR i = 0 to i < (n - gap_size)
Step 6:              IF Arr[i + gap_size] > Arr[i], then
                           SWAP Arr[I + gap_size], Arr[i]
                           SET Flag = 0
Step 7: END
```

Figure 14.8 Algorithm for shell sort

PROGRAMMING EXAMPLE

9. Write a program to implement shell sort algorithm.

```
#include<stdio.h>
main()
{
        int arr[10];
        int i, j, n, flag = 1, gap_size, temp;
        printf("\n ENter the number of elements in the array : ");
        scanf("%d", &n);
        printf("\n Enter %d numbers : ");
        for(i=0;i<n;i++)
                scanf("%d", &arr[i]);
        while(flag == 1 || gap_size > 1)
        {
                flag = 0;
                gap_size = (gap_size + 1) / 2;
                for(i=0; i< (n - gap_size); i++)
                {
                        if( arr[i+gap_size] > arr[i])
                        {
                                temp = arr[i+gap_size];
                                arr[i+gap_size] = arr[i];
                                arr[i] = temp;
                        }
                }
        }
        printf("\n The sorted array is : \n");
        for(i=0;i<n;i++)
                printf("%d", arr[i]);
}
```

14.10 COMPARISON OF SORTING ALGORITHMS

Table 14.1 compares the average-case and worst-case time complexities of different sorting algorithms discussed so far.

Table 14.1 Comparison of algorithms

Algorithm	Average Case	Worst Case
Bubble sort	$O(n^2)$	$O(n^2)$
Bucket sort	$O(n.k)$	$O(n^2.k)$
Selection sort	$O(n^2)$	$O(n^2)$
Insertion sort	$O(n^2)$	$O(n^2)$
Shell sort	–	$O(n \log^2 n)$
Merge sort	$O(n \log n)$	$O(n \log n)$
Heap sort	$O(n \log n)$	$O(n \log n)$
Quick sort	$O(n \log n)$	$O(n^2)$

SUMMARY

- Internal sorting deals with sorting the data stored in the memory, whereas external sorting deals with sorting the data stored in files.
- In bubble sorting, consecutive adjacent pairs of elements in the array are compared with each other.
- For insertion sort, the best case occurs when the array is already sorted. In this case, the running time of the algorithm has a linear running time (i.e. $O(n)$). Similarly, the worst case occurs when the array is sorted in reverse order. Therefore, in the worst case, insertion sort has a quadratic running time (i.e. $O(n^2)$).
- Selection sort is a sorting algorithm that has a quadratic running time complexity given as $O(n^2)$, thereby making it inefficient to be used on large lists.
- Merge sort is a sorting algorithm that uses the divide, conquer, and combine algorithmic paradigm. *Divide* means partitioning the n-element array to be sorted into two sub-arrays of n/2 elements in each sub-array. *Conquer* means sorting the two sub-arrays recursively using merge sort. *Combine* means merging the two sorted sub-arrays of size n/2 each to produce a sorted array of n elements. The running time of merge sort in average case and worst case can be given as $O(n \log n)$.
- Quick sort has an average case running time complexity of $O(n \log n)$. In the worst case, it has a quadratic running time complexity given by $O(n^2)$.
- Radix sort is a linear sorting algorithm that uses the concept of sorting names in alphabetical order.
- Heap sort has a running time complexity of $O(n \log n)$. It sorts an array in two phases. In the first phase, it builds a heap of the given array. In the second phase, the root element is deleted.
- Shell sort is considered as an improvement over insertion sort, as it compares elements separated by a gap of several positions.

GLOSSARY

Bubble sort A sorting algorithm which sorts by comparing each adjacent pair of values in a list in turn, swapping the values if necessary, and repeating the pass through the list until there are no more swaps.

Heap sort A sorting algorithm that builds a heap and then repeatedly extracts the maximum value (in case of max-heap). The execution time of heap sort is $O(n \log n)$.

Insertion sort A sorting algorithm that sorts by repeatedly taking the next value and inserting it into the final sorted list. The execution time of insertion sort is `O(n2)`.

Internal sort A variant of sorting algorithm that exclusively uses the main memory during the sort. This assumes high-speed random access to all memory.

Merge sort A sorting algorithm that splits the values to be sorted into two groups, recursively sorts each group, and merges them into a final sorted sequence. The execution time of merge sort is `O(n log n)`.

Quick sort A sorting algorithm that selects an element from the array (the pivot), partitions the remaining elements into those greater than and less than this pivot, and recursively sorts the partitions.

Radix sort A multiple pass sorting algorithm that distributes each value to a bucket according to part of the value's key beginning with the least significant digit of the key. After each pass, values are collected from the buckets, keeping the values in order, then redistributed according to the next most significant digit of the key.

Selection sort A sorting algorithm that repeatedly looks through the remaining values to search for the lowest value in order to move it to its final location. The execution time of selection sort is `O(n²)` and the number of swaps performed is `O(n)`, where `n` is the number of elements.

Shell sort The sorting algorithm in which `i` sets of `n/i` items are sorted on each pass, typically with insertion sort. On each succeeding pass, `i` is reduced until it is 1 for the last pass.

Sort The process of arranging values in a predetermined order. There are dozens of algorithms, the choice of which depends on factors such as the number of values relative to the working memory, the cost of comparing keys versus the cost of moving items, etc.

EXERCISES

Review Questions

1. Define sorting. What is the importance of sorting?
2. What are the different types of sorting techniques? Which sorting technique has the least worst case?
3. Explain the difference between bubble sort and quick sort. Which one is more efficient?
4. Sort the elements 77, 49, 25, 12, 9, 33, 56, 81 using
   (a) insertion sort (b) selection sort
   (c) bubble sort (d) merge sort
   (e) quick sort (f) radix sort
   (g) shell sort
5. Compare heap sort and quick sort.
6. Quick sort shows quadratic behaviour in certain situations. Justify.
7. If the following sequence of numbers is to be sorted using quick sort, then show the iterations of the sorting process.
   42, 34, 75, 23, 21, 18, 90, 67, 78
8. Sort the following sequence of numbers in descending order using heap sort.
   42, 34, 75, 23, 21, 18, 90, 67, 78
9. A certain sorting technique was applied to the following data set,
   45, 1, 27, 36, 54, 90
   After two passes, the rearrangement of the data set is given as below:
   1, 27, 45, 36, 54, 90
   Identify the sorting algorithm that was applied.
10. A certain sorting technique was applied to the following data set,
    81, 72, 63, 45, 27, 36
    After two passes, the rearrangement of the data set is given as below:
    27, 36, 80, 72, 63, 45
    Identify the sorting algorithm that was applied.
11. A certain sorting technique was applied to the following data set,
    45, 1, 63, 36, 54, 90
    After two passes, the rearrangement of the data set is given as below:
    1, 45, 63, 36, 54, 90
    Identify the sorting algorithm that was applied.

12. Write a recursive function to perform selection sort.
13. Compare the running time complexity of different sorting algorithms.
14. Discuss the advantages of insertion sort.

Programming Exercises

1. How many swaps will be performed to sort the following numbers using bubble sort?
   7, 1, 4, 12, 67, 33, 45
2. Write a program to implement bubble sort. Analyse its strengths and weaknesses.
3. Write a program to implement insertion sort.
4. Write a program to implement selection sort.
5. Write a program to implement merge sort.
6. Write a program to implement quick sort.
7. Write a program to implement radix sort.
8. Write a program to sort an array of integers in descending order using the following sorting techniques:
   (a) insertion sort (b) selection sort
   (c) bubble sort (d) merge sort
   (e) quick sort (f) radix sort
   (g) shell sort
9. Write a program to sort an array of floating point numbers in descending order using the following sorting techniques:
   (a) insertion sort (b) selection sort
   (c) bubble sort (d) merge sort
   (e) quick sort (f) radix sort
   (g) shell sort
10. Write a program to sort an array of names using bucket sort.

Multiple Choice Questions

1. A card game player arranges his cards and picks them one by one. To which sorting technique you can compare this example?
   (a) Bubble sort (b) Selection sort
   (c) Merge sort (d) Radix sort
2. Which sorting deals with sorting the data stored in the computer's memory?
   (a) Insertion sort (b) Internal sort
   (c) External sort (d) Radix sort
3. In which sorting, consecutive adjacent pairs of elements in the array are compared with each other?
   (a) Bubble sort (b) Selection sort
   (c) Merge sort (d) Radix sort
4. Which term means sorting the two sub-arrays recursively using merge sort?
   (a) Divide (b) Conquer
   (c) Combine (d) All of these
5. Which sorting algorithm sorts by moving the current data element past the already sorted values and repeatedly interchanging it with the preceding value until it is in its correct place?
   (a) Insertion sort (b) Internal sort
   (c) External sort (d) Radix sort
6. Which algorithm uses the divide, conquer, and combine algorithmic paradigm?
   (a) Selection sort (b) Insertion sort
   (c) Merge sort (d) Radix sort
7. Quick sort is faster than
   (a) Selection sort (b) Insertion sort
   (c) Bubble sort (d) All of these
8. Which sorting algorithm is also known as tournament sort?
   (a) Selection sort (b) Insertion sort
   (c) Bubble sort (d) Heap sort

True or False

1. For insertion sort, the best case occurs when the array is already sorted.
2. Selection sort has a linear running time complexity.
3. The running time of merge sort in the average case and the worst case is `O(n log n)`.
4. The worst case running time complexity of quick sort is `O(n log n)`.
5. Heap sort is an efficient and a stable sorting algorithm.
6. External sorting deals with sorting the data stored in the computer's memory.
7. Insertion sort is less efficient than quick sort, heap sort, and merge sort.
8. The average case of insertion sort has a quadratic running time.

9. The partitioning of the array in quick sort is done in O(n) time.
10. Heap sort is a stable sorting algorithm.

Fill in the Blanks

1. Sorting means ______.
2. ______ sort shows the best average-case behaviour.
3. ______ deals with sorting the data stored in files.
4. O(n^2) is the running time complexity of ______ algorithm.
5. In the worst case, insertion sort has a ______ running time.
6. ______ sort uses the divide, conquer, and combine algorithmic paradigm.
7. In the average case, quick sort has a running time complexity of ______.
8. The execution time of bucket sort in average case is ______.
9. The running time of merge sort in the average and the worst case is ______.
10. The efficiency of quick sort depends on ______.

15 Hashing and Collision

Learning Objective

In this chapter, we will discuss another data structure known as hash table. We will see what a hash table is and why do we prefer hash tables over simple arrays. We will also discuss hash functions, collisions, and the techniques to resolve collisions.

15.1 INTRODUCTION

In Chapter 12, we discussed two search algorithms: *linear search* and *binary search*. Linear search has a running time proportional to `O(n)`, while binary search takes time proportional to `O(log n)`, where `n` is the number of elements in the array. Binary search and binary search trees are efficient algorithms to search for an element. But what if we want to perform the search operation in time given as `O(1)`? In other words, is there a way to search an array in constant time, irrespective of its size?

There are two solutions to this problem. To analyse the first solution, let us take an example. In a small company of 100 employees, each employee is assigned an `Emp_ID` in the range 0–99. To store the records in an array, each employee's `Emp_ID` acts as an index into the array where this employee's record will be stored as shown in Fig. 15.1.

Key		Array of Employee's Record
Key 0 →	[0]	Employee record with Emp_ID 0
Key 1 →	[1]	Employee record with Emp_ID 1
Key 2 →	[2]	Employee record with Emp_ID 2
....		....
....		....
Key 98 →	[98]	Employee record with Emp_ID 98
Key 99 →	[99]	Employee record with Emp_ID 99

Figure 15.1 Record of employees

In this case, we can directly access the record of any employee, once we know his `Emp_ID`, because the array index is the same as that of the `Emp_ID` number. But practically, this implementation is hardly feasible.

Let us assume that the same company uses a five digit `Emp_ID` as the primary key. In this case, key values will range from 00000 to 99999. If we want to use the same technique as above, we need an array of size 100,000, of which only 100 elements will be used. This is illustrated in Fig. 15.2.

Key		Array of Employee's Record
Key 00000 →	[0]	Employee record with Emp_ID 00000
....		....
Key n →	[n]	Employee record with Emp_ID n
....		....
Key 99998 →	[99998]	Employee record with Emp_ID 99998
Key 99999 →	[99999]	Employee record with Emp_ID 99999

Figure 15.2 Record of employees with a five digit Emp_ID

Observe how impractical it is to waste so much storage space just to ensure that each employee' record is in a unique and predictable location.

Whether we use a two digit primary key (`Emp_ID`) or a five digit key, there are just 100 employees in the company. Thus, we will be using only 100 locations in the array. Therefore, in order to keep the array size down to the size that we will actually be using (100 elements), another good option is to use just the last two digits of the key to identify each employee. For example, the employee with `Emp_ID` 79439 will be stored in the element of the array with index 39. Similarly, the employee with `Emp_ID` 12345 will have his record stored in the array at the 45th location.

In the second solution, we see that the elements are not stored according to the *value* of the key. So in this case, we need a way to convert a five-digit key number to a two-digit array index. We need a function which will do the transformation. In this case, we will use the term *hash table* for an array and the function that will carry out the transformation will be called a *hash function.*

15.2 HASH TABLE

Hash table is a data structure in which keys are mapped to array positions by a hash function. In our example we use a hash function that extracts the last two digits of the key. Therefore, we map the keys to array locations or array indexes. A value stored in the hash table can be searched in `O(1)` time using a hash function to generate an address from the key (by producing the index of the array where the value is stored).

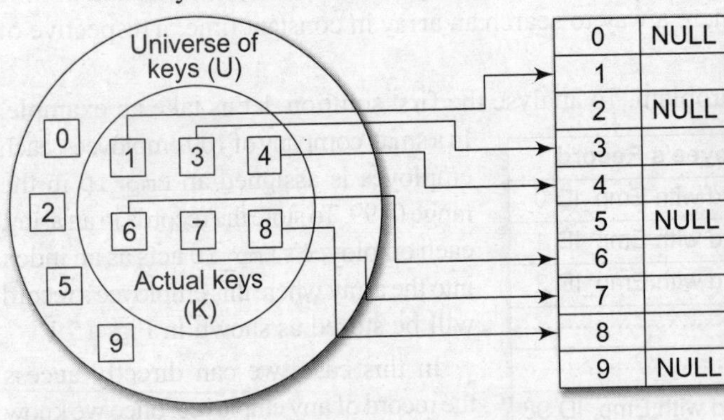

Figure 15.3 Direct relationship between key and index in the array

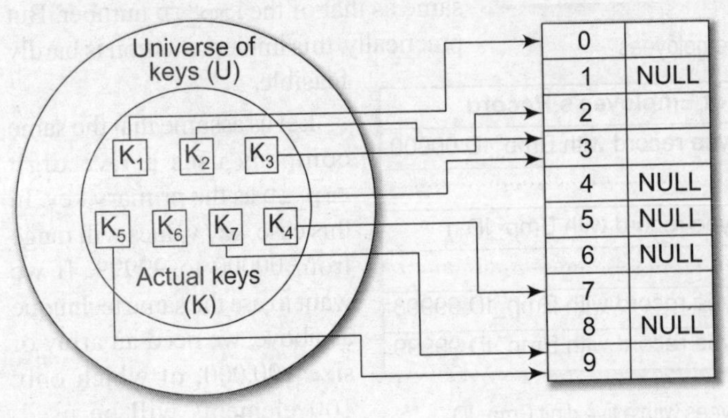

Figure 15.4 Relationship between keys and hash table index

Figure 15.3 shows a direct correspondence between the key and the index of the array. This concept is useful when the total universe of keys is small and when most of the keys are actually used from the whole set of keys. This is equivalent to our first example, where there are 100 keys for 100 employees.

However, when the set `K` of keys that are actually used is smaller than that of `U`, a hash table consumes less storage space. The storage requirement for a hash table is just `O(k)`, where `k` is the number of keys actually used.

In a hash table, an element with key `k` is stored at index `h(k)` and not `k`. It means a hash function `h` is used to calculate the index at which the element with key `k` will be stored. Thus, the process of mapping the keys to appropriate locations (or indexes) in a hash table is called *hashing.*

Figure 15.4 shows a hash table in which each key from the set `K` is mapped to locations generated by using a hash function. Note that keys k_2 and k_6 point to the same memory location. This is known as *collision*. That is, when two or more keys map to the same memory location, a collision is said to occur. Similarly, keys k_5 and k_7 also collide. Therefore, the main goal of using a hash function is to reduce the range of array indices that have to be handled. Therefore, instead of having U values, we just need K values, thereby reducing the amount of storage space required.

15.3 HASH FUNCTION

A hash function is simply a mathematical formula which when applied to a key, produces an integer which can be used as an index for the key in the hash table. The main aim of a hash function is that elements should be relatively, randomly, and uniformly distributed. It produces a unique set of integers within some suitable range. Hash functions are used to reduce the number of collisions. In practice, there is no hash function that eliminates collision completely. A good hash function can only minimize the number of collisions by spreading the elements uniformly throughout the array.

There is no magic formula to create a hash function but in this section, we will discuss the popular hash functions which help to minimize collisions. But before that, let us first look at the properties of a good hash function.

Properties of a Good Hash Function

Low cost The cost of computing a hash function must be small, so that using the hashing technique becomes preferable over other approaches. For example, if binary search algorithm can search an element from a sorted table of n items with $\log_2$ n key comparisons, then the hash function must cost less than performing $\log_2$ n key comparisons.

Determinism A hash procedure must be deterministic. This means that the same hash value must be generated for a given input value. However, this criteria excludes hash functions that depend on external variable parameters (such as the time of day) and on the memory address of the object being hashed (because address of the object may change during processing).

Uniformity A good hash function must map the keys as evenly as possible over its output range. This means that the probability of generating every hash value in the output range should roughly be the same. The property of uniformity also minimizes the number of collisions.

15.4 DIFFERENT HASH FUNCTIONS

In this section, we will discuss the hash functions using numeric keys. However, there can be cases in real-world applications where you have alphanumeric keys rather than simple numeric keys. In such a case, the ASCII values of the characters can be added to transform it into its equivalent numeric key. Once this transformation is done, any of the hash functions given below can be applied to generate the hash value.

15.4.1 Division Method

It is the most simple method of hashing an integer x. The method divides x by M and then uses the remainder thus obtained. In this case, the hash function can be given as

```
h(z) = z mod M
```

The division method is quite good for just about any value of M and since it requires only a single division operation, the method works very fast. However, extra care should be taken to select a suitable value for M.

For example, suppose M is an even number, then h(x) is even if x is even; and h(x) is odd if x is odd. If all possible keys are equi-probable, then this is not a problem. But if even keys are more likely than odd keys, then the division method will not spread the hashed values uniformly.

Generally, it is best to choose M to be a prime number because making M a prime number increases the likelihood that the keys are mapped with a uniformity in the output range of values. Then, M should also be not too close to the exact powers of 2. If we have

$$h(k) = x \bmod 2^k$$

then, the function will simply extract the lowest k bits of the binary representation of x.

The division method is extremely simple to implement. The following code segment illustrates how to do this:

```
int const M = 97; // a prime
int h (int x)
{ return (x % M); }
```

A potential drawback of the division method is that while using this method, consecutive keys map to consecutive hash values. On one hand, this is good as it ensures that consecutive keys do not collide, but on the other, it also means that consecutive array locations will be occupied. This may lead to degradation in performance.

EXAMPLE 15.1: Calculate the hash values of keys 1234 and 5462.

Setting m = 97, hash values can be calculated as:

```
h(1234) = 1234 % 97 = 70
h(5642) = 5642 % 97 = 16
```

15.4.2 Multiplication Method

The steps involved in the multiplication method are as follows:

Step 1: Choose a constant A such that 0 < A < 1.

Step 2: Multiply the key k by A.

Step 3: Extract the fractional part of kA.

Step 4: Multiply the result of Step 3 by m and take the floor.

Hence, the hash function can be given as:

$$h(x) = \lfloor m \, (kA \bmod 1) \rfloor$$

where, (kA mod 1) gives the fractional part of kA and m is the total number of indices in the hash table.

The greatest advantage of this method is that it works practically with any value of A. Although the algorithm works better with some values, but the optimal choice depends on the characteristics of the data being hashed. Knuth has suggested that the best choice of A is:

$$» (\text{sqrt}5 - 1) / 2 = 0.6180339887$$

EXAMPLE 15.2: Given a hash table of size 1000, map the key 12345 to an appropriate location in the hash table.

We will use A = 0.618033, m = 1000, and k = 12345

```
h(12345) = ⌊ 1000 (12345 × 0.618033 mod 1) ⌋
         = ⌊ 1000 (7629.617385 mod 1) ⌋
         = ⌊ 1000 (0.617385) ⌋
         = ⌊ 617.385 ⌋
         = 617
```

15.4.3 Mid-Square Method

The mid-square method is a good hash function which works in two steps.

Step 1: Square the value of the key. That is, find k^2.

Step 2: Extract the middle r bits of the result obtained in Step 1.

The algorithm works well because most or all bits of the key value contribute to the result. This is because all the digits in the original key value contribute to produce the middle two digits of the squared value. Therefore, the result is not dominated by the distribution of the bottom digit or the top digit of the original key value.

In the mid-square method, the same r bits must be chosen from all the keys. Therefore, the hash function is given as:

```
h(k) = s
```

where, s is obtained by selecting r bits from k^2

EXAMPLE 15.3: Calculate the hash value for keys 1234 and 5642 using the mid-square method. The hash table has 100 memory locations.

Note that the hash table has 100 memory locations whose indices vary from 0–99. This means that only two digits are needed to map the key to a location in the hash table, so r = 2.

When k = 1234, k^2 = 1522756, h (k) = 27

When k = 5642, k^2 = 31832164, h (k) = 21

Observe that the 3rd and 4th digits starting from the right are chosen.

15.4.4 Folding Method

The folding method works in the following two steps:

Step 1: Divide the key value into a number of parts. That is, divide k into parts $k_1, k_2, \ldots, k_n$, where each part has the same number of digits except the last part which may have lesser digits than the other parts.

Step 2: Add the individual parts. That is, obtain the sum of $k_1 + k_2 + \ldots + k_n$. The hash value is produced by ignoring the last carry, if any.

Note that the number of digits in each part of the key will vary depending upon the size of the hash table. For example, if the hash table has a size of 1000, then there are 1000 locations in the hash table. To address these 1000 locations, we need at least three digits, therefore, each part of the key must have three digits except the last part which may have lesser digits.

EXAMPLE 15.4: Given a hash table of 100 locations, calculate the hash value using folding method for keys 5678, 321, and 34567.

Since there are 100 memory locations to address, we will break the key into parts where each part (except the last) will contain two digits. Therefore,

Key	5678	321	34567
Parts	56 and 78	32 and 1	34, 56 and 7
Sum	134	33	97
Hash value	34 (ignore the last carry)	33	97

15.5 COLLISIONS

As discussed earlier in this chapter, collisions occur when the hash function maps two different keys to the same location. Obviously, two records cannot be stored in the same location. Therefore, a method used to solve the problem of collision, also called *collision resolution technique*, is applied. The two most popular method of resolving a collision are:

(a) open addressing, and
(b) chaining.

In this section, we will discuss both these techniques in detail.

15.5.1 Collision Resolution by Open Addressing

Once a collision takes place, open addressing computes new positions using a probe sequence and the next record is stored in that position. In this technique, all the values are stored in the hash table. The hash table contains two types of values: *sentinel values* (for example, –1) and *data values*. The presence of a sentinel value indicates that the location contains no data value at present but can be used to hold a value.

When a key is mapped to a particular memory location, then the value it holds is checked. If it contains a sentinel value, then the location is free and the data value can be stored in it. However, if the location already has some data value stored in it, then other slots are examined systematically in the forward direction to find a free slot. If not even a single free location is found, then we have an `OVERFLOW` condition.

The process of examining memory locations in the hash table is called *probing*. Open addressing technique can be implemented using linear probing, quadratic probing, and double hashing.

Linear Probing

The simplest approach to resolve a collision is linear probing. In this technique, if a value is already stored at a location generated by `h(k)`, then the following hash function is used to resolve the collision:

```
h(k, i) = [h'(k) + i] mod m
```

Here, `m` is the size of the hash table, `h'(k) = (k mod m)`, and `i` is the probe number and varies from 0 to `m-1`.

Therefore, for a given key `k`, first the location generated by `[h'(k) mod m]` is probed; because for the first time `i=0`. If the location is free, the value is stored in it, else the second probe generates the address of the location given by `[h'(k) + 1]mod m`. Similarly, if the location is occupied, then subsequent probes generate the address as `[h'(k) + 2]mod m`, `[h'(k) + 3]mod m`, `[h'(k) + 4]mod`, `[h'(k) + 5]mod m`, and so on, until a free location is found.

The same procedure is applied while storing a value as well as searching a value. In case of searching a value, the value stored at the location generated by the hash function is compared with the value to be searched. If the two values match, then the search operation is successful. Otherwise, if the values do not match, then we examine other locations in sequence until the value is found.

Note

Linear probing is known for its simplicity. When we have to store a value, we try the slots: `[h'(k)]mod m`, `[h'(k) + 1]mod m`, `[h'(k) + 2]mod m`, `[h'(k) + 3]mod m`, `[h'(k) + 4]mod m`, `[h'(k) + 5]mod m`, and so no, until a vacant location is found.

EXAMPLE 15.5: Consider a hash table of size 10. Using linear probing, insert the keys 72, 27, 36, 24, 63, 81, 92, and 101 into the table.

Let `h'(k) = k mod m, m = 10`

Initially, the hash table can be given as:

0	1	2	3	4	5	6	7	8	9
–1	–1	–1	–1	–1	–1	–1	–1	–1	–1

Step 1:

```
            Key = 72
h(72, 0) = (72 mod 10 + 0) mod 10
         = (2) mod 10
         = 2
```

Since `T[2]` is vacant, insert key 72 at this location.

0	1	2	3	4	5	6	7	8	9
–1	–1	72	–1	–1	–1	–1	–1	–1	–1

Step 2:

```
            Key = 27
h(27, 0) = (27 mod 10 + 0) mod 10
         = (7) mod 10
         = 7
```

Since `T[7]` is vacant, insert key 27 at this location.

0	1	2	3	4	5	6	7	8	9
–1	–1	72	–1	–1	–1	–1	27	–1	–1

Step 3:

```
            Key = 36
h(36, 0) = (36 mod 10 + 0) mod 10
         = (6) mod 10
         = 6
```

Since `T[6]` is vacant, insert key 36 at this location.

0	1	2	3	4	5	6	7	8	9
–1	–1	72	–1	–1	–1	36	27	–1	–1

Step 4:

```
            Key = 24
h(24, 0) = (24 mod 10 + 0) mod 10
         = (4) mod 10
         = 4
```

Since T[4] is vacant, insert key 24 at this location.

0	1	2	3	4	5	6	7	8	9
–1	–1	72	–1	24	–1	36	27	–1	–1

Step 5: Key = 63

h(63, 0) = (63 mod 10 + 0) mod 10
= (3) mod 10
= 3

Since T[3] is vacant, insert key 63 at this location.

0	1	2	3	4	5	6	7	8	9
–1	–1	72	63	24	–1	36	27	–1	–1

Step 6: Key = 81

h(81, 0) = (81 mod 10 + 0) mod 10
= (1) mod 10
= 1

Since T[1] is vacant, insert key 81 at this location.

0	1	2	3	4	5	6	7	8	9
0	81	72	63	24	–1	36	27	–1	–1

Step 7: Key = 92

h(92, 0) = (92 mod 10 + 0) mod 10
= (2) mod 10
= 2

Now T[2] is occupied, so we cannot store the key 92 in T[2]. Therefore, try again for the next location. Thus probe, i = 1, this time.

Key = 92

h(92, 1) = (92 mod 10 + 1) mod 10
= (2 + 1) mod 10
= 3

Now T[3] is occupied, so we cannot store the key 92 in T[3]. Therefore, try again for the next location. Thus probe, i = 2, this time.

Key = 92

h(92, 2) = (92 mod 10 + 2) mod 10
= (2 + 2) mod 10
= 4

Now T[4] is occupied, so we cannot store the key 92 in T[4]. Therefore, try again for the next location. Thus probe, i = 3, this time.

Key = 92

h(92, 3) = (92 mod 10 + 3) mod 10
= (2 + 3) mod 10
= 5

Since `T[5]` is vacant, insert key 92 at this location.

0	1	2	3	4	5	6	7	8	9
–1	81	72	63	24	92	36	27	–1	–1

Step 8:

```
            Key = 101
     h(101, 0) = (101 mod 10 + 0) mod 10
               = (1) mod 10
               = 1
```

Now `T[1]` is occupied, so we cannot store the key 101 in `T[1]`. Therefore, try again for the next location. Thus probe, `i = 1`, this time.

```
            Key = 101
     h(101, 1) = (101 mod 10 + 1) mod 10
               = (1 + 1) mod 10
               = 2
```

Now `T[2]` is occupied, so we cannot store the key 101 in `T[2]`. Therefore, try again for the next location. Thus probe, `i = 2`, this time.

```
            Key = 101
     h(101, 2) = (101 mod 10 + 2) mod 10
               = (1 + 2) mod 10
               = 3 mod 10
               = 3
```

Now `T[3]` is occupied, so we cannot store the key 101 in `T[3]`. Therefore, try again for the next location. Thus probe, `i = 3`, this time.

```
            Key = 101
     h(101, 3) = (101 mod 10 + 3) mod 10
               = (1 + 3) mod 10
               = 4 mod 10
               = 4
```

Note that this procedure will be repeated until the hash function generates the address of location 8 which is vacant and can be used to store the value in it.

Searching a value using Linear Probing

While searching for a value in a hash table, the array index is re-computed and the key of the element stored at that location is checked with the value that has to be searched. If a match is found, then the search operation is successful. The search time, in this case, is given as `O(1)`. If the key does not match, then the search function begins a sequential search of the array that continues until:

- the value is found, or
- the search function encounters a vacant location in the array, indicating that the value is not present, or
- the search function terminates because the table is full and the value is not present.

In the worst case, the search operation may have to make `(n-1)` comparisons, and the running time of the search algorithm may take `O(n)` time. The worst case will be encountered when the table is full and after scanning all the `(n-1)` elements, the value is either present at the last location or not present in the table.

Thus, we see that with the increase in the number of collisions, the distance from the array index computed by the hash function and the actual location of the element increases, thereby increasing the search time.

Pros and Cons

Linear probing finds an empty location by doing a linear search in the array beginning from position `h(k)`. Although the algorithm provides good memory caching through good locality of reference, but the drawback of this algorithm is that it results in clustering, and thus a higher risk of more collision where one collision has already taken place. The performance of linear probing is sensitive to the distribution of input values.

As the hash table fills, clusters of consecutive cells are formed and the time required for a search increases with the size of the cluster. In addition to this, when a new value has to be inserted into the table at a position which is already occupied, that value is inserted at the end of the cluster, which again increases the length of the cluster. Generally, an insertion is made between two clusters that are separated by one vacant location. But with linear probing, there are more chances that subsequent insertions will also end up in one of the clusters, thereby potentially increasing the cluster length by an amount much greater than one. More the number of collisions, higher the probes that are required to find a free location and lesser is the performance. This phenomenon is called *primary clustering*. To avoid primary clustering, other techniques like quadratic probing and double hashing are used.

Quadratic Probing

In this technique, if a value is already stored at a location generated by `h(k)`, then the following hash function is used to resolve the collision:

$$h(k, i) = [h'(k) + c_1 i + c_2 i^2] \bmod m$$

where, `m` is the size of the hash table, $h'(k) = (k \bmod m)$, `i` is the probe number that varies from 0 to `m-1`, and c_1 and c_2 are constants such that c_1 and $c_2 \neq 0$.

Quadratic probing eliminates the primary clustering phenomenon of linear probing because instead of doing a linear search, it does a quadratic search. For a given key `k`, first the location generated by `h'(k) mod m` is probed. If the location is free, the value is stored in it, else subsequent locations probed are offset by factors that depend in a quadratic manner on the probe number `i`. Although quadratic probing performs better than linear probing, but to maximize the utilization of the hash table, the values of c_1, c_2, and `m` need to be constrained.

EXAMPLE 15.6: Consider a hash table of size 10. Using quadratic probing, insert the keys 72, 27, 36, 24, 63, 81, and 101 into the table. Take $c_1 = 1$ and $c_2 = 3$.

Let $h'(k) = k \bmod m$, $m = 10$

Initially, the hash table can be given as:

0	1	2	3	4	5	6	7	8	9
–1	–1	–1	–1	–1	–1	–1	–1	–1	–1

We have,

$$h(k, i) = [h'(k) + c_1 i + c_2 i^2] \bmod m$$

Step 1: Key = 72

h(72) = [72 mod 10 + 1 × 0 + 3 × 0] mod 10
= [72 mod 10] mod 10
= 2 mod 10
= 2

Since T[2] is vacant, insert the key 72 in T[2]. The hash table now becomes:

0	1	2	3	4	5	6	7	8	9
–1	–1	72	–1	–1	–1	–1	–1	–1	–1

Step 2: Key = 27

h(27) = [27 mod 10 + 1 × 0 + 3 × 0] mod 10
= [27 mod 10] mod 10
= 7 mod 10
= 7

Since T[7] is vacant, insert the key 27 in T[7]. The hash table now becomes:

0	1	2	3	4	5	6	7	8	9
–1	–1	72	–1	–1	–1	–1	27	–1	–1

Step 3: Key = 36

h(36) = [36 mod 10 + 1 × 0 + 3 × 0] mod 10
= [36 mod 10] mod 10
= 6 mod 10
= 6

Since T[6] is vacant, insert the key 36 in T[6]. The hash table now becomes:

0	1	2	3	4	5	6	7	8	9
–1	–1	72	–1	–1	–1	36	27	–1	–1

Step 4: Key = 24

h(24) = [24 mod 10 + 1 × 0 + 3 × 0] mod 10
= [24 mod 10] mod 10
= 4 mod 10
= 4

Since T[4] is vacant, insert the key 24 in T[4]. The hash table now becomes:

0	1	2	3	4	5	6	7	8	9
–1	–1	72	–1	24	–1	36	27	–1	–1

Step 5: Key = 63

h(63) = [63 mod 10 + 1 × 0 + 3 × 0] mod 10
= [63 mod 10] mod 10
= 3 mod 10
= 3

Since `T[3]` is vacant, insert the key 63 in `T[3]`. The hash table now becomes:

0	1	2	3	4	5	6	7	8	9
–1	–1	72	63	24	–1	36	27	–1	–1

Step 6:

```
        Key = 81
    h(81) = [81 mod 10 + 1 × 0 + 3 × 0] mod 10
          = [81 mod 10] mod 10
          = 81 mod 10
          = 1
```

Since `T[1]` is vacant, insert the key 81 in `T[1]`. The hash table now becomes:

0	1	2	3	4	5	6	7	8	9
–1	81	72	63	24	–1	36	27	–1	–1

Step 7:

```
        Key = 101
    h(101) = [101 mod 10 + 1 × 0 + 3 × 0] mod 10
           = [101 mod 10 + 0] mod 10
           = 1 mod 10
           = 1
```

Since `T[1]` is already occupied, the key 101 can not be stored in `T[1]`. Therefore, try again for next location. Thus probe, `i = 1`, this time.

```
        Key = 101
    h(101) = [101 mod 10 + 1 × 1 + 3 × 1] mod 10
           = [101 mod 10 + 1 + 3] mod 10
           = [101 mod 10 + 4] mod 10
           = [1 + 4] mod 10
           = 5 mod 10
           = 5
```

Since `T[5]` is vacant, insert the key 101 in `T[5]`. The hash table now becomes:

0	1	2	3	4	5	6	7	8	9
81	–1	72	63	24	101	36	27	–1	–1

Pros and Cons

Quadratic probing caters to the primary clustering problem that exists in the linear probing technique. Quadratic probing provides good memory caching because it preserves some locality of reference. But linear probing does this task better and gives a better cache performance.

One of the major drawbacks with quadratic probing is that a sequence of successive probes may only explore a fraction of the table, and this fraction may be quite small. If it happens, then we will not be able to find an empty location in the table despite the fact that the table is by no means full. In the example given above, try to insert the key 92 and you will encounter this problem.

Although quadratic probing is free from primary clustering, it is still liable to what is known as *secondary clustering*. It means that if there is a collision between two keys, then the same probe sequence will be followed for both. With quadratic probing, the potential for multiple collisions increases as the table becomes full. This situation is usually encountered when the hash table is more than full.

Quadratic probing is widely applied in the Berkeley Fast File System to allocate free blocks.

Searching a value using Quadratic Probing

While searching a value using the quadratic probing technique, the array index is re-computed and the key of the element stored at that location is checked with the value that has to be searched. If the desired key value matches the key value at that location, then the element is present in the hash table and the search is said to be successful. In this case, the search time is given as `O(1)`. However, if the value does not match, then the search function begins a sequential search of the array that continues until:

- the value is found, or
- the search function encounters a vacant location in the array, indicating that the value is not present, or
- the search function terminates because the table is full and the value is not present.

In the worst case, the search operation may take `(n-1)` comparisons, and the running time of the search algorithm may be `O(n)`. The worst case will be encountered when the table is full and after scanning all the `(n-1)` elements, the value is either present at the last location or not present in the table.

Thus, we see that with the increase in the number of collisions, the distance from the array index computed by the hash function and the actual location of the element increases, thereby increasing the search time.

Double Hashing

To start with, double hashing uses one hash value and then repeatedly steps forward an interval until an empty location is reached. The interval is decided using a second, independent hash function, hence the name *double hashing*. In double hashing, we use two hash functions rather than a single function. The hash function in the case of double hashing can be given as:

$$h(k, i) = [h_1(k) + ih_2(k)] \bmod m$$

where, m is the size of the hash table, $h_1(k)$ and $h_2(k)$ are two hash functions given as $h_1(k) = k \bmod m$, $h_2(k) = k \bmod m'$, i is the probe number that varies from 0 to m-1, and m′ is chosen to be less than m. We can choose m′ = m-1 or m-2.

When we have to insert a key k in the hash table, we first probe the location given by applying $[h_1(k) \bmod m]$ because during the first probe, i = 0. If the location is vacant, the key is inserted into it, else subsequent probes generate locations that are at an offset of $[h_2(k) \bmod m]$ from the previous location. Since the offset may vary with every probe depending on the value generated by the second hash function, the performance of double hashing is very close to the performance of the ideal scheme of uniform hashing.

Pros and Cons

Double hashing minimizes repeated collisions and the effects of clustering. That is, double hashing is free from problems associated with primary clustering as well as secondary clustering.

Example 15.7: Consider a hash table of size = 11. Using double hashing, insert the keys 72, 27, 36, 24, 63, 81, 92, and 101 into the table. Take `h1 = (k mod 10)` and `h2 = (k mod 8)`.

Let `m = 11`

Initially, the hash table can be given as:

0	1	2	3	4	5	6	7	8	9
–1	–1	–1	–1	–1	–1	–1	–1	–1	–1

We have,

$$h(k, i) = [h_1(k) + ih_2(k)] \bmod m$$

Step 1: Key = 72

$$h(72, 0) = [72 \bmod 10 + (0 \times 72 \bmod 8)] \bmod 11$$
$$= [2 + (0 \times 0)] \bmod 10$$
$$= 2 \bmod 11$$
$$= 2$$

Since `T[2]` is vacant, insert the key 72 in `T[2]`. The hash table now becomes:

0	1	2	3	4	5	6	7	8	9
–1	–1	72	–1	–1	–1	–1	–1	–1	–1

Step 2: Key = 27

$$h(27, 0) = [27 \bmod 10 + (0 \times 27 \bmod 8)] \bmod 10$$
$$= [7 + (0 \times 3)] \bmod 10$$
$$= 7 \bmod 10$$
$$= 7$$

Since `T[7]` is vacant, insert the key 27 in `T[7]`. The hash table now becomes:

0	1	2	3	4	5	6	7	8	9
–1	–1	72	–1	–1	–1	–1	27	–1	–1

Step 3: Key = 36

$$h(36, 0) = [36 \bmod 10 + (0 \times 36 \bmod 8)] \bmod 10$$
$$= [6 + (0 \times 4)] \bmod 10$$
$$= 6 \bmod 10$$
$$= 6$$

Since `T[6]` is vacant, insert the key 36 in `T[6]`. The hash table now becomes:

0	1	2	3	4	5	6	7	8	9
–1	–1	72	–1	–1	–1	36	27	–1	–1

Step 4: Key = 24

$$h(24, 0) = [24 \bmod 10 + (0 \times 24 \bmod 8)] \bmod 10$$
$$= [4 + (0 \times 0)] \bmod 10$$
$$= 4 \bmod 10$$
$$= 4$$

Since `T[4]` is vacant, insert the key 24 in `T[4]`. The hash table now becomes:

0	1	2	3	4	5	6	7	8	9
–1	–1	72	–1	24	–1	36	27	–1	–1

Step 5:

```
         Key = 63
h(63, 0) = [63 mod 10 + (0 × 63 mod 8)] mod 10
         = [3 + (0 × 7)] mod 10
         = 3 mod 10
         = 3
```

Since `T[3]` is vacant, insert the key 63 in `T[3]`. The hash table now becomes:

0	1	2	3	4	5	6	7	8	9
–1	–1	72	63	24	–1	36	27	–1	–1

Step 6:

```
         Key = 81
h(81, 0) = [81 mod 10 + (0 × 81 mod 8)] mod 10
         = [1 + (0 × 1)] mod 10
         = 1 mod 10
         = 1
```

Since `T[1]` is vacant, insert the key 81 in `T[1]`. The hash table now becomes:

0	1	2	3	4	5	6	7	8	9
–1	81	72	63	24	–1	36	27	–1	–1

Step 7:

```
         Key = 92
h(92, 0) = [92 mod 10 + (0 × 92 mod 8)] mod 10
         = [2 + (0 × 4)] mod 10
         = 2 mod 10
         = 2
```

Now `T[2]` is occupied, so we cannot store the key 92 in `T[2]`. Therefore, try again for the next location. Thus probe, `i = 1`, this time.

```
         Key = 92
h(92, 1) = [92 mod 10 + (1 × 92 mod 8)] mod 10
         = [2 + (1 × 4)] mod 10
         = (2 + 4) mod 10
         = 6 mod 10
         = 6
```

Now `T[6]` is occupied, so we cannot store the key 92 in `T[6]`. Therefore, try again for the next location. Thus probe, `i = 2`, this time.

```
       Key = 92
h(92) = [92 mod 10 + (2 × 92 mod 8)] mod 10
      = [2 + (2 × 4)] mod 10
      = [2 + 8] mod 10
      = 10 mod 10
      = 0
```

Since `T[1]` is vacant, insert the key 81 in `T[1]`. The hash table now becomes:

0	1	2	3	4	5	6	7	8	9
92	81	72	63	24	–1	36	27	–1	–1

Step 8: `Key = 101`

```
h(101, 0) = [101 mod 10 + (0×101 mod 8)] mod 10
          = [1 + (0×5)] mod 10
          = 1 mod 10
          = 1
```

Now `T[1]` is occupied, so we cannot store the key 101 in `T[1]`. Therefore, try again for the next location. Thus probe, `i = 1`, this time.

```
Key = 101
h(101, 1) = [101 mod 10 + (1×101 mod 8)] mod 10
          = [1 + (1×5)] mod 10
          = [1 + 5] mod 10
          = 6
```

Now `T[6]` is occupied, so we cannot store the key 101 in `T[6]`. Therefore, try again for the next location. Thus probe, `i = 2`, this time.

```
Key = 101
h(101, 2) = [101 mod 10 + (2×101 mod 8)] mod 10
          = [1 + (2×5)] mod 10
          = [1 + 10] mod 10
          = 11 mod 10
          = 1
```

Now `T[1]` is occupied, so we cannot store the key 101 in `T[1]`. Therefore, try again for the next location. Thus probe, `i = 3`, this time.

```
Key = 101
h(101, 3) = [101 mod 10 + (3×101 mod 8)] mod 10
          = [1 + (3×5)] mod 10
          = [1 + 15] mod 10
          = 16 mod 10
          = 16
```

Now `T[6]` is occupied, so we cannot store the key 101 in `T[6]`. Therefore, try again for the next location. Thus probe, `i = 4`, this time.

```
Key = 101
h(101, 4) = [101 mod 10 + (4×101 mod 8)] mod 10
          = [1 + (4×5)] mod 10
          = [1 + 20] mod 10
          = 21 mod 10
          = 1
```

Now `T[1]` is occupied, so we cannot store the key 101 in `T[1]`. Therefore, try again for the next location. Thus, probe `i = 5`. Repeat the entire process until a vacant location is found. Here, you see that we have probed so many times to insert the key 101 in the hash table. Although double hashing is a very efficient algorithm, it always requires `m` to be a prime number. In our case `m=10`, which is not a prime number; hence, the degradation in performance. Had `m` been equal to 11, the algorithm would have worked very efficiently. Thus, we can say that the performance of the technique is sensitive to the value of `m`.

If the hash table becomes nearly full, then the running time for insertion, deletion, and lookup operations takes longer time to execute. So in such a situation, create a new hash table of double size than that of the original hash table.

For each key in the original hash table, compute the new hash value and insert it into the new hash table. Finally, free the memory occupied by the original hash table.

15.6 COLLISION RESOLUTION BY CHAINING

In chaining, each location in the hash table stores a pointer to a linked list that contains all the key values that were hashed to the same location. That is, location l in the hash table points to the head of the linked list of all the key values that hashed to l. However, if no key value hashes to l, then location l in the hash table contains `NULL`. Figure 15.5 shows how the key values are mapped to location l in the hash table and stored in a linked list that corresponds to l.

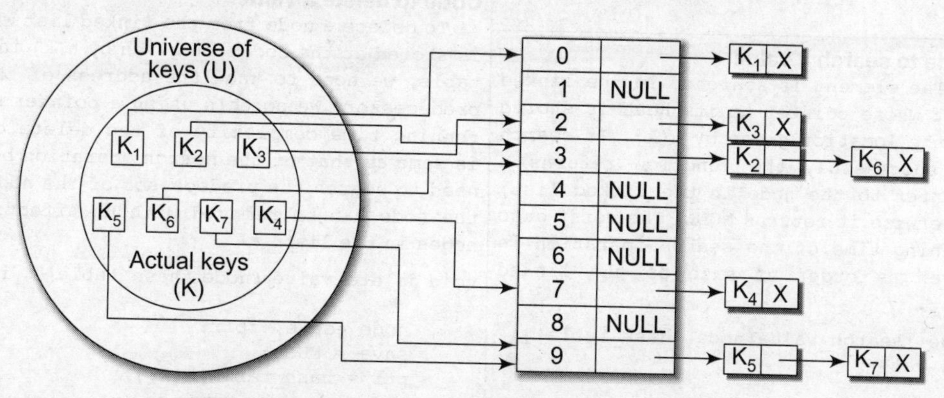

Figure 15.5 Keys being hashed to a chained hash table

15.6.1 Operations on a Chained Hash Table

Searching for a value is as simple as scanning the linked list for an entry with the given key. Insertion operation appends the key to the end of the linked list pointed by the hashed location. Deleting a key requires searching the list and removing the element.

Chained hash tables with linked lists are widely used due to the simplicity of the algorithms to insert, delete, and search a key. The code for these algorithms is exactly the same as that of inserting, deleting, and searching a value in a single linked list that we have already studied in Chapter 8.

While the cost of inserting a key in a chained hash table is `O(1)`, the cost for deleting and searching a value is given as `O(l)` where `l` is the number of elements in the list of that location. Searching and deleting takes more time because these operations scan the entries of the selected location for the desired key.

In the worst case, searching a value may take a running time of `O(n)`, where n is the number of key values stored in the chained hash table. This case arises when all the key values are inserted into the linked list of the same location (of the hash table). In this case, the hash table is ineffective.

Table 15.1 gives the code to initialize a hash table as well as the codes to insert, delete, and search a value in a chained hash table.

Table 15.1 Codes to initialize, insert, delete, and search a value in a chained hash table

Structure of the node

```
typedef struct node_HT
{
   int value;
   struct node *next;
}node;
```

Code to initialize a chained hash table

```
/* Initializes m location in the
chained hash table.
The operation takes a running time of
O(m) */
void initializeHashTable (node
*hash_table[], int m)
{
   int i;
   for(i=0i<=m;i++)
          hash_table[i]=NULL;
}
```

Code to search a value

```
/* The element is searched in the linked
list whose pointer to its head is stored
in the location given by h(k). If search
is successful, the function returns a
pointer to the node in the linked list;
otherwise it returns NULL. The worst case
running time of the search operation is
given as order of size of the linked
list. */
node *search_value(node *hash_table[],
int val)
{
      node *ptr;
      ptr = hash_table[h(x)];
      while ( (ptr!=NULL) &&
(ptr->value != val))
             ptr = ptr->next;
      if (ptr->value == val)
             return ptr;
      else
             return NULL;
}
```

Code to insert a value

```
/* The element is inserted at the beginning of the
linked list whose pointer to its head is stored in
the location given by h(k). The running time of
the insert operation is O(1), as the new key value
is always added as the first element of the list
irrespective of the size of the linked list as
well as that of the chained hash table. */
node *insert_value( node *hash_table[], int val)
{
      node *new_node;
      new_node = (node *)malloc(sizeof(node));
      new_node->value = val;
      new_node->next = hash_table[h(x)];
      hash_table[h(x)] = new_node;
}
```

Code to delete a value

```
/* To delete a node from the linked list whose head
is stored at the location given by h(k) in the hash
table, we need to know the address of the node's
predecessor. We do this using a pointer save. The
running time complexity of the delete operation
is same as that of the search operation because we
need to search the predecessor of the node so that
the node can be removed without affecting other
nodes in the list. */
void delete_value (node *hash_table[], int val)
{
      node *save, *ptr;
      save = NULL;
      ptr = hash_table[h(x)];
      while ((ptr != NULL) && (ptr->value != val))
      {
            save = ptr;
            ptr = ptr->next;
      }
      if (ptr != NULL)
      {
            save->next = ptr->next;
            free (ptr);
      }
      else
            printf("\n VALUE NOT FOUND");
}
```

Example 15.8: Insert the keys 7, 24, 18, 52, 36, 54, 11, 23 and 60 in a chained hash table of 9 memory locations. Use `h(k) = k mod m`.

In this case, `m=9`. Initially, the hash table can be given as:

0	NULL
1	NULL
2	NULL
3	NULL
4	NULL
5	NULL
6	NULL
7	NULL
8	NULL
9	NULL

Step 1: Key = 7

`h(k) = 7 mod 9`

= 7

Create a linked list for location 7 and store the key value 7 in it as its only node.

0	NULL
1	NULL
2	NULL
3	NULL
4	NULL
5	NULL
6	NULL
7	→ 7 X
8	NULL
9	NULL

Step 2: Key = 24

`h(k) = 24 mod 9`

= 6

Create a linked list for location 6 and store the key value 24 in it as its only node.

0	NULL
1	NULL
2	NULL
3	NULL
4	NULL
5	NULL
6	→ 24 X
7	→ 7 X
8	NULL
9	NULL

Step 3: Key = 18

`h(k) = 18 mod 9`

= 0

Create a linked list for location 0 and store the key value 18 in it as its only node.

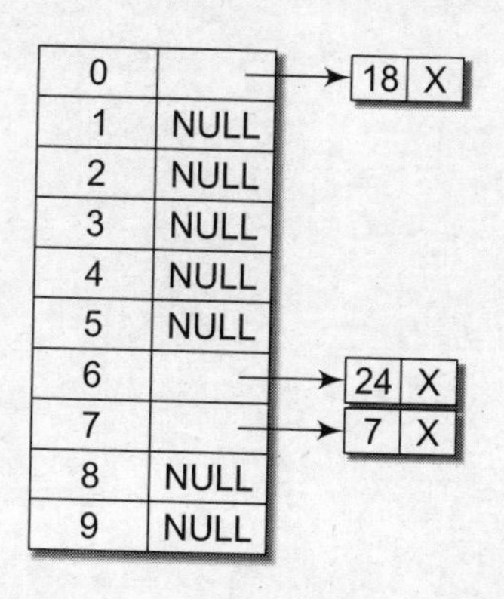

Step 4: Key = 52

`h(k) = 52 mod 9`

= 7

Insert 52 at the beginning of the linked list of location 7.

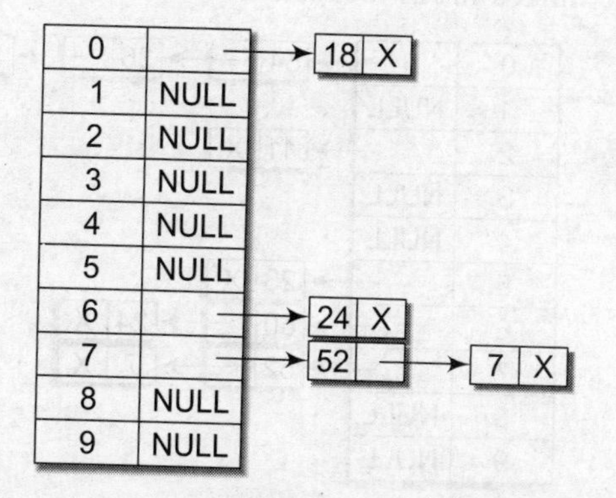

Step 5: Key = 36

`h(k) = 36 mod 9` = 0

Insert 36 at the beginning of the linked list of location 0.

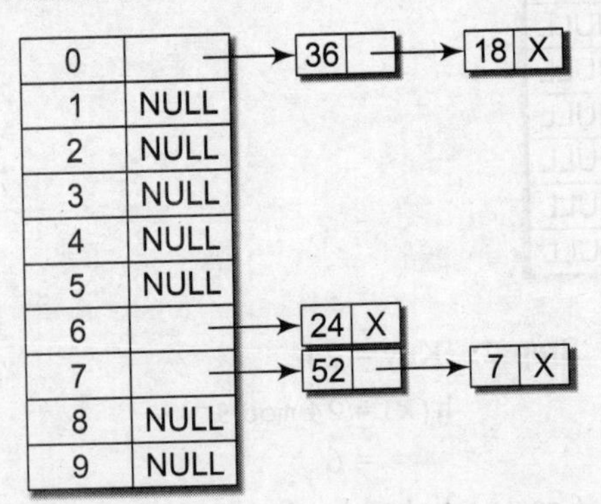

Step 6: Key = 54

`h(k) = 54 mod 9` = 0

Insert 54 at the beginning of the linked list of location 0.

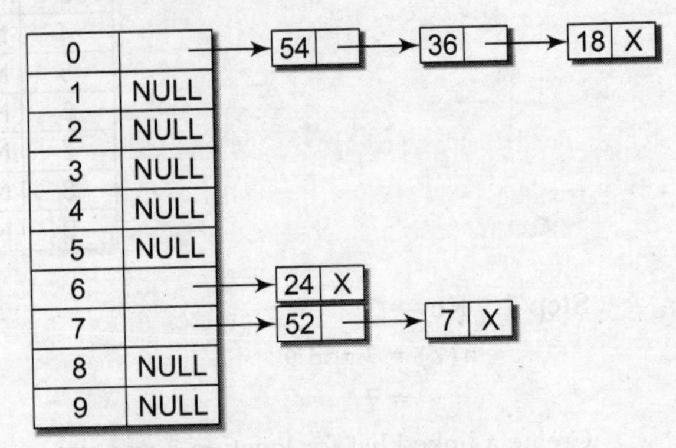

Step 7: Key = 11

`h(k) = 11 mod 9` = 2

Create a linked list for location 2 and store the key value 11 in it as its only node.

Step 8: Key = 23

`h(k) = 23 mod 9` = 5

Create a linked list for location 5 and store the key value 23 in it as its only node.

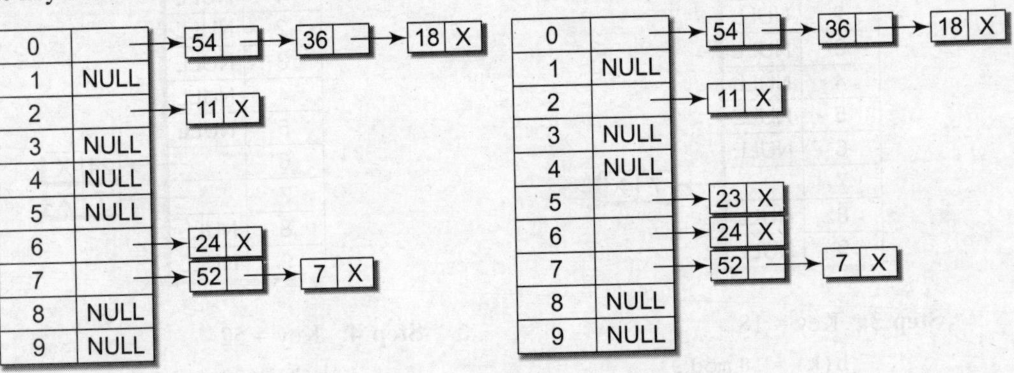

Step 9: Key = 60

`h(k) = 60 mod 9` = 6

Insert 60 at the beginning of the linked list of location 6.

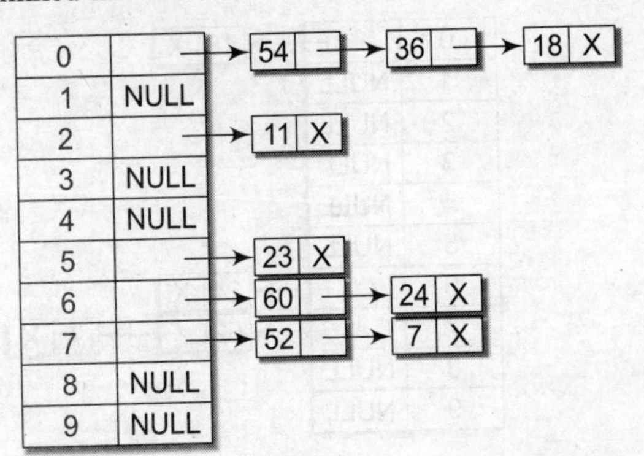

15.6.2 Pros and Cons

The main advantage of using a chained hash table is that it remains effective even when the number of key values to be stored is much higher than the number of locations in the hash table. However, with the increase in the number of keys to be stored, the performance of a chained hash table does degrade gracefully (linearly). For example, a chained hash table with 1000 memory locations and 10,000 stored keys will give 5 to 10 times less performance as compared to the performance of a chained hash table with 10,000 locations. But the conclusion is that, a chained hash table is still 1000 times faster than a simple hash table.

The other advantage of using chaining for collision resolution is that its performance, unlike quadratic probing, does not degrade when the table is more than half full. This technique is absolutely free from clustering problems and thus provides an efficient mechanism to handle collisions.

However, chained hash tables inherit the disadvantages of linked lists. First, to store even a key value, the space overhead of the next pointer in each entry can be significant. Second, traversing a linked list has poor cache performance, making the processor cache ineffective.

15.7 PROS AND CONS OF HASHING

One advantage of hashing is that no extra space is required to store the index as in the case of other data structures. In addition, a hash table provides fast data access and an added advantage of rapid updates.

On the other hand, the primary drawback of using the hashing technique for inserting and retrieving data values is that it usually lacks locality and sequential retrieval by key. This makes insertion and retrieval of data values even more random.

All the more, choosing an effective hash function is more an art than a science. It is not uncommon to (in open-addressed hash tables) to create a poor hash function.

15.8 APPLICATIONS OF HASHING

Hash tables are widely used in situations where enormous amounts of data have to be accessed to quickly search and retrieve information. A few typical examples where hashing is used are given here.

Hashing is used for database indexing. Some DBMSs store a separate file known as indexes. When data has to be retrieved from a file, the key information is first found in the appropriate index file which references the exact record location of the data in the database file. This key information in the index file is often stored as a hashed value.

Hashing is used as symbol tables. For example, in Fortran language to store variable names. Hash tables speeds up execution of the program as the references to variables can be looked up quickly.

In many database systems, File and Directory hashing is used in high-performance file systems. Such systems use two complementary techniques to improve the performance of file access. While one of theses techniques is caching which saves information in the memory, the other is hashing which makes looking up the file location in the memory much quicker than most other methods.

Hash tables can be used to store massive amount of information for example, to store driver's license records. Given the driver's license number, hash tables help to quickly get information about the driver (i.e. name, address, age).

Hashing technique is used for compiler symbol tables in C++. The compiler uses a symbol table to keep a record of the user-defined symbols in a C++ program. Hashing facilitates the compiler to quickly look up variable names and other attributes associated with symbols. Hashing is also widely being used for Internet search engines.

Miscellaneous Applications Hashing is used in telephone book databases to quickly find a person's telephone number.

Hashing is used in electronic library catalogs for quickly searching among millions of materials stored in the library. Hash tables are used to store passwords for systems with multiple users. Hashing enables fast retrieval of the password which corresponds to a given username.

Uses of Hashing in the Real World

CD Databases For CD's, it is desirable to have a world-wide CD database so that when users put their disk in the CD player, they get a full table of contents on their own computer's screen. The main thing is that these tables are not stored on the disks themselves. That is, the CD does not store any information about the songs, rather this information is downloaded from the database. The critical issue to solve here is that CD's have no ID numbers stored on them, so how will the computer know which CD has been put in the player? The only information that can be used is the track length, and every CD is different.

Basically, a big number is cooked up from the track lengths, also known as a 'signature'. This signature is used to identify any particular CD. The signature is nothing but a value obtained by hashing. For example, a number of length of 8 or 10 hex. digits is cooked up; the number is then sent to the database, and that database just looks for the closest match. The reason being that track length may not be measured exactly.

Drivers Licenses/insurance Cards Like our CD example, even the driver's license numbers or insurance card numbers are cooked up using hashing from things that never change: date of birth, name, etc.

Sparse Matrix A sparse matrix is a two-dimensional array in which most of the entries contains a 0. That is, in a sparse array there are very few non-zero entries. Of course, we can store 2D array as it is, but this would lead to sheer wastage of valuable memory. So another possibility is to store the non-zero elements of the spare matrix as elements in a 1D array. That is by using hashing, we can store a two-dimensional array in a one-dimensional array. There is a one to one correspondence between the elements in the sparse matrix and the elements in the array. This concept is clearly visible in Fig. 15.6.

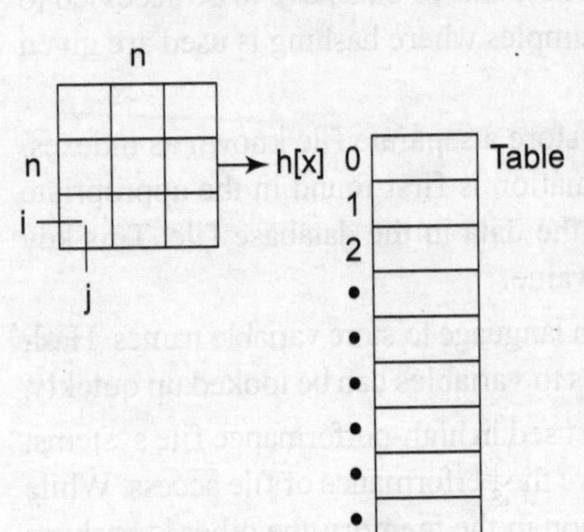

Figure 15.6 Sparse matrix

If the size of the sparse matrix is `n X n`, and there are `N` non-zero entries in it, then from the coordinates `(i,j)` of a matrix, we determine an index `k` in an array by a simple calculation. Thus, we have `k=h(i,j)` for some function `h`, called a hash function.

The size of the 1D array is proportional to `N`. This is far better from the size of the sparse matrix that required storage proportional to `nXn`. For example, if we have a triangular sparse matrix `A`, then an entry `A[i, j]` can be mapped to an entry in the 1D array by calculating the index using the hash function `h(i, j) = i(i-1)/2 + j`.

File Signatures File signatures provide a compact means of identifying files. We use a function, `h[x]`, the file signature, which is a property of the file. Although we can store files by name but signature provides a compact identity to the file.

Since signature depends on the contents of the file, if any change is made to the file, then the signature will change. In this way, the signature of a file can be used as a quick verification to see if anyone has altered the file, or if it has lost a bit during transmission. Signatures are widely used for files that stores marks of the students.

Game Boards In the game board for tic-tac-toe or chess, a position in a game may be stored using a hash function.

Graphics In graphics, a central problem is the storage of objects in a scene or view. For this, we organize our data by hashing. Hashing can be used to make a grid of appropriate size, an ordinary vertical-horizontal grid. (Note that a grid is nothing but a 2D array, and there is a one to one correspondence when we move from a 2D array to a single array).

So, we store the grid as a 1D array as we did in the case of sparse matrix. All points that fall in one cell will be stored in the same place. If a cell contains three points, then these three points will be stored in the same entry. The mapping from the grid cell to the memory location is done by using a hash function. The key advantage of this method of storage is fast execution of operations like the nearest neighbor search.

SUMMARY

- Hash table is a data structure in which keys are mapped to array positions by a hash function. A value stored in the hash table can be searched in `O(1)` time using a hash function to generate an address from the key.
- The storage requirement for a hash table is just `O(k)`, where `k` is the number of keys actually used. In a hash table, an element with key `k` is stored at index `h(k)`, not `k`. This means that a hash function `h` is used to calculate the index at which the element with key `k` will be stored. Thus, the process of mapping keys to appropriate locations (or indexes) in a hash table is called hashing. When two or more keys maps to the same memory location, a collision is said to occur.
- Division method divides `x` by `M` and then use the remainder thus obtained. A potential drawback of the division method is that using this method, consecutive keys map to consecutive hash values.
- Multiplication method applies the hash function given as $h(x) = \lfloor m (kA \bmod 1) \rfloor$
- Mid square method is a good hash function which works in two steps. First, it finds k^2 and then extracts the middle `r` bits of the result.
- Once a collision takes place, open addressing computes new positions using a probe sequence and the next record is stored in that position. In this technique of collision resolution, all the values are stored in the hash table. The hash table will contain two types of values- either sentinel value (for example, –1) or a data value.
- Linear probing enables good memory caching, through good locality of reference, but the drawback of this algorithm is that it results in clustering. More the number of collisions, higher the probes that are required to find a free location and lesser is the performance. This phenomenon is called primary clustering. To avoid primary clustering, other techniques like quadratic probing and double hashing are used.
- Quadratic probing eliminates the primary clustering and provides good memory caching. But, it is still liable to what is known as secondary clustering. This means that if there is a collision between two keys, then the same probe sequence will be followed for both.
- In double hashing, we use two hash functions rather a single function. The performance of double

hashing is very close to the performance of the ideal scheme of uniform hashing. It minimizes repeated collisions and the effects of clustering.

- In chaining, each location in the hash table stores a pointer to a linked list that contains all the key values that were hashed to the same location. While the cost of inserting a key in a chained hash table is `O(1)`, the cost for deleting and searching a value is given as `O(l)`, where l is the number of elements in the list of that location. However, in the worst case, searching for a value may take a running time of `O(n)`.

GLOSSARY

Clustering free The situation in which a collision resolution technique spreads out entries in a hash table.

Clustering The tendency for elements in a hash table using open addressing to be stored together, even when the table has enough empty space to spread them out.

Collision resolution technique The mechanism to handle collisions, that is, when two or more values should be kept in the same location of a hash table.

Collision When two or different keys maps/hash to the same location, then a collision is said to occur. That is, it is the situation in which one or more values should be kept in the same location of the hash tables.

Direct chaining A collision resolution technique in which the hash table is an array of links to lists, where each list holds all the values having the same hash value.

Double hashing A method of open addressing for a hash table in which a collision is resolved by searching the table for an empty location at intervals specified by a different hash function, thereby minimizing clustering.

Hash function A function that is applied to map keys to integers in-order to get an even distribution on a smaller set of values.

Hash table A table in which keys are mapped to locations as calculated by application of hash functions.

Linear probing A hash table in which a collision is resolved by storing the value in the next empty location in the array following the occupied place.

Multiplication method A hash function that makes use of the first `p` bits of the key times an irrational number.

Open addressing A collision resolution technique in which all the values are stored within the hash table. Whenever collision occurs, new positions are computed, giving a probe sequence, and checked until an empty position is found.

Primary clustering The tendency of some collision resolution techniques to create long runs of occupied slots near the hash function position of keys.

Probe sequence In case of collision, probe sequence is the list of locations which a method for open addressing produces as alternative location for the key.

Quadratic probing A method of open addressing for a hash table in which a collision is resolved by storing the values in the next empty location given by a probe sequence. The space between locations in the sequence increases quadratically.

Second clustering The tendency of some collision resolution technique to create long runs of occupied slots near the hash function position of keys.

EXERCISES

Review Questions

1. Define a hash table.
2. What do you understand by a hash function? Give the properties of a good hash function.
3. How is a hash table better than a direct access table (array)?
4. Write a short note on the different hash functions. Give suitable examples.
5. Calculate hash values of keys: 1892, 1921, 2007, 3456 using different methods of hashing.

6. What is collision? Explain the various techniques to resolve a collision. Which technique do you think is better and why?
7. Consider a hash table with size = 10. Using linear probing, insert the keys 27, 72, 63, 42, 36, 18, 29, and 101 into the table.
8. Consider a hash table with size = 10. Using quadratic probing, insert the keys 27, 72, 63, 42, 36, 18, 29, and 101 into the table. Take $c_1 = 1$ and $c_2 = 3$.
9. Consider a hash table with size = 11. Using double hashing, insert the keys 27, 72, 63, 42, 36, 18, 29, and 101 into the table. Take $h_1 = k \bmod 10$ and $h_2 = k \bmod 8$.
10. What is hashing? Give its applications. Also, discuss the pros and cons of hashing.
11. Explain chaining with examples.
12. Write short notes on:

    Linear probing

    Quadratic probing

    Double hashing

Multiple Choice Questions

1. In a hash table, an element with key `k` is stored at index

   (a) `k` (b) `log k`

   (c) `h(k)` (d) k^2
2. In any hash function, `M` should be a

   (a) prime number (b) composite number

   (c) even number (d) odd number
3. Consecutive keys map to consecutive hash values is a property of which hash function?

   (a) division method

   (b) multiplication method

   (c) folding method

   (d) mid-square method
4. The process of examining memory locations in the hash table is called

   (a) hashing (b) collision

   (c) probing (d) Addressing
5. Which probing is applied in the Berkeley Fast File System to allocate free blocks?

   (a) linear probing (b) quadratic probing

   (c) double hashing (d) hashing
6. Which open addressing technique is free from clustering problems?

   (a) linear probing (b) quadratic probing

   (c) double hashing (d) hashing

True or False

1. Hash table is based on the property of locality of reference.
2. Binary search takes `O(n log n)` time to execute.
3. The storage requirement for a hash table is $O(k^2)$, where `k` is the number of keys.
4. Hashing takes place when two or more keys maps to the same memory location.
5. A good hash function completely eliminates collision.
6. `M` should not be too close to exact the powers of 2.
7. A sentinel value indicates that the location contains valid data.
8. Linear probing is sensitive to the distribution of input values.
9. A chained hash table is faster than a simple hash table.

Fill in the Blanks

1. Linear search takes ______ time to execute.
2. In a hash table, keys are mapped to array positions by a ______.
3. ______ is the process of mapping keys to appropriate locations in a hash table.
4. In open addressing, hash tables stores two values ______ and ______.
5. When there is no free location in the hash table then ______ occurs.
6. More the number of collisions, higher is the number of ______ to find free location ______ which eliminates primary clustering but not secondary clustering.

16 Files and Their Organization

Learning Objective

In this chapter, we will discuss the basic attributes of a file and see the different ways in which files are organized in the secondary memory. Then, we will learn different indexing strategies that allow efficient and faster access to these files. Finally, we will discuss external sorting. We already know that there are two types of sorting: internal sorting and external sorting. Chapter 12 covered internal sorting in detail. In this chapter, we will deal with the concept of external sorting.

16.1 INTRODUCTION

Most of the applications today collect a huge amount of data. A lot of data is collected everywhere in one form or the other. For example, when we seek admission in a college, a lot of data such as our name, address, phone number, the course in which to seek admission, aggregate of marks obtained in the last examination, and so on, are collected. Similarly, to open a bank account, you need to provide a lot of input. All these data were traditionally stored on paper documents, but handling these documents had always been a mess.

Similarly, scientific experiments and satellites also generate enormous amounts of data. Therefore, in order to efficiently analyse all the data that has been collected from different sources, it has become a necessity to store the data in computers in the form of files.

In computer terminology, a file is a block of useful data which is available to a computer program and is usually stored on a persistent storage medium. Storing a file on a persistent storage medium like hard disk ensures the availability of the file for future use. These days, files stored on computers are a good alternative to store documents that were once stored in offices and libraries.

16.2 DATA HIERARCHY

Every file contains data which can be organized in a hierarchy to present a systematic organization. The data hierarchy includes terms such as fields, records, files, and database. These terms are defined below.

- A *data field* is an elementary unit that stores a single fact. A data field is usually characterized by its type and size. For example, student's name is a data field that stores the name of students. This field is of type *characters* and its size can be set to be of maximum 20 characters or 30 characters depending on the requirement.
- A *record* is a collection of related data fields that is seen as a single unit from the application point of view. For example, the student's record may contain data fields such as name, address, phone number, roll number, marks obtained, and so on.

- A *file* is a collection of related records. For example, if there are 60 students in a class, then there are 60 records. All these related records are stored in a file. Similarly, we can have a file of all the employees working in an organization, a file of all the customers of a company, a file of all the suppliers, so on and so forth.
- A *directory* stores information of related files. A directory organizes information so that users can find it easily. For example, consider Fig. 16.1 that shows how multiple related files are stored in a student directory.

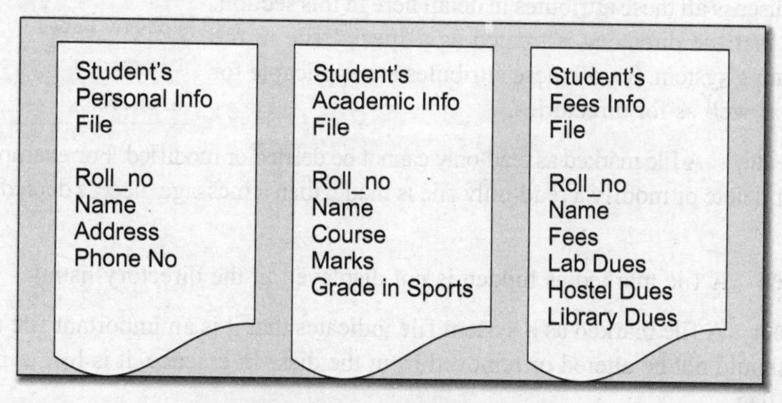

Figure 16.1 Student directory

16.3 FILE ATTRIBUTES

Every file in a computer system is stored in a directory. Each file has a list of attributes associated with it that gives the operating system and the application software more information about the file and how it is intended to be used.

Any software program looks up the directory entry to discern the attributes of a file, to take decisions about how to treat the file. For example, if a program attempts to write to a file that has been marked as a read-only file, then the program prints an appropriate message to notify the user that he is trying to write to a file that is meant only for reading.

Similarly, there is an attribute called *hidden*. When you attempt to execute the `DIR` command in DOS, then the files whose hidden attribute is set will not be displayed. These attributes are explained in this section.

File name It is a string of characters that stores the name of files. File naming conventions vary from one operating system to the other.

File position It is a pointer that points to the position at which the next read/write operation will be performed.

File structure It indicates whether the file is a text file or a binary file. In the text file, the numbers (integers or floating point) are stored as a string of characters. A binary file, on the other hand, stores the numbers in the same way as they are represented in the main memory.

File access method It indicates whether the records in a file will be accessed sequentially or randomly. In sequential access mode, records are read one by one. That is, if 60 records of students are stored in the `STUDENT` file, then to read the record of 39th student, you have to go through the record of the first 38 students. However, in random access, records can be accessed in any order.

Attributes flag A file can have six additional attributes attached to it. These attributes are usually stored in a single byte, with each bit representing a specific attribute. Each bit that is set to '1' means that the file has that attribute turned on. Table 16.1 shows a list of attributes and their position in the attribute flag or attribute byte.

Therefore, if a system file is set as hidden and read-only, then its attribute byte can be given as 00000111. We will discuss all these attributes in detail here in this section. Note that the directory is treated as a special file in the operating system. So, all these attributes are applicable for files as well as for directories.

Table 16.1 Attribute flag

Attribute	Attribute Byte
Read-Only	00000001
Hidden	00000010
System	00000100
Volume Label	00001000
Directory	00010000
Archive	00100000

Read-only A file marked as read-only cannot be deleted or modified. For example, if an attempt to either delete or modify a read-only file is made, then a message 'access denied' is displayed to the user.

Hidden A file marked as hidden is not displayed in the directory listing.

System A file marked as a system file indicates that it is an important file used by the system and should not be altered or removed from the disk. In essence, it is like a 'more serious' read-only flag.

Volume label Every disk volume is assigned a label for identification. The label can be assigned at the time of formatting the disk or later through various tools such as the DOS command `LABEL`.

Directory In directory listing, the files and sub-directories of the current directory are differentiated by a directory-bit. This means that the files that have the directory-bit turned on are actually sub-directories containing one or more files. Theoretically, one can convert a file to a directory by changing this bit. But practically, trying to do this would result in a mess.

Archive The archive bit is used as a communication link between programs that modify files and those that are used for backup. Most backup programs allow the user to do an incremental backup. Incremental backup selects for backup only those files which have been modified since the last backup.

When the backup program takes the backup of a file, or in other words, when the program archives the file, it clears the archive bit (set it to zero). Subsequently, if any program modifies the file, it turns on the archive bit (set it to 1). Then, whenever the backup program is run, the backup program would check the archive bit of each file to know which files have been modified since its last run. The backup program will then archive only those files which were modified.

Thus, the archive attribute is used for backup. When a file is created, the archive bit is turned on. When a backup program archives the file, it sets the archive bit to zero. When any program modifies the file, the attribute bit is turned on, so that the modified file can be backed up.

16.4 TEXT FILES AND BINARY FILES

A *text file,* also known as a flat file or an ASCII file, is structured as a sequence of lines. The data in a text file, whether it is numeric or non-numeric, is stored using its corresponding ASCII codes. The end of a text file is often denoted by placing a special character, called an end-of-file marker, after the last line in a text file.

A *binary file* contains any type of data encoded in binary form for computer storage and processing purposes. A binary file can contain text that is not broken up into lines. A binary file stores data in a format that is similar to the format in which the data is stored in the main memory. Therefore, a binary file is not readable by humans and it is up to the program reading the file to make sense of the data that is stored in the binary file and convert it into something meaningful (e.g. a fixed length of record).

Binary files contain formatting information that only certain applications or processors can understand. It is possible for humans to read text files which contain only ASCII text, while binary files must be run on the appropriate software or processor so that the software or processor can transform the data in order to make it readable by humans. For example, only Microsoft Word can interpret the formatting information in a Word document.

Although text files can be manipulated by any text editor, they do not provide efficient storage. In contrast, binary files provide efficient storage of data, but they can be read only through an appropriate program.

16.5 BASIC FILE OPERATIONS

The basic operations that can be performed on a file are given in Fig. 16.2.

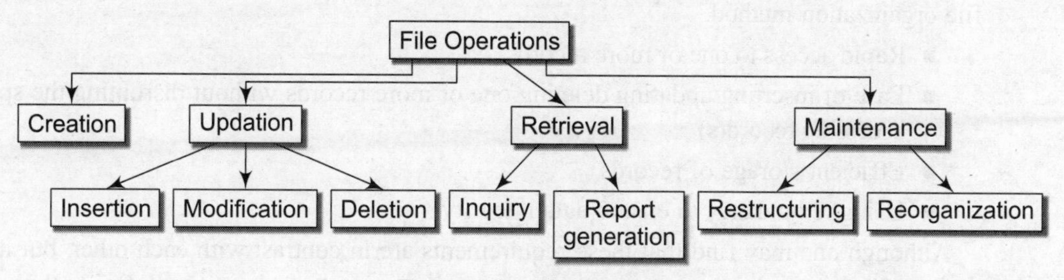

Figure 16.2 File operations

Creating a file A file is created by specifying its name and mode. Then, the file is opened for writing records that are read from an input device. Once all the records have been written into the file, the file is closed. The file is now available for future read/write operations by any program that has been designed to use it in some way or the other.

Updating a file Updating a file means changing the contents of the file to reflect a recent snapshot of the real world. That is, a file is updated to reflect a more current picture of reality. A file can be updated in the following ways:

- Inserting a new record in the file. For example, if a new student joins the course, we need to add his record in the STUDENT file.
- Deleting an existing record. For example, if a student quits a course in the middle of the session, his record to be deleted from the STUDENT file.
- Modifying an existing record. For example, if the name of a student was spelt incorrectly, then correcting the name will be a modification of the existing record.

Retrieving from a file It means extracting useful data from a given file. Information can be retrieved from a file either for an inquiry or for report generation. An inquiry for some data retrieves low volumes of data, while report generation may retrieve a large volume of data from the file.

Maintaining a file It involves restructuring or re-organization of the file to improve the performance of the programs that access those files. Restructuring a file keeps the file organization unchanged and changes only the structural aspects of the file (for example, changing the field width or adding/deleting fields). On the other hand, file reorganization may involve changing the entire organization of the file. We will discuss file organization in detail in the next section.

16.6 FILE ORGANIZATION

We know that a file is a collection of related records. The main issue in file management is the way in which the records are organized inside the file because it has a significant effect on the system performance. Organization of records means the *logical* arrangement of records in the file (for example, based on their ordering or the placement of related records close to each other in the file), and not the physical layout of the file as stored on a storage media.

Choosing an appropriate file organization is a design decision, hence it must be done keeping the priority of achieving good performance with respect to the most likely usage of the file. Therefore, the following considerations should be kept in mind before selecting an appropriate file organization method:

- Rapid access to one or more records
- Ease of inserting/updating/deleting one or more records without disrupting the speed of accessing record(s)
- Efficient storage of records
- Using redundancy to ensure data integrity

Although one may find that these requirements are in contrast with each other, but it is the designer's job to find a good compromise among them, to get an adequate solution to the problem at hand. For example, the ease of addition of records can be compromised to get fast access to data.

In this section, we will discuss some of the file organization techniques that can be considered for a particular situation at hand.

16.6.1 Sequential Organization

A sequentially organized file stores the records in the order in which they were entered. That is, the first record that was entered is written as the first record in the file, the second record entered is written as the second record in the file, and so on. As a result, new records are added only at the end of the file.

Sequential files can be read only sequentially, starting with the first record in the file. Sequential file organization is the most basic way to organize a large collection of records in a file. Figure 16.3 shows `n` records numbered from 0 to `n-1` stored in a sequential file.

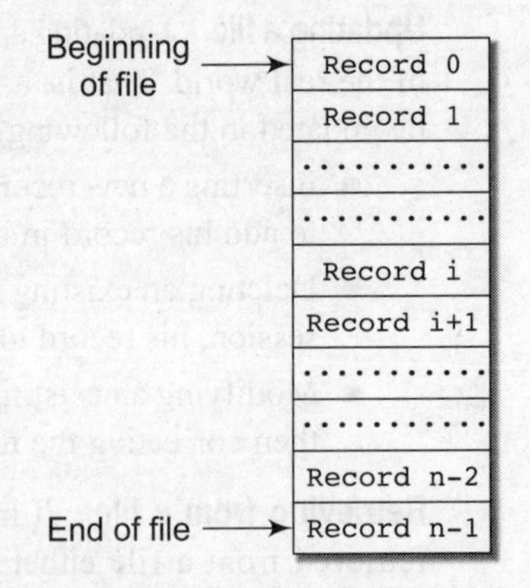

Figure 16.3 Sequential file organization

Once we store the records in a file, we cannot make these records shorter or longer. We cannot even delete the records from a sequential file. However, a file can be updated only if the length does not change. In case we need to delete or update one or more records, we have to replace the records by creating a new file.

In sequential file organization, all the records have the same size and the same field format, and every field has a fixed size. The records are sorted based on the value of one field or a combination of two or more fields. This field is known as the *key*. Every key uniquely identifies the records in a file. Thus, every record has a different value for the key field. Records can be sorted in either ascending or descending order.

Sequential files are generally used to generate reports or to perform sequential reads of large amount of data which some programs prefer to do such as payroll processing of all the employees of the organization. Sequential files can be easily stored both on the disk and on the tape. Table 16.2 summarizes the features, advantages, and disadvantages of sequential file organization.

Table 16.2 Sequential file organization

Features	Advantages	Disadvantages
• Records are written in the order in which they were entered. • Records are read and written sequentially. • Deletion or updation of one or more records calls for replacing the original file with a new file that contains the desired changes. • Records have the same size and the same field format. • Records are sorted on a key value. • Generally used for report printing or sequential reads.	• Simple and easy to handle. • No extra overheads involved. • Sequential files can be stored on magnetic disks as well as magnetic tapes. • Well suited for batch-oriented applications.	• Records can be read only sequentially. If i^{th} record has to be read, then all the i-1 records must be read. • Does not support update operation. A new file has to be created and the original file has to be replaced with the new file that contains the desired changes. • Cannot be used for interactive applications.

16.6.2 Relative File Organization

Relative file organization provides an effective way to access individual records directly. In a relative file organization, records are ordered by their *relative key*. It means the record number represents the location of the record relative to the beginning of the file. The record numbers range from 0 to `n-1`, where `n` is the number of records in the file. For example, the record with Record Number 0 is basically the first record in the file. The records in a relative file are of fixed length.

Therefore, in relative files, records are organized in ascending *relative record number*. A relative file can better be thought of as a single dimension table stored on a disk, in which the relative record number is the index into the table. Relative files can be used for both random as well as sequential access. For sequential access, records are simply read one after another.

Relative files provide support for only one key, that is, the relative record number. This key must be numeric and must take a value between 0 and the current highest Relative Record Number – 1. This means that enough space must be allocated for the file to contain the records with relative record numbers between 0 and the highest record number – 1. For example, if the highest relative record number is 1,000, then space must be allocated to store 1,000 records in the file.

Figure 16.4 shows a schematic representation of a relative file which has been allocated enough space to store 100 records. Although it has space to accommodate 100 records, not all the locations are occupied. The locations marked as 'free' are yet to store records in them. Therefore, every location in the table either stores a record or is marked as `FREE`.

Relative record number	Records stored in memory
0	Record 0
1	Record 1
2	FREE
3	FREE
4	Record 4
....	....
98	FREE
99	Record 99

Figure 16.4 Relative file organization

Relative file organization provides random access by directly jumping to the record which has to be accessed. For example, if the records are of fixed length and we know that each record occupies 20 bytes and the base address of the file is 1000; then any record `i` can be accessed using the following formula.

```
Address of ith record = base_address + (i-1) * record_length
```

Therefore, if we have to access the 5th record, then the address of the 5th record can be given as:

```
1000 + (5-1) * 20
= 1000 + 80
= 1080
```

Note that the base address of the file refers to the starting address of the file. We took `i-1` in the formula because record numbers start from 0 rather than 1. So, the 5th record is actually the 4th record. Table 16.3 summarizes the features, advantages, and disadvantages of relative file organization.

Table 16.3 Relative file organization

Features	Advantages	Disadvantages
• Provides an effective way to access individual records. • The record number represents the location of the record relative to the beginning of the file. • Records in a relative file are of fixed length. • Relative files can be used for both random as well as sequential access. • Every location in the table either stores a record or is marked as FREE.	• Ease of processing. • If the relative record number of the record that has to be accessed is known, then the record can be accessed instantaneously. • Random access of records makes access to relative files fast. • Allows deletions and updates in the same file. • Provides random as well as sequential access of records with low overhead. • New records can be easily added in the free locations based on the relative record number of the record to be inserted. • Well suited for interactive applications.	• Use of relative files is restricted to disk devices. • Records can be of fixed length only. • For random access of records, the relative record number must be known in advance.

16.6.3 Indexed Sequential File Organization

Indexed sequential file organization stores data for fast retrieval. The records in an indexed sequential file are of fixed length and every record is uniquely identified by a key field. We maintain a table known as the *index table* which stores the record number and the address of all the records. That is for every file, we have an index table. This type of file organization is called as indexed sequential file organization because physically the records may be stored anywhere, but the index table stores the address of those records.

The ith entry in the index table points to the ith record of the file. Initially, when the file is created, each entry in the index table contains `NULL`. When the ith record of the file is written, free space is obtained from the free space manager and its address is stored in the ith location of the index table. Figure 16.5 shows this scheme.

Now, if one has to read the 4th record, then there is no need to access the first three records. Address of the 4th record can be obtained from the index table and the record can be straightaway read from the specified address (742, in our example). Conceptually, the index sequential file organization can be visualized as shown in Fig. 16.6.

Record number	Address of the Record
1	765
2	27
3	876
4	742
5	NULL
6	NULL
7	NULL
8	NULL
9	NULL

Figure 16.5 Indexed sequential file organization

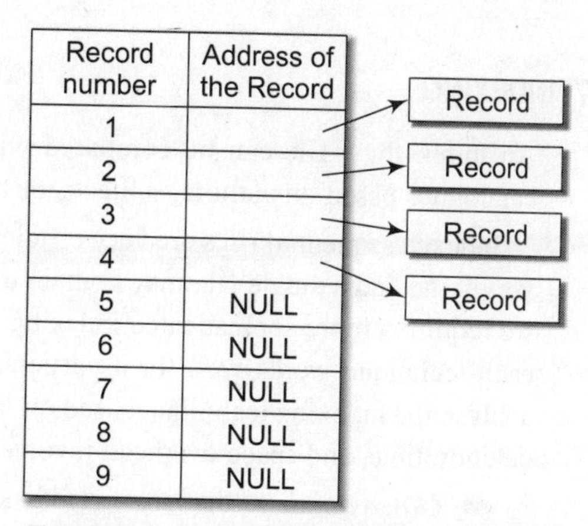

Figure 16.6 Indexed sequential file organization

An indexed sequential file uses the concept of both sequential files as well as relative files. While the index table is read sequentially to find the address of the desired record, a direct access is made to the address of the specified record in order to access it randomly.

Indexed sequential files perform well in situations where sequential access as well as random access is made to the data. Indexed sequential files can be stored only on devices that support random access, for example, magnetic disks.

For example, take an example of a college where the details of students are stored as an indexed sequential file. This file can be accessed in two ways:

- *Sequentially*, to print the aggregate marks obtained by each student in a particular course, or
- *Randomly*, for example, to modify the name of a particular student.

Table 16.4 summarizes the features, advantages, and disadvantages of indexed sequential file organization.

Table 16.4 Indexed sequential file organization

Features	Advantages	Disadvantages
• Provides fast data retrieval. • The records are of fixed length. • Index table stores the address of the records in the file. • The ith entry in the index table points to the ith record of the file. • While the index table is read sequentially to find the address of the desired record, a direct access is made to the address of the specified record in order to access it randomly. • Indexed sequential files perform well in situations where sequential access as well as random access is made to the data.	• The key improvement is that the indexes are small and can be searched quickly, allowing the database to then access only the records it needs. • Supports applications that require both batch and interactive processing. • Records can be accessed sequentially as well as randomly. • Updates the records in the same file.	• Indexed sequential files can be stored only on disks. • Needs extra space and overhead to store indices. • Handling these files is more complicated than handling sequential files. • Supports only fixed length records.

16.7 INDEXING

An index for a file can be compared with a catalogue in a library. Like a library has card catalogues based on authors, subjects, or titles, a file can also have one or more indices.

Indexed sequential files are very efficient to use, but in real-world applications, these files are very large and a single file may contain even millions of records. Therefore, in such situations, we require a more sophisticated indexing technique. There are several indexing techniques and each technique works well for a particular application. For a particular situation at hand, we analyse the indexing technique based on factors such as access type, access time, insertion time, deletion time, and space overhead involved. Basically, there are two kinds of indices:

- *Ordered indices* that are sorted based on the key values.
- *Hash indices* that are based on the values generated by applying a ***hash function***.

16.7.1 Ordered Indices

Indexes are used to provide fast random access to records. As stated above, a file may have multiple indices based on different key fields. An index of a file may be a primary index or a secondary index.

Primary index In a sequentially ordered file, the index whose search key specifies the sequential order of the file is defined as the primary index. For example, if the records of students are stored in a `STUDENT` file in a sequential order starting from roll number 1 to roll number 60, and if the search key is roll number, that is, we want to search record for, say, roll number 10, then the student's roll number is the primary index. Indexed sequential files are a common example where a primary index is associated with the file.

Secondary index An index whose search key specifies an order different from the sequential order of the file is called as the secondary index. For example, if the record of a student is searched by his name, then the name is a secondary index. Secondary indices are used to improve the performance of queries on non-primary keys.

16.7.2 Dense and Sparse Indices

In a dense index, the index table stores the address of every record in the file. However, in a sparse index, the index table stores the address of only some of the records in the file. Although sparse indices are easy to fit in the main memory, but if a dense index fits in the memory, it is more efficient to use than a sparse index. Figure 16.7 shows a dense index and a sparse index for an indexed sequential file.

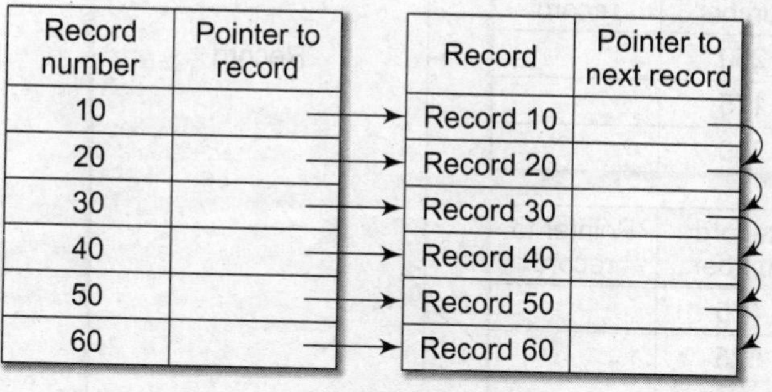

(a) Dense index

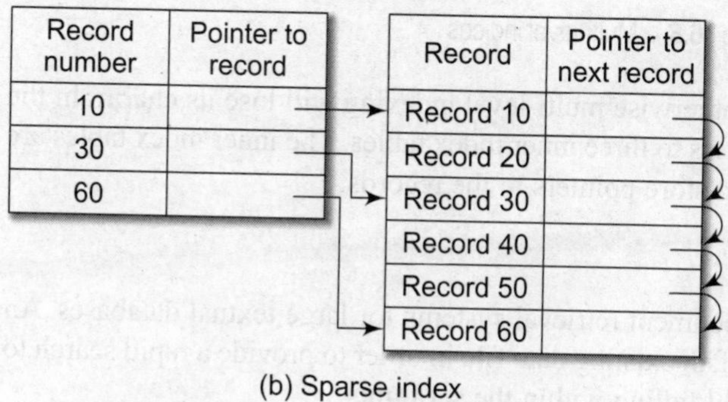

(b) Sparse index

Figure 16.7 Dense index and sparse index

Note that these records need not be stored in consecutive memory locations. The pointer to next field stores the address of the next record.

By looking at the dense index, it can be concluded directly whether the record exists in the file or not. This is not the case in sparse index. In sparse index, to locate a record, we find an entry in the index table with the largest search key value that is either less than or equal to the search key value of the desired record. Then, we start at that record pointed to by that entry in the index table and then proceed searching the record using the sequential pointers in the file, until the desired record is obtained. For example, if we need to access record number 40, then record number 30 is the largest key value that is less than 40. So jump to the record pointed by record number 30 and move along the sequential pointer to reach record number 40.

Thus we see that sparse index takes more time to find a record with the given key. Dense indices are faster to use, while sparse indices require less space and impose less maintenance for insertions and deletions.

16.7.3 Multi-Level Indices

In real-world applications, we have very large files that may contain millions of records. For such files, a simple indexing technique will not suffice. In such a situation, we use multi-level indices. To understand this concept, consider a file that has 10,000 records. If we use simple indexing, then we need an index table that can contain at least 10,000 entries to point to 10,000 records. If each entry in the index table occupies 4 bytes, then we need an index table of 4 × 10000 bytes = 40000 bytes. Finding such a big space consecutively is not always easy. So, a better scheme is to index the index table. Figure 16.8 which shows this technique.

Figure 16.8 shows a two-level multi-indexing. We can continue further by having a 3-level indexing and so on. But practically, we use second-level indexing. Note that second and higher-

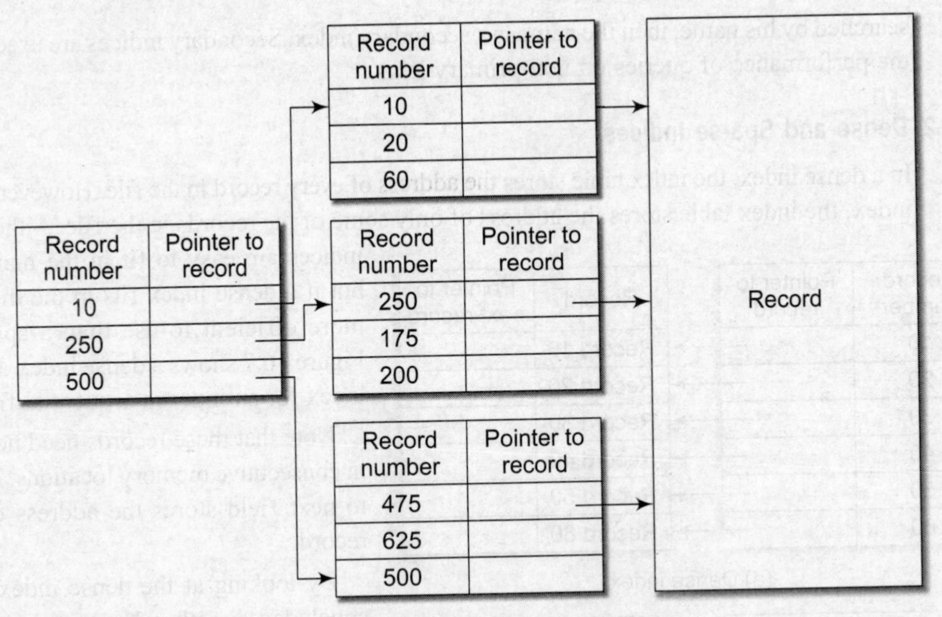

Figure 16.8 Multi-level indices

level indexing must always be sparse, otherwise multi-level indexing will lose its charm. In the figure, the main index table stores pointers to three inner index tables. The inner index tables are basically sparse index tables that in turn store pointers to the records.

16.7.4 Inverted files

Inverted files are commonly used in document retrieval systems for large textual databases. An inverted file reorganizes the structure of an existing data file in order to provide a rapid search to be made for all records having one field falling within the set limits.

For example, inverted files are widely used by bibliographic databases that may store author names, title words, journal names, etc. When a term or keyword specified in the inverted file is identified, the record number is given and a set of records corresponding to the search criteria are created.

Thus, for each keyword, an inverted file contains an inverted list that stores a list of pointers to all occurrences of that term in the main text. Therefore, given a keyword, the addresses of all the documents containing that keyword can easily be located very fast.

There are two main variants of inverted indexes:

- A record-level inverted index (also known as *inverted file index* or *inverted file*) stores a list of references to documents for each word.
- A word-level inverted index (also known as *full inverted index* or *inverted list*) in addition to a list of references to documents for each word also contains the positions of each word within a document. Although this technique needs more time and space, it offers more functionality (like phrase searches).

Therefore, the inverted file system consists of an index file in addition to a document file (also known as *text file*). It is this index file that contains all the keywords which may be used as search terms. For each keyword, an address or reference to each location in the document where that word occurs is stored. There is no restriction on the number of pointers associated with each word.

For efficiently retrieving a certain word from the index file, the keywords are sorted in a specific order (usually alphabetically).

However, the main drawback with this structure is that when new words are added in the documents or text files, the whole file must be reorganized. Therefore, a better technique is to use B-trees.

16.7.5 B-Tree Index

A database is defined as a collection of data organized in a fashion that facilitates updating, retrieving, and managing the data (that may include anything, such as names, addresses, pictures, and numbers). Databases are very commonly used everyday. For example, an airline reservation system maintains a database of flights, customers, and tickets issued. A university maintains a database of all its students. These real-world databases may contain millions of records that may occupy gigabytes of storage space.

For a database to be useful, it must support fast retrieval and storage of data. Since it is impractical to maintain the entire database in the memory, B-trees are used to index the data in order to provide fast access.

For example, to search data from an un-indexed and unsorted database containing `n` key values may take a running time of `O(n)` in the worst case, but if the same database is indexed with a B-tree, the search operation will run in `O(log n)` time.

Majority of the DBMS today use the B-tree index technique as the default indexing method. This technique supersedes other techniques of creating indexes, mainly due to its data retrieval speed, ease of maintenance, and simplicity. Figure 16.9 shows a B-tree index.

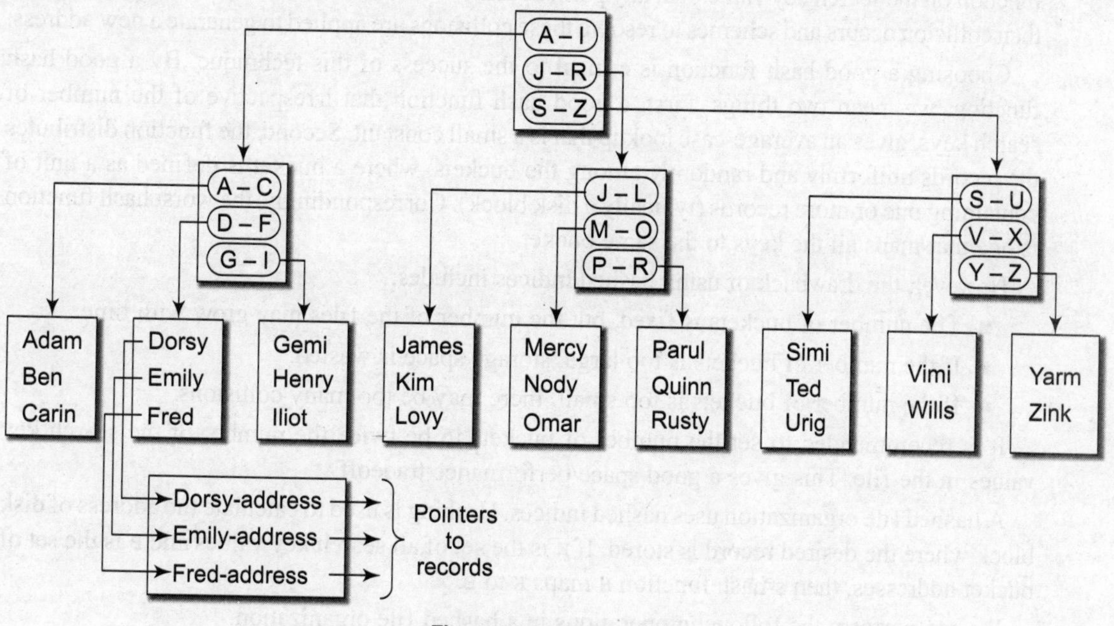

Figure 16.9 B-tree index

It forms a tree structure with the root at the top. The index consists of a B-tree (balanced tree) structure based on the values of the indexed column. In this example, the indexed column is *name* and the B-tree is created using all the existing names that are the values of the indexed column. The upper blocks of the tree contains index data pointing to the next lower block, thus forming a hierarchical structure. The lowest level blocks, also known as leaf blocks, contain pointers to the data rows stored in the table.

If a table has a column that has many unique values, then the selectivity of the column is said to be high. B-tree indexes are most suitable for highly selective columns, but it causes a sharp increase in the size when the indexes contain concatenation of multiple columns.

The B-tree structure has the following advantages:

- Since the leaf nodes of the tree are at the same depth, retrieval of any record from anywhere in the index takes approximately the same time.
- B-trees improve the performance of a wide range of queries that either search a value having an exact match or range searches.
- It provides fast and efficient algorithms to insert, update, and delete records that maintain the key order.
- B-trees perform well for small as well as large tables. Its performance does not degrade as the size of a table grows.
- B-trees optimize costly disk access.

16.7.6 Hashed Indices

In the last chapter, we have discussed hashing in detail. The same concept of hashing can be used to create hashed indexes.

So far, we have studied that hashing is used to compute the address of a record by using a hash function on the search key value. If at any point of time, the hashed values map to the same address, then collision occurs and schemes to resolve these collisions are applied to generate a new address.

Choosing a good hash function is critical to the success of this technique. By a good hash function, we mean two things. First, a good hash function that irrespective of the number of search keys, gives an average-case lookup that is a small constant. Second, the function distributes the records uniformly and randomly among the buckets, where a bucket is defined as a unit of containing one or more records (typically a disk block). Correspondingly, the worst hash function is one that maps all the keys to the same bucket.

However, the drawback of using hashed indices includes:

- The number of buckets is fixed, but the number of the files may grow with time.
- If the number of buckets is too large, storage space is wasted.
- If the number of buckets is too small, there may be too many collisions.

It is recommended to set the number of buckets to be twice the number of the search key values in the file. This gives a good space/performance tradeoff.

A hashed file organization uses hashed indices. Hashing is used to calculate the address of disk block where the desired record is stored. If `K` is the set of all search key values and `B` is the set of bucket addresses, then a hash function `H` maps `K` to `B`.

We can perform the following operations in a hashed file organization.

Insert To insert a record that has k_i as its search value, use the hash function $h(k_i)$ to compute the address of the bucket for that record. If the bucket is free store the record, else use chaining to store the record.

Search To search a record having the key value k_i, use $h(k_i)$ to compute the address of the bucket where the record is stored. The bucket may contain just one or several records, so check for every record in the bucket (by comparing k_i with the key of every record) to finally retrieve the desired record with the given key value.

Delete To delete a record with key value k_i, use $h(k_i)$ to compute the address of the bucket where the record is stored. The bucket may contain just one or several records so check for every record in the bucket (by comparing k_i with the key of every record). Then, delete the record as we delete a node from linear linked list. We have already studied how to delete a record from a chained hash table in Chapter 13.

Note that in a hashed file organization, basically the secondary indices need to be organized using hashing.

16.8 EXTERNAL SORTING

As discussed in Chapter 14, external sorting is a sorting technique that can handle massive amounts of data. It is usually applied when the data being sorted does not fit into the main memory (RAM) and therefore, a slower memory (usually a magnetic disk or even a magnetic tape) needs to be used.

EXAMPLE 16.1: Let us consider external merge sort algorithm for sorting 700 MB data using only 100 MB of RAM. Follow the steps given below.

Step 1: Read 100 MB of the data in RAM and sort this data using any conventional sorting algorithm like quick sort.

Step 2: Write the sorted data back to the magnetic disk.

Step 3: Repeat Steps 1 and 2 until all the data (in 100 MB chunks) is sorted. All these 7 chunks, that are sorted needs to be merged into one single output file.

Step 4: Read the first 10 MB of each of the sorted chunks and call them input buffers. So, now we have 70MB of data in the RAM. Allocate the remaining RAM for output buffer.

Step 5: Perform 7-way merging and store the result in the output buffer. If at any point of time, the output buffer becomes full, then write its contents to the final sorted file. However, if any of the 7 input buffers gets empty, fill it with the next 10 MB of its associated 100 MB sorted chunk or else mark the input buffer (sorted chunk) as exhausted if it does not has any more left with it. Make sure that this chunk is not used for further merging of data.

The external merge sort can be visualized as given in Fig. 16.10.

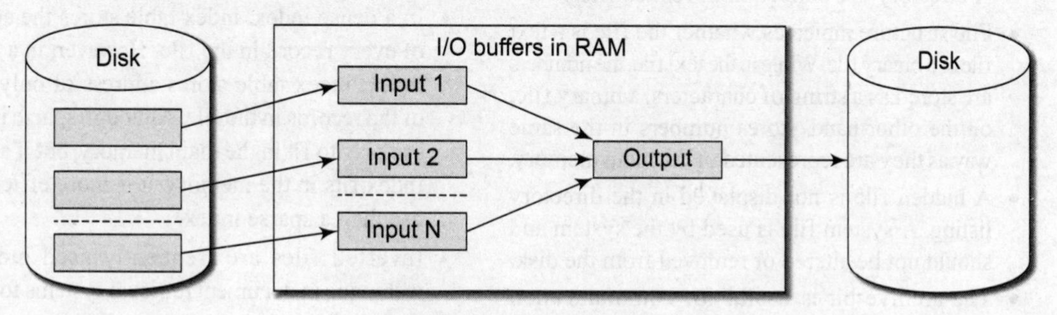

Figure 16.10 External merge sort

16.8.1 Generalized External Merge Sort Algorithm

From the example above, we can now present a generalized merge sort algorithm for external sorting. If the amount of data to be sorted exceeds the available memory by a factor of K, then K chunks (also known as K run lists) of data are created. These K chunks are sorted and then a K-way merge is performed. If the amount of RAM available is given as X, then there will be K input buffers and 1 output buffer.

In the above example, a single-pass merge was used. But if the ratio of data to be sorted and available RAM is particularly large, a multi-pass sorting is used. We can first merge only the first half of the sorted chunks, then the other half, and finally merge the two sorted chunks. The exact number of passes needed depends on the following factors:

- Size of the data to be sorted when compared with the available RAM
- Physical characteristics of the magnetic disk such as transfer rate, seek time, etc.

16.8.2 Applications of External Sorting

External sorting is used to update a master file from a transaction file. For example, updating the `EMPLOYEES` file based on new hires, promotions, increments, and dismissals.

It is also used in database applications for performing operations like *Projections* and *Joins*. Projection means selecting a subset of fields and join means joining two files on a common field to create a new file whose fields are the union of the fields of the two files. External sorting is used to remove duplicate records.

SUMMARY

- A file is a block of useful information which is available to a computer program and is usually stored on a persistent storage medium.
- Every file contains data. This data can be organized in a hierarchy to present a systematic organization. The data hierarchy includes terms such as fields, records, files, and database.
- A data field is an elementary unit that stores a single fact. A record is a collection of related data fields that is seen as a single unit from the application point of view. A file is a collection of related records. A directory is a collection of related files.
- File structure indicates whether the file is a text file or a binary file. While in the text file, the numbers are stored as a string of characters, a binary file, on the other hand, stores numbers in the same way as they are represented in the main memory.
- A hidden file is not displayed in the directory listing. A system file is used by the system and should not be altered or removed from the disk.
- The archive bit is useful for communication between programs that modify files and programs that are used for backup.
- A sequentially organized file stores records in the order in which they were entered.
- In relative file organization, records in a file are ordered by their relative key. Relative files can be used for both random accesses of data as well as for sequential access.
- In indexed sequential file, every record is uniquely identified by a key field. We maintain a table known as the index table that stores record number and the address of the record in the file.
- In a dense index, index table stores the address of every record in the file. However, in a sparse index, index table stores address of only some of the records in the file. Although sparse indices are easy to fit in the main memory, but if a dense index fits in the memory, it is more efficient to use than a sparse index.
- Inverted files are frequently used indexing technique in document retrieval systems for large textual databases. An inverted file reorganizes the structure of an existing data file in order to

provide a rapid search for all records having one field falling within set limits.

- A database is defined as a collection of data organized in a fashion that facilitates updating, retrieving, and managing the data.
- External sorting is a sorting technique that can handle massive amounts of data. It is usually applied when the data being sorted does not fit into the main memory (RAM) and therefore, a slower memory (usually a magnetic disk or even a magnetic tape) needs to be used.

GLOSSARY

External sort A sort algorithm that uses external memory (like magnetic tape or disk) during the sort. Since internal sorting algorithms assume high-speed random access to all intermediate memory, they are unsuitable if the values to be sorted do not fit in the main memory.

Index file A file that stores keys and an index into another file.

Inverted index An index into a set of texts of the words in the texts. The index is accessed by applying some search technique. Each index entry gives the word and a list of texts, along with locations within the text, where the word occurs.

Key The part of data by which it is sorted, indexed, cross referenced, etc.

EXERCISES

Review Questions

1. Why do we need files?
2. Explain the terms field, record, file organization, key, and index.
3. Define file. Explain all the file attributes.
4. How is archive attribute useful?
5. Differentiate between a binary file and a text file.
6. Explain the basic file operations.
7. What do you understand by the term file organization? Briefly summarize the different file organizations that are widely used today.
8. Give a brief note on indexing.
9. Differentiate between sparse index and dense index.
10. Explain the significance of multi-level indexing with an appropriate example.
11. What are inverted files? Why are they needed?
12. Give the merits and demerits of a B-tree index.
13. Explain the external merge sort algorithm with the help of an example.

Multiple Choice Questions

1. The data hierarchy can be given as
   (a) Primary index (b) Secondary index
   (c) Hashed index
2. Which is an important file used by the system and should not be altered or removed from the disk?
   (a) Hidden file (b) Archived file
   (c) System file
3. Updating a file includes
   (a) Fields, records, files and database
   (b) Records, files, fields and database
   (c) Database, files, records and fields
4. Searching a student record by his name is an example of
   (a) Inserting (b) Deleting
   (c) Modifying (d) All of these

True or False

1. When a backup program archives the file, it sets the archive bit to one.
2. In a text files, data is stored using ASCII codes.
3. A binary file is more efficient than a text file.
4. Maintenance of a file involves re-structuring or re-organization of the file.
5. Relative files can be used for both random access of data as well as for sequential access.
6. In a sparse index, index table stores the address of every record in the file.

7. Higher level indexing must always be sparse.
8. B-tree indexes are most suitable for highly selective columns.

Fill in the Blanks

1. ______ is a block of useful information.
2. A data field is usually characterized by its ______ and ______.
3. ______ is a collection of related data fields.
4. A file is a collection of related ______.
5. ______ is a pointer that points to the position at which next read/write operation will be performed.
6. ______ indicates whether the file is a text file or a binary file.
7. Index table stores ______ and ______ of the record in the file.
8. If the file stores all its records in a sequential manner and the index whose search key specifies the sequential order of the file is defined as the ______ index.
9. ______ files are frequently used indexing technique in document retrieval systems for large textual databases.
10. ______ is a collection of data organized in a fashion that facilitates updating, retrieving, and managing the data.

Index